The Cambridge Dictionary of Probability and Its Applications

Probability comes of age with this, the first dictionary of probability and its applications in English, which supplies a guide to the concepts and vocabulary of this rapidly expanding field. Besides the basic theory of probability and random processes, applications covered here include financial and insurance mathematics, operations research (including queueing, reliability, and inventories), decision and game theory, optimization, time series, networks, and communication theory, as well as classic problems and paradoxes.

The dictionary is reliable, stable, concise, and cohesive. Each entry provides a rigorous definition, a sketch of the context, and a reference pointing the reader to the wider literature. As the only dictionary on the market, this will be a guiding reference for all those working in, or learning, probability together with its applications.

DAVID STIRZAKER was an Associate Professor in the Mathematical Institute at the University of Oxford, and is now an Emeritus Research Fellow at St John's College, Oxford. His previous titles include *Probability and Random Processes* (with Geoffrey Grimmett) and *Elementary Probability*.

The Cambridge Dictionary of Probability and Its Applications

DAVID STIRZAKER
St John's College, Oxford

CAMBRIDGE
UNIVERSITY PRESS

CAMBRIDGE
UNIVERSITY PRESS

University Printing House, Cambridge CB2 8BS, United Kingdom

Cambridge University Press is part of the University of Cambridge.

It furthers the University's mission by disseminating knowledge in the pursuit of education, learning and research at the highest international levels of excellence.

www.cambridge.org
Information on this title: www.cambridge.org/9781107075160

First published 2015

Printed in the United Kingdom by TJ International Ltd. Padstow Cornwall

A catalogue record for this publication is available from the British Library

Library of Congress Cataloguing in Publication data
The Cambridge dictionary of probability and its applications / [edited by] David Stirzaker, St John's College, Oxford.
pages cm
Includes bibliographical references and index.
ISBN 978-1-107-07516-0 (Hardback : alk. paper) 1. Probabilities–Dictionaries. 2. Mathematical statistics–Dictionaries. I. Stirzaker, David. II. Title: Dictionary of probability and its applications.
QA273.A3C36 2015
519.2–dc23 2015009562

ISBN 978-1-107-07516-0 Hardback

This book is dedicated to the benefactors of St John's College, Oxford

Preface

This is a dictionary of technical terms that commonly arise in probability and its applications – except statistical terms, for which the reader should consult a dictionary of statistics. Besides the basic theory of probability and random processes, applications covered here include financial and insurance mathematics, operations research (including, e.g. queueing, reliability, and inventories), decision and game theory, optimization, time series, point processes, networks, branching processes, communication theory, classic problems and paradoxes, and several other stochastic models for physical and biological processes that have attracted the analytical attention of probabilists. There is, of course, some overlap with the probabilistic aspects of statistical vocabulary. However, the book is not an encyclopaedia, and the applications of probability are many and increasing at an increasing rate; so rapidly indeed, that it might be said that probability is the new calculus. Universal coverage is therefore neither intended nor possible, but it is hoped that enough terms are defined to make the book useful to a large number of those working in, or learning, the theory of probability together with its applications. There are more than 3100 headwords, which, together with subsidiary definitions and aliases, takes the total number of terms defined to somewhere near 5000. Almost all entries also include at least one reference, which may be to the original paper, a survey paper, an important and useful consequent paper, a textbook, or a monograph. It should be noted that there has been a necessary element of randomness in choosing which to cite of several equally meritorious works.

Note

In the sample space of all possible dictionaries, there is an event on which this book contains no serious errors or omissions; but that is almost surely not the way to bet. Corrections to the former and suggestions for the latter will be gratefully received by the author and the Press, who will also be pleased to see ideas for better references.

Remark

Entries are in strict lexicographical order of their headings, ignoring spaces, punctuation, and the possessive ‘s’; for this ordering numerals and Greek letters are almost always spelt out, and terms are never inverted: e.g. see ‘stochastic matrix’, not ‘matrix, stochastic’.

Abbreviations

a.e.	almost everywhere
a.s.	almost surely
arcsine	inverse sine function
arg z	argument θ of $z = \lvert z\rvert e^{i\theta}$
Ber(p)	Bernoulli distribution, parameter p
Beta(a, b)	beta distribution, parameters a, b
$\beta(a, b)$	beta function
B.M.	Brownian motion
B(n, p)	binomial distribution, parameters n, p
cdf	cumulative distribution function
cgf	cumulant generating function
chf	characteristic function
$\chi^2(n)$	chi-squared distribution
CLT	central limit theorem
cos, cosh	cosine, hyperbolic cosine
cov	covariance
cts	continuous
det	determinant
distn	distribution
edn	edition
eqn	equation
et al.	and others
Exp(λ)	exponential distribution, parameter λ
exp	exponential function
fBm	fractional Brownian motion
fdds	finite-dimensional distributions
fn	function
$\Gamma(m, \lambda)$	gamma distribution, parameters m, λ
Geo(p)	geometric distribution, parameter p
iff	if and only if
i.i.d.	independent and identically distributed
Im(z)	imaginary part of z
inequ.	inequality
inf	infimum
LHS	left-hand side
LIL	law of the iterated logarithm
lim	limit
log	logarithm
max	maximum
mgf	moment generating function
min	minimum
m.s.	mean square
$N(\mu, \sigma^2)$	normal distribution, mean μ, variance σ^2
pde	partial differential equation
pdf	probability density function
pgf	probability generating function
pmf	probability mass function
Poi(λ)	Poisson distribution, parameter λ
rcll	right continuous with left limits
Re(z)	real part of z
resp.	respectively
$\rho(X, Y)$	correlation function
RHS	right-hand side
r.v., r.vs	random variable, random variables
r.vr, r.vrs	random vector, random vectors
r.w.	random walk
SDE	stochastic differential equation
sec, sech	secant, hyperbolic secant
sin, sinh	sine, hyperbolic sine
SLLN	strong law of large numbers
sup	supremum
tan	tangent
thm	theorem
U(a, b)	uniform distribution, parameters (i.e. end points) a, b
u.i.	uniformly integrable
var	variance
v.v.	vice versa
WLLN	weak law of large numbers
wlog	without loss of generality
w.p. 1	with probability one
wrt	with respect to

Symbols and notation

(a, b), $[a, b]$	open, closed interval
$\|A\|$	cardinality of A
$A \cup B$	union
$A \cap B$	intersection
$A \backslash B = A \cap B^c$	difference
A^T or A'	transpose
A^c	complement
$\mathcal{A}, \mathcal{B}, \mathcal{F}, \mathcal{G}, \mathcal{H}, \dots$	algebras and σ-fields
$\mathrm{B}(n, p)$	binomial distribution
$\mathbb{C}$	complex plane
cov	covariance
$c(h), c(t)$	autocovariance
$d_{\mathrm{TV}}, d_{\mathrm{KL}}$	total variation distance, Kullback–Leibler divergence
e	base of natural logarithms
$\mathbb{E}$	expectation
$\mathbb{E}(X\|Y)$	conditional expectation
$f(x), f_X(x)$	density
$f(x\|y), f_{X\|Y}(x\|y)$	conditional density
$f(x, y)$	joint density
$F(x), F_X(x)$	cumulative distribution
$F(x, y)$	joint distribution
$F(x\|y), F_{X\|Y}(x\|y)$	conditional distribution
$G(s), G_X(s)$	generating function
$\dot{g}(x) = g'(x) = \frac{\mathrm{d}g}{\mathrm{d}x}$	derivative
$H(X)$, $H(X,Y)$, $H(X\|Y)$	Shannon entropies
i	$\sqrt{-1}$
$i, j, k, l, m, n, s, t, u, v, \dots$	indices and time parameters
$\mathrm{I}(A)$	indicator function
$\mathrm{Im}(z), \mathrm{Re}(z)$	imaginary part of z, real part of z
$\log_2$	logarithm with base 2
max ($\vee$), min ($\wedge$)	maximum, minimum
$M(t), M_X(t)$	moment generating function
$n!$	factorial
$\binom{n}{r}$	binomial coefficient
$\mathbb{N}$	positive integers
$N(t)$	Poisson, or renewal, or counting process
$N(\mu, \sigma^2)$	normal distribution
$p(x), p_X(x)$	mass function, probability
$p(x, y)$	joint mass function
$p(x\|y), p_{X\|Y}(x\|y)$	conditional mass function
$p_{jk}, p(j, k), p_{jk}(n)$	transition probabilities
$\mathbb{P}(\cdot), \mathbb{Q}(\cdot)$	probability functions, or measures
$\mathbb{R}$	reals
var	variance
$W(t)$	Wiener process
$X, Y, Z, W, X(\omega), \dots$	random variables
$\mathbf{X}, \mathbf{Y}, \mathbf{Z}, \dots$	random vectors
$\|X\| = \det \mathrm{X}$	determinant of the matrix X
$\|x\|$	modulus or absolute value of x
$x^+ = x_+$	$\max\{0, x\}$
$x^- = x_-$	$-\min\{0, x\}$
$[x]$	integer part of x
$x^{\overline{n}}, x^{\underline{n}}$	rising, falling factorial
$\bar{z}$	complex conjugate
$\mathbb{Z}$	integers
$\mathbb{Z}^+$	non-negative integers
$\beta(a, b)$	beta function
γ	Euler's constant
$\Gamma(m, \lambda)$	gamma distribution
$\Gamma(t)$	gamma function
$\delta_{ij} = \delta(i-j) = I(i=j)$	Kronecker delta
$\delta(t)$	Dirac delta
μ	mean, or other parameter
$\rho(X, Y)$	correlation
$\rho(t)$	autocorrelation
$\phi(t), \phi_X(t)$	characteristic function
$\phi(x), \Phi(x)$	standard normal density, distribution function
$\boldsymbol{\pi}$	stationary distribution
$\chi^2(n)$	chi-squared distribution
ω	elementary event, or sample point
Ω	sample space
$\emptyset$	empty set, or impossible event
$\|\| \cdot \|\|$	norm
∇, ∇^2	gradient, Laplacian
$\propto$	proportional

A

1-randomness *see* Martin-Löf randomness

Abelian sandpile (alias: **Bak–Tang–Wiesenfeld model; sandpile automaton; lattice avalanche model; Dhar's Abelian sandpile**) A classic interacting particle system, where the particles are envisaged as grains of sand that form piles at the points of a subset D of the d-dimensional integer lattice. Each vertex can support grains up to some finite limit; at integer times a new grain is added to a randomly chosen point $v \in D$. If this addition causes the pile to exceed its limit, it will topple onto its neighbours according to some rule of redistribution. This **toppling** may induce further topplings, and a complete set of topplings is called an **avalanche**. The process is called Abelian when the final stable result of an avalanche is independent of the order of execution of the topplings. No grains are created in avalanches, but sand may be lost at a boundary. Other sandpile models exist, e.g. **Zhang's sandpile** in which a random quantity of sand, continuously distributed on an interval $[a, b]$, is added to a randomly chosen vertex. If its capacity is exceeded, the pile is equally divided among all the vertex's neighbours. Related models include the Bak–Sneppen model, the chip-firing game, and the Dirichlet game. [G. Pruessner, *Self-Organized Criticality*. Cambridge University Press, 2012]

absolute continuity A strengthening of the idea of continuity that arises in the context of real-valued functions and measures. One definition asserts that a real function $F(x), x \in \mathbb{R}$ is absolutely continuous if it can be written as the Lebesgue integral of a function $f(x)$; $F(x) = \int_a^x f(u)\mathrm{d}u$. Thus the distribution of a random variable having a density on $\mathbb{R}$ (or $\mathbb{R}^d$), is absolutely continuous. Equivalently, $F(x)$ is absolutely continuous on $[a, b]$ if, for any $\varepsilon > 0$, there is $\delta > 0$ such that $\sum_k |F(b_k) - F(a_k)| < \varepsilon$, for any collection of disjoint intervals $(a_k, b_k) \in [a, b]$, such that $\sum_k (b_k - a_k) < \delta$.

A classic example of a continuous, yet not absolutely continuous, distribution function is the Cantor distribution. Similarly, but more generally, a measure (or distribution) $\mu(\cdot)$ is said to be absolutely continuous with respect to a measure (or distribution) $\nu(\cdot)$ if $\mu(A) = \int_A f(x)\nu(\mathrm{d}x)$ for some $f(\cdot)$. When μ and f are probability distributions, f is called a **density** of μ with respect to ν. More generally, f is called a Radon–Nikodym derivative of μ with respect to ν.

The Radon–Nikodym theorem (sometimes called the Lebesgue–Nikodym thm) shows that μ is absolutely continuous wrt ν iff $\mu(A) = 0$ whenever $\nu(A) = 0$. In this case one says that μ is dominated by the **dominating measure** ν, and writes $\mu \ll \nu$. If also $\nu \ll \mu$, then ν and μ are said to be **equivalent measures**.

The concepts coincide for a distribution function $F(x)$ corresponding to a probability measure $\mathbb{P}(\cdot)$ on $\mathbb{R}$ such that $\mathbb{P}((-\infty, x]) = F(x)$; $\mathbb{P}(\cdot)$ is absolutely continuous wrt the Lebesgue measure iff $F(\cdot)$ is an absolutely continuous function. [D. Pollard, *A User's Guide to Measure Theoretic Probability*. Cambridge University Press, 2002]

absolute deviation The absolute value $|X - q|$ of the difference $X - q$ between a random variable X and some given q, commonly called the mean, the median, or the 'true value' of some quantity, of which X is a sample or realization. It is sometimes called the absolute error; and $|X - q|/q$ may be called the **relative deviation** (or error). [M. Bean, *Probability: The Science of Uncertainty*. American Mathematical Society, 2001]

absolute moment The rth absolute moment of the r.v. X is $\mathbb{E}(|X|^r) = \mathbb{E}(|X^r|)$. If r is not an integer then this is the absolute fractional moment; note that for z complex and r real, $z^r = |z|^r \mathrm{e}^{\mathrm{i}r\theta}$, where θ is the argument of z. The first absolute moment may be called the **mean (absolute) deviation**. [K. L. Chung, *Course in Probability Theory*, 2nd edn revised. Academic Press, 2001]

absolute monotonicity The probability generating function $G(x)$ of a non-negative integer-valued r.v. has a power series expansion with positive coefficients for $0 \le x < 1$; any function with this property is said to be absolutely monotone. From a theorem of S. Bernstein, this holds iff

$$\sum_{r=0}^{k} \binom{k}{r} (-1)^{k-r} G(rh) \geq 0, \;\; 0 \leq k \leq n-1, h = n^{-1}, \quad n \geq 1.$$

[W. Feller, *Introduction to Probability Theory and its Applications*, Vol. 2, 2nd edn. Wiley, 1971]

absolute probability This is used to denote the unconditional probability of an event, where the lack of conditioning is to be stressed. For example, for a Markov chain X_n with initial distribution a_r and n-step transition probabilities $p_{ij}(n)$, the absolute probability of the event $X_n = j$ is $\sum_r a_r p_{rj}(n)$; in contrast to the conditional $p_{ij}(n)$. Similarly, the term **absolute probability fn** is used in an axiomatization of probability defined on the statements of a propositional language; where the use of 'absolute' stresses the distinction from **relative probability fns**. These are functions of two propositions, one of which is a conditioning statement seen as evidence or support for the other. [W. Feller, *Introduction to Probability Theory and its Applications*, Vol. 1, 3rd edn. Wiley, 1968]

absolute regularity *see* mixing

absorbing A term arising in the context of random processes whose evolution can be seen as the movement of a particle in some space. A point is absorbing if the particle remains there for all time after its first arrival there. Such a point (or state of the process) may be called a **sink** or **trap**.

A collection of such points may be called an **absorbing barrier**, or absorbing boundary. They are important not only as models of reality, but also because all first-passage problems can be formulated in terms of the original process together with additional (artificial) absorbing barriers.

A related concept is that of the **absorbing set** S, which the process can never leave after entering for the first time, but may continue its evolution within S. [S. Karlin & H. Taylor, *A First Course in Stochastic Processes*, 2nd edn. Academic Press, 1975; *A Second Course in Stochastic Processes*. Academic Press, 1981]

absorption-time distribution *see* phase-type distribution

accelerated life model Arises in reliability theory (and other contexts) when the failure-time distribution has a heavy tail (or a large mean) so that empirical results are delayed and expensive. The core of the technique is the assumption that the parameters of the model depend on some measure of stress (e.g. temperature or loading). Data may be obtained quickly and cheaply under high stress and used, via the model, to predict low-stress lifetimes. [V. Bagdonavičius & M. Nikulin, *Accelerated Life Models*. Chapman & Hall, 2002]

acceptability A finite set $X_1, \ldots, X_n$ of r.vs is called acceptable if for any $t \in \mathbb{R}$, $\mathbb{E}\exp\,[t\sum_1^n X_r] \leq \prod_1^n \mathbb{E}\exp(tX_r)$; while a sequence $X_1, X_2, \ldots$ is acceptable if any finite subset is acceptable. [R. Giuliano Antonini *et al.*, Convergence of series of dependent φ-subgaussian random variables, *J. Math. Anal. Appl.*, 338(2), 1188–1203, 2008]

acceptance-complement A method for sampling from a density $f_X(x)$, as follows. Choose non-negative functions $f_1(x)$ and $f_2(x)$, and a proper density $f_Y(x)$, such that for $x \in \mathbb{R}$ $f_X(x) = f_1(x) + f_2(x)$, and $f_1(x) \leq f_Y(x)$. Then:

1. Generate Y with density $f_Y(y)$, and independent U uniform on $[0, 1]$.
2. If $U \leq f_1(Y)/f_Y(Y)$, then set $X = Y$; otherwise generate Z with density $f_2(z)/\int_{\mathbb{R}} f_2(x)\,dx$, and set $X = Z$. The output X has density $f_X(x)$.

[R. Kronmal & A. Peterson, Jr, An acceptance-complement analogue of the mixture-plus-acceptance-rejection method for generating random variables, *ACM Trans. Math. Software*, 10(3), 271–281, 1984]

acceptance-rejection (alias: **rejection method; Von Neumann's method**) A method for sampling from a density $f_X(x)$, as follows. Choose a density $f_Y(y)$ such that for some constant a, and all $x, 0 \leq f_X(x) \leq af_Y(x)$.

1. Generate independent U and Y, where Y has density $f_Y(y)$, and U is uniform on $(0, 1)$.
2. If $aUf_Y(Y) \leq f_X(Y)$, then set $X = Y$. Otherwise reject Y and go to step 1.

[R. Rubinstein & D. Kroese, *Simulation and the Monte Carlo Method*, 2nd edn. Wiley, 2008]

acceptance sampling A class of quality control methods that decide whether to accept the output of some industrial (or other) process as satisfactory (or reject it as unsatisfactory) on the basis of a random sample from each lot. There are two main types of sentencing rule (i.e. the decision function for acceptance or rejection of the lot):

1. A plan (or rule) that depends on the number of defective (or non-conforming) items is said to be acceptance by attributes.
2. A plan that depends on measurement of some variable property of the items is acceptance by variables.

More complicated procedures are typically sequential, rather than simply lot-based.

[E. Schilling & D. Neubauer, *Acceptance Sampling in Quality Control*, 2nd edn. Chapman & Hall, 2009]

accessibility A concept arising in the context of Markov chains. Let $X_n \in S$ be a chain with discrete state space S, with $i, j \in S$. Then j is accessible from i if, for some finite n, $\mathbb{P}(X_n = j \,|\, X_0 = i) > 0$. Equivalently, if T_{ij} is the hitting time of j from i, then j is accessible from i if $\mathbb{P}(T_{ij} < \infty) > 0$.

We write $i \to j$ for this property, and synonyms are: i **leads to** (or communicates with) j; j can be reached (or is **reachable**) from i; j is a **consequent** of i.

If $i \to j$ and $j \to i$, then i and j are said to **communicate** (or inter-communicate); and we write $i \leftrightarrow j$ for this property, which is an equivalence relation. If the state space of a chain comprises exactly one such equivalence class, then it is called **irreducible**.

For a Markov chain in continuous time with standard transition probabilities $p_{ij}(t)$, the **Lévy dichotomy** asserts that either $p_{ij}(t) > 0$ for all $t > 0$, or $p_{ij}(t) = 0$ for all t. If $p_{ij}(t)$ have a stable Q-matrix Q, and $\hat{p}_{ij}(t)$ are the transition probabilities for the associated Feller-minimal process $\hat{X}(t)$, then $i \to j$ for $\hat{X}(t)$ iff $i \to j$ for the jump chain defined by Q. For diffusions complications arise, but it is often convenient to say that a state (or set of states) is accessible from x_0 if its occupation measure is greater than zero when started from x_0.

The term also arises in the general theory of processes, where a stopping time T is called accessible if there is a sequence $T_1, T_2, \ldots$ of predictable times such that $\mathbb{P}(\cup_{r=1}^{\infty}\{T_r = T < \infty\}) = \mathbb{P}(T < \infty)$, in which case the sequence is said to catch (or envelop) T. [O. Kallenberg, *Foundations of Modern Probability*, 2nd edn. Springer, 2002]

access time Arises in the context of Markov chains as a generalization of hitting times. Let a finite irreducible Markov chain be started with distribution p, and consider all rules that stop the chain at a time when it has distribution q. The minimum over all such rules of the expected stopping time of the chain is called the **access time** from p to q, denoted by $A(p, q)$. If p and q are degenerate on states i and j respectively, then the access time is just the hitting time of j from i.

If π is the stationary distribution of the chain, and $\delta(i)$ is the degenerate distribution on the state i, then $\max_i A(\delta(i), \pi)$ is called the **mixing time**. Also the quantity $\sum_i \pi_i A(\delta(i), \pi)$ is called the **reset time**. Finally, the **forget time** is $\min_\tau \max_i A(\delta(i), \tau)$, and the distribution τ that achieves this minimum is the forget distribution. [L. Lovász & P. Winkler, Reversal of Markov chains and the forget time, *Combin. Probab. Comp.*, 7, 189–204, 1998]

activated random walk An interacting particle system on the vertices of a suitable graph. Initially a number of particles are distributed (possibly randomly) on the vertices of the graph; they are all active, which is to say that each performs a symmetric r.w. on the vertices independently of the others. Each particle becomes inactive (falls asleep) at rate λ, and once inactive (sleeping) it stops walking until another (necessarily active) particle is present at the same vertex, whereupon it resumes the random walk. In particular, no particle can fall asleep at a vertex with two or more particles present. If there is a time T after which a vertex V is never again visited by an active particle, then V is said to **fixate** at fixating time T. *See also* frog model. [R. Dickman *et al.*, Activated random walkers: Facts, conjectures and challenges, *J. Stat. Phys.*, 138, 126–142, 2010]

actuarial table *see* life table

adapted process (alias: **non-anticipating process**) It is natural to assume that for a random process evolving in time, its distribution at time t (given its past) should not depend on its future values. The adapted process makes this idea formal, thus: given a filtration $(\mathcal{F}_t)$, a random process $X(t)$ is adapted to $(\mathcal{F}_t)$ if, for all t, $X(t)$ is $\mathcal{F}_t$-measurable. Any process is adapted to its natural or induced filtration $(\mathcal{F}_t) = \sigma\{X(s) : s \leq t\}$.

In continuous time, a right-(or left-)continuous adapted process $X(t)$ has the stronger property of being progressively measurable. In discrete time an adapted process is both progressively measurable and an optional process. [D. Applebaum, *Lévy Processes and Stochastic Calculus*, 2nd edn. Cambridge University Press, 2009]

adaptive rejection A method for sampling from a density $f_X(x)$ that is a computationally efficient modification of the acceptance-rejection method. Whenever a rejection occurs, the bounding function $af_Y(x)$ is modified to reduce the number of future rejections. [W. Gilks *et al.*, Adaptive rejection Metropolis sampling, *Appl. Stat.*, 44, 455–472, 1995]

adaptive stochastic control Broadly, that part of control theory that seeks to optimize some function of the realizations of a random process with uncertain parameters, in which the control policy is adapted in the light of updated estimates of the state variables and the unknown parameters.

addition rule (alias: **additive property**) The axiom of classical probability theories which asserts that the probability of the union of disjoint events is the sum of their probabilities. Formally, if $A \cap B = \emptyset$, then $\mathbb{P}(A \cup B) = \mathbb{P}(A) + \mathbb{P}(B)$.

This rule is natural when probability is interpreted as frequency, or proportion, or an expression of symmetry, but its necessity has been questioned for interpretations such as degree of belief, or degree of support.

The special form of the rule, written for arbitrary A and B as $\mathbb{P}(B) = \mathbb{P}(A \cap B) + \mathbb{P}(A^c \cap B)$, is often called the **partition rule** or the **law of total** (or **complete**) **probability**. More strongly, Kolmogorov's additivity axiom requires that for a countable pairwise disjoint collection of events $A_i, \mathbb{P}(\bigcup A_i) = \sum \mathbb{P}(A_i)$, and $\mathbb{P}(\cdot)$ is said to be **countably additive** or **sigma-additive**. *See* inclusion-exclusion bounds. [H. Tijms, *Understanding Probability*, 3rd edn. Cambridge University Press, 2012]

additive functional Arises in the context of a strong Markov process $X(t) = X(t, \omega)$. A continuous additive functional of (or associated with) $X(t, \omega)$ is a random process $A(t, \omega)$ such that $A(0, \omega) = 0$, and for $s, t \geq 0$, $A(t+s, \omega) = A(t, \omega) + A(s, \omega_t^+) = A(t, \omega) + A(s, \omega) \circ \theta_t$; where ω_t^+ is the sample path shifted by t (i.e. seen after t) and θ_t is the shift operator. Furthermore, $A(t, \omega)$ is required to be continuous, non-decreasing, and adapted to the filtration generated by $X(t, \omega)$. A classic example is $A(t) = \int_0^t f(X(u))\mathrm{d}u$, where $f(\cdot)$ is a bounded non-negative function; another example is the local time of a diffusion at x. Some writers do not require the functional to be non-decreasing. A **multiplicative functional** is defined analogously by $M(t+s) = M(s)\,M(t) \circ \theta_s$. [S. Karlin & H. Taylor, *A Second Course in Stochastic Processes*. Academic Press, 1981]

additive process (alias: **independent (or orthogonal) increments process**) If, for any n, and $t_1 < t_2 < \cdots < t_n$, the increments $X(t_2) - X(t_1), \ldots, X(t_n) - X(t_{n-1})$ of $X(t)$ are independent, then $X(t)$ is said to be an additive (or strongly additive) process. If the increments are only uncorrelated, then the process is said to be weakly additive, or to have orthogonal increments. The Wiener, Lévy, and Poisson processes are classic examples of additive processes. [K. Itô, *Essentials of Stochastic Processes*, English translation. American Mathematical Society, 2006]

additive property *see* addition rule

additive set functional *see* valuation

Adelson recursion A formula for the computation of the terms of a compound Poisson distribution, recursively. Thus, if $X = \sum_{r=1}^{N} Y_r$ where N is Poisson (λ), and the Y_r are i.i.d.r.v. with $\mathbb{P}(Y_r = k) = p_k, k \geq 0$, then $\mathbb{P}(X = 0) = \exp\{-\lambda(1 - p_0)\}$, and $x\mathbb{P}(X = x) = \lambda \sum_{k=0}^{x-1} (x-k) p_{x-k}\, \mathbb{P}(X = k), \quad x \geq 1$. The case $p_1 = 1$ is the classic Poisson recursion., and a generalization is called the **Panjer** (or **Adelson–Panjer**) **recursion**. [R. Adelson, Compound Poisson distributions, *Oper. Res. Q.*, 17(1), 73–75, 1966]

adiabatic effect Arises in the context of Monte Carlo simulations; it refers to the possibility that an over-hasty annealing schedule may leave the system (process) in a locally extreme (metastable) state, rather than the desired global extreme (ground state). [B. Gidas, Metropolis-Type Monte Carlo simulation algorithms and simulated annealing, in *Topics in Contemporary Probability and its Applications*, L. Snell (ed.). CRC Press, 1995]

adjoint source In communication theory, that memoryless discrete source which has the same distribution over its symbols as a given ergodic Markov source in equilibrium. [D. Welsh, *Codes and Cryptography*. Oxford University Press, 1988]

Adleman's theorem Addresses randomized algorithms in the context of computational complexity. It asserts that the set of decision problems solvable by a probabilistic Turing machine in polynomial time (with error probability less than $p \in (0, 1/2)$ is included in the set of problems decidable by a Turing machine whose circuits and running time are polynomially bounded as functions of the length of the input, denoted by P/poly. [R. Motwani & P. Raghavan, *Randomized Algorithms*. Cambridge University Press, 1995]

admission control Arises in queueing theory, or related systems, when the arrival process of customers (or items) depends on the number, and possibly types, of those in service. The control may be random, as e.g. in balking and reneging, or it may be imposed with the aim of optimizing some measure of performance. [S. Stidham & R. Weber, A survey of Markov decision models for control of networks of queues, *Queueing Sys.*, 13(1–3), 291–314, 1993]

adversarial input (alias: **worst-case**) A concept arising in game theory that has applications in areas as disparate as cryptography and sequential job scheduling in OR. Broadly, it is the assumption that the random inputs to the system are in some sense worst-case; that is, as if chosen by an adversary.

affine Markov process A term arising in two contexts:

1. Given a Markov chain X_n, if there exist suitable functions f and g such that, for all $n \geq 0, X_{n+1} = g(X_n) + f(X_n)Z_{n+1}$, where the Z_n are i.i.d.r.vs independent of X_0, then the chain is said to be affine.
2. If the transition semi-group of a Markov process in continuous time is such that the logarithm of its characteristic function is an affine transformation of the initial state, then the process is called affine.

[T. Hurd & A. Kuznetsov, Affine Markov chain model of multifirm credit migration, *J. Credit Risk*, 3(1), 3–29, 2007; D. Dawson & Z. Li, Skew convolution semigroups and affine Markov processes, *Ann. Probab.*, 34(3), 1103–1142, 2006]

affine process *see* harness

age (alias: **current life; spent waiting time; backward recurrence time**) In a renewal process at time t, the time elapsed since the most recent prior event is called the age of the process. If there is no prior event the age is taken to be t, for a process started at 0. If the interevent time is not arithmetic, with distn fn $F(x)$, then the limiting distn of the age as $t \to \infty$ is $A(x) = (\mathbb{E}X)^{-1}\int_0^x (1 - F(u))\mathrm{d}u$, provided $\mathbb{E}X = \int_0^\infty (1 - F(u))\mathrm{d}u$ is finite. [G. R. Grimmett & D. R. Stirzaker, *Probability and Random Processes*, 3rd edn. Oxford University Press, 2001]

age-dependent process (alias: **Bellman–Harris process**) Generalizes the ordinary Galton–Watson process by requiring each individual to complete a random lifetime before branching. In the simplest model, lifetimes are independent of each other and the family sizes. More generally, a family-size distribution may depend on the age of the individual giving rise to it. [P. Haccou *et al.*, *Branching Processes*. Cambridge University Press, 2005]

ageing Arises in the analysis of reliability and maintenance of systems (in the most general sense), as well as in demographic and actuarial theories. In the simplest classical model, a system comprises n components (usually satisfying some dependency relations), each of which has a random lifetime. For a component aged t, whose lifetime distribution is $F(x)$, for a lifetime T with density $f(x)$, we define $r(t) = f(t)/\{1 - F(t)\}$, which is called either the **failure rate fn** or the **hazard rate fn** (or the **force of mortality** in actuarial contexts).

There is a considerable taxonomy based on the properties of $r(t)$, and suitable functions of $r(t)$. The **cumulative hazard fn** is $H(t) = \int_0^t r(u)\mathrm{d}u = -\log(1 - F(t))$(also, and confusingly, sometimes called the hazard function), and for a random lifetime T, the residual lifetime $R(t)$ at t is the random variable $T - t$, conditional on the event $T > t$. Then these classifications are commonplace:

(A) If $H(t)$ is convex (resp. concave) in t, then T is said to have increasing (resp. decreasing) failure rate, denoted by IFR (resp. DFR).
(B) If $t^{-1}H(t)$ is non-decreasing (resp. non-increasing) then T is said to have increasing (resp. decreasing) failure rate on average, denoted by IFRA (resp. DFRA).
(C) If for all $u, v \geq 0, H(u+v) \geq H(u) + H(v)$(resp. $\leq$), then T is said to be new better (resp. worse) than used, denoted by NBU (resp. NWU).
(D) If $\mathbb{E}T \geq \mathbb{E}R(t)$(resp $\leq$), then T is said to be new better (resp. worse) than used in expectation, denoted by NBUE (resp. NWUE).
(E) If $\mathbb{E}R(t)$ is non-increasing (resp. non-decreasing) in t, then T has decreasing (resp. increasing) mean residual life, denoted by DMRL (resp. IMRL).

It can be shown that IFR $\Rightarrow$ DMRL $\Rightarrow$ NBUE and that IFR $\Rightarrow$ IFRA $\Rightarrow$ NBU $\Rightarrow$ NBUE.

The equivalent expression, $1 - F(t) = \mathrm{e}^{-H(t)} = \exp\{-\int_0^t r(u)\mathrm{d}u\}$, may be called the **exponential formula for reliability** in appropriate contexts. More generally, when F has no density, the hazard rate over $(t, t+a], a > 0$, is

$$r(t,a) = \frac{F(t+a) - F(t)}{1 - F(t)} = \mathbb{P}(t < X \leq t + a \,|\, X > t).$$

More narrowly, if X is integer-valued, then $r(t) = \mathbb{P}(X = t \,|\, X \geq t)$. [M. Finkelstein, *Failure Rate Modelling for Reliability and Risk*. Springer, 2008]

age replacement model (or policy) Any one of a large number of procedures for maintaining the reliability of a system by replacing components according to rules based on their current lifetime, or age. [T. Nakagawa, *Maintenance Theory of Reliability*. Springer, 2005]

age-structured process *see* age-dependent process

aggregation *see* lumpability

aggregation-disaggregation A method sometimes used in the analysis of Markov chains arising in the context of very large networks (of queues, or processors). If the state space can be partitioned into subsets within which transitions are very likely, but transitions into other subsets are very unlikely, then one may proceed by first considering inter-set transitions, and then intra-set transitions, to yield good approximate results. [P. Schweitzer, A survey of aggregation-disaggregation in large Markov chains, in *Numerical Solution of Markov Chains*, W. Stewart (ed.). Marcel Dekker, 1991]

aggregation paradox Refers to the fact that it is possible that a number of distinct and disparate (but commensurate) data-sets may all point towards some common conclusion, but when combined (i.e. pooled) the aggregated data-sets suggest the opposite conclusion. *See* Yule–Simpson paradox. [D. Saari, A chaotic exploration of aggregation paradoxes, *SIAM Rev.*, 37(1), 37–52, 1995]

Agrawal–Biswas test A randomized method for testing the primality of a given integer n. It relies on the fact that for any $c > 1$, such that $\gcd(n, c) = 1$, n is prime iff $(x - c)^n = x^n - c$, modulo n. Broadly, the method chooses a random monic polynomial $R(x)$ of degree $\lceil \log n \rceil$ = smallest integer not less than $\log n$. If $(x + 1)^n \neq x^n + 1$, modulo $(R(x), n)$, then n is declared composite; otherwise n may be prime. [L. Rempe-Gillen & R. Waldecker, *Primality Testing for Beginners*. American Mathematical Society, 2014]

agreeing function Arises in a number of contexts, e.g. decision theory and theories of probability, where a family T of objects are subject to some ordering induced by, e.g., preferences or beliefs. The real-valued function $a(\cdot)$ is said to agree with this ordering if, for all $x, y \in T, a(x) \leq a(y)$ iff y is preferred to x (or belief in y is greater than belief in x). *See* utility functions. [J. Halpern, A counterexample to theorems of Cox and Fine, *J. Artif. Intel. Res.*, 10, 67–85, 1999]

Ahlswede–Daykin inequality (alias: **four functions theorem**) This has applications in random graphs, and other areas of probabilistic combinatorics. One form is this: Let π be the set of all subsets of $\{1, 2, \ldots, n\}$ (i.e. its power set), and let $a(\cdot), b(\cdot), c(\cdot)$, and $d(\cdot)$ be four non-negative functions on π. For a family ϕ of subsets of π, the sum $\sum_{S \in \phi} a(S)$ is denoted by $A(\phi)$, and likewise for b, c, and d.

If, for any two subsets S and T of π, we have $a(S)b(T) \leq c(S \cup T)\, d(S \cap T)$, then for any two families ϕ and θ of subsets of π, $A(\phi)B(\theta) \leq C(\phi \cup \theta)D(\phi \cap \theta)$, where $\phi \cup \theta = \{A \cup B : A \in \phi, B \in \theta\}$, and similarly for $\phi \cap \theta$.

It is an example of a correlation inequality, such as those of Harris, FKG, Holley and Fishburn–Shepp, all of which it implies. [R. Ahlswede & D. Daykin, An inequality for the weights of two families of sets, their unions and intersections, *Probab. Theor. Rel. Fields*, 43(3), 183–185, 1978]

Airy distribution Usually denotes one of two distributions:

1. The **map-Airy** distribution, so called because it arises in the study of random maps (i.e. planar graphs), which has density on $\mathbb{R}$ given by $f(x) = 2[xA(x^2) - A'(x^2)]\exp\left[-\frac{2}{3}x^3\right]$, where $A(x) = \pi^{-1} \int_0^\infty \cos\left(xt + \frac{1}{3}t^3\right)dt$, is the Airy function. It has zero mean, but no higher moments.
2. The **area-Airy** distribution, so called because it is the distribution of the area (suitably scaled) of the positive Brownian excursion, which may be defined on $\mathbb{R}^+$ by its moments m_r given by $m_r = A_r 2\sqrt{\pi}/\Gamma\left(\frac{1}{2}(3r - 1)\right),\quad r \geq 1$, where A_r are the Airy constants $\{A_1 = 1/2, A_2 = 5/4, A_3 = 45/4, \ldots\}$. It arises as the limiting distribution of a number of random structures.

[C. Banderier & G Louchard, *Philippe Flajolet and The Airy Function*, online, www.stat.purdue.edu/~mdw/ChapterIntroductions/banderier-louchard.pdf

Aldous–Broder method A procedure for generating a uniform spanning tree on a finite connected graph G (i.e. one selected uniformly at random from all spanning trees). From any vertex v run a symmetric random walk on G until all vertices have been visited. For any vertex $u \neq v$, let $e(u)$ be the edge used to make the first visit to u. The collection of such edges is the required random tree. [S. Evans *et al.*, Rayleigh processes, real trees, and root growth with regrafting, *Probab. Theor. Rel. Fields*, 134(1), 81–126, 2006]

Aldous's condition Arises in considering the weak convergence of a sequence of random processes, where it supplies a sufficient condition for tightness in terms of behaviour after stopping times; a later weaker condition is similarly denoted. [D. Aldous, Stopping times and tightness

I & II, *Ann. Probab.*, 6(2), 335–340, 1978; 17(2), 586–595, 1989]

Aldous's integrated superBrownian excursion (or ISE) Introduced by D. Aldous as a description for random distribution of masses, this is a random probability measure arising as a continuous rescaled limit of random trees. [D. Aldous, Tree-based models for random distribution of mass, *J. Stat. Phys.*, 73(3–4), 625–641, 1993]

aleatory Random. From the Latin *aleator*, meaning one who throws a die; the Latin word for die being *alea* (which was also the name of a popular Roman board game).

The term 'aleatory probability' may be applied in the context of uncertain physical events lying in the future, such as the decay of subatomic particles or radioactive atoms; it is to be contrasted with epistemic probability, which arises simply from a lack of knowledge (see probability). [I. Hacking, *The Emergence of Probability*. Cambridge University Press, 1975]

Alexandrov's theorem *see* portmanteau theorem

algebraic probability *see* quantum probability

algorithmic complexity *see* Kolmogorov complexity

algorithmic probability (alias: **Solomonoff probability; universal probability**) Commonly denotes a probability distribution assigned to binary strings (finite or infinite) on an a priori basis, i.e. axiomatically not empirically (as the uniform distribution is similarly assigned by the principle of insufficient reason). The concept was introduced by R. Solomonoff for the purposes of inductive inference, and is closely related to Kolmogorov complexity. In the basic case of a finite binary string b the so-called universal (algorithmic) probability $\mathbb{P}(b)$ of b is defined in terms of a universal (Turing) computer with random (prefix-free) input p having length $l(p)$ yielding output $T(p)$.

Then $\mathbb{P}(b) = \sum_{p:T(p)=b} 2^{-l(p)}$, where $\sum_p 2^{-l(p)} = \Omega_c < 1$ is Chaitin's constant. Note that the term algorithmic probability also has its standard interpretation. [M. Hutter, *Universal Artificial Intelligence*. Springer, 2005; M. Neuts, *Algorithmic Probability*. Chapman & Hall, 1995]

algorithmic randomness Arises in considering the extent to which a given infinite sequence (typically binary) can be seen as random. Broadly speaking, the issue is decided by putting the sequence through a number of suitable tests (essentially statistical and probabilistic); then sequences that pass the test may be called algorithmically random.

The field of such tests is large and varied, but e.g. a random sequence should not be readily compressible (in the sense of having a short encoding); it should be essentially unpredictable (so that no gambler can expect to win by betting on its unrevealed future); and it should pass all statistical tests for randomness (such as obeying the appropriate laws of large numbers). See Martin-Löf randomness. [R. Downey & D. Hirschfeldt, *Algorithmic Randomness and Complexity*. Springer, 2010]

aliasing (alias: **alias method; Walker's method**) A technique for sampling from the distribution $p = (p_1, \ldots, p_n)$ of a simple random variable X. It relies on the fact that X can be represented as a composite (or compound) random variable in the form $X = Y_U$, where U is uniform on $[1, \ldots, n-1]$, and $Y_k \in \{x_k, x_{a(k)}\}$, for a suitable function $a(\cdot) : \{1, \ldots, n-1\} \to [1, \ldots, n]$, called the alias function. Thus sampling from p is equivalent to sampling from a randomly chosen Bernoulli distribution. Put simply, one can always set

$$p = \frac{1}{n-1}\sum_{r=1}^{n-1} V_r,$$

where each V_r is a vector of length n with at most two non-zero entries, one of which can be forced to lie in the rth place of V_r by renumbering. [W. Hörmann *et al.*, *Automatic Nonuniform Random Variate Generation*. Springer, 2004]

Ali–Silvey divergence *see* f-divergence

Allais paradox (alias: **Allais phenomenon; Allais problem**) Arises in decision theory and utility theory. The Von Neumann–Morgenstern model for choice under uncertainty assumes that choices are independent of irrelevant alternatives. M. Allais constructed uncertain choices such that the irrelevant alternatives affected the choices of real decision makers. Behavioural decision theory accounts for this in a number of ways, including the certainty effect, or by postulating regret as a factor in decision-making. The paradox can be seen either as a failure of people to be rational, or as a failure of classical utility to account for people's behaviour. [I. Hacking, *Introduction to Probability and Inductive Logic*. Cambridge University Press, 2001]

Allison mixture Refers to the fact that two independent white noise processes can be mixed to yield a process that is not white noise. Formally, let $X_n^{(1)}$ and $X_n^{(2)}$ be independent stationary processes with means λ and $\mu, \lambda \neq \mu$, whose autocorrelation functions are both given by $\rho(X_n^{(i)}, X_{n+m}^{(i)}) = \delta(m)$, where $\delta(m)$ is the Kronecker delta. Let Z_n be a two-state Markov chain with transition matrix $\begin{pmatrix} p & 1-p \\ 1-q & q \end{pmatrix}$, with $p+q \neq 1$, and $pq \neq 0$, in equilibrium. Then the process $W_n = X_n^{(Z_n)}$ is stationary, but is not white noise, in that $\rho(W_n, W_{n+m}) \neq \delta(m)$. [D. Abbott, Developments in Parrondo's paradox, in *Applications of Nonlinear Dynamics*, V. Longhini *et al.* (eds). Springer, 2009]

allocation problem (alias: **scheduling problem; sequencing problem**) A large class of problems in operations research that require the allocation of some resource (or vector of resources), which may be random, to satisfy various (possibly random) constraints, according to some criterion of optimality. For example: bin-packing; portfolio allocation, knapsack problems; bandwidth allocation; network routing; buffering; makespan scheduling; bandits; job-shop problems; repairman scheduling; etc. [S. Sarin *et al.*, *Stochastic Scheduling*, Cambridge University Press, 2010]

almost surely (alias: **with probability 1; almost everywhere; almost certainly; a.s.; a.e.**) An event A such that $\mathbb{P}(A) = 1$ is said to occur almost surely; by extension the same is said of any probability statement that holds on an event of probability 1. Thus e.g. for r.vs defined on the same space, to say that '$X = Y$, a.s.' is to say that $\mathbb{P}(\{\omega : X(\omega) = Y(\omega)\}) = 1$. Note that an almost sure event A need not be the certain event Ω; the complement of an almost sure event A is a **null event**. The term 'almost everywhere' is commonly used in the more general context of measure theory, where it is said of a property (or assertion) that holds everywhere except on a set of measure zero. [J. Stoyanov, *Counterexamples in Probability*, 3rd edn. Dover, 2013]

alpha-mixing (alias: **strong mixing**) A concept that formalizes the idea that a sequence $X_n, n \in \mathbb{Z}$, of random variables may not be independent, but widely separated members of the sequence may be negligibly dependent on each other. Formally, let A be any event that depends only on $\{X_r : r \leq k\}$, and B any event that depends only on $\{X_s : s \geq n+k\}$, and define

$$\alpha(n) = \sup_{k \geq 1} |\mathbb{P}(A \cap B) - \mathbb{P}(A)\mathbb{P}(B)|.$$

If, for any A and B, $\alpha(n) \to 0$ as $n \to \infty$, then the sequence X_n is said to be α-mixing. It is the weakest of a number of so-called strong-mixing conditions, of which independence is the strongest. [R. Bradley, Basic properties of strong mixing conditions, *Probab. Surv.*, 2, 107–144, 2005]

alternating renewal A random process that arises as a model for a system that works (or is on, or is up) for a random time to failure, and is then off (or down) for a random time until repair; this cycle is independently repeated ad infinitum. Formally, let $X_0, X_1, \ldots$ be a sequence of independent non-negative r.vs such that $X_{2r-1}, r \geq 1$ have distn $F(x)$, and $X_{2r}, r \geq 1$, have distn $G(x)$. This, together with the counting function $N(t) = \max\{n : \sum_{r=0}^{n} X_r \leq t\}$, comprises an alternating renewal process. It is ordinary if X_0 has the distn $G(x)$; otherwise it is delayed. [D. R. Cox, *Renewal Theory*. Methuen, 1962]

amalgamation paradox *see* Yule–Simpson paradox

ambit process Arises in modelling the evolution of suitable spatial processes in time. Broadly, for $t \in \mathbb{R}$ and $\mathbf{r} \in \mathbb{R}^d$, one defines a so-called ambit field as an integral $Y(t, \mathbf{r}) = \int k(s, t; \mathbf{x}, \mathbf{r}) M(\mathrm{d}s, \mathrm{d}\mathbf{x})$, where M is an appropriate random measure and k some (possibly random) kernel. Then the values of the field along a path in $\mathbb{R}^{d+1}$ constitute an ambit process. [O. Barndorff-Nielsen *et al.*, Ambit processes and SDES, in *Advanced Mathematical Methods for Finance*, G. Di Nunno & B. Øksendal (eds). Springer, 2011]

American option Arises in financial mathematics, where it refers to contracts in which the holder of an option can exercise it at any time up to the expiry (or maturity) date T. [M. Musiela & M. Rutkowski, *Martingale Methods in Financial Modelling*, 2nd edn. Springer, 2005]

Ammeter process Arises in insurance mathematics, where it denotes a Cox (i.e. doubly stochastic Poisson) process such that the random intensity $\lambda(t)$ is given for $t \geq 0$ by $\lambda(t) = A_n, n-1 \leq t < n, n \geq 1$, where $A_n, n \geq 1$, is a sequence of i.i.d.r.vs. [J. Grandell, *Mixed Poisson Processes*. Chapman & Hall, 1997]

Andersen *see* Sparre Andersen (model, theorem, etc.)

Anderson's theorem (or inequality) In its simplest form this asserts that if X has an origin-symmetric unimodal density, and Y is independent of X, then $\mathbb{P}(|X| \leq a) \geq \mathbb{P}(|X+Y| \leq a)$. More generally, if the r.vr $\mathbf{X} \in \mathbb{R}^d$ has an origin-symmetric unimodal density, and A is an origin-symmetric convex set in $\mathbb{R}^d$, then $\mathbb{P}(\mathbf{X} \in A) \geq \mathbb{P}(\mathbf{X}+\mathbf{Y} \in A)$ for any $\mathbf{Y} \in \mathbb{R}^d$ that is independent of $\mathbf{X}$. [T. Anderson, The integral of a symmetric convex set and some probability inequalities, *Proc. Amer. Math. Soc.*, 6, 170–176, 1955]

annealing method (alias: **simulated annealing**) Arises in optimization. It is a probabilistic procedure for finding the global extremes of a function $f(x), x \in S$, that may have many widely dispersed local extremes in the state space S, which is commonly a lattice or a graph. In such cases, elementary optimization methods such as exhaustion, or gradient (hill-climbing) methods (possibly stochastic) are often useless. The method is a type of Markov chain Monte Carlo technique, and one version is implemented as follows.

Assume we seek a global minimum; simulate a Markov chain $X_n, n \geq 0$, on S, whose stationary distribution is $\pi(x) \propto \exp\{f(x)T^{-1}\}$ (being a Gibbs distn), where the arbitrary parameter T is called the temperature. As $T \downarrow 0$, this distribution is supported by the set M of global minima of $f(x)$. Therefore $T = T(n)$ is set as a non-increasing function of time, called the annealing schedule, and when T is small enough the chain is very likely to be in M. The nomenclature arises from the analogy with the creation of physical objects with low internal energy (and hence stress) by first heating and then cooling according to a good schedule. A related (mixture) method, in which a chain is run in equilibrium on a state space augmented by values of T, is called **simulated tempering**. *See* Gibbs sampling. [B. Gidas, Metropolis-Type Monte Carlo simulation algorithms and simulated annealing, in *Topics in Contemporary Probability and its Applications*, L. Snell (ed.). CRC Press, 1995]

annihilating particle system A type of interacting particle system in which particles may disappear on encountering another particle. An example is the annihilating random walk on $\mathbb{Z}^d$, in which all particles present at $t = 0$ perform independent simple random walks on the lattice, and any particles that meet at a vertex (or use the same edge), are instantly removed. Interest centres, e.g., on the asymptotic behaviour of the probability $p(t)$ that the origin (or other given site) is occupied at time t. More generally, particles may be of two (or more) types, with only different types (say) annihilated on meeting. Related processes include those where particles destroy themselves (death process), kill neighbours (coalescence), or reproduce (branching annihilation). [T. Liggett, *Interacting Particle Systems*. Springer, 2005]

announceable time *see* predictability

annuity Arises in actuarial maths. A periodic (classically, annual) payment, which may vary, and may run for a fixed term, or for the annuitant's lifetime, or forever (in which last case it may be called a perpetuity). Formally, the present value of an annuity (equal to the fair purchase price) is $V = \sum_{j=1}^{T} A_j \prod_{k=1}^{j-1} D_k$, where T is the duration of the annuity (lifetime), A_j are the amounts paid, and D_j are the discount factors subsuming the effects of inflation and interest rates. [D. Dickson *et al.*, *Actuarial Mathematics for Life Contingent Risks*. Cambridge University Press, 2009]

anomalous diffusion (alias: **enhanced diffusion, sub- and super diffusion**) Denotes a class of real-world processes that are essentially diffusive in nature, but whose detailed properties are different from those of the classic Brown–Einstein–Fick motion (BEF). If $X(t)$ is a BEF process, then its mean-square displacement $\mathbb{E}\big(X(t)^2\big)$ is proportional to the elapsed time t; in an anomalous diffusion $A(t), \mathbb{E}\big(A(t)^2\big) \propto t^a$, where $a < 1$, $a > 1$, and $a = 2$ correspond respectively to **subdiffusion**, **superdiffusion**, and **ballistic diffusion**. Examples include tracers in turbulence, transient currents in photocopiers, and diffusive percolation. A classic anomalous diffusion model is the Montroll–Weiss process, in which a particle makes i.i.d. random jumps in $\mathbb{R}^d$ at the instant of a renewal process. Another class of models comprises the fractional diffusion processes. These may be variously derived, but typically their probability distributions satisfy fractional diffusion equations, i.e. of the form $\partial^a f / \partial t^a = \Delta^\beta f$, where $\partial^a / \partial t^a$ is a time-fractional derivative, and Δ^β is a space-fractional differential operator. [D. Ben-Avraham & S. Havlin, *Diffusion and Reactions in Fractals and Disordered Systems*. Cambridge University Press, 2000]

Anscombe's condition For a sequence of r.vs $X_n, n \geq 0$, this condition amounts to requiring uniform continuity in probability. Formally, for any $\varepsilon > 0$ and $\delta > 0$, there exist $c > 0$ and $m < \infty$, such that for all $n > m$

$$\mathbb{P}\left(\max_{(1-c)n\leq k\leq(1+c)n}|X_k - X_n| > \varepsilon\right) < \delta.$$

[P. Berti *et al.*, An Anscombe-type theorem, *J. Math. Sci.*, 196(1), 15–22, 2014]

Anscombe's quartet Comprises four artificial sets of bivariate data (X, Y), having the same marginal means and variances, the same covariance, and the same regression of Y upon X. However, when plotted they exhibit entirely different apparent functional relationships between their respective X and Y. [F. Anscombe, Graphs in statistical analysis, *Amer. Stat.*, 27(1), 17–21, 1973]

Anscombe's theorem This is a limit theorem for randomly indexed sequences of random variables. Formally, let $X_n, n \geq 0$ satisfy Anscombe's condition and converge in distribution as $n \to \infty$; we write $X_n \to^D X$. If $M(n), n \geq 0$, is an integer-valued random process, and $a(n), n \geq 0$ a non-random sequence of positive numbers increasing to ∞ as $n \to \infty$, where $M(n)/a(n)$ converges in probability to 1, then $X_{M(n)}$ converges in distribution (with the same limit as X_n), as $n \to \infty$. Note that it is not assumed that $M(n)$ is independent of X_n. [A. Gut, Anscombe's theorem 60 years later, *Seq. Anal.*, 31(3), 368–396, 2012]

anticipating integral A type of stochastic integral in which the integral is not necessarily adapted to the σ-field generated by the integrator. That is to say, it fails to be predictable, i.e. non-anticipating. Examples include the **Skorokhod integral**, when the integrator is the Wiener process; and the **Ma–Protter–San Martin** integral, when the integrator is a normal martingale. [C. Tudor, Martingale-type stochastic calculus for anticipating integral processes, *Bernoulli*, 10(2), 313–325, 2004]

antithetic variables Arise in Monte Carlo methods, where they were introduced by J. Hammersley and K. Morton to improve efficiency by reducing the variance of outcomes. For example, if r.vs X and Y, with the same distn F, are simulated to estimate their common mean, the variance of the elementary estimator $\frac{1}{2}(X+Y)$ is $\frac{1}{2}\operatorname{var} X + \operatorname{cov}(X, Y)$. Independent simulations thus yield a larger variance than simulations such that $\operatorname{cov}(X+Y) < 0$ (i.e. antithetic, from the Greek word meaning opposite). In this example, one may simulate a uniform r.v. U on $(0,1)$, and set $X = F^{-1}(U)$ and $Y = F^{-1}(1-U)$; it was shown by P. Moran that this minimizes the correlation between X and Y having the same distn. [S. Ross, *Simulation*, 5th edn. Academic Press, 2013]

a priori probability A probability determined independently of experience or empirical evidence (as it relates to the problem at hand). Such probabilities may be fixed subjectively, or by logical argument, or by applying some principle, such as insufficient reason or maximum entropy. Such a probability may be used as a prior in Bayes methods; but note that not all priors are determined a priori. [H. Jeffreys, *Theory of Probability*, 3rd edn. Oxford University Press, 1961]

arbitrage Arising in financial mathematics, this denotes any collection of financial transactions, or betting scheme, that yields a gain with probability 1. (A weak arbitrage yields no loss with probability 1, and a positive probability of positive gain.) A set of offered bets, or offered prices for assets and derivatives, is arbitrage-free (or a no-arbitrage market), if no such collection (or portfolio) exists. Standard market models for pricing derivatives and options require that the result must be arbitrage-free; in the simpler models this is sufficient to determine a unique fair price. More general models may require other (stronger) principles to constrain uniqueness of prices. A set of betting odds that offers an arbitrage opportunity is called a **Dutch book**. [T. Björk, *Arbitrage Theory in Continuous Time*, 3rd edn. Oxford University Press, 2009]

arbitrage theorem (alias: **no-arbitrage theorem; first fundamental theorem of finance**) Arises in the context of betting, with natural applications to financial mathematics. Informally, suppose that n betting opportunities are available on a game (or experiment) with m possible outcomes, and the amounts to be bet on each opportunity are free to choose. Then the theorem asserts that either there is a probability distribution on the outcomes such that all bets are fair (in the sense of expected return), or there is a betting scheme which wins with probability 1. Formally, let the betting scheme (amounts bet on each of the n opportunities) be $(b_1, \ldots, b_n) \in \mathbb{R}^n$, with the total return, when the outcome of the game is j, denoted by $r(\mathbf{b}, j) = \sum_{k=1}^{n} b_k r(k,j)$, where $r(k,j)$ is the return on b_k for outcome j. Then either there is a probability distn $(p_1, \ldots, p_m)$ such that, for all k, $\sum_{j=1}^{m} p_j r(k,j) = 0$, or there is a betting scheme $(a_1, \ldots, a_n)$ such that, for all j, $\sum_{k=1}^{n} a_k r(k,j) > 0$; yielding an arbitrage. This theorem may be derived from the duality theorem of linear programming. [S. Le Roy & J. Werner, *Principles of Financial Economics*. Cambridge University Press, 2001]

ARCH process An acronym for autoregressive conditionally heteroscedastic process; the word

heteroscedastic indicates that the innovations of the A-R process have time dependent variance (and is little used outside this context). Generalizations of this may be denoted by GARCH process. [C. Gouriéroux, *ARCH Models and Financial Applications*. Springer, 1997]

archery contest A classic example of game theory. Two independent archers of equal skill possess m and n arrows respectively. They approach their targets, in view of each other, from the same distance, at the same speed, while the probability of a hit increases to 1 as they near their targets. The first to hit the target wins, and the problem is to devise an optimal strategy. [D. Blackwell & M. Girshick, *Theory of Games and Statistical Decisions*. Wiley, 1954; reprinted 1979, Dover]

Archimedean copula A type of copula that is useful and important, because of its ease of construction and convenient properties. In two dimensions the Archimedean copula is given by $C(x, y) = g^{(-1)}(g(x)+ g(y))$, where the generator $g(z) \geq 0$, $z \in [0, 1]$, is a strictly decreasing convex continuous function, with $g(1) = 0$. Here, $g^{(-1)}(\cdot)$ is the ordinary inverse if $g(0) = \infty$, or else

$$g^{(-1)}(x) = \begin{cases} 0, & x \geq g(0) \\ g^{-1}(x), & 0 \leq x < g(0) \end{cases}.$$

They are called 'Archimedean' because they satisfy a form of the Archimedean property of real numbers, with multiplication defined by $x * y = C(x, y)$. Examples include the **Gumbel**, where $g(x) = (-\log x)^a, a \geq 1$; and the **Clayton**, where $g(x) = (x^{-a} - 1)/a, a \in [-1, \infty)\backslash\{0\}$. [R. Nelsen, *Introduction to Copulas*, 2nd edn. Springer, 2006]

arcsine distribution A r.v. X has the arcsine distribution on [0, 1] if its density is $f(x) = \{x(1-x)\}^{-\frac{1}{2}}/\pi$, with distn fn $F(x) = (2/\pi)\text{arcsine}\sqrt{x} = \frac{1}{2} + \pi^{-1}\text{arcsine}(2x - 1)$. This arises in the *arcsine laws*.

A r.v. Y has the arcsine distribution on $[-1, 1]$ if its density is $f(y) = (1 - y^2)^{-\frac{1}{2}}/\pi$, with distn fn $F(y) = \frac{1}{2} + \pi^{-1} \sin^{-1} y$.

This arises as the distribution of the amplitude of a sine wave with random phase uniform on $[0, 2\pi]$. For such Y, both Y^2 and $(1 + Y)/2$ have the density of X, above.

A r.v. Z has the **generalized arcsine distn** on [0, 1] if its density is a type of beta density $f_a(z) = \pi^{-1} \sin(\pi a) z^{-a}(1 - z)^{a-1}$, $0 < a < 1$. Such distributions arise in limit theorems for Lévy processes, renewal processes, diffusions, and random walks (see arcsine law). [W. Feller, *Introduction to Probability Theory and its Applications*, Vol. 2, 2nd edn. Wiley, 1971]

arcsine law Arises in the contexts of random walk and the Wiener process. The basic arcsine law stems from the simple symmetric r.w. on the integers, starting from the origin O and observed up to its 2nth step. Consider these random variables: L the location of the last visit to O before the 2nth step; V the number of visits of the walk to positive integers before the 2nth step; and Y the step at which the walk first achieves its maximum before the 2nth step. Then, for large enough n, the distribution fn of all three of L/n, V/n, and Y/n is approximately $\frac{2}{\pi} \sin^{-1}\sqrt{x}$.

Lévy's arcsine laws make the same assertion about the analogous three r.vs defined for the Wiener process observed up to time t. **Spitzer's arcsine law** generalizes the basic case to a general random walk $S_n = \sum_{r=1}^{n} X_r$. With V and Y defined as above, it asserts that V/n and Y/n have the generalized arcsine density $f_a(x)$ in the limit as $n \to \infty$, iff $\lim_{n\to\infty} \frac{1}{n} \sum_{r=1}^{n} \mathbb{P}(S_r > 0) = a$; which is called **Spitzer's condition**. R. Doney has shown that it is equivalent to the condition $\lim_{n\to\infty} \mathbb{P}(S_n > 0) = a$.

More generally, the **intermediate arcsine laws** assert that for the Wiener process tied down at $u > 1$ (i.e. the Brownian bridge over $[0, u]$), both V (the time spent in $\mathbb{R}^+$ during [0, 1]) and Y (the time of first reaching the maximum in [0, 1]) have the density

$$f(x) = \frac{1}{\pi u}\left\{\pi + \left[\frac{u(u-1)}{x(1-x)}\right]^{1/2} - \tan^{-1}\left[\frac{u(u-1)}{x(1-x)}\right]^{1/2}\right\}, \quad 0 < x < 1.$$

Finally, the time of the last zero in [0, 1] has distribution function

$$G(x) = \frac{2}{\pi} \sin^{-1}\left\{\frac{x(u-1)}{u-x}\right\}^{1/2}, \quad 0 < x < 1.$$

[S. Watanabe, Generalized arc-sine laws for one-dimensional diffusions and random walks, in *Proceedings of Symposia in Pure Mathematics #57, Stochastic Analysis*, M. Cranston & M. Pinsky (eds). American Mathematical Society, 157–172, 1995; J. Bertoin & R. Doney, Spitzer's condition for random walks and Lévy processes, *Ann. Inst. H. Poincaré*, 33(2), 167–178, 1997]

arctic circle theorem Arises in the context of uniformly random domino (i.e. 2 × 1 rectangle) tilings of a diamond-shaped region (the Aztec diamond), where it asserts that as the size of the diamond increases, the tiles outside its inscribed circle have a similarly oriented (brickwork) pattern with probability arbitrarily close to 1. They are said to be frozen in the arctic (or polar) region outside the so-called arctic circle. [D. Romik, Arctic circles, domino tilings, and square Young tableaux, *Ann. Probab.*, 40(2), 611–647, 2012]

area-interaction process A type of point process that is either more clustered, or more scattered, than points of the Poisson process; according to the values of a scalar parameter. In its simplest form, the density of the process (with respect to the reference distribution of a Poisson process of unit rate) is given, for some finite (compact) set A, containing n points of the reference process, by $f(x) \propto a^n \exp\{-bV(x)\}$. Here, $V(x)$ is the volume of the intersection with A of the union of spheres of radius $r > 0$ centred at the points $(x_1, \ldots, x_n) = x$ of the reference process. The case $b = 0$ yields a Poisson process; the case $b > 0$ exhibits clustering (attraction), and is called the Widom–Rowlinson model (or penetrable spheres model); the case $b < 0$ exhibits scattering (repulsion or dispersion). [A. Baddeley & M. van Lieshout, Area-interaction point processes, *Ann. Inst. Stat. Math.*, 47(4), 601–619, 1995]

arithmetic distribution A r.v. X has an arithmetic distn if X is a multiple of some $a > 0$ with probability 1. When a is the largest number with this property, it is called the **span** (or **step size**) **of X**. *See* lattice distribution. [R. Gallager, *Stochastic Processes*. Cambridge University Press, 2013]

Arnold's distribution *see* Fisher's distribution

arrival theorem (alias: **ASTA problem; random observer property**) Arises in queueing theory (and related fields) where one seeks sufficient and/or necessary conditions such that, in equilibrium, the stationary distribution of the system is the same as the stationary distribution of the imbedded process comprising the system at the instants when customers arrive. If this holds, one says that Arrivals See Time Averages ≡ ASTA. One necessary and sufficient condition is that the conditional intensity of the arrival point process at any arrival time T, given the state $S(T)$ of the system, is independent of $S(T)$. Under weaker conditions one has that Poisson Arrivals See Time Averages ≡ PASTA. [R. Wolff, Poisson arrivals see time averages, *Oper. Res.*, 30(2), 223–231, 1982]

Arrow–Debreu security Arises in financial mathematics, where it is a very simple derivative (or instrument) that pays the holder one unit (of the numeraire) if the underlying asset (or set of assets) is in some given state s at the fixed time T; otherwise it pays nothing. Suitable derivatives can be written as linear combinations of a set of Arrow–Debreu securities. [A. Etheridge, *A Course in Financial Calculus*. Cambridge University Press, 2002]

Arrow–Pratt measure Arises in the mathematics of finance, economics, and decision theory, where it seeks to quantify risk-aversion. For an individual with utility function $U(x)$, the Arrow–Pratt measure (or index, or coefficient) of absolute risk aversion is $a(x) = -U''(x)/U'(x)$, where $U'(x)$ is called the **marginal utility**. The A-P index of relative risk aversion (a measure weighted by the actual 'wealth' x), is $r(x) = -xU''(x)/U'(x)$. **Arrow's hypothesis** is that $a'(x) \leq 0$ and $r'(x) \geq$. For examples, where $b > 0$:

$u(x) = -\exp(-bx)$, with $a = b$, and $r = bx$,

$u(x) = (1 - x^{-b})/b$, with $a = (1 + b)/x$, and $r = 1 + b$,

$u(x) = \log(b + x)$, with $a = (b + x)^{-1}$, and $r = x/(b + x)$.

[S. A. Ross, Some stronger measures of risk aversion in the small and the large with applications, *Econometrica*, 49(3), 621–638, 1981]

asian option (alias: **average option; mean value option**) Arises in financial mathematics, where it denotes an option whose pay-off is a function of the mean value of the underlying asset over the period of the contract. Note that the 'mean' may be arithmetic, geometric, or weighted. [L. C. G. Rogers & Z. Shi, The value of an asian option, *J. Appl. Probab.*, 32(4), 1077–1088, 1995]

asset In financial mathematics, anything that has a measurable monetary value, and can be owned. If it can be bought and sold, it is a tradable asset, and a set of tradable assets constitutes a market. A time-series (or index, or stock price) is not itself an asset, but derivative contracts on such random processes are assets. [T. Höglund, *Mathematical Asset Management*. Wiley, 2008]

asset-selling problem *see* house-selling problem

associated distribution *see* Esscher transform

associated queue Let $Z_n = \sum_{r=1}^{n} X_r$ be a random walk with i.i.d. steps $X_r, r \geq 1$. Then the queue associated to Z_n is that with service times $S_n = \max\{X_n, 0\}$ and interarrival times $T_n = \max\{-X_n, 0\}$. The association is expressed in the fact that the waiting time W_n of the nth customer of the queue has the distribution of $\max[0, Z_1, \ldots, Z_{n-1}]$; this is Pollaczek's thm. Other properties of the walk also correspond to queueing quantities of interest. Conversely, the sequence $W_0 = 0, W_{n+1} = (W_n + X_{n+1})^+$, is called the queue induced by the random walk. [R. Wolff, *Stochastic Modelling and the Theory of Queues*. Prentice-Hall, 1989]

associated random walk (alias: **conjugate random walk**) Let $S_n = \sum_{r=1}^n X_r$ be a random walk whose steps have mgf $M(\theta) = \mathbb{E}\exp(\theta X_1)$ such that $M'(0) \neq 0$, and $M(a) = 1$ for some $a \neq 0$; with distn fn $F(x)$. Let $Y_r, r \geq 1$, be i.i.d. with the tilted distn fn $F_Y(y) = \int_{-\infty}^y e^{ax} dF(x)$. Then the r.w. $S_n^a = \sum_{r=1}^n Y_r$ is said to be associated with (or conjugate to) S_n. [W. Feller, *Introduction to Probability Theory and its Applications*, Vol. 2, 2nd edn. Wiley, 1971]

association inequality (alias: **Chebyshov's association inequality**) Let $f(X)$ and $g(X)$ be non-decreasing real-valued functions of the random variable X. Then $\mathbb{E}[f(X)g(X)] \geq \mathbb{E}f(X)\mathbb{E}g(X)$. If f is non-decreasing and g is non-increasing then $\mathbb{E}[f(X)g(X)] \leq \mathbb{E}f(X)\mathbb{E}g(X)$; which is to say that $\mathrm{cov}(f(X), g(X)) \leq 0$. For real-valued functions $f(\mathbf{x})$ and $g(\mathbf{x})$ that are non-decreasing in each component of a vector $\mathbf{X}$ of independent r.vs, it remains true that $E(fg) \geq EfEg$, a result that may be called Harris's inequality. [S. Boucheron *et al.*, *Concentration Inequalities*. Oxford University Press, 2013]

association of random variables A concept (or type) of dependence for the components of a random vector $(X_1, \ldots, X_n) = \mathbf{X}$. The vector $\mathbf{X}$ is said to be associated if, for all functions $f(\mathbf{x})$ and $g(\mathbf{x})$ that are non-decreasing in each coordinate, we have $\mathrm{cov}[f(\mathbf{X}), g(\mathbf{X})] \geq 0$, whenever the covariance exists. If the inequality is reversed, then $\mathbf{X}$ may be called negatively associated. The property of being associated is preserved by all the following operations on random vectors: (i) taking subsets; (ii) forming unions of independent vectors (sets); (iii) forming sets of non-decreasing functions; and (iv) taking limits in distribution. If $\mathbf{X}$ is associated, then it is weakly associated, which is to say that for all non-decreasing functions $f_r, 1 \leq r \leq n$, $\mathbb{E}\prod_{r=1}^n f_r(X_r) \geq \prod_{r=1}^n \mathbb{E}f_r(X_r)$. Also, $\mathbf{X}$ is weakly associated iff it is positively (upper) orthant dependent, i.e. $\mathbb{P}(\cap_1^n \{X_r > x_r\}) \geq \prod_1^n \mathbb{P}(X_r > x_r)$, for all $\mathbf{x}$.

In two dimensions, more strongly, the positive quadrant dependence of X and Y implies that they are associated. [P. Oliveira, *Asymptotics for Associated Random Variables*. Springer, 2012]

asymptotic entropy *see* entropy function; entropy rate

asymptotic equipartition property (alias: **AEP; entropy ergodic theorem; typicality property**) An important theorem in the theory of communication. According to C. Shannon, a message of length n from a discrete memoryless source is a sequence $(X_1, \ldots, X_n) = X_n \in A^n$ where the X_r are i.i.d.r.vs taking values in the alphabet A, and having entropy $H(X)$. Then, for any $\varepsilon > 0$ and $\delta > 0$, there exists an $m < \infty$ such that for $n > m$ there is a set $T_n \in A^n$, called the **typical set**, such that $\mathbb{P}(\mathbf{X}_n \in Y_n) > 1 - \varepsilon$, and for any message $\mathbf{x} \in T_n$ we have $2^{-n(H(X)+\delta)} < \mathbb{P}(\mathbf{X}_n = \mathbf{x}) < 2^{-n(H(X) - \delta)}$. Shannon's observation is the corollary that for many practical purposes it is therefore possible to treat messages of length $n > m$ as though there were only $2^{nH(X)}$ of them, each having approximately the same probability, 2^{-nH}. *See* Shannon–McMillan–Breiman–Chung theorem. [T. Cover & J. Thomas, *Elements of Information Theory*, 2nd edn. Wiley, 2006]

Athreya–Ney embedding A technique for embedding an appropriate discrete-time process X_n in a continuous-time process $X(t)$ (where X_n and $X(t)$ may be random vectors), so that the process $X(T_n)$, being $X(t)$ observed at its jump times T_n, has the same distribution as X_n. There is a more general version of the method called **Rubin's construction**. [B. Davis, Reinforced random walk, *Probab. Theor. Rel. Fields*, 84, 203–229, 1990]

Athreya–Ney–Nummelin pseudo-atom construction (or splitting) A technique that allows the ideas of renewal theory to be used in developing the limit theory of Harris-recurrent Markov chains. Broadly, the state-space S of the original chain X is augmented by a point $\boldsymbol{\pi}$, which may be called a pseudo-atom, such that the chain X^* on the augmented space visits $\boldsymbol{\pi}$ infinitely often with probability 1 (so that $\boldsymbol{\pi}$ is a recurrence point), and X^* is essentially equivalent to the original chain X. The idea was developed independently by Nummelin at the same time as Athreya and Ney. [E. Nummelin, A splitting technique for Harris-recurrent Markov chains, *Probab. Theor. Rel. Fields*, 43, 309–318, 1978]

atom Broadly, an atom of probability $\mathbb{P}(A) > 0$ is such that it cannot be expressed as a sum of positive probabilities. In particular, the distribution of a random variable X has an atom at a if $\mathbb{P}(X = a) > 0$; the distribution function $F_X(x)$ thus has a jump (discontinuity) at a, and the number of atoms of X is at most countable. More generally, an atom A of a probability space is an event A, with $\mathbb{P}(A) > 0$, such that for any event $B \subset A$, either $\mathbb{P}(B) = 0$ or $\mathbb{P}(B) = \mathbb{P}(A)$. If C is the union of the atoms, then $\Omega \backslash C$ is called the diffuse (or atomless) part of the space, and C is said to be the atomic part. Discrete random variables have an atomic distn. Thus, a space is atomless if any event can be subdivided into events of positive probability.

Similarly an atom of a σ-algebra $\mathcal{F}$ is a set $A \in \mathcal{F}$ such that the only proper subset of A that is still in $\mathcal{F}$ is the empty set $\emptyset$. And if $A_1, \ldots, A_n$ are disjoint sets whose union is Ω, then each A_r is called an atom of the partition. [O. Kallenberg, *Foundations of Modern Probability*, 2nd edn. Springer, 2002]

attainable claim *see* contingent claim

attractive event (alias: **positive dependence**) Events A and B are attracted to each other if $\mathbb{P}(A \mid B) > \mathbb{P}(A)$. If the inequality is reversed, then A and B are said to repel each other (or to be negatively dependent). If A and B attract each other, then A and B^c repel each other.

attribute Arises in decision theory, when the rewards (or costs) consequent on a decision are of more than one type, and are represented by a vector. The vector's components are referred to as attributes. [J. Figueira *et al.* (eds), *Multiple Criteria Decision Analysis*. Springer, 2009]

augmentation Arises in the rigorous definition of random processes. Given a filtration $\mathcal{F}_t, t \geq 0$, on a probability space $(\Omega, \mathcal{F}, \mathbb{P})$ (where $\mathcal{F}_t$ may not be complete and right-continuous), there exists a smallest complete and right-continuous extension of $\mathcal{F}_t$. This is called the (usual) augmentation, or enlargement, and the filtration thus obtained satisfies the so-called usual conditions. [R. Bass, *Stochastic Processes*. Cambridge University Press, 2011]

augmented state space (alias: **supplementary variable**) Arises most often in the context of random processes; where, typically, a process that does not have the Markov property can have its state space expanded in such a way that the process on the larger state space is Markovian. For example, let $S_n = \sum_{r=1}^{n} X_r$ be a random walk, and consider $Z_m = \sum_{r=1}^{m} S_r$. Then Z_m is not a Markov process, but the bivariate (augmented) process (S_n, Z_n) is Markov. It can be seen as a form of embedding. [D. R. Cox, The analysis of non-Markovian stochastic processes by the inclusion of supplementary variables, *Proc. Camb. Phil. Soc.*, 51(3), 433–441, 1955]

Aumann expectation (alias: **selection expectation**) The first, and most frequently seen, of many methods for defining the expectation of a random closed (or compact) set X, i.e. X is a random element in the space of all closed (or compact) subsets of some metric space such as $\mathbb{R}^d$. For such X, a selection of X is a random vector V, such that $V \in X$ with probability 1. Then the Aumann expectation of X is $\mathbb{E}X = \{\mathbb{E}V : V \text{ is a selection of } X \text{ and } \mathbb{E}V \text{ exists}\}$; where the expectation is the Bochner expectation, taken coordinatewise. The condition $\mathbb{E}(\sup\|x\| : x \in X) < \infty$ is sufficient for $\mathbb{E}X$ to be non-empty. [I. Molchanov, *Theory of Random Sets*. Springer, 2005]

autobinomial model (alias: **Besag's autobinomial**) A Markov random field introduced by J. Besag as a model describing the 'texture' of plane images. The sites, on a square lattice, may each be in any one of $n+1$ states. The conditional distribution of the state of a vertex V, given the states of neighbour of V, is binomial $\mathrm{B}(n,p)$, where p is a function of the states of neighbours. Analogously, replacing the binomial distn by another, one may define autoPoisson, autogamma, and autonormal (or autoGaussian) models. [J. Besag, Spatial interaction and the statistical analysis of lattice systems, *J. Roy. Stat. Soc. B*, 36(2), 192–236, 1974]

autocorrelation (function, or coefficient) (alias: **correlation function**) Arises in the theory of random processes, and is important in many applications. Let $X(t)$ be a finite variance random process, where time is commonly either discrete $t \in \mathbb{Z}$, or continuous $t \in \mathbb{R}$. Then the correlation $\rho(X(s), X(t))$ is called the autocorrelation (function), or acf; and likewise, the covariance $\mathrm{cov}(X(s), X(t))$ is called the **autocovariance** (function). These are of most use and importance when $X(t)$ is a weakly (or second-order) stationary process so that $\rho(X(s), X(t)) = \rho(t-s)$, is a function of $h = t - s$ only, as is the covariance $\mathrm{cov}(X(s), X(t)) = c(h)$; the argument h is often called the **lag**. Then $\rho(h)$ has these properties:

(i) When real, it is symmetric: $\rho(h) = \rho(-h)$.

(ii) When X, and hence ρ, may take complex values, $\overline{\rho(h)} = \overline{\rho(-h)}$, where $\overline{\rho}$ is the complex conjugate.

(iii) It is non-negative definite, which is to say that $\sum_{j,k} \rho(t_k - t_j) z_j \bar{z}_k \geq 0$ for real t_r and complex z_s.

(iv) It has the so-called **spectral representation** $\rho(h) = \int_{-\infty}^{\infty} \mathrm{e}^{\mathrm{i}h\lambda} \mathrm{d}F(\lambda)$ (provided ρ is continuous at $h = 0$), where $F(\lambda)$ is a distn fn called the spectral distn. When $t \in \mathbb{Z}$ one has more simply $\rho(h) = \int_{-\pi}^{\pi} \cos(h\lambda) \mathrm{d}G(\lambda)$, where G is a distn symmetric on $[-\pi, \pi]$. By the Fourier inversion theorem, if $F'(\lambda) = f(\lambda), f(\lambda) = \frac{1}{2\pi} \int_{-\infty}^{\infty} \mathrm{e}^{-\mathrm{i}t\lambda} \rho(t) \mathrm{d}t$, and $f(\lambda)$ is called the spectral density.

The pair of results in (iv) is often called the **Wiener–Khinchin thm**, though some writers use that term for the corollary

$$f(\lambda)\operatorname{var}X = \lim_{T\to\infty}\frac{1}{2T}\mathbb{E}\left|\int_{-T}^{T} X(t)e^{-i\lambda t}dt\right|^2.$$

(v) The **Cramér representation** asserts that, when $\rho(t)$ is continuous, there is a random (spectral) process $S(\lambda)$, having orthogonal increments, such that,

$$X(t) = \int_{-\infty}^{\infty} e^{it\lambda} dS(\lambda),$$

where $\mathbb{E}\mid S(v) - S(u)\mid^2 = F(v) - F(u), \quad u \le v.$

Finally, note that some authors, especially when writing for engineers, define a quantity called the (sample) autocorrelation of $X(t)$ as

$$r_T(s) = \frac{1}{T}\int_0^T X(t+s)\overline{X(t)}\,dt.$$

For a nice, ergodic, unit-variance, zero-mean process $X(t)$, as $T \to \infty$, $r_T(s) \to \rho(s)$ with probability 1. The term **serial correlation** is also used for this, especially in discrete time. [G. Box *et al.*, *Time Series Analysis*, 4th edn. Wiley, 2008]

automaton Arises in the context of theories of control, reliability, organisms, computing, etc., as a general mathematical description of such systems. Broadly, it can be seen as an abstract machine whose responses to all feasible inputs are described by a set of states with a set of rules for moving between states and supplying outputs. If the rules or inputs are random, then it is a probabilistic (or stochastic) automaton. **Von Neumann's problem** is to decide whether (and if so, how) it is possible for automatons (or automata) to have unreliable components, but nevertheless to be reliable to a known degree. [M. Rabin, Probabilistic automata, *Inf. Control*, 6(3), 230–245, 1963]

autoregressive process (or model) Arises in the theory of time series, being used for modeling and prediction of real world series (or indices). The process $X_n, n \in \mathbb{Z}$, is said to be a (linear) autoregressive (or AR) process of order p, denoted by AR(p), if for constants $\phi_1, \ldots, \phi_p$, and all n, $X_n - \phi_1 X_{n-1} - \phi_2 X_{n-2} - \cdots - \phi_p X_{n-p} = Z_n$, where the Z_n are (usually) independent (or, more weakly, orthogonal) r.vs with common finite variance and zero mean (thus constituting a white noise process). More generally, X_n is a non-linear AR process of order p if for some function f, and all n, $X_n - f(X_{n-1}, \ldots, X_{n-p}) = Z_n$.

Most authors require X_n to be a (weakly) stationary sequence, which imposes some conditions on $\phi_1, \ldots, \phi_p$ (or f). Defining the backward shift operator B such that $BX_m = X_{m-1}$, the above equation is written as $\boldsymbol{\phi}(B)X_n = Z_n$. More generally, if for some polynomial $\boldsymbol{\theta}$ of order q, we have for all n that $\boldsymbol{\phi}(B)X_n = \boldsymbol{\theta}(B)Z_n$, the process is an autoregressive-moving average process, denoted by ARMA (p, q). If for some integer $d \ge 1$, we have $\boldsymbol{\phi}(B)\,(1-B)^d X_n = \boldsymbol{\theta}(B)Z_n$, then the process is autoregressive integrated moving average, denoted by ARIMA(p, q, d). If $d \in \left(-\frac{1}{2}, \frac{1}{2}\right)$ is not zero, then the process is autoregressive fractionally integrated moving average, denoted by ARFIMA(p, q, d). If $X(t), t \in \mathbb{R}$ is such a process run in continuous time, so that the weighted sums of $X(t)$ and $Z(t)$ are replaced by integrals, then one obtains a CARMA process (and CAR, CMA, etc., likewise). [P. Brockwell & R. Davis, *Time Series*, 2nd edn. Springer, 1991]

availability Arises in the theory of the reliability of systems, broadly as a measure of the degree to which the system S can function or perform its task. More formally, if $\boldsymbol{\phi}(t)$ is the indicator of the event that S is working at time t ($\boldsymbol{\phi} = 1$ if it is, $\boldsymbol{\phi} = 0$ if it is not), then the (point) availability of S is $a(t) = \mathbb{P}(\boldsymbol{\phi} = 1)$. This is also called the **Barlow–Proschan availability**.

For a system that is totally non-repairable, $a(t)$ is the same as the reliability $R(t) = \mathbb{P}(T > t)$, where $T = \inf\{t : \boldsymbol{\phi}(t) = 0\}$ is the time of first failure. (It is therefore not uncommon for the availability function of a component, whether repairable or not, to be called its reliability, blurring the distinction.)

For repairable systems the simplest model assumes that S is either available or unavailable (on or off; up or down) and then for any subset B of $\mathbb{R}^+$ the set-availability (or interval-availability if B is an interval) is the expected proportion of B during which S works, so $|B|a(B) = \int_{t\in B} a(t)dt$. If the limit of this exists as $|B| \to \infty$, it may be called the limiting availability. [A. Høyland & M. Rausand, *System Reliability Theory*, 2nd edn. Wiley, 2004]

average Most commonly, when used without qualification, this denotes the arithmetic mean $n^{-1}(x_1 + \cdots + x_n)$ of a set of numbers $x_1, \ldots, x_n$. More generally, the weighted average of $x_1, \ldots, x_n$ with weights $w_1, \ldots, w_n$ is $\sum_{r=1}^n w_r x_r$, where $\sum_i^n w_i = 1$. Yet more generally, for any increasing continuous function f, the generalized mean, or average, is $f^{-1}\left(\sum_1^n w_r f(x_r)\right)$.

In other contexts, average may refer to the expected value, or mean of a random variable; or, in very loose usage, it may refer to any one of various other quantities associated with centrality, location, typicality, etc. *See* Kolmogorov–Nagumo mean. [J. Aczél, On mean values, *Bull. Amer. Math. Soc.*, 54, 392–400, 1948]

average cost (criterion) Arises in sequential decision procedures, where it is natural to see a policy as good if it minimizes the expected average cost over time. The minimization may be over a finite period (finite horizon), or over $[0, t]$ in the limit as $t \longrightarrow \infty$ (infinite horizon); the latter may be called the long-run expected average cost. Future costs may be discounted, leading to the average discounted cost criterion. [X. Guo & O. Hernández-Lerma, *Continuous-time Markov Decision Processes*. Springer, 2009]

average option *see* asian option

averaging property A term of wide natural application in mathematics. In probability it (perhaps) most often arises thus:

1. Denoting the basic property of conditional expectation (and conditional probability), e.g. when X and Y are discrete r.vs, $\mathbb{E}X = \sum_y \mathbb{E}(X \mid Y = y)\mathbb{P}(Y = y)$.

   In the same context, if X is a r.v. defined on the probability space $[\Omega, \mathcal{F}, \mathbb{P}]$ and $\mathcal{G}$ is a sub-σ-field of $\mathcal{F}$, the fact that $\int_A \mathbb{E}(X \mid G)\mathrm{d}\mathbb{P} = \int_A X\mathrm{d}\mathbb{P}$, for all $A \in \mathcal{G}$, is sometimes called the **partial averaging property** of conditional expectation.
2. Denoting the harmonic property of random walk on a lattice (or graph), or more generally of Markov chains (and likewise, of the Wiener process).

[O. Kallenberg, *Foundations of Modern Probability*, 2nd edn. Springer, 2002]

avoidance function Arises in random point (or random set) processes. If $\sum$ is a random set (in $\mathbb{R}^d$, say), and A $\subseteq \mathbb{R}^d$ is arbitrary, then the avoidance fn is $a(A) = \mathbb{P}(\sum \cap A = \emptyset)$, where A ranges over a suitable family of sets. When $\sum$ is a point process, and $N(A)$ is the number of points of $\sum$ in A, $a(A) = \mathbb{P}(N(A) = 0)$. If the family of sets A is sufficiently rich in membership, many random set (or point) processes can be characterized by their avoidance fn. [D. Daley & D. Vere-Jones, *Introduction to the Theory of Point Processes*, 2nd edn. Springer, Vol. I, 2003; Vol. II, 2008]

axial distribution Arises in the context of random vectors. A probability distn on the sample space of unoriented lines (in $\mathbb{R}^d, d \geq 2$) through the origin is said to be axial. Such a distn may be implemented either as an antipodally symmetric distribution on the unit sphere, or as any distn on the unit hemisphere. If the lines are oriented (i.e. vectors) one has a directional (or vectorial) distn. If such a distn is concentrated around a great circle it is called **equatorial**, if concentrated around any circle (great or small) it is a **girdle distn**; if concentrated at the ends of a diameter of the sphere, it is called **bipolar**. [N. Fisher *et al.*, *Statistical Analysis of Spherical Data*. Cambridge University Press, 1987]

axioms of probability Arise in formalizing any interpretation (or view) of probability, so that a mathematical theory can be developed. Several axiomatizations exist; that most commonly seen in mathematical probability postulates a probability space $[\Omega, \mathcal{F}, \mathbb{P}]$, where Ω is a set, $\mathcal{F}$ a field of subsets of Ω, and $\mathbb{P}$ a function defined on the members of $\mathcal{F}$. The structure is then developed by defining the concepts (and properties) of independence, conditioning, random variables, expectation, and so on. The best known (by far) rules are called Kolmogorov's axioms. [P. Suppes & M. Zanotti, *Foundations of Probability with Applications*. Cambridge University Press, 1996]

Azéma's martingale A number of martingales have this name. The first, also called the **excursions martingale**, is $Y(t) = (Z(t))^{\frac{1}{2}}\mathrm{sign}\, W(t)$, where $W(t)$ is the Wiener process, and $Z(t)$ is the current duration (at time t) of the excursion in progress of $W(t)$ away from the origin. A second Azéma martingale is $A(t) = \left[\frac{1}{2}\pi Z(t)\right]^{\frac{1}{2}} - L(t)$, where $L(t), t > 0$, is the local time of $W(t)$ at 0, and $Z(t)$ is as above. [J. Obłój & M. Yor, An explicit Skorokhod embedding for the age of Brownian excursions and Azéma martingale, *Stoch. Proc. Appl.*, 110(1), 83–110, 2004]

Azéma–Yor martingale A class of martingales defined in terms of the Wiener process $W(t)$, the supremum function, $S(t) = \sup_{s \leq t} W(t)$, and a non-negative fn $f(x)$ such that, $\int_0^\infty f(x)\mathrm{d}x = 1$ (and f is Borel-measurable). Then $M(t) = -f(S(t))(S(t) - W(t)) + \int_0^{S(t)} f(x)\, \mathrm{d}x$ is an Azéma–Yor martingale. It may be equivalently defined by $M(t) = \int_0^t f(S(t))\, \mathrm{d}W(t)$. [L. Carraro *et al.*, On Azéma–Yor processes, their optimal properties, and the Bachelier-drawdown equation, *Ann. Probab.*, 40(1), 372–400, 2012]

Azuma's inequality *see* Hoeffding's inequality

B

Bachelier's equations Arise in the context of financial mathematics. For a random process $X(t)$, with transition density function $f(x,t)$, having independent stationary (i.e. time-homogenous) increments, L. Bachelier showed that f satisfies

$$f(x, s+t) = \int_{-\infty}^{\infty} f(y,s)f(x-y,t)\mathrm{d}y.$$

Similar equations, derived in other contexts, are called the Chapman–Kolmogorov eqn, or the Einstein–Bachelier–Smoluchowski eqn, or other combinations of these names. On the assumption of continuity, Bachelier obtained the solution $f(x,t) = (2\pi t)^{-\frac{1}{2}}\exp\left(-\frac{x^2}{2t}\right)$ amounting to an informal derivation of the Wiener process $W(t)$. Postulating a model for a stock price $S(t)$ as $S(t) = S(0)(1 + \sigma W(t))$, Bachelier's (second) equation gives the formula for the value of a call option on $S(t)$, at time T, with strike price K, as $\mathbb{E}[(S(T) - K)^+]$. [L. Bachelier, *Theory of Speculation*, trans. by M. Davis & A. Etheridge. Princeton University Press, 2006]

Bachelier–Lévy formula Arises in the context of the Wiener process $W(t)$, where it gives the density of the first-passage time T of $W(t)$ to the boundary $a + bt, a > 0$, in the form $f(t) = at^{-3/2}\phi((a+bt)/t^{\frac{1}{2}})$, with distn fn $F(t) = 1 - \Phi\left((a+bt)/t^{\frac{1}{2}}\right) + \mathrm{e}^{-2ab}\Phi\left((bt-a)/t^{\frac{1}{2}}\right)$, where ϕ and Φ are the standard normal density and distn fn. [H. Lerche, *Boundary Crossing of Brownian Motion*. Springer, 1986]

backward and forward equations (alias: **Kolmogorov equations; retrospective and prospective equations**) Two related sets of differential (or integral, or integro-differential) equations that arise in the context of Markov processes in continuous time. The differential eqn forms are often called Kolmogorov's eqns. Informally, and very roughly speaking, for a Markov process $X(t)$, backward equations arise by considering (conditioning on) the development of the process immediately following its initiation at time $t = 0$; whereas forward equations arise by considering the most recent development of the process prior to time t. These equations take different forms depending on the state space of the process. For example a Markov chain with finite matrix of transition probabilities $p_{jk}(t) = P$, and Q-matrix Q has backward and forward equations $\frac{\partial P}{\partial t} = QP$, and $\frac{\partial P}{\partial t} = PQ$, respectively; the forward equation(s) are often called the **master eqn(s)** in applications. There is an integral eqn formulation involving P, Q, the holding time parameters, and the routing matrix of the jump chain.

By contrast, a suitable (pure) diffusion process with parameters $a(t,x)$ and $b(t,x)$ has backward and forward eqns of the respective forms

$$\frac{\partial f}{\partial s} = -a(s,x)\frac{\partial f}{\partial x} + \frac{1}{2}b(s,x)\frac{\partial^2 f}{\partial x^2},$$

$$\frac{\partial f}{\partial t} = -\frac{\partial}{\partial y}(a(t,y)f) + \frac{1}{2}\frac{\partial^2 f}{\partial y^2}(b(t,y)f).$$

The forward equation is often called the **Fokker–Planck** or Smoluchowski eqn in applications. Such equations may also be formed for suitable functions of a Markov process $X(t)$, and of its sample paths; as in, e.g. the Feynman–Kac formula (or method). [W. Feller, *Introduction to Probability Theory and its Applications*, Vol. 2, 2nd edn. Wiley, 1971]

backward coupling *see* Propp–Wilson method

backward induction (alias: **Zermelo's method**) That form of mathematical induction in which the initial hypothesis is $h(N), N > 0$, and the recursive step establishes $h(n)$, given $h(n+1), n < N$. Use of the method is natural in sequential decision theory (e.g. finite-horizon Markov decision processes), the valuation of derivatives by finite differences, establishing the value-iteration method of dynamic programming, and in the solution of suitable games. [U. Schwalbe & P. Walker, Zermelo and the early history of game theory, *Games Econ. Behav.*, 34(1), 123–137, 2001]

backward martingale (alias: **reverse martingale**) A sequence $(X_n, n \geq 0)$ of finite-mean random variables constitutes a backward martingale if, for all n, $\mathbb{E}(X_n \mid X_{n+1}, X_{n+2}, \ldots) = X_{n+1}$. The terminology arises because X_n is a backward martingale iff the sequence $Z_n \to X_{-n}$ is a martingale. As $n \to \infty$, the backward martingale converges both in mean and with probability one. [J. Neveu, *Discrete-parameter Martingales*. North-Holland, 1975]

backward recurrence time *see* age

Baconian probability Denotes, and arises in, a theory of probability based on its interpretation as a degree of logical support; that is, the Baconian probability of an event A is a measure of the weight of evidence, or inductive support for A. When formalized, such theories usually differ significantly from mathematical probability, because evidence which fails to support the truth of A is not necessarily (or even usually) therefore evidence for the falsity of A. Thus it is not the case that $\mathbb{P}(A \cup B) = \mathbb{P}(A) + \mathbb{P}(B) - \mathbb{P}(A \cup B)$ in general. In one formulation, the Baconian probability satisfies $\mathbb{P}(A \cup B) = \max\{\mathbb{P}(A), \mathbb{P}(B)\}$. [L. J. Cohen, Historical remarks on the Baconian conception of probability, *J. Hist. Ideas*, 41(2), 219–235, 1980]

Bahadur–Kiefer process Arises in the context of the empirical process and is used in proving an important relationship with the quantile process. Formally, let $U_1, U_2, \ldots$ be i.i.d. r.vs, each uniform on (0, 1) and let $F_n(x) = n^{-1}\sum_{r=1}^{n} I(U_r \le x)$ be the (right-continuous) empirical distribution function; where $I(U_r \le x)$ is the indicator of the event $\{U_n \le x\}$. The inverse, or quantile, function is $Q_n(x) = \inf[t : F_n(t) \ge x]$, with $Q_n(0) = 0$. Then $a_n(x) = n^{1/2}[F_n(x) - x]$ is the uniform empirical process, and $b_n(x) = n^{\frac{1}{2}}(Q_n(x) - x)$ is the uniform quantile process, and the Bahadur–Kiefer process is $R_n(x) = a_n(x) + b_n(x)$. The Bahadur–Kiefer thm asserts that as $n \to \infty$ $\frac{n^{1/4}}{(\log n)^{1/2}} \frac{\|R_n\|}{\|a_n\|^{1/2}} \to 1$, with probability 1, where $\|u(x)\| = \sup_{0 \le x \le 1} |u(x)|$. [J. Einmahl, A short and elementary proof of the main Bahadur-Kiefer theorem, *Ann. Probab.*, 24(1), 526–531, 1996]

Baik–Deift–Johansson theorem Arises in the theory of random permutations and random matrices. Let L_n be the longest increasing subsequence of a random permutation of $\{1, \ldots, n\}$ Then the theorem asserts that, for positive constants a and b, $\mathbb{E}L_n = 2n^{1/2} - an^{1/6} + \mathrm{o}(n^{1/6})$, and $\{\mathrm{var}\ L_n\}^{1/2} = bn^{1/6} + \mathrm{o}(n^{1/6})$. [J. Baik *et al.*, The distribution of the length of the longest increasing subsequence of random permutations, *J. Amer. Math. Soc.*, 12(4), 1119–1178, 1999]

Bak–Sneppen model Arises as a simple model in the theory of evolution, that exhibits self-organized criticality. It may be defined on a finite connected graph by attaching i.i.d.r.vs that are uniform on (0, 1) to each vertex, with this updating rule: for the vertex with the smallest r.v. attached, replace this and the r.vs at each of its neighbours with fresh i.i.d.r.vs uniform on (0, 1). In the original model the graph was circular (i.e. a regular polygon). [R. Meester & D. Znamenski, Limit behaviour of the Bak–Sneppen evolution model, *Ann. Probab.*, 31(4), 1986–2002, 2003]

Bak–Tang–Wiesenfeld model *see* Abelian sandpile

balance equations (alias: **balance conditions; weak symmetry**) Arise in the context of reversible Markov chains. Given a Markov chain X, with transition matrix p_{ij}, and a measure λ_i on the state space S of X, the detailed balance equations are $\lambda_j p_{jk} = \lambda_k p_{kj}$, for all j and k in S. Summing over k yields the **full balance eqns,** $\sum_{k \in S} \lambda_k p_{kj} = \sum_{k \in S} \lambda_j p_{jk} = \lambda_j$, since p_{jk} is a stochastic matrix. If $\sum_{j \in S} \lambda_j = 1$ then $(\lambda_1, \ldots, \lambda_{|S|})$ is a stationary distn of (or an invariant distn for) the Markov chain and an important theorem is this: an irreducible chain having p_{ij} as its transition matrix is reversible in equilibrium iff there is a stationary distn satisfying the detailed balance eqns.

In continuous time, similar results are expressed in terms of the Q-matrix, provided that the chain is irreducible and non-explosive. Then $\lambda = (\lambda_1, \ldots, \lambda_{|S|})$ and $Q = (q_{ij})$ are in detailed balance if $\lambda_j q_{jk} = \lambda_k q_{kj}$ for all k and j. If A is a proper subset of the state space S and $\lambda_j \sum_{k \in A} p_{jk} = \sum_{k \in A} \lambda_k p_{kj}$, these are the **partial (or local) balance eqns**. If a Markov chain on S is truncated to A, then the equilibrium distribution of the truncated chain is that of the original chain conditioned on its being in A, iff the partial balance equations hold. Finally, some authors say that p_{ij} is **weakly symmetric** with symmetrizing measure λ_i, if $\lambda_i p_{ij} = \lambda_j p_{ji}$ for all i and j, instead of the above. [F. Kelly, *Reversibility and Stochastic Networks*. Wiley, 1979; reprinted by Cambridge University Press, 2011]

balayage (alias: **sweeping (swept) measure**) A concept used in solving Laplace's eqn in a regular domain $D \subseteq \mathbb{R}^d$ with given continuous value on the boundary of D; the so-called Dirichlet problem. The term appears in probability because of the link between classical potential theory and Brownian motion. Let T_D be the hitting time of D by a Wiener process. Then the balayage, $b(x, D)$, of the Dirac measure $\delta(x)$ on D is the distribution of the position of the Wiener process started from x, at time T_D. Similarly, if W^* is the Wiener process started at a point with probability distribution $P_0(x)$, the balayage of P_0 on D is the distribution of the position of W^* at its hitting time of D, i.e. $\int b(x, D) \mathrm{d}P_0(x)$. [S. Port & C. Stone, *Brownian Motion and Classical Potential Theory*. Academic Press, 1978]

Balkema–De Haan–Pickands theorem (alias: **second theorem of extreme value theory**) In extreme value theory, this arises when considering observations that exceed a high threshold value t; this is often called the peak-over-threshold (or POT), method. Consider a random variable X having an extreme-value distn, of one of the types specified by the **Fisher–Tippett–Gnedenko thm** (first theorem of extreme value theory). Then the exceedance (or excess) of X over t, conditional on $X > t$, has a limiting distribution as $t \to \infty$ that is of generalized Pareto form. Formally, for some $b > 0$, and $a \in \mathbb{R}$, either $\mathbb{P}(X - t > y \,|\, X > t) \to \left(1 + \frac{a}{b}y\right)^{-\frac{1}{a}}$, $a \neq 0$, where $y \geq 0$ for $a > 0$, and $0 \leq y \leq -\frac{b}{a}$ for $a < 0$; or, when $a = 0$, the limiting distn is exponential with mean b. *See also* extreme value distributions. [R.-D. Reiss & M. Thomas, *Statistical Analysis of Extreme Values*, 3rd edn. Birkhäuser, 2007]

balking A queue discipline in which a customer who arrives, and could join the queue, may immediately (on inspection of the system, including, e.g. the queue length and the possibly different types of customer in it) refuse to queue, and leaves. By contrast, **reneging** occurs when a customer who has joined the queue leaves (or is ejected) at some point of the waiting time before service. [D. Gross *et al.*, *Fundamentals of Queueing Theory*, 4th edn. Wiley, 2008]

ballistic point (or path) Arises in stochastic flow. Typically, such a flow $X(t)$ is mixing (or diffusive) and distances travelled are of order $\sqrt{t}$; that is, $X(t) \sim \sqrt{t}$ with probability 1. More generally, particles may move linearly; such particles (and their paths) are called ballistic, and the set of ballistic points is $B = \{X(t) \in \mathbb{R} : \liminf_{t\to\infty} ||X(t)||/t > 0\}$. [M. Scheutzow & D. Steinsaltz, Chasing balls through martingale fields, *Ann. Probab.*, 30(4), 2046–2080, 2002]

ballot theorem The classical ballot theorem, first proved by W. Whitworth in 1878. Suppose when the final votes are all counted, in an election with just two candidates A and B, it is found that A has a votes, and B has b votes, where $a > b$. Then the probability that, throughout the counting, A was always ahead of B, is $\frac{a-b}{a+b}$. The theorem was proved again by Bertrand in 1887, and is sometimes called **Bertrand's ballot thm**. *See* hitting-time theorem. [W. Whitworth, Arrangements of m things of one sort and n things of another sort, under certain conditions of priority, *Messenger Math.*, 8(1878–1879), 105–114, 1879]

Ball–Torous model Arises in financial mathematics as a model for the price process $Z(t)$ of a zero-coupon bond that pays 1 at time T, and nothing else. It sets $Z(t) = Z(0)\exp[rt + \sigma B(t)], \sigma > 0$, where $B(t)$ is the Brownian bridge over $[0, T]$. [S. Cheng, On the feasibility of arbitrage-based option pricing when stochastic bond prices are involved, *J. Econ. Theor.*, 53(1), 185–198, 1991]

ball walk Arises as a sampling method for calculating volumes; at each step of the walk inside a continuous convex body K, the process either steps to a point Z picked uniformly at random in the ball, radius b, centred at its current position, if $Z \in K$, or when $Z \notin K$ it stays put. [M. Jerrum, *Counting, Sampling, and Integrating: Algorithms and Complexity*. Birkhäuser, 2003]

Banach's matchbox problem A classic problem of elementary probability. A mathematician initially has two matchboxes, each of N matches, one in each of the right and left coat pockets. When a match is required, one or the other pocket is selected uniformly at random to supply it. After how many such removals is it first discovered that a box is empty? [W. Feller, *Introduction to Probability Theory and its Applications*, Vol. 1, 3rd edn. Wiley, 1968]

bandit process Arises in sequential Markov decision procedures, as a model for the following common real-world problem. At a sequence of times, a decision maker (called an agent) is faced with a number of choices, each yielding random rewards (or costs), called a system. The dependence of the rewards on the decision is initially unknown, but choosing a given action yields information about its rewards. A classic example is supplied by medical trials, where a number of treatments are available, whose relative merits are unknown; choosing a given treatment supplies a reward/cost, together with information on its efficacy. In all such problems there is a trade-off between finding out more about the system or making the best choice on the basis of what is already known.

Bandit processes model this dilemma in terms of m (usually independent) Markov processes $X_1, \ldots, X_m$, called projects. Choosing project r at time t yields a reward $g_r(X_r(t))$, and the project changes its state according to its transition matrix. Only one project can be chosen, and unchosen projects neither change their state, nor yield a reward. The name arises because of the analogy with a casino gambling machine, called a bandit, and the above model may be called an m-armed bandit process. When the object is to maximize the total expected discounted reward, under very mild assumptions, it turns out that there is a vector of functions $G_1(X_1(t)), \ldots, G_m(X_m(t))$, called **Gittins indices**, such that the optimal policy at any time is to choose the project (pull the arm) with largest index at time t; this is the **Gittins–Jones thm**. Extensions of the problem include the so-called **restless bandit problem**, in which the projects may change their state (perhaps according to different rules) even when not chosen by the agent. [J. Gittins et. al, *Multi-armed Bandit Allocation Indices*, 2nd edn. Wiley, 2011]

band-limited A finite-variance stationary process $X(t), t \in \mathbb{R}$, is said to be band-limited if its spectral distribution vanishes outside a finite interval. In this case, by the Nyquist–Shannon sampling theorem, the process is determined by its values at a sequence of lattice times. [S. Miller & D. Childers, *Probability and Random Processes*, 2nd edn. Academic Press, 2012]

bandwidth A concept arising in the theory of second-order (weakly) stationary processes, where it denotes the interval that supports the spectral distribution; or, more informally, the interval that supports the significant part of the spectrum. Because a larger bandwidth enables a communication channel to transmit more information, the term is sometimes used figuratively to mean channel capacity. [S. Miller & D. Childers, *Probability and Random Processes*, 2nd edn. Academic Press, 2012]

Barabási–Albert model (or method) (alias: **preferential attachment model**) A procedure for producing random graphs that arises as a model to describe the scale-free power-law distribution of vertex degrees that is observed in real-world networks.
A Barabási–Albert tree is generated recursively from an initial edge. At each subsequent step a new edge is added, with one end a new vertex; the other end is attached to the existing tree by selecting an edge uniformly at random, and attaching the new edge to either end with equal probability. Thus the vertex for attachment is selected with probability proportional to its vertex-degree, which is an example of preferential attachment. If each new vertex may be attached to $m > 1$ vertices, one obtains a Barabási–Albert random graph. *See* Yule–Simon distribution. [S. Bornholdt & H. Ebel, World Wide Web scaling exponent from Simon's 1955 model, *Phys Rev. E*, 64, 035104, 2001]

Barbour–Hall inequality Arises in Poisson approximation, thus: let $S = \sum_{r=1}^{n} X_r$ be a sum of independent Bernoulli r.vs, where $\mathbb{P}(X_r = 1) = p_r = 1 - \mathbb{P}(X_r = 0)$ and let $\lambda = \sum_{r=1}^{n} p_r$. Then $\frac{1}{2}\sum_{r\geq 0}|\mathbb{P}(S = r) - e^{-\lambda}\lambda^r/r!| \leq (1 - e^{-\lambda})\lambda^{-1} \sum_{r=1}^{n} {p_r}^2$. [A. Barbour & P. Hall, On the rate of Poisson convergence, *Math. Proc. Camb. Phil. Soc.*, 95(3), 473–480, 1984]

Barlow–Yor inequality Gives bounds for moments of local times of suitable martingales. For example, let $L(t, a), t \geq 0, a \in \mathbb{R}$, be the local time at a of the Wiener process, and define $L_t = \sup_a L(t, a)$. Then, if T is a stopping time for the Wiener process, there are constants c_p and C_p such that for $p > 0$, $c_p \mathbb{E}T^{p/2} \leq \mathbb{E}L_T^p \leq C_p \mathbb{E}T^{p/2}$. [B. Davis, On the Barlow-Yor inequalities for local time, *Séminaire de Probabilités de Strasbourg*, 21, 218–220, 1987]

Barndorff-Neilsen distribution *see* hyperbolic distribution

Barnes–Allan fBm *see* fractional Brownian motion

barrier Broadly, for a random process evolving in a state space S, a barrier B is some subset of S such that the development of the process is in some way altered on arriving at (or passing through) the frontier of B. There are several different types of such barriers, depending on the various real-world effects that they seek to describe: e.g. (i) reflecting barrier, which returns the particle analogously to reflected light; (ii) absorbing barrier, at which the particle stops; (iii) elastic barrier, at which the particle undergoes some immediate return from B that is not analogous to optical reflexion; (iv) sticky barrier, which may trap the particle for a (possibly random) time; (v) osmotic barrier, which may permit a crossing or reflection, at random; (vi) interface, which the particle crosses without hindrance, but behaves differently on each side; and (vii) resetting barrier, which entails jumps to a set of base-states. Typically, these properties are formulated in the shape of boundary conditions, depending on the nature of the barrier and the process. (The frontier, or boundary, of the state space S is also a barrier, of course.) [S. Karlin & H. Taylor,

A Second Course in Stochastic Processes. Academic Press, 1981]

barrier options A class of exotic options that depend on the behaviour of the underlying asset $X(t)$ with respect to some barrier B. The simplest depend merely on whether or not $X(t)$ reaches B, and there are two types: a knock-in option is activated if $X(t)$ reaches B before the exercise time T; a knock-out option is annulled if $X(t)$ reaches B before T. More precisely, if $B < X(0)$ then the knock-in option is called down-and-in, the knock-out option is called down-and-out. For $B > X(0)$ one defines up-and-in and up-and-out options similarly. These can be used to modify most vanilla options, so that for example a European down-and-out option with strike price K has pay-off $\left(X(T) - K\right)^{+}\mathrm{I}\{X(t) > B; 0 \leq t \leq T\}$, where I is the usual indicator. [M. Yor *et al.*, Some combinations of Asian, Parisian and barrier options, in *Mathematics of Derivative Securities*, M. Dempster & S. Pliska (eds). Cambridge University Press, 61–87, 1997]

Bartky's sampling scheme This is a stopped random walk arising in a simple sequential decision procedure. One form is this: to test the quality of a large batch of items, one takes a sequence of (independent) samples each of size m, and the number of defective items in the rth sample is denoted by $X_r + 1, r \geq 1$. If the cumulative total $S_n = \sum_{r=1}^{n} X_r$ (a random walk) ever falls below zero, then the entire batch is accepted; if S_n ever rises above a level R, the batch is rejected. (The threshold R is chosen to minimize the cost of inspection, accepted defectives, and rejection.) [W. Bartky, Multiple sampling with constant probability, *Ann. Math. Stat.*, 14(4), 363–377, 1943]

Bartlett's equation A forward equation for the moment generating function $M(\theta, t) = \mathbb{E}e^{\theta X(t)}$, of a suitable diffusion process $X(t)$. Formally, if $X(t)$ has parameters $a(x,t) = \sum_n a_n(t)x^n$ and $b(x,t) = \sum_n b_n(t)x^n$, then the equation may be written as

$$\frac{\partial M}{\partial t} = \left[\theta b\left(\frac{\partial}{\partial \theta}, t\right) + \frac{1}{2}\theta^2 a\left(\frac{\partial}{\partial \theta}, t\right)\right] M,$$

with the usual conventions for expressing the operator $\frac{\partial}{\partial \theta}$. [D. Cox & H. Miller, *Theory of Stochastic Processes*. Methuen, 1965; reprinted by Chapman & Hall, 2001]

Bartlett–Lewis model (alias: **Poisson cluster process**) Arises as a description of spatial and spatio-temporal processes. One form is this: There is an underlying Poisson process in $\mathbb{R}^d$, and from each point of this process there originates an independent random walk having a finite number of steps, whose endpoints comprise the process. Originally formulated by M. Bartlett as a model for road-traffic, and used by P. Lewis to describe computer failures, it has been much used to model rainfall. [D. R. Cox & V. Isham, *Point Processes*. Chapman & Hall, 1980]

Bartlett's spectrum The spectral measure (i.e. essentially the Fourier transform of the covariance measure) associated with a second-order stationary random measure on $\mathbb{R}^d$. [D. Daley & D. Vere-Jones, *Introduction to the Theory of Point Processes*, 2nd edn. Springer, Vol. I, 2003]

Bartlett's theorem Arises in the theory of queueing systems such that customers move around the system independently of each other. Asserts that if the arrival process of customers is a Poisson process of rate $\lambda(t)$, and a customer arriving at time s is in some state (or set of states) A, at time $s + t$, with probability $P(s,t)$, then the total number of customers in A at time t is Poisson, with parameter $\int_{-\infty}^{t} \lambda(s)P(s, t-s)\mathrm{d}s$. [J. F. C. Kingman, *Poisson Processes*. Oxford University Press, 1993]

base-rate fallacy (alias: **test paradox; false positive paradox**) Arises when conditional probabilities are misinterpreted as absolute probabilities (as psychologists have shown people often do). There are several classic examples, of which the best known (and most hackneyed) is the Medical Test. This infallibly makes the correct diagnosis if you have the disease D; but, with probability 10^{-2}, wrongly says that you do have D when, in fact, you do not. When the test asserts that you have the disease, what is the probability that the diagnosis is correct? Without knowing the probability, $\mathbb{P}(D)$ that a random individual (you) has the disease, one cannot say; to say that it is necessarily high is to commit the base-rate fallacy. [E. Poulton, *Behavioural Decision Theory*. Cambridge University Press, 1994]

basic coupling (alias: **Wasserstein coupling; Vasershtein coupling**) A coupling method for a Markov chain on a state space S that was originally formulated in the context of interacting particle systems. If the transition probabilities of the chain are p_{ij}, the basic coupling for the bivariate Markov chain on $S \times S$ has transition probabilities given (in an obvious notation) by $\mathbb{P}\left((i,j) \to (k,k)\right) = \min\{p_{ik}, p_{jk}\} = p_{ik} \wedge p_{jk}$ and for $i \neq j, k \neq l$

$$\mathbb{P}((i,j) \to (k,l)) = (p_{ik} - p_{jk})^+(p_{jl} - p_{il})^+ / \left\{1 - \sum_k p_{ik} \wedge p_{jk}\right\}.$$

[T. Lindvall, *Lectures on the Coupling Method.* Wiley, 1992; reprinted by Dover, 2002]

basket option Arising in financial mathematics, this is an option (derivative contract) defined in terms of a weighted average of underlying assets, which may be of any form. [M. Musiela & M. Rutkowski, *Martingale Methods in Financial Modelling*, 2nd edn. Springer, 2005]

Bass–Pyke strong law A strong law of large numbers for a family (or an array) of random variables indexed by sets. Let $X_r, r \in \mathbb{N}^d$, be a family of i.i.d.r.vs having finite mean $\mathbb{E}X$. Let C be a collection of Borel-measurable subsets of $[0,1]^d$, and let $|A(\varepsilon)|$ be the volume (i.e. Lebesgue measure) of the region that is within a (Euclidean) distance ε of the frontier (boundary) of a set $A \in C$. If, as $\varepsilon \to 0$, $\sup_{A\in C}|A(\varepsilon)| \to 0$, then, as $n \to \infty$, $\sup_{A\in C}|n^{-d}\sum_{j\in nA}X_j - \mathbb{E}X|A|| \to 0$, where $nA = (nx : x \in A)$, and $|A|$ is the volume of A. [R. Bass & R. Pyke, A strong law of large numbers for partial-sum processes indexed by sets, *Ann. Probab.*, 12(1), 268–271, 1984]

batch discipline Arises in queueing theory, where it denotes a rule in which customers arrive in groups or are served in groups, or both. In the last case, the departing groups need not be the same as the arriving groups. [R. Wolff, *Stochastic Modelling and the Theory of Queues.* Prentice-Hall, 1989]

Bates–Neyman model This seeks to describe accident proneness in a population. A randomly selected individual is assumed to have X and Y accidents respectively, in two disjoint time periods (e.g. successive years), where X and Y have Poisson distributions with respective parameters Z and cZ; here $c > 0$, Z has the gamma density $f(z) = b^a z^{a-1} e^{-bz}/\Gamma(a)$, and X and Y are conditionally independent given Z. Then their joint distn is the bivariate negative binomial

$$f(x,y) = \frac{\Gamma(a+x+y)}{x!y!\Gamma(a)} b^a c^y (1+b+c)^{-(a+x+y)}.$$

[G. Bates & J. Neyman, Contributions to the theory of accident proneness, *Univ. Calif. Publ. Stat.*, 215–276, 1952]

battle of the sexes A classic problem of game theory, comprising a two-person game where the pay-off matrix has, e.g., the form

$$\begin{pmatrix} (a,d) & (b,b) \\ (c,c) & (d,a) \end{pmatrix}, \quad b \le c < a < d.$$

Players maximize joint rewards by cooperating, but separately prefer incompatible strategies. For a single engagement, randomization offers an equilibrium strategy. [K. Binmore, *Playing for Real.* Oxford University Press, 2007]

Baum–Katz Laws Arise in considering the behaviour of the tail of the distribution of the partial sums $S_n = \sum_{r=1}^n X_r$ of a sequence of i.i.d.r.vs X_r, in the case when X_r does not have finite moments of all orders. One formulation is this: for $r \ge s < 2$, $t > 0$, $\sum_{n=1}^\infty n^{\frac{r}{s}-2}\mathbb{P}(|S_n| \ge tn^{\frac{1}{s}}) < \infty$ iff $\mathbb{E}|X|^r < \infty$ and, for the case $r \ge 1$, $\mathbb{E}X = 0$. [H. Lanzinger & U. Stadtmüller, Refined Baum-Katz laws for weighted sums of iid random variables, *Stat. Probab. Lett.*, 69(3), 357–368, 2004]

Baxter–Donsker formula Arises in the context of a random process $X(t)$ with stationary independent increments, where it supplies an expression for the double Laplace transform of the function $S(x,t) = \mathbb{P}[\sup_{0\le u\le t} X(u) \le x]$. For such a process $X(t)$, there is a function $L(\theta)$ such that $\mathbb{E}\exp\{i\theta X(t)\} = \exp\{tL(\theta)\}$. If $L(\theta)$ is real, then the Baxter–Donsker formula is

$$u\int_0^\infty \int_0^\infty e^{-ut-vx} d_x s(x,t) dt = \exp\left\{\frac{1}{2\pi}\int_u^\infty \int_{-\infty}^\infty \frac{v}{v^2+\theta^2} \frac{L(\theta)}{y(y-L(\theta))} d\theta dy\right\}$$

(with a related expression when $L(\theta)$ is complex), valid for $u > 0, v > 0$. [G. Baxter & M Donsker, On the distribution of the supremum functional for process with stationary independent increments, *Trans. Amer. Math. Soc.*, 85(1), 73–87, 1957]

Baxter's identities A set of operator identities that find applications in the fluctuation theory of partial sums of sequences of i.i.d.rvs; in particular, they yield the Pollaczek–Spitzer (or Spitzer's) identity, among other similar relationships. [G. Baxter, An analytic problem whose solution follows from a simple algebraic identity, *Pacific J. Math.*, 10(3), 731–742, 1960]

Baxter processes *see* Gaussian process

Bayesian decision theory A branch of decision theory in which the outcomes (rewards or losses) consequent upon an action (or decision) depend on an unknown parameter $w \in W$, where W is the set of all possible states of nature (or states of the world, or parameter space). The set of actions is A, and if action a is taken when the world is in state w, then a loss $L(a, w)$ is incurred. The distinguishing feature of Bayesian decisions is the assumption that there is a probability distn $\pi(w)$, on W. Then the expected loss $\sum_w L(a, w)\pi(w)$ (or $\int L(a, w)\pi(\mathrm{d}w)$), is denoted by $r(a, \pi)$ and called the risk of a under π. The Bayes risk is the infimum over $a \in A$, $r_B(\pi) = \inf_{a \in A} r(a, \pi)$, and if it is attained by some a_B, then a_B is called the Bayes decision (or rule, or solution) against π. If the infimum is not attained, but it is possible to take an action with risk arbitrarily close to r_B, this is an extended Bayes decision. [J. Berger, *Statistical Decision Theory and Bayesian Analysis*, 2nd edn. Springer, 1985]

Bayes network (alias: **belief (or credal) network; Bayes net; directed graphical model**.) Arises in contexts such as machine learning and expert systems. It is a directed acyclic graph $G = (E, V)$, where each vertex $v \in V$ is labelled with both a random variable X_v, and the conditional distribution of X_v given its parents (i.e. the random variables at the other end of all edges incoming to v). Any variable labelling a vertex that is not descendent from v is conditionally independent of X_v, given the labels of parents of X_v. The **Markov blanket** of v is the smallest subset B of v such that X_v is conditionally independent of all labels in $V \backslash B$, given all the labels in $B \backslash \{v\}$. It comprises the parents, children, and spouses of V (where a spouse is a vertex with which V shares parenthood of some other $w \in V$).

If the tree has no branches (a bare trunk), and the edges are all oriented in the same direction (seen as time), then the process is a Markov chain. In general, it can be regarded as a directed Markov tree, having a local directed Markov property. [R. Cowell *et al.*, *Probabilistic Networks and Expert Systems*. Springer, 2007]

Bayes probability *see* inverse probability

Bayes strategy Broadly, any rule for making decisions that minimizes risk (or maximizes expected utility) where the expectation is taken over some prior (or posterior, following an experiment) distribution of states of the world. For example, in game theory when a player has a subjective (or other prior) distribution on the other players' strategies. [J. Berger, *Statistical Decision Theory and Bayesian Analysis*, 2nd edn. Springer, 1985]

Bayes theorem (alias: **Bayes theorem of Inverse Probability**) This is a straightforward consequence of the definition of conditional probability, that is of great importance because of its interpretations and applications. Formally, it is the statement that

$$\begin{aligned}\mathbb{P}(A|B) &= \mathbb{P}(A \cap B)/\mathbb{P}(B) = \mathbb{P}(B|A)\mathbb{P}(A)/\mathbb{P}(B)\\ &= \frac{\mathbb{P}(B|A)\mathbb{P}(A)}{\mathbb{P}(B|A)\mathbb{P}(A) + \mathbb{P}(B|A^C)\mathbb{P}(A^C)},\end{aligned}$$

where the expanded version is often called Bayes formula, or Bayes lemma, etc. (This was proposition 5 of T. Bayes's famous essay.) In one important interpretation, A is a hypothesis about the state of nature (or the value of a parameter) and B is an observed event (which it is assumed depends on the state of nature). In another interpretation, A is a possible cause of the observed effects B. Since A is envisaged as preceding B, temporally or causally, the statement of the theorem is one of inverse probability; that is, the probability of causes (or states), given their effects (or consequences). The direct equivalent is $\mathbb{P}(B \mid A) = \mathbb{P}(A \cap B)/\mathbb{P}(A)$.

Bayes used this theorem to consider the following problem in drawing inferences from observations. Let X be binomial $\mathrm{B}(n, \theta)$, where θ is an unknown parameter (or state of nature). It is natural to treat θ as a random variable, though it may well not be. Assuming that θ has a (prior) distribution that is uniform on $[0, 1]$ (which is an application of the indifference principle), he showed that for $0 \le c < d \le 1$

$$\mathbb{P}(c < \theta \le d, X = x) = \binom{n}{x} \int_c^d \theta^x (1-\theta)^{n-x} \mathrm{d}\theta,$$

$$\mathbb{P}(c < \theta \le d \mid X = x) = \frac{(n+1)!}{x!(n-x)!} \int_c^d \theta^x (1-\theta)^{n-x} \mathrm{d}\theta.$$

Both of these have also been referred to as 'Bayes theorem'. In considering the conditional densities $f(x \mid y)$ and $f(y \mid x)$ of jointly continuous r.vs X and Y, the analogous fact that $f(y \mid x) = f(x \mid y) f_Y(y)/f_X(x), f_X(x) > 0$, is sometimes called the generalized Bayes theorem. [T. Bayes & R. Price, An essay towards solving a problem in the doctrine of chances, *Phil. Trans.*, 53, 370–418, 1763; A. Dale, *A History of Inverse Probability*, 2nd edn. Springer, 1991]

BBGKY hierarchy Arises in interacting particle systems (statistical mechanics), where it comprises a family of integro-differential equations such that the distribution function for n particles is expressed in terms of integrals of distributions for $n + 1$ particles, $n \ge 1$. [D. Ya. Petrina, *Stochastic Dynamics and Boltzmann hierarchy*. de Gruyter, 2009]

BCMP theorem (alias: **Baskett–Chandy–Muntz–Palacios theorem**) Arises in queueing theory, where it identifies a class of networks having the property that the equilibrium distribution of queue lengths has a product form over nodes of the network. This is effectively a generalization of Jackson's network theorem. In a BCMP network arrivals must be Poisson, and independent for different types of customer, and routing of customers between queues must be independent of queue lengths. Each queue may be of classic M/M/1 type, but they may also (subject to some conditions on the nature of service times) be of three further types: shared service (each customer present gets an equal share of the server's attention); instant service (infinitely many servers); or last-come-first-served, with preemption (but no loss of elapsed service time if preempted). [F. Baskett *et al.*, Open, closed, and mixed networks of queues with different classes of customers, *J. ACM*, 22(2), 248–260, 1975]

Beardwood–Halton–Hammersley theorem Arises in the Travelling Salesman Problem, which seeks the shortest route R_n through n points in $\mathbb{R}^d$. It asserts that if the points are i.i.d. in $\mathbb{R}^d$, having a density $f(x)$ that is zero outside a finite d-dimensional box, then as $n \longrightarrow \infty$

$$R_n/n^{(d-1)/d} \longrightarrow c(d)\int_{\mathbb{R}^d} f(x)^{(d-1)/d}\,\mathrm{d}x,$$

with probability 1, where $c(d)$ depends on d, but not $f(x)$. [J. Steele, *Probability Theory and Combinatorial Optimization*. SIAM, 1997]

behavioural decision theory (alias: **behavioural models**) Classical decision theory, as developed in the hands of A. Wald and his successors, is concerned with identifying optimal choices on the assumption that the agents involved have mathematically rational preferences. There is evidence that real individuals do not meet these criteria, so behavioural decision theory seeks to characterize empirical preferences, and hence describe the nature of decisions under risk or uncertainty that are actually made. (The classical theory is said to be normative, the behavioural theory is descriptive.) [E. Poulton, *Behavioural Decision Theory*. Cambridge University Press, 1994]

Behrens–Fisher distribution Let X and Y be independent r.vs having t-distributions with respective parameters λ and μ, and let $\theta \in (0, \frac{\pi}{2})$. Then the r.v. $Z = X\cos\theta - Y\sin\theta$ is said to have a Behrens–Fisher distn, with parameters λ, μ, and θ. [Fco. J. Girón & C. del Castillo, The Multivariate Behrens–Fisher distribution, *J. Multivariate Anal.*, 101(9), 2091–2102, 2010]

Belgian progression (alias: **le montant Belge; breadwinner system**) A betting system resembling the better-known Labouchere system. A player engages in a series of bets, and in the course of so doing maintains a sequence S of numbers that guide the amounts wagered:

1. Fix an integer $B > 0$.
2. Initially S is empty; the first wager is of size 1, and this is the first member of S.
3. After a loss of size L, place L at the right of the sequence; if the same loss is incurred in B consecutive bets then the size of the next wager is increased by 1.
4. After a win of size W, delete consecutive terms from the left of S whose sum is W, if possible; if impossible then delete the minimum to exceed W and insert the (positive) difference on the left in replacement.
5. Bet sizes are the same after a win, unless the resulting wager is greater than the sum σ of all terms in S, in which case the next bet is $\sigma + 1$.
6. When S is empty, the gambler is in profit by 1, and can either stop with that gain, or begin afresh.

[S. Ethier, *The Doctrine of Chances*. Springer, 2010]

belief function Arises in the quest to measure how far one accepts the truth of propositions such as, e.g., statements of fact ('The event A occurred'), or the statements of a propositional language. A subjective interpretation of probability asserts that $\mathbb{P}(A)$ simply measures our degree of belief in the occurrence of A, so in this case the probability $\mathbb{P}(\cdot)$ can be called a belief function. If one further asserts that a rational believer must be happy to use all these beliefs as fair odds for bets, but would not be happy to lose money for certain, then this belief function obeys the usual (mathematical) axioms for probability. If these constraints on belief are rejected, then belief functions do not necessarily obey the rules for probability functions; *see* Dempster–Shafer theory. [R. Yager & L. Liu (eds) *Classic Works of the Dempster–Shafer Theory of Belief Functions*. Springer, 2008]

belief propagation (alias: **sum-product method; Lauritzen–Spiegelhalter–Pearl procedure**) Arises in Bayes networks and similar probabilistic structures, where it gives a method for calculating marginal distributions in terms of suitable sums of products of conditional probabilities. [R. Cowell *et al.*, *Probabilistic Networks and Expert Systems*. Springer, 2007]

Bell–Anderson model (alias: **size-dependent branching**) Arises as a model for the growth and division of cells, in which the probability of a cell dividing at any time is proportional to its size at that time, and the mass of the cell at splitting is divided among its daughter cells. [M. Aout, A branching cell population model allowing for cell death, *J. Math. Stat.*, 2(2), 373–377, 2006]

Bell inequality (alias: **Bell–Boole inequality**) A basic inequality that has important implications in quantum mechanics. J. Bell's form asserts that for jointly distributed random variables X, Y, and Z, each taking values in $\{-1, +1\}$, we have $1 - \mathbb{E}(XY) \geq \mathbb{E}|XZ - YZ|$; in fact this holds more generally for X, Y, and Z taking values in $[-1, 1]$. A version expressed in terms of probabilities is Wigner's inequality; more generally for any three events A, B, and C, $\mathbb{P}(A \cap B^c) + \mathbb{P}(B \cap C^c) \geq \mathbb{P}(A \cap C^c)$, in which form it was known to G. Boole. Similarly, for any four random variables W, X, Y, and Z, taking values in $[-1, 1]$, $\mathbb{E}|XZ - YZ| + \mathbb{E}|XW + YW| \leq 2$. [J. Bell, *Speakable and Unspeakable in Quantum Mechanics*, 2nd edn. Cambridge University Press, 2004]

Bellman equation (alias: **optimality equation; dynamic programming equation; Wald–Bellman equation**) Arises in the context of optimal control, or more broadly in a large class of problems such that one seeks to make a sequence of decisions that are optimal (according to a suitable criterion). The verbal statement of the essential idea is Bellman's (optimality) principle; the mathematical formulation is Bellman's (optimality) eqn; and the whole process of finding the optimal solution by this method is called dynamic programming. The underlying framework comprises a process $X(t) \in S$, which may be random; for simplicity we assume that $t \in \mathbb{N}$. At each time t there is a set A of feasible (allowable) actions (or decisions), depending on $X(t)$ (and possibly t). Then, depending on the state $X(t)$ and the action a selected, a reward $R(X, a)$ is accrued (or a cost incurred) and the distribution of the next value $X(t+1)$ of the process is fixed. The problem is to choose a sequence of actions to maximize a function of the rewards (such as total expected reward, or expected discounted reward).

Bellman's principle asserts that: 'An optimal policy has the property that, whatever the initial state and initial decisions are, the remaining decisions must constitute an optimal policy with regard to the state resulting from the first decision.' For definiteness, assume that $S = \mathbb{Z}$, that rewards are simply added, and that the process $X(t)$ runs over $0, \ldots, T$. Define the so-called value function $V_T(x_0)$ to be the maximum possible reward when $X(0) = x_0$. Denote the probability that $X(t+1)$ is k, given $X(t) = j$ and action a is taken by $p(j, k;\ a)$. Then the maximum reward following the first step is $R(x_0, a) + \sum_k p(x_0, k; a) V_{T-1}(k)$. Hence, by Bellman's principle, Bellman's eqn is $V_T(x_0) = \max_{a \in A}\{R(x_0, a) + \sum_k p(x_0, k; a) V_{T-1}(k)\}$.

A similar argument can be applied to suitable processes in continuous time, yielding differential equations that may be called the Hamilton–Jacobi–(Bellman) eqn. Such problems become much harder to solve as the size of S, or dimensionality of $X(t)$, increases. This is called the **curse of dimensionality**, and was so named by R. Bellman in 1961. [M. Puterman, *Markov Decision Processes*. Wiley, 1994]

Bellman–Harris process *see* age-dependent process

bell-shaped density In general, a term denoting a density $f(x)$ resembling the cross-section of a bell, being continuous and symmetric about a single mode m, where $f(x)$ decreases for $x > m$ and has just two points of inflection where $f'' = 0$. More narrowly, bell curve refers to the normal density.

Benders' decomposition (alias: **Benders' partitioning; L-shaped decomposition**) Arises in stochastic programming in the presence of many constraints, where it gives an iterative method that performs a sequence of simpler optimizations, checking each for feasibility and optimality at each stage. [A. Conejo *et al.*, *Decomposition Techniques in Mathematical Programming*. Springer, 2006]

Benford's distribution (alias: **law of anomalous numbers; first-(or leading-)digit law; significant-digit law**) Arises empirically in collections of numbers derived from unrelated or disparate sources. It is often found that the distribution of the first significant digit of the (decimal) numbers therein is given approximately by Benford's distribution:

$$f(k) = \log_{10}(1 + k^{-1}), \quad 1 \leq k \leq 9.$$

This discovery arose from the observation, made by S. Newcomb in 1881, and again by F. Benford in 1938, that well-used tables of logarithms are more heavily thumbed (and grubbier) towards the beginning. (The result holds independently of the base of the number, in the appropriate reformulation.) This phenomenon has been explained theoretically by **Hill's**

thm, which (broadly) asserts that if random samples are taken from randomly selected distributions, in such a way that no particular scale of measurement is favoured over another in the long run, then limiting distn of significant digits is that of Benford's Law. [A. Berger & T. Hill, A basic theory of Benford's Law, *Probab. Surv.*, 8, 1–126, 2011]

Benini distribution The distribution function $F(x)$ is called a Benini distribution if it satisfies, for positive a and b,

$$\log(1 - F(x)) = -a\log\left(\frac{x}{c}\right) - b\left\{\log\left(\frac{x}{c}\right)\right\}^2, x \geq c > 0.$$

[C. Kleiber & S. Kotz, *Statistical Size Distributions in Economics and Actuarial Sciences*. Wiley, 2003]

Benktander distributions Arise in insurance mathematics, where they are used when heavy-tailed distributions are called for. There are two types, whose distributions $F(x)$ are given, for $x > 1$, and suitable $c > 0$, by

$$1 - F(x) = cx^{a-1}\exp\{-b(x^a - 1)/a\}, 0 < a < 1, b > 0,$$

and

$$1 - F(x) = c\left(1 + 2\left(\frac{b}{a}\right)\log x\right)x^{-a-1}\exp\{-b(\log x)^2\},$$

$$a > 0, a(a+1) > 2b.$$

[C. Kleiber & S. Kotz, *Statistical Size Distributions in Economics and Actuarial Sciences*. Wiley, 2003]

Bennett's inequality Arises in the context of bounding the probability distribution of a sum of independent bounded random variables. It may denote any one of three improvements of Bernstein's inequality, given by G. Bennett, but usually refers to this one: let $X_1, \ldots, X_n$ be independent random variables that are bounded, i.e. $|X_r - \mathbb{E}X_r| \leq b_r$ and (without loss of generality) assumed to have zero mean. Write $b = \max_r b_r$ and $S = \sum_{r=1}^{n} X_r$, and $\sigma^2 = \text{var}S$. Then, for $t > 0$,

$$\mathbb{P}(S \geq \sigma t) < e^{\sigma t/b}(1 + tb/\sigma)^{-\left(t\sigma/b + (\sigma/b)^2\right)}$$
$$= \exp\{-(\sigma/b)^2 g(tb/\sigma)\},$$

where $g(x) = (1+x)\log(1+x) - x$. [S. Boucheron *et al.*, *Concentration Inequalities*. Oxford University Press, 2013]

Beppo Levi theorem (or lemma) (alias: **B. L. monotone convergence theorem; Fatou–B.L. theorem**) Arises in the theory of integration. In the context of probability theory, one form is this: let X_n be a non-decreasing sequence of random variables with finite expectations $\mathbb{E}X_n$, such that $\sup_n \mathbb{E}X_n < \infty$; then, as $n \to \infty$, X_n converges a.s. to a finite random variable X such that $\mathbb{E}X = \lim_{n\to\infty}\mathbb{E}X_n < \infty$. [M. Capinski & E. Kopp, *Measure, Integral and Probability*, 2nd edn. Springer, 2004]

Beran distribution A family of rotationally invariant directional distributions in $\mathbb{R}^d$, $d \geq 2$, whose mth member has a density proportional to $\exp\{\sum_{r=1}^{m} a_r H_r\}$, where H_r is the (spherical) surface harmonic of order r (i.e. H_r is homogeneous of degree r in $x_1, \ldots, x_d$ and satisfies Laplace's equation), for suitable constants a_r. [R. Beran, Exponential models for directional data, *Ann. Stat.*, 7(6), 1162–1178, 1979]

Berge's inequality A bivariate inequality of Chebyshov type for jointly distributed r.vs X and Y, having correlation ρ, and standard deviations σ_X and σ_Y. For $t > 0$, $\mathbb{P}(|X - \mathbb{E}X| \geq t\sigma_X$ or $|Y - \mathbb{E}Y| \geq t\sigma_Y) \leq t^{-2}\left[1 + (1-\rho^2)^{\frac{1}{2}}\right]$. Lal's inequality extends this to assert that for s, $t > 0$, $\mathbb{P}(|X - \mathbb{E}X| \geq s\sigma_X$ or $|Y - \mathbb{E}Y| \geq t\sigma_Y) \leq \frac{1}{2s^2t^2}\left[s^2 + t^2 + [(s^2+t^2)^2 - 4\rho^2 s^2 t^2]^{\frac{1}{2}}\right]$. [I. Savage, Probability inequalities of the Tchebycheff type, *J. Res. Nat. Bus. Stand. B, Math. Math. Phys.*, 65, 211–222, 1961]

Berkson's paradox (or fallacy, or bias) Denotes the fact that attributes that are independent in a population, and which are sampled independently from the population (at different rates) need not be independent in the resulting selected subpopulation. This can be seen as an example of the Yule–Simpson paradox, where the conditioning event is selection. [J. Berkson, Limitations of the application of fourfold table analysis to hospital data, *Biometrics Bull.*, 2(3), 47–53, 1946]

Bermuda option Arises in financial mathematics, where it denotes a derivative contract that can be exercised only at one of a predetermined set B of times up to the expiry date T. The set B may be discrete (e.g. any Monday), or an interval, e.g. $[\tau, T]$ for $\tau > 0$. [M. Musiela & M. Rutkowski, *Martingale Methods in Financial Modelling*, 2nd edn. Springer, 2005]

Bernoulli distribution The r.v. X is said to have the Bernoulli distribution with parameter p if $X \in \{0, 1\}$, and

$\mathbb{P}(X = 1) = p = 1 - \mathbb{P}(X = 0)$. Then X may be called a Bernoulli r.v. (or an indicator r.v. in several contexts) (*see* Bernoulli trial). More generally, a vector $X \in \{0, 1\}^d$ may be said to have a d-dimensional multivariate Bernoulli distribution, which may be parameterized in various different ways. [J. Teugels, Some representations of the multivariate Bernoulli and binomial distributions, *J. Multivariate Anal.*, 32(3), 256–268, 1990]

Bernoulli ensemble Broadly, a random matrix whose entries X_{ij} are i.i.d. binary random variables. More strictly, a matrix whose entries are i.i.d. Bernoulli r.vs (so that $X_{ij} \in \{0, 1\}$); the random matrix whose entries $X_{ij} \in \{-1, 1\}$ may be called a **Rademacher ensemble.** [T. Tao, *Topics in Random Matrix Theory*. American Mathematical Society, 2012]

Bernoulli–Laplace (urn) model Arises as an elementary model for diffusion. Two urns each contain n balls (particles), and of these $2n$ balls, n are of one type and n are of another type. At each stage of the process two balls are selected, one from each urn, independently at random, and exchanged between their respective urns. More generally, there may be unequal numbers of each type. The term also denotes an interacting particle system that is otherwise known as the symmetric exclusion process. [P. Donnelly *et al.*, Approach to stationarity of the Bernoulli-Laplace diffusion model, *Adv. Appl. Probab.*, 26 (3),715–727, 1994]

Bernoulli process (alias: **binary process; Bernoulli scheme (or shift)**) In its basic form, this comprises a sequence of i.i.d.r.vs $X_n, n \in I$, where $X_n \in \{0, 1\}$ and the index set I is commonly $\mathbb{N}$ or $\mathbb{Z}$. Such a process arises naturally as the indicators of success in a sequence of Bernoulli trials. More generally, a sequence of i.i.d. simple r.vs (or of r.vs taking values in a compact set) may be called a Bernoulli process; similarly, for a probability measure $\mathbb{P}$ on a standard probability space (isomorphic to a mixture of Lebesgue measure on [0,1] and discrete distn) the product measure $\mathbb{P}^{\mathbb{Z}}$ on the usual product space may be called a Bernoulli measure. In ergodic theory, a suitable 1–1 invertible measure-preserving transformation T on a probability space S such that the family $T^n S, n \in \mathbb{Z}$ is independent, may be called a Bernoulli shift; some authors use the terms Bernoulliness and Bernoullicity in this context. [R. Gallager, *Stochastic Processes*. Cambridge University Press, 2013; D. Ornstein, *Ergodic Theory, Randomness, and Dynamical Systems*. Yale, 1974]

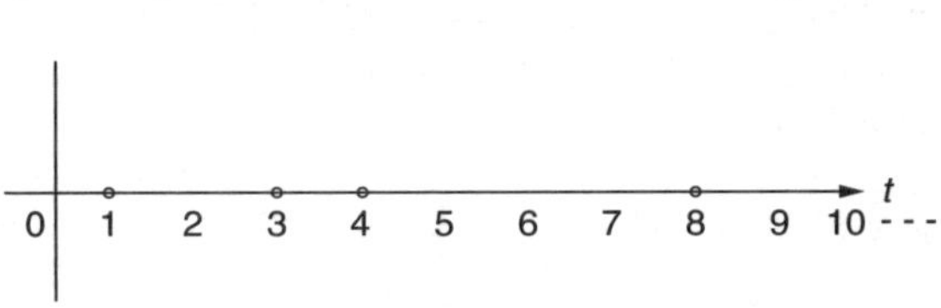

Simple Bernoulli process whose points (occurrences) are denoted by ∘.

Bernoulli random walk (alias: **simple random walk; binomial r.w.**) The random walk $S_n = \sum_{r=1}^{n} X_r$, where the sequence X_r, comprises i.i.d.r.vs having a Bernoulli (two-point) distribution. Usually, it is assumed that $X_r \in \{-1, 1\}$. Occasionally the term is used for the monotone walk when $X_r \in \{0, 1\}$. [F. Spitzer, *Principles of Random Walk*, 2nd edn. Springer, 1976]

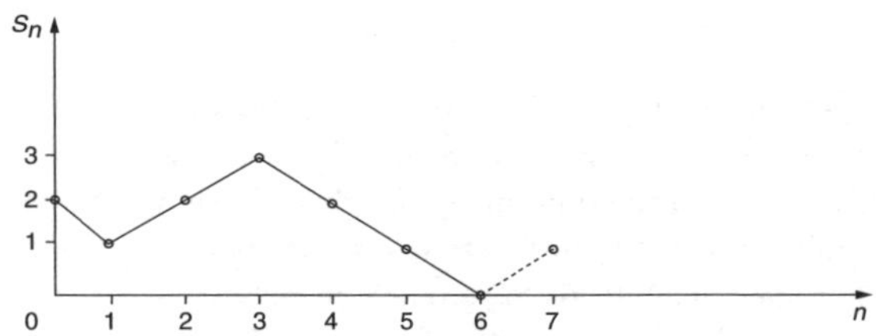

Bernoulli, or simple, random walk. Here $S_0 = 2$, and the walk makes a first passage to the origin at step 6.

Bernoulli sieve A procedure for generating a random ordered partition (or composition) of an integer n. It can be visualized as the random placement of n balls in a sequence of boxes, according to a suitable distribution, or as the following game for n players: there is an infinite sequence $C_1, C_2, \ldots$ of i.i.d.r.vs taking values in $(0, 1)$. At the first round, given C_1, all n players each flip an independent coin with probability C_1 of showing heads, and any player whose coin yields a head is eliminated. The survivors repeat the procedure with coins for which the probability of heads is C_2, and so on. Ignoring rounds that all players survive, the number dropping out in each round supplies the decomposition. [A. Gnedin, The Bernoulli sieve, *Bernoulli*, 10(1), 79–96, 2004]

Bernoulli's theorem (alias: **weak law of large numbers for Bernoulli trials**) Consider n independent trials, each yielding success with probability p, or failure with probability $q = 1 - p$. (This is thus a Bernoulli process: in Bernoulli's original proof $p = r/(r + s)$, where

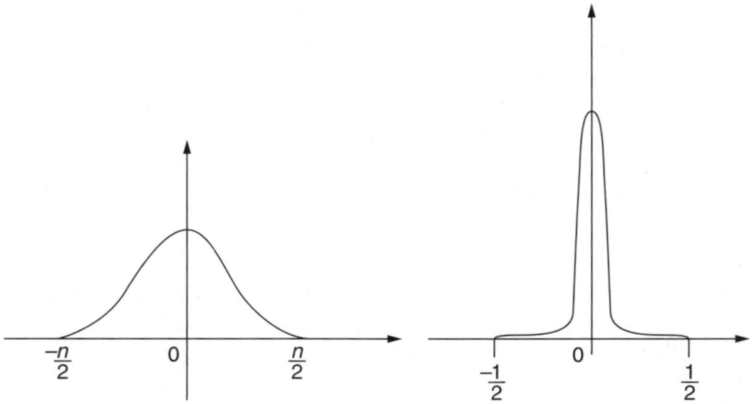

***Left:* Distribution of probability mass for $X_n - \frac{1}{2}n$, where X_n has B $(n, \frac{1}{2})$ distribution for large n. *Right:* Distribution of probability mass for $\frac{1}{n}(X_n - \frac{1}{2}n)$; as n increases the diagram approaches a spike shape. Histograms have been smoothed for simplicity.**

r and s are integers.) Let the total number of successes be S_n, so that the relative frequency of successes is $R_n = S_n/n$. Then (informally) the theorem shows that the probability that R_n differs from p by any positive amount, however small, tends to zero as n increases.

Formally, for any $\varepsilon > 0$, as $n \to \infty$, $\mathbb{P}(|R_n - p| > \varepsilon) \to 0$. This was first proved by James Bernoulli during 1688–1690, and published posthumously in *Ars Conjectandi* in 1713. In that same year another Bernoulli, Nicholas, proved a markedly stronger version of the theorem (tighter bounds on the tail of the binomial distribution), so it could also be called Bernoullis' theorem. It is of great importance in the history of probability. [A. Hald, *History of Probability and Statistics and their Applications Before 1750*. Wiley, 1990]

Bernoulli trial A probability space with a sample space $\Omega = \{S, F\}$, comprising two points. A random variable X defined on Ω may be called a Bernoulli r.v.; or an indicator if $X \in \{0, 1\}$. A probability space in which Ω comprises $t \geq 3$ points may be called a **de Moivre trial**; the joint distribution of the numbers of outcomes in n independent trials is multinomial. [W. Feller, *Introduction to Probability Theory and its Applications*, Vol. 1, 3rd edn. Wiley, 1968]

Bernstein–Feller central limit theorem (alias: **generalized CLT; GCLT**) Let $X_r, r \geq 1$, be a sequence of independent r.vs having respective distribution functions $F_r(x)$. The GCLT holds for this sequence if there exist sequences of constants a_r and $b_r > 0, r \geq 1$, such that (as $n \to \infty$) the array X_r/b_n, is null (i.e. u.a.n., or infinitesimal) and $\lim_{n\to\infty}\mathbb{P}(b_n{}^{-1}\sum_{r=1}^{n} X_r - a_n \leq x) \to \Phi(x)$. The Bernstein–Feller CLT asserts that this GCLT holds iff there are constants $C_r > 0$ such that $\lim_{n\to\infty}\sum_{r=1}^{n}\mathbb{P}(|X_r| \geq C_n) = 0$, and $\lim_{n\to\infty} C_n^{-2} \sum_{r=1}^{n} \mathrm{var}[X_r I(X_r \leq C_n)] = \infty$, where $I(\cdot)$ is the usual indicator function. [H. Tucker, *A Graduate Course in Probability*. Academic Press, 1967; reprinted by Dover, 2014]

Bernstein function *see* completely monotone

Bernstein's inequality Arises in the context of bounding the probability distn of a sum of independent random variables whose absolute moments satisfy a so-called Bernstein condition. Let $X_1, \ldots, X_n$ be independent random variables such that $\mathrm{var}\, X_r = \sigma_r^2$, and whose means, without loss of generality, are assumed to be zero. Set $S = \sum_{r=1}^{n} X_r$, and $\sigma^2 = \mathrm{var}\, S$. If, for $q \geq 2$, and some constant C, $\mathbb{E}|X_r|^q \leq \frac{1}{2}\sigma_r^2 C^{q-2} q!$, which is Bernstein's condition, then for $t > 0$

$$\mathbb{P}(S \geq t\sigma) \leq \exp\left\{\frac{-t^2}{2 + 2ct/\sigma}\right\}.$$

If, more simply, with probability one we have $\max_r |X_r| \leq m$, then

$$\mathbb{P}(S \geq t\sigma) \leq \exp\left\{\frac{-t^2}{2 + \frac{2}{3}mt/\sigma}\right\}.$$

A related inequality, for such uniformly bounded r.vs, also called Bernstein's inequality, is

$$\mathbb{P}\left(\left|\frac{1}{n}S\right| > t\right) \leq 2\exp\left\{-an\min\left\{\frac{t}{b}, \frac{t^2}{b^2}\right\}\right\},$$

for some absolute constant a, and suitable constant b. [S. Boucheron *et al.*, *Concentration Inequalities*. Oxford University Press, 2013]

Bernstein–Kac theorem *see* Skitovich–Darmois theorem

Bernstein's lemma In probability theory, this denotes a Slutsky-type theorem that is commonly stated thus: let $Z_n = X_n + Y_n$ for a sequence of random vectors $(X_n, Y_n), n \geq 1$. If X_n converges in distribution (as $n \to \infty$) to X with distribution $F(x)$, and Y_n converges in probability to 0, then the distribution of Z_n converges to $F(x)$ as $n \to \infty$. [H. Kobayashi *et al.*, *Probability, Random Processes, and Statistical Analysis*. Cambridge University Press, 2012]

Bernstein's method A term used to denote several ideas; in probability it most often arises in considering limit laws for sums of weakly dependent r.vs by dividing the sum into blocks that are asymptotically independent, or whose contribution is asymptotically negligible; this is also called the **blocking** (or **cross-section**) **method**. [E. Bolthausen, On the central limit theorem for stationary mixing random fields, *Ann. Probab.*, 10(4), 1047–1050, 1982]

Bernstein (reciprocal) process *see* reciprocal process

Bernstein's theorem Any of several results, but often this: A function $M(\theta), \theta \geq 0$, is the Laplace transform of a probability distribution iff $M(0) = 1$, and it has derivatives $M^{(n)}(\theta)$ of all orders such that $(-1)^n M^{(n)}(\theta) \geq 0, \theta > 0$, for all n (which is to say that $M(\theta)$ is completely monotone). [W. Feller, *Introduction to Probability Theory and its Applications*, Vol. 2, 2nd edn. Wiley, 1971]

Bernstein–Weierstrass theorem Let $g(p)$ be a continuous function, $0 \leq p \leq 1$. Then as $n \to \infty$, and uniformly in $0 \leq p \leq 1$,

$$\sum_{k=0}^{n} g\left(\frac{k}{n}\right)\binom{n}{k}p^k(1-p)^{n-k} \to g(p).$$

[W. Feller, *Introduction to Probability Theory and its Applications*, Vol. 2, 2nd edn. Wiley, 1971]

Berry–Esseen inequality *see* smoothing inequality

Berry–Esseen theorem Arises in the context of the central limit theorem, where it supplies a bound for the accuracy of the normal approximation to the distribution of the suitably scaled sum of n independent random variables $X_1, \ldots, X_n$. For the simplest case, these are i.i.d. zero-mean with var $X_r = \sigma^2$, $\mathbb{E}|X_r|^3 = \tau < \infty$, and we set $S_n = \sum_{r=1}^n X_r/(\sigma\sqrt{n})$, with distn $F_n(x)$. Then there is an absolute (or numerical) constant C (which is to say that C does not depend on any other parameters of the problem) such that

$$\sup_x |F_n(x) - \Phi(x)| \leq C\tau\sigma^{-3}n^{-\frac{1}{2}},$$

where $\Phi(x)$ is the standard normal distn fn. And, for all values of x, there is a numerical constant D such that $|F_n(x) - \Phi(x)| \leq \dfrac{D\tau}{(1+|x|^3)\sigma^3\sqrt{n}}$, which may be called the non-uniform Berry–Esseen bound. More generally, when the r.vs are not identically distributed, so that var$X_r = \sigma_r^2$, and $\mathbb{E}|X_r|^3 = \tau_r$, we set $S_n = \sum_{r=1}^n X_r/(\sum_{r=1}^n \sigma_r^2)^{\frac{1}{2}}$ and we have Esseen's bound

$$\sup_x |F_n(x) - \Phi(x)| \leq B\sum_{r=1}^{n}\tau_r \Big/ \left(\sum_{r=1}^{n}\sigma_r^2\right)^{3/2},$$

for some numerical constant B. [W. Feller, *Introduction to Probability Theory and its Applications*, Vol. 2, 2nd edn. Wiley, 1971]

Bertrand's box paradox A classic problem constructed to show that elementary uninformed reasoning about questions in probability, especially those involving conditional probability, often leads to errors. The problem is this: there are three boxes B_1, B_2, and B_3, with two compartments in each box, and one coin in each compartment. B_1 contains two gold coins, B_2 contains two silver coins, and B_3 contains a coin of each metal. You select a box at random, and open a compartment at random, to find a gold coin. What is the probability γ that the other compartment of your box contains a gold coin? A standard use of conditional probability proves that

$\gamma = \frac{2}{3}$. (By contrast, the following fallacious argument suggests that $\gamma = \frac{1}{2}$: 'The selected box is either B_1 or B_3, so $\gamma = \frac{1}{2}$ because you were equally likely to have picked either of these'.) [P. Gorroochurn, *Classic Problems of Probability*. Wiley, 2012]

Bertrand's circle paradox (alias: **Bertrand's random chord paradox**) A classic problem constructed to show the difficulty of defining a natural probability measure to describe the distribution of a randomly selected geometrical object. The problem is this: an equilateral triangle T is inscribed in a circle C, and a chord R of the circle is drawn at random. What is the probability λ that the length of a side of T exceeds the length of R? Since the distribution of R is undefined, the problem is ill-posed and has no unique solution. Here are five:

(a) Pick a point P uniformly at random in C, and take P as the midpoint of R. Then $\lambda = \frac{1}{4}$.
(b) Pick the ends of R independently and uniformly at random on the perimeter of C. Then $\lambda = \frac{1}{3}$.
(c) Pick a radius D of C uniformly at random, and pick a point P uniformly on D to be the midpoint of R. Then $\lambda = \frac{1}{2}$.
(d) Pick a point P uniformly at random in C and choose R to pass through P with direction selected uniformly at random. Then $\lambda = \frac{2}{3} - \sqrt{3}/(2\pi)$.
(e) Pick a point P uniformly at random on a radius D selected uniformly at random, and choose R to pass through P with direction selected uniformly at random. Then $\lambda = \frac{2}{3} + \frac{1}{\pi} \log \tan \frac{\pi}{12}$.

The difficulty is resolved if it is agreed that the distribution of the line should be invariant under translations and rotations. In that case, the appropriate interpretation of 'at random' is (c), and $\lambda = \frac{1}{2}$.
[E. Jaynes, The well-posed problem, *Foundations Phys.*, 3(4), 477–492, 1973]

Bessel distribution A variety of distributions are known under this name. Those most commonly seen include these densities:

(a) $f_1(x) = e^{-x} x^{-1} r I_r(x), \quad x \geq 0, \quad r \geq 1$, where $I_r(x)$ is the modified Bessel function of the first kind. It arises naturally in the context of random walks: if T is the first-passage time to r of a simple symmetric random walk, and $X_n, n > 1$, are i.i.d.r.vs having the exponential density with parameter 1, then the density of $\sum_{r=1}^{T} X_r$ is $f_1(x)$.
(b) For positive constants a, b, and c,

$f_2(x) =$

$$2^{(c-3)/2} x^{(c-1)/2} a^c b^{(1-c)} \exp\left\{-\frac{a}{2}x - \frac{b^2}{4a}\right\} I_{c-1}\left\{b\left(\frac{x}{2}\right)^{\frac{1}{2}}\right\}, x \geq 0.$$

(c) For $a > 0$, $b > -\frac{1}{2}$, and $x \in \mathbb{R}$

$$f_3(x) = a^{b+1} |x|^b K_b(a|x|) \Big/ \left\{2^b \sqrt{\pi}\Gamma\left(b + \frac{1}{2}\right)\right\},$$

where K_b is the modified Bessel function of the second kind, and Γ is the gamma function. Densities of this type are sometimes called (Bessel) K form densities.
In addition, a discrete r.v. X on the non-negative integers is said to have the Bessel distribution if

$$\mathbb{P}(X = n) = \left(\frac{a}{2}\right)^{2n+b} \Big/ \{n!\, I_b\,(a)\Gamma\,(n + b + 1)\}, \quad n \geq 0$$

where $a > 0$ and $b > -1$. Finally, the transition density of the Bessel process is sometimes referred to as the Bessel density. [L. Yuan & J. Kalbfleisch, On the Bessel distribution and related problems, *Ann. Inst. Stat. Math.*, 52(3), 438–447, 2000]

Bessel process (alias: **radial Brownian motion**) A diffusion (i.e. continuous Markov) process, that arises naturally in the context of Brownian motion (the Wiener process) in d dimensions, which we denote by $W(t) = (W_1(t), \ldots, W_d(t))$. Then $R_d(t) = (W_1{}^2 + \cdots + W_d{}^2)^{\frac{1}{2}}$, $t \geq 0$, is the d-dimensional Bessel process, being the radial distance of $W(t)$ from the origin. The transition (Bessel) density is

$$f(y,\ t;\ x) = t^{-1}\, y^{d-1}\, (xy)^{1-d/2}\, I_{\frac{d}{2}-1}\left(\frac{xy}{t}\right) \exp\left(-\frac{x^2 + y^2}{2t}\right),$$

where I is a modified Bessel function of the first kind. Sometimes one writes $v = \frac{d}{2} - 1$, and refers to a Bessel process of order (or index) v. More generally, a Bessel process can be defined for non-integer d as the solution of the SDE

$$dR_d = \frac{d-1}{2R_d} dt + dW(t),$$

where $W(t)$ is the (standard) Wiener process. The process $S_d(t) = R_d^2(t)$ is the **squared Bessel process**, which is sometimes called Feller's diffusion. It may be otherwise defined as the solution of $dS_d = d\,dt + 2\sqrt{S_d}dW(t)$. Processes related to the Bessel process include:

(a) The **Bessel bridge**, which is a Bessel process $R_d(t), 0 \leq t \leq s$, conditional on its taking some value v at time $s > 0$.
(b) The **Bessel clock** $C(t)$, which is defined as the integral

$$C(t) = \int_0^t \frac{ds}{|B_d(s)|^2}, \; t \geq 0,$$

where $|B_d(t)|$ is the radial distance from the origin of a Brownian motion, in $d \geq 2$ dimensions such that $B_d(0) \neq 0$. [A. Gőing-Jaeschke & M. Yor, A survey and some generalizations of Bessel processes, *Bernoulli*, 9(2), 313–349, 2003]

beta distribution A beta random variable X has density $f(x) = x^{a-1}(1-x)^{b-1}/\beta(a,b), 0 < x < 1$, where $a > 0$, $b > 0$, and $\beta(a, b)$ is the beta function, which is written in terms of gamma functions as $\beta(a,b) = \Gamma(a)\Gamma(b)/\Gamma(a+b)$. The beta distn arises naturally as the distns of the order statistics of a collection of independent r.vs that are uniformly distributed as $(0,1)$. It also arises as the distribution of $X/(X+Y)$, where X and Y are independent gamma random variables, with parameters (λ, a) and (λ, b) respectively. The random variable Y having the binomial distribution $B(n, X)$, where X has a beta distn, is called **beta-binomial.** [A. Gupta & S. Nadarajah (eds), *Handbook of Beta Distribution and its Applications*. Dekker, 2004]

beta ensemble *see* random matrix

beta-prime distribution (alias: **inverted beta; Pearson type VI**) Has density $f(x) = \{\beta(p,q)\}^{-1}x^{p-1}(1+x)^{(-p-q)}$, $x>0, p>0, q>0$, where $\beta(p,q)$ is the beta function. [E. Keeping, *Introduction to Statistical Inference*. Van Nostrand, 1962]

betting system (alias: **gambling system, or strategy**) Broadly, any set of rules followed by a gambler in placing bets. The term usually refers to a rule for placing a sequence, and generally includes a stopping (or restarting) rule. For example, the classic linear system is the so-called linear progression, or **D'Alembert's system**, in which the initial stake (at evens) is 1 (without loss of generality), and then at any winning play the subsequent stake is decreased by 1, while at any losing play the subsequent stake is increased by 1. Thus, whenever the rule calls for a stake of 0, the gambler is up by 1 and may quit or restart. All mathematical analysis of such systems must include the assumption that the sum wagered can depend only on past wagers and outcomes. Formally, in terms of the natural filtration $\mathcal{F}_n$ generated by wagers and outcomes, the next bet at the nth wager is $\mathcal{F}_{n-1}$-measurable, i.e. predictable. Some better known systems are: Belgian; bold play; Labouchere (Condorcet); Kelly; Oscar; martingale; timid play; (*see* martingale transform). [S. Ethier, *The Doctrine of Chances*. Springer, 2010]

Beurling's projection theorem (alias: **Beurling–Nevanlinna (circular) projection theorem**) Arises as a classic theorem about harmonic measure in a disc. The connexion between Brownian motion and potential theory permits this formulation: let $\mathbf{Z}(\mathbf{a}, t)$ be two-dimensional (planar) Brownian motion started from $\mathbf{a}$, for $|\mathbf{a}| < 1$; and let $C \subseteq D$ be a compact set in the unit disc D. Define the interval $I \subseteq [-1, 0]$ to be the set of points $\{-(x^2+y^2)^{1/2} : (x,y) \in C\}$ (which is called either the radial, or the circular, projection of C on to the negative real axis). Then the theorem asserts that the probability $P(\mathbf{a}, c)$ that $\mathbf{Z}(\mathbf{a}, t)$ hits C before it hits the unit circle is not less than the probability $P(|\mathbf{a}|, I)$ that $\mathbf{Z}((|\mathbf{a}|, 0), t)$ hits I before it hits the unit circle. [R. Bass, *Probabilistic Techniques in Analysis*. Springer, 1995]

Bhatia–Davis inequality This asserts that for a random variable X such that $m \leq X \leq M$, $\operatorname{var} X \leq (M - \mathbb{E}X)(\mathbb{E}X - m)$, with equality iff the distribution is the binary (two-point) distribution on $\{m, M\}$. (R. Bhatia & C. Davis, A better bound on the variance, *Amer. Math. Mon.*, 107(4), 353–357, 2000]

Bhattacharyya distance Any one of several measures of the extent to which two probability distributions p and q differ from each other. When $p(x)$ and $q(x)$ are discrete, we define the Hellinger affinity (or Bhattacharyya coefficient, or fidelity similarity, and other names) as $a(p,q) = \sum_x \{p(x)\,q(x)\}^{1/2}$. Then the function $d_1 = -\log a(p,q)$ is the Bhattacharyya distance between p and q (the functions $d_2 = \cos^{-1} a(p,q)$, and $d_3 = \{\cos^{-1} a(p,q)\}^2$, have also been called Bhattacharyya distances). More generally, when p and q are densities (or Radon–Nikodym derivatives with respect to a common dominating measure μ) the same applies with the Hellinger affinity function $a(p,q) = \int (pq)^{1/2}\, d\mu = \int (pq)^{1/2}\, dx$, when μ is Lebesgue measure. It is related to *f*-divergence, and is a special case ($\alpha = 1/2$) of Chernoff divergence. [A. Bhattacharyya, On a measure of divergence between two multinomial populations, *Sankhyâ*, 7(4), 401–406,

1946; S.-H. Cha, A comprehensive survey on distance/similarity measures between probability density functions, *Int. J. Math. Model. Method. Appl. Sci.*, 1(4), 300–307, 2007]

Biane–Imhof identity Arises in the context of Brownian motion $D(t)$ having positive drift μ, that is $D(t) = W(t) + \mu t$, where $W(t)$ is the Wiener process. Let X be a random variable independent of $D(t)$, having exponential density with parameter 2μ. Then the identity asserts that the total time spent by $D(t)$ in the negative real axis has the same distribution as the first time at which $D(t)$ takes the value X. [P. Salminen & M. Yor, Perpetual integral functionals as hitting and occupation times, *Electr. J. Probab.,* 10(11), 371–419, 2005]

Bichteler–Dellacherie theorem (alias **B–D–Meyer–Mokobodzky theorem; fundamental theorem of stochastic integration**) Arises in the theory of stochastic integrals, where it asserts (broadly speaking) that if a stochastic integral of the form $\int f dX$ is to have reasonable properties, such as obeying the bounded convergence theorem, then the most general form that the integrator X can take is to be a semi-martingale. [P. Protter, *Stochastic Integration and Differential Equations*, 2nd edn. Springer, 2005]

Bickel–Lehmann dispersive order *see* dispersion order

Bienaymé inequality *see* Chebyshov's inequality

bifractional Brownian motion (alias: **two-parameter fractional Brownian motion**) Arises as a generalization of fractional Brownian motion, as an attempt at a better description of, e.g., turbulence. It is a continuous, zero-mean, Gaussian process, $X(t), t \geq 0$, with $X(0) = 0$, having covariance function $c(s,t) = \frac{1}{2^K}\{[t^{2H} + s^{2H}]K - |t-s|^{2HK}\}$, where $0 < H < 1$, and $0 < K \leq 1$. If $K = 1$, it is an ordinary fBm with Hurst parameter H. [C. Tudor, *Analysis of Variations for Self-similar Processes*. Springer, 2013]

Biggins (martingale convergence) theorem Arises in the context of branching random walk, where it supplies conditions under which an associated martingale converges in mean to a non-trivial limit. It is a generalization of the Kesten–Stigum theorem. [K. Athreya & P. Jagers (eds), *Classical and Modern Branching Processes*. Springer, 1997]

Biham–Middleton–Levine traffic model An interacting particle system that displays phase transitions and self-organizing properties. One form is this: each vertex of the square lattice $\mathbb{Z}^2$ independently receives a car with probability p; each car is independently equally likely to be facing north or east (in the conventional orientations). At odd times east-facing cars move one step east if possible; at even times north-facing cars move one step if possible. (All cars retain their initial direction of travel throughout.) For p near 1, the system jams; for p near 0, it flows freely. At intermediate values, jams and free flowing zones coexist in simulations. [I. Benjamini *et al.*, The Biham–Middleton–Levine traffic model for a single junction, *Progr. Probab.,* 60, 97–116, 2007]

bilateral exponential distribution *see* double exponential distribution

Billard's theorem Addresses the convergence of random Fourier series. One form is this: let $S(\theta) = \sum_{n=1}^{\infty} X_n e^{i\theta n}, 0 \leq \theta \leq 2\pi$ where X_n is a sequence of independent random variables that are symmetric about 0. Then the theorem asserts that these are equivalent statements:

(i) S is a bounded function with probability 1.
(ii) S is a continuous fn with probability 1.
(iii) The series converges uniformly on $[0, 2\pi]$ with probability 1.
(iv) The series converges almost everywhere in $[0, 2\pi]$ with probability 1.

[J.-P. Kahane, *Some Random Series of Functions*, 2nd edn. Cambridge University Press, 1985]

Billingsley's metric *see* Skorokhod metric

bi-Markov process A two-parameter process that arises as (broadly speaking) the tensor product of two ordinary Markov processes each having one time parameter. A classic example is bi-Brownian motion, $B(s,t) = (W_1(s), W_2(t))$, where W_1 and W_2 are independent d-dimensional Wiener processes. A bi-Markov process is not to be confused with a bivariate (vector-valued) Markov process $Z(t) = (X(t), Y(t))$. [B. Davis & T. Salisbury, Connecting Brownian paths, *Ann. Probab.*, 16(4), 1428–1457, 1988]

binary channel Strictly, a channel in communication theory both of whose input and output alphabets each comprise just two symbols, in particular when the chance of error is the same for each input symbol one has the binary symmetric channel (BSC), or more generally the binary asymmetric channel. More loosely, the term is used for channels whose input is binary, such as:

1. binary erasure, in which any input symbol is deleted with some probability r; this deletion is known to the recipient (BEC);
2. binary output-symmetric, in which the input is real and binary (e.g. +1 or -1) and the output is some r.v. whose distribution is symmetric about the value of the input.

[T. Richardson & R. Urbanke, *Modern Coding Theory*. Cambridge University Press, 2007]

binary (or digital) option A term used in financial mathematics to denote a class of contingent claims (derivative contracts) whose pay-offs depend on binary r. vs that are typically indicator functions of the asset price $S(T)$ at the exercise date T. Examples include cash or nothing, with pay-off $CI(S(T) > K)$; asset or nothing, with pay-off $S(T)I(S(T) > K)$; gap or nothing, with pay-off $(S(T) - C)I(S(T) > K)$; etc. [N. Bingham & R. Kiesel, *Risk-neutral Valuation*, 2nd edn. Springer, 2004]

Bingham distribution An axial, or antipodally symmetric, distribution on the sphere whose density is proportional to $\exp\{x^T Ax\}$ for some symmetric matrix A (where x^T is the transpose of the vector x). It arises constructively by conditioning a zero-mean multinormal r.v. to lie on the sphere. More generally, the **Fisher–Bingham**, or **Kent, distn** has density proportional to $\exp(a^T x + x^T Ax)$ for a suitable constant vector a, and symmetric A. [C. Bingham, An antipodally symmetric distribution on the sphere, *Ann. Stat.*, 2(6), 1201–1225, 1974]

binomial central limit theorem *see* De Moivre's theorem

binomial distribution The probability distribution of a binomial random variable X, taking the form $\mathbb{P}(X = r) = \frac{n!}{r!(n-r)!} p^r (1-p)^{n-r}$, $0 \le r \le n, 0 \le p \le 1$, and denoted by $\mathrm{B}(n, p)$. It has mean $\mathbb{E}X = np$ and variance $\mathrm{var}\, X = np(1-p)$. It arises naturally as the distribution of the number of successes in n independent Bernoulli trials; or equivalently as the distribution of the sum of n i.i.d. indicator r.vs Y_r, with $\mathbb{E}Y_r = p, 1 \le r \le n$; or in other

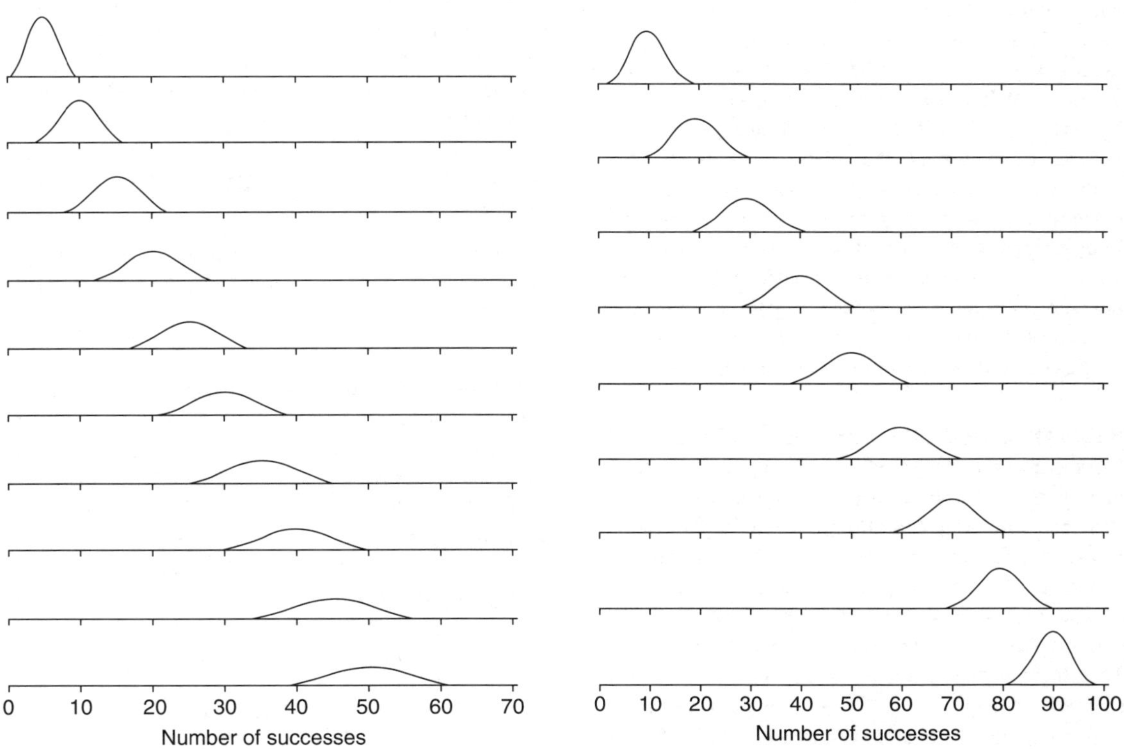

Binomial distributions. ***Left:*** **Fixed** $\boldsymbol{p}$ **and varying** $\boldsymbol{n}$**; from top to bottom,** $\boldsymbol{n = 10, 20, \ldots, 100}$**.** ***Right:*** **Fixed** $\boldsymbol{n}$ **and varying** $\boldsymbol{p}$**; from top to bottom,** $\boldsymbol{p = 10\%, 20\%, \ldots, 90\%}$**. Histograms have been smoothed for simplicity.**

phraseology it is the distribution of the sum of a random sample of size n from the Bernoulli distribution. It is ubiquitous, and important in the development of probability, with both Bernoulli's thm and De Moivre's thm proved initially for such a sequence of Bernoulli trials. More generally, the **multivariate binomial distn** is the joint distribution of the sums of the components of a sample of size n from a multivariate Bernoulli distn. Thus, e.g., a bivariate binomial distn has pgf $G(s,t) = (\alpha + \beta s + \gamma t + \delta st)^n$ for $\alpha + \beta + \gamma + \delta = 1$, and $\alpha, \beta, \gamma, \delta$ all non-negative. The inverse binomial distn may refer either to the negative binomial distribution or to the lost-games distribution. [J. Teugels, Some representations of the multivariate Bernoulli and binomial distributions, *J. Multivariate Anal.*, 32(2), 256–268, 1990]

binomial moment A non-negative, integer-valued r.v. X has rth binomial moment given by $\beta_r = \mathbb{E}\binom{X}{r}$. [V. N. Sachkov, *Probabilistic Methods in Combinatorial Analysis*, Cambridge University Press, 1997]

bin-packing A classic problem of operations research, which typically presents as follows: there are a number of bins (usually of the same size) and a number of objects of random sizes $X_1, X_2, \ldots$; the objective is to place objects in bins, so that the sum of their sizes does not exceed the capacity of their assigned bin, while minimizing the number of bins used. If the objects are presented in sequence and must be placed in order of presentation, this is the on-line bin-packing problem. In the closely related **knapsack problem**, each object also has a random value $V_1, V_2, \ldots$, and the problem is to place the objects in a bin (subject to the capacity constraint) so as to maximize the expected total value of the objects so packed. [H. Kellerer *et al.*, *Knapsack Problems*. Springer, 2004]

Birkhoff's ergodic theorem (alias: **strong ergodic theorem; pointwise ergodic theorem; a.e. ergodic theorem**) Arises in the context of physical (dynamical) systems evolving in equilibrium, where it gives the precise relationships between averages taken over time, and system averages (expected values). In the framework of random processes, one form is this: Let $X_n, n \geq 1$, be a strongly stationary sequence of r.vs with $\mathbb{E}|X_1| < \infty$. Then there exists a random variable Y, with $\mathbb{E}Y = \mathbb{E}X_1$, such that as $n \to \infty$, $\frac{1}{n}\sum_{r=1}^{n} X_r \to Y$ with probability 1. More precisely, the random variable $Y = \mathbb{E}(X_1|\mathcal{J})$, where $\mathcal{J}$ is the σ-field of invariant events of the sequence. If $\mathcal{J}$ is trivial, containing only null or certain events, then $Y = \mathbb{E}X_1$, and the process is called ergodic.

In an alternative framework, let T be a measure-preserving transformation on a measure space $\{\Omega, \mathcal{F}, \mathbb{P}\}$. Then for any r.v. such that $\mathbb{E}|X| < \infty$, as $n \to \infty$, $\frac{1}{n}\sum_{r=0}^{n-1} X(T^r\omega) \to \mathbb{E}(X|\mathcal{J})$, for $\omega \in \Omega$. If the transformation is ergodic then $\mathbb{E}(X|\mathcal{J}) = \mathbb{E}X$; typically this is established by mixing conditions. [G. D. Birkhoff, Proof of the ergodic theorem, *Proc. Nat. Acad. Sci.*, 17(12), 656–660, 1931]

Birkhoff's theorem Asserts that any doubly stochastic $n \times n$ matrix P can be written as a convex combination of at most $(n-1)^2 + 1$ permutation matrices. It is sometimes called the **Birkhoff–Von Neumann thm**, as J. Von Neumann later provided a more direct proof that is more widely applicable. [R. Bapat & T. Raghavan, *Nonnegative Matrices and Applications*. Cambridge University Press, 1997]

Birnbaum–Marshall inequality A maximal inequality that generalizes the inequalities of Chebyshov, Doob, Hájek, Rényi, and Kolmogorov. Let $0 = X_0, X_1, X_2, \ldots$ be a sequence of random variables with finite rth moment, $r \geq 1$, and such that with probability 1, $\mathbb{E}(|X_n||X_1, \ldots, X_{n-1}) \geq c_n|X_{n-1}|$, $n \geq 2$, for constants $c_n \geq 0$. For any sequence $a_k > 0$, define $b_k = \max\{a_k, a_{k+1}c_{k+1}, \ldots, a_n c_{k+1}, \ldots, c_n\}$. Then

$$\mathbb{P}(\max_{1\leq k\leq n} a_k|X_k| \geq 1) \leq \sum_{k=1}^{n} b_k^r(\mathbb{E}|X_k|^r - c_k^r\mathbb{E}|X_{k-1}|^r).$$

[E. Slud, Generalization of an inequality of Birnbaum and Marshall, with applications to growth rates for submartingales, *Stoch. Proc. Appl.*, 24(1), 51–60, 1987]

Birnbaum–Orlicz distance A measure of separation between two distribution functions $F(x)$ and $G(x)$, given in terms of a function $u(x) \geq 0, x \geq 0$, that is continuous, and non-decreasing, and satisfies $u(2x) \leq cu(x)$ for all x and some finite c (this being Orlicz's condition). Then the Birnbaum–Orlicz uniform distance is $\sup_{x\in\mathbb{R}} u(|F(x) - G(x)|)$, and the Birnbaum–Orlicz average distance is (for suitable F and G) $\int_{-\infty}^{\infty} u(|F(x) - G(x)|)\mathrm{d}x$. [S. Rachev *et al.*, *The Methods of Distances in the Theory of Probability and Statistics*. Springer, 2013]

Birnbaum–Raymond–Zuckerman inequality In its basic form, for sufficiently large t, this asserts for independent r.vs $X_r, r \geq 1$, having common variance σ^2, that $\mathbb{P}\left(\sum_{r=1}^{n}(X_r-\mathbb{E}X_r)^2\geq t^2\right)\leq(n\sigma^2/t^2)(1-n\sigma^2/4t^2)$; there are improved versions. [R. Savage, Probability inequalities of the Tchebycheff type, *J. Res. Nat. Bur. Stand. B*, 65B(3), 211–226, 1961; Z. W. Birnbaum *et al.*, A generalization of Tshebychev's inequality to two dimensions, *Ann. Math. Stat.*, 18(1), 70–79, 1947]

Birnbaum–Saunders (or fatigue life) distribution Arises in reliability theory as a model for lifetime distributions. There are two parameters $a > 0$ and $b > 0$, and a r.v. T with this distribution is such that

$$\frac{1}{a}\left\{\left(\frac{T}{b}\right)^{\frac{1}{2}}-\left(\frac{b}{T}\right)^{\frac{1}{2}}\right\},$$

has a standard normal density, with distn fn $\Phi(t)$. Equivalently,

$$F_T(x)=\Phi\left\{\left(\frac{1}{a}\right)\left(\sqrt{\frac{x}{b}}\right)-\left(\sqrt{\frac{b}{x}}\right)\right\},$$

for $x > 0$. [Z. Birnbaum & S. Saunders, A new family of life distributions, *J. Appl. Probab.*, 6(2), 319–327, 1969]

birth–assassination process A variant of the classic branching process in continuous time. From the instant of their birth, particles produce offspring at the instants of (conditionally) independent Poisson processes, until they are assassinated. The initial particle is assassinated after a time X with distribution F. Its offspring are free from risk of assassination until their parent is removed, after which time their remaining lifetimes until assassination are i.i.d. r.vs with the distribution F of X. This process continues through successive generations unless it ceases on the event of extinction E. If $\mathbb{P}(E) = 1$, the process is called stable. [D. Aldous & W. Krebs, The birth-and-assassination process, *Stat. Probab. Lett.*, 10(5), 427–430, 1990]

birthday problem A classic problem of elementary probability usually posed as follows: it is assumed that individuals are independently equally likely to have their birthday on any day of a year that has 365 days. The original problem asks for the probability p that at least two individuals in a group of n share the same birthday, and then asks for the smallest n such that $p > ½$; the answer to the latter question being $n = 23$. The **strong birthday problem** asks for the probability q that everyone in the group shares their birthday, and for the smallest n such that $q > ½$; the answer here is $n = 3064$. The related **first-coincidence problem** considers the group arriving one by one, and asks which of them is most likely to be the first arrival to share a birthday with one of those already present; the answer is the 20th arrival. Finally, one may ask for the expected size of the group needed until two people first share a birthday; the ceiling of this number is 25. [A. Das Gupta, The matching, birthday, and the strong birthday problem: a contemporary review, *J. Stat. Planning Inference*, 130(1–2), 377–389, 2005]

birth–death process This is a Markov process $X(t)$, in continuous time, taking values in the non-negative integers, that arises in many fields of application. Informally, it can be seen as the path of a particle making steps of sizes ± 1 on the non-negative integers, whose positive and negative jumps from the point n occur at rates λ_n and μ_n respectively, where λ_n is the birth-rate and μ_n is the death-rate. More formally, the Q-matrix of $X(t)$ is tri-diagonal, thus:

$$\begin{pmatrix} -\lambda_0 & \lambda_0 & 0 & 0 & \dots \\ \mu_1 & -(\lambda_1+\mu_1) & \lambda_1 & 0 & \dots \\ 0 & \mu_2 & -(\lambda_2+\mu_2) & \lambda_2 & \dots \\ \vdots & \vdots & \vdots & \ddots & \end{pmatrix}.$$

If $\lambda_0 = 0$, then the process is absorbed at 0; all other rates are assumed positive, so $\lambda_n\mu_n > 0$. Special cases include:
(a) **Simple birth–death**: $\lambda_n = n\lambda, \mu_n = n\mu$.
(b) **Random walk** reflected at the origin: $\lambda_n = \lambda, \mu_n = \mu$.
(c) **Immigration–death**: $\lambda_n = \lambda, \mu_n = n\mu$.
Note that the simple birth–death process is sometimes called the **Feller–Arley process**, including the case when λ and μ depend on t. The behaviour of the process is to a considerable extent determined by the so-called **potential coefficients,** $c_n, n \geq 0$, where $c_0 = 1$ and

$$c_n=\frac{\lambda_0\lambda_1\dots\lambda_{n-1}}{\mu_1\dots\mu_n}, n\geq 1.$$

For example, the process is uniquely determined by its birth and death rates if $\sum_{n=0}^{\infty}(\lambda_n c_n)^{-1} = \infty$; and there is a stationary distribution π if $s = \sum_{n=0}^{\infty} c_n < \infty$, and then $\pi_n = c_n/s$. [M. Kijima, *Markov Processes for Stochastic Modeling*. Chapman & Hall, 1997]

birth process Broadly, a random process on the integers that increases with time. More commonly, it usually refers to the Markov process $X(t)$ in continuous time, taking values in the non-negative integers, in which the time spent in state n is exponential with parameter λ_n, and such that at the end of this interval, $X(t)$ jumps to $n + 1$. In particular, if $\lambda_n = n\lambda$, then $X(t)$ is the simple birth process, also called the Yule process (or Yule–Furry process). If $\lambda_n = \lambda$, then $X(t)$ is the Poisson process. The birth process is honest (i.e. non-explosive) if and only if $\sum_{n=0}^{\infty} \lambda_n^{-1} = \infty$. [J. Norris, *Markov Chains*. Cambridge University Press, 1997]

birth time Arises in the context of path decompositions of a Markov chain $X_n, n \geq 0$, having stationary transition probabilities. Let T be a random time for X_n, which is to say that T is a non-negative integer-valued function of the process $X_n, n \geq 0$. Then T is a birth time if the post-T process $A(T) = [X_T, X_{T+1}, \ldots]$ is Markov with stationary transition probabilities (that are not necessarily the same as those of X_n). Similarly T is a death time if the pre-T process $B(T) = [X_0, \ldots, X_{T-1}]$ is likewise Markov. If $A(T)$ and $B(T)$ are conditionally independent given X_{T-1} (or X_T), and T is a birth time or a death time, then $\{A(T), B(T)\}$ is said to be a path decomposition, and T is said to be a regular birth time or a regular death time accordingly. [A. Pittenger & M. Sharpe, Regular birth and death times, *Probab. Theor. Rel. Fields*, 58(2), 231–246, 1981]

Bismut–Elworthy–Li formula (alias: **Bismut's formula**) Arises in the context of stochastic differential equations on Riemannian manifolds. The term is commonly used to refer to results of the following type: let $X(x,t)$ be an Itô diffusion arising from the SDE $\mathrm{d}X = a(X)\mathrm{d}t + \sigma(X)\mathrm{d}W$, where $X(x,0) = x$, and $W(t)$ is the Wiener process in d dimensions. For fixed t, a Malliavin weight M is a random variable such that for all suitable functions $g(X) \in \mathbb{R}$ one can write $\nabla \mathbb{E} g(X(x,t)) = \mathbb{E}(g(X(x,t))M)$, where ∇ denotes the gradient operator.

The Bismut–Elworthy–Li formula supplies an explicit representation for M in the form of a stochastic integral $M = \frac{1}{t}\int_0^t R(s)Y(x,s)\mathrm{d}W(s)$, where $R(s)$ is the right-inverse of σ, and $Y(x,s)$ is the first variation process of $X(x,t)$. [G. Di Nunno *et al.*, *Malliavin Calculus for Lévy Processes with Applications to Finance*. Springer, 2009]

Bissinger's distribution (alias: **STER distribution**) A random variable X has a Bissinger distribution on the non-negative integers if it can be written in the form $\mathbb{P}(X = r) = \sum_{k=r+1}^{\infty} k^{-1}\mathbb{P}(Y = k)$, where Y is a random variable distributed on the positive integers. If U is a r.v. having the uniform density on (0, 1), then $[UY]$, i.e. the integer part of UY, has the distribution of X. [J. Mačutek, Discrete probability distributions generated by the generalised STER summation, *Braz. J. Probab. Stat.*, 16, 1–6, 2002]

bivariate normal distribution (alias: **binormal distribution**) Random variables X and Y have a standard binormal distn if their joint density f is

$$f(x,y) = \frac{1}{2\pi(1-\rho^2)^{1/2}} \exp\left[\frac{-1}{1-\rho^2}(x^2 - 2\rho xy + y^2)\right],$$

$$(x,y) \in \mathbb{R}^2,$$

where $\rho \in (-1,1)$ is the correlation coefficient of X and Y. It is denoted either by $N(0,V)$, where V is the covariance matrix $\begin{pmatrix} 1 & \rho \\ \rho & 1 \end{pmatrix}$, or by binormal ($\rho$). The joint characteristic function of the pair (X,Y) is

$$\phi(s,t) = \exp\left[-\frac{1}{2}(s^2 + 2\rho st + t^2)\right],$$

and they are independent iff uncorrelated. More generally, X and Y are said to be jointly binormal if it is the case that for all real a and b, $aX + bY$ has a normal distribution. This includes singular cases, e.g. $\rho^2 = 1$, when the density $f(x,y)$ cannot exist in the form above. The name binormal has also been used to denote the **bilateral normal-tail density**

$$f(x) = \begin{cases} \sqrt{\frac{2}{\pi}}\left(\frac{1}{\sigma+\tau}\right)\exp\left(-x^2/(2\sigma^2)\right), & x \leq 0 \\ \sqrt{\frac{2}{\pi}}\left(\frac{1}{\sigma+\tau}\right)\exp\left(-x^2/(2\tau^2)\right), & x > 0, \end{cases}$$

(alias the joined half-Gaussian density), and also to denote the mixed (or compound) normal density, which for $0 < p < 1$ is

$$f(x) = \frac{p}{\sqrt{2\pi}\sigma}\exp\left(-\frac{x^2}{2\sigma^2}\right) + \frac{1-p}{\sqrt{2\pi}\tau}\exp\left(-\frac{x^2}{2\tau^2}\right).$$

[N. Balakrishanan & C.-D. Lai, *Continuous Bivariate Distributions*, 2nd edn. Springer, 2009]

bivariate Poisson distribution The random variables X and Y have a bivariate Poisson distn if, for non-negative λ, μ, and ν,

$$\mathbb{P}(X = x, Y = y) = \frac{\mathrm{e}^{-\lambda}\lambda^x}{x!}\mathrm{e}^{-\mu}\frac{\mu^y}{y!}\mathrm{e}^{-\nu}\sum_{r=0}^{x\wedge y}\binom{x}{r}\binom{y}{r}r!\left(\frac{\nu}{\lambda\mu}\right)^r,$$

for $x \geq 0, y \geq 0$, and where $x \wedge y = \min\{x,y\}$. It has Poisson marginals, and $\mathrm{cov}(x,y) = \nu$, and its joint pgf is $\exp\{\lambda(s-1) + \mu(t-1) + \nu(st-1)\}$. [S. & K. Kocherlakota, *Bivariate Discrete Distributions*. Dekker, 1992]

bivariate probability integral transform *see* Kendall distribution function

Björk–Grandell process Arises in insurance mathematics as a model for a claims process $C(t)$. It is a type of Cox process, defined thus: let (A_r, X_r), be a sequence of i.i.d.r.vrs, where $X_r > 0$ and $A_r \geq 0$. Define $S_n = \sum_{r=1}^{n} X_r$, and $\lambda(t) = A_n$, for $S_{n-1} \leq t < S_n$, and $\Lambda(t) = \int_0^t \lambda(s)\mathrm{d}s$. If $N(t)$ is a Poisson process of rate 1, the B-G process is the point process $C(t) = N[\Lambda(t)]$. [J. Grandell, *Aspects of Risk Theory*. Springer, 2011]

Black–Derman–Toy model Arises in the mathematical finance as a model to describe the evolution of interest rates in terms of the short (instantaneous) rate $r(t)$. When formalized in continuous time, it is written in terms of an SDE for $X(t) = \log r(t)$ as

$$dX(t) = (a(t) + b(t)X(t))\mathrm{d}t + c(t)\mathrm{d}W(t),$$

where $W(t)$ is the Wiener process, and $b(t)$ depends on the volatility of $r(t)$ and is therefore not a free parameter. A simplified model sets $b(t) = 0$, and the more general **Black–Karasinski** process, also called the extended B-D-T model, sets $b(t)$ as a free parameter. [M. Musiela & M. Rutkowski, *Martingale Methods in Financial Modelling*, 2nd edn. Springer, 2005]

Black–Scholes–Merton theory Arises in mathematical finance, where it supplies a model for financial markets, and, within the model, a rationale for pricing options, together with a technique (dynamic hedging) intended to make the trading of such options risk-free. The basic fair prices for the standard put and call options are called the **Black–Scholes formulas**, or (generally) the Black–Scholes–Merton formulas.

In its barest essentials, the model comprises a bond $B(t)$ and a stock $S(t)$, where $t \geq 0$ is the time. The bond models a risk free asset, and $B(t)$ is assumed to satisfy $dB(t) = rB(t)dt$, where r is the (short) interest rate. The stock is assumed to satisfy the SDE (called lognormal dynamics), $dS = aSdt + bSdW$, where $W(t)$ is the Wiener process, for suitable constants a, called the drift (or local mean rate of return), and b, called the volatility (or variance parameter). The market is assumed to be unconstrained, to operate continuously with no fees or taxes, and the stock (which can be shorted) pays no dividends. A fair (or market) price for a suitable option should have the property that no trading strategy can achieve an arbitrage, and this requirement yields the various Black–Scholes–Merton formulas. Furthermore, there exists a tradable portfolio of bond and stock which has the value of the option at the exercise date, thus rendering the fair price risk-free; this is known as dynamic hedging. The portfolio is called a hedge, and is said to replicate the option. [M. Capinski & E. Kopp, *The Black–Scholes Model*. Cambridge University Press, 2012]

Blackwell's approachability theorem Arises in the context of game theory. where it can be seen as an extension to games with vector pay-offs, of Von Neumann's minimax theorem for games with scalar pay-offs. It gives conditions under which a closed subset C of the pay-off space, called the approachable set, contains a possible long-run average pay-off for player A, no matter what B's strategy is. [D. Blackwell, An analog of the minimax theorem for vector pay-offs, *Pacific J. Math.*, 6(1), 1–8, 1956]

Blackwell's comparison of experiments theorem *see* equivalence theorem

Blackwell–Dubins maximal inequality (or theorem) This supplies a stochastic bound (i.e. a stochastically dominating r.v.) for the maximum of any regular martingale $X(t), t \geq 0$, that is closed by Z (where Z may be called the last element). Let the distribution of Z be F, and denote the left-continuous inverse of F by $F^{-1}(y), 0 \leq y \leq 1$. Define the Hardy–Littlewood maximal function $H_F(t) = \frac{1}{1-t}\int_t^1 F^{-1}(y)dy, 0 \leq t < 1$, and set $M = H_F(U)$ where U is a r.v. uniform on (0,1). Then the theorem asserts that for any martingale $X(t)$ closed by Z, $\mathbb{P}(\max_t X(t) \geq x) \leq \mathbb{P}(M \geq x)$. [D. Gilat & I. Meilijson, A simple proof of a theorem of Blackwell and Dubins on the maximum of a uniformly integrable martingale, *Séminaire de Probabilités*, 22, 214–216, 1988]

Blackwell–Dubins theorem *see* Hunt's lemma

Blackwell–MacQueen urn scheme This generalizes the classical Pólya urn scheme in which balls are of two colours (or possibly a finite number of colours) to a scheme allowing a continuum of colours. The limiting distribution (as the number of draws increases) is the discrete Ferguson–Dirichlet distn. Formally, let $m(\cdot)$ be a finite positive measure on the (Borel) subsets of a complete separable metric space S. A sequence of random variables $X_n \in S, n \geq 1$, is a Pólya sequence with parameter m if for any $A \subseteq S$, $\mathbb{P}(X_1 \in A) = m(A)/m(S)$, and

$\mathbb{P}(X_{n+1} \in A | X_1, \ldots, X_n) = m_n(B)/m_n(S)$, where $m_n = m + \sum_{r=1}^{n} \delta(X_r)$, and $\delta(x)$ denotes unit measure concentrated at x. Then the Blackwell–MacQueen theorem asserts that, as $n \to \infty, m_n/m_n(S)$ converges a.s. to a Ferguson–Dirichlet distn (among other assertions). [D. Blackwell & J. MacQueen, Ferguson distributions via Pólya urn schemes, *Ann. Stat.*, 1(2), 353–355, 1973]

Blackwell's renewal theorem Arises in the context of a renewal process $N(t) = \max[n : \sum_{r=1}^{n} X_r \leq t], t \geq 0$, where $X_r, r \geq 1$, is a sequence of i.i.d. positive r.vs with mean μ. The theorem asserts that if X_r is not a lattice r.v. then as $t \to \infty, h > 0$,

$$\mathbb{E}(N(t+h) - N(t)) = \sum_{k=1}^{\infty} \mathbb{P}(t < \sum_{r=1}^{k} X_r \leq t+h) \to h/\mu.$$

(By convention $1/\mu = 0$ when $\mu = \infty$.) If X_r is a lattice with period d, then for integer m, $\mathbb{E}(N(t+md) - N(t)) \to md/\mu$. This extends the equivalent theorem for arithmetic X_r, known as the Erdös–Feller–Pollard theorem. The equivalent theorem for two-sided renewal, or random walk, is sometimes called the **Blackwell–Feller–Orey** (or **Feller–Orey**) theorem. [S. Asmussen, *Applied Probability and Queues*, 2nd edn. Springer, 2003]

Blackwell space (alias: **Lusin space**) Arises in the measure-theoretic foundations of probability where it comprises a measurable space $(\Omega, \mathcal{F})$, such that $\mathcal{F}$ is a separable σ-algebra, and for every real-valued, $\mathcal{F}$-measurable function f, the range of f is an analytic (or Souslin) set. It was pointed out by D. Blackwell that any probability measure defined on $(\Omega, \mathcal{F})$ yields a perfect probability space. [D. Blackwell, On a class of probability spaces, *Proc. 3rd Berkeley Symp. Math. Stat. Probab.*, 2, 1–6, 1956]

Blackwell's theorem Arises in decision theory, in an application to the comparison of experiments. Informally, it asserts that a decision-maker seeking to maximize expected utility will prefer to draw on the results of experiments that yield more information (rather than less), for a suitable interpretation of the expression 'more information.' [M. Leshno & Y. Spector, An elementary proof of Blackwell's theorem, *Math. Social Sci.*, 25(1), 95–98, 1992]

blocking Arises in queueing theory and its applications in network flows. For example, if a queue has a finite waiting room that is full, customers arriving to see a full waiting room are said to be blocked, and the proportion of customers that are so blocked is the blocking probability. The random process of such blocked customers is called the overflow; its members may either be lost, or they may join an overflow queue, or they may attempt to enter the waiting room later on. This latter action is called a **retrial**, and such customers are said to be in **orbit**. For another example, if queues are in tandem (i.e. in series), or in a network, and intermediate queues between stations are limited, or not permitted, then a customer denied service at the next station may slow, or even stop, the operation of the server at the station just quitted. This server (or station) is said to be blocked. [S. Balsamo *et al.*, *Analysis of Queueing Networks with Blocking*. Kluwer, 2001]

blocking probability Arises in the context of random graph models for a variety of real-world systems. Let G be a connected graph with one vertex called the source, and another the sink. If each edge of the graph is blocked with some probability (not necessarily the same) then the probability that every path in G from the source to the sink includes at least one blocked edge is the blocking probability.

Bloom filter A hash-coding (data-storage) method that achieves a reduction in space (storage) requirements by allowing a small probability of error in identification; specifically, a small number of objects (messages) may be wrongly identified as members of the stored set. [H. Bloom, Space/time trade-offs in hash coding with allowable errors, *Commun. ACM*, 13(7), 422–426, 1970]

Blum–Blum–Shub generator A non-linear recursive generator of pseudo-random numbers, which is popular in the context of cryptography. [L. Blum *et al.*, A simple unpredictable pseudorandom number generator, *SIAM J. Comp.*, 15(2), 364–383, 1986]

Blumenthal–Getoor index Any of several indices that describe some sample path properties of a suitable Lévy process $X(t), t \geq 0$, e.g. the intensity of jumps and the rate of growth of $X(t)$. The upper index b is defined in terms of the Lévy measure $\lambda(x)$ as $b = \inf\{t > 0 : \int_{|x|<1} |x|^t \lambda(\mathrm{d}x) < \infty\} \in [0, 2]$. When the process is stable of order (or index) σ, then $b = \sigma$. The lower indices a and a' are defined in terms of the Lévy exponent $\phi(x)$, e.g. $a' = \sup[t \geq 0 : |x|^{-t} \mathrm{Re}\phi(x) \to \infty \text{ as} |x| \to \infty]$. [R. Blumenthal & R. Getoor, Sample functions of stochastic processes with stationary independent increments, *J. Math. Mech*, 10(3), 493–516, 1961]

Blumenthal's zero-one (or 0–1) law Arises in the context of the Wiener process, and other suitable diffusions, where it is useful in studying the properties of sample-paths. The basic form is this: let $W(t)$ be the Wiener process, and let $\sigma\{W(s) : 0 \leq s \leq t\}$ denote the sigma-field generated by the process over $[0, t]$. Define $F_0^+ = \bigcap_{t>0} \sigma\{W(s) : 0 \leq s \leq t\}$. Then any event $A \in F_0^+$ has probability 0 or 1. [P. Mörters & Y. Peres, *Brownian Motion*. Cambridge University Press, 2010]

Bobkov–Houdré conjecture Asserts that if X and Y are i.i.d, symmetric r.vs such that, for $z > 0$, $\mathbb{P}(|X + Y|/\sqrt{2} \geq z) \leq \mathbb{P}(|X| \geq z)$, then X and Y are normally distributed. S. Kwapien, M. Pycia, and W. Schachermayer, and independently, K. Olestkiewicz, have shown that the conjecture is true. [S. Kwapien *et al.*, Proof of a conjecture of Bobkov & Houdré, *Electr. Commun. Probab.*, 1, 7–10, 1996]

Bobkov's inequality Arises in considering Boolean functions, where it yields a proof of the Gaussian isoperimetric inequality. [S. Bobkov, An isoperimetric inequality on the discrete cube, and an elementary proof of the isoperimetric inequality in Gauss space. *Ann. Probab.*, 25(1), 206–214, 1997]

Bochner expectation A generalization of the ordinary (Lebesgue) expectation to yield the expected value of random elements taking values in a real Banach space that may be infinite-dimensional. If the space is finite-dimensional, then the Bochner expectation is essentially the coordinate-wise Lebesgue expectation, i.e. $\mathbb{E}\mathbf{V} = (\mathbb{E}V_1, \ldots, \mathbb{E}V_d)$, in d dimensions. If the space is infinite-dimensional and separable, then the Bochner expectation can be defined, for any element $\mathbf{V}$ with $\mathbb{E}\|\mathbf{V}\| < \infty$, by taking the limit of a suitable sequence $\mathbf{V}_n$, of simple elements having well-defined Bochner expectation, converging to $\mathbf{V}$. If the space is not separable, the condition $\mathbb{E}\|\mathbf{V}\| < \infty$ is no longer sufficient for the existence of the Bochner expectation of $\mathbf{V}$. [R. Dudley, *Uniform Central Limit Theorems*, 2nd edn. Cambridge University Press, 2014]

Bochner's theorem (alias: **Bochner–Khinchin theorem**) Arises in characterizing those functions $\phi(t)$ that can be the characteristic fn of a real random variable X. It asserts that $\phi(t)$ is such a characteristic fn iff it is uniformly continuous on $\mathbb{R}$, with $\phi(0) = 1 \geq |\phi(t)|$, and non-negative definite, which is to say that $\sum_{j,k} \phi(t_j - t_k) z_j \bar{z}_k \geq 0$, for all real $t_1, \ldots, t_n$, and complex $z_1, \ldots, z_n$ and $n \in \mathbb{N}$. A similar result holds for joint characteristic functions of finite-dimensional distributions. The equivalent characterization in infinitely many dimensions is the **Bochner–Minlos thm**. [K. Itô, *Introduction to Probability Theory*. Cambridge University Press, 1984]

bold play (or strategy) Arises when a gambler is betting against the odds with a fixed goal, i.e. if the gambler's initial fortune is $f \in (0, 1)$, no further bets will be made if the gambler's fortune is ever greater than or equal to 1, which is the goal. Then for a fortune f the bold gamble is to bet either all you have, or as much as would yield a total fortune of 1, whichever is the smaller. For example suppose the game is **Red-and-Black**, in which the casino returns the stake s together with winnings s, with probability $P \leq ½$, or takes the stake with probability $1 - P$. If the gambler's objective is to maximize the probability of achieving the goal of a fortune of size 1, then the famous **Dubins–Savage thm** asserts that bold play is optimal. The result holds quite widely, but by no means universally, when playing such unfavourable games. [S. Ethier, *The Doctrine of Chances*. Springer, 2010]

Boltzmann distribution *see* Gibbs distribution

Boltzmann machine Arises in the theory of neural networks as a particular example of a Markov chain Monte Carlo procedure, for a binary random field whose equilibrium distribution is a special type of Gibbs distribution. There are applications to machine learning and combinatorial optimization. [G. Winkler, *Image Analysis, Random Fields, and Markov Chain Monte Carlo Methods*, 2nd edn. Springer, 2003]

Bondesson class This is the smallest class of distributions on $(0, \infty)$ that includes all mixtures of exponential densities, and is closed under the operations of convolution and taking limits in distribution on $(0, \infty)$. The extension to d dimensions is called the **Goldie–Steutel–Bondesson class**. [O. Barndorff-Nielsen *et al.*, Some classes of multivariate infinitely divisible distributions admitting stochastic integral representations, *Bernoulli*, 12(1), 1–33, 2006]

bond percolation *see* Broadbent–Hammersley percolation

bond pricing Arises in financial mathematics, where markets seek to trade in bonds, i.e. securities that pay a one unit dividend at a fixed time $T > 0$. If the bond yields no other payment, it is said to be a zero-coupon bond; intermediate interest payments are called coupons. It is assumed that for every $T > 0$, there is a market for zero-coupon bonds maturing at T, and that there is a bank (money) account that pays interest at rate $r(t)$ at time $t > 0$ (called the short rate). A classic model assumes that $r(t)$ is a diffusion process (arising from any one of a number of SDEs called interest rate models). If the price at t of a bond maturing at T is $p(t, T, r(t))$ and the market is taken to be arbitrage-free, the bond pricing problem is to determine $p(t, T, r(t))$. Under weak conditions it can be shown that $p(t, T, r)$ satisfies a partial differential equation, called the **term structure eqn**, whose solution may be written as the expected value (with respect to an appropriate martingale measure) of a functional of the interest-rate process $r(t)$. [Y.-K. Kwok, *Mathematical Models of Financial Derivatives*, 2nd edn. Springer, 2008]

bond process (alias: **edge process**) Arises in models for physical systems that may be represented as a graph G, with vertex set V and edge set E. To any edge $e \in E$ we may attach a random variable $X(e)$, and the collection is called a bond (or edge) process. The term is especially used when $X(e)$ is the indicator of the edge being blocked (to the flow of some notional fluid), or not. In this case one speaks of the bond percolation process. Similarly, to any vertex $v \in V$ we may attach a r.v. $Y(v)$, yielding a vertex (or site) process in general, and a site percolation process when the $Y(v)$ are indicator r.vs. [G. Grimmett, *Percolation*, 2nd edn. Springer, 1999]

Bonferroni inequalities Arise as a generalization of Boole's inequalities, thus: for events $A_1, \ldots, A_n$ define $P_a = \mathbb{P}(\cup_1^n A_i)$ to be the probability that any of the n events occur, and $S_r = \sum_I \mathbb{P}(\cap_{i \in I} A_i)$, where the summation is over the subsets I of $[1, \ldots, n]$ such that $|I| = r$. Then $P_a \geq S_1 - S_2$; $P_a \leq S_1 - S_2 + S_3$; $P_a \geq S_1 - S_2 + S_3 - S_4$, and so on. Related bounds for the probability that exactly m (or at least m) of the n events occur may be called Bonferroni-type bounds, or inclusion–exclusion bounds, e.g. (among many improvements) the **Hunter–Worsley inequality** asserts that for any spanning tree T on $\{1, \ldots, n\}$, $P_a \leq S_1 - \sum_{r,s:r \sim s} \mathbb{P}(A_r \cap A_s)$, where $r \sim s$ if r and s are joined by an edge of T. [J. Galambos & I. Simonelli, *Bonferroni-type Inequalities with Applications*. Springer, 1996]

Boolean function Arises in optimization, control, reliability, percolation, cryptography, games, voting theories, and decision theory, among other areas. A Boolean variable takes values in $\{0, 1\}$, and a function $f(x_1, \ldots, x_n) \in \{0, 1\}$, from $\{0, 1\}^n$ to $\{0, 1\}$, is called a Boolean function. If, more generally, such an f takes values in $\mathbb{R}$, it is called pseudo-Boolean. Typically, the arguments are Bernoulli (indicator) r.vs. Inequalities that arise in such cases (very often in combinatorial optimization) are Boolean inequalities (no confusion with Boole's inequalities can arise), or pseudo-Boolean inequalities in the more general case. [C. Garban & J. Steif, *Noise Sensitivity and Percolation*. Cambridge University Press, 2015; R. O'Donnell, *Analysis of Boolean Functions*. Cambridge University Press, 2014]

Boolean model (alias: **Poisson particle process; germ–grain model; centre-satellite process**) Arises in geometrical probability as a model for many spatial processes, such as randomly scattered objects or heterogeneous materials. It comprises a Poisson process N of points in $\mathbb{R}^d$, $d \geq 1$, together with a collection of i.i.d random compact sets, each of which is attached to a point of N, e.g. each point may be the centre of a randomly sized circle, or of a cluster of points. The final realization of the process is the union of the sets (possibly overlapping, unlike the hard-sphere model). The points of N may be called germs, and the sets grains, leading to the name germ–grain model. If each attached set is a finite point set, this is the Poisson cluster (or Bartlett–Lewis) model. If the union of grains has an unbounded connected component, the process is called supercritical; if the germs can move around as time passes, it is a dynamic Boolean model. Despite the terminology, such processes were unknown to Boole, the name being coined by G. Matheron. [S. Chiu *et al.*, *Stochastic Geometry and its Applications*, 3rd edn. Wiley, 2013]

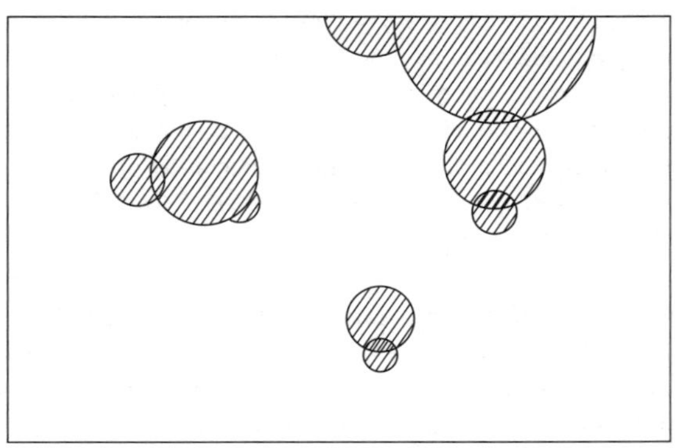

Boolean model formed from randomly selected compact sets centred at the points of a Poisson process.

Boolean network A graph in which the nodes or the edges (or both) are labelled with a Boolean variable taking values in $\{0,1\}$, that is in many cases of interest a realization of an indicator r.v. In applications, such a network may evolve randomly over time according to rules defined in terms of the current configuration. [I. Shmulevich & E. Dougherty, *Probabilistic Boolean Networks*. SIAM, 2010]

Boole's inequalities For events $A_l, \ldots, A_n$, we have

$$\mathbb{P}(\bigcup_{r=1}^{n} A_r) \leq \sum_{r=1}^{n} \mathbb{P}(A_r), \text{ and}$$

$$\mathbb{P}(\bigcap_{r=1}^{n} A_r) \geq 1 - \sum_{r=1}^{n} \mathbb{P}(A_r^c) = \sum_{r=1}^{n} P(A_r) - (n-1).$$

This is generalized by the Bonferroni inequalities, and the inclusion–exclusion bounds. [E. Seneta, On the history of the strong law of large numbers and Boole's equality, *Historia Mathematica*, 19, 24–39, 1992]

Borel–Cantelli lemmas Arise in the context of an infinite sequence of events $A_1, A_2, \ldots$, that are from the same space. It is often important to know the probability that only finitely many of them occur, or that infinitely many do, an event denoted by $A = [A_n \text{ i.o.}]$. The Borel–Cantelli lemmas assert that
(a) $\mathbb{P}(A) = 0$ if $\sum_{r=1}^{\infty} \mathbb{P}(A_r) < \infty$;
(b) $\mathbb{P}(A) = 1$ if $\sum_{r=1}^{\infty} \mathbb{P}(A_r) = \infty$;
and the events $A_1, A_2, \ldots$, are independent. There are a number of extended B-C lemmas that seek to weaken the strong condition of independence in (b). For example, define

$$a = \liminf_{n\to\infty} \left[\sum_{j=1}^{n} \sum_{k=1}^{n} \mathbb{P}(A_j \cap A_k) / \left(\sum_{r=1}^{n} \mathbb{P}(A_r) \right)^2 \right],$$

then if $\sum_1^\infty P(A_r) = \infty$, the **Erdös–Rényi B–C lemma** asserts that $\mathbb{P}(A) = 1$ if $a = 1$; this holds if, e.g., $\mathbb{P}(A_j \cap A_k) \leq \mathbb{P}(A_j)\mathbb{P}(A_k)$ for all $j \neq k$, which is implied by pairwise independence, for example. More strongly, the **Kochen–Stone–Spitzer B–C lemma** asserts that $\mathbb{P}(A) \geq a^{-1}$, when $\sum_r \mathbb{P}(A_r) = \infty$. Among other extensions, we have the **Dubins–Freedman–Lévy B-C lemma**. Let A_n and $I(A_n), n \geq 1$, be a sequence of events and their indicators, adapted to the filtration $\mathcal{F}_n, n \geq 1$, where $0 \leq P(A_1) < 1$. Then the lemma asserts that as $n \to \infty$, with probability 1,

$$r_n = \sum_{k=1}^{n} I(A_k) / \sum_{k=1}^{n} \mathbb{P}(A_k \mid \mathcal{F}_{k-1}).$$

converges to a finite limit r, and $r = 1$ a. s. whenever $\sum_{n=1}^{\infty} \mathbb{P}(A_n \mid \mathcal{F}_{n-1}) = \infty$. [T. K. Chandra, *The Borel–Cantelli Lemma*. Springer, 2012]

Borel field *see* sigma-field

Borell's inequality (alias: **Borell–CIS inequality**) Arises in the theory of Gaussian processes; a common form is this: let $X(t), t \in T$, be a zero-mean Gaussian process such that $\mathbb{P}\left(\sup_{t\in T}|X(t)| < \infty\right) = 1$, and define $s^2 = \sup_{t\in T}\mathbb{E}[X(t)^2]$. Then $m = \mathbb{E}[\sup_{t\in T}X(t)] < \infty$, and for $t > 0$,

$$\mathbb{P}\left(|\sup_{t\in T} X(t) - m| > t\right) \leq 2\exp\left[-\frac{1}{2}t^2/s^2\right].$$

This was shown independently by Borell and Tsirelson–Ibragimov–Sudakov, which explains the alias when we note that Cirelson is an alternative transliteration for Tsirelson. [R. Adler & J. Taylor, *Random Fields and Geometry*. Springer, 2007]

Borell–Sudakov–Tsirelson inequality *see* Gaussian isoperimetric inequality

Borel measure Arises in measure theory, and has important uses in probability. On the real line, a Borel set is one obtained by countable unions or intersections of open intervals in $\mathbb{R}$; the Borel field is the smallest σ-field including all the open intervals of $\mathbb{R}$. Any countably additive (measure) function $m(\cdot) \geq 0$ defined on the Borel sets may be called a Borel measure; in particular, if $m(\mathbb{R}) = 1$ it is a (Borel) probability measure. Borel sets and measures are defined similarly (in terms of the sigma-field generated by the open sets) for $\mathbb{R}^d, d \geq 2$, or any other suitable space Ω. A real-valued function $g(\cdot)$ on $\mathbb{R}$ is called Borel-measurable (or said to be a Borel function) if the inverse $g^{-1}(B)$ of any Borel set B is also a Borel set. More generally, for a measurable space $(\Omega, \mathbb{F}), g(\cdot): \Omega \to \mathbb{R}$ is a Borel function if $g^{-1}(B) \in \mathcal{F}$ for any Borel set B. When $(\Omega, \mathcal{F}, \mathbb{P})$ is a probability space a Borel function from Ω to $\mathbb{R}$ is a random variable, or a random vector if from Ω to $\mathbb{R}^d$. The class of Borel sets in $\mathbb{R}^d$ is sometimes called the **Borel tribe.** [R. Ash & C. Doléans-Dade, *Probability and Measure Theory*, 2nd edn. Academic Press, 2000]

Borel's normal number theorem (alias: **Borel's strong law of large numbers**) Arises naturally in the context of number theory, and also in flipping a sequence of fair coins. A number $x \in [0,1]$ is said to be normal (or absolutely or simply normal) if its binary (dyadic) expansion $x = 0, b_1, b_2, b_3, \ldots$ has the property that $\frac{1}{n}\sum_{r=1}^{n} b_r \to \frac{1}{2}$, as $n \to \infty$. The theorem asserts that the Lebesgue measure of the normal numbers in $[0, 1]$ is 1.

In its probabilistic interpretation, this asserts that a random variable U that is uniform in $[0, 1]$ is a.s. normal, which is equivalent to the assertion that a sequence of i.i.d.r.vs $X_r, r \geq 1$, each having the Bernoulli distn with parameter 1/2 (i.e. a sequence of fair coins), obeys the strong law of large numbers, $\frac{1}{n}\sum_{r=1}^{n} X_r \to \frac{1}{2}$, a.s., which is Borel's SLLN. [P. Billingsley, *Probability and Measure*, 3rd edn. Wiley, 1995]

Borel's paradox (alias: **Borel–Kolmogorov paradox**) Arises as an illustration of the inadmissibility of defining conditional probability with respect to an isolated null event. Let a point P picked uniformly at random on the surface of a sphere have latitude X and longitude Y. Then the conditional distribution of the longitude given the latitude is uniform on $[0, 2\pi]$. But the conditional distribution of the latitude given the longitude, distributed on $[-\pi/2, \pi/2]$, is not uniform; the density is $\frac{1}{2}|\cos x|$. The 'paradox' becomes perhaps more striking if the given latitude in the first case is that of the equator, being a great circle also. [D. Pollard, *A User's Guide to Measure Theoretic Probability*. Cambridge University Press, 2002]

Borel–Tanner distribution Arises in queueing theory, as the distribution of the number of customers served in a busy period of the M/D/1 queue, with r customers waiting as the period starts. It has the form $\mathbb{P}(X = n) = re^{-an}a^{n-r}n^{n-r-1}/(n-r)!$, for $n \geq r$, and some $a \in [0, 1)$. The case $r = 1$ is called the Borel distribution. [F. Haight & M. Breuer, The Borel–Tanner distribution, *Biometrika*, 47(1–2), 143–150, 1960]

Bose–Einstein distribution Arises in statistical mechanics where it describes the distribution of the n indistinguishable particles of a perfect gas among their possible energy levels. Probabilists use the term to denote the distribution of n indistinguishable objects among r boxes, which attaches equal probability to each of the $\binom{n+r-1}{n}$ distinguishable assignments of the objects (where any number of objects may be in any box). [W. Feller, *Introduction to Probability Theory and its Applications*, Vol. 1, 3rd edn. Wiley, 1968]

Bougerol identity Arises in the context of exponential functionals of the Wiener process $W(t)$. If $B(t)$ is a second standard Wiener process, independent of $W(t)$, then the identity asserts that, for any fixed $t > 0$, $\sinh B(t)$ has the same distribution as $W\left(\int_0^t \exp\left(-2B(s)\right)\mathrm{d}s\right)$. [S. Vakeroudis, Bougerol's identity in law and extensions, *Probab. Surv.*, 9, 411–437, 2012]

bounded differences inequality *see* McDiarmid's inequality

Bourgain's entropy criterion Arises in ergodic theory, where it shows that the pointwise (that is, a.s.) convergence of a suitable sequence of operators entails that the metrical entropy of the sequence is uniformly bounded. [R. Jones, Ergodic theory and connections with analysis and probability, *New York J. Math.*, 3A, 31–67, 1997]

Box–Muller transformation Arises in sampling from the normal distn (i.e. simulating normal random variables) thus: if U and V are i.i.d.r.vs that are uniformly distributed on $[0, 1]$, then X and Y given by $(X, Y) = (\{-2\log U\} \cos\{2\pi V\}, \{-2\log U\} \sin\{2\pi V\})$ are independent standard normal r.vs. [J. Gentle, *Random Number Generation and Monte Carlo Methods*, 2nd edn. Springer, 2003]

brackets process Arises in the theory of martingales, in several contexts, where there are two so-called brackets processes. Let $M_n, n \geq 0$, be a martingale such that $M_0 = 0$ and $\mathbb{E}M_n^2 < \infty$ for all n. Define the process A_n recursively by $A_0 = 0$ and $A_n - A_{n-1} = \mathbb{E}(M_n^2 - M_{n-1}^2|\mathcal{F}_{n-1}) = \mathbb{E}((M_n - M_{n-1})^2|\mathcal{F}_{n-1})$, where $\mathcal{F}_{n-1}$ is the filtration on which the martingale is defined (to which it is adapted). Then $A_n, n \geq 0$, is denoted by $\langle M\rangle$ (or sometimes $\langle M, M\rangle_n$, or $\langle M\rangle_n$), and called the (predictable) quadratic variation process, or the angle brackets process (or the sharp, or pointed brackets process). It may also be defined as the unique increasing predictable process such that $\langle M\rangle_0 = 0$ and $M_n^2 - \langle M\rangle_n$ is a martingale.

Secondly, define the square-bracket process $[M]$ as

$$[M]_n = [M, M]_n = \sum_{r=0}^{n}(M_r - M_{r-1})^2$$
$$= 2\sum_{r=0}^{n} M_{r-1}(M_r - M_{r-1}),$$

which is also called the optional quadratic variation process. The process $M_n^2 - [M]_n$ is also a martingale, that is uniformly integrable if $\sup_n \mathbb{E}M_n^2 < \infty$. The analogous process $[M]_t$ may be defined for any local martingale $M(t)$ in continuous time, and $\langle M \rangle_t$ also if $M(t)$ is locally square integrable. The cross-variation processes between martingales X_n and Y_n are defined for angle-brackets by $\langle X, Y \rangle_n = \frac{1}{4}\{\langle X+Y \rangle_n - \langle X-Y \rangle_n\}$, and similarly for $[X,Y]_n$. *See* quadratic variation. [M. H. A. Davis, Martingale representation and all that, in: *Advances in Control, Communication Networks, and Transportation Systems.* Birkhäuser, 2005, 57–69]

branching Erlang distribution *see* Coxian distribution

branching process Arises naturally as a model for reproducing, or otherwise proliferating, entities. The simplest (and classical) example is the Galton–Watson process (also called the G-W-Bienaymé branching, or ordinary branching, or just the branching, process), which we denote by $Z_n(x)$, $n \geq 0$, and describe thus: the process starts at time 0 with $Z_0 = x$ particles. At any time $n \geq 0$, each of the $Z_n(x)$ particles in existence is replaced by a random number of replacements (called its family) whose sizes are the i.i.d.r.vs $X(n,1), \ldots, X(n, Z_n)$, where $G(s) = \mathbb{E}s^{X(n,z)}$. Then $Z_{n+1}(x) = X(n,1) + \cdots + X(n, Z_n)$. It is essential to the analysis of $Z_n(x)$ that it has the so-called **branching property**, viz $Z_n(x+y)$ has the same distribution as $Z_n(x) + \widetilde{Z}_n(y)$, where $\widetilde{Z}_n(y)$ has the same distribution as $Z_n(y)$ and is is independent of $Z_n(x)$. Then, more generally, a branching process is a family of random elements (which may be continuous r.vs, or measure-valued, etc.) that has the branching property in some suitable form. Elementary examples include: age-dependent branching processes, in which particles complete random lifetimes before replacement; spatial branching processes, in which particles perform random walks, or follow a diffusion process before replacement; and multitype branching, where particles are of various kinds.

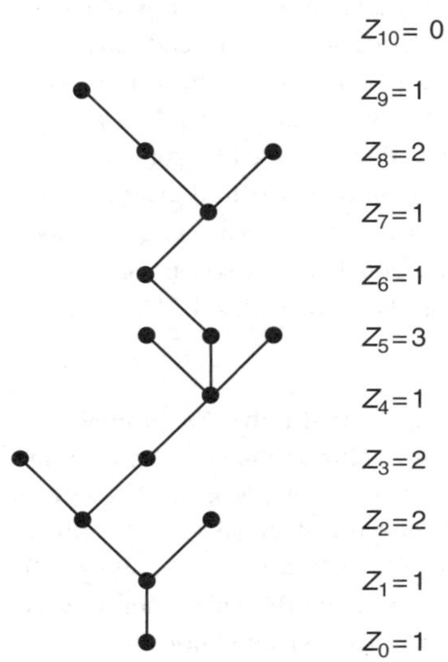

Branching process.

For branching processes in which the family-size may be zero, there is a possibility of extinction. In the case of the Galton–Watson process this yields the basic **criticality thm:** let $Z_0 = 1$, and $G'(1) = \mathbb{E}X(n,1) = \mu$, and consider $\mathbb{P}(Z_n = 0) = q_n$, which is the probability of extinction at or before the nth generation. Then the theorem asserts that as $n \to \infty, q_n \to \eta$, where η is the smallest positive root of $G(s) = s$, and (a) $\eta = 1$ if $\mu < 1$; (b) $\eta < 1$ if $\mu > 1$; (c) if $\mu = 1$, then $\eta = 1$ provided that var $X(n,1) > 0$. This final case is called a **critical process**; the cases $\mu > 1$ and $\mu < 1$ are called **supercritical** and **subcritical** respectively. *See* multiplicative process and Kesten–Stigum thm. [P. Haccou *et al.*, *Branching Processes.* Cambridge University Press, 2005; K. Athreya & P Jagers (eds), *Classical and Modern Branching Processes.* Springer, 1997]

breadwinner system *see* Belgian progression

Bregman divergence Any of a family of dissimilarity measures that can supply a measure of difference between two probability distributions $p_1, \ldots, p_n$, and $q_1, \ldots, q_n$, in terms of a convex function $g(\cdot)$ from $\mathbb{R}^n$ to $\mathbb{R}$ whose gradient is $\nabla g(x) = \left(\frac{\partial g}{\partial x_1}, \ldots, \frac{\partial g}{\partial x_n}\right)$. Its basic form is $d(\mathbf{p},\mathbf{q}) = g(\mathbf{p}) - g(\mathbf{q}) + (\mathbf{q}-\mathbf{p}) \cdot \nabla g(\mathbf{q})$ (where $\mathbf{a} \cdot \mathbf{b}$ is the scalar product of $\mathbf{a}$ and $\mathbf{b}$); more narrowly, for a convex function $h(\cdot)$ from $\mathbb{R}$ to $\mathbb{R}$, $s(p,q) = \sum_{r=1}^{n} \{h(p_r) - h(q_r) + (q_r - p_r)h'(q_r)\}$ is called the separable version. [W. Stummer & I. Vajda, On Bregman distances and divergences of probability measures, *Trans. IEEE Inf. Theor.*, 58(3), 1277–1288, 2012]

Breiman's extra-sensory perception problem *see* Chow–Robbins problem

Breiman's theorem Any of several results in various fields, but perhaps most often a lemma arising in the context of Breiman's type of arcsine law: let X and Y be independent and non-negative, where Y is regularly varying with exponent $a \geq 0$, while $\mathbb{E}X^{a+\delta} < \infty$ for some $\delta > 0$. Then, as $z \to \infty, \mathbb{P}(XY > z) \sim \mathbb{P}(Y > z)\mathbb{E}X^a$.

Elsewhere, **Breiman's realistic gambling thm** arises in the context of a realistic casino, thus: let $X_n, n \geq 1$, be the sequence of fortunes of some gambler. It is realistic to require that for some $b > 0$, $|X_{n+1} - X_n| \geq b$, i.e. the casino has a minimum stake. Furthermore, X_n must realistically be a non-negative supermartingale, i.e. only subfair bets are available, with no credit. The theorem asserts that a gambler who always plays when $X_n \geq b$ will be ruined with probability 1 (in the sense that $X_N < b$ for some finite N). **Freedman's thm** asserts that a gambler who wishes to make N as large as possible (in the sense that $\mathbb{P}(N > n)$ is maximized for all n) should always bet the minimum stake; timid play is best. Breiman's optimal system theorem for favourable games is often called the **Kelly–Breiman thm.** *See also* Shannon–McMillan–Breiman–Chung theorem. [L. Breiman, On some limit theorems similar to the arc-sin law, *Theor. Probab. Appl.*, 10(2), 323–331, 1965; L. Breiman, *Probability*. Addison-Wesley, 1968; reprinted by SIAM, 1992; L. Breiman, Optimal gambling systems for favourable games, *Proc. Berk. Symp. Math. Stat. Probab.*, 1, 65–78, 1961]

Breit–Wigner distribution *see* Cauchy distribution

Brennan–Schwartz model Arises in financial mathematics, where it postulates an SDE to describe the joint behaviour of the short-term interest rate $r(t)$ (the instantaneous rate), and the long-term rate $l(t)$ (that is the consol rate, i.e. the rate implied by the price of a bond with fixed coupon (interest) and infinite maturity date). There are several formulations; one is this:

$$\mathrm{d}r = [a + b(l - r)]\mathrm{d}t + cr\mathrm{d}W_1;$$
$$\mathrm{d}X = [\alpha + \beta l + \gamma r]\mathrm{d}t + \sigma \mathrm{d}W_2,$$

where $X = \log l(t)$, and (W_1, W_2) comprises a pair of correlated Wiener processes. [M. Brennan & E. Schwartz, Continuous time approach to the pricing of bonds, *J. Banking Finan.*, 3(2), 133–155, 1979]

Brier score Arises as a measure of the accuracy of a set of predicted probabilities, e.g. the stated chances of rain in a sequence of weather forecasts. If p_r is the predicted probability for an event A_r, and I_r is the indicator of the event's occurrence, the Brier score is $\frac{1}{n}\sum_{r=1}^{n}(p_r - I_r)^2$, where n is the number of predictions. [G. Candille & O. Talagrand, Evaluation of probabilistic prediction systems for a scalar variable, *Q. J. Roy. Metereol. Soc.*, 131, 2131–2150, 2005]

Brillouin's information Arises as a measure of the information content of a given message of length n, i.e. a sequence of symbols from an alphabet A of size a. If the rth letter makes n_r appearances, the Brillouin information is $B = \log \frac{n!}{n_1! \ldots n_a!}$, in appropriate units. As the length of message increases the average B/n converges to the Shannon information, provided that the letter-frequencies converge to fixed proportions. The Brillouin information is also used as an index of diversity, in which context it is called the B. function, B. measure, B. index, etc. [L. Brillouin, *Science and Information Theory*, 2nd edn. Academic Press, 1962; reprinted by Dover, 2004]

Broadbent–Hammersley model *see* percolation process

Broadbent–Hammersley percolation (alias: **bond percolation**) Arises in the context of the following question: a stone is porous and immersed in water. Is the middle of the stone wet? The model considers the square (or cubic) lattice whose coordinates are integers. It is envisaged that water may flow along the lattice edges between adjacent points, except that each edge is independently blocked with probability $1 - p$; otherwise it is open, and water flows. The classic questions of percolation ask for what values of p is the set of wet atoms finite, or infinite? [G. Grimmett, *Percolation*, 2nd edn. Springer, 1999]

Brook's factorization theorem Arises in extending the simple Markov property in one dimension to Markov random fields. Let $f(x_1, \ldots, x_n)$ be a joint probability distribution (on the product state-space S^n). If, for given x_r and x_s, there is a factorization ϕ of f (as a product of functions of the x_i), such that no factor depends on both x_r and x_s, then x_r and x_s are said to be separated in ϕ. The theorem asserts that there is a factorization Φ of f such that every such pair (x_r, x_s), separated by some ϕ, is separated in Φ. *See* Hammersley–Clifford theorem. [D. Brook, On the distinction between the conditional probability and the joint probability approaches in the specification of nearest-neighbour systems, *Biometrika*, 51(3–4), 481–483, 1964]

Brownian bridge (alias: **tied-down (or pinned) Wiener process (or Brownian motion)**) Arises in the study of empirical distribution functions, and has various applications, e.g. in financial mathematics. Its intuitively natural definition is this: Fix $0 \leq x \leq y < \infty$, and $s > 0$, and define $S(t), 0 \leq t \leq s$, to be Brownian motion $B(t)$ (i.e. the standard Wiener process), conditional on $B(0) = x$ and $B(s) = y$. Then $S(t)$ is a Brownian bridge from x to y with span s; if $x = y = 0$ and $s = 1$ we have the (standard) Brownian bridge. It may be alternatively defined as a continuous Gaussian process such that

$\mathbb{E}S(t) = x + (y - x)t/s$, and

$\text{cov}(S(u), S(v)) = \min(u, v) - uv/s.$

Further, it may be defined as the continuous diffusion X started at x with parameters (instantaneous mean and variance) $a(t, X) = -X/(s - t)$, and $b(t, X) = 1$; $X \in \mathbb{R}, 0 \leq t \leq s$. It has several other representations in terms of $B(t)$ itself, thus all these are Brownian bridges:
(a) $S_1(t) = (1 - t)B(t/(1 - t)), S_1(1) = 0$;
(b) $S_2(t) = B(t) - tB(1)$;
(c) $S_3(t) = (1 - t)\int_0^t \frac{1}{1-s} \mathrm{d}B(s)$, $S_3(1) = 0$.
Also, if $T > 0$ is fixed, and U is the last zero of $B(t)$ before time T, then conditional on $U = u$ the restriction of $B(t)$ to the interval $[0, U]$ is a Brownian bridge from 0 to 0 with span u. Finally $S_4(t) = \frac{\sqrt{2}}{\pi}\sum_{r=1}^{\infty}\frac{1}{r}\sin(r\pi t)Z_r$, is a Brownian bridge when Z_r are i.i.d. normal $N(0, 1)$ r.vs. Analogous bridges in $d > 1$ dimensions are likewise most naturally defined by conditioning on the end points. [A. Borodin & P. Salminen, *Handbook of Brownian Motion*, 2nd edn. Birkhäuser, 2002]

Brownian coefficient (alias: **Gaussian coefficient (not to be confused with the q-binomial coefficients of quantum theory)**) Arises in the context of a Lévy process $X(t), t \geq 0$, with characteristic exponent $\psi(\theta)$ such that $\mathbb{E}e^{i\theta X(t)} = e^{-t\psi(\theta)}$. Then $\lim_{|\theta|\to\infty}\theta^{-2}\psi(\theta) = \frac{1}{2}\sigma^2$, where $\sigma^2 \geq 0$ is called the Brownian coefficient. [R. Doney, Introduction to Lévy processes, in Fluctuation theory for Lévy processes, in: *Ecole d'Eté de Probabilités de Saint-Flour XXXV-2005*, 1–8, 2007]

Brownian disconnexion exponent Arises in two-dimensional Brownian motion started at $(1, 0)$. If any segment of the path of $B(s), 0 \leq s \leq t$, forms a closed loop around the origin, then the origin is said to have been disconnected from ∞. Let $T_r = \inf\{t \geq 0 : |B(t)| = r\}$ be the first hitting time of the circle radius r, and let p_r denote the probability that the origin is not disconnected from ∞ before time T_r. Then the disconnexion exponent is $d = \lim_{r\to\infty} -\frac{\log p_r}{\log r}$; it is conjectured that $d = 1/4$. Similarly, if S_r is the first hitting time of the circle radius r by an independent Brownian motion $W(t)$ started from $(-1, 0)$, and q_r is the probability that $B(t)$ on $[0, T_r]$ does not meet $W(t)$ on $[0, S_r]$, then the intersection exponent is $\lim_{r\to\infty}\frac{-\log q_r}{\log r}$. Similar exponents for random walk in the plane are defined likewise. [G. Lawler & E. Puckette, The disconnection exponent for simple random walk, *Israel J. Math.*, 99(1), 109–121, 1997]

Brownian distance covariance (alias: **Wiener covariance**) A measure of dependence between random variables X and Y defined thus: Let X' and Y' be i.i.d. copies of X and Y respectively, and let $W(t)$ and $V(t)$, $t \in \mathbb{R}$, be i.i.d. Wiener processes that are independent of X, Y, X', Y'. Write $X(W) = W(X) - \mathbb{E}(W(X)|W)$, and likewise for Y, V, and primed quantities. Then the Brownian covariance $B(X, Y) > 0$ is given by $[B(X,Y)]^2 = \mathbb{E}(X(W)X'(W)Y(V)Y'(V))$. The Brownian distance correlation is then given by $\rho = B(X, Y)/\{B(X,X)B(Y,Y)\}^{1/2}$. [G. Székely & M. Rizzo, Brownian distance covariance, *Ann.. Appl. Stat.*, 3(4), 1236–1265, 2009]

Brownian excursion Arises naturally as the continuous limit of an excursion of a random walk, i.e. its path (or graph) until it first repeats its initial value. If the continuous limit of the walk is Brownian motion, the distribution of the size $X(t) \geq 0$ at time $t \in (0, 1)$, of the excursion from 0 (and back to 0) over the interval $[0, 1]$ has density

$$f(x) = 2x^2[2\pi t^3(1 - t)^3]^{-\frac{1}{2}}\exp\left[-\frac{x^2}{2t(1 - t)}\right].$$

Informally, if $B(t)$ is the standard Brownian motion, then the standard (normalized) Brownian excursion $X(t)$ can be defined as $B(t)$ on $[0, 1]$ conditioned on $B(1) = 0$, and $B(t) > 0$ for $0 < t < 1$; an idea which can be made precise. From another point of view, at any time $t > 0$, Brownian motion $B(t)$ is almost surely enjoying an excursion (possibly negative), called the excursion straddling t. The length of this excursion has density

$$f(x) = (t^{\frac{1}{2}} - \max\{t - x, 0\}^{\frac{1}{2}})/\left(\pi x^{\frac{3}{2}}\right), x > 0.$$

If U is the last zero of $B(t)$ before $t = 1$, V the first zero of $B(t)$ after $t = 1$, then conditioned on $U = u$ and $V = v$, the process $|X(t)|$ is a standard Brownian excursion, where

$$X(t) = \frac{1}{\sqrt{v - u}}B(u + t(v - u)), \quad 0 \leq t \leq 1; \text{this is}$$

Lévy's representation of the Brownian excursion. [F. Knight, *Essentials of Brownian Motion and Diffusion*. American Mathematical Society, 1981]

Brownian frontier Arises in two-dimensional Brownian motion, the path of which will almost surely cross itself, thus creating finite regions in $\mathbb{R}^2$ enclosed by the path. The union of these regions and the path is the hull, and the boundary of the hull is the Brownian frontier. It has Hausdorff dimension 4/3. [G. Lawler *et al.*, The dimension of the planar Brownian frontier is 4/3, *Math. Res. Lett.*, 8(4), 401–411, 2001]

Brownian (or Wiener) functional A random variable (or more generally, a random element), arising as a suitable function of the path of a Wiener process; more formally, a suitably measurable function on the Wiener space of continuous functions, equipped with Wiener measure. Examples include: integrals, first-passage times, occupation times, etc. [I. Karatzas & S. Shreve, *Brownian Motion and Stochastic Calculus*, 2nd edn. Springer, 1998]

Brownian map Arises as the continuous limit (with increasing size), of a type of large connected planar random graph. (A planar graph is drawn on the surface of a sphere, or equivalently the plane, and edges meet only at vertices of the graph.) Each such graph (of the type called a *p*-angulation), has a distance function supplied by its edges, so the resulting continuous limit, the Brownian map, can be seen as a random metric space. It is homeomorphic to the sphere, and has Hausdorff dimension 4. [J.-F. Le Gall, Uniqueness and universality of the Brownian map, *Ann. Probab.*, 41(4), 2880–2960, 2011]

Brownian meander Arises as the continuous limit of a sequence of suitable random walks started from the origin that are conditioned to stay positive thereafter. Its most natural definition is as a Brownian motion $B(t)$ conditional on the event that $B(s) > 0, 0 < s \leq t$; if the interval of positivity is $(0, 1]$, this yields the standard meander $B^+(t), 0 \leq t \leq 1$, where $B^+(1)$ has the *Rayleigh density* $f(x) = x\exp\left[-\frac{1}{2}x^2\right], x > 0$. Alternatively, let U be the time of the last visit to zero of Brownian motion before $t = 1$; then conditional on $U = u$, the Brownian meander is $B^+(t) = |B(u + t(1 - u))|/(1 - u)^{1/2}, 0 \leq t \leq 1$.

Furthermore, it may also be defined as a three-dimensional Bessel bridge from $(0, 0, 0)$ to $(0, 0, R)$, where R has the Rayleigh density above. Finally, note that the process $B(t) - \min_{0\leq s\leq 1} B(s)$ is called **Brownian double meander**. [R. Mansuy & M. Yor, *Aspects of Brownian Motion*. Springer, 2008]

Brownian motion (alias: **Wiener process; Bachelier–Wiener process.**) Arises as a model for the Brownian movement of particles suspended in a fluid (first observed by Robert Brown in 1827). It serves as a driving process for other models, and is thus of widespread application in the physical sciences and financial mathematics. It has several definitions and characterizations, including:

1. The almost surely continuous process $W(t), t \geq 0$, having independent increments that have a stationary normal distn, where $W(0) = 0$ and $W(t)$ has the normal density with mean 0 and variance t. More formally, for all n and choice of $0 \leq t_1 \leq \ldots t_n$, the random variables $W(t_r) - W(t_{r-1}), 2 \leq r \leq n$, are independent zero-mean normal random variables with respective variances $t_r - t_{r-1}, 2 \leq r \leq n$.
2. The continuous homogeneous Markov process whose transition function (or density or kernel) satisfies the standard diffusion equation $f_t = \frac{1}{2}f_{xx}$; or, more indirectly, whose infinitesimal increments (i.e. $X = W(t + h) - W(t)$ for small h), satisfy a number of technical conditions in addition to $\mathbb{E}X = 0$ and $\operatorname{var} X = h + o(h)$.
3. The continuous zero-mean Gaussian process with $W(0) = 0$ and covariance function $\mathbb{E}(W(s)W(t)) = \min\{s,t\}$.
4. The process $W(t)$, adapted to the filtration $\mathcal{F}_t$, with respect to which both $W(t)$ and $W(t)^2 - t$ are martingales having continuous sample paths starting from zero.

For such definitions to be non-empty, it is necessary to produce both a probability space on which $W(t)$ can exist, and a demonstration that it does so. This is Wiener's theorem of 1923, in which the construction is in terms of a limit (which may for example be based on a suitable sequence of scaled symmetric simple random walks). The resulting process is self-similar, with fractal properties, and while the sample paths are indeed continuous they are also almost surely nowhere differentiable.

There are many extensions of B.M., some of which follow. The name 'Wiener process' often replaces 'Brownian motion'.

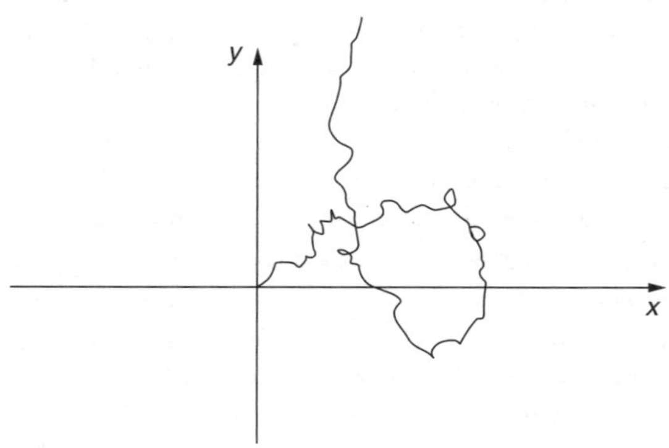

Brownian motion in the plane.

circular B.M. This is the B.M. $W(t)$ wrapped around a circle; if the circle has circumference 1, it may be written as $X(t) = W(t)$ modulo 1. On a circle of radius 1, it can be represented in the complex plane as $Z(t) = \mathrm{e}^{iW(t)}$.

(This circular B.M. should not be confused with Lévy's multiparameter B.M. evaluated at the points of a circle in the parameter space.)

conditional B.M. This is usually interpreted to mean B.M. conditioned on some event determined by the process itself, such as B.M. conditional on its not having exited the interval $[-a, a]$ before time t.

d-dimensional B.M. This is a vector of $d \geq 1$ independent Brownian motions; if $d = 2$, this is **planar B.M.**

drifting B.M. (alias: **arithmetic B.M.**) Arises naturally as a model for a particle undergoing Brownian movement in fluid that is itself moving. More formally, the drifting B.M. is defined as $D = at + bW(t)$, for real constants a and b, where $W(t)$ is the Wiener process. More generally, at is replaced by a **drift fn** $a(t)$.

elastic B.M. Arises in extending the idea of an elastic barrier from discrete processes to the Wiener process. In a simple random walk an elastic barrier either absorbs the walk on arrival with probability α, or reflects it with probability $1 - \alpha$; the number of visits before absorption (conditional on at least one) is geometric. If the origin is deemed to be an elastic barrier for B.M, the set $\{t : W(t) = 0\}$ has Lebesgue measure zero, so elasticity is defined using the Lévy local time at zero. Thus, elastic (or reflecting elastic) B.M. is B.M. that is killed (absorbed at 0) when its Lévy local time at 0 first exceeds X, where X is an independent exponentially distributed r.v.

geometric B.M. (alias: **exponential B.M.**) Arises in financial mathematics as a model for the evolution of an asset price over time. It is defined in terms of the Wiener process $W(t)$ as $S(t) = \exp[at + bW(t)], t \geq 0$, for constants a and b. Alternatively it may be defined as a a solution of the SDE $\mathrm{d}S = (a + \frac{1}{2}b^2)S\mathrm{d}t + bS\mathrm{d}W(t)$.

hyperbolic B.M. A diffusion process in the upper half plane that is the solution of the pair of SDEs $\mathrm{d}X = Y\mathrm{d}W_1$, and $\mathrm{d}Y = Y\mathrm{d}W_2$, where W_1 and W_2 are independent standard Wiener processes. The solution, started from $(0, 1)$, is $Y(t) = \exp\left[W_2 - \frac{1}{2}t\right]$, and $X(t) = \int_0^t \exp\left[W_2(s) - \frac{1}{2}s\right]\mathrm{d}W_1(s)$.

The transition density satisfies $\frac{\partial f}{\partial t} = \frac{1}{2}y^2\nabla^2 f$, where ∇^2 is the usual Laplacian operator.

invariance of B.M. A number of operations on Brownian motion $B(t)$ leave it unchanged, including:

1. Scaling. For $a > 0$ the process $X(t) = a^{-1}B(a^2t)$ is Brownian motion.
2. Inversion. The process $Y(t) = tB(t^{-1})$ for $t > 0$, with $Y(0) = 0$, is Brownian motion.
3. Conformal invariance.

iterated B.M. Informally, this is a d-dimensional B. M. where time is run by a standard Wiener process, with a necessary modification to exclude negative times. Formally, let $W^+(t), W^-(t)$ and $W(t)$ be independent standard Wiener process (where $W(t)$ is 1-dimensional, and W^+ and W^- are d-dimensional processes) and define

$$Z(t) = \begin{cases} W^+(t), & t \geq 0, \\ W^-(-t), & t < 0. \end{cases}$$

Then the iterated B.M. started at zero is $Z(W(t))$. (The process $Z(t)$ is sometimes called **two-sided B.M.)** Another iterated B.M., which is also called **Brownian-time B.M.,** is simply $B(|W(t)|)$, where B is a (possibly d-dimensional) B.M., and $|W(t)|$ is the standard Wiener process reflected at zero.

killed B.M. Broadly, a process that is B.M. until some time T, which may be random, and need not be finite with probability 1, at which time the process enters a special state, which may be either adjoined to $\mathbb{R}^d$ (in which case it may be called a coffin or cemetery state) or some point of $\mathbb{R}^d$. For example, if T is the first-passage time to some set B, then the new process is B.M. killed in B.

Brownian pillow (alias: **completely tucked Brownian sheet**) Arises naturally in two dimensions as a two-parameter Gaussian process with a.s. continuous realizations, whose covariance function is $(s_1 \wedge t_1 - s_1t_1)(s_2 \wedge t_2 - s_2t_2)$ for $0 \leq s_i, t_i \leq 1, i = 1, 2$; it can be seen as a Brownian sheet tied down (or sewn) along the perimeter of the unit square. In n dimensions the covariance fn is $\prod_1^n(s_i \wedge t_i - s_it_i)$.

reflected B.M. Broadly, any B.M. in the presence of a reflecting (or retaining) barrier. Very often, more narrowly in one dimension, it simply means $|W(t)|$, where $W(t)$ is the standard Wiener process.

Brownian scenery The process $S(t) = \int_{\mathbb{R}} L(t, x)\mathrm{d}W(x), t \geq 0$, where $L(t, x)$ is the local time process of a Brownian motion independent of the two-sided Wiener process $W(x), x \in \mathbb{R}$, is called B.M. in Brownian Scenery.

skew B.M. Arises as a model for diffusion in the presence of an asymmetric (osmotic) barrier at the origin (say); a particle arriving at the barrier is expelled (immediately) on the positive side with probability p, or on the negative side with probability $q = 1 - p$. One way of formalizing this is to consider the standard Wiener process reflected at 0, whose path comprises a sequence of (positive) excursions from 0 to 0. Now change the sign of each excursion (to negative) independently with probability q.

spherical B.M. Arises as a model for the position $S(t)$ of a particle executing Brownian movement on the surface of a sphere. It may be formalized (in three dimensions) as the (unique) spatially and temporally homogenous diffusion on the sphere, whose transition density $f(t, \theta, \phi)$ (as a function of the spherical polar coordinates θ and ϕ) satisfies $f_t = \frac{1}{2}\nabla^2 f$, where ∇^2 is the (spherical) Laplacian (or Laplace–Beltrami operator). Alternatively, its polar coordinates $\theta(t) \in (0, \pi)$ and $\phi(t) \in (0, 2\pi)$ solve the pair of SDEs: $\mathrm{d}\phi = \sigma(\sin\theta)^{-1}\mathrm{d}Y$, and $\mathrm{d}\theta = \frac{1}{2}\sigma^2(\tan\theta)^{-1}\mathrm{d}t + \sigma \mathrm{d}X$, where X and Y are independent standard Wiener processes, and $\sigma > 0$.

The process can also be represented in terms of three-dimensional B.M. $B(t)$ by using a time-change. Let $|B(t)| = R(t)$, and define $C(t) = \int_o^t R(s)^{-2}\mathrm{d}s$. Then $S(t)$ is the process defined by $S(C(t)) = B(t)/R(t)$. Formalizations in $d > 3$ dimensions exist similarly.

subordinated B.M. A process of the form $W(T(t))$, where $W(t)$ is a Wiener process (in d dimensions) and $T(t) \in \mathbb{R}^+$ is a suitable non-decreasing random process called the subordinator (or directing process, or operational time). If $T(t)$ has independent increments, the subordinated process has the Markov property.

super-Brownian motion Arises as a continuous limit of the empirical count over areas (suitably rescaled), of a large population of particles enjoying a branching random walk. If the two random mechanisms (branching and dispersion of offspring) are nicely balanced, the limiting process is a continuous-time Markov process taking values in a set of finite measures; this result may be called **Watanabe's thm**. The line of descent of any particle inspected at time t will be a Brownian motion executed from the time and place of its birth.

The extended random process that takes values in the set of finite measures on such paths is called historical super-Brownian motion. This process arises as the limit of several other similar spatially branching processes, and is a special case of a larger family called superprocesses.

two-sided B.M. This is a B.M. $X(t)$ defined for $t \in \mathbb{R}$, with $X(0) = 0$. It may be realized from two independent B.M.s $U(t)$ and $V(t)$ by setting $X(t) = U(t)$ for $t \geq 0$, and $X(t) = V(-t)$ for $t < 0$.

Brownian quartet Denotes the four processes: Brownian motion; Brownian bridge; Brownian excursion; and Brownian meander. [J. Pitman, Brownian motion, bridge, excursion, and meander characterized by sampling at independent uniform times, *Electr. J. Probab.*, 4(11), 1–33, 1999]

Brownian ratchet (alias: **Brownian motor; stochastic resonance**) Arises in several branches of the natural sciences, where the term broadly describes systems in which particles executing ordinary Brownian motion are subject to some other occasional intervention (which may be periodic, irregular, or random) that has the effect of inducing drift, i.e. directed transport of particles which may be interpreted as getting useful work out of noise. The concept of such rectified Brownian motion needs care in interpretation, as exemplified by the **Lippmann–Smoluchowski–Feynman ratchet**; this has a vane executing circular B.M. in one box (due to the erratic battering by molecules in the box) attached by an axle to a ratchet and pawl that permits only clockwise motion enclosed in another box. A careful analysis is required to avoid a 'solution' that violates the laws of thermodynamics. [J. Parrondo & L. Dines, Brownian motion and gambling: from ratchets to paradoxical games, *Contemporary Phys.*, 45(2), 147–157, 2004; R. Ait-Haddou & W. Herzog, Brownian ratchet models of molecular motors, *Cell Biochem. Biophy.*, 38(2), 191–213, 2003]

Brownian sausage (alias: **Wiener sausage**) This is derived from a standard Wiener process $W(t)$, in d dimensions. Denote the d-dimensional ball (or sphere) of radius r, centred at x, by $S_r(x)$. Then the Wiener sausage of radius r and length t is $\bigcup_{0 \leq s \leq t} S_r(W(s))$. More generally, one can replace $S_r(x)$ by the translation to x of any compact set $K(0)$ that includes the origin, and branching Brownian sausages, and super-Brownian sausages, are defined analogously. [M. Donsker & S. Varadhan, Asymptotics for the Wiener sausage, *Commun. Pure. Appl. Math.*, 28(4), 525–565, 1975]

Brownian sheet (alias: **two-parameter Wiener (or Brownian) process; Wiener–Chentsov (or Chentsov) field**) The standard Brownian sheet $S(u, v)$ is the zero-mean, continuous Gaussian process, for $u \in \mathbb{R}^+, v \in \mathbb{R}^+$, with covariance function $\mathbb{E}\{S(u, v)S(x, y)\} = \min\{u, x\} \min\{v, y\}$. It may alternatively be specified as the process on $u, v \geq 0$ that has independent normally distributed increments (the 'increments' being defined over disjoint rectangles). Similarly, one defines the Ornstein–Uhlenbeck sheet, which can be represented as $S(e^{-x}, e^{-y})e^{(x+y)/2}, x, y \geq 0$. More generally, the fractional Brownian sheet X is zero-mean and Gaussian, with covariance

$$\mathbb{E}\{X(s_1,s_2)X(t_1,t_2)\}=\frac{1}{4}\prod_{r=1}^{2}\left(|s_r|^{2H_r}+|t_r|^{2H_r}-|s_r-t_r|^{2H_r}\right),$$
$$0\leq H_r\leq 1.$$

Sheets in $d > 2$ dimensions are defined similarly. (The Brownian sheet should not be confused with the Brownian surface.) [M. Lifshits, *Lectures on Gaussian Processes*. Springer, 2012]

Brownian snake (alias: **Le Gall's Brownian Snake**) A Markov process taking values in the set of Brownian motion paths (in $\mathbb{R}^d$), stopped at a point $L(t)$ called the lifetime process, $L(t) \in \mathbb{R}^+$. The snake is initially a Brownian (Wiener) process on $[0, L(0)]$. At later times, if $L(t)$ decreases the existing snake is erased to run only over its new (shorter) lifetime, and as $L(t)$ increases the snake grows by a Wiener process that is independent of the existing segment. The lifetime process is often specified as Brownian motion reflected at the origin, in which case the snake is linked with super-Brownian motion. [J.-F. Le Gall, *Spatial Branching Processes, Random Snakes, and Partial Differential Equations*. Birkhäuser, 1999]

Brownian spider (alias: **Walsh's Brownian motion (or spider)**) This is ordinary Brownian motion executed on a finite number of lines radiating from the origin, in which the particle picks a line at random on which to continue whenever it is at the origin. [S. Vakeroudis & M. Yor, A Scaling proof for Walsh's Brownian motion extended arc-sin law, *Electron. Commun. Probab.*, 17(63), 1–9, 2012]

Brownian surface (alias: **Lévy's multiparameter Brownian field**) The standard Brownian surface is the continuous, zero-mean Gaussian process $X(t_1 \ldots, t_d)$, $t_r \in \mathbb{R}$, such that $X(0, \ldots, 0) = 0$, and the covariance is $\mathbb{E}[(X(s)X(t)] = \frac{1}{2}\{|s| + |t| - |t - s|\}$, where $|t| = [\sum_{r=1}^{d} t_r{}^2]^{1/2}$ is the usual Euclidean distance. The case $d = 2$ was considered by B. Mandelbrot as a fractal model for landscape. More generally, the fractional Brownian surface, with index H, has $2\mathbb{E}(X(s)X(t)) = |s|^{2H} + |t|^{2H} - |t - s|^{2H}, 0 \leq H \leq 1$. (The Brownian surface should not be confused with the Brownian sheet.) [K. Falconer, *Fractal Geometry*, 3rd edn. Wiley, 2014]

Brownian web Arises as the continuous (diffusion) limit of coalescing random walks in the plane; since the limit of the random walks yield Brownian motions, the limit can be loosely described as a collection of coalescing Brownian motions from all possible starting points. If the random walks (or Brownian motions) are allowed to branch suitably, one obtains a limiting object called the **Brownian net**. [L. Fontes *et al.*, The Brownian web: characterization and convergence, *Ann. Probab.*, 32(4), 2857–2883, 2004]

Brownian zeros The set of points $Z = \{t : B(t) = 0\}$, where $B(t)$ is Brownian motion. With probability 1, Z is closed with no isolated zeros, yet Z has zero Lebesgue measure. Furthermore, its Hausdorff dimension is 1/2. [P. Mörters & Y. Peres, *Brownian Motion*. Cambridge University Press, 2010]

Brox's diffusion (alias: **Brox–Schumacher diffusion**) Arises as a model for diffusion in a random medium (i.e. environment), in particular when the environment is a (two-sided) Wiener process. For a suitable function $w(x) \in \mathbb{R}, x \in \mathbb{R}$, a diffusion $X(t)$ in the environment $w(x)$ has generator $\frac{1}{2}e^{w(x)} \frac{d}{dx} \left(e^{-w(x)} \frac{d}{dx}\right)$ (where, if $w(x)$ is differentiable, $X(t)$ is a solution of the SDE $dX = -\frac{1}{2}w'(X(t))dt + dB(t)$, and $B(t)$ is an independent Wiener process). For Brox's diffusion $w(x)$ is a realization of a two-sided Wiener process (which may be drifting). [T. Brox, A one-dimensional diffusion process in a Wiener medium, *Ann. Probab.*, 14(4), 1206–1218, 1986]

Brunet–Derrida particle system A branching random walk in which the number of particles in $\mathbb{R}$ is kept fixed by deleting the leftmost particle at any instant when a new particle appears. Under suitable conditions on the distribution of family locations, the system of particles ultimately moves rightwards at a constant speed, and one speaks of a travelling wave. [R. Durrett & D. Remenik, Brunet–Derrida particle systems, free boundary problems, and Wiener-Hopf equations, *Ann. Probab.*, 39(6), 2043–2078, 2011]

Brunk–Chung strong law Let $X_n, n \geq 1$, be zero-mean r.vs with $\sum_{n=1}^{\infty} n^{-(1+r)}\mathbb{E}(|X_n|^{2r}) < \infty$, for some $r \geq 1$. Then $n^{-1}\sum_{k=1}^{n} X_k \to 0$ a.s. and in $2r$th mean as $n \to \infty$. The case $r = 1$ is better-known as Kolmogorov's strong law. D. Brunk established the case for any integer r. [Y. Chow & H. Teicher, *Probability Theory*, 3rd edn. Springer, 1997]

Brun's sieve In a probabilistic context, this term refers to this theorem: let $X_n, n > 1$, be a sequence of non-negative integer-valued r.vs such that for each $r \geq 1$, and some $\lambda > 0$, $\lim_{n\to\infty} \binom{X_n}{r} = \frac{\lambda^r}{r!}$. Then as $n \to \infty$, X_n converges in distribution to a Poisson distn with parameter λ. That is to say, convergence of factorial moments implies convergence in distribution. [N. Alon & J. Spencer, *The Probabilistic Method*, 3rd edn. Wiley, 2008]

Bruss odds rule (alias: **sum-the-odds theorem, e^{-1} law**) Arises in the optimal stopping problem which seeks to maximize the probability of stopping at the last success in a finite sequence of independent (Bernoulli) trials, $X_1, \ldots, X_n$. The Bruss odds theorem asserts that the objective is achieved by this rule: stop at the first success (i.e. the smallest r such that $X_r = 1$) such that the sum of the odds for success at any later trial is less than or equal to 1. More formally, stop at

$$\min\left\{ r : X_r = 1 \text{ and } \sum_{k=r+1}^{n} \mathbb{P}(X_k = 1)/(1 - \mathbb{P}(X_k = 1)) \leq 1 \right\};$$

where the kth odds ratio is deemed to be ∞ if $\mathbb{P}(X_k = 1) = 1$. Furthermore, the probability that this rule succeeds, exceeds e^{-1}, whatever the values of $\mathbb{P}(X_k = 1), 1 \leq k \leq n$. [R. Dendievel, New developments of the odds theorem, *Math. Sci.*, 38(2), 2013]

B-stability (alias: **branching stability**) Arises in applications of branching processes. Let X be a r.v. with characteristic function $\phi(t)$, and let $G(s)$ be the probability generating function of a positive integer-valued r.v. Y. Then X is B-stable with respect to Y if $G(\phi(t)) = \phi(ct)$ for some scalar c (which depends on ϕ and G). [C. Mallows & L. Shepp, B-stability, *J. Appl. Probab.*, 42(2), 581–586, 2005]

Buffon's needle (problem) The celebrated problem that initiated geometrical probability in 1733: a needle is dropped at random onto a horizontal plane that is ruled with a family of parallel lines, distance a from each other. If the length L of the needle satisfies $L \leq a$, then the probability that the needle intersects a line is $(2L)/(\pi a)$ (with the natural interpretation of 'at random'). If $a < L$, the probability of at least one intersection is

$$\frac{2}{\pi}\cos^{-1}\frac{a}{L} + \frac{2L}{\pi a}\left\{1 - \left(1 - \frac{a^2}{L^2}\right)^{\frac{1}{2}}\right\}.$$

In any case, the expected number of intersections of the needle with lines is $(2L)/(\pi a)$, and this does not depend on the straightness of the needle. Thus in Buffon's noodle problem, if a limp noodle of length L (or length L of string) is dropped on the plane, the expected number of intersections is likewise $(2L)/(\pi a)$; likewise one may speak of Buffon's bread-stick, which may break. If the plane is further ruled with a family of lines distance b apart, all perpendicular to the first family, then the probability of an intersection by a straight needle of length $L < \min[a, b]$ is $(2La + 2Lb - L^2)/(\pi ab)$. There are several extensions of the idea. For example, for a bounded subset B of the plane the Buffon needle probability $\mathbb{P}(B)$ of B is the probability that a random line of the plane (i.e. a randomly dropped needle of infinite length) intersects B. It is also called the **Favard length**, or **Crofton measure**, of B. [Y. Peres *et al.*, Fractals with positive length and zero Buffon needle probability, *Amer. Math. Mon.*, 110(4), 314–325, 2003; M. Crofton, On the theory of local probability, applied to straight lines drawn at random in a plane, *Phil. Trans. Roy. Soc.*, 158, 181–199, 1868; H. Solomon, *Geometric Probability*. SIAM, 1978]

Bunyakovskii inequality *see* Cauchy–Schwarz inequality

Bunyakovsky inequality *see* Cauchy–Schwarz inequality

Burg entropy *see* entropy

Burgers superprocess Arises as a continuous limit of a model for a branching interacting particle system. It is defined as a solution $Y(t, x)$ of the stochastic partial differential equation

$$\frac{\partial y}{\partial t} = \frac{1}{2}\frac{\partial^2 y}{\partial x^2} - ay\frac{\partial y}{\partial x} + by^{1/2}W(\mathrm{d}t, \mathrm{d}x),$$

in one space dimension, where W is white noise in two dimensions, $t \geq 0, x \in \mathbb{R}$. [G. Bonnet & R. Adler, The Burgers superprocess, *Stoch. Process. Appl.*, 117(2), 143–164, 2007]

burglar problem (alias: **stopping a success run**) Arises in optimal stopping, thus; let $X_1, X_2, \ldots$ be i.i.d.r.vs, and let $Z_1, Z_2, \ldots$ be a sequence of i.i.d. Bernoulli trials. On the rth day, your gain is X_r if Z_r is a success, but on any day s when Z_s is a failure, you lose your entire accumulated fortune. The problem asks: what is the optimal strategy for retirement? [T. Ferguson, Optimal stopping and applications, available online at: www.math.ucla.edu/~tom]

Burke's theorem (alias: **Burke–Reich theorem**) Arises in queueing theory, when the arriving customers form a Poisson process of rate λ and the queue is in equilibrium. In the original form, suppose also that there is a single server, whose service times are exponential with parameter μ, yielding an M/M/1 queue. Then Burke's theorem asserts that the departing customers form a Poisson process of rate λ, and the size of the queue at time t is independent of the Poisson process of departures before t. It follows that in a sequence of queues (in tandem) having this property, the times in the queues are independent for any customer passing through the system. Various other queues have this property, such as M/M/s and M/G/∞. The usual short proof is an elegant appeal to reversibility, due to E. Reich, hence the alias. [P. Burke, The output of a queueing system, *Oper. Res.*, 4(6), 699–704, 1956; E. Reich, Waiting times when queues are in tandem, *Ann. Math. Stat.*, 28(3), 768–773, 1957]

Burkholder–Davis–Gundy inequality Any of a large family of inequalities arising in the context of a suitable martingale $M_t, t \geq 0$, in discrete or continuous time, that relate the long-run properties of the running maximum of $|M_t|$ to its difference process, i.e. the quadratic variation process $[M]_t$ in continuous time, or the sum of squared martingale differences $[M]_n = \sum_{r=1}^{n}(M_r - M_{r-1})^2$ in discrete time. Broadly speaking, it asserts that in $L_p, p > 0$, $M_t^* = \sup_{0 \leq s \leq t}|M_s|$ grows at the same rate as $([M_t])^{1/2}$. Formally, in discrete time, there exist universal constants (depending only on p, not on the martingale) a_p and b_p such that

$$a_p \mathbb{E}\left\{[M]_n{}^{p/2}\right\} \leq \mathbb{E}(M_n{}^{*p}) \leq b_p \mathbb{E}\left\{[M]_n{}^{p/2}\right\}.$$

More generally, a similar inequality holds with a suitable non-decreasing convex function $\Phi(x)$ replacing the power function x^p.

Burkholder's inequality arises in the same context, and takes a similar form, broadly asserting that the martingale itself grows at the same rate as its quadratic variation process. Formally, for $p > 1$, there exist universal constants A_p and B_p such that

$$A_p \mathbb{E}\left([M]_n{}^{p/2}\right) \leq \mathbb{E}[|M_n|^p] \leq B_p \mathbb{E}\left([M]_n{}^{p/2}\right).$$

See Marcinkiewicz–Zygmund inequality. [Y. Chow & H Teicher, *Probability Theory*, 3rd edn. Springer, 1997]

burn-in time (alias: **transient time**) A term arising in Markov chain Monte Carlo methods to denote the iterations required until the distribution of the chain is close enough (in some sense) to the desired equilibrium for the purposes of the simulation. The expression 'burn-in' also arises in reliability theory, when the failure-rate of some lifetime distribution may be initially high for small values of time. The interval over which components with such a distribution are run before being confidently installed, or the first period of time that systems are run with such components, is called burn-in. [W. Gilks *et al.*, *Markov Chain Monte Carlo in Practice.* Chapman & Hall, 1995]

Burr distribution (type III) *see* Dagum distribution

Burr distribution (type XII) In insurance mathematics, and economics, this often denotes the distribution of the simple power law transformation of a Pareto r.v. that has density $f(x) = \frac{abc^a x^{b-1}}{(c+x^b)^{a+1}}, x > 0$, in one parameterization, where a, b, and c are all positive constants. It is one of many distribution functions discussed by I. Burr, and is called the **Singh–Maddala distn** by some authors. *See* Dagum distribution. [I. Burr, Cumulative frequency functions, *Ann. Math. Stat.*, 13(2), 215–232, 1942]

busy period Arises in queueing theory, where it denotes the interval of time from any moment when an arriving customer finds the server (or system of queues) entirely idle (or free), until the next moment at which the server (or system) is idle. The time between successive busy, periods is the **idle period**, and the interval between successive moments when an arriving customer finds the server (or system) free is a **cycle.** [S. Asmussen, *Applied Probability and Queues*, 2nd edn. Springer, 2003]

C

Cameron–Martin formula Any of a number of expressions evaluating suitable functionals of the Wiener process $W(t), t \geq 0$. A classic example is

$$\mathbb{E}\exp\left\{-\frac{1}{2}\lambda^2\int_0^t (W(s))^2 ds\right\} = \{\cosh(\lambda t)\}^{-\frac{1}{2}}.$$

[R. Cameron & W. Martin, Evaluation of Various Wiener integrals by use of certain Sturm-Liouville differential equations, *Bull. Amer. Math. Soc.*, 51(2), 73–90, 1945]

Cameron–Martin–Girsanov theorem (alias: **Girsanov theorem**) Arising originally in the context of the Wiener process (and then, more generally, suitable diffusions and martingales), the term denotes any of several results relating transformations of the process by means of suitable translations (Cameron–Martin thm) to transformations of the process, by means of a substitution of suitable measures (Girsanov's thm); there are important applications in financial mathematics. (The ultimate extension to local martingales may be called **Lenglart's thm.**)

One often-seen basic version for the Wiener process defined on a probability space (Ω, $\mathcal{F}$, $\mathbb{P}$) is this: let $h(t)$ be bounded on $0 \leq t \leq T$ (or, more generally, an $\mathcal{F}_t$-predictable function such that $\mathbb{E}\exp\left(\frac{1}{2}\int_0^T h'(t)^2 dt\right) < \infty$), and define the r.v.

$$(^*)\ L = \exp\left\{\int_0^T h'(t)dW(t) - \frac{1}{2}\int_0^T h'(t)^2 dt\right\},$$

and the measure $\mathbb{Q}$ on (Ω, $\mathcal{F}$) by

$$(\ddagger)\mathbb{Q}(A) = \mathbb{E}(LI(A)),$$

for $A \in \mathcal{F}_T$, where $I(A)$ is the usual indicator. Then $W(t) - \int_0^t h(s)ds$ is a Wiener process on the space $(\Omega, \mathcal{F}, \mathbb{Q})$ (or, equivalently, on this space $W(t)$ is a drifting Wiener process with drift coefficient $-h$). In this context (*) is often called the **Cameron–Martin formula**, and ($\ddagger$) is often called the **Girsanov change of measure** (or transformation). The space of continuous functions $h(t), 0 \leq t \leq T$, with $h(0) = 0$, such that $\int_o^T h'(t)^2 dt < \infty$ is called **Cameron–Martin space**. *See* change of measure. [K. L. Chung & R. Williams, *Introduction to Stochastic Integration*, 2nd edn. Birkhäuser, 1990; I. Karatzas & S. Shreve, *Brownian Motion and Stochastic Calculus*, 2nd edn. Springer, 1998]

Campbell's theorem (alias: **Campbell's formulas; Campbell–Hardy theorem**) Arises in considering sums $S = \sum g(X_i)$, where the sum is taken over all the points X_i of a Poisson process in $\mathbb{R}$ (or in $\mathbb{R}^d$), for suitable functions $g(\cdot)$. Examples include the **shot-noise process**; and, more particularly, assuming stars comprise a Poisson process in space it follows that the gravitational field at the (arbitrary) origin is of the form $S = \sum m_i/|X_i|$. Denoting the mean (intensity) measure of the process by $\lambda(x)$, the term Campbell's theorem is used for expressions such as these (in one dimension): $\mathbb{E}S = \int_{\mathbb{R}} g(x)\lambda(dx)$; $\operatorname{var} S = \int_{\mathbb{R}} g(x)^2\lambda(dx)$; $\mathbb{E}e^{\theta S} = \exp[\int_{\mathbb{R}}(e^{\theta g(x)} - 1)\lambda(dx)]$; and, for two such sums S_1 and S_2, $\operatorname{cov}(S_1, S_2) = \int_{\mathbb{R}} g_1(x)g_2(x)\lambda(dx)$. (In his paper of 1909 giving such results, N. Campbell noted that the main principle of the argument was supplied by G. Hardy.) Analogous results in the theory of Palm processes are termed **Mecke's thm** (or Campbell–Mecke); there is a tendency to use the term 'Campbell's theorem' for a wide range of similar identities in the general theory of point processes. [J. F. C. Kingman, *Poisson Processes.* Oxford University Press, 1993]

Camp–Meidell–Narumi inequality (alias: **Camp–Meidell inequality**) Any of a number of inequalities for the tail $\mathbb{P}(X > t)$, (or tails $\mathbb{P}(|X| > t)$), of a continuous random variable X, whose density is required to satisfy certain conditions; most conspicuously conditions of monotonicity. They are thus seen as generalizations of Chebyshov's inequality and Gauss' inequality. [B. Camp, A new generalization of Tchebycheff's statistical inequality, *Bull. Amer. Math. Soc.*, 28(9), 427–432, 1922; M. Meidell, Sur un problème du calcul des probabilités, *Comptes Rendus*, 173, 806–808, 1922; S. Narumi, On further inequalities with possible applications to problems in the theory of probability, *Biometrika*, 15, 245–253, 1923]

cancellation condition *see* independence axiom

canonical random process (alias: **coordinate representation process; directly given process**) Arises in considering a random process $X_n(\omega)$, $n \geq 1, \omega \in \Omega$, defined on a probability space $(\Omega, \mathcal{F}, \mathbb{P})$, when – as is usually the case – the object of interest is the distribution of $\mathbf{X}(\omega) = (X_1(\omega), X_2(\omega), \ldots)$, denoted by $\mathbb{P}_X(B) = \mathbb{P}(\mathbf{X} \in B)$, where B is any element of a suitable σ-field $\mathcal{B}^\infty$ of subsets of $\mathbb{R}^\infty$; that is, no distinction is made between two versions of $\mathbf{X}$ defined on different spaces but having the same distribution. Then a convenient – and arguably the simplest – version to take as representing all of them is the process $\hat{X}_n(x)$, $x \in \mathbb{R}^\infty$, defined on the probability space $(\mathbb{R}^\infty, \mathcal{B}^\infty, \mathbb{P}_X)$ by the so-called coordinate mapping (or projection) $\hat{X}_n(x_1, x_2, \ldots) = x_n$. This has the same joint distributions as $X_n, n \geq 1$, and is called the canonical (or coordinate) representation (or version). The idea extends to processes in continuous time, where one commonly chooses versions whose sample paths have useful regularity properties, e.g. the canonical Wiener process takes values in Wiener space equipped with the Wiener measure. [D. Revuz, & M. Yor, *Continuous Martingales and Brownian Motion*, 3rd edn. Springer, 1999; P. Billingsley, *Probability and Measure*, 3rd edn. Wiley, 1995; L. Breiman, *Probability*. Addison-Wesley, 1968, reprinted by SIAM, 1992]

canonical random variable A canonical form of some set of mathematical objects is their simplest or most natural representation, in a standard framework, without loss of generality. The term canonical r.v. is used to denote several such objects:

1. Any r.v. X defined on a probability space $(\Omega, \mathcal{F}, \mathbb{P})$ has a distribution function $F(x) = \mathbb{P}(\{\omega : X(\omega) \leq x\})$. Defining the probability measure $\mathbb{P}_X(B) = \mathbb{P}(\{\omega : X(\omega) \in B\})$, for any Borel set B in the Borel field $\mathcal{B}$ of $\mathbb{R}$, yields a probability space $(\mathbb{R}, \mathcal{B}, \mathbb{P}_X)$, with completion $(\mathbb{R}, \overline{\mathcal{B}}, \mathbb{P}_X)$, and the r.v. $\hat{X}$ defined on this space by $\hat{X}(x) = x$ has distribution $F(x)$; $\hat{X}$ may be called the canonical form of X.
2. More loosely, for any probability space $(\Omega, \mathcal{F}, \mathbb{P})$, the Ω-valued random element (or object) X defined by $X(\omega) = \omega, \omega \in \Omega$, has distribution $\mathbb{P}$, and is called a canonical random variable.
3. Let U be the canonical uniform r.v. and let $F(x)$ have inverse function (generalized as necessary) $F^{-1}(\cdot)$, so that $F^{-1}(U)$ has distribution function $F(x)$; in this context U is sometimes called the canonical r.v.
4. The term is differently formalized in free probability (see references).

[D. Voiculescu *et al.*, *Free Random Variables*. American Mathematical Society, 1992; N. Krylov *et al.*, *Stochastic PDEs and Kolmogorov Equations in Infinite Dimensions*. Springer, 1999]

canonical scale *see* scale function

Cantelli's inequality Let X be a zero-mean random variable with absolute moments $a_r = \mathbb{E}(|X|^r)$, $r \geq 1$. Then for $t^r \geq a_{2r}/a_r > 0$, we have $\mathbb{P}(|X| \geq t) \leq (a_{2r} - a_r^2)/\{(t^r - a_r)^2 + a_{2r} - a_r^2\}$. The term 'Cantelli's inequality' also often denotes the one-sided Chebyshov inequality. [M. Fréchet, *Recherches Théoriques Modernes sur la Théorie des Probabilités*. Paris, 1937, 123–126; I. Savage, Probability inequalities of the Tchebycheff type, *J. Res. Nat. Bureau Stand. B*, 65B(3), 211–226, 1961]

Cantelli's theorem (alias: **Cantelli's SLLN**) Asserts that i.i.d.r.vs $X_1, X_2, X_3, \ldots$ with finite fourth moment obey the strong law of large numbers; formally, if $\mathbb{E}X_1^4 < \infty$, then $\mathbb{P}(\lim_{n\to\infty}\frac{1}{n}\sum_{r=1}^n (X_r - \mathbb{E}X_r) \to 0) = 1$. More generally, for independent r.vs, it asserts that the above statement holds if $\sum_{n=1}^\infty n^{-2}(S_2^2(n) + S_4(n)) < \infty$, where $S_r(n) = \sum_{k=1}^n \mathbb{E}(X_k - \mathbb{E}X_k)^r, r = 2,4$. [J. Walsh, *Knowing the Odds*. American Mathematical Society, 2012]

Cantor distribution (alias: **Lebesgue's function**) A classic example of a purely singular distribution function; it is derived from the Cantor ternary set $T \subseteq [0, 1]$. The set T is obtained by deleting the open middle third $(\frac{1}{3}, \frac{2}{3})$ of $[0, 1]$, then deleting the middle thirds of the surviving intervals, and so on. Any point x remaining in the limiting set T has ternary expansion $x = \sum_{r=1}^\infty t_r 3^{-r}$, where $t_r \in \{0, 2\}$. The Cantor distribution fn $F(x)$ is defined for $x \in T$ by $F_1(x) = \sum_{r=1}^\infty t_r 2^{-(r+1)}$, and then for $y \in T^c, 0 \leq y \leq 1$, by $F(y) = \sup_{t \in T}\{F_1(t) : t < y\}$. It follows that $F'(x) = 0$ for almost all $x \in [0, 1]$, though F is continuous and non-decreasing with $F(0) = 0$ and $F(1) = 1$. The characteristic function of this distn is $\phi(t) = e^{it/2}\prod_{r=1}^\infty \cos(t\,3^{-r})$. [G. Grimmett & D. Welsh, *Probability: An Introduction*, 2nd edn. Oxford University Press, 2014]

capacitance (alias: **capacity**) A term from electrostatics that arises in the theory of Markov processes because the same equations govern both certain electrical quantities and various probabilistic quantities. Very informally, this happens because (in suitable cases) the partition theorem of conditioning can be seen as analogous to some conservation laws of physics. Consequently, the way in which probabilities dispose themselves over the state-space of a Markov process corresponds to the distribution of charge or potential in an associated electrostatic problem. For example, suppose an electrical network has nodes (each with a capacity and charge, and consequent potential) joined by wires of given conductivities. Then, in equilibrium, the laws of Kirchoff and Ohm yield equations that are equivalent to the balance equations of a Markov chain whose transition probabilities are determined by the wire conductivities. The capacitances of the nodes correspond to an invariant measure for the chain. Similar connexions appear for similar reasons for martingales and Brownian motion. [F. Kelly, *Reversibility and Stochastic Networks*, Wiley, 1979; reprinted by Cambridge University Press, 2011]

capacity (alias: **performance rate**) Arises in multi-state reliability theory, in which components can be in one of several states indicating their level of availability, or ability to perform a task, called their capacity; this will generally be a vector random process $X(t)$. Then the overall capacity, or performance rate, may be assumed to be a function $\eta(X(t))$ of the vector of component capacities. [B. Natvig, *Multistate Systems Reliability Theory with Applications*. Wiley, 2011]

cap and floor In financial mathematics, the terms denote derivative instruments depending on some underlying interest rate $L(t)$, which is assumed to be a random process. Such contracts pay out a suitable function of the excursions of $L(t)$, above a cap C, or below a floor F, during a future time-interval $[s, t]$. [M. Musiela & M. Rutkowski, *Martingale Methods in Financial Modelling*, 2nd edn. Springer, 2005]

Carathéodory's extension theorem Arises in measure theory, with natural applications in probability. Let $m(A)$ be a measure defined on an algebra $\mathcal{A}$ of sets in Ω (which is to say that $\mathcal{A}$ is closed under finite unions and intersections and complements), where $m(A) \geq 0$ for all $A \in \mathcal{A}$, and $m(\Omega) = 1$, and $m(A) = \sum_i m(S_i)$ for any countable collection of disjoint sets S_i whose union is $A \in \mathcal{A}$. Then the theorem asserts that there is a unique extension of $m(\cdot)$ to a measure on the smallest σ-algebra (σ-field) of subsets of Ω that includes $\mathcal{A}$. As a typical application, this enables the definition of a product probability space for independent random variables by extension from the elementary product of probabilities on rectangles. [R. Durrett, *Probability: Theory and Examples*, 4th edn. Cambridge University Press, 2010]

cardioid distribution (alias: **cosine distribution**) A member of the class of circular distns, having density $f(x) = (2\pi)^{-1}\{1 + 2c\cos(x - m)\}, 0 \leq x < 2\pi$ for $0 \leq c \leq 1/2$. Note that the cardioid curve itself, with polar equation $r = a(1 - \cos\theta)$, is actually a degenerate limaçon: $r = a + b\cos\theta$. [N. Fisher, *Statistical Analysis of Circular Data*. Cambridge University Press, 1993]

Cardy's formula (alias: **Cardy–Smirnov formula**) Arises in two-dimensional critical percolation models. It supplies, for a suitable lattice, a general explicit expression for the limiting probability p that there exists an open path of edges between a given pair of opposite sides of a rectangle, as the size of the rectangle increases. There are various versions, but in one convenient parameterization the rectangle is similar to $\{\pm e^{\pm i\theta}\}$, $0 \leq \theta \leq \frac{1}{2}\pi$ (in an obvious notation in the complex plane), and then

$$p = \frac{3\Gamma(\frac{2}{3})}{\Gamma(\frac{1}{3})^2}(\sin\theta)^{2/3}\, {}_2F_1\left(\frac{1}{3}, \frac{2}{3}, \frac{4}{3}; \sin^2\theta\right),$$

where ${}_2F_1$ is Gauss's hypergeometric function. The proof of this on a triangular lattice is called **Smirnov's thm.** [G. Grimmett, *Probability on Graphs*. Cambridge University Press, 2010]

Carleman's criterion (or condition) Arises in the context of the moment problem, where it asserts that a probability distribution on $\mathbb{R}$ is uniquely determined by its moments $m_r = \mathbb{E}X^r, r \geq 1$, if $\sum_{r=0}^{\infty} m_{2r}^{-r/2}$ diverges. If the distribution is confined to $\mathbb{R}^+$, then it is sufficient that $\sum_{r=0}^{\infty} m_r^{-r/2}$ diverges. Neither condition is necessary. [J. Stoyanov, *Counterexamples in Probability*, 3rd edn. Dover, 2013]

carousel problem A classic problem of operations research in which various objects are stored in bins on a carousel that rotates (possibly in either direction) in front of a server; the object is to maximise the efficiency of the system given a sequence of random demands. [M. Vlasiou, A Lindley-type equation arising from a carousel problem. *J. Appl. Probab.*, 41(4), 1171–1181, 2004]

Carroll's paradox Any of several superficially plausible logical arguments that lead to self-evidently false conclusions as constructed by the logician C. L. Dodgson, writing also as Lewis Carroll. In the context of probability he proposed this problem: a bag contains two counters, as to which nothing is known except that each is either black or white. Ascertain their colours without taking them out of the bag. The 'answer' provided purports to prove that necessarily one was black and one was white (Carroll's sense of humour getting the better of him here, he refrains from explaining the flaw in the reasoning). [C. L. Dodgson, *Pillow Problems*, 4th edn. Macmillan & Co., 1895; reprinted by Dover, 1958]

cascade process (alias: **multiplicative (cascade) process**) The concept of a cascade arises in a number of applications, and the term therefore covers a broad field of models. The classic example is the particle cascade, in which each particle carries an energy (or mass), and may branch into a family of lesser particles among which the energy (or mass) is divided. A related example is polymer degradation. Such processes are often isomorphic to a suitable branching random walk. Another classic example is in turbulence, where the branching is in scales of motion, and energy is randomly distributed from larger to smaller scale motions. This was summarized by L. F. Richardson thus: 'Big whirls have little whirls that feed on their velocity, Little whirls have lesser whirls and so on to viscosity…'. Classic models for such phenomena include the Kolmogorov rock-crushing process. These ideas lead to one formalization of a cascade process as a sequence of transformations $T_n, n \geq 1$, of some suitable set Ω (i.e. a dynamical system) having the property that for $\omega \in \Omega$, $T_n(T_m(\omega)) = T_{n+m}(\omega)$. In particular, stochastic cascade processes obeying a random multiplicative and self-similar recurrence are often called Mandelbrot (or multiplicative) cascades and the resulting objects often have fractal properties in the limit. More broadly, any physical process (or its model) of events (or operations) giving rise to objects that are subject to a sequence of similar events (or operations) may be called a cascade. [D. Harte, *Multifractals*. Chapman & Hall, 2001; Q. Liu, On generalized multiplicative cascades, *Stoch. Proc. Appl.*, 86(2), 263–286, 2000]

casting lots *see* lottery

Catalan numbers Arise in many counting problems, including examples in probability. The nth Catalan number is $c_n = \frac{1}{n+1}\binom{2n}{n}, n \geq 0$, with generating function $c(x) = \sum x^n c_n = \frac{1}{2x}\left(1 - (1-4x)^{\frac{1}{2}}\right)$. For example, if T is the first-passage time, of the simple symmetric r.w. on the integers, from 0 to 1, then $\mathbb{E}s^T = \frac{s}{2}c\left(\frac{s^2}{4}\right) = \frac{1}{s}\left[1 - (1-s^2)^{\frac{1}{2}}\right]$. The generalized Catalan numbers have generating function $c(x,a) = \frac{2a-1-(1-4ax)^{1/2}}{2(x+a-1)}$, and arise (for example) as the moment sequence of the (generalized beta) density $f(x) = \frac{(4a-x)^{1/2}}{2\pi(1+(a-1)x)x^{1/2}}$. The **Fuss–Catalan numbers** f_n are given by $f_n = \frac{1}{1+nr}\binom{n(r+1)}{n}$, $r = 1, 2, 3, \ldots, n > 0$; and arise as moment sequences of members of a family of densities (free Bessel distns) of interest in free probability and random matrices. [A. Nica & R. Speicher (eds), *Lectures on the Combinatorics of Free Probability*. Cambridge University Press, 2006; T. Koshy, *Catalan Numbers with Applications*. Oxford University Press, 2009]

catalytic branching (alias: **contact action process**) Classical catalysis involves two substances, the first a reactant which evolves in the presence of the second, a catalyst, which is not itself permanently altered thereby. A basic model describes the evolution of the reactant as a spatial branching process (e.g. branching r.w. or diffusion) with non-constant branching rate (or mechanism) whose possibly random variation over space and time is seen as reflecting changes in the concentration of the catalyst. If the catalyst's distribution is independent of the reactant, this is one-sided catalysis (and amounts to branching in a random environment). If the catalyst's evolution depends on the reactant, then this is mutual catalysis. By taking a suitable continuous limit one obtains catalytic superprocesses, which may be one-sided or mutual. [L. Gorostiza & B. Ivanoff (eds), Stochastic models, *CMS Conference Proc.*, vol. 26, American Mathematical Society, 2000]

Cauchy class Arises as a stationary Gaussian process $X(t), t \in \mathbb{R}^n$, with (isotropic) correlation function $c(h) = (1 + |h|^a)^{-b/a}$, where $|h| = \left(\sum_1^n h_r^2\right)^{1/2}$, $0 < a \leq 2$, and $b > 0$. The case when $a = 2$ has been called the Cauchy model, and the Cauchy class is therefore also known as the generalized Cauchy (correlation) model. [T. Gneiting & M. Schlather, Stochastic models that separate fractal dimension and the Hurst effect, *SIAM Rev.*, 46(2), 269–282, 2004]

Cauchy distribution (alias: **Lorentz distribution; Breit–Wigner distribution**) A standard Cauchy r.v., centred at the origin, has density $f(x) = \frac{1}{\pi(1+x^2)}$, $-\infty < x < \infty$, with distn fn $F(x) = \pi^{-1}\tan^{-1}x + \frac{1}{2}$, and characteristic fn $\phi(t) = e^{|t|}$. If a random line passing through the origin makes an angle θ with the x-axis, where θ is uniformly distributed on $[0, 2\pi]$, then the x-coordinate of the point of intersection with the line $y = 1$ has a Cauchy density. It is a stable distribution with (characteristic) exponent (or index) 1. More generally, a random vector $X = X_1, \ldots, X_n$ has a multivariate Cauchy distn iff $\sum_{r=1}^{n} a_r X_r$ has a Cauchy distn for every real vector $(a_1, \ldots, a_n)$. A random variable having the density

$$f(x) = \frac{\Gamma(m)}{\Gamma\left(m - \frac{1}{2}\right)\Gamma\left(\frac{1}{2}\right)} \frac{1}{(1+x^2)^m}, m > \frac{1}{2}, x \in \mathbb{R},$$

is said to have a Cauchy–type distribution. For its characteristic fn; see *Kendall's Advanced Statistics*, Example 3.13. [T. Ferguson, A representation of the symmetric bivariate Cauchy distribution, *Ann. Math. Stat.*, 33(4), 1256–1266, 1962]

Cauchy process *see* Lévy process

Cauchy product *see* convolution

Cauchy–Schwarz inequality (alias: **Bunyakovsky inequality; Schwarz inequality**) The standard formulation in probability theory asserts that if random variables X and Y have finite variance then $\mathbb{E}(XY) < \infty$, and $[\mathbb{E}(XY)]^2 \leq \mathbb{E}(X^2)\mathbb{E}(Y^2)$, with equality iff $\mathbb{P}(aX + bY = 0) = 1$ for some real a and b that are not both zero. It is a special case of Hölder's inequality, and holds more generally for inner products in a suitable space. [G. Hardy *et al.*, *Inequalities*, 2nd edn. Cambridge University Press, 1952; J. M. Steele, *Cauchy–Schwarz Masterclass*. Cambridge University Press, 2004]

causal (alias: **future-independent; stable**) A concept arising from our perception of time as having a direction, and of causes preceding effects. This is formalized as a property of any time series that arises in describing processes evolving in time, or seeks to explain or predict one sequence in terms of another. Thus, for $t \in \mathbb{Z}$, let Z_t be a sequence of zero-mean i.i.d.r.vs (called noise), and X_t a weakly stationary sequence of r.vs, such that for constants ϕ_t and θ_t

$$X_t = \sum_{r=1}^{p} \phi_r X_{t-r} + \sum_{r=1}^{q} \theta_r Z_{t-r} + Z_t.$$

That is to say, X_t is an autoregressive moving-average process. Then X_t is causal if there are constants c_t such that $\sum_0^\infty |c_r| < \infty$ and $X_t = \sum_{r=0}^\infty c_r Z_{t-r}$. If $\phi(z) = 1 - \phi_1 z - \cdots - \phi_p z^p$ and $\theta(z) = 1 + \theta_1 z + \cdots + \theta_q z^q$ have no zeros in common, then the above condition for X_t to be causal is equivalent to either of the following:
(a) Z_t is independent of X_r for all $r < t$;
(b) $\phi(z)$ has no zeros in $|z| \leq 1$.
Inversely, if there are constants b_t such that $\sum_0^\infty |b_t| < \infty$ and $Z_t = \sum_{r=0}^\infty b_r X_{t-r}$, then X_t is said to be invertible. [P. Brockwell & R. Davis, *Time Series*, 2nd edn. Springer, 1991]

causality (alias: **causation**) Denotes the concepts of cause and effect, i.e. one event bringing about another, which are essential to the development of mathematics, science, and reasoning in general. (But the question whether causality is a feeling of the mind or a law governing the external universe seems to be unresolved.) In probability the concept is extended to the idea that one event may tend to bring about another, without necessarily doing so in every case, called probabilistic causality. Applications are extensive in statistics, which seeks to distinguish factors that are (probabilistically) causal from those that are not; or, more weakly, to identify factors that are predictive without necessarily being causal in the strict sense. *See* Granger causal and Wiener causal. [J. Pearl, *Causality*. Cambridge University Press, 2000]

causal Markov property (alias: **local Markov property**) Arises in network models describing relationships between families of random variables. Let $G = (V, E)$ be a directed graph on a set V of vertices joined by the edges of E that are directed from parent vertices to their offspring vertices. A collection of r.vs $X_v, v \in V$, has the causal (or local) Markov property if every X_v is conditionally independent of the random variables indexed by its non-descendant vertices, given the values of the random variables indexed by its parent vertices. Here a directed edge is interpreted as a direct causal relation (in that direction) between the variables at those vertices. [P. J. Green *et al.* (eds), *Highly Structured Stochastic Systems*. Oxford University Press, 2003]

causative Markov chain Denotes a type of Markov chain that is non-homogeneous in time, as follows: let $X(n), n \geq 0$ be a countable state-space Markov chain matrix of transition probabilities with $p_{jk}(n,m) = \mathbb{P}\big(X(m) = k | X(n) = j\big)$ denoted by $P(n,m)$. Such a chain is a constant causative chain if $P(n, n+1) = P(n-1, n)C$ for some matrix C, which is called a causative matrix with respect to $P(0,1)$. The concept extends to continuous time, where in the regular case the Q-matrix is given by $Q = At + B$, and A and B are homogeneous Q-matrices. [J. Johnson, Continuous-time constant causative Markov chains, *Stoch. Proc. Appl.*, 26,161–171,1987]

Cayley–Moser problem (alias: **house-selling problem**) A classic problem of optimal stopping, in the finite-horizon case. One version is L. Moser's: you will receive m offers $X_1, \ldots, X_m$ that are i.i.d.r.vs, each uniformly distributed on (0,1). They arrive in sequence, and you may stop at any time to accept the current offer. You must accept some offer, and you must accept X_m if no previous offer has been accepted. In A. Cayley's version, the offers are chosen randomly without replacement from a set of known numbers $x_1, \ldots, x_n, n \geq m$. [T. Ferguson, *Optimal Stopping and Applications*, electronic text at: www.math.ucla.edu/~tom/]

cell model (alias: **cell process**) Broadly, a process (commonly random) of non-overlapping regions of the plane or higher-dimensional space. A natural application arises in wireless communication, where a cell is that region surrounding an antenna where its signal is more powerful than any other, and powerful enough to be received. If the cells exhaust space, it is a random tessellation. More generally, it can be seen as a type of non-overlapping coverage process. For the related process in which one considers the possibly overlapping regions in which signals from antennas can be received, one may speak of a **footprint process**. (Cells do not overlap, but footprints may.) [F. Baccelli & B. Blaszczyszyn, On a coverage process ranging from the Boolean model to the Poisson–Voronoi tessellation, *Adv. Appl. Pub.*, 33, 293–323, 2001]

cemetery (alias: **coffin state**) Arises in the context of a dishonest Markov process $X(t)$ (i.e. one that is non-conservative, or sub-Markovian). In this case, for some t, and some $x \in S$, where S is the state space of $X(t)$ we have $\mathbb{P}(X(t) \in S \mid X(0) = x) < 1$. This may arise because the process is explosive, or it may be killed by some agency. In either case, it is convenient to add (or adjoin) an extra absorbing state Δ to S, where Δ is called the cemetery, yielding a new state space $\Delta \cup S$ on which one defines transition probabilities for an honest Markov process; $\mathbb{P}(X(t) \in \Delta \cup S) = 1$ for all t. [L. Rogers & D. Williams, *Diffusions, Markov Processes and Martingales*, Vol. 1, 2nd edn. Wiley, 1994; reprinted by Cambridge University Press, 2000]

censored process (alias: **watched process**) Broadly, for a random process $X(t)$, with state-space S, if the process $X^A(t)$ is obtained by deleting those segments of the sample paths of $X(t)$ that lie in $A^c \subseteq S$, and connecting the remaining segments that lie in $A \subseteq S$, then $X^A(t)$ is called $X(t)$ censored in A^c (or $X(t)$ watched in A). Such processes arise naturally in considering Markov chains on an infinite state space, that are approximated by truncating the state space. Let P be the transition matrix of such a chain in discrete time, and partition it by states membership of A or A^c, thus:

$$P = \begin{matrix} & A & A^c \\ A & T & U \\ A^c & R & Q \end{matrix} \quad \text{i.e.} \quad P = \begin{pmatrix} T & U \\ R & Q \end{pmatrix}.$$

Defining the process $X^A(n)$ whose nth term is the value of $X(t)$ on its nth visit to A, we have that the censored process $X^A(n)$ is also a Markov chain with transition matrix $P^A = T + U(\sum_{r=0}^{\infty} Q^r)R$ (with the convention that $0 \cdot \infty = 0$, should any entries in the sum be infinite). [D. Freedman, *Approximating Countable Markov Chains*. Holden-Day, 1972; reprinted by Springer, 1983]

censored random variable *see* truncation

centering constant A random variable X is said to be centred at c if we replace X by $X - c$, where c is constant. If $\mathbb{E}X < \infty$, then $X - \mathbb{E}X$ is said to be centred at (or on) its expectation; more concisely, the term centred is often used as a synonym for zero-mean. If X is proper, with median m, then $X - m$ is centred at (or on) its median. *See* location parameter. [M. Loève, *Probability Theory*, 4th edn. Springer, 1978]

centering function Arises in the context of a random process $X(t), t \in T$, having independent increments. It was shown by P. Lévy that there is a non-random centering function $c(t)$ such that the process $Z(t) = X(t) - c(t)$ has desirable smoothness properties, in which case $Z(t)$ is called a centred process. This may be called the **centering lemma**. It was further shown by R. Pyke that a Lévy process $Y(t)$ can be centred by a function $c(t)$ such that $Y(t) - c(t)$ also has stationary, as well as independent, increments. [J. Doob, *Stochastic Processes*. Wiley, 1953]

central asymptotic theorem (alias: **generalized central limit theorem**) Arises as a generalization of the classical central limit theorem to the case when summands are dependent. In practice, authors usually use more specific terms that indicate the nature of the dependence, e.g. **martingale CLT**; **Markov chain CLT**; **mixing CLT**; etc. [M. Loève, *Probability Theory*, 4th edn. Springer, 1978, p.361]

central convergence theorem (alias: **generalized (or extended) central limit theorem**) Arises as a natural generalization of the classical central limit theorem to the case when the summands are not necessarily identically distributed. Specifically, let (X_{nr}) be a null array of random variables (or vectors such that, as $n \to \infty$, the nth row sum $\sum_r X_{nr}$ converges in distribution). Then this limiting distribution is infinitely divisible. Yet more precisely the so-called **central convergence criteria** supply conditions detailing exactly which infinitely divisible distribution arises in the limit. When the limit is normal, it is usual to speak of the central limit theorem. When the limit is not normal, one sometimes refers to non-central limit theorems. [O. Kallenberg, *Foundations of Modern Probability*, 2nd edn. Springer, 2002, p. 303; M. Loève, *Probability Theory*, 4th edn. Springer, 1978, p. 311; R. Laha & V. Rohatgi, *Probability Theory*. Wiley, 1979, p. 307]

central limit theorem (alias: **CLT; normal approximation theorem**) In probability theory the central limit problem is to find what happens to the distribution of a sum $S_n = X_1 + \cdots + X_n$ of independent random variables as n increases (where the X_r are often identically distributed, but not necessarily so). A central limit theorem gives conditions under which the distn of S_n is approximately normal for large n, i.e. more formally, that there are sequences of constants a_n and b_n such that, as $n \to \infty$, for $x, y \in \mathbb{R}$,

$$(^*)\ \mathbb{P}(x \le (S_n - a_n)/b_n \le y) \to \Phi(y) - \Phi(x),$$

where Φ denotes the standard normal distribution function. The first and simplest such theorem arises when each X_r is a Bernoulli (or indicator) random variable such that $\mathbb{P}(X_r = 1) = p = 1 - q = 1 - \mathbb{P}(X_r = 0)$. Then A. De Moivre's CLT (of 1738), in the form given later by P. Laplace, is $\mathbb{P}\big(x < (S_n - np)/(npq)^{1/2} \le y\big) \to \Phi(y) - \Phi(x)$, where $a_n = +np$, and $b_n = (npq)^{1/2}$. This is called the **normal approximation to the binomial distn**. More generally, if the X_r are i.i.d.r.v.s with mean μ and variance $\sigma^2 \in (0, \infty)$ then $\mathbb{P}\big(x \le (S_n - n\mu)/(n\sigma^2)^{1/2} \le y\big) \to \Phi(y) - \Phi(x)$; this is perhaps the result most often denoted by the term 'the central limit theorem'. More generally, in this i.i.d. case there exist sequences a_n and b_n such that (*) holds iff $\lim_{x\to\infty}\{x^2\mathbb{P}(|X_1| > x)/\mathbb{E}\{X_1^2 I[|X_1| \le x]\} = 0$ where $I\{|X| \le x\}$ is the indicator of $\{|X| \le x\}$. For yet more CLTs *see* Lyapunov's theorem and the Lindeberg–Lévy–Feller theorem. [W. Adams, *The Life and Times of the CLT*, 2nd edn. American Mathematical Society, 2009; P. Moran, *Introduction to Probability Theory*. Oxford University Press, 1968; H. Fischer, *History of the CLT*. Springer, 2010]

centre-satellites process *see* Neyman–Scott process

certain event (alias: **sure event**) *see* almost surely; event

certainty effect A property of people that complicates the application of decision theory. According to classical utility theory, the utility of money is effectively unchanged by non-decreasing linear transformations. However, individuals actually tend to overvalue sure gains (and undervalue sure losses) in comparison with an uncertain gamble yielding the same expected outcome. For example, consider the choice between: (a) a 90% chance of losing £10 000, with a 10% chance of losing nothing; or (b) a certain loss of £9000. The great majority of people prefer (a), though this increases their risk. *See* Allais paradox. [P. Gärdenfors & N.-E. Sahlin (eds), *Decision, Probability, and Utility*. Cambridge University Press, 1988]

certainty equivalence principle Arises in control theory, where – broadly speaking – it asserts that in suitable linear models with quadratic cost functions the optimal control is achieved by using linear least-square estimates of unknown parameters (unobservables). [P. Whittle, *Risk-sensitive Optimal Control*. Wiley, 1990]

certainty equivalent An important concept in utility theory. Suppose you are offered an uncertain reward X, or random prize; this is classically represented by the possession of a lottery ticket with pay-off X. If there is a fixed (non-random) sum C that you would exchange indifferently for the ticket, or vice-versa, then that is your certainty equivalent of X. Since you are indifferent between rewards having the same utility, it follows that $C = u^{-1}\big(\mathbb{E}u(X)\big)$, where $u(\cdot)$ is your utility function. If your increasing utility function is risk averse, then the so-called **Pratt** (or **Arrow–Pratt**) **risk premium** is $\mathbb{E}X - C \ge 0$. [R. Keeney & H. Raiffa, *Decisions with Multiple Objectives*. Cambridge University Press, 1993]

chain rule Broadly, a type of formula in which some quantity is evaluated (or expanded) by performing a sequence of essentially similar operations; the result may be a product, or sum, or nested. Several arise in probability, e.g.

(i) **Chain rule for probabilities** (also called the **multiplication rule**): for events A, B, and C, $\mathbb{P}(A \cap B \cap C) = \mathbb{P}(A)\mathbb{P}(B|A)\mathbb{P}(C|A \cap B)$, and more generally, for events $A_1, \ldots, A_n$, $\mathbb{P}(\cap_{r=1}^n A_r) = \mathbb{P}(A_1)\mathbb{P}(A_2|A_1)\cdots\mathbb{P}(A_n|\cap_{r=1}^{n-1} A_r)$.

(ii) **Chain rule for conditional expectation** (or **tower rule**): for random variables X, Y, and Z, $\mathbb{E}X = \mathbb{E}(\mathbb{E}(X|Y)) = \mathbb{E}(\mathbb{E}(\mathbb{E}(X|Y, Z)|Y))$. Or, more abstractly, for σ-fields $\mathcal{F}$ and $\mathcal{G}$, $\mathcal{F} \subseteq \mathcal{G}$,

$\mathbb{E}X = \mathbb{E}(\mathbb{E}(X \mid \mathcal{F})) = \mathbb{E}(\mathbb{E}(\mathbb{E}(X \mid \mathcal{G}) \mid \mathcal{F}))$. There are natural extensions to more r.vs, and longer increasing sequences of σ-fields, and to densities of continuous r.vs, and to Radon–Nikodym derivatives.

(iii) **Chain rule for entropy:** the Shannon entropy $H(X_1, \ldots, X_n)$ of the random vector $X_1, \ldots, X_n$ may be written in terms of conditional entropies as $H(X_1,\ldots,X_n) = \sum_{r=2}^{n} H(X_r|X_1,\ldots,X_{r-1}) + H(X_1)$.

(iv) **Chain rule for relative entropy:** let $p(x, y)$ and $q(x, y)$ be joint distributions, with marginal distributions $p(x)$ and $q(x)$ for the first variable, and conditional distributions $p(y \mid x)$ and $q(y \mid x)$ given the first variable. Then the relative entropy $d(p, q)$, between p and q, satisfies $d(p(x, y), q(x, y) = d(p(x), q(x) + d(p(y \mid x), q(y \mid x))$.

(v) **Chain rule for mutual information:** the mutual information I between Y and the r.vr $X_1, \ldots, X_n$ can be written in terms of conditional mutual informations as

$$I(Y;X_1,\ldots,X_n) = I(Y;X_1) + \sum_{r=2}^{n} I(Y;X_r|X_1,\ldots,X_{r-1}).$$

[T. Cover & J. Thomas, *Elements of Information Theory*, 2nd edn. Wiley, 2006]

chain rule for conditional probabilities *see* multiplication rule for probabilities

Chaitin number (or constant) *see* algorithmic probability

Champernowne's distribution Arises in econometrics, where D. Champernowne used it to fit the observed distribution of incomes. It is given in several forms, including this one (due to Champernowne) for the distribution function

$$F(x) = 1 - \theta^{-1} \tan^{-1}\left\{\frac{\sin\theta}{\cos\theta + (x/m)^a}\right\},$$
$$x \geq 0, \text{ for } 0 < \theta < \pi,\ a > 0, \text{ and } m > 0.$$

For large x this is approximately a Pareto distribution, and Champernowne developed a model for the evolution of income distributions yielding a Pareto-type equilibrium distribution. The distn can be seen as one of Perks's generalized logistic distns; in particular as $\theta \to 0$, $F(x) \to \frac{(x/m)^a}{1+(x/m)^a}$, and the transformation $(x/m)^a = \mathrm{e}^y$ yields the logistic density $f(y) = \left(\frac{1}{2}\operatorname{sech}\frac{y}{2}\right)^2$, $y \in \mathbb{R}$.

[D. Champernowne, A model of income distribution, *Econ. J.*, 63(250), 318–351, 1953]

change of measure Broadly, this arises when the probability measure $\mathbb{P}(\cdot)$, implicit in the definition of some random variable (or process) in the probability space $(\Omega, \mathcal{F}, \mathbb{P})$, is exchanged for (i.e. replaced by) a different probability measure $\mathbb{Q}(\cdot)$; the sample space Ω and σ-field $\mathcal{F}$ remaining unchanged. There are extensive applications in financial mathematics where the stock price process $S(t)$ is equipped with two measures: $\mathbb{P}(\cdot)$ which describes the evolution of the stock price over time; and the martingale (or risk-neutral) measure $\mathbb{Q}(\cdot)$ under which the discounted value of $S(t)$ is a martingale and thus determines fair (i.e. no arbitrage) prices for options. There are also applications in large deviation theory which employs the so-called tilted measures. The basic mathematical example is supplied by a discrete random variable X with probability measure $p(x) = \mathbb{P}(X = x)$. If this is replaced by $q(x)$, we call the resulting random variable Y, so that $\mathbb{Q}(Y = y) = q(y)$ (though it should be noted that $X(\cdot)$ and $Y(\cdot)$ are the same measurable function on the sample space Ω so that $X(\omega) = Y(\omega)$). By construction, the distribution of Y is obtained by multiplying $p(x)$ by the so-called likelihood ratio $L(x) = q(x)/p(x)$ (where it is required that p and q have the same support). As a function of X, this is a random variable and there is a change-of-measure formula $\mathbb{Q}(Y \in A) = \mathbb{E}\{L(X)I(X \in A)\}$. For continuous r.vs, and more generally, the likelihood ratio is replaced by the Radon–Nikodym derivative.

Now consider a random process X_n, adapted to the filtration $\mathcal{F}_n$, where $\mathbb{P}_n$ is the restriction of $\mathbb{P}$ to $\mathcal{F}_n, n \geq 1$; and let $\mathbb{Q}$ be another measure whose restriction to $\mathcal{F}_n$ is $\mathbb{Q}_n$ and such that $\mathbb{Q}_n(A) = 0 \Leftrightarrow \mathbb{P}_n(A) = 0$. Then there is a random process $L_n, n \geq 1$, on $(\Omega, \mathcal{F}, \mathbb{P})$, that is a martingale wrt $\mathcal{F}_n$ (called the likelihood ratio martingale), such that for any $A \in \mathcal{F}_n$, whose indicator under $\mathbb{P}$ is $I(A)$, we have $\mathbb{Q}(A) = \mathbb{E}(L_n I_{\mathbb{P}}(A))$. If the random processes under $\mathbb{P}(\cdot)$ and $\mathbb{Q}(\cdot)$ have respective joint densities $f(x_1, \ldots, x_n)$ and $g(x_1, \ldots, x_n)$ then $L_n = g(X_1, \ldots, X_n)/f(X_1, \ldots, X_n)$. The further generalization to Wiener processes and diffusions is the Cameron–Martin–Girsanov theorem.

[F. Klebaner, *Introduction to Stochastic Calculus with Applications*, 3rd edn. Imperial College Press, 2012;
O. Barndorff-Nielsen & A. Shiryaev, *Change of Time and Change of Measure*. World Scientific, 2010]

change of measure identity *see* importance sampling identity

change of numeraire Arises in financial mathematics when one seeks to price claims in some other unit of value, e.g. crowns instead of dollars. More generally, one seeks a valuation in terms of units of one tradable asset (e.g. treasury bonds) rather than another. In each case the valuations should be the same at any time (on conversion) but the risk-neutral measures are different for each numeraire. Formally, then, a change of numeraire is a change of measure between two martingale measures. [T. Björk, *Arbitrage Theory in Continuous Time*, 3rd edn. Oxford University Press, 2009]

change of scale *see* scale function

change of time (alias: **time-change; time substitution; subordination; transformation of time; operational time.**) Broadly if $X(t)$ is a random process, and $T(t)$ is a suitable function of the time-parameter t, then $T(t)$ is a change of time, and $X\big(T(t)\big)$ is a time-changed process. The function $T(t)$ may be a random variable for each t. For example, let $N(t)$ be a Poisson process of unit rate, and define $M(t) = \int_0^t m(s)\mathrm{d}s$, where $0 < m(s) < \infty$. Then the time-changed process $X(t) = N\big(M(t)\big)$ is a non-homogeneous Poisson process with intensity function $m(t)$, and any such process can thus be represented as a time-change of the standard Poisson process.

More generally, if $X(t)$ is a Markov process, and $T(t)$ is a random process with non-negative independent increments, then $X\big(T(t)\big)$ is also a Markov process (under mild conditions of good behaviour); $T(t)$ is called the directing process, or subordinator, or leading process, or operational time, and $X(T)$ is said to be directed by, or subordinate to $T(t)$. For example, if $a > 0$ and $W(t)$ is the standard Wiener process, then $X(t) = \mathrm{e}^{-at}W(\mathrm{e}^{2at} - 1)$ is an Ornstein–Uhlenbeck process, and $Y(t) = (1 - t)$ $W(\frac{t}{1-t})$ is a standard Brownian bridge. More generally, if $a(t)$ is a non-negative function for $t > 0$, such that $T(t) = \int_0^t (a(s))^2\mathrm{d}s < \infty$ for $t \geq 0$, then $W\big(T(t)\big)$ has the same distribution as the diffusion $Z(t) = \int_0^t a(s)\mathrm{d}W(s)$. Considerable development of these concepts leads to the result that any regular continuous diffusion on its natural scale can be represented by a time-changed version of the Wiener process (and hence, by a change of scale, so too for any continuous local martingale, under technical conditions). [O. Barndorff-Nielsen & A. Shiryaev, *Change of Time and Measure*. World Scientific, 2010]

change-of-variable formula Arises in considering transformations and functions of r.vs, r.vrs, and random processes. In the basic case, X is a continuous r.v. with density $f(x)$, and $Y = h(X)$, where $h(\cdot)$ is strictly increasing and continuously differentiable; with differentiable inverse $k(\cdot)$. Then Y has density $g(y) = f(k(y))k'(y))$ (the so-called change of variable formula); further, $\mathbb{E}Y = \int_{\mathbb{R}} h(x)f(x)\mathrm{d}x$, which is sometimes called the change of variable formula for expectations. Similar results hold in the multivariate case. For example, let X and Y have joint density $f(x,y)$ and suppose that $U = u(X,Y)$, $V = v(X,Y)$ is a continuously differentiable, one–one map to $\mathbb{R}^2$. Then $\mathbb{E}U = \int\int u(x,y)$ $f(x,y)\mathrm{d}x\mathrm{d}y$, with an analogous formula for V; while U and V have joint density $g(u,v) = f\big(x(u,v),\ y(u,v)\big)J$, where J is the Jacobian of the transformation.

In the context of diffusion processes, Itō's change of variable formula supplies the SDE for continuously twice differentiable functions of a diffusion process. [H. Tuckwell, *Elementary Applications of Probability*, 2nd edn. Chapman & Hall, 1995]

change-point problem (alias: **disruption problem; disorder problem; quickest detection problem**) A classic problem of optimal stopping, arising in the context of monitoring systems for potentially damaging changes. In its simplest form, you are presented with a sequence of i.i.d.r.vs (X_r) having distribution F, which at a time T will change to the presentation of a sequence of i.i.d.r.vs (Y_s) with distribution $G \neq F$. The change-time T is unknown, but it is desired to assert that it has occurred as soon as possible after the event (and not before). [A. Shiryaev, *Optimal Stopping Rules*. Springer, 2008]

Chan–Karolyi–Longstaff–Sanders model (alias: **CKLS family**) Arises in financial mathematics as a family of SDEs used to describe interest rates and asset prices. It has the form $\mathrm{d}X(t) = \big(a + bX(t)\big)\mathrm{d}t + c\big(X(t)\big)^\alpha \mathrm{d}W(t)$, where a, b, c are non-negative constants $\alpha \geq 1/2$, and $W(t)$ is the Wiener process. It includes several other models as special cases (e.g. Vasicek, CIR, Brennan–Schwartz, etc.). [S. Iacus, *Simulation and Inference for SDEs*. Springer, 2008]

channel (alias: **communication channel**) Any of a number of models for the various devices that accept an input (called a message, word, or string) at one point, and supply at another point an output from which it is intended to reconstruct the input message as accurately as possible. The basic case is the discrete memoryless noisy channel, defined thus: There is a finite set A of symbols for input, and a finite set B of outputs, these are often termed alphabets; not uncommonly it is the case that

$A \equiv B$. In a noisy channel the output Y is a random function of the input X, and it is thus characterized by the family of conditional distributions $p(b \mid a) = \mathbb{P}(Y = b \mid X = a), a \in A, b \in B$; the array $p(b \mid a)$ is often called the **channel matrix**. If the outputs are conditionally independent given the inputs for a message of any length, the channel is called **memoryless**; formally this is the condition $p(b_1, \ldots, b_r \mid a_1, \ldots, a_r) = \prod_{j=1}^{r} p(b_j \mid a_j),\ r \geq 1$, on the conditional distribution of the output sequence $(b_1, \ldots, b_r)$ given input $(a_1, \ldots, a_r)$.

Otherwise, the channel is said to have a memory. Very often $A = (a_1, a_2)$ and $B = (b_1, b_2)$, in which case one has a **binary channel**. Many different types of channel exist, classified in various ways, e.g.

- The input and output alphabets may be countable or uncountable, corresponding to discrete or continuous r.vs respectively. If they are jointly normally distributed one speaks of a Gaussian channel.
- The channel may operate in continuous time, with input and output comprising jointly distributed random processes.
- There may be several inputs; this is a **multi-access channel**.
- There may be several outputs; this is a **broadcast channel**.

(*See also* Shannon's theorem; symmetric channel.) Analogously, in the context of quantum communication, a quantum channel is a suitable affine mapping of the (convex) set of all quantum states (where a quantum state is a positive matrix operator of unit trace, called a density operator, in a separable Hilbert space H). [T. Cover & J. Thomas, *Elements of Information Theory*, 2nd edn. Wiley, 2006; M. Wilde, *Quantum Information Theory*. Cambridge University Press, 2013]

channel capacity Broadly, the capacity of a channel is a measure of its ability to communicate information by accepting messages at one point and reproducing them at another, with arbitrarily small probability of error in the reproduction; the units of measurement being bits per symbol, or bits per second, according to circumstances. Thus in the ideal simplest case of a noiseless binary channel, the capacity is one bit per symbol, achieved when the input distribution is uniform on the two symbols of the alphabet; or equivalently, when for any $n \geq 1$ the 2^n possible messages of length n are equally likely. More generally, the capacity of any discrete noiseless channel is $C = \lim_{n \to \infty} \frac{1}{n} \log_2 M(n)$, where $M(n)$ is the number messages of length n permitted as input. In the case of a noisy discrete memoryless channel, Shannon's noisy coding theorem shows that the maximum rate of transmission in bits per symbol, with arbitrary accuracy, is $C = \sup I(X;Y)$, where $I(X;Y)$ is the mutual information between input X and output Y, and the supremum is over possible distributions of X. Similar theorems yield equivalent capacity in e.g. continuous channels, and channels, whose input and output are jointly stationary, etc. Analogously, the capacity of a quantum channel is the supremum of the achievable rates of transmission (of qubits) in n uses of the channel, as n increases.

A related concept is that of **channel capacity-cost,** in which transmission entails a penalty, and the capacity is achieved subject to constraints on the average penalty permitted per symbol. [R. McEliece, *Information and Coding*. Cambridge University Press, 2002; M. Wilde, *Quantum Information Theory*. Cambridge University Press, 2013]

chaos expansion (alias: **Wiener–Itō expansion (or representation, or decomposition); Itō expansion**) Arises in stochastic calculus and the theory of SDEs, where it expresses any suitable (square integrable) functional X of the Wiener process over $[0, t]$ as a sum of multiple Wiener–Itō integrals. Formally, it asserts the existence of unique non-random functions f_n, determined by X, such that $X = \mathbb{E}X + \sum_{n=1}^{\infty} \int_0^t \int_0^{t_1} \cdots \int_0^{t_{n-1}} f_n(t_1, \ldots, t_n) \mathrm{d}W(t_1) \cdots \mathrm{d}W(t_n)$, where $0 < t_1 < \cdots < t_n < t$, and $W(\cdot)$ denotes the Wiener process. Note that the smallest closed linear space of zero-mean finite-variance normal r.vs that contains the nth term for all suitable (measurable) functions f_n, is called the nth Wiener homogeneous chaos, and the space comprising the direct sum of the first n chaoses is the nth polynomial chaos. The first chaos is the smallest closed space of normal r.vs containing $W(t)$ itself. Similar chaoses may be defined in terms of Rademacher r.vs (Rademacher chaos), or the Poisson process (Poisson chaos).

Lévy chaos is an expansion in terms of multiple stochastic integrals with respect to an infinitely divisible random measure. [C. Houdré & V. Pérez-Abreu (eds), *Chaos Expansions, Multiple Wiener-Itō Integrals and Their Applications*. CRC Press, 1994]

Chapman–Kolmogorov equation(s) Arises in the context of Markov processes, where it follows from an application of the Markov property to the transition probabilities. It takes its simplest form for a Markov chain $X(n)$, in discrete time $n \geq 0$, with discrete state space S in which case the equation is

$$\mathbb{P}(X_{m+n}=k|X_0=i)=\sum_{j\in S}\mathbb{P}(X_m=j|X_0=i)\mathbb{P}(X_{m+n}=k|X_m=j), i,k\in S;$$
$$=\sum_{j\in S}\mathbb{P}(X_m=j|X_0=i)\mathbb{P}(X_n=k|X_0=j)$$

for a time-homogeneous chain, i.e. having stationary transition probabilities. In this case it is more concisely written in terms of the matrix $P(n)$ of n-step transition probabilities as $P(m+n) = P(m)P(n)$; which is essentially the same in continuous time for such chains, i.e. $P(s+t) = P(s)P(t), s, t \geq 0$. The more general integral form of the equations for chains on more general spaces may be called the Chapman–Kolmogorov identity. The equation was first given by S. Chapman in 1928, in generalizing Einstein's theory of Brownian motion to particles moving in a non-uniform medium. In 1931 A. Kolmogorov extended the identity to cover a more general class of what are now (since 1934) called Markov processes, and showed that in suitably regular cases there is an equivalence with Kolmogorov backward and forward eqns in continuous time. The term generalizal Chapman–Kolomogorov eqn may be used for the analogous equation arising when the times between jumps are not exponentially distributed, so that the process is semi-Markov, rather than Markov. (Note that the truth of the equation is not sufficient for a process to be Markov.) [R. Chapman, On the Brownian displacement and thermal diffusion of grains suspended in a non-uniform fluid, *Proc. Roy. Soc. Lond. A.*, 119, 34–54, 1928; W. Feller, *Introduction to Probability Theory and its Applications*, Vol. 1, 3rd edn. Wiley, 1968; Vol. 2, 2nd edn. Wiley, 1971]

characteristic function For a r.v. X the characteristic function, also written as chf, ch. fn., or c.f., is defined for all real t by, $\phi(t) = \mathbb{E}\exp(\mathrm{i}tX)$; and knowledge of $\phi(t)$ uniquely determines the distribution function $F(x)$ of X. Thus $\phi(t) = \int_{\mathbb{R}}\exp(\mathrm{i}tx)\mathrm{d}F(x)$, the Fourier–Stieltjes transform of F; and when X has density $f(x), \phi(t)$ is the Fourier transform of $f(x)$; $F(x)$ and $f(x)$ being given as integral transforms of $\phi(t)$ by their corresponding Fourier inversion formulas. The chf $\phi(t)$ is positive definite, and uniformly continuous on $\mathbb{R}$, with $\phi(0) = 1$. If $\mathbb{E}|X|^k < \infty$, then $\phi^{(k)}(0) = \mathrm{i}^k\,\mathbb{E}X^k$. Furthermore, if $\phi(t)$ can be extended from the real line to a holomorphic (analytic) function $\phi(z)$ that exists in a strip $\{z : |\mathrm{Im}(z)|\langle a\rangle 0\}$ then all moments of X are finite, and as $k \to \infty$, lim sup $\{|\mathbb{E}X^k|/k!\}^{1/k} \leq C < \infty$, and X is uniquely determined by its moments. These properties, together with the continuity theorem and the fact that the chf of a sum of independent r.vs is the product of their separate chfs, account for the great importance of chfs in probability.

More generally, any random vector $X_1, \ldots, X_d$ has joint chf $\phi(t_1, \ldots, t_d)$ defined for $t \in \mathbb{R}^d$ by $\phi(t) = \mathbb{E}\exp[\mathrm{i}(t_1X_1 + \cdots + t_dX_d)]$. This is a type of characteristic functional. [E. Lukacs, *Characteristic Functions*, 2nd edn. Griffin, 1970]

characteristic functional Arises in considering e.g. random processes $X(t), t \in \mathbb{R}$; or random point processes in $\mathbb{R}^n$, where it plays a role analogous to that of the characteristic function $\phi(\theta)$ of a r.v. X (of which it can be seen as an infinite-dimensional equivalent). In a commonly seen basic form, a suitable random process $X(t), t \in \mathbb{R}$, has characteristic functional $\phi(\theta) = \mathbb{E}(\exp\{\mathrm{i}\int_{\mathbb{R}}\theta(t)X(t)\mathrm{d}t\})$, for an appropriate class of so-called test functions $\theta(t), t \in \mathbb{R}$. More generally, when X is a random variable (or element) in a suitable linear space S, the characteristic functional is $\phi(\theta) = \mathbb{E}\exp\{\mathrm{i}\theta X\}$, where θ is an element in an appropriate space of linear functionals on S. Laplace functionals are defined analogously to the Laplace transform. [A. N. Kolmogorov, La transformation de Laplace dans les espaces linéaires, *C. R. Acad. Sci. Paris*, 200, 1717–1718, 1935; D. Daley & D. Vere-Jones, *Introduction to the Theory of Point Processes*, 2nd edn. Springer, Vol. I, 2003; Vol. II, 2008]

Chebyshov's inequality (alias: **Bienaymé inequality; Chebyshev inequality, and numerous other transliterations of Уебышёв**) Gives a bound on the probability in the tail of the distribution of a r.v. X in terms of its variance $\sigma^2 = \mathrm{var}\,X < \infty$, thus $\mathbb{P}(|X - \mathbb{E}X| \geq t) \leq \sigma^2/t^2, t > 0$. A related inequality asserts that for any non-negative even function $g(x)$, that increases for $x > 0, \mathbb{P}(|X| \geq t) \leq \mathbb{E}g(X)/g(t)$. This may be called a Chebyshov-type inequality; the case $g(X) = |X|$ is often called Markov's inequality. Many improvements exist, e.g. the **Mallows–Richter improvement** of Chebyshov is $\mathbb{P}(|X - \mathbb{E}X| \geq t) \leq \sigma^2/\{\mathbb{E}(|X - \mathbb{E}X|||X - \mathbb{E}X| \geq t)\}^2$. The **one-sided Chebyshov inequality**, often called *Cantelli's inequality*, is $\mathbb{P}(|X - \mathbb{E}X| \geq t) \leq \sigma^2/(\sigma^2 + t^2)$, $t > 0$, with Mallows–Richter improvement given by $\mathbb{P}(X - \mathbb{E}X \geq t) \leq \sigma^2/(\sigma^2 + \{\mathbb{E}(X - \mathbb{E}X|X - \mathbb{E}X \geq t)\}^2)$. [C. Mallows & D. Richter, Inequalities of Chebyshov type involving conditional expectations, *Ann. Math. Stat.*, 40(6), 1922–1932, 1969]

Note: In transliterating the Cyrillic version of Chebyshov into the Roman alphabet, these forms (at least) have been given for the seven phonemes, in order: {č, ch, tch, tsch}, {e}, {b}, {i,y}, {š, sh, ch, sch}, {e, o}, {v, f, ff, w}, yielding 256 possibilities, of which at least 36 have been used. The transliteration used here is optimized for anglophone readers; for them the Russian pronunciation of the penultimate Cyrillic letter 'ë' lies somewhere between 'o' and 'yo', but closer to the former. We quote:

But there is one I leave to you,
Whatever you may choose to do:
Put letters in or leave them out,
Garnish with accents round about,
Finish the name with eff or off:
There is no way to spell Уебышёв.

[R. P. Boas, *Lion Hunting and Other Mathematical Pursuits*. MAA, 1995, 204–205 (reprinted with permission)]

Cheeger's inequality Arose originally as a lower bound for eigenvalues λ of the Laplacian operator, in the form $\lambda \geq h^2/4$, where h is a constant that may be called Cheeger's constant. It may be adapted to various probabilistic contexts, e.g. suppose that the transition matrix of a finite ergodic reversible Markov chain has second-largest eigenvalue λ and conductance Φ. Then Cheeger's inequality takes the form $\frac{1}{2}\Phi^2 \leq 1 - \lambda \leq 2\Phi$. [P. Diaconis & D. Stroock, Geometric bounds for eigenvalues of Markov Chains, *Ann. Appl. Probab.*, 1(1), 36–61, 1991]

Chen's inequality In probability theory this denotes a multivariate generalization of an inequality due to H. Chernoff; one form is this: Let $X_1, \ldots, X_k$ be i.i.d. standard normal r.vs, and let $g(x_1, \ldots, x_k) \in \mathbb{R}$ have all its first-order partial derivatives (in fact, absolutely continuous in each coordinate), then:

$$\operatorname{var} g(X_1, \ldots, X_k) \leq \sum_{r=1}^{k} \mathbb{E}\left\{\frac{\partial g}{\partial x_r}(X_1, \ldots, X_k)^2\right\},$$

where equality holds with finite RHS iff there are constants $c, c_1, \ldots, c_k$ such that $g(x_1, \ldots, x_k) = c + \sum_{r=1}^{k} c_r x_r$, a.s. [L. Chen, An inequality for the multivariate normal distribution, *J. Multivariate Anal.*, 12(2), 306–315, 1982]

Chernoff's bounds (alias: **Cramér–Chernoff bounds (or method)**) Inequalities that supply geometric (i.e. exponential) bounds for the probability in the tail of a distribution, and are of particular use when applied to sums of independent r.vs. The bounds are members of a family of related bounds (e.g. Hoeffding's) that apply the so-called Bernstein method. The classic form is this: let $S_n = \sum_{r=1}^{n} X_r$, where X_r are i.i.d.r.vs, and denote the minimum of the mgf $M(t) = \mathbb{E}\,e^{t(X-a)}$, for $t \in \mathbb{R}$, by $m(a)$. Then for $a < \mathbb{E}X_1 > -\infty, \mathbb{P}(S_n < na)$ behaves roughly like $m(a)^n$. Formally:

(i) If $\mathbb{E}X_1 > -\infty$, and $a < \mathbb{E}X_1$, then $\mathbb{P}(S_n \leq na) \leq m(a)^n$.
(ii) If $\mathbb{E}X_1 < \infty$, and $a \geq \mathbb{E}X_1$, then $\mathbb{P}(S_n \geq na) \leq m(a)^n$.
(iii) If $0 < t < m(a)$, whether $\mathbb{E}X_1$ exists or not,

$$\lim_{n\to\infty} \frac{(m(a)-t)^n}{\mathbb{P}(S_n \geq na)} = 0 = \lim_{n\to\infty} \frac{(m(a)-t)^n}{\mathbb{P}(S_n \leq na)}.$$

Note that it is not required that $M(t)$ exists for all $t \in \mathbb{R}$. Similar and related bounds are often said to be of Chernoff type. [H. Chernoff, A measure of asymptotic efficiency for tests of a hypothesis, *Ann. Math. Stat.*, 23(4), 493–507, 1952]

Chernoff's divergence (alias: **Chernoff (relative) information; Chernoff alpha distance**) A measure of divergence between probability distributions. For any discrete distns $(p_1, p_2, \ldots, p_n)$ and $(q_1, \ldots, q_n)$ it is defined for $0 \leq \alpha \leq 1$ as $C(p, q; \alpha) = -\log \sum_r p_r^\alpha q_r^{1-\alpha}$. It is a concave function of $\alpha \in [0, 1]$, and the maximum $C(p, q) = \max_{\alpha\in[0,1]} C(p, q; \alpha)$ may be called the **Chernoff information**. The Chernoff divergence is defined for all $\alpha \in \mathbb{R}$, for distributions p and q such that $p_r > 0 \Leftrightarrow q_r > 0$. It is related to the Rényi divergence $R(p, q; \alpha)$ by $R = C(p, q; \alpha)/(1-\alpha)$; sometimes called the normalized Chernoff divergence. Further, C is a Bhattacharyya distance when $\alpha = 1/2$, and it is also related to the family of f-divergences $D(p, q; f)$ by $C(p, q; \alpha) = -\log[D(p, q; f) + 1]$ when $f(x) = x^{1-\alpha} - 1$. The divergence for densities and other probability measures are defined analogously. [H. Chernoff, A measure of asymptotic efficiency for tests of a hypothesis, *Ann. Math. Stat.* 23, 493–507, 1952]

Chernoff's inequality Asserts that for a standard normal $N(0, 1)$ r.v. X, and any absolutely continuous function $g(x)$ having derivative $g'(x), x \in \mathbb{R}$, we have $\operatorname{var} g(X) \leq \mathbb{E}[g'(X)^2]$ whenever the LHS is finite. The basic form has been extended so that e.g. **Cacoullos' inequality** asserts that $\{\mathbb{E}g'(X)\}^2 \leq \operatorname{var} g(X)$; **Chen's inequality** extends it to a function $g(x_1, \ldots, x_k)$ of several standard normal r.vs; and the **Olkin–Shepp inequality** asserts than when $g_1(x), \ldots, g_k(x)$ are suitable absolutely continuous functions the matrix with (j, k)-th entry $\mathbb{E}\{g_j'(X)g_k'(X)\} - \operatorname{cov}\{g_j(X), g_k(X)\}$ is non-negative definite. **Klassen's inequality** is a generalization to arbitrary r.vs, X. [I. Olkin & L. Shepp, A matrix variance inequality, *J. Stat. Plan. Inference*, 130(1–2), 351–358, 2005]

chinese restaurant process Arises in the theory of random partitions, and Dirichlet processes, where it has applications in genetics. It is a random process that generates a random partition of $\{1, \ldots, n\}$ into blocks, and can otherwise be interpreted as a Pólya urn process with an unlimited number of colours, or as a preferential attachment process. It is described thus: n individuals enter a room with an unlimited number of tables, and each sits at an empty table or an already occupied table (with no choice for the first individual) as follows. If individuals $1, 2, \ldots, r$ have sat at k tables with $s_1, \ldots, s_k$ respectively at each table, then the $(k+1)$-th arrival sits at an empty table with probability $\frac{ak+b}{r+b}$, or at table t with probability $\frac{s_t-a}{r+b}$, $1 \leq t \leq k$, for suitable constants a and b. [D. Aldous, *Exchangeability and Related Topics*, St.-Flour, Summer School, No. 13. Springer, 1985]

chip-firing game (alias: **Dirichlet game**) A model arising in graph theory (that can be seen as a type of Abelian sandpile) in which C chips are distributed among the vertices of a connected graph. At each stage, one of the vertices that has at least as many chips as its vertex degree is chosen (by some rule) and then transfers a chip to each one of its neighbours – called firing the vertex. If the game terminates when no vertex has more chips than its degree, the final configuration and the number of steps to reach it are independent of the rule for choosing vertices. [A. Björner *et al.*, Chip-firing games on graphs, *Euro. J. Combinatorics*, 12(4), 283–291, 1991]

chi-squared distribution (alias: **chi-square distribution; χ^2 distribution; gamma $\{\frac{1}{2}, \frac{1}{2}d\}$**) A chi-squared random variable X has density

$$f(x) = \left(\frac{1}{2}\right)^{d/2} \frac{1}{\Gamma\left(\frac{1}{2}d\right)} x^{\frac{1}{2}d-1} e^{-\frac{1}{2}x}, x \geq 0$$

where d is a positive integer called the number of degrees of freedom. The density $f(x)$ arises naturally as that of the sum of the squares of d independent standard normal random variables; it therefore has mean d, variance $2d$, and characteristic function $\phi(t) = (1-2it)^{-\frac{1}{2}d}$. For positive non-integer $d, f(x)$ is a gamma density with parameters $\frac{1}{2}$ and $\frac{1}{2}d$. The square root of X is sometimes said to have a **chi distn** (or a **Maxwell distn** when $d = 3$, or a Rayleigh distribution when $d = 2$. [H. Lancaster, *The Chi-squared Distribution*. Wiley, 1969]

chi-squared divergence (alias: **Pearson's χ^2 divergence**) A measure of divergence between probability distributions. For discrete distributions $(p_1, \ldots, p_n)$ and $(q_1, \ldots, q_n)$ it is defined (provided that $q_r = 0 \Rightarrow p_r = 0$ by $S(p,q) = \sum_r p_r^2/q_r - 1 = D(p,q;f)$, $f(x) = x^2 - 1$, where $D(p,q;f)$ is the f divergence. It is never less than the Kullback–Leibler divergence. More generally, for a measure $\mathbb{P}$ that is absolutely continuous wrt a measure $\mathbb{Q}$,

$$S(\mathbb{P}, \mathbb{Q}) = \int \left(\frac{d\mathbb{P} - d\mathbb{Q}}{d\mathbb{Q}}\right)^2 d\mathbb{Q}.$$

[S.-H. Cha, Comprehensive survey on distance/similarity measures between probability density functions, *Int. J. Math. Model. Method. Appl. Sci.*, 1(4), 300–307, 2007]

chooser option Arises in financial mathematics where it denotes a derivative contract (on some asset) that includes the right to choose (at or before some fixed date) whether the option shall be a put or a call. [I. Nelken, *Handbook of Exotic Options*. McGraw-Hill, 1996]

Choquet capacity (alias: **fuzzy measure**) Denotes a non-negative set function $C(\cdot)$ on a suitable collection $\mathcal{F}$ of subsets of a space Ω, e.g. the power set when Ω is finite, or a σ-field of subsets more generally. The capacity $C(\cdot)$ can be seen as a generalization of the probability set function $\mathbb{P}(\cdot)$, in which the axiom of additivity is replaced by monotonicity. That is to say, if $A \subseteq B$ then $C(A) \leq C(B)$, where $C(\emptyset) = 0$. A natural example arises in the framework of irreducible Markov chains, thus: if A is any set of states, let $C(A)$ be the probability that the chain ever enters A. There are other applications in probabilistic counterparts of potential theory, and such capacities arise in models for importance in Boolean systems, and other areas of decision theory and game theory. If $C(A \cup B) \leq C(A) + C(B)$ for any A and B, then $C(\cdot)$ is subadditive; more generally, if

$$C\left(\bigcap_{r=1}^{n} A_r\right) \leq \sum_r C(A_r) - \sum_{r<s} C(A_r \cup A_s) + \cdots + (-1)^{n-1} C\left(\bigcup_{1}^{n} A_r\right),$$

then it is said to be a Choquet capacity of order n (or n-alternating). (A probability measure is a countably additive Choquet capacity.) An integral wrt $C(\cdot)$ is a Choquet integral. [G. Choquet, Theory of capacities, *Annal. Institut Fourier*, 5, 131–295, 1954]

Choquet–Deny theorem Arises in describing the solutions of a type of integral equation that occurs in probability theory in a number of contexts; one form is this: let $F(x)$ be a distribution function on $(0, \infty)$ (i.e. not a lattice distn), and $g(x)$ a continuous bounded function on $\mathbb{R}$ such that $g(x) = \int_0^\infty g(x-y)\, dF(y)$, then $g(x)$ must be a constant. The theorem holds for distributions (measures) and functions on suitable groups. The function g may be said to be harmonic with respect to F (or F-harmonic). [W. Feller, *Introduction to Probability Theory and its Applications*, Vol. 2, 2nd edn. Wiley, 1971]

Choquet–Kendall–Matheron theorem Arises in the theory of random sets, where it asserts that the distribution of a random closed subset $\sum$ of a suitable space Ω is given by the collection $C(K) = \mathbb{P}(\sum \cap K \neq \emptyset)$, where K is any compact subset of the space Ω, and $C(K)$ is a Choquet capacity (that is infinitely alternating). [G. Matheron, *Random Sets and Integral Geometry*. Wiley, 1975]

Choquet representation Choquet theory is a branch of functional analysis that has several links to, and applications in, probability. Broadly, let C be a convex compact subset of a suitable space, then the Choquet representation theorem asserts that for any $z \in C$ there is a probability measure $\mathbb{P}(\cdot)$ on the extreme points of C such that z has the integral representation $z = \int x \mathrm{d}\,\mathbb{P}(x)$. [R. Phelps, *Lectures on Choquet's Theorem*. Springer, 2001]

Chover's law of the iterated logarithm Describes the large deviations of the partial sums $\sum_{r=1}^{n} X_r$ of a sequence of i.i.d.r.vs that have a symmetric stable distribution, i.e. $\mathbb{E}e^{itX_r} = \exp(-|t|^a)$, $0 < a < 2$. Then Chover's LIL asserts that with probability 1,

$$\limsup_{n\to\infty} \left| n^{-1/a} \sum_{r=1}^{n} X_r \right|^{(\log\log n)^{-1}} = e^{1/a}.$$

[J. Chover, A law of the iterated logarithm for stable summands, *Proc. Amer. Math. Soc.*, 17(2), 441–443; 1966]

Chow's inequality *see* Hájek–Rényi–Chow inequality

Chow–Robbins problem (alias: **S_n/n problem; Breiman's extra-sensory perception problem**)
A classic problem of optimal stopping, arising in the context of a sequence $X_1, X_2, \ldots,$, observed term by term, where the X_r are i.i.d.r.vs with $\mathbb{P}(X_1 = 1) = \mathbb{P}(X_1 = -1) = 1/2$, and $S_n = \sum_{r=1}^{n} X_r$. The problem is to find the stopping rule (finite stopping time) $T \in \{1, 2, \ldots\}$ that maximizes $\mathbb{E}\left\{\frac{a+S_T}{b+T}\right\}$, where a, and $b \geq 0$, are fixed integers. Y. Chow and H. Robbins found such a rule, a result later generalized by A. Dvoretzky to the case when the X_r are any i.i.d.r.vs with finite variance. [Y. Chow & H. Robbins, On optimal stopping rules for S_n/n, *Illinois J. Math.*, 9(3), 444–454, 1965; A. Dvoretzky, Existence and properties of certain optimal stopping rules, *Proc. 5th Berk. Symp. Math. Stat. Probab.*, 1, 441–452, 1967.]

Chow–Robbins theorem Arises in considering sums $S_n = \sum_{r=1}^{n} X_r$ of i.i.d.r.vs $X_1, X_2, \ldots$ such that $\mathbb{E}|X_1| = \infty$. It asserts that for any sequence of constants, $C_n, n \geq 1$, either (or both) of the following must hold:
(a) $\limsup_{n\to\infty} |S_n/C_n| = \infty$ with probability 1,
(b) $\liminf_{n\to\infty} |S_n/C_n| = 0$ with probability 1.
Thus the probability that S_n/C_n converges to a non-zero finite constant is zero. Note that the solution to the Chow–Robbins problem, or several other results in optimal stopping, may also be referred to as a 'Chow–Robbins theorem'. [Y. Chow & H. Robbins, Sums of independent random variables with infinite moments, and 'fair' games, *Proc. Natl. Acad. Sci. USA*, 47(3), 330–335, 1961]

Chow's theorem An extension of the Hsu–Robbins–Erdös theorem on complete convergence. Thus, let $X_1, X_2, \ldots$ be i.i.d.r.vs with zero mean, and define $S_n = \sum_{r=1}^{n} a_{nr} X_r$, where a_{nr} is a triangular array (i.e. $a_{nr} = 0$ for $r > n$). If for some $0 < C < \infty$ and $0 < p \leq 1$, we have $|a_{nr}| \leq C \sum_{r=1}^{n} a_{nr}^2 \leq C^2 n^{-p}$ and $\mathbb{E}|X_1|^{2/p} \leq C$, then for $t > 0$, $\sum_{n=1}^{\infty} \mathbb{P}(|S_n| \geq t) < \infty$, which is to say that S_n converges completely to 0. An analogous extension of the Hsu–Robbins–Erdös strong law holds similarly. [Y. Chow, Some convergence theorems for independent random variables, *Ann. Math. Stat.* 37(6), 1482–1493, 1966]

Chung–Diaconis–Graham process A type of random walk on the integers modulo p (denoted by $\mathbb{Z}_p$) that arises in the context of generating pseudo random numbers. In particular, those authors considered the random sequence $X_n, n \geq 0$, of integers satisfying $X_{n+1} = 2X_n + B_n$ (modulo p), where $X_0 = 0$ and the B_n are i.i.d.r.vs taking values in $\mathbb{Z}_p$, and p is odd. More generally, one considers $X_{n+1} = A_n X_n + B_n$ (modulo p), where the A_n are i.i.d.r.vs. The problem of interest is to determine how long it takes for the distribution of X_n to be close to uniform on $\mathbb{Z}_p$. [F. Chung *et al.*, Random walks arising in random number generation, *Ann. Probab.*, 15(3), 1148–1165, 1987]

Chung–Erdös inequality Asserts that for arbitrary events $A_1, \ldots, A_n$,

$$\left(\sum_{r=1}^{n} \mathbb{P}(A_r) \right)^2 \leq \mathbb{P}\left(\bigcup_{r=1}^{n} A_r \right) \sum_{i=1}^{n} \sum_{j=1}^{n} \mathbb{P}(A_i \cap A_j).$$

It arises in the context of the Borel–Cantelli lemma. [K. L. Chung & P. Erdös, On the application of the Borel–Cantelli lemma, *Trans. Amer. Math. Soc.*, 72(1), 179–186, 1952]

Chung–Erdös–Sirao test Arises in the context of uniform continuity of the Wiener process $W(t)$. Define the modulus of continuity

$$S(\delta,t)=\sup_{\substack{0\le u\le v\le t\\ v-u\le\delta}}|W(u)-W(v)| \text{ and for}$$

$s\in(0,t]$, let $g(s)>0$ be continuous and non-decreasing in s, with $g(s)s^{-\frac{1}{2}}$ decreasing for small enough s. Then $S(\delta,t)\le g(\delta)$ for all small enough δ, with probability 1 or with probability 0, according as $\int_{0+}g(s)^3s^{-7/2}\exp\left(-\frac{1}{2}g(s)^2/s\right)\mathrm{d}s$ converges or diverges. For the function $g(s)=c\left(2s\log\frac{1}{s}\right)^{\frac{1}{2}}$, the dichotomy is equivalent to $c>1$ or $c\le 1$, a result known as **Lévy's Hölder condition**. [K. Itô & H. McKean, *Diffusion Processes and Their Sample Paths*. Springer, 1965]

Chung–Erdös test Arises as an application of the Borel–Cantelli lemma to the simple symmetric r.w. $S_n=\sum_{r=1}^n X_r$ where $\mathbb{P}(X_r=+1)=\mathbb{P}(X_r=-1)=\frac{1}{2}$, and the X_r are independent r.vs. Let $g(n)>0$ be increasing in n, and N_n be the number of positive terms in $\{S_1,\ldots,S_n\}$. Then $\mathbb{P}(N_n\le n/g(n)$ infinitely often) is 0 or 1 according as $\sum_n\frac{1}{n\sqrt{g(n)}}$ is finite or infinite. It is often seen in the context of the Wiener process. [K. Chung & P. Erdös, Application of the Borel–Cantelli lemma, *Trans. Amer. Math. Soc.*, 72, 179–186, 1952]

Chung–Fuchs theorem Arises in considering the recurrence or transience of a random walk in $\mathbb{R}^d$, and thus generalizes Pólya's theorem for simple random walks in $\mathbb{Z}^d$. Denote the walk by $S_n=\sum_{r=1}^n X_r$, where the X_r are i.i.d.r.vs when $d=1$, or i.i.d.r.vrs when $d\ge 2$, having chfs $\phi(\theta)$ or $\phi(\boldsymbol{\theta})$ respectively. Then the **Chung–Fuchs condition** (or criterion) says the walk is recurrent iff, for $b>0$, $\sup_{0<a<1}\int_{|\theta|<b}\mathrm{Re}\frac{1}{1-a\phi(\theta)}\mathrm{d}\theta=\infty$. It follows that for $d\ge 3$ the walk is transient; for $d=1$ the walk is recurrent iff $\mathbb{E}|X_1|$ exists and $\mathrm{E}X_1=0$; for $d=2$ the walk is recurrent iff $\mathbb{E}|X_1|^2$ exists and $\mathbb{E}\mathbf{X}_1=0$. The Chung–Ornstein theorem is a simple proof of the cases $d=1$ and $d=2$. Note that the usual form of Chung–Fuchs condition for the recurrence of a Lévy process with chf $\exp(-t\phi(\theta))$ is the divergence of $\int_{|\theta|<b}\mathrm{Re}\phi(\theta)^{-1}\mathrm{d}\theta$, $b>0$. [R. Bhattacharya & E. Waymire, *Basic Course in Probability Theory*. Springer, 2007]

Chung–Kennedy formula Let M be the maximum of the standard Brownian excursion on $[0,1]$. Then the distribution function of M is given by the formula

$$F(x)=1-2\sum_{r=1}^{\infty}(-1)^{r-1}\exp\left(-\frac{1}{2}r^2x^2\right)$$
$$=\frac{\sqrt{8\pi}}{x}\sum_{k=1}^{\infty}\exp\left(\frac{-(2k-1)^2\pi^2}{2x^2}\right).$$

The range of the standard Brownian bridge has the same distribution, an observation that may be called the Chung–Kennedy identity. A probabilistic explanation of this fact is supplied by **Vervaat's decomposition**. [K. Chung, Excursions in Brownian motion, *Ark. Math.*, 14(2), 155–177, 1976; D. Kennedy, Distribution of the maximum Brownian excursion, *J. Appl. Probab.*, 13(2), 371–376, 1976; W. Vervaat, A relation between Brownian bridge and Brownian excursion, *Ann. Probab.*, 7(1), 143–149, 1979]

Chung's law of the iterated logarithm Arises in considering large deviations of the running maximum $M_n=\max_{1\le r\le n}|S_r|$ of the absolute value of the partial sums $S_n=\sum_{r=1}^n X_r$ of a sequence X_r of independent zero-mean r.vs. Denoting $s_n^2=\sum_{r=1}^n\mathrm{var}\,X_r$, with a mild condition on the third moment of X_r, it asserts that if, as $n\to\infty$, $g_n\downarrow 0$ and $g_n^{-1}=o\left\{(\log\log s_n)^{\frac{1}{2}}\right\}$, then $\mathbb{P}(M_n<g_ns_n)=(1+o(1))\exp(-\frac{\pi^2}{8g_n^2})$; and, for $x>0$, $\mathbb{P}(M_n<xs_n)\to\frac{4}{\pi}\sum_{r=0}^{\infty}\frac{(-1)^r}{2r+1}\exp\left\{-\frac{(2r+1)^2\pi^2}{8x^2}\right\}$. In the context of the Wiener process, where $M(t)=\sup_{0\le s\le t}|W(s)|$, Chung's L.I.L. usually denotes the assertion that with probability 1, $\liminf_{t\to\infty}(t^{-1}\log\log t)^{\frac{1}{2}}M(t)=\pi/\sqrt{8}$; and the corollary that $\mathbb{P}(M(t)\le x)\le\frac{4}{\pi}\exp\left(\frac{-\pi^2t}{8x^2}\right)$ is Chung's bound. [K. L. Chung, On the maximum partial sums of sequences of independent random variables, *Trans. Amer. Math. Soc.*, 64, 205–233, 1948]

Chung–Smirnov law of the iterated logarithm Arises in considering the deviations of empirical distribution functions, i.e. the rate of convergence in the Glivenko–Cantelli theorem. Let $F(x)$ be a continuous distribution function, and $F_n(x)$ the corresponding empirical distribution function. (That is, $F_n(x)=\frac{1}{n}\sum_{r=1}^n H(x-X_r)$, where $X_1,\ldots,X_n$ are i.i.d. with distn fn $F(x)$, and $H(y)=I([0,\infty))$ is the distn fn of the distribution degenerate at 0, i.e. the Heaviside unit step fn.) Then the Chung–Smirnov law asserts that as $n\to\infty$, with probability 1, lim sup $\left[[2n/\log\log n]^{\frac{1}{2}}\sup_{x\in\mathbb{R}}|F_n(x)-F(x)|\right]=1$. [K. L. Chung, An estimate concerning the Kolmogoroff limit distribution,

Trans. Amer. Math. Soc., 67, 36–50, 1949; N. V. Smirnov, *Uspekhi Mat. Nauk.*, 10, 179–206, 1944 (in Russian)]

Chung's strong law of large numbers Generalizes the SLLNs of W. Feller and Marcinkiewicz–Zygmund to the sums $S_n = \sum_{r=1}^{n} X_r$ of zero-mean independent r.vs $X_r, r \geq 1$, that are not necessarily identically distributed. Let $a_n > 0$ be a sequence increasing to ∞ as $n \to \infty$, and $g(x) > 0$ an even continuous function on $\mathbb{R}$ such that – as $|x|$ increases $-g(x)/x$ increases and $g(x)/x^2$ decreases. The theorem asserts that whenever $\sum_n \mathbb{E}g(X_n)/g(a_n) < \infty, \sum_n X_n/a_n$, converges a.s. Conversely, if $\sum_n \mathbb{E}g(X_n)/g(x_n) = \infty$, then there is a sequence of zero-mean independent r.vs X_n' with the distributions of X_n such that $\sum_n X_n'/a_n$ diverges a.s. [K. L. Chung, Notes on some strong laws of large numbers, *Amer. J. Math.*, 69(1), 189–192, 1947; K. L. Chung, *Course in Probability Theory*, 2nd edn revised. Academic Press, 2001]

Chung's test Arises in considering the running absolute maxima $M_n = \max_{1\leq r\leq n}|S_n|$ of the partial sums $S_n = \sum_{r=1}^{n} X_r$ of a sequence $X_r, r \geq 1$, of independent zero-mean r.vs (having mild conditions on their moments). Writing $V_n^2 = \sum_{r=1}^{n} \operatorname{var} X_r$ with $g(\cdot)$ a function such that $\lim_{n\to\infty} g(V_n^2) = \infty$, the test asserts that $\mathbb{P}(M_n < \pi V_n/\big(g(V_n^2)\sqrt{8}\big)$ infinitely often) is zero or one according as $\int^{\infty} y^{-1} g^2(y) \exp\{-g^2(y)\} dy$ converges or diverges. The term often denotes a special case applicable to the Wiener process $W(t)$: for a non-decreasing function $g(t) > 0$ such that $g(t)/\sqrt{t}$ is non-increasing, $\mathbb{P}\big(\sup_{0\leq s\leq t}|W(s)| < \sqrt{t}/g(t) \text{ i.o}\big)$ is 0 or 1 according as $\int^{\infty} y^{-1} g^2(y) \exp\left\{\frac{-\pi^2}{8} g^2(y)\right\} dy$ converges or diverges. [K. Chung, On the Maximum partial sums of sequences of independent random variables, *Trans. Amer. Math. Soc.*, 64(2), 205–233, 1948]

Chung's theorem Often refers to the several results shown above, or to any of several other theorems, e.g. (not exhaustively):

1. A uniform version of the SLLN (Chung, 1951).
2. Characterizing conditional independence relative to a σ-field (Chung, 2002, Thm 9.2.1).
3. Any one of several theorems on Markov processes (Chung, 1967).

[K. L. Chung, in *Proc. 2nd Berkeley Symp. Math. Stat. Prob.* University of California Press, 1951, pp. 341–352; K. L. Chung, *Markov Chains with Stationary Transition Probabilities.* Springer, 1967; K. L. Chung, *Course in Probability Theory*, 2nd edn revised. Academic Press, 2001]

Ciesielski–Taylor theorem States that the first-passage time to the unit sphere in $\mathbb{R}^d$, by d-dimensional Brownian motion, has the same distribution as the total time spent inside the unit sphere in $\mathbb{R}^{d+2}$ by $(d+2)$-dimensional Brownian motion (this latter is the occupation (or sojourn) time, which is finite because the process is transient). [Z. Ciesielski & S. Taylor, First passage times and sojourn times for Brownian motion in space, and the exact Hausdorff measure of the sample path, *Trans. Amer. Math. Soc.*, 103, 434–450,1962]

circular distribution Broadly, a distribution (usually a density) of a random variable Θ confined to the interval $[0, 2\pi]$. Such distributions arise naturally in the context of random directions in two dimensions, in which case the probability density is envisaged as distributed on the perimeter of the unit circle. The variable is naturally interpreted as the angle made with some reference radius. A significant subset of these densities may be obtained from probability distributions on $\mathbb{R}$ or $\mathbb{R}^2$ by various transformations, including:

1. Wrapping a univariate density of some r.v. X around the circle; then $\Theta = X$, modulo 2π.
2. Stereographic projection of a density on the line to the circle.
3. Transformation of a bivariate density to plane polars.
4. Inverse trigonometric functions of suitably supported densities.

Popular examples of such densities include the wrapped normal, the cardioid, the von Mises, wrapped Cauchy, and the uniform. The **Jones–Pewsey family** $f(\theta) \propto \{1 + \tanh(ka)\cos(\theta - m)\}^{1/a}$, $m - \pi < \theta \leq m + \pi$ includes many of these as special cases: $a = 0$ yields von Mises; $a = -1$ yields the wrapped Cauchy; $a = +1$ yields the cardioid; $k = 0$ yields the uniform density. [M. Jones & A. Pewsey, A family of symmetric distributions on the circle, *J. Amer. Stat. Ass.*, 100, 1422–1428, 2005]

circular law (alias: **Girko's law**) This describes the empirical distribution (in $\mathbb{C}$) of the eigenvalues $\lambda_1, \ldots, \lambda_n$ of a class of random $n \times n$ matrices. Let X_n be an $n \times n$ matrix having entries $n^{-1/2} Z_{jk}, 1 \leq j, k \leq n$, where the Z_{jk} comprise an array of i.i.d. zero-mean complex-valued random variables having unit variance. Define the empirical distribution of the eigenvalues $\lambda_1, \ldots, \lambda_n$ in $\mathbb{C}$ through the function (empirical measure) $m_n(A) = \frac{1}{n}\sum_{k=1}^{n} I_k, A \subseteq \mathbb{C}$, where I_k is the indicator of the event that $\lambda_k \in A$ (A being any suitable test set in $\mathbb{C}$). The circular law asserts that as $n \to \infty$ the empirical measure converges with probability 1 to the uniform measure on the unit disc. (Furthermore, the real eigenvalues converge to the uniform density on $[-1, 1]$.) Sufficient conditions for the law to hold are (for example) that the real and imaginary parts of Z_{jk} have finite 6th moment and a bounded joint density. [V. L. Girko, Circular law, *Theor. Probab. Appl.*, 29(4), 694–706, 1984]

circular normal distribution *see* von Mises distribution

circular triad Arises when preferences, rankings, or paired comparisons are not transitive; most commonly used in statistics. In the probabilistic context, one may have random variables X, Y, Z such that $\mathbb{P}(X > Y) > \frac{1}{2}$, $\mathbb{P}(Y > Z) > \frac{1}{2}, \mathbb{P}(Z > X) > \frac{1}{2}$, and one speaks of a non-transitivity paradox, as in e.g. Gale's roulette or Efron's dice. And one may, likewise, have circular tetrads, pentads, hexads, heptads, octads, enneads, dekads, etc., and, generally, polyads. [R. Savage, The paradox of nontransitive dice, *Amer. Math. Mon.*, 101(5), 429–456, 1994]

claims process *see* risk process and ruin problem

Clark–Ocone formula (alias: **Clark–Haussmann–Ocone formula**) Arises in stochastic analysis where, very broadly speaking, it supplies a form of the fundamental theorem of calculus in terms of appropriate stochastic integrals, with a suitable type of derivative. In its simplest form, developed by J. Clark, one considers a martingale $M(t)$ defined on the stochastic basis $\{\Omega, \mathcal{F}, \mathcal{F}_t, \mathbb{P}\}$ generated by the Wiener process $W(t), t \geq 0$; so that $M(t)$ is adapted to $\mathcal{F}_t$ (and also assumed square-integrable). Then the Itô calculus (Itô–Kunita–Watanabe theorem) asserts that there is an adapted process $Y(t)$ such that $M(t) = M(0) + \int_0^t Y(s)\mathrm{d}W(s)$. Then Clark's form of the theorem asserts that (with suitable conditions) $Y(t)$ can be explicitly represented as $Y(t) = \mathbb{E}(D_t M|\mathcal{F}_t)$, where D_t is a Fréchet derivative. U. Haussmann extended this formula to include functionals $M(t)$ of suitable Itô diffusions; and D. Ocone generalized the results in terms of the Malliavin derivative. The expression $\mathbb{E}(D_t X|\mathcal{F}_t)$ is referred to as the (optional or predictable), projection of the derivative (of whatever type) $D_t X$. The theorem has been further extended to e.g. fractional B.M. and Lévy processes. [J. Clark, *Ann. Math. Stat.*, 41, 1282–1295, 1970; U. Haussmann, *Stochastics*, 3, 17–27, 1979; D. Ocone, *Stochastics*, 12(3–4), 161–185, 1984]

class In the context of Markov chains, this term is almost universally used to denote the subsets of the state space that form equivalence classes of communicating states. Several properties of states of a Markov chain are called class properties, because they must be shared by all states in the class, or none. The proof of such a result may be called a class solidarity theorem. Besides the essential class property of communicating, other class properties include those of being essential, periodic, transient, recurrent, ergodic, null-recurrent, etc. In particular, a closed class C of a chain is such that the probability of ever making a transition out of C, into C^c, is zero. [J. Norris, *Markov Chains*. Cambridge University Press, 1997]

class (D) process Arises in the context of decomposition theorems of the Doob–Meyer type. Let $X(t), t \geq 0$, be defined on the stochastic basis $[\Omega, \mathcal{F}, \mathcal{F}_t, \mathbb{P}], t \geq 0$ (which satisfies the usual conditions), where $X(t)$ is $\mathcal{F}_t$-measurable and progressively measurable. (By convention, set $\mathcal{F}_\infty = \mathcal{F}$.) Then $X(t)$ is said to be of class (D) (or of Dirichlet class) if, for the family of stopping times T, the collection $X_T I(T < \infty)$ is uniformly integrable. Any uniformly integrable martingale is of class (D); the Doob–Meyer theorem asserts that any submartingale of class (D) is uniquely decomposed as the sum of a u.i. martingale and a predictable increasing process. Further, a process $X(t)$ is of class (DL) if for every $t > 0$, the family X_T, where T runs over stopping times less than t, is uniformly integrable. A local martingale is a martingale iff it is of class (DL). [D. Revuz, & M. Yor, *Continuous Martingales and Brownian Motion*, 3rd edn. Springer, 1999]

classical probability (alias: **a priori probability**) Denotes an interpretation of probability that allows numerical values of probabilities to be assigned to suitable events in advance of the results of any experiments; it can be seen as essentially a type of symmetry principle. In its basic form (implicit in the work of early probabilists, and made explicit by P. Laplace), it asserts that for an experiment where there is no reason to suppose any of its n outcomes to be more likely than another, an event A that occurs as a result of r of these outcomes has probability $\mathbb{P}(A) = r/n$. (*See* indifference principle.) This is relatively uncontentious in the context of flipped coins, and well-made dice or roulette wheels; and also fairly plausible in considering cards dealt from a well-shuffled pack. But the principle fails when extended to infinite sample spaces, or to experiments that do not have some intrinsic symmetry (other than symmetry of ignorance, which is insufficient); though something of the idea can be saved by invoking a maximum entropy principle, or an appropriate invariance principle. The term classical probability space is sometimes used to denote one having equally likely elementary events. [I. Hacking, *The Emergence of Probability*. Cambridge University Press, 1975; R. Carnap, *An Introduction to the Philosophy of Science*, M. Gardner (ed.). Dover, 1995]

classic coupling *see* Doeblin coupling

Clayton copula *see* copula

Clement matrix *see* Kac matrix

closed Jackson network *see* Gordon–Newell network

closed martingale, (alias: **complete, regular, Doob's, or Lévy martingale**) If $X_n, n \geq 0$, is a martingale (with natural filtration $\mathcal{F}_n$) and there exists a r.v. Z such that $\mathbb{E}|Z| < \infty$ and $X_n = \mathbb{E}(Z \mid \mathcal{F}_n)$, then the martingale is said to be closed (or closable) by Z; this is equivalent to the uniform integrability of $X_n, n \geq 0$, and also to the convergence in mean of X_n as $n \to \infty$. The definition of a u.i. martingale by $X_n = \mathbb{E}(Z \mid F_n)$ for some r.v. Z and a filtration $\mathcal{F}_n$ (on a suitable space) is called Doob's, Lévy's, or the **Doob–Lévy construction**. [F. Klebaner, *Introduction to Stochastic Calculus with Applications*, 3rd edn. Imperial College Press, 2012]

cluster process (alias: **clustering (point) process**) Broadly, cluster processes are defined on the basis of an underlying point process in a suitable space, which may be $\mathbb{R}$, or $\mathbb{R}^d$, or some more general space. To each such point, which is regarded as a cluster centre, there is associated a subsidiary point process. Usually such processes are independent and identically distributed about their given centre (seen as the origin), and the superposition of all such points is the cluster process, provided that it is finite on finite regions. The centres of the underlying process may or may not be included. The special case when the underlying process is a Poisson process is of the greatest importance, yielding Poisson cluster processes. Of these, two classic examples are the **Neyman–Scott process**, in which a random number of points are independently and identically distributed about the cluster centre; and the **Bartlett–Lewis process**, in which the points of the cluster are the steps of a finite random walk beginning at the cluster centre. A commonly seen special case of the Bartlett–Lewis process arises when the process is defined in $\mathbb{R}$, and the random walk is a terminating renewal process. [D. Daley & D. Vere-Jones, *Introduction to the Theory of Point Processes*, 2nd edn. Springer, Vol. I, 2003; Vol. II, 2008]

coalescence (alias: **agglomeration; aggregation; coagulation; flocculation; accretion; linking; gelation; etc.**) Stochastic coalescence models (or coalescents) seek to describe real-world phenomena arising in a wide range of physical situations, e.g. aerosols, snowflakes, polymers, astronomy, genetics, etc. In its simplest form, a coalescent is a random process (often having the Markov property) whose state space comprises all feasible groupings of a set M of basic entities, where M may be constant (in the conservative case), or it may grow or decrease. Groups of sizes x and y coalesce at random, at a rate proportional to a function $c(x, y)$ called the coalescent rate kernel. A well-known general model of this type is the Marcus–Lushnikov process; the choices $c(x, y) = 1$ and $c(x, y) = xy$ are popular. More generally, the groups may be moving in some space, and perhaps aggregating at some suitable point, e.g. or in diffusion limited aggregation, as preferential attachment models. More narrowly, the term coalescent often arises in the analysis of genetic variation, as exemplified by Kingman's coalescent; other coalescents include those of Bolthausen–Sznitman and Pitman–Sagitov. [D. Aldous, Deterministic and stochastic models for coalescence (aggregation and coagulation) a review of the mean-field theory for probabilists, *Bernoulli*, 5(1), 3–48, 1999]

coalescing random walk An interacting particle system in which, broadly, one envisages particles performing independent random walks in some state space, where any particles occupying the same location immediately coalesce and thenceforth move as one particle. A classic example arises on the square lattice of points with integer coordinates in $\mathbb{R}^d, d \geq 1$. Initially, each point supports a particle, which then perform independent simple symmetric random walks, coalescing when they meet. One may ask for the probability $V_d(t)$ that the origin (say) carries a particle at time $t > 0$. It has been shown that, as $t \to \infty$, asymptotically, $V_1(t) \sim (\pi t)^{-\frac{1}{2}}; V_2(t) \sim \frac{\log t}{\pi t}; V_d(t) \sim (tp_d)^{-1}, d \geq 3$, where p_d is the probability that the walk in d dimensions ever returns to its starting point. [T. Liggett, *Interacting Particle Systems*. Springer, 2005]

coalescent theory Arises in population genetics, where it seeks to describe the history of a random (genetic) sample of a population, instead of describing the evolution of the genetic characteristics of the entire population. See e.g. Kingman's coalescent. [J. F. C. Kingman, Origins of the coalescent: 1974–1982, *Genetics*, 156 (4), 1461–1463, 2000]

Cobham's formulas Supply the expected waiting time for customers of different types entering a non-preemptive priority scheduling scheme, in a queueing system in equilibrium, that may have one or more servers. For example, suppose there are m levels of priority, and customers of priority p arrive according to a Poisson process of rate λ_p (independent of other types), and have a service distribution $F_p(x)$ with mean $\mu^{-1}{}_p$. Then for the single server queue with non-preemptive priority service, the expected waiting time for a customer having priority level p, is given by

$$W_p = \frac{\frac{1}{2}\lambda \int_0^\infty x^2 \, \mathrm{d}F(x)}{[1 - \sum_{r=1}^{p-1} (\lambda_r/\mu_r)][1 - \sum_{r=1}^{p} (\lambda_r/\mu_r)]},$$

where $F(x) = (\sum_{p=1}^{m} \lambda_r)^{-1} \sum_{r=1}^{m} \lambda_r F_r(x)$ and $0 < \lambda_r < \mu_r < \infty$; $1 \leq r \leq m$. [F. Baccelli & P. Brémaud, *Elements of Queueing Theory*, 2nd edn. Springer, 2003]

Cochran's theorem alias Cochran–Fisher (or Fisher–Cochran) theorem Arises in considering the distribution of quadratic forms in a vector $\mathbf{X} = (X_1, \ldots, X_n)$ of i.i.d. normal r.vs. One form of the theorem is this: suppose that $\sum_{r=1}^{n} X_r^2 = \sum_{j=1}^{m} Q_j$, where $Q_j = \mathbf{X}^T A_j \mathbf{X}$, and the matrix A_j has rank r_j. Then these are equivalent:

(i) $r_1 + \cdots + r_m = n$
(ii) each Q_j has a chi-squared distribution
(iii) the Q_j are independent.

[W. Cochran, The distribution of quadratic forms in a normal system, *Math. Proc. Camb. Phil. Soc.*, 30(2), 178–191, 1934]

codifference Arises in seeking measures of dependence between random variables with infinite variance, particularly in the case of stationary time-series with infinite variance where it supplies a function that may replace the role of the autocovariance function. Formally, if X and Y are random variables, their codifference is $\log \mathbb{E}e^{i(X-Y)} - \log \mathbb{E}e^{iX} - \log \mathbb{E}e^{-iY}$. It finds particular application in stationary sequences of stable random variables. If X and Y have a standard bivariate normal distn, then the codifference is their correlation. [P. Kokoszka & M. Taqqu, Infinite variance stable ARMA processes, *J. Time Series Anal.*, 15(2), 203–220, 1994]

coefficient of coherence Arises in considering the association, or level of dependence, between two stationary time series, with respective spectral densities $f_1(\omega)$ and $f_2(\omega)$, where the cross-spectral densities are $f_{12}(\omega) = \overline{f_{21}(\omega)}$. The coefficient of coherence is $C(\omega)$ (or, sometimes, $C(\omega)^2$), where $C(\omega) = |f_{12}(\omega)| / (f_1(\omega) f_2(\omega))^{1/2}$. Essentially, it is the correlation coefficient of the (random) spectral representations of the two stationary processes. [L. Koopmans, *Spectral Analysis of Time Series*, 2nd edn. Academic Press, 1995]

coefficient of ergodicity (alias: **ergodic (or ergodicity) coefficient; contraction coefficient**) Arises in the context of the theory of Markov chains, notably in connexion with the question of the rate of convergence to equilibrium of an ergodic chain. The classic example is Dobrushin's ergodicity coefficient. In general, if P is an $n \times n$ stochastic matrix, and $\|\cdot\|$ is a norm on $\mathbb{R}^n$, then an ergodic coefficient $\tau(P)$ with respect that norm is $\tau(P) = \sup \|\mathbf{x}P\|$, where the supremum is taken over vectors $\mathbf{x}$ such that $\|\mathbf{x}\| = 1$, and $\mathbf{x1} = 0$. The function so defined satisfies:

1. $\tau(PQ) \leq \tau(P)\tau(Q)$, for stochastic matrices P and Q; which is submultiplicativity.
2. $\tau(P) = 0$ iff the rank of P is 1.
3. If $\lambda \neq 1$ is any eigenvalue of P, then $|\lambda| \leq \tau(P)$. In particular, if $\tau(P) < 1$, then P is said to be a scrambling matrix; if, for some $\lambda > 1, \tau(P^r) < 1$, then P is said to be weakly scrambling with scrambling integer (or index) r.

[E. Seneta, Coefficients of ergodicity: structure and applications, *Adv. Appl. Probab.*, 11(3), 576–590, 1979]

coefficient of probability A term used by J. W. Gibbs to denote the joint density f of the phase of a physical system in his theory of statistical mechanics; the natural logarithm of f was called the **index of probability** of the phase. Logicians and philosophers (and others) also use the term 'coefficient of probability' to mean simply probability. (Statisticians and economists sometimes use this term to denote other concepts.) [J. W. Gibbs, *Elementary Principles in Statistical Mechanics*, 1902, reprinted by Cambridge University Press, 2010]

coefficient of variation *see* variation coefficient

coherence (alias: **consistency**) Arises in seeking to develop the axioms of probability from a subjective (or Bayesian) interpretation, in which probability is seen as a degree of belief. Natural methods of measuring beliefs include ascertaining the lowest odds at which an individual will accept a bet, or determining a utility function by means of preferences. These methods were proposed by F. P. Ramsey, who also noted that the probability functions so obtained obey the usual laws of probability (including the additivity axiom) provided that those probabilities are required to exclude the possibility of a Dutch book against the individual. That is to say, there is no set of bets using the given probabilities such that the individual must lose whatever the outcome. This is called a requirement of rationality, and a set of probabilities (or beliefs, or preferences) having this property is said to be coherent (or, by Ramsey, **consistent**). [F. P. Ramsey, *The Foundations of Mathematics*. R & K.P., 1931; reprinted by Routledge, 2000]

coherent In reliability theory, the structure function $\phi(\mathbf{X}) \in \{0,1\}$ of a system of components with (possibly random) states $X_1, \ldots, X_n$ (where $X_r \in \{0,1\}$), is coherent if it is monotone (which is to say that $\phi(\mathbf{x}) \geq \phi(\mathbf{y})$ whenever $x_r \geq y_r, 1 \leq r \leq n$), with $\phi(\mathbf{0}) = 0$ and $\phi(\mathbf{1}) = 1$, and also has no irrelevant components (where the rth component is irrelevant if $\phi(\cdot)$ is a constant function of the state variable X_r). Some authors do not impose the final condition of all-relevant components for coherence. Informally, a monotone structure will not deteriorate if the performance of any component C is improved, and such improvement of C must improve the system in at least one configuration of states. [R. Barlow, F. Proschan, & L. Hunter, *Mathematical Theory of Reliability*. Wiley, 1965; reprinted by SIAM, 1996]

cointegration Arises in the theory of time series. A time series $X(n), n \in Z^+$, is said to be integrated of order d if, when it is differenced $d \geq 1$ times, the result is a stationary (and invertible) ARMA process. If $X(n)$ and $Y(n)$ are both integrated of order d, and for some a and b with $ab \neq 0$ the series $Z(n) = aX(n) + bY(n)$ is integrated of order $c < d$, then $X(n)$ and $Y(n)$ are said to be cointegrated. A classic toy example arises when $X(n)$ is a random walk, and $Y(n) = X(n) + W(n)$(with $W(n)$ a white noise process) is jittered random walk. (The pair has been labelled: 'A drunk and their dog'.) [R. Engle & C. Granger, Co-integration and error correction, *Econometrica*, 55 (2), 251–276, 1987]

collapsibility *see* lumpability

collar option Arises in financial mathematics, where it is a derivative option on an underlying asset $S(t), t \geq 0$, with pay-off $V_T = \min\{\max\{S(T), a\}, b\}$, for some $0 < a < b$. [A. Etheridge, *A Course in Financial Calculus*. Cambridge University Press, 2002]

collective (alias: **Von Mises collective**) A concept introduced by Richard Von Mises in an attempt to define probability. His definition was:

> A collective is a mass phenomenon or a repetitive event, or, simply a long sequence of observations for which there are sufficient reasons to believe that the relative frequency of the observed attribute would tend to a fixed limit if the observations were indefinitely continued. This limit will be called the probability of the attribute considered within the given collective.

This cumbersome (Von Mises' own word) description was adopted to avoid the more obvious failures in the frequently definition of probability. But the appearance of the phrase 'sufficient reasons to believe' attracts the conclusion that this is simply an over-complicated version of a subjective definition of probability. The original concept has been refined by later writers, but the objections remain in force. [R. Von Mises, *Wahrscheinlichkeit, Statistik und Warheit*, 1928 (in German), later English editions (*Probability, Statistics and Truth*) in 1939, 1951, 1957, and then 1981 by Dover]

collective marks (alias: **Van Dantzig's collective marks**) An artifice (or device) that may be used in the context of generating functions. For example, suppose that an urn U contains a random number X of balls, each of which is independently unmarked with probability u, or marked with probability $1 - u$. Then evaluating the pgf $G_X(s)$ of X at $s = u$ gives $G_X(u) = \mathbb{E}u^X = \mathbb{P}$ (all balls in U are unmarked). Now consider N independent and identically distributed urns like U, containing a total $B(N) = \sum_{r=1}^{N} X_r$ of balls. Noting that the contents of any urn are entirely unmarked with probability $G_X(u)$, by the above, and then using the same argument with X replaced by N, yields the identity $\mathbb{E}u^{B(N)} = G_{B(N)}(u) = G_N(G_X(u))$ immediately, without the usual calculation using conditioning or summation. By interpreting λ as the parameter of an independent exponential r.v. T, similar identities are likewise proved for the mgf

$$M_X(\lambda) = \mathbb{E}e^{-\lambda X} = \int_0^\infty e^{-\lambda x} dF_X(x).$$

The variable T is traditionally referred to as the **time of catastrophe**. It may be seen as an example of embedding. [D. Van Dantzig, Sur la méthode des fonctions génératrices, *Colloques Internationaux du CNRS*, 13, 29–46, 1949]

collective risk *see* Cramér–Lundberg model

collector problem (alias: **coupon collecting; cigarette card problem; dixie cup problem**) A classic problem of elementary probability: there are C different types of object (called coupons) of which you seek to collect a set. At each independent trial, in which you obtain something including a coupon, you are equally likely to find any one of the C types. Natural questions include the distribution and moments of the random number T of trials necessary to achieve the required set of all C types. One has $\mathbb{E}T = C\sum_{r=1}^{C} r^{-1}$; and, as $C \to \infty$, the distribution of $C^{-1}(T - C\log C)$ converges to the extreme value density $f(x) = e^{-x}\exp(-e^{-x}), x \in \mathbb{R}$. The problem is often reformulated in terms of an urn model; there are applications in ecology, error-detection and optimization problems. More generally, if the probability of obtaining a coupon of the nth type at any trial is p_n, then

$$\mathbb{E}T = \sum_{r=1}^{C} (-1)^{r+1} \sum_{n_1 < \cdots < n_r} (p_{n_1} + \cdots + p_{n_r})^{-1}$$
$$= \int_0^{\infty} 1 - \prod_{n=1}^{C} (1 - \exp(-p_n x))\mathrm{d}x.$$

[L. Holst, Birthday, collectors, occupancy, and other problems, *Int. Stat. Rev.*, 54(1), 15–27, 1986]

collision probability (alias: **matching probability**) Arises from the application of combinatorial probability to many real-world problems, e.g. height allocation in air traffic control, random choice of identifiers and network slots by packets in internet protocols, hashing, DNA mapping, lotteries. The applications explain the choice of vocabulary to denote the concept, which may be canonically formulated in terms of random variables or urns:

1 Suppose that m blue balls and n red balls are thrown independently and uniformly at random into u urns (or bins). Any urn containing balls of both colours indicates a collision, and the collision probability q is the chance of this occurring; $p = 1 - q$ is called the success probability in many applications. Where collision occurs with B blue balls and R red balls in the same urn, min $\{B, R\}$ may be called the matching number.
2 Alternatively, one envisages two random vectors $\{B_1, \ldots, B_m\}$ and $\{R_1, R_2, \ldots, R_n\}$ each term of which may take one of u values; a collision, or a match, occurs if at least one element of B and one of R takes the same value.

[S. Johnson & S. Kotz, *Urn Models*. Wiley, 1977]

colouring theorem *see* thinning; marking theorem

combinatorial central limit theorem (alias: **Hoeffding's CLT; Motoo's CLT**) Let $A_n = a_n(i,j)$, $1 \le i,j \le n < \infty$ be an $n \times n$ matrix of real numbers, for each n; and for all n let $\pi_n = (\pi_{(1)}, \ldots, \pi_{(n)})$ be a uniformly chosen random permutation of $(1, \ldots, n)$. Then a combinatorial central limit theorem supplies sufficient conditions for a suitably scaled and centred form of $\sum_{i=1}^{n} a_n(i, \pi_{(i)})$ to have an asymptotically normal distribution as $n \to \infty$. Such theorems were proved by W. Hoeffding and M. Motoo, generalizing an earlier result of A. Wald & J. Wolfowitz, and G. Noether. The corresponding Berry–Esseen type rate-of-convergence result is **Bolthausen's thm.** [E. Bolthausen, An estimate of the remainder in a combinatorial CLT, *Probab. Theor. Rel. Fields*, 66(3), 379–386, 1984; M. Motoo, On Hoeffding's combinatorial CLT, *Ann. Inst. Stat. Math.*, 8, 145–154, 1957]

combinatorial probability A term used to denote that part of probability in which problems are solved by enumeration, or by the other methods of combinatorial theory. It is not to be confused with probabilistic combinatorics, which means either (i) the study of random combinatorial objects (such as random graphs), or (ii) the use of random objects to solve combinatorial problems (as in P. Erdös's probabilistic method). [J. Steele, *Probability Theory and Combinatorial Optimization*. SIAM, 1997]

coming down from infinity A Markov process $X(t), t \ge 0$, in continuous time, is said to come down from infinity if $\lim_{t\downarrow 0} X(t) = \infty$ (with probability 1), but $\mathbb{P}(X(t) < \infty \text{ for } t > 0) = 1$. In discrete time, for $X_n \in \mathbb{Z}^+$, the corresponding property is that $\lim_{j\uparrow\infty} \mathbb{P}(X_n = k | X_0 = j)$ exists for all $k \ge 0$ and $n > 0$. The concept arises naturally in seeking non-trivial entrance distributions for Markov processes (i.e. not the stationary distn), and it has applications to coalescents. [J. T. Cox, Entrance Laws for Markov Chains, *Ann. Probab.*,

5(3), 533–549, 1977; J. Berestycki *et al.*, The Λ-coalescent speed of coming down from infinity. *Ann. Probab.*, 38(1), 207–233, 2010]

communication channel *see* channel

communication theory (alias: **information theory**) In mathematical and probabilistic contexts, this usually denotes C. Shannon's theory of communication, and its extensive development since, with applications to data-compression, cryptography, and the reliable transmission of messages in many contexts.

Shannon's original model envisages a source or a sender A (who selects a message from a set of possible messages), and a channel (which may be very generally interpreted) that conveys a possibly corrupted form of the message to a receiver B. The participants A and B may wish to communicate as swiftly or accurately or as privately as possible, giving rise to various distinct branches of the theory. A central concept is that of the information content of the message, which is determined by the amount of freedom A has to choose it, and related to the reduction in B's uncertainty about the message when the channel delivers its output. The messages themselves may be e.g. random variables taking values in some alphabet, quantum states, suitable computer programs, etc. Other important concepts include channel capacity and entropy. [C. Shannon, A mathematical theory of communication, *Bell Syst. Tech. J.*, 27, 379–423 and 623–656, 1948]

commute time (alias: **commuting time**) Let j and k be two states of an irreducible Markov chain on a countable state space. For a chain started in j, the commute time between j and k, denoted by $C(j,k)$, is the number of steps until the first occasion at which the chain revisits j having made a visit to k at some intervening time. Denoting the first-passage time from j to k by T_{jk}, it follows by construction (and the Markov property) that $C(j,k) = T_{jk} + T_{kj} = C(k,j)$. Further, for any three states i, j, and k, $C(i,k) \leq C(i,j) + C(j,k)$. Note that $D(j,k) = T_{jk} - T_{kj}$ has been called the **difference time**. [D. Levin, Y. Peres, & E. Wilmer, *Markov Chains*. American Mathematical Society, 2009]

comonotonic (alias: **comonotone**) Random variables $X(\omega)$ and $Y(\omega)$, defined on the same probability space, are said to be comonotonic if for all $\omega_1, \omega_2 \in A$, $[X(\omega_1) - X(\omega_2)][Y(\omega_1) - Y(\omega_2)] \geq 0$, where $\mathbb{P}(A) = 1$. This holds iff the joint distribution $F_{X,Y}(x,y)$ has the property $F_{X,Y}(x,y) = \min\{F_X(x), F_Y(y)\}$. Furthermore, X and Y are comonotonic iff there is a random variable Z and two non-decreasing functions $u(\cdot)$ and $v(\cdot)$ such that (X,Y) has the same distribution as $(u(Z), v(Z))$. Other characterizations exist, and more generally, a probability measure on $\mathbb{R}^n$ (or r.vr) is comonotonic iff its support is a.s. a totally ordered set.

Random process $X(t)$ and $Y(t), t \geq 0$, defined on the same (filtered) probability space are comonotonic if for all t the random variables $X(t)$ and $Y(t)$ are comonotonic. [G. Deelstra *et al.*, An overview of comonotonicity and its applications in finance and insurance, in *Advanced Mathematical Methods for Finance*, J. Di Nunno & B. Øksendal (eds). Springer, 2011, pp. 155–180]

compensator (alias: **dual-predictable projection**) Arises in the context of martingale theory. Let $X_n, n \geq 0$, be a random process adapted to the filtration $\mathcal{F}_n$. If it is possible to write $X_n = M_n + A_n$, where M_n is a martingale for $\mathcal{F}_n$, and A_n is an increasing predictable (i.e. $\mathcal{F}_n$ measurable, or adapted) process, then $A_n, n \geq 0$ is called a compensator (or 'the', if it is unique a.s.), or a compensating process. A similar definition holds in continuous time; thus e.g.:

1. Let $W(t)$ be the Wiener process; then $A(t) = t$ is the compensator of $W^2(t)$, as $W^2(t) - t$ is a martingale.
2. Let $N(t)$ be the Poisson process; then λt is the compensator both of $N(t)$ and of $(N(t) - \lambda t)^2$. The fact that a unique compensator exists for any submartingale is Doob's theorem; the same result in continuous time is the Doob–Meyer theorem. The idea of compensation extends to e.g. semi-martingales and random measure processes.

[M. Jeanblanc *et al.*, *Mathematical Methods for Financial Markets*. Springer, 2009]

competition theorem (or lemma) A property of independent exponentially distributed r.vs that is of importance for Markov processes in continuous time. Let $X_k, k \geq 1$ be independent r.vs, where X_k is exponential with parameter λ_k, and $0 < \lambda = \sum_{r=1}^{\infty} \lambda_r < \infty$. Define $Z = \inf_i X_i$. Then Z is attained by a unique member X_M of the sequence $X_1, X_2, \ldots$, with probability 1. Furthermore, Z is exponential with parameter λ, and $\mathbb{P}(M = m) = \lambda_m / \lambda$, and Z and M are independent. [P. Brémaud, *Markov Chains*. Springer, 1999]

complementary event If A is an event in the sample space Ω, then $\Omega \backslash A = A^c$ is the complementary event, and $\mathbb{P}(A^c) = 1 - \mathbb{P}(A)$. There are several equivalent notations for complement, e.g. $A'; \bar{A}; \sim A; \Omega - A; cA; \Omega \backslash A$.

complete class Arises in decision theory, as initiated by A. Wald, in which the risk function $r(\theta, d(x))$ of a decision rule $d(x)$, based on the realization x of a random variable X, is the expected loss incurred by the rule $d(X)$ when the true state of nature is θ. A rule $b(x)$ is r-better than $d(x)$ if $r(\theta, b) \leq r(\theta, d)$ for all θ, with strict inequality for at least one θ; rules b and d are equivalent if $r(\theta, b) = r(\theta, d)$ for all θ. If there is no r-better rule than d, then d is **admissible**; otherwise d is inadmissible.

A class C of decision rules is then complete if, given any rule d not in C, there is a rule $b \in C$ that is r-better than d. If, given $d \notin C$, there is a rule $b \in C$ that is equivalent to, or r-better than d, then C is essentially complete. If no proper subset of C is complete, then C is minimal complete. [J. Berger, *Statistical Decision Theory and Bayesian Analysis*, 2nd edn. Springer, 1985]

complete convergence

1. A sequence $X_n, n \geq 1$, of random variables is said to converge completely to X if, for all $\varepsilon > 0$, $\sum_{r=1}^{\infty} \mathbb{P}(|X_r - X| > \varepsilon) < \infty$. This implies the a.s. convergence of X_n to X, as $n \to \infty$, and the converse is true if the X_n are independent.
2. A sequence $F_n(x), n \geq 1$, of distribution functions is said to converge completely to $F(x)$ if, as $n \to \infty, F_n(x)$ converges to $F(x)$ at all continuity points of $F(x)$ (i.e. $F_n(x)$ converges weakly), and $F(x)$ is a proper distn fn. This is equivalent to convergence in the Lévy metric for distns. (Note that some authors call this weak convergence, while referring to weak convergence as vague convergence.)

[P. Hsu and H. Robbins, Complete convergence and the law of large numbers, *Proc. Nat. Acad. Sci.*, 33(2), 25–31, 1947; P. Moran, *Probability Theory*. Oxford University Press, 1968]

complete filtration A filtration $\mathcal{F}_t$, defined on some basis $\{\Omega, \mathcal{F}, \mathbb{P}\}$ is said to be complete if $\mathcal{F}$ itself is complete (which is to say that it contains all the subsets of sets A such that $\mathbb{P}(A) = 0$, i.e. $\mathbb{P}$-null sets), and each $\mathcal{F}_t$ includes all the $\mathbb{P}$-null sets of $\mathcal{F}$. This process of inclusion may be called completion. [O. Kallenberg, *Foundations of Modern Probability*, 2nd edn. Springer, 2002]

complete intensity function A random process that arises naturally in the context of a point process having counting function $N(s, t)$ that is finite and orderly, i.e. $N(s, t)$ is the number of points of the process in $[s, t]$, where $\mathbb{E}N(s, t) < \infty$, and $\mathbb{P}(N(t, t+h) > 1) = o(h)$. Then the complete intensity fn $\lambda(t|H(t))$ is given by $\lambda(t|H(t)) = \lim_{h \downarrow 0+} h^{-1}\mathbb{P}(N(t, t+h) > 0|H(t))$, where $H(t)$ is the history of the process up to time t. If the process is confined to $\mathbb{R}^+$, then the analogous function $\lambda^+(t|H(t))$ may be called the **conditional intensity fn**. [D. R. Cox & V. Isham, *Point Processes*. Chapman & Hall, 1980]

completely monotone

1. Let $m_k, k \geq 0$, be real numbers such that for $r \geq 0$,

$$\sum_{i=0}^{r} (-1)^i \binom{r}{i} m_{k+i} \geq 0.$$

Then $m_k, k \geq 0$, is said to be a completely monotone sequence. A theorem of F. Hausdorff asserts that this is so (when $m_0 = 1$), iff they are the moments of a probability distribution.

2. A function $m(t)$ on $[0, \infty)$ is completely monotone if $(-1)^r(\mathrm{d}/\mathrm{d}t)^r m(t) \geq 0$ for all $r \geq 0$. When $m(0) = 1$, a theorem of S. Bernstein asserts that this is so iff $m(t)$ is the mgf (i.e. $m(t) = \mathbb{E}e^{-tX}$), of a random variable X.

[W. Feller, *Introduction to Probability Theory and its Applications*, Vol. 2, 2nd edn. Wiley, 1971]

completely random measure Arises when one seeks to extend the idea and property of independent increments (as enjoyed by the Poisson process $N(t), t \in \mathbb{R}$), to processes defined on more general spaces, for example $\mathbb{R}^d, d > 1$. In general, if S is some suitable set (with a σ-field $\mathcal{F}$ of subsets of S), and $X(\cdot)$ is a random measure on S (i.e. a random variable taking values in a set of σ-additive functions from $\mathcal{F}$ to $[0, \infty]$), then X is a completely random measure if $X(A_1), \ldots, X(A_n)$ are independent whenever $A_1, \ldots, A_n$ are disjoint members of $\mathcal{F}$. More broadly, if X is some random object (or element) having this property, one may speak of a completely random process. [J. F. C. Kingman, Completely random measures, *Pacific J. Math.*, 21(1), 59–78, 1967]

complete market Arises in mathematical finance, where a market is a vector of asset prices over time, $\boldsymbol{S}(t)$, jointly distributed according to a probability measure $\mathbb{P}$, and generating a natural filtration $\mathcal{F}_t$. If every t-option (i.e. $\mathcal{F}_t$-measurable claim, or derivative, based on the underlying asset process $\boldsymbol{S}$), can be hedged (or replicated, or duplicated, or reached, or attained) by a self-financing portfolio of assets (with probability 1), then the market is said to be complete. The **completeness thm** asserts that an arbitrage-free market is complete iff there is a measure $\mathbb{Q}$, equivalent to $\mathbb{P}$, under which the asset prices in $\boldsymbol{S}$ are martingales; this holds generally in discrete time, and with technical conditions in continuous time. [N. Bingham & R. Kiesel, *Risk-neutral Valuation*, 2nd edn. Springer, 2004]

completeness Let the family of random variables X_θ have distributions $F(x\,|\,\theta), \theta \in \Theta$. If the condition $\mathbb{E}g(X_\theta) = 0$ for all θ, entails that $g(x) = 0$ (except possibly on a null event), then the family of distributions $F(x\,|\,\theta)$ is said to be complete. [E. Lehmann & J. Romano, *Testing Statistical Hypotheses*, 3rd edn. Springer, 2005]

complete probability space (alias: **completion of a space**) A probability space (Ω, $\mathcal{F}$, $\mathbb{P}$) is called complete if all subsets of null events are also events. (A null event A has $\mathbb{P}(A) = 0$.) This nomenclature is inherited from measure theory, in which a complete measure has the property that all subsets of any null set are measurable. For example, on $\mathbb{R}$ Lebesgue measure is complete, whereas Borel measure is not. A non-decreasing family $\mathcal{F}_0, \mathcal{F}_1, \ldots$ of sub-sigma-fields of $\mathcal{F}$ is called complete if $\mathcal{F}_0$ contains all the $\mathbb{P}$-null events in $\mathcal{F}$. [P. Billingsley, *Probability and Measure*, 3rd edn. Wiley, 1995]

component Broadly, a subsidiary part of anything; most commonly it means one or more of the variables of a random vector. In particular, in the context of reliability theory it refers to any constituent part of the system, and hence also (by extension) to the random variable that describes the state of the component. Thus one may speak of a binary (or Boolean) system with components $(X_1, \ldots, X_n)$, where each X_r is an indicator of the events working/not working. More generally, the components X_r may supply a more detailed description of functionality. [R. Barlow, F. Proschan, & L. Hunter, *Mathematical Theory of Reliability*. Wiley, 1965; reprinted by SIAM, 1996]

composition method (alias: **mixture method**) A procedure for sampling from a density *f*(*x*) that can be expressed in the form $f(x) = \sum_r p_r f_r(x)$, where p_r is the probability distribution of an integer-valued r.v. *N*, and each $f_r(x)$ is a density. One proceeds by first simulating *N*, and then sampling from $f_N(x)$. If at least one of the $f_r(x)$ is sampled using acceptance-rejection, then one speaks of the MAR (or CAR) method. [R. Rubinstein & D. Kroese, *Simulation and the Monte Carlo Method*, 2nd edn. Wiley, 2008]

compounding (alias: **generalized distributions; contagion; composition; contamination**) Broadly, a random variable (or process) is said to have a compound (or generalized) distn when it is obtained from a family of r.vs (or processes) depending on a parameter $\lambda \in S$, by assigning a probability distribution to *S*, i.e. randomizing the parameter λ as Λ. The variable λ may be a natural parameter such as the mean or variance; or it may be a simple index; or it may be time. The idea is readily defined by examples:

(a) *Compounding with respect to a parameter*. Let B (n, p) be the Binomial distribution, and suppose that the parameter p has the beta $\mathrm{B}(\alpha, \beta)$ distribution. This yields the so-called beta-binomial (or Pólya) distribution (also called the negative hypergeometric distribution when α and β are integers). For another example, suppose that the parameter λ in a Poisson distribution has a gamma density with parameters μ and m. Then the resulting compound distribution is negative binomial.

(b) *Compounding with respect to an index*. Let $X_r, r \in \mathbb{N}$ be a collection of random variables and N a non-negative integer-valued r.v. Then X_N is a compound r.v.; when N is independent of the X_r, it is often called a **mixture**, or a **contaminated distn**.

(a) *Compounding with respect to time*. If S_n is a random process, with N as above, then S_N is a compounded process, that may be seen as the sequence $S_1, S_2, \ldots$ stopped (or sampled) at time $n = N$. The case when N is a stopping time for the sequence is important. A classic example is when $S_n = \sum_{r=1}^{n} X_r$ arises as a partial sum; *see* Wald's identity.

[P. Moran, *Introduction to Probability Theory*. Oxford University Press, 1968; S. Ross, *Introduction to Probability Models*, 11th edn. Academic Press, 2014]

compound option In financial mathematics, an option contract written in terms of an asset (or assets), that is itself an option. [Y.-K. Kwok, *Mathematical Models of Financial Derivatives*, 2nd edn. Springer, 2008]

compound Poisson (alias: **generalized Poisson; composed Poisson**) Arises in the context of Poisson processes and counts. Let $Y_1, Y_2, \ldots$ be i.i.d.r.vs, with chf ϕ_Y, and N an independent Poisson random variable with parameter λ. Then the r.v. $C = \sum_{r=1}^{N} Y_r$, with chf $\phi(s) = \exp\{\lambda(\phi_Y(s) - 1)\}$, is compound Poisson. If $N(t)$ is a Poisson process, then $C(t) = \sum_{r=1}^{N(t)} Y_r$ is a compound Poisson process. The only infinitely divisible distributions on the non-negative integers are compound Poisson in which the summands Y_r are i.i.d. integer-valued r.vs (that are non-negative).

The compound Poisson r.v. is not to be confused with the mixed Poisson r.v. having distribution $\mathbb{P}(M = n) = \int_0^\infty \frac{\lambda^n e^{-\lambda}}{n!} \mathrm{d}F(\lambda)$, for some distn fn F, i.e. the Poisson distn compounded with respect to its parameter. The term doubly compound Poisson process is sometimes used to denote a compound Cox process.

[T. Rolski *et al.*, *Stochastic Processes for Insurance and Finance*. Wiley, 1999]

concentration Broadly, the extent to which a probability measure is a apportioned principally among its most likely values, or allocated near them or some other values, such as the mean (when the measure is on a metric space). There are several coefficients, indices, inequalities, and so forth, that quantify the degree of concentration of a distribution, reflecting the importance of the concept. The elementary classic example of such a measure is the variance σ^2, where σ is the standard deviation, which is likewise such a measure and may be called a concentration parameter. More importantly, a concentration inequality supplies a bound for the deviation of a r.v. from its mean or median (say).

These coefficients of concentration work for measures on a metric space, by considering the spread of the measure, i.e. by how far the r.v. deviates from its mean (say). More generally, for probability measure on a measure-space, coefficients of concentration work by considering how dispersed the measure is (and may be called diversity coefficients). Classic examples, for a discrete distribution $p_r, r \geq 1$, include the Shannon entropy $H(p) = -\sum_r p_r \log p_r$, and the Gini–Simpson index $1 - \sum_r p_r^2$, where $\sum_r p_r^2$ is the quadratic concentration or coincidence index.

(It would have been useful if the distinction between deviation (from the mean, say) and dispersion (among the elements of the space) had been preserved in the literature, but it has not.) [M. Ledoux, *The Concentration of Measure Phenomenon*. American Mathematical Society, 2001; S. Boucheron *et al.*, *Concentration Inequalities*. Oxford University Press, 2013]

concentration ellipse *see* ellipse of concentration

concentration function (alias: **Lévy concentration function; maximal concentration function**) Informally, this is a measure of the spread of the distribution of a random variable X about its domain in $\mathbb{R}$. Formally, for any $\varepsilon > 0$, the Lévy (or ε-) concentration $L_\varepsilon(X)$ of X is given by $L_{2\varepsilon}(X) = \sup_{x\in\mathbb{R}} \mathbb{P}(|X - x| \leq \varepsilon)$. It is decreased by convolution, so that for independent X and $Y, L_\varepsilon(X + Y) \leq \min[L_\varepsilon(X), L_\varepsilon(Y)]$. For example, if X is normal, $N(0, 1)$, then as $\varepsilon \to 0$ $L_\varepsilon(X) = \sqrt{\frac{2}{\pi}}\varepsilon(1 + o(1))$. If X has variance $\sigma^2 < \infty$, then by Chebyshov's inequality $L_\varepsilon(X) \geq 1 - 4\sigma^2/\varepsilon^2$.

More precisely, Lévy's concentration-variance inequality is $L_\varepsilon(X) \geq \frac{3}{2n+1}\left(1 - \frac{4}{n(n+1)}\frac{\sigma^2}{\varepsilon^2}\right)$, where $n > 0$ is the integer such that $\varepsilon^2(n^2 - 1) < 12\sigma^2 \leq \varepsilon^2(n^2 + 2n)$. There are several other concentration functions, e.g. Kawata's $M(X) = \frac{1}{2\varepsilon}\int_{-\infty}^{\infty} \{\mathbb{P}(|X - x| \leq \varepsilon)\}^2 dx$, which is related to the quadratic entropy; and Kunisawa's $K(X) = \varepsilon\int_0^\infty e^{-\varepsilon t}|\phi(t)|^2 dt$, where $\phi(t)$ is the characteristic function of X. [P. Lévy, *Théorie de l'Addition des Variables Aléatoires*, 2nd edn. Gauthier-Villars, 1954; W. Hengartner & R. Theodorescu, *Concentration Functions*. Academic Press, 1973]

concentration inequality (alias: **deviation inequality; tail inequality**) Broadly, this supplies a bound on the probability that a random variable X differs (or deviates) by more than $t > 0$ from some value m; typically m is the mean or median of X. That is, a function $b(t)$ such that for $t > 0, \mathbb{P}(|X - m| \geq t) \leq b(t)$. One may also speak of a two-sided tail inequality, or bound. Note that such bounds are customarily required to hold for all t, in contrast to large deviation inequalities that hold asymptotically for large enough t. Classic examples include Chebyshov's inequality, where $b(t) = \text{var}(X)/t^2$; and stronger results such as those of S. Bernstein, H. Chernoff, and W. Hoeffding yield exponential bounds, e.g. when X is binomial $\mathrm{B}(n, p)$ then $b(t) = 2\exp(-2nt^2)$. Typically, such inequalities are sought for functions of random vectors, and especially for sums of independent random variables; one has the inequalities of Efron–Stein, C. McDiarmid, and M. Talagrand. [S. Boucheron *et al.*, *Concentration Inequalities; a Non-asymptotic Theory of Independence*. Oxford University Press, 2013; N. Berestycki & R. Nickl, *Concentration of Measure*, www.statslab.cam.ac.uk/~beresty/teach/cm10.pdf]

concomitant (alias: **induced order statistic**) Let $(X_r, Y_r), 1 \leq r \leq n$, be a bivariate random sample from the joint distn $F(x, y)$, and denote the order statistics of the X_s by $X_{(i,n)}$. Then the second component of the vector whose first component is $X_{(i,n)}$ is $Y_{[i,n]}$, called the concomitant of the ith-order statistic of the X_s. [H. David & H. Nagaraja, *Order Statistics*, 3rd edn. Wiley, 2003]

concordance function Let (X, Y) and (U, V) be independent vectors, having respective joint densities $f(x, y)$ and $g(u, v)$, such that $X \overset{d}{=} U$, and $Y \overset{d}{=} V$, i.e. they have common marginal distns. They are said to be concordant if $(X - U)(Y - V) > 0$, and discordant otherwise. The concordance function (or coefficient) is $Q = \mathbb{P}((X - U)\ (Y - V) > 0) - \mathbb{P}((X - U)$ $(Y - V) < 0)$. See also *dependence*. [W. Hoeffding, On the distribution of the rank correlation coefficient when the variates are not independent, *Biometrika*, 34(314), 183–196, 1947]

concordance order A type of stochastic ordering. Let $F(x,y)$ and $G(u,v)$ be the joint distribution functions of the random vectors (X,Y) and (U,V) respectively, where $X \stackrel{d}{=} U$, and $Y \stackrel{d}{=} V$, i.e. they have common fixed marginals. If, for all (s,t), we have $F(s,t) \le G(s,t)$, then (X,Y) is said to be less concordant than (U,V). One writes $(X,Y) \le_{conc} (U,V)$. It is equivalent to the following statement: For all increasing functions $h(\cdot)$ and $k(\cdot)$ such that the covariances exist, $\mathrm{cov}(h(X),k(Y)) \le \mathrm{cov}(h(U),k(V))$. It follows that the order is invariant under monotone transformations of the arguments, and can hence be expressed in terms of their copulas. Thus, copulas $C(s,t)$ and $K(s,t)$ are in concordance order $C < K$ if $C(s,t) \le K(s,t)$ for all $(s,t) \in (0,1)^2$. [R. Nelsen, *Introduction to Copulas*, 2nd edn. Springer, 2006]

conditional density The density of a suitable conditional distribution. In its typical elementary form, it arises when the r.vs X and Y have joint density $f(x,y)$, in which case the conditional density of Y given that $X = x$ is given by $f_{Y|X}(y \mid X = x) = f(y \mid x) = f(x,y)/f_X(x)$, where $f_X(x) > 0$ is the marginal density of X. Conversely, if the conditional density of Y is well-defined as a function of X with X having density $f_X(x)$, then $f(x,y)$ is defined as $f(y \mid x) f_X(x)$.

More generally, suppose Y is a random variable on the space $(\Omega, \mathcal{F}, \mathbb{P})$, and $\mathcal{G}$ is a sub-σ-field of $\mathcal{F}$. If the conditional probability distribution function of Y given $\mathcal{G}$ (or with respect to $\mathcal{G}$), denoted by $F(y \mid \mathcal{G})$, can be written as $F(y \mid \mathcal{G}) = \int_{-\infty}^{y} f(t \mid \mathcal{G})\mathrm{d}t$ for some function f, then f is the conditional density of Y given $\mathcal{G}$. [W. Feller, *Introduction to Probability Theory and its Applications*, Vol. 2, 2nd edn. Wiley, 1971]

conditional determinism *see* pull-through property

conditional distribution Commonly denotes the probability distribution of some r.v. Y, given the value of another r.v. X that is jointly distributed with Y; it can be seen as a special case of conditional probability. Thus when X is discrete the conditional distribution of Y given X is given by $\mathbb{P}(Y \le y \mid X = x) = \mathbb{P}(\{Y \le y\} \cap \{X = x\}) / \mathbb{P}(X = x)$, whenever $\mathbb{P}(X = x) \ne 0$. This yields a conditional density if differentiable in y. When X takes uncountably many values (e.g. if X is continuous) the elementary definition fails. In one resolution of the difficulty, for X and Y defined on the probability space $(\Omega, \mathcal{F}, \mathbb{P})$, the conditional distribution of Y given X is any function $\mathbb{P}(Y \le y \mid X(\omega))$ on Ω that is measurable wrt the σ-field generated by X, and such that for all $y \in \mathbb{R}$, $\mathbb{P}(Y \le y \mid X \in B) = \int_{\omega: X \in B} \mathbb{P}(Y \le y \mid X(\omega)\mathrm{d}\mathbb{P}(\omega)$, for any Borel set $B \in \mathbb{R}$. Any two members of this class of functions take the same values a.s. Alternatively, the conditional distribution may be defined as a suitable regular conditional probability function. More generally, one likewise defines the conditional distribution of a r.vr given another. [P. Billingsley, *Probability and Measure*, 3rd edn. Wiley, 1995; L. Breiman, *Probability*. Addison-Wesley, 1968; reprinted by SIAM, 1992]

conditional entropy A concept arising naturally in C. Shannon's theory of communication, where it quantifies the expected remaining uncertainty about the value of an unknown r.v. Y (the input to a noisy channel), when we learn the value of X (the output of the channel). Formally, if X and Y are simple random variables, with joint probability distn $f(x,y); x \in A, y \in B$, and conditional distn $f(y \mid x)$ of Y given X, then the conditional entropy of Y given X is denoted by $H(Y \mid X)$, where

$$\begin{aligned} 0 \le H(Y \mid X) &= -\sum_{x,y} f(x,y) \log_2 f(y \mid x) \\ &= H(X,Y) - H(X) \le H(Y). \end{aligned}$$

Here $H(X), H(Y)$, and $H(X,Y)$ are the entropies of X, Y, and (X,Y), respectively. See also equivocation. [C. Shannon, A mathematical theory of communication, *Bell Syst. Tech. J.*, 27, 379–423 and 623–656, 1948]

conditional expectation Arises naturally in the context of jointly distributed random variables X and Y, where one seeks the expected value of Y conditional on X, i.e. given that $X = x$. When X and Y are discrete, the definition of conditional probability yields

$$\mathbb{E}(Y \mid X = x) = \sum_{y} y \mathbb{P}(Y = y, X = x)/\mathbb{P}(X = x).$$

When X and Y are jointly continuous, the defn of conditional density yields

$$\mathbb{E}(Y \mid X = x) = \int y f(x,y)/f_X(x)\mathrm{d}y.$$

When only X is discrete, denoting the probability space on which X and Y are defined by $(\Omega, \mathcal{F}, \mathbb{P})$, we have $\mathbb{P}(Y \in A \mid X = x) = \mathbb{P}(Y \in A, X = x)/\mathbb{P}(X = x)$; it follows from the formal definition of expectation in general that in this case (when the expectation exists),

$$\mathbb{E}(Y|X = x) = \int_{\omega:X=x} Y(\omega)\mathbb{P}(d\omega)/\mathbb{P}(X = x).$$

Allowing x to run over all possible values of X yields the conditional expectation of Y given X, which is a random variable (since it depends on X) denoted by $\mathbb{E}(Y \mid X)$. It has the property that for all functions $g(X)$ such that the expectations exist

$$\mathbb{E}(Yg(X)) = \mathbb{E}(g(X)\mathbb{E}(Y \mid X)).$$

When X is not discrete, these elementary arguments fail, since $\mathbb{P}(X = x) = 0$ for at least one x in the range of X. The difficulty is resolved by defining $\mathbb{E}(Y \mid X)$ through the two key properties above: the conditional expectation of Y given X is a function $\Psi(X)$ of X such that for any Borel set B,

$$\mathbb{E}(YI(X \in B)) = \mathbb{E}(I(X \in B)\Psi(X)),$$

where $I(X \in B)$ is the indicator of the event that X lies in an arbitrary Borel set B of $\mathbb{R}$. The resulting function exists and is almost surely unique (i.e. there is an equivalence class of such functions that differ on events of probability zero).

More formally and generally, one likewise defines the conditional expectation of Y with respect to a sub σ-field $\mathcal{G}$ of $\mathcal{F}$; $\Psi(X)$ is recovered by letting $\mathcal{G}$ be the σ-field generated by X. [L. Breiman, *Probability*. Addison-Wesley, 1968; reprinted by SIAM, 1992]

conditional independence If events A, B, and C are such that $\mathbb{P}(A \cap B \mid C) = \mathbb{P}(A \mid C)\mathbb{P}(B \mid C)$, then A and B are said to be conditionally independent given C. This does not imply the independence of A and B (unless $C = \Omega$), nor does the independence of A and B imply their conditional independence, in general. Random variables X and Y are conditionally independent given C if, for any intervals A and B in $\mathbb{R}$,

$$\mathbb{P}(X \in A, Y \in B \mid C) = \mathbb{P}(X \in A \mid C)\mathbb{P}(Y \in B \mid C),$$

where C is an event in the space on which X and Y are defined. If $C = \{W \in \mathrm{I}\}$, where W is a random variable on the same space, and I an arbitrary interval of $\mathbb{R}$, then X and Y are conditionally independent given W. The conditional independence of X and Y given a sub-σ-field $\mathcal{G}$ is defined similarly. [K. L. Chung, *Course in Probability Theory*, 2nd edn revised. Academic Press, 2001]

conditional Poisson process *see* mixed Poisson process

conditional probability Arises in considering the probability of an event A, given the further condition that same event B, with probability $\mathbb{P}(B) > 0$, is known, or assumed, to occur. Then the so-called conditional probability of A given B is defined to be $\mathbb{P}(A \mid B) = \mathbb{P}(A \cap B)/\mathbb{P}(B)$. It follows that for a countable family of disjoint events $B_i, i \in I$, whose union is B, $\mathbb{P}(A \cap B) = \sum_{i\in I}\mathbb{P}(A \mid B_i)\mathbb{P}(B_i)$. The concept of conditioning is often applied in the context of jointly distributed discrete r.vs X and Y, so that $\mathbb{P}(Y = y \mid X = x) = \mathbb{P}(X = x, Y = y)/\mathbb{P}(X = x)$ whenever $\mathbb{P}(X = x) \neq 0$. In practice, given a probability space $(\Omega, \mathcal{F}, \mathbb{P})$, it is often natural and desirable to require the conditional probability of an event A given an outcome in Ω whose probability is zero; it is therefore necessary to consider the conditional probability measure $\mathbb{P}(A \mid \mathcal{G})$ where $\mathcal{G} \subseteq \mathcal{F}$ is the appropriate sub-σ-field of $\mathcal{F}$.

Then the conditional probability wrt the sub-σ-field $\mathcal{G}$ of $\mathcal{F}$ is defined as a function $\mathbb{P}(A \mid \mathcal{G})$ on Ω that is measurable wrt $\mathcal{G}$ (i.e. is a r.v. on $(\Omega, \mathcal{G})$), and such that for any $B \in \mathcal{G}$

$$\mathbb{P}(A \cap B) = \int_{\omega\in B} \mathbb{P}(A|\mathcal{G})d\mathbb{P}(\omega).$$

Any two such functions differ only on a null event. Alternatively, this conditional probability may be defined in terms of the equivalent conditional expectation given $\mathcal{G}, \mathbb{E}(X \mid \mathcal{G})$, by setting, $\mathbb{P}(A \mid \mathcal{G}) = \mathbb{E}(I(A) \mid \mathcal{G})$, where $I(A)$ is the usual indicator.

In seeking a converse definition of conditional expectation in terms of conditional probability, one requires that there is a version of $\mathbb{P}(A \mid \mathcal{G})$ satisfying the extra condition that for every fixed ω, $\mathbb{P}(\cdot \mid \omega)$ is a probability measure on $\mathcal{F}$; in this case it is said to be a regular conditional probability, and it is the case that with probability 1,

$$\mathbb{E}(X \mid \mathcal{G}) = \int X(\omega)\mathbb{P}(d\omega \mid \mathcal{G}).$$

[T. Seidenfeld *et al.*, Improper regular conditional distributions, *Ann. Probab.*, 29(4), 1612–1624, 2001; P. Billingsley, *Probability and Measure*, 3rd edn. Wiley, 1995; L. Breiman, *Probability*. Addison-Wesley, 1968; reprinted by SIAM, 1992]

conditional variance theorem (alias: **total variance formula; variance decomposition**) For any finite-variance, jointly distributed r.vs X and Z, $\operatorname{var} X = \mathbb{E}\operatorname{var}(X \mid Z) + \operatorname{var}(\mathbb{E}(X \mid Z))$, where $\operatorname{var}(X \mid Z)$ is the variance of the conditional distribution of X, given Z. It is a special case of the **conditional covariance thm**: for any finite-variance r.vs X, Y, and Z,

$$\text{cov}(X, Y) = \mathbb{E}\text{cov}(X, Y|Z) + \text{cov}(\mathbb{E}(X|Z), \mathbb{E}(Y|Z)).$$

An important special case of the conditional variance thm is often called **Burgess's variance thm**: let $X_1, \ldots, X_N$ be a random number N of finite variance i.i.d.r.vs that are conditionally independent given N; then

$$\text{var}\left(\sum_{r=1}^{N} X_r\right) = \mathbb{E}N\,\text{var}\,X + (\mathbb{E}X)^2 \text{var}\,N.$$

[R. Durrett, *Essentials of Probability.* Duxbury, 1994]

conductance Arises in the context of the rate of mixing of a time-reversible irreducible Markov chain X_n, i.e. the rate at which its n-step distn $p_{ij}(n)$ converges to its stationary distribution $\pi_i, i \in S$, where S is the state space of X. Consider the collection of proper subsets L of S, such that $0 < \pi(L) = \sum_{i \in L} \pi_i \leq \frac{1}{2}$, with complements L^c. Then the conductance of the Markov chain is $\Phi = \min_L \left\{ \sum_{i \in L, j \in L^c} \pi_i p_{ij} / \sum_{i \in L} \pi_i \right\}$. Roughly speaking Φ measures the tendency of the distn of the chain to spread about its state space, or mix. Formally, if X has the property that $p_{ii} \geq \frac{1}{2}$ for all i, then

$$\max_{i,j \in S} \{\pi_j^{-1} |p_{ij}(n) - \pi_j|\} \leq \left(1 - \frac{1}{2}\Phi^2\right)^n / \min_{i \in S} \pi_i.$$

This resembles **Cheeger's inequality** as applied to Markov chains, so Φ is sometimes called a Cheeger-type constant. Further, the quantity $\pi(L)/\sum_{i \notin L, j \in L} \pi_i p_{ij}$ has been called the **amplification factor** of L. [M. Jerrum & A. Sinclair, Conductance and the rapid mixing property, in *Proc. 20th ACM Symp. Theory of Computing*, 1988, pp. 235–244; A. Sinclair, *Algorithms for Random Generation and Counting*. Birkhäuser, 1993]

confirmation function Arises in that branch of confirmation theory which seeks to use probability theory in formalizing the scientific method of induction. Formally, let $\mathbb{P}(\cdot)$ be a probability fn on a field $\mathcal{F}$ of propositions, and consider a hypothesis H, evidence E, and background knowledge K; all elements of $\mathcal{F}$. Then a function $C(E, H \mid K)$ is called a (Bayesian) confirmation measure if for all probability spaces $(\Omega, \mathcal{F}, \mathbb{P})$:

(i) $C(E, H \mid K) = 0$ iff E and H are conditionally independent given K

(ii) $C(E, H \mid K) > 0$ iff $\mathbb{P}(E \cap H \mid K) > \mathbb{P}(E \mid K)\mathbb{P}(H \mid K)$, which is to say that E and H are conditionally attractive given K

(iii) $C(E, H|K) < 0$ iff $\mathbb{P}(E \cap H|K) < \mathbb{P}(E|K)\mathbb{P}(H|K)$ which is to say that E and H are conditionally repellent given K.

[J. Earman, *Bayes or Bust?* MIT Press, 1992]

conformal invariance of Brownian motion (alias: **Lévy's planar B.M. theorem**) Let $Z(t)$ be two-dimensional B.M. in the plane, which we take as the complex plane so that $Z = Z_0 + (W_1, W_2)$, where $W_1(t)$ and $W_2(t)$ are independent Wiener processes. Let $Z_0 \in D$, where D is a domain in $\mathbb{C}$, and Z is stopped on leaving D. If $f(z)$ is an analytic function of the complex variable z, then $f(Z(t))$ is a time-changed B.M. More formally, $f(Z(t))$ has the same distribution as $Z(T(t))$, where $T(t) = \int_0^t |f'(Z(u))|^2 du$. [G. Lawler, *Conformally Invariant Processes in the Plane*. American Mathematical Society, 2005]

conformal martingale Let $X(t)$ and $Y(t)$ be real continuous local martingales. Then $Z(t) = X(t) + iY(t)$ is a complex(-valued) local martingale. Such a martingale is called conformal if it satisfies the following equivalent conditions:

(i) Z^2 is a local martingale

(ii) $f(Z)$ is a martingale for all holomorphic functions $f(z)$ of the complex variable z

(iii) $\langle X, Y \rangle = 0 = \langle X, X \rangle - \langle Y, Y \rangle$, where $\langle X, X \rangle$ is the quadratic variation such that $X^2 - \langle X, X \rangle$ is a local martingale.

[R. Getoor & M. Sharpe, Conformal martingales, *Inventiones Math.*, 16(4), 271–308, 1972]

congestion *see* Erlang loss system; queueing theory

conglomerability Arises in the context of finitely additive probability measures, where, in particular, B. de Finetti constructs the 'Paradox of non-conglomerability'. Essentially, this asserts that one can have a finitely (but not countably) additive probability function $\mathbb{P}(\cdot)$, such that $a \leq \mathbb{P}(A \mid B_r) \leq b$ for every set B_r in a countable partition $B_r, r \geq 1$, of the sample space Ω; but, nevertheless, $\mathbb{P}(A) \notin [a, b]$. In such a case $\mathbb{P}(\cdot)$ is said to be non-conglomerable. Formally, let $\mathcal{F}$ be an algebra of subsets of Ω, and define a full conditional probability function $\mathbb{P}(A \mid B)$ as a function of the elements A of $\mathcal{F}$, given any non-empty element B of $\mathcal{F}$, such that:

(i) $\mathbb{P}(A \mid A) = 1$.

(ii) $\mathbb{P}(\cdot \mid B)$ is a finitely additive probability measure on $\mathcal{F}$.

(iii) $\mathbb{P}(A \mid C) = \mathbb{P}(A \mid B)\mathbb{P}(B \mid C)$ for any $A \subseteq B \subseteq C \subseteq \Omega$.

The function $\mathbb{P}(A) = \mathbb{P}(A \mid \Omega)$ is called the associated unconditional probability measure. Then $\mathbb{P}(\cdot \mid \cdot)$ is said to be conglomerable if for every partition $B_r, r \geq 1$, of Ω, and every A, $\inf_r \mathbb{P}(A \mid B_r) \leq \mathbb{P}(A) \leq \sup_r \mathbb{P}(A \mid B_r)$. If Ω is countable, then $\mathbb{P}(\cdot \mid \cdot)$ is conglomerable iff $\mathbb{P}$ is countably additive; this was de Finetti's conjecture. [B. de Finetti, *Probability, Induction, and Statistics*. Wiley, 1972; T. Armstrong, Conglomerability of probability measures on Boolean algebras, *J. Math. Anal. Appl.*, 150(2), 335–358, 1990]

conjugate distribution *see* Esscher transform

conjugate random walk *see* associated random walk

connective constant Arises in the context of self-avoiding random walk on a lattice L. Let $S(n)$ be the number of self-avoiding walks on the lattice that start at the origin and have length n. Then the connective constant is $C = \lim_{n\to\infty}(S(n))^{1/n}$. For the hexagonal lattice, $C = (2 + \sqrt{2})^{1/2}$. [N. Madras & G Slade, *The Self-avoiding Walk.* Birkhäuser, 1993]

conservation law *see* invariance relation

conservative In the natural sciences a conservation law asserts the constancy of some measurable quantity q, as the system in which q is embedded evolves, e.g. the conservation of mass. By analogy, the term 'conservative' is applied in various instances where probability mass is conserved. Thus:

1. In population and genetic models, a vector-valued random process of non-negative entries $(X_1(t), \ldots, X_{N(t)}(t))$ is called conservative if

$$\sum_{r=1}^{N(t)} X_r(t) = \text{constant}, \; t \geq 0.$$

2. Similarly, a measure-valued branching process is conservative if the total measure remains constant over time.
3. A regular diffusion $X(t), t \geq 0$, with killing time T may be called conservative on its set I of possible initial values if $\mathbb{P}(T = \infty \mid X(0) = x) = 1$, for all $x \in I$.
4. The Q-matrix $Q = (q_{ij})$ of a continuous-time Markov chain taking values in $\mathbb{Z}$ is conservative if, for all $i \in \mathbb{Z}$, $\sum_{j\neq i} q_{ij} = -q_{ii} = q_i < \infty$. This is seen informally as a requirement that the chain cannot disappear instantaneously (to infinity). Such a process may still explode (to the boundary of its state space) in finite time, but augmenting the state space by adjoining the boundary (compactification) yields a conservative process.
5. The semi-group T_t arising from the transition kernel $k(x, A)$ of a Markov process (via the definition $Tf(x) = \int k(x, dy) f(y)$ for suitable functions f) is said to be a conservative semi-group if $T_t 1 = 1$.

[S. Karlin & H. Taylor, *A Second Course in Stochastic Processes*. Academic Press, 1981]

contact process (alias: **Harris contact process; Harris epidemic model**) An interacting particle system that can be seen as a model for the spread of infection among individuals located at the points of $\mathbb{Z}^d$, the d-dimensional integers. Any point is labelled 1 if the individual is infected, or 0 if the individual is susceptible. The contact process $X(t)$ is a Markov process whose values are subsets of $\mathbb{Z}^d$, where $X(t)$ is the set of points labelled 1 at time t. Infected sites recover, to become susceptible, independently of all other sites, at a time after first infection that is Exp(1), i.e. exponentially distributed with parameter 1. Any susceptible (uninfected) site becomes infected during $(t, t+h)$ with probability $\lambda ih + o(h)$, where i is the number of infected neighbours of the site at time t; $0 \leq i \leq 2d$. The process may also be seen as a percolation process in which one axis (time) is continuous and directed. There is a critical value λ_c such that for $\lambda \leq \lambda_c$ the infection dies out, but for $\lambda > \lambda_c$ the complete convergence theorem asserts that $X(t)$ either dies out or converges to a stationary distn that assigns probability 0 to extinction. See **shape thm** in Eden–Richardson model. [T. Harris, Contact interaction on a lattice, *Ann. Probab.*, 2, 969–988, 1974]

contiguity Arises in the context of sequences of probability measures, where it supplies a characterization of proximity (or closeness) of two such sequences, $\mathbb{P}_n$ and $\mathbb{Q}_n$, both defined on the corresponding sequence of measure spaces $(\Omega_n, \mathcal{F}_n)$. Then $\mathbb{P}_n$ is contiguous to (or with respect to) $\mathbb{Q}_n$ if for any sequence of measurable sets A_n such that $\mathbb{Q}_n(A_n) \to 0$ it follows that $\mathbb{P}_n(A_n) \to 0$, as $n \to \infty$. If conversely $\mathbb{Q}_n$ is contiguous to $\mathbb{P}_n$, then they are contiguous (or mutually contiguous). The property is transitive, so that the contiguity of $\mathbb{P}_n$ and $\mathbb{Q}_n$, and of $\mathbb{Q}_n$ and $\mathbb{M}_n$, implies the contiguity of $\mathbb{P}_n$ and $\mathbb{M}_n$. [L. LeCam, Locally asymptotically normal families of distribution, *Univ. Calif. Publ. Stat.*, 3, 37–98, 1960]

contingency-type distribution *see* Plackett's distribution

contingent claim (alias: **derivative security (or asset, or contract); option**) Arises in financial mathematics, where it denotes a security (or contract, or asset) whose value at its exercise (or contract, or claim) date T, in the future, is a function of the values of one or more underlying assets (such as a stock price, or other index or asset); more generally it may also depend on the values of the underlying before T. Broadly, the principal types of contingent claim comprise:

1. Options, where the owner has the right (but not the duty) to make some purchase, sale, or other similar transaction. The right to buy is a call option, the right to sell is a put option, and there are many combinations, variants, and extensions of these simple so-called **vanilla** options. *See* exotic options.
2. Futures and forwards contracts (the former being exchange traded, the latter not), where the ultimate transaction at T is not optional.
3. Swaps, which are essentially non-cash forwards.

In the usual model, the underlying is a component of an arbitrage-free market; if there is a self-financing trading strategy (or market portfolio) whose value at T is a.s. the same as the claim, then the claim is said to be replicable, or hedgeable, or reachable, or attainable, or marketable, and is replicated by the strategy. If every contingent claim is attainable, the market is called complete.
[R. Elliott & P. Kopp, *Mathematics of Financial Markets*, 2nd edn. Springer, 2005]

continuity correction Many discrete random variables and processes (indexed by a parameter n, say), have a limit (as $n \to \infty$) that is continuous and analytically more tractable than the exact discrete version. However, the accuracy of the approximation may be compromised for moderate values of n.

A continuity correction is a systematic adjustment made to deal with this problem. For example, consider a binomial $\mathrm{B}(n,p)$ random variable X. The central limit theorem tells us that as $n \to \infty$, for integer b,

$$\mathbb{P}(X \le b) \sim \Phi\left(\frac{b-np}{\sqrt{npq}}\right).$$

However, for moderate values of n, a more accurate estimate of $\mathbb{P}(a \le X \le b)$ is given by

$$\Phi\left(\frac{b+\frac{1}{2}-np}{\sqrt{npq}}\right) - \Phi\left(\frac{a-\frac{1}{2}-np}{\sqrt{npq}}\right),$$

for integer a and b. Inspection of a sketch of the normal curve shows why this is so. [W. Feller, On the normal approximation to the binomial, *Ann. Math. Stat.*, 16(4), 319–329, 1945]

continuity theorem (alias: **Lévy's continuity theorem; Lévy–Cramér continuity theorem**) Relates the convergence of a sequence of distribution functions $F_n(x), n \ge 1$, to the convergence of their respective characteristic functions $\phi_n(t)$, as $n \to \infty$. One form asserts that the following are equivalent:

(a) $\lim_{n\to\infty} F_n(x) = F(x)$ at every point x where the (proper) distribution function $F(x)$ is continuous.
(b) $\lim_{n\to\infty} \phi_n(t) = \phi(t)$ for all real t, where $\phi(t)$ is the chf of the (proper) distribution function $F(x)$.
(c) For every bounded continuous function $u(x)$ on $\mathbb{R}$,

$$\lim_{n\to\infty} \int_{\mathbb{R}} u(x)\mathrm{d}F_n(x) = \int_{\mathbb{R}} u(x)\mathrm{d}F(x).$$

More strongly, it is sufficient for (*a*) that $\psi(t) = \lim_{n\to\infty}\phi_n(t)$ exists for all real t, where $\psi(t)$ is continuous at $t = 0$. Then $\psi(t)$ is the chf of $F(x)$.

Similar theorems consider the convergence of mgfs; among them being **Curtiss's thm**: suppose that $M_n(t), n \ge 1$, are the mgfs of $F_n(x)$, and that $M_n(t)$, exists for all n and $|t| < a > 0$. If $\lim_{n\to\infty} M_n(t) = M(t)$ exists and is finite for all $t \in [-b, b], |b| < |a|$, then there exists a distribution function $F(x)$ such that $\lim_{n\to\infty} F_n(x) = F(x)$ wherever $F(x)$ is continuous, and $F(x)$ has mgf $M(t)$ that exists for $|t| \le b$.

Likewise, there is a continuity theorem for pgfs, thus: let $p_n(x) = \mathbb{P}(X_n = x), x \ge 0; n \ge 1$, be discrete (integer-valued x) distributions having respective probability generating functions $G_n(s)$, where $G_n(1) = 1$. Then $\lim_{n\to\infty} p_n(k) = p(k)$ exists for all k iff $\lim_{n\to\infty} G_n(s) = G(s)$ exists for $0 < s < 1$, and then $G(s)$ is the pgf of the discrete (possibly defective) distribution $p(k)$. [J. Curtiss, A note on moment generating functions, *Ann. Math. Stat.*, 13(4), 430–433, 1942; E. Lukacs, *Characteristic Functions*, 2nd edn. Griffin, 1970]

continuous mapping theorem *see* Mann–Wald theorem

continuous random process The property of continuity, with respect to the time-parameter t, arises naturally in the context of a continuous-time random process $X_t = X(t,\omega), t \in \mathbb{R}, \omega \in \Omega$, defined on a probability space $(\Omega, \mathcal{F}, \mathbb{P})$. Note that the process X_t itself may comprise random variables that are either discrete or continuous (in the other meaning of that word). There are various concepts of continuity in t for such a process, depending on the characteristic under consideration; these are often seen:

1. X_t is continuous in distribution at t if $\lim_{s\to t} F(s,x) = \lim_{s\to t}\mathbb{P}(X_s \le x) = F(t,x)$, at all points X such that the distribution function $F(t,x)$ of X_t is continuous.
2. X_t is continuous in the rth mean at t if $\lim_{s\to t}\mathbb{E}\{|X(t) - X(s)|^r\} = 0$. In particular, it is continuous in mean-square for $r = 2$, and this is equivalent to the continuity in the ordinary sense of its covariance function.
3. X_t is continuous in probability at t, or stochastically continuous, if for $\varepsilon > 0$ $\lim_{s\to t}\mathbb{P}(|X(s) - X(t)| > \varepsilon) = 0$.
4. X_t is continuous with probability 1 (or a.s. continuous), at t if $X(t,\omega)$ is a continuous function of t (in the ordinary sense) at t, on an event that has probability 1. (That is, the event that $X(t,\omega)$ is discontinuous at t is null.)
5. X_t is sample continuous, or pathwise continuous, on $[a,b]$ if $X(t,\omega)$ is a continuous function of t over the interval $t \in [a,b]$, on an event that has probability 1.

Right-(or left-) continuous random processes may be defined likewise in each case. [J. Doob, *Stochastic Processes*. Wiley, 1953]

continuous random variable (alias: **absolutely continuous r.v.**) A term often used to denote a random variable X whose distribution function $F(x)$ can be written as the integral of a function $f(x)$ called the density of X: $F(x) = \int_{-\infty}^{x} f(y)dy$. For a so-called jointly continuous random vector $\mathbf{X}$, one has $F(\mathbf{x}) = \int_{-\infty}^{x_1}\cdots\int_{-\infty}^{x_n} f(\mathbf{y})\mathbf{dy}$, likewise. In general, this property of F is called absolute continuity (with respect to the measure implicit in the integral). Strictly speaking, therefore, 'continuous random variable' is shorthand for 'random variable whose probability measure is absolutely continuous with respect to Lebesgue measure'. The use of an abbreviation is natural, but this one is unfortunate since X, seen as a function, need have no ordinary properties of continuity. Note that continuous r.vs have continuous distribution functions, but not all distribution functions that are continuous have a density; there may be a singular part. [D. Pollard, *A User's Guide to Measure Theoretic Probability*. Cambridge University Press, 2002; J. Kingman & S. Taylor, *Introduction to Measure and Probability*. Cambridge University Press, 1966]

continuous-state branching process (alias: **Jiřina process**) This is an analogue of the ordinary Galton–Watson branching process, which is continuous in the sense that it takes values in $\mathbb{R}^+$. It may be obtained by rescaling the state and time parameter of a discrete branching process, suitably, and taking an appropriate limit. More directly, it may be defined as a Markov process $(X(t), t \ge 0)$, that is continuous in probability (but not necessarily a.s. continuous), that is characterized by a function $b(x), x \ge 0$, called the **branching mechanism** as follows: The transition probabilities of $X(t)$ are determined by the branching property for $s, t \ge 0$, $\mathbb{E}\left(e^{-\lambda X(s+t)}|X(s)\right) = \exp(-m(\lambda,t)X(s))$, where $m(\lambda,t)$ is the unique non-negative solution of $\frac{\partial m}{\partial t} = -b(m)$, with $m(\lambda, 0) = \lambda \ge 0$. [M. Jiřina, Stochastic branching processes with continuous state space, *Czech Math. J.*, 8, 292–313, 1958]

continuous-time random walk (alias: **CTRW**) A term used very much in the physical sciences literature to denote the random walk in $\mathbb{R}^d$, or on a d-dimensional lattice, defined thus: The i.i.d.r.vs $T_r > 0$, $r \ge 1$, denote the intervals between steps of the walk (also called holding, waiting, or pausing, etc., times) having mgf $M(\theta)$; the i.i.d.r.vs $X_r, r \ge 1$, denote the successive step sizes at the step-times $S_n = \sum_{r=1}^{n} T_r$, starting from the origin, where X_r has chf $\phi(s)$; the joint Laplace–Fourier transform of (T_r, X_r) is $L(\theta, s)$. Then the density $P(w,t)$ of the position W of the particle at time t has Laplace–Fourier transform $\int_o^\infty\int_{-\infty}^{\infty} e^{-\theta t}e^{isw}P(w,t)dwdt = \{1 - m(\theta)\}\{\theta(1 - L(\theta,s))\}^{-1}$, which may be called the **Scher–Lax eqn**, or the generalized **Montroll–Weiss eqn**; in the special case when step sizes are independent of holding times, so that $L(\theta,s) = M(\theta)\phi(s)$, one speaks of the **Montroll–Weiss eqn**. [E. Montroll & G Weiss, Random walks on lattices, *J. Math. Phys.*, 6(2), 167–181, 1965]

continuum random tree (alias: **Aldous's random tree**) A type of random metric space that is defined in terms of a continuum tree, which is a metric space such that between any two points of the space, x and y, there is unique path (this is the tree property) whose length is determined by the metric; the points of the space are then equipped with a probability distribution. Finally, a continuum random tree is obtained by assigning a probability measure to the family of continuum trees. There are a number of ways of specifying such a measure; one arises from a suitable scaling limit, as $n \to \infty$, of the Galton–Watson process conditioned to have total population n, prior to extinction. [D. Aldous,

The Continuum Random Tree II, London Math. Soc. Lecture Notes Series, 167. Cambridge University Press, 1991]

contour process (alias: **search-depth process; height process; exploration process**) Arises in the context of branching processes, and related processes that generate a random planar tree. The contour process of such a tree, denoted by $C(t)$, can be seen as the distance from the root at time t of a particle that starts from the root at time $t = 0$, and traverses the edges of the tree at constant speed, according to a given lexicographic (i.e. linear) order. For a branching process, the natural ordering is by birth-order of offspring: if the process is simple Galton–Watson, then the edges between parents and offspring are of length 1; whereas for age-dependent branching the lengths correspond to times between births. The particle explores the tree starting at the root. On arriving at the first birth it turns to the edge generated by that offspring, and so on. There are two possible rules for the particle on reaching a terminal node: either it may retrace its path until it first reaches a node with an untraversed branch (yielding a continuous contour) or it can simply jump there. The contour uniquely encodes the tree (provided that individuals are not required have exactly one offspring). An important special case is the critical binary Galton–Watson process conditioned to have total population size n, whose contour process, when suitably scaled, converges to a Brownian excursion as $n \to \infty$; this can be seen as the contour process of a continuum tree. [J.-F. Le Gall, *Spatial Branching Processes, Random Snakes, and Partial Differential Equations*. Birkhäuser, 1999]

control variate A method of reducing the variance in Monte Carlo integration, and related methods. The broad idea of the technique is that something should be known about a random variable Y that is correlated with the random variable X of primary interest. Then Y is said to be a control variate for X.

Specifically, suppose we seek to estimate $\mathbb{E}X$ by simulation of X, and Y is a random variable with known mean $\mathbb{E}Y$, and Y is not independent of X. Then, for any constant c, the random variable $Z = X + c(Y - \mathbb{E}Y)$ is also an unbiased estimator of X. But var $Z =$ var $X + c^2$ var $Y + 2c\,\mathrm{cov}(X, Y)$, which is minimized when c is $\hat{C} = -\mathrm{cov}(X, Y)/\mathrm{var}\, Y$, and this minimum variance is $\mathrm{var}_{\min} Z = [1 - \rho^2(X, Y)]$ var X, where $\rho(X, Y)$ is the correlation coefficient. [R. Rubinstein & D. Kroese, *Simulation and the Monte Carlo Method*, 2nd edn. Wiley, 2008]

convergence Questions of convergence arise throughout probability theory, and there are several differently defined types of convergence arising in various contexts. Classic examples of naturally arising problems of convergence include: the stability of statistical ratios; the central limit theorem; and the laws of large numbers. The principal forms of convergence for a sequence $X, X_1, X_2, \ldots$ of r.vs are listed:

1. **Almost sure convergence** (alias: convergence with probability one, or convergence almost surely, or convergence almost everywhere, or strong convergence of random variables). If $X, X_1, X_2, \ldots$ are defined on the same probability space $(\Omega, \mathcal{F}, \mathbb{P})$, and $\mathbb{P}(\lim_{n\to\infty} X_n = X) = 1$, then we say X_r converges to X almost surely, or a.s., or a.e., or w.p. 1, and write $X_r \to^{a.s.} X$, or $X_r \to^{a.e.} X$. More explicitly, $\mathbb{P}(\{\omega : X_n(\omega) \to X(\omega) \text{ as } n \to \infty\}) = 1$. More strongly, if for any $\varepsilon > 0$, $\sum_{r=1}^{\infty} \mathbb{P}(|X_r - X| > \varepsilon) < \infty$, then X_n is said to converge completely to X; this implies convergence a.s., and is equivalent to it if the X_r are independent. Also more strongly, if $\lim_{n\to\infty} X_n(\omega) = X(\omega)$ for all $\omega \in \Omega$, then X_n is said to converge pointwise, or everywhere, or surely, to X.
2. **Convergence in probability** (alias: convergence in measure). If, for any $\varepsilon > 0$, $\lim_{r\to\infty} \mathbb{P}(|X_r - X| > \varepsilon) = 0$ then we say X_r converges in probability to X, or i.p., and write $X_r \to^{\mathbb{P}} X$. Note that this requires only the successive pairs (X_r, X) to be jointly distributed. It is implied by almost sure convergence, and is equivalent to it if the sequence X_n is monotone (on some probability space).
3. **Convergence in rth mean.** If for some $r \geq 1$ $\lim_{n\to\infty} \mathbb{E}(|X_n - X|^r) = 0$, then we say that X_n converges in rth mean to X, and write $X_n \to^r X$. In particular, when $r = 2$, X_n is said to converge in mean square, and we may write $X_n \to^{m.s.} X$. Convergence in rth mean implies convergence in probability, but not conversely; it does not imply almost sure convergence. More strongly, if for any $\varepsilon > 0$ and some $r > 0$, $\mathbb{E}\{(\sup\{n : |X_n - X| \geq \varepsilon\})^r\} < \infty$ then X_n is said to converge to X r-quickly (where sup $\emptyset = 0$). This does imply almost sure convergence. More generally, if $f(\cdot)$ is a real-valued function and $\lim_{n\to\infty} f(X_n - X) = 0$, then X_n is said to converge in f to X.
4. **Convergence in distribution** (alias: convergence in law). Let $X, X_1, \ldots$ have distribution functions $F(x), F_1(x), F_2(x)$, where these distribution functions are all proper (i.e. not defective) in the sense that $\lim_{x\to-\infty} F(x) = 0$ and $\lim_{x\to\infty} F(x) = 1$. If

$\lim_{n\to\infty} F_n(x) = F(x)$ at all points of the set $C(F)$ of points at which $F(x)$ is continuous, then we say X_n converges to X in distribution (or in law) and write $X_n \xrightarrow{D} X$, or $X_n \xrightarrow{L} X$ (but note that this is a slight abuse of notation, as the r.vs X_n and X need not be defined on the same probability space). This is implied by convergence in probability, but not in general conversely; though convergence in distribution to a constant C does imply convergence in probability to C. There are several equivalent characterizations of convergence in distribution (this fact may be called the portmanteau theorem); commonly seen are these:

(i) $\mathbb{E}g(X_n) \to \mathbb{E}g(X)$ for all bounded continuous fns $g(\cdot)$;

(ii) $\mathbb{E}g(X_n) \to \mathbb{E}g(X)$ for all functions continuous on the interval (a, b) (*or* $[a, b]$), for any a and b in $C(F)$. (Such an interval is called an interval of continuity.)

5. **Convergence of distributions**. In cases of interest it is necessary to allow the possibility that $F(x)$ is defective (or improper), entailing the following definitions (but note that some writers differ in their terminology):

(i) If for all a and b in $C(F)$, with $a < b$ (i.e. for every interval of continuity) $\lim_{n\to\infty}(F_n(b) - F_n(a)) = F(b) - F(a)$, then $F_n(x)$ is said to converge vaguely to F, and we may write $F_n \xrightarrow{v} F$. If in addition F is proper, then F_n is said to converge properly to F.

(ii) If $\lim_{n\to\infty} F_n(x) = F(x)$ for all $x \in C(F)$, i.e. at all the continuity points of F, then F_n is said to converge weakly to F and we write $F_n \xrightarrow{w} F$. If in addition F is proper, then F_n is said to converge completely to F.

In fact, all these cases are equivalent when $F(x)$ is proper and authors simply speak of weak convergence and write $F_n \xrightarrow{w} F$. This and the previous definitions extend naturally to sequences of random vectors in $\mathbb{R}^d$.

6. **Convergence of probability measures**. Arises naturally in considering a sequence $X, X_1, X_2, \ldots$ of random elements (or objects) taking values in a metric space S that is not necessarily $\mathbb{R}^d$. If $\mathbb{P}$, $\mathbb{P}_1, \ldots$ is a sequence of probability measures such that for every bounded continuous function $g \in \mathbb{R}$,

$$\lim_{n\to\infty} \int_S g\mathrm{d}\mathbb{P}_n = \int_S g\mathrm{d}\mathbb{P},$$

then $\mathbb{P}_n$ is said to converge weakly to $\mathbb{P}$, and we write $\mathbb{P}_n \xrightarrow{w} \mathbb{P}$; the notation $\mathbb{P}_n \Rightarrow \mathbb{P}$ is also used. Related forms of convergence measures include these:

(i) If $\int_S g\mathrm{d}\mathbb{P}_n \to \int_S g\mathrm{d}\mathbb{P}$ for all continuous fns g that vanish at infinity, then $\mathbb{P}_n$ is said to converge vaguely, we write $\mathbb{P}_n \xrightarrow{v} \mathbb{P}$.

(ii) If $\int_S g\mathrm{d}\mathbb{P}_n \to \int_S g\mathrm{d}\mathbb{P}$ for all bounded (Borel-) measurable fns g then $\mathbb{P}_n$ converges setwise; we write $\mathbb{P}_n \xrightarrow{s} \mathbb{P}$.

(iii) If $\mathbb{P}_n$ converges to $\mathbb{P}$ in total variation norm then it is said to converge strongly; we write $\mathbb{P}_n \xrightarrow{\mathrm{TV}} \mathbb{P}$.

Thus, if $X, X_1, \ldots$ is a sequence of random elements in S having corresponding measures $\mathbb{P}, \mathbb{P}_1, \ldots$, then X_n is said to converge weakly (to X) if $\mathbb{P}_n \xrightarrow{w} \mathbb{P}$; and likewise for the other modes if applicable.

[P. Billingsley, *Convergence of Probability Measures*, 2nd edn. Wiley, 1999; S. Resnick, *A Probability Path.* Birkhäuser, 1999; D. Stroock, *Probability Theory*, 2nd edn. Cambridge University Press, 2011]

convergence to (or of) types theorem (alias: **Khinchin's convergence to types theorem**) Arises when a sequence of r.vs, $X_n, n \geq 1$, converges in distribution to X having non-degenerate distribution function $F_X(x)$. The theorem asserts that if there is a non-degenerate distribution function $G(x)$, and sequences of constants $a_n > 0$ and $b_n \in \mathbb{R}, n \geq 1$, such that $\lim_{n\to\infty}\mathbb{P}(a_nX_n + b_n \leq x) = G(x)$ (at continuity points of $G(x)$) then there exist constants a and b such that $a_n \to a, b_n \to b$, and $G(x)$ is the distribution of $aX + b$. That is to say, the convergence in distribution of X_n determines $G(x)$ up to location and scale and also determines a_n and b_n up to asymptotic equivalence as $n \to \infty$.

Note that r.vs X and Y are said to be of the same type if for some constants c and d, $cX + d$ has the same distribution as Y. *See* extreme value distributions.

[S. Resnick, *A Probability Path.* Birkhäuser, 1999; L. Breiman, *Probability.* Addison-Wesley, 1968; reprinted by SIAM, 1992]

convex order A type of stochastic ordering of random variables X and Y. If $\mathbb{E}u(X) \leq \mathbb{E}u(Y)$ for any convex function $u(\cdot)$ such that the expectations exist, then X is smaller than Y in convex order, denoted by $X \leq_{\mathrm{cx}} Y$. More weakly, if $\mathbb{E}u(X) \leq \mathbb{E}u(Y)$ for all increasing convex functions $u(\cdot)$, then X is smaller than Y in increasing convex order, denoted by $X \leq_{\mathrm{icx}} Y$. The latter is equivalent to the stop-loss ordering. The former is equivalent to stop-loss order for r.vs such that $\mathbb{E}X = \mathbb{E}Y$. Both types constitute what is called a **variability order**. [T. Rolski *et al.*, *Stochastic Processes for Insurance and Finance.* Wiley, 1999; M. Shaked & J. Shanthikumar, *Stochastic Orders.* Springer, 2007]

convolution (alias: **Cauchy product; composition**) An operation on functions (or series) that arises naturally in considering sums of independent random variables. Thus if $p_n = \mathbb{P}(X = n)$ and $q_n = \mathbb{P}(Y = n), n \geq 0$, are the probability distributions of independent non-negative integer-valued r.vs X and Y, then the distribution $f_n = \mathbb{P}(X + Y = n)$ of their sum is given by $f_n = p_0q_n + p_1q_{n-1} + \cdots + p_nq_0$. The sum on the right is the convolution (or Cauchy product) of p_n and q_n, which may be denoted by $p_n * q_n$, or $(p * q)_n$. More generally, for independent r.vs X and Y with respective distn fns $F(x)$ and $G(y)$, the distn fn of their sum is $H(x) = \int F(x - y)\mathrm{d}G(y) = (F * G)(x) = F(x) * G(x)$. If X, Y, and $X + Y$ have densities f, g, and h, then one has $h(x) = \int f(x - y)g(y)\mathrm{d}y = (f * g)(y)$. Here $f*g$ is the convolution (or composition) of f and g. [W. Feller, *Introduction to Probability Theory and its Applications*, Vol. 2, 2nd edn. Wiley, 1971; G. Högnäs & A. Mukherjea, *Probability Measures on Semigroups*, 2nd edn. Springer, 2011]

convolution equivalence Arises in relating the behaviour of tails of sums of independent r.vs to the behaviour of tails of the summands. Let $F(x) < 1, x \in \mathbb{R}$, be a distribution function, and let $F * F$ be its convolution with itself. It has an exponential tail with parameter (or rate) $a \geq 0$ if

$$\lim_{x\to\infty} \frac{1 - F(x - y)}{1 - F(x)} = \mathrm{e}^{ay}, y \in \mathbb{R}.$$

Such a distribution is further said to be convolution equivalent if $\lim_{x\to\infty} \frac{1-F*F(x)}{1-F(x)} = 2m < \infty$ exists. If it does, then $m = \int_{-\infty}^{\infty} \mathrm{e}^{ax}\mathrm{d}F(x)$. [A. Pakes, Convolution equivalence and infinite divisibility, *J. Appl. Probab.*, 44(2), 295–305, 2007]

Conway's leading number formula (or algorithm) This arises when waiting for the first to appear of two possible patterns of lengths m and n, in a sequence of rolls of a q-sided die. Let $A = a_1a_2 \ldots a_n$ and $B = b_1 \ldots b_m$ be two such patterns, i.e. two strings whose symbols are taken from the q symbols labeling the q sides of the die. Then the **leading number** of A over B is (A, B), where $(A, B) = \sum_{r=0}^{n-1} \phi(n - r)q^r$, and $\phi(\cdot)$ is this indicator function:

$$\phi(k) = \begin{cases} 1 & \text{if the string } a_ka_{k+1}\ldots a_n \text{ is a prefix of the string } b_1b_2\ldots b_m \\ 0 & \text{otherwise.} \end{cases}$$

That is to say, $\phi(k) = 1$ if the final $n - k + 1$ symbols of A are the same as the first $n - k + 1$ symbols of B. When q is replaced by an arbitrary variable z, this yields a generating function for the $\phi(n - r)$, which is called the correlation polynomial $(A, B)(z)$. If now the die is rolled repeatedly, the odds that the string B appears as a consecutive sequence of rolls, before A so appears, is given by $\frac{(A,B)-(A,A)}{(B,A)-(B,B)}$, which is Conway's formula.

When $m = n$, and two players take turns in choosing a sequence, the game is called **Penney-Ante**. [S.-Y. Li, A martingale approach to the study of occurrence of sequence patterns in repeated experiments, *Ann. Probab.*, 8(6), 1171–1176, 1980]

copula (alias: **dependence function**) Arises in describing and categorizing the possible dependence relations of the components of a random vector, given the marginal distribution of each component. Broadly, an n-dimensional copula is a joint distribution function $C(x_1, \ldots, x_n)$ all of whose marginal distributions $C_{X_r}(x_r), 1 \leq r \leq n$, are uniform on $[0, 1]$. Formally, in two dimensions $C(x, y)$ is a copula, $0 \leq x, y \leq 1$, if:

1. $C(x, 0) = C(0, y) = 0, 0 \leq x, y \leq 1$; this is called grounding; $C(x, y)$ is grounded.
2. $C(x, 1) = x, 0 \leq x \leq 1; C(1, y) = y, 0 \leq y \leq 1$; thus the marginals of $C(x, y)$ are uniform on $(0, 1)$.
3. For any $0 \leq u \leq x \leq 1$ and $0 \leq v \leq y \leq 1, C\ (x, y) - C(u, y) - C(x, v) + C(u, v) \geq 0$; in this case $C(x, y)$ is called 2-increasing.

The connexion of copulas with general multivariate distributions is supplied by **Sklar's thm**, which in two dimensions takes this form: let X and Y have joint distribution function $F(x, y)$ with marginal distribution functions $F_X(x)$ and $F_Y(y)$. Then there is a copula $C(u, v)$ such that for $(x, y) \in \mathbb{R}^2, F(x, y) = C(F_X(x), F_Y(y))$, where $C(u, v)$ is unique if the marginals are continuous. Conversely, for distributions F_X and F_Y and any copula C, the function $C(F_X(x), F_Y(y))$ is a joint distribution having F_X and F_Y as its marginals. The set (equivalence class) of all joint distributions having F_X and F_Y as marginals is the **Fréchet–Hoeffding class**. It follows that X and Y, having copula $C(u, v)$ are independent iff $C(u, v) = uv, 0 \leq u, v \leq 1$. Constraints on copulas are supplied by the **Fréchet–Hoeffding bounds**, i.e. for any copula C, $\max\{0, u + v - 1\} \leq C(u, v) \leq \min\{u, v\}$, where each bound is itself a copula.

Notable examples of copulas include:

1. Archimedean copula.
2. **Clayton copula** which has the form $C_a(u, v) = (u^{-a} + v^{-a} - 1)^{-1/a}, a \in [-1, \infty)\backslash\{0\}, 0 \leq u, v \leq 1$.
3. **Gaussian or normal** copula, which in two dimensions is obtained from the distribution function $F(x, y)$ of the standard binormal distribution as

$C(u,v) = F(\Phi^{-1}(u), \Phi^{-1}(v))$, where Φ is the distn fn of the standard normal r.v.

4. **Marshall–Olkin** copula; has the form $C(u,v) = \min\{uv^{1-a}, u^{1-b}v\}, 0 \le a, b \le 1$.

[R. Nelsen, *Introduction to Copulas*, 2nd edn. Springer, 2006]

Cornish–Fisher inversion Supplies an expansion (or approximation) for the quantile function of an absolutely continuous r.v. X in terms of its cumulants; this is of particular use when X is the standardized sum of i.i.d.r.vs. When X has zero mean and unit variance, the first two terms are

$$F^{-1}(y) = \Phi^{-1}(y) + \frac{K_3}{6}\{\Phi^{-1}(y)^2 - 1\} + \cdots$$

where $\Phi(\cdot)$ is the standard normal distribution fn, and K_3 is the third cumulant of X. [E. Cornish & R. Fisher, Moments and cumulants in the specification of distributions, *Rev. I.S.I.*, 5, 307–320, 1937]

correlation A concept that arises in considering the degree of dependence, or association, between random variables. The term is used both loosely, to denote an apparent lack of independence, and also precisely, in the definition of correlation coefficient, etc. Correlation is useful in prediction, but unreliable when it comes to inferring causation. [E. Eells, *Probabilistic Causality*. Cambridge University Press, 1991]

correlation coefficient (alias: **Pearson's product moment coefficient**) Arises from the joint distribution of random variables X and Y, whose variances are finite and non-zero. The correlation $\rho(X,Y)$ between X and Y is $\rho(X,Y) = \mathrm{cov}(X,Y)\{\mathrm{var}\,X\,\mathrm{var}\,Y\}^{\frac{-1}{2}}$, where $\mathrm{cov}(X,Y)$ is the covariance of X and Y. If $\rho = 0$, then X and Y are uncorrelated; this is implied by their independence, but does not imply it. The number ρ^2 is sometimes called the coefficient of determination. The correlation ρ satisfies $|\rho| \le 1$, with equality iff $\mathbb{P}(aX + bY = c) = 1$, for some constants a, b, and c. If $\rho > 0$, one may speak of direct correlation, if $\rho < 0$ the correlation is said to be inverse or negative. (Note that a number of empirical statistics may be referred to as correlation coefficients.) The coefficient ρ arises naturally in the context of minimum mean squared error linear prediction, i.e. in choosing a and b to minimise $\mathbb{E}\{(Y-(aX+b))^2\}$ to obtain a predictor $\hat{Y}$ of Y that is linear in X. Then the minimizing values $\hat{a}$ and $\hat{b}$ are given by $\hat{Y} = \hat{a}X + \hat{b} = \rho(X - \mathbb{E}X)\mathrm{var}\,Y/\mathrm{var}\,X + \mathbb{E}Y$. More generally, given a suitable random vector $(X_1, \ldots, X_n)$, one may find the analogous predictors $\hat{X}_1$ and $\hat{X}_2$ of X_1 and X_2 that are each linear in $X_3, \ldots, X_n$. Then $\rho(X_1 - \hat{X}_1, X_2 - \hat{X}_2))$ is called the **partial correlation coefficient**. Note that the conditional correlation coefficient $\rho(X_1, X_2 | X_3, \ldots, X_n)$ (defined in the natural way by the conditional joint distribution of X_1 and X_2), has also been called the **partial correlation**. [P. Whittle, *Probability via Expectation*, 4th edn. Springer, 2000]

correlation function *see* autocorrelation function

correlation inequality A term occasionally applied to theorems that establish results of the form $\mathbb{P}(\cap_{r=1}^n A_r) \ge \prod_{r=1}^n \mathbb{P}(A_r)$ for a collection of events A_r; or similar inequalities. Examples of inequalities include those of Bell, Harris, Janson, Suen, and the FKG inequality. [G. Brightwell & W. Trotter, A combinatorial approach to correlation inequalities, *Discrete Math.*, 257(2–3), 311–327, 2002; B. Bollobás & O. Riordan, *Percolation*. Cambridge University Press, 2006]

correlation ratio Arises in considering the nature of dependence between random variables X and Y; the correlation ratio of X with respect to (or relative to) Y is denoted by $r_{X|Y}$, where $r_{X|Y}^2 = \mathrm{var}\{\mathbb{E}(X \mid Y)\}/\mathrm{var}\,X = 1 - \mathbb{E}\,\mathrm{var}(X \mid Y)/\mathrm{var}\,X$; and the correlation ratio of Y to X likewise is given by $r_{Y|X}^2 = \mathrm{var}\{\mathbb{E}(Y \mid \mathrm{X})\}/\mathrm{var}\,Y = 1 - \mathbb{E}\,\mathrm{var}(Y \mid X)/\mathrm{var}\,Y$. Denoting the correlation between X and Y by ρ, we have $0 \le \rho^2 \le r_{X|Y}^2 \le 1$, and $0 \le \rho^2 \le r_{Y|X}^2 \le 1$, where the central inequality follows from the Cauchy–Schwarz inequality, and the rightmost inequality follows from the conditional variance theorem. For $r_{Y|X}$ the former is an equality if $\mathbb{E}(Y \mid X) = aX + b$ with probability 1, for some a and b, and the latter is an equality if Y is a non-random function of X; so that $\mathrm{var}(Y \mid X) = 0$. [H. Frank & S. Althoen, *Statistics: Concepts and Applications*. Cambridge University Press, 1994; K. Pearson, *The Grammar of Science*, 1911; reprinted by Cosimo, 2007]

cosine distribution *see* cardioid distribution

countable state space process *see* discrete random process

countably additive A probability (or other measure) function is said to have this property if, for any countable collection $A_1, A_2, \ldots$, of disjoint events (or measurable sets), we have $\mathbb{P}(\cup_{r=1}^\infty A_r) = \sum_{r=1}^\infty \mathbb{P}(A_r)$. Note that a probability measure $\mathbb{P}(\cdot)$ is countably additive iff it is a continuous set function, which is to say that $\lim_{n\to\infty}\mathbb{P}(A_n) = \mathbb{P}(\lim_{n\to\infty}A_n)$, for any sequence of events A_n having a limit $A = \lim_{n\to\infty}A_n$. [D. Pollard, *A User's Guide to Measure Theoretic Probability*. Cambridge University Press, 2002]

counterfeit coin problem A classic puzzle; in its simplest form it requires the identification of a single counterfeit coin *C* (which is either heavy or light) from a set of coins that are otherwise identical, and all (except *C*) weigh the same. The only means of discrimination is a balance, and one requires the fewest uses of the balance. It can be seen as a problem in information theory; and there are many variants. [R. Guy & R. Nowakowski, Coin-weighing, *Amer. Math. Mon.*, 102(2), 164–167, 1995]

counter model A broad class of random processes arising in the theories of queueing, reliability, and inventories, as well as natural physical applications. There is an underlying point process (events, or particles) together with a mechanism that responds to the events (the counter). Classically, the arrival of a particle locks the counter for a dead-time (or dead period) *D*, during which other particles have no effect and are not counted. For so-called type-I counters, the mechanism is freed at the end of the dead-time, and resumes counting. For type-II counters, particles arriving during *D* are not counted but initiate their own dead-times; the mechanism is freed at the expiry of the last of these. [R. Pyke, Counter models, *Ann. Math. Stat.*, 29, 737–754, 1985]

counting process Arises naturally whenever random phenomena can be represented by points in time or space, yielding a point process (to which a counting process can be seen as a type of dual process). Let $N(A)$ be the number of points of a point process in a subset *A* of the space *S* in which points occur. Then $N(A)$ is said to count the points in *A*, and yields a counting process as *A* varies. In many cases *S* is $\mathbb{R}$, or $\mathbb{R}^+$, interpreted as time, and $N(A) = N([0, t]) = N(t)$. Then in this case a counting process $N(t)$ is formally defined as an integer-valued, right-continuous, non-decreasing, adapted process on $\mathbb{R}^+$, with $N(0) = 0$; it is called simple if all jumps are of size 1. More generally, a point process is a random countable subset of some suitable (e.g. completely separable metric) space, and the corresponding counting process is formalized in terms of a counting measure $m(\cdot)$ (or a family *B* of Borel sets), such that for any $A \in B, m(A)$ takes non-negative integer values that are finite for bounded *A*. [D. Daley & D. Vere-Jones, *Introduction to the Theory of Point Processes*, 2nd edn. Springer, Vol. II, 2008]

coupling A concept, and an analytical technique, that arises when it is necessary to compare two probability distributions, e.g. the distributions of random variables (or random vectors, or processes) defined on their respective probability spaces $(\Omega, \mathcal{F}, \mathbb{P})$ and $(\Omega', \mathcal{F}', \mathbb{P}',)$. The term is used most often in two senses, one broad, one narrow.

Broadly (or in the wide sense) a coupling of the two probability spaces denoted above is a (joint) probability distribution $\mathbb{Q}$ on the product space $(\Omega \times \Omega', \mathcal{F} \times \mathcal{F}')$, whose marginal distributions are $\mathbb{P}$ and $\mathbb{P}'$, i.e. $\mathbb{Q}(A \times \Omega') = \mathbb{P}(A)$ and $\mathbb{Q}(\Omega \times A') = \mathbb{P}'(A')$. By extension, two random variables *X* and *X'* on these two

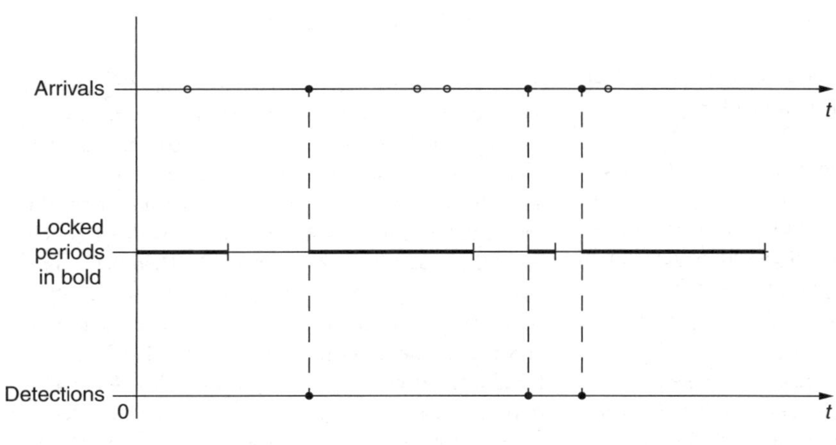

Undetected arrivals during locked (or dead) periods are denoted by ○, yielding a process of detected arrivals denoted by •.

spaces can be coupled by forming a random vector (Y, Y') defined on the product space such that Y has the distribution of X, and Y' has the distribution of X'. The most elementary coupling is the independence coupling, when Y and Y' are independent; but dependent couplings are commonly seen.

More narrowly, a coupling of two random process $X(t)$ and $X'(t)$ having the same state space S, is a bivariate process $(Y(t), Y'(t))$, defined on the natural product space, such that $Y(t)$ has the same distribution as $X(t)$, and $Y'(t)$ the same distribution as $X'(t)$; together with a time T (usually random), such that $Y(t) = Y'(t), t \geq T$. In this case T is called the coupling time; if $\mathbb{P}(T < \infty) = 1$ then one has a successful coupling, and two processes are said to admit coupling if there is a coupling which is successful for any initial distribution of $X(t)$ and $Y(t)$ at $t = 0$. If there is a non-zero probability that the processes never meet (or are always distant at least $d > 0$ from each other) one speaks of a shy coupling, and the processes exhibit shyness. More generally, one has an ε-coupling if the process are such that for two a.s. finite random times S and $T, X(S) = Y(T)$, where $|S - T| < \varepsilon$ and $X(t) = Y(t)$ for $t > \max\{S, T\}$.

There are many ways of arranging a coupling of random processes; examples include:

1. **Basic coupling** (alias: Vasershtein, or Wasserstein, coupling).
2. Co-adapted coupling. A coupling of $(X(t), Y(t))$ is co-adapted if there is a common filtration $F_t, t \geq 0$, such that $X(t)$ and $Y(t)$ are both adapted to F_t, and each is separately a Markov process with respect to F_t, i.e. the distribution of $X(t)$ given F_s is the same as that of $X(t)$ given $X(s)$, with the equivalent property for $Y(t)$. (This does not entail that $(X(t), Y(t))$ is Markov.)
3. **Doeblin coupling** (alias: classical coupling). When processes are run independently until they meet, at the coupling time.
4. Efficient coupling. Arises in the context of a coupled pair $(X_1(t), X_2(t))$ of copies of a Markov process $X(t)$, with coupling time T. Denote the eigenvalue of second-largest modulus of the matrix of transition probabilities (or of the generator, more generally) by λ. Then the coupling is said to be efficient if $\mathbb{P}(T \geq t) = \mathrm{O}(\mathrm{e}^{-\lambda t})$; and said to be super-efficient if $\mathbb{P}(T \geq t) = \mathrm{o}(\mathrm{e}^{-\lambda t})$.
5. Exact coupling. Sometimes used to denote a standard form of coupling in which processes meet at time T, to distinguish it from an ε-coupling defined above.
6. Maximal coupling. One that achieves the upper bound in the coupling inequality.
7. Mineka coupling. Arises as a coupling designed for sums of independent r.vs that are not necessarily identically distributed.
8. Ornstein coupling. Arises in coupling random walks on the integers.
9. Reflection coupling. Arises in coupling diffusions in $\mathbb{R}^d, d \geq 2$. For example, given Brownian motion $X(t), t \geq 0$, started at x, one creates a Brownian motion starting at $y \neq x$ whose path is the reflexion of X in the hyperplane perpendicularly bisecting the line joining x to y; coupling occurs on this plane.
10. Shift coupling. A form of coupling in which the paths of two process $X(t)$ and $Y(t)$ coincide up to a random time shift; it can be seen as an ε-coupling (defined above) in which one abandons the requirement that $|S - T| < \varepsilon$.

Finally, so-called **weak couplings** are accomplished in terms of distributions, rather than sample paths. See also Kendall–Cranston coupling. [T. Lindvall, *Lectures on the Coupling Method.* Wiley, 1992; reprinted by Dover, 2002; W. Doeblin & H. Cohn (eds), Doeblin and modern probability, in which are: *The Coupling Technique in Interacting Particle Systems*, T. Liggett; H. Thorisson, *Coupling and Shift Coupling.* American Mathematical Society, 1993]

coupling inequality (alias: **coupling bound; coupling lemma**) A term used in both a broad and a narrow sense. Broadly, let X and Y be r.vs defined on a common probability space and taking values in the same state space. Then the coupling inequality bounds the total variation distance thus:

$$d_{TV}(X, Y) = 2 \sup_A |\mathbb{P}(X \in A) - \mathbb{P}(Y \in A)| \leq 2\mathbb{P}(X \neq Y).$$

(Note that some writers define d_{TV} as half the above, in which case the factor 2 is omitted in two terms.) If equality holds, then one refers to a maximal coupling; such a coupling always exists. More narrowly, let $X(t)$ and $Y(t)$ be processes defined on a common probability space, and having the same state space, with T a coupling time such that $X(t) = Y(t)$ for $t \geq T$. Then the coupling inequality is $d_{TV}(X(t), Y(t)) \leq \mathbb{P}(T > t)$. If equality holds, then the coupling is maximal. The existence of a maximal coupling for homogeneous Markov chains is **Griffeath's thm**; for general processes it is **Goldstein's thm**. [H. Thorisson, *Coupling, Stationarity, and Regeneration.* Springer, 2000; S. Goldstein, Maximal coupling, *Probab. Theor. Relat. Fields*, 46, 193–201, 1979]

coupon collecting problem *see* collector problem

Cournot's principle One of the two ways in which mathematical (axiomatic) probability is applied to the real world; the other being the stability of statistical ratios. It was expressed thus by A. Kolmogorov, in terms of an event A that arises from an experiment E with probability $\mathbb{P}(A)$: 'If $\mathbb{P}(A)$ is very small, one can be practically certain that in only one realization of E the event A would not occur at all.' [A. N. Kolmogorov, *Foundations of the Theory of Probability*, 2nd English edn. Chelsea, 1956]

covariance Real-valued random variables X and Y with finite variance, have covariance

$$\begin{aligned}\mathrm{cov}(X,Y) &= \mathbb{E}\{(X-\mathbb{E}X)(Y-\mathbb{E}Y)\}\\ &= \mathbb{E}(XY)-\mathbb{E}X\mathbb{E}Y = \mathrm{cov}(Y,X).\end{aligned}$$

If $\mathrm{cov}(X,Y)=0$, then X and Y are said to be uncorrelated. Independence of X and Y entails $\mathrm{cov}(X,Y)=0$, but not conversely. For complex-valued X and Y, $\mathrm{cov}(X,Y)=\mathbb{E}\{(X-\mathbb{E}X)(\overline{Y-\mathbb{E}Y})\} = \overline{\mathrm{cov}(Y,X)}$, where the overbar denotes the complex conjugate. For random vectors $\mathbf{X}=(X_1,\ldots,X_m)$ and $\mathbf{Y}=(Y_1,\ldots,Y_n)$, the matrix V whose (i,j)-th entry is $V_{ij}=\mathrm{cov}(X_i,Y_j), 1\le i\le m, 1\le j\le n$, is called the **covariance matrix** (and also the dispersion matrix). It is most often seen when $\mathbf{X}=\mathbf{Y}$, in which case $V_{ii}=\mathrm{var}\,X_i$ and it may therefore be called the variance–covariance matrix of $\mathbf{X}$, or simply the variance (matrix) of $\mathbf{X}$; var $\mathbf{X}$. In this case $V=\mathrm{var}\,\mathbf{X}$ is positive semi-definite (or non-negative-definite); if it is not strictly positive-definite then the components of $\mathbf{X}$ are a.s. linearly dependent. Replacing each V_{ij} by the corresponding correlation coefficient $\rho(X_i,X_j)$ yields the **correlation matrix** (having 1s on the diagonal), when $0<\mathrm{var}\,X_i<\infty$ for all i. (V is sometimes called the moment matrix, but this term has other uses.) [W. Feller, *Introduction to Probability Theory and its Applications*, Vol. 2, 2nd edn. Wiley, 1971]

covariance density Arises naturally in the context of a suitable point process X on $\mathbb{R}$, say; let $N(s,t)$ be the number of points of X in (s,t). Then if $\lim_{h,k\to 0}\mathrm{cov}(N(s,s+h),N(t,t+k))/(hk)=f(s,t)$, the function f is called the covariance density. More generally, when $M(\cdot)$ is a suitable random measure and $\mathrm{cov}(M(A),M(B)$ is absolutely continuous with respect to Lebesgue measure, a covariance density arises similarly. [D. R. Cox & V. Isham, *Point Processes*. Chapman & Hall, 1980]

covariance representation Let $\mathbf{X}$ and $\mathbf{Y}$ be independent standard multivariate normal random variables in $\mathbb{R}^n$. Then for suitably differentiable functions f and g, from $\mathbb{R}^n$ to $\mathbb{R}$

$$\mathrm{cov}(f(\mathbf{X}),g(\mathbf{Y}))=\int_0^1 \mathbb{E}\langle\nabla f(\mathbf{X}),\nabla g(u\mathbf{X}+\sqrt{1-u^2}\mathbf{Y})\rangle \mathrm{d}u,$$

where ∇ denotes the usual gradient differential operator and $\langle .,. \rangle$ is the usual inner product (dot product) on $\mathbb{R}^n$. An analogous identity holds for functions f and g on $\{-1,1\}^n$, called the **Bernoulli covariance representation**. [S. Bobkov *et al.*, On Gaussian and Bernoulli covariance representations, *Bernoulli*, 7(3), 439–451, 2001]

covariance stationary (alias: **weakly, or wide-sense, or second-order stationary**) A random process $X(t), t\in T$, is said to be covariance stationary if $\mathbb{E}|X(t)|^2<\infty, t\in T$, and $c(t)=\mathrm{cov}(X(s+t),X(s))$ does not depend on s. For complex-valued processes, equivalently, $\mathrm{cov}(X(s+t),\overline{X(s)})$ does not depend on s. It is customary, but not essential, to require also that $\mathbb{E}X(t)$ does not depend on t. [J. Doob, *Stochastic Processes*. Wiley, 1953]

covariation Arises in seeking to characterize dependence among random variables and processes that do not have a covariance. In particular, for r.vs X and Y that are jointly stable of order (or index) $\alpha\in(1,2]$. If the spectral (Feldheim) measure of X and Y is $S(\cdot,\cdot)$, then the covariation of X on Y is $\mathrm{covar}[X,Y]=\iint uv^{\langle\alpha-1\rangle}S(\mathrm{d}u,\mathrm{d}v)$, where $x^{\langle y\rangle}=|x|^y$ sign $x, x, y\in\mathbb{R}$. When X and Y are binormal, with $\alpha=2$, $\mathrm{covar}[X,Y]=\frac{1}{2}\mathrm{cov}(X,Y)$. [G. Samorodnitsky & M. Taqqu, *Stable Non-Gaussian Process*. Chapman & Hall, 1994]

covariation process (alias: **cross-variation; quadratic covariation process**) Arises in the context of continuous local martingales $X(t)$ and $Y(t)$ defined on the same filtration. Then the covariation process of X and Y is given in terms of the ordinary quadratic variation by $\langle X,Y\rangle=\frac{1}{4}(\langle X+Y,X+Y\rangle-\langle X-Y,X-Y\rangle)$. This formula is sometimes called the **polarization identity**, and $\langle X,Y\rangle$ may be called the **brackets process**. Its importance resides in the fact that $XY-\langle X,Y\rangle$ is a continuous local martingale. [O. Kallenberg, *Foundations of Modern Probability*, 2nd edn. Springer, 2002]

coverage problem Broadly, coverage problems consider the properties of the unions and intersections of a family of random subsets of some set Ω. For a classic elementary example, let n intervals in $\mathbb{R}$ each of length a have their midpoints placed independently and uniformly at random in an interval I of $\mathbb{R}$ of length 1. What is the probability that their union includes I? Or covers it twice over? What is the expected size of the set of points included in at least r intervals? When $\Omega = \mathbb{R}^n, n \geq 3$, these become space filling problems. More generally, the sets may be of random size and shape, and centred at the points of some point process in Ω. And in a variant, the sets may be disjoint, as in packing and parking problems. [P. Hall, *Introduction to the Theory of Coverage Problems*. Wiley, 1988]

covering function Arises in stochastic geometry. Suppose X is a random closed subset of $\mathbb{R}^n$ (say), defined on a suitable space $(\Omega, \mathcal{F}, \mathbb{P})$. Then the covering function of X defined for $\mathbf{r} \in \mathbb{R}^n$, is the probability $P(\mathbf{r})$ that $\mathbf{r}$ lies in X, i.e. $P(\mathbf{r}) = \mathbb{P}(\{\omega : \mathbf{r} \in X(\omega)\}$. [D. Stoyan, Random sets, *Int. Stat. Rev.*, 6(1), 1–27, 1998]

cover time (alias: **exploration time**) Arises in Markov (and semi-Markov) processes on a finite state space S. For each $i \in S$, the cover time starting from i is the number of transitions (steps) until the chain has visited every other state in S at least once. That is, defining T_{ij} to be the hitting time of j from i, $C_i = \max_{j \in S} T_{ij}$. Then the cover time for the chain is $C = \max_{i \in S} \mathbb{E} C_i$.

The definition extends in the natural way to chains in continuous time. A related quantity is the **cover and return time** C_i^+ defined for $i \in S$ by $C_i^+ = \min\{n \geq C_i : X_n = i\}$, with $C^+ = \max_{i \in S} \mathbb{E} C_i^+$. [D. Levin *et al.*, *Markov Chains and Mixing Times*. American Mathematical Society, 2009]

Cover's universal portfolio *see* universal portfolio

Cox–Grimmett coefficient Arises in the Cox–Grimmett CLT for random variables indexed by $\mathbb{Z}^d, d \geq 1$. In its elementary form, when $d = 1$, it is this: let $X_1, X_2, \ldots$ be a sequence of zero mean random variables that are *associated*. If $\inf_{r \geq 1} \operatorname{var} X_r > 0$, and $\sup_{r \geq 1} \mathbb{E}|X_r|^3 < \infty$, and $\infty > u(n) = \sup_{r \geq 1} \sum_{|j-r| \geq n \geq 0} \operatorname{cov}(X_j, X_r) \to 0$ as $n \to \infty$, then $\sum_{r=1}^n X_r \{\operatorname{var} \sum_{r=1}^n X_r\}^{-\frac{1}{2}}$ converges in distribution to $N(0,1)$ as $n \to \infty$, and $u(n)$ is called the Cox–Grimmett coefficient. [J. T. Cox & G. Grimmett, CLTs for associated random variables and the percolation model, *Ann. Probab.*, 12(2), 514–528, 1984]

Coxian distribution (alias: **phase-type distribution; branching Erlang distribution**) Arises naturally in the theory of queueing networks; the family of Coxian distributions is useful because they have mgfs that are rational functions of the argument t (where $M(t) = \mathbb{E}e^{-tX}$ is the mgf of X), and they are dense in the set of distributions of positive r.vs (so that any positive r.v. can be approximated in distribution by a Coxian r.v., as closely as we please). Explicitly, let $X_1, X_2, \ldots$ be a sequence of independent exponentially distributed r.vs with respective parameters $\lambda_1, \lambda_2, \ldots$, such that $\lambda_r \neq \lambda_s$ for $r \neq s$. Let $0 \leq p_r \leq 1$ for $1 \leq r \leq n$, with $p_n = 1$, set $q_r = 1 - p_r$, and define the random variable N independent of the X_s, having distribution $\mathbb{P}(N = r) = q_1 q_2 \ldots q_{r-1} p_r$. Then the r.v. $Z = \sum_{r=1}^N X_r$ has a Coxian distn; if $M < N$ is another r.v. independent of the X_s then $Y = \sum_{r=M}^N X_r$ is a generalized Coxian distn.

If $\mathbb{P}(N = n) > 0$, then one has a Coxian-n (or Cox-n) distn. The distribution can be seen as a mixture of hypoexponential (or Erlang, or phase) distns. It can also be seen as the time to absorption of a suitable Markov chain in continuous time. *See* method of phases; phase-type distribution. [S. Asmussen *et al.*, Fitting phase-type distns, *Scand. J. Stat.*, 23(4), 419–441, 1996]

Cox–Ingersoll–Ross model (alias: **CIR process; mean-reverting square-root diffusion process**) In financial mathematics this is an interest rate model for the instantaneous short rate $r(t)$ that postulates the SDE $dr(t) = \mathcal{K}(\theta - r)dt + \sigma\sqrt{r}dW(t)$, where $\mathcal{K}, \theta > 0$, and $W(t)$ is a Wiener process. [J. C. Cox *et al.*, Theory of the term structure of interest rates, *Econometrica*, 53(2), 385–407, 1985]

Cox process (alias: **doubly stochastic Poisson process; mixed Poisson process; Poisson process in a random environment**) Arises as a non-homogeneous Poisson process whose intensity $\lambda(t)$ is itself a random process. Formally, let $\Lambda(t)$ be a non-negative non-decreasing random process for $t \geq 0$, independent of an ordinary Poisson process of constant intensity $N(t), t \geq 0$. Then $C(t) = N(\Lambda(t))$ is a Cox process. By the conditional variance theorem, $\operatorname{var} C(t) = \int_0^t \mathbb{E}\Lambda(y)dy + \operatorname{var} \int_0^t \Lambda(y)dy \geq \mathbb{E}C(t)$, and $C(t)$ is said to be overdispersed, unless equality holds when $\int_0^t \Lambda(y)dy$ is non-random. Some authors, especially in the actuarial literature, call $\Lambda(t)$ the structure function. If $N(t)$ is compound, one may speak of a doubly compound Poisson process. [D. R. Cox, Some statistical methods connected with series of events, *J. Roy. Stat. Soc. B*, 17(2), 129–164, 1955]

Cox's theorem R. T. Cox addressed the problem of logical inference under uncertainty, i.e. the question of how far one should believe statements that are supported by uncertain evidence. The question is to prove that any measure of belief (a belief function) satisfying a number of assumptions about what constitutes rational and consistent behaviour must obey essentially the same rules as a probability measure, up to isomorphism. Such a theorem would be significant for the interpretation of probability as a degree of belief (or logical support). The required theorem holds under a number of different sets of assumptions. [J. Halpern, Cox's theorem revisited, *J. Artif. Intel. Res.*, 11, 429–435, 1999]

C_p inequality (alias: **C_r inequality**) For any random variables X and Y, and $p > 0, \mathbb{E}(|X+Y|^p) \leq C_p\{\mathbb{E}(|X|^p) + \mathbb{E}(|Y|^p)\}$, where $C_p = 1$ for $0 < p \leq 1$, and $C_p = 2^{p-1}$ for $p > 1$. More generally, $\mathbb{E}|X_1 + \cdots + X_n|^p \leq C_p \sum_{r=1}^{n} \mathbb{E}|X_r|^p$, where $C_p = 1$ for $0 < p \leq 1$, and $C_p = n^{p-1}$ for $p > 1$. [M. Loéve, *Probability Theory*, 4th edn. Springer, 1978]

Craig–Hotelling theorem (alias: **Craig's theorem**) In the original form, this asserts that for a vector $\mathbf{X} = (X_1, \ldots, X_n)$ of independent standard normal r.vs, and real symmetric $n \times n$ matrices A and B, the quadratic forms $\mathbf{X}'A\mathbf{X}$ and $\mathbf{X}'B\mathbf{X}$ are independent iff $AB = 0$, (where $\mathbf{X}'$ is the transpose). The following more general result, due to several later authors, is often called Craig's thm: let A and B be real symmetric matrices, and let $\mathbf{X} = (X_1, \ldots, X_n)$ have a multivariate normal distribution with mean vector μ, and covariance matrix V, that may be singular. Then $\mathbf{X}'A\mathbf{X}$ and $\mathbf{X}'B\mathbf{X}$ are independent iff $VAVBV = 0 = VAVB\mu$ and $VBVA\mu = 0 = \mu'AVB\mu$. If V is not singular, then the condition $AVB = 0$ is necessary and sufficient. [H. Hotelling, A note on a matric theorem of A. T. Craig, *Ann. Math. Stat.*, 15, 427–429, 1944]

Cramér's condition A term that may refer to any one of several conditions on a random variable, most commonly one of these:

1. The strongly non-lattice condition (also called Cramér's condition (C)). This holds for a r.v. X with characteristic function $\phi(t)$ if $\lim_{|t|\to\infty}\sup\phi(t) < 1$. No discrete r.v. can be strongly non-lattice.
2. Cramér's moment condition (also called the light tail condition). This holds for a r.v. X if for some $a > 0, \mathbb{E}e^{t|X|} < \infty$ for $|t| < a$ (or, in variant form, for some $a < \infty$ and $r \geq 2$, we have $\mathbb{E}|X|^r \leq r!a^r\mathbb{E}X^2$).
3. Regularity conditions (i.e. suitable differentiability) required of density functions for the validity of certain theorems in statistics, e.g. Cramér–Rao lower bound, and consistency of MLE estimators.

[H. Cramér, *Collected Works*. Springer, 1993]

Cramér's criterion Asserts that a continuous and bounded function $\phi(t), t \in \mathbb{R}$, is the characteristic fn of a r.v. X iff for all $a > 0$, and $x \in \mathbb{R}$, the integral $\int_0^a \int_0^a \exp\{i(s-t)x\}\phi(s-t)\mathrm{d}s\mathrm{d}t$ is real and non-negative, and $\phi(0) = 1$. [E. Lukacs, *Characteristic Functions*, 2nd edn. Griffin, 1970]

Cramér's formula *see* Pollaczek–Khinchin formula

Cramér–Leadbetter theorem Supplies a formula for the factorial moments of the number of upcrossings of a level u during $0 \leq t \leq T$ by a stationary Gaussian process, whose sample path has a continuous derivative with probability 1 on $[0, T]$. [H. Cramér & M. Leadbetter, Moments of the number of crossings of a level, *Ann. Math. Stat.*, 36(6), 1656–1663, 1965]

Cramér–Lévy theorem (alias: **Cramér's theorem**) Supplies this characterization of the normal distribution: If random variables X and Y are independent, and $X+Y$ is normally distributed, then X and Y are both normal random variables. That is to say, a normal r.v. Z cannot be decomposed as the sum of two independent r.vs; a property sometimes referred to as primality under convolution. [W. Feller, *Introduction to Probability Theory and its Applications*, Vol. 2, 2nd edn. Wiley, 1971]

Cramér–Lundberg model (alias: **compound Poisson model**) *see* risk process

Cramér's model A random sequence suggested by H. Cramér as a plausible analogue for generating conjectures about the actual distribution of prime numbers. Thus, let $X_n, n \geq 2$ be a sequence of independent random variables such that

$$\mathbb{P}(X_n = 1) = (\log n)^{-1} = 1 - \mathbb{P}(X_n = 0), n \geq 3,$$

and consider the random sequence $S = \{n \geq 2 : X_n = 1\}$ (where $X_2 = 1$). Then any property that holds for S with probability 1 may be conjectured as being true for the actual sequence of primes. For example, **Cramér's conjecture** is the statement that as $x \to \infty$, the largest gap between pairs of consecutive primes, both less than x, is of order $(\log x)^2$. [H. Cramér, On the order of magnitude of the difference between consecutive primes, *Acta Arithmetica*, 2, 23–46, 1937]

Cramér's theorem (alias: **Cramér's expansion; Cramér's large deviations theorem**) A classic limit theorem for the probability of large deviations of the partial sums $S_n = \sum_{r=1}^{n} X_r$ of a sequence of i.i.d.r.vs X_r with finite variance σ^2. Let $F_n(x) = \mathbb{P}(S_n/(\sigma\sqrt{n}) \leq x)$, and $\Phi(x)$ denote the standard normal distribution, and assume Cramér's condition of light tails, i.e. $\mathbb{E}e^{tX_1} < \infty$ for $|t| < a > 0$. Then for $x = o(\sqrt{n})$ as $n \to \infty$,

$$\frac{1 - F_n(x)}{1 - \Phi(x)} = \exp\left\{\frac{x^3}{\sqrt{n}}\lambda\left(\frac{x}{\sqrt{n}}\right)\left(1 + O\left(\frac{x}{\sqrt{n}}\right)\right)\right\},$$

where $\lambda(y) = \sum_{r=0}^{\infty} C_r y^r$ is the Cramér series, whose coefficients depend only on the moments of X_1 (with a similar statement for $F_n(-x)/\Phi(-x)$). Subsequent sharper large deviation theorems for sums of i.i.d.r.vs, proved by different methods, may also be called Cramér's thm, or the Cramér–Chernoff thm. [D. Stroock, *Probability Theory: An Analytic View*. Cambridge University Press, 1993]

Cramér's transform (alias: **Legendre transform; Fenchel–Legendre transform; rate function; Chernoff transform; Cramér–Chernoff transform**) Arises in the theory of large deviations. Let $M(t) = \mathbb{E}e^{tX_1} < \infty, \text{for}|t| < a > 0$, with cumulant generating function $K(t) = \log M(t)$. Then the Cramér transform (of the distribution $F(x)$ of X_1) is $I(a) = \sup_{x\in\mathbb{R}}\{ax - K(x)\}, a \in \mathbb{R}$. For example, if $X_1 isN(0, \sigma^2)$, then $I(a) = \frac{1}{2}\frac{a^2}{\sigma^2}$. Its importance is illustrated by the Cramér–Chernoff thm (alias Cramér's thm), which asserts that for $S_n = \sum_{r=1}^{n} X_r$ where the X_r are i.i.d.r.vs with the mgf $M(t)$ as defined above, $\mathbb{P}(S_n \geq nx) \leq e^{-nI(x)}, x > \mathbb{E}X_1$, and furthermore that this bound is tight in the sense that

$$\lim_{n\to\infty}\left(\frac{-1}{n}\log\,\mathbb{P}(S_n \geq nx)\right) = I(x).$$

Likewise $\mathbb{P}(S_n \leq nx) \leq e^{-nI(x)}, x \leq \mathbb{E}X_1$. Yet further, the Bahadur–Rao thm asserts that $\mathbb{P}(S_n \geq nx)$ is asymptotic to $e^{-nI(x)}/\{2\pi n\hat{x}^2 K''(\hat{x})\}^{1/2}$, where $\hat{x}$ is the value of x yielding the supremum in the definition of $I(a)$, and K'' is the second derivative of $K(x)$. [O. Kallenberg, *Foundations of Modern Probability*, 2nd edn. Springer, 2002]

Cramér–Wold device (alias: **C-W theorem**) Arises in considering the convergence in distribution of a sequence of random vectors in $\mathbb{R}^d$, where it supplies a method (or device) for reducing the problem to convergence in one dimension. Thus, let $X_n, n \geq 1$, be a sequence of random vectors; then X_n converges in distribution to X iff $t^T X_n$ converges in distribution to $t^T X$ (as $n \to \infty$) for every $t \in \mathbb{R}^d$. Part of the theorem entails the assertion that if $F(\mathbf{x})$ and $G(\mathbf{x})$ are two distributions that agree on every half-space of $\mathbb{R}^d$ of the form $\sum_{r=1}^{d} t_r x_r \leq z$, then $F(\mathbf{x}) \equiv G(\mathbf{x})$. [H. Cramér & H. Wold, Some theorems on distribution functions, *J. Lond. Math. Soc.*, 11, 290–295, 1936]

credal set Arises in decision-making (or inference) under uncertainty, in a particular model, that can be seen as an extension of the methods of Bayesian decision theory in which belief is described by a prior distribution that is updated (by Bayes theorem) as new evidence appears. In the extended form, belief is modelled by a convex set of probability distributions, called a credal set. [F. Cozman, Credal networks, *Artif. Intell.*, 120, 199–233, 2000]

credence *see* degree of belief

Cressie–Read divergence A measure of divergence between probability distributions $(p_1, \ldots, p_n)$ and $(q_1, \ldots, q_n)$ given for $a \in \mathbb{R}$ by

$$d(p, q) = \frac{2}{a(a+1)}\sum_{r=1}^{n} p_r\left(\left(\frac{p_r}{q_r}\right)^a - 1\right).$$

[N. Cressie & T. Read, Multinomial goodness-of-fit-tests, *J. Roy. Stat. Soc. B*, 46(3), 440–464, 1984]

critical parameter Arises in many probability models of real-world phenomena. Commonly, such systems exhibit qualitatively different behaviour for different values of some parameter p. If there is a value p_c of p separating regimes in which the system acts differently, then p_c is the critical value (or point) of the parameter p. When p is a vector, there may be a critical line or a critical surface separating different behaviours. A classic example is the Galton–Watson branding process for which the critical value of the mean family size μ is $\mu = \mu_c = 1$, separating certain extinction from possible extinction. More generally, in physical systems, the change of regimes may be called a phase transition. If some measurable property of the system, that depends on p and is denoted by $f(p)$, is given approximately near p_c by a relationship of the form $f = |p - p_c|^a, p > p_c$ (and $f = |p - p_c|^b, p < p_c$), then a may be called a **critical exponent** (as may b). [G. Grimmett, *Percolation*, 2nd edn, 1999; and *The Random-cluster Model*. Springer, 2006]

Croft–Kingman lemma Arises in the context of random processes in continuous time, when one seeks to derive theorems about $X(t), t \geq 0$, by applying known theorems about random processes in discrete time to the discrete h-skeletons $X(nh), n \geq 0, h > 0$, of $X(t)$. The most commonly seen lemma asserts that for a continuous function $f(t) \in \mathbb{R}, 0 < t < \infty$, such that $\lambda(h) = \lim_{n\to\infty} f(nh)$ exists and is finite for $h > 0$, the limit $L = \lambda(h)$ is independent of h and $\lim_{t\to\infty} f(t) = L$. [J. F. C. Kingman, Ergodic properties of continuous-time Markov processes and their discrete skeletons, *Proc. Lond.. Math. Soc.*, 13, 593–604, 1963]

Crofton's theorem (alias: **Crofton's formula**) Any of a number of results in geometrical probability (stochastic geometry) and various generalizations in integral geometry. In their simplest form, two such formulas are these:

1. Let $Q_1, \ldots, Q_n$ be n points independently and uniformly distributed in some domain S, whose volume (or measure) is s, and let $X(s) = X(Q_1, \ldots, Q_n)$ be a random variable that depends on the positions of the points, where $m(s) = \mathbb{E}X(s)$. Denote the boundary of S by B. Then, subject to some regularity conditions on S, Crofton's formula asserts that

$$\frac{\mathrm{d}m(s)}{\mathrm{d}s} = \frac{n}{s}\{\mathbb{E}(X|\mathbb{Q}_1 \in B) - m(s)\}.$$

2. The concept of picking a line at random can be made precise in terms of the uniform measure $\mathrm{d}\mu = \mathrm{d}p\,\mathrm{d}\theta$ on the space L of all straight lines in the plane, where p is the perpendicular distance from the origin, and θ the orientation, of a such a line λ. If S is any rectifiable curve in the plane, having length $L(S)$, and $n(S, \lambda)$ is the number of intersections of S with a line λ, then Crofton's formula is

$$\int_{\lambda \in L} n(S, \lambda)\mathrm{d}p\,\mathrm{d}\theta = 2L(S).$$

In the particular case when S is a convex region, and K is any convex subregion of S, Crofton's formula yields the probability that a random line intersects K, conditional on its intersecting S, as $L(K)/L(S)$. Similarly, denoting the intersection of the compact interior I of S with λ by $d(S, \lambda)$, $\int_{\lambda \in L} d(S, \lambda)\mathrm{d}p\,\mathrm{d}\theta = \pi a(S)$ where $a(S)$ is the area of I. [D. Stoyan *et al.*, *Stochastic Geometry and its Applications*, 2nd edn. Wiley, 1995; R. Coleman, Intersections of random lines, *Rend. Circolo Mat. Palermo*, II(65), 67–77, 2000]

Cromwell's rule Arises in Bayesian decision theory and inference where it asserts the inadvisability of assigning prior probabilities of 0 (or of 1), to any uncertain event. [D. Lindley, *Making Decisions*, 2nd edn. Wiley, 1985]

cross-covariance Arises in the context of a stationary multivariate time series $\mathbf{X}(t)$, where the (auto) covariance matrix of the process, with lag h, is

$$A(h) = E\{(\mathbf{X}(t+h) - \mathbf{m})(\mathbf{X}(t) - \mathbf{m})^{\mathrm{T}}\} = (a_{ij})$$

where $\mathbf{m} = \mathbb{E}\mathbf{X}(t)$, and T denotes transpose. Then each diagonal entry $a_{ii}(h)$ is the autocovariance of $X_i(t)$, and each off-diagonal entry $a_{ij}(h)$ is the cross covariance of $X_i(t)$ and $X_j(t)$, The cross-correlation function is obtained on dividing by $[a_{ii}(0)a_{jj}(0)]^{1/2}$. [P. Brockwell & R. Davis, *Time Series*, 2nd edn. Springer, 1991]

cross-entropy *see* Shannon entropy; Kerridge's inaccuracy

crossover probability Arises in a number of contexts, thus:

1. In communication theory, the crossover probability of a symbol entering a binary symmetric channel is the probability that it is transmitted incorrectly.
2. In the context of mathematical genetics, a single model for a chromosome envisages an ordered set of sites (originating from the centromere) each of which has a value drawn from one or other parent. At any site, the probability that the next value is from the other parent is the crossover probability.

crudely stationary (alias: **simple stationarity**) Arises in considering a random point process $N(\cdot)$ on the line, i.e. such that for each interval $I \in \mathbb{R}, N(I)$ is a non-negative integer-valued r.v. that is finite when I is finite. Such a process is crudely (or simply) stationary if $N(I_1)$ and $N(I_2)$ have the same distribution for any two intervals I_1 and I_2 of equal length. [K. L. Chung, Crudely stationary counting processes, *Amer. Math. Mon.*, 79(8), 867–877, 1972]

crude Monte Carlo (alias: **crude experiment**) In Monte Carlo methods (of integration, and related simulation techniques) this denotes the estimation or evaluation of same quantity θ through the sample mean of a number of (simulated) i.i.d.r.vs each having mean θ. Such crude methods are refined by using e.g. control or antithetic variates. [J. Hammersley & D. Handscomb, *Monte Carlo Methods*. Methuen, 1964]

Crump–Mode–Jagers process A branching process in which individuals produce offspring at random during their lifetime (and, more generally, after their lifetime). Formally any individual I born into the population at time A_I has a random lifetime L_I and is removed at time $A_I + L_I$. The offspring born to I comprise a point process N_I on $[A_I, \infty)$, which will (in most cases) terminate at $A_I + L_I$. The joint distribution of L_I and N_I is the same for all individuals, and individuals behave independently with respect to lifetime and family. The special case when offspring all appear at $A_I + L_I$ is the Bellman–Harris process. [K Crump & C. Mode, A general age-dependent branching process I and II, *J. Math. Anal. Appl.*, 24, 494–508, 1968; 25, 8–17, 1969]

crypto-deterministic A term introduced by E. T. Whittaker to describe systems whose evolution is deterministic, but whose initial conditions are random. [J. Moyal, Causality, determinism, and probability, *Philosophy*, 24 (91), 310–317, 1949]

cryptography Broadly, the art of communicating privately and securely, using channels that are themselves not necessarily private or secure. Effectively, this entails the encryption (or encoding) of the original message (or datastream) in such a way that an unauthorized interceptor of the cryptogram cannot recover the original message (the secrecy problem), or substitute a message that purports to come from the sender (the authentication problem), or convince the recipient that the message is from a different sender (the signature problem). These tasks of the interceptor, i.e. breaking the code, are referred to as cryptanalysis, and the entire field of study is cryptology. Cryptographers seek to accomplish their ends by constructing an effective secrecy system, in which the concepts of randomness and complexity are key. [O. Goldreich, *Foundations of Cryptography*. Cambridge University Press, Vol. I, 2001; Vol. II, 2004]

Csáki's formula Arises in the standard Brownian bridge $V(t), 0 \le t \le 1$. Let $M(t)$ be the supremum, and $m(t)$ the infimum of $V(t)$ for $0 \le t \le 1$. Then

$$\mathbb{P}(M(t)/(M(t) - m(t)) \le v) = (1 - v)(1 - \pi v \cot(\pi v)), \quad 0 < v < 1.$$

[D. Revuz, & M. Yor, *Continuous Martingales and Brownian Motion*, 3rd edn. Springer, 1999]

Csiszár divergence *see* f-divergence

cumulant function *see* Laplace exponent

cumulants (alias: **semi-invariants; half-invariants**) A sequence of numbers associated with a random variable X having characteristic function $\phi(t)$, that may be defined as follows. Since $\phi(t)$ is continuous, and $\phi(0) = 1$, there exists $a > 0$ such that $\phi(t) \neq 0, |t| \le a$; and so we can define $K(t) = \log\, \phi(t), |t| \le a$ (where $\log\, \phi = 0$ for $\phi = 1$). Then $K(t)$ is called the **cumulant generating function** of X (or the second characteristic function of X). If $\mathbb{E}(|X|^n) < \infty$, then there exist $K_1, K_2, \ldots, K_n$ such that $K(t) = \sum_{r=1}^{n} (it)^r K_r/r! + o(t^n)$, and K_r is called the rth cumulant of X. They have the property that, for independent X and Y, $K_r(X + Y) = K_r(X) + K_r(Y)$, in an obvious notation. The so-called joint cumulants of $X_1, \ldots, X_n$ are obtained by a similar expansion of $\log\{\mathbb{E}\, \exp(\mathrm{i}\sum_{r=1}^{n} t_r X_r)\}$, and may also be called cross-cumulants. It is a famous theorem of J. Marcinkiewicz that if, for some positive integer k, $K_r(X) = 0$ for $r \ge k$, then X is a normal r.v.

If the moment generating function $M(t) = \mathbb{E}\mathrm{e}^{tX}$ is finite for $|t| < a > 0$, the function

$$\hat{K}(t) = \log M(t) = \sum_{r=1}^{\infty} \frac{1}{r!} K_r(X) t^r$$

also generates the cumulants, and is often called the kumulant function to avoid confusion with $K(t)$. When $X \ge 0$, the log Laplace transform $L(t)$ is used: $L(t) = \log\, \mathbb{E}\mathrm{e}^{-tX}$. [J. Marcinkiewicz, Sur un propriété de large loi de Gauss, *Math. Zeit.*, 44, 612–618, 1939]

cumulative damage model *see* shock model

current life *see* age

cut-off phenomenon (alias: **threshold phenomenon**) Arises in the context of a type of ergodic Markov chain whose distribution exhibits an abrupt transition in its convergence to equilibrium. Broadly, the chain is nowhere near equilibrium before this transition interval, but essentially in its steady state afterwards. Formally, consider a family of Markov chains indexed by a parameter $n \ge 1$, where each is ergodic and takes values in a finite state space. Let $d_n(t)$ be the total variation distance between the distribution of the nth chain at time t and its stationary distribution, and define the a-mixing time for the nth chain by $T_n(a) = \min[t : d_n(t) < a]$, $0 < a < 1$. Then the sequence displays (or has, or exhibits) cut-off if $\lim_{n\to\infty} T_n(a)/T_n(1 - a) = 1$, $0 < a < 1$; or, equivalently, as $n \to \infty$,

$$\lim d_n(cT_n(a)) = \begin{cases} 1, & \text{if} c < 1; \\ 0, & \text{if} c > 1. \end{cases}.$$

The cut-off at T_n has cut-off window W_n if $W_n = o(T_n)$ and both of these hold: $\lim_{b\to\infty}\lim\inf_{n\to\infty} d_n(T_n - bW_n) = 1; \lim_{b\to\infty}\lim\inf_{n\to\infty} d_n(T_n + bW_n) = 0$. Informally, one refers to the cut-off time as the time for the chain to reach equilibrium, or mix. [D. Levin *et al.*, *Markov Chains and Mixing Times*. American Mathematical Society, 2009]

cut time Given a random process $X(t) \in \mathbb{R}^d, 0 \le t \le T$, a time K is called a cut time over $[0, T]$ for X if $[X(u) : 0 \le u < K] \cap [X(v) : K < v \le T] = \emptyset$. That is, the path of X up to time K has no point in common with path after K up to T; then $X(K)$ may be called the cut point. For example, when $X(t)$ is Brownian motion, the Hausdorff dimension of the set of cut times over $[0, 1]$ equals $1 - \mathrm{b}_d$, where b_d is the intersection exponent. [G. Lawler, Hausdorff dimension of cut points of Brownian motion, *Elec. J. Probab.*, 1, 1–20, 1996]

cycle (alias: **cycle time**) Arises in the theory of random processes in describing the recurrent completions of some round of events. For example, in a single server queue, the time between successive instants at which an arriving customer finds the server free is a busy cycle. More generally, the interval between the regeneration points (or times) of a regenerative system may be called a cycle, e.g. the interval between the successive entries to some given state of a Markov process. [S. Asmussen, *Applied Probability and Queues*, 2nd edn. Springer, 2003]

cycle formula Arises in the classic construction of a positive invariant measure for an irreducible persistent stochastic Markov chain transition matrix P. For any state k, let T_k be the time of first return of the Markov chain X_n to k, which may be called a cycle. Then the vector ρ of expected times spent in each state $i \ne k$ during $[0, T_k - 1]$ supplies the required invariant measure, where

$$\rho_i = \mathbb{E}\left[\sum_{r=0}^{T_k-1} I(X_r = i) | X_0 = k\right], i \in S,$$

is called the cycle formula, where $I\ (X_r = i)$ is the indicator of the event $X_r = i$, and $\rho_k = 1$.

cycle-stationary Let $X(t), t \in \mathbb{R}$, be a suitable (measurable) random process, jointly distributed with a simple point process $N(t), t \in \mathbb{R}$, having a point at zero, i.e. $N(\{0\}) = 1$. The pair is cycle-stationary if the joint distribution of (X, N) is unchanged by a shift of origin to any of the points of $N(t)$. [H. Thorisson, *Coupling, Stationarity, and Regeneration.* Springer, 2000]

cyclically interchangeable (alias: **cyclically exchangeable**) Random variables $X_1, \ldots, X_n$ are cyclically interchangeable if their n cyclic permutations each have the same joint distribution. It is interesting that if the X_r are integer-valued, with maximum value 1, they satisfy Dwass's ballot theorem, for $0 \le k \le n, k$ an integer

$$\mathbb{P}\left(\sum_{r=1}^{i} X_r > 0 \text{ for all } 1 \le i \le n \middle| \sum_{r=1}^{n} X_r = k\right) = \frac{k}{n}.$$

[L. Addario-Bery & B. Reed, Ballot theorems old and new, in *Horizons of Combinatorics*, E. Györi *et al.* (eds). Springer, 2008]

cyclic class Arises in describing the behaviour of Markov chains, when, for an irreducible chain with period d, the states are partitioned into d disjoint sets $C_0, \ldots, C_{d-1}$, such that the chain being in C_r always moves to C_{r+1} at its next step, $0 \le r \le d - 2$ (or from C_{d-1} to C_0), the C_r are called cyclic classes. [J. Doob, *Stochastic Processes*. Wiley, 1953]

cyclic queue May denote either of two types of queue:

1. One where customers queue in any of several queues, and the server moves cyclically around these.
2. A queuing network in which customers move between stations cyclically.

[F. Kelly, *Reversibility and Stochastic Networks*, Wiley, 1979; reprinted by Cambridge University Press, 2011]

cylinder set Arises in considering the infinite product spaces entailed in the analysis of random processes defined on an infinite parameter set, e.g. $\mathbb{Z}$ or $\mathbb{R}$. For example, let $(\Omega_r, \mathcal{F}_r)$ be a sequence of measurable spaces, with Ω denoting the set of all sequences $(\omega_1, \omega_2, \ldots.)$ such that $\omega_r \in \Omega_r, r \ge 1$. If C_n is the set of sequences in Ω such that $(\omega_1, \ldots, \omega_n) \in B(n)$, where $B(n)$ is a set in the product space $\prod_{r=1}^{n}\Omega_r$, then C_n is called the cylinder with base $B(n)$. It is measurable if $B(n) \in \prod_{r=1}^{n}\mathcal{F}_r$, and it is a rectangle if $B(n) = B_1 \times B_2 \times \cdots \times B_n$, where $B_r \in \Omega_r, 1 \le r \le n$. [R. Ash & C. Doléans-Dade, *Probability and Measure Theory*, 2nd edn. Academic Press, 2000]

D

Dagum distribution (alias: **Burr type III distribution; inverse Burr type XII distribution; kappa distribution; generalized log-logistic distribution**) Arising in actuarial mathematics and economics, the basic form of the distribution function is $F(x) = \{1+(\frac{x}{a})^b\}^{-c}$, for $a,\ b,\ c,\ x > 0$. [C. Kleiber & S. Kotz, *Statistical Size Distributions in Economics and Actuarial Sciences*. Wiley, 2003]

Dalang–Morton–Willinger theorem Arises in financial mathematics, where it supplies a quite general form of the first fundamental theorem of asset pricing. Broadly, it asserts that for a sequence of asset-price vectors $S_1, S_2, \ldots, S_T$, each being random vectors taking values in $\mathbb{R}^d$, the condition of no-arbitrage on this sequence entails the existence of an equivalent probability measure, under which the sequence $S_1, S_2, \ldots, S_T$ is a martingale. [F. Delbaen, & W. Schachermayer, *The Mathematics of Arbitrage*. Springer, 2006]

Dambis–Dubins–Schwarz theorem Supplies an important connection between continuous martingales and the Wiener process. Let $M(t), t \geq 0$ be a continuous local martingale with $M(0) = 0$, whose quadratic variation satisfies $\langle M, M\rangle_t \to \infty$ as $t \to \infty$. Then there is a time-change process $T(t)$ such that $B(t) = M(T(t))$ is a Brownian motion (Wiener process), and $M(t) = B(\langle M, M\rangle_t), t \geq 0$. A partial extension to a vector of continuous local martingales is called Knight's theorem. [D. Revuz & M. Yor, *Continuous Martingales and Brownian Motion*, 3rd edn. Springer, 1999]

dam process (alias: **fluid queue**) A classic example of a storage process, characterized by the fact that the dam (or buffer in computing applications) has capacity K, say, and unreleased input that would raise the content of the pool above this level is spilled. Models for such processes may be formulated in discrete time; e.g. the recurrence

$$Z_{n+1} = \begin{cases} K & \text{for } Z_n + W_n - R_n \geq K, \\ 0 & \text{for } Z_n + W_n - R_n \leq 0, \\ Z_n + W_n - R_n & \text{otherwise,} \end{cases}$$

where Z_n denotes the content of the dam, and W_n and R_n represent the random inflow and release at the nth stage. *See also* Moran gamma process. [N. Prabhu, *Stochastic Storage Processes*, 2nd edn. Springer, 1980]

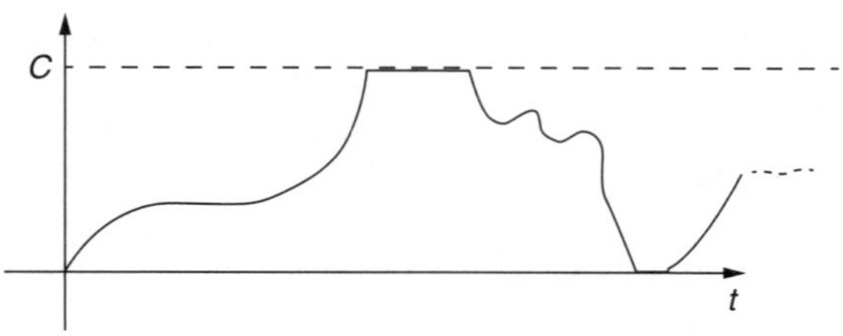

Dam process with capacity *C*.

Daniel Bernoulli's marriage problem A classic (1768) problem in probability. Suppose that $2n$ individuals form n married pairs at time $t = 0$. At succeeding times, each individual may be removed (independently of all the others) with probability given by the appropriate life-table. (In the elementary problem, individuals are of the same age at $t = 0$; more generally, they may be of different ages.) What is the expected number of years until a given pair is reduced to a singleton? And what is the expected number of surviving pairs at $t = r$? [D. Bernoulli, De usu algorithmi infinitesimalis & De duratione matrimoniorum, *Novi Commentarii Acad. Petrop*, 1766–7(XII), 87–98 & 99–126, 1768; D. Steinsaltz, Random time changes for sock-sorting and other stochastic process limit theorems, *Elec. J. Probab.*, 4(14), 1999]

Daniell–Kolmogorov theorem (alias: **Kolmogorov's extension (or existence) theorem**) In its basic form, this supplies conditions for the existence of a random process $X(t), t \in T$, whose finite dimensioned distributions are those of a specified set $F(t; x) = \mathbb{P}(X(t) \leq x_1, \ldots, X(t_n) \leq x_n)$ as t_i and $x_i, 1 \leq i \leq n$, range over all possible values for any given finite n. These so-called consistency conditions (or **Kolmogorov's compatibility conditions**) are:

1. $F(t_1, \ldots, t_{n-1}, t_n; x_1, \ldots, x_{n-1}, x_n) \to F(t_1, \ldots, t_{n-1}; x_1, \ldots, x_{n-1})$, as $x_{n+1} \to \infty$.

2. Let πx denote $(x_{\pi(1)}, \ldots, x_{\pi(n)})$ for any vector x and permutation π of $(1, \ldots, n)$, then $F(t;x) = F(\pi t; \pi x)$ for any n, x, t, and π.

[P. Billingsley, *Probability and Measure*, 3rd edn. Wiley, 1995; G. Shafer & V. Vovk, The sources of Kolmogorov's Grundbegriffe, *Stat. Sci.*, 21(1), 70–98, 3.2, 2006]

Daniels system Arises in the analysis of the reliability of systems. The classical form of the model is presented as a bundle of parallel threads clamped at each end, and subject to stress (tension). The strength (i.e. the level of stress at which it fails) of each thread is a random variable; the threads may be assumed i.i.d., and in more general models strength may be assumed to vary over time. What is the overall strength of the bundle when placed under stress, i.e. the load at which the bundle fails? The model has applications beyond this narrow formulation. [H. Daniels, The statistical theory of the strength of bundles of threads, *Proc. Roy. Soc. A*, 183, 405–435, 1945]

Darling–Erdős theorem A limit theorem for the maximum of normalized partial sums $S_n = \sum_{r=1}^{n} X_r$ of a sequence of i.i.d.r.vs $X_1, X_2, \ldots$, such that $\mathbb{E}X_1 = 0$, var $X_1 = 1$, and $\mathbb{E}(|X_1|^3) < \infty$. In this case, for $x \in \mathbb{R}$, $\lim_{n\to\infty} \mathbb{P}\big(\max_{1\le r\le n} r^{-1/2} S_r \le xa(n) + b(n)\big) = \exp(-e^{-x})$, where $a(n) = (2\log\log n)^{-1/2}$, and $b(n)/a(n) = 2\log\log n + \frac{1}{2}\log\log\log n - \frac{1}{2}\log(4\pi)$. [D. Darling & P. Erdös, A limit theorem for the maximum of normalized sums of i.i.d. random variables, *Duke Math. J.*, 1956, 23, 143–154]

Darling–Kac theorem A limit theorem for suitable functionals (including the occupation times) of a Markov process $X(t)$ with stationary transition probabilities. Let $V(X(t)) \ge 0$ be a non-negative function of $X(t)$, and assume the following Darling–Kac condition, sometimes called condition (A), or condition (D-K), i.e. there exists a function $h(s) \to \infty$, as $s \to 0$, and a constant $C > 0$, such that, as $s \to 0$, $\big(h(s)\big)^{-1}\int_0^\infty e^{-st}\mathbb{E}\big(V\big(X(t)\big)|X(0) = x\big)dt - C \to 0$, uniformly in those x such that $V(x) > 0$. Then the Darling–Kac thm asserts: if for some $0 \le a \le 1$, $h(s) = s^{-a}L(s^{-1})$, where $L(s^{-1})$ is slowly varying as $s \to 0$, then $[Ch(t^{-1})]^{-1}\int_0^t V\big(X(y)\big)dy$ converges in distribution as $t \to \infty$; the limit being a Mittag–Leffler distn. Conversely, if $b(t)\int_0^t V\big(X(y)\big)dy$ converges in distn as $t \to \infty$, for some scaling function $b(t)$, then the scaling is of the above type, and the limiting distn is a scaled Mittag–Leffler distn. [D. Darling & M. Kac, On occupation times for Markoff processes, *Trans. Amer. Math. Soc.*, 84(2), 444–458, 1957]

data processing theorem Asserts that processing information (i.e. applying a transformation or function to a source, or random variable), destroys (or reduces) information; where by information we mean Shannon information. Formally, let X be a random variable (source), and let Y have a distribution conditional on X and Z a distribution conditionally independent of X given Y; denoted by $p(y|x)$ and $p(z|y)$ respectively. Thus X, Y, and Z form a Markov chain, often represented by $X \to Y \to Z$. We may interpret the arrows as a data-processing step. Now let $I(X;Y)$ be the mutual information between X and Y, and $I(X;Y|Z)$ the conditional mutual information. Then $I(X;Y) \ge I(X;Z)$, and $I(X;Y) \ge I(X;Y|Z)$. That is to say, informally, no manipulation of data can increase the Shannon information content. Of course, that is not to say that manipulation of data cannot make it easier to interpret, which is quite a different thing. [R. Gray, *Entropy and Information Theory*, 2nd edn. Springer, 2011; M. Pardo & I. Vajda, Distances of discrete distributions satisfying the data processing theorem of information theory, *IEEE Trans. Inf. Theor.*, 43(4), 1288–1293, 1997]

Davis distribution Proposed as a model for income distribution, this has density $f(x) = \dfrac{b^d}{\zeta(d)\Gamma(d)x^{d+1}}\dfrac{1}{\exp(b/x) - 1}$, $x \ge 0$, for $d > 1$ and $b > 0$. [C. Kleiber & S. Kotz, *Statistical Size Distributions in Economics and Actuarial Sciences*. Wiley, 2003]

Davydov's inequality Let X and Y be jointly distributed real random variables and suppose that $p^{-1} + q^{-1} + r^{-1} = 1$, for positive p, q, and r. Then for some constant $c \le 8$, the covariance cov(X, Y) satisfies $|\text{cov}(X,Y)| \le c\alpha^{1/p}(\mathbb{E}|X|^q)^{1/q}(\mathbb{E}|Y|^r)^{1/r}$, where α is the strong mixing coefficient: $\alpha = \sup_{A,B}|\mathbb{P}(A \cap B) - \mathbb{P}(A)\mathbb{P}(B)| \le \frac{1}{4}$, and the supremum is taken over all events A and B such that A is in the sigma-field generated by X and B is in the sigma-field generated by Y. [Y. Davydov, Convergence of distributions generated by stationary stochastic processes, *Theor. Probab. Appl.*, 13, 691–696, 1968]

Dawid's prequential principle Arises in the context of prediction and forecasting. Of necessity, any forecaster must have a rule for formulating predictions; and it is natural to seek to compare the efficacy of competing rules. The (weak) prequential principle asserts that any such assessment of merits should be based only on the predictions issued and the outcomes actually observed,

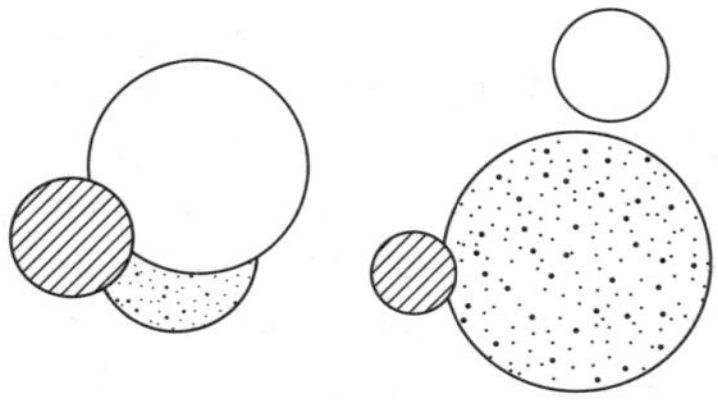

Dead leaves model with occlusion.

but not on any internal aspects of the model underlying the rule. [A. P. Dawid & V. Vovk, Prequential probability, *Bernoulli*, 5(1), 125–162, 1999]

dead leaves model (alias: ***Matheron dead leaves model***; ***falling leaves model***; ***occlusion model***) A classic random tessellation of $\mathbb{R}^d, d \geq 2$, arising as a model for the statistical features of images (in the case $d = 2$), or for morphology of materials (when $d = 3$). A standard construction in the plane proceeds thus: Let π be a Poisson process of rate λ in $\mathbb{R}^2 \times (-\infty, 0]$, where the third dimension is time, up to time $t = 0$. Each point $Q(x, y, t)$ of the process is assigned a closed set $S(Q)$; typically a circle centred at Q. If $\hat{Q}(u, v, s)$ is another such point, with $s < t$, then $S(Q)$ obliterates the intersection (if non-empty) $S(Q) \cap S(\hat{Q})$. At $t = 0$, one has a stationary tessellation of the plane, and an object of particular interest is the random process of sets $S(Q)$ no part of which has been obliterated, which forms a non-overlapping (hard-sphere model) collection. In d dimensions, with spherical 'leaves', the fraction of the plane in this set is 2^{-d}. [C. Bordenave *et al.*, The dead leaves model, *Adv. Appl. Probab.*, 38(1), 31–46, 2006]

dead time (or dead period) *see* counter model

death process (alias: **pure death process**) The non-increasing Markov process on the non-negative integers, in which the holding time in any state k is exponential with parameter μ_k, and the ensuing transition is to the state $k - 1$. Essentially, this is a birth–death process with zero birth-rates. When $\mu_k = k\mu$, one has the simple death process. More generally, $\mu_k(t)$ may be a function of time in the non-homogeneous death process. [K. Itô, *Essentials of Stochastic Processes*, English translation. American Mathematical Society, 2006]

de Bruijn's identity Supplies a relationship between Shannon's differential entropy and Fisher's information; in various forms. Let $X = Y + Z$, where the continuous r.v. Y has unit variance, and is independent of the normal random variable Z with mean zero and variance t. Denote the Shannon differential entropy and Fisher information of X by $H(X)$ and $J(X)$ respectively. Then $\frac{\partial H}{\partial t} = \frac{1}{2}J$ (or $\frac{\partial H}{\partial t} = \frac{1}{2}\log_2 eJ$, if logarithms to base 2 are used in the definition of entropy). There is also an integral form of de Bruijn's identity: let X be any finite variance r.v. with differential entropy $H(X)$. Then

$$H(X) = \int_0^1 J(\sqrt{t}X + \sqrt{1-t}Z)/(2t)^{-1}\mathrm{d}t,$$

where Z is a normal random variable that is independent of X, with the same mean and variance as X. [A Barron, Entropy and the central limit theorem, *Ann. Probab.*, 14(1), 336–342, 1986]

début theorem Arises in establishing conditions such that hitting times (of random processes in continuous time) are measurable and hence also stopping times. Let $(\Omega, \mathcal{F}, \mathbb{P})$ be a probability space, and $\mathcal{F}_t, t \geq 0$, a filtration of $\mathcal{F}$ that satisfies the usual conditions (i.e., it is right-continuous, and $\mathcal{F}_t$ contains all the null sets of $\mathbb{P}$). Let A be a subset of $[0, \infty) \times \Omega$, and define the début of A thus: $D_A(\omega) = \inf[t \geq 0 : (t, \omega) \in A]$. Then the début theorem asserts that if the indicator $I(A)$ is a *progressively measurable* process (for $\mathcal{F}_t$), then D_A is a stopping time for $\mathcal{F}_t$. [R. Bass, The measurability of hitting times, *Elec. Commun. Probab.*, 5, 99–105, 2010]

decay parameter Let $p_{ij}(t)$ be the stationary transition probabilities of a standard conservative Markov chain in continuous time with a countable state space. If for each i, j, and some $\lambda > 0$, as $t \to \infty$, there are constants π_j such that $p_{ij}(t) - \pi_j = \mathrm{O}(\exp(-\lambda t))$, then the chain is said to be exponentially ergodic with decay parameter λ. In discrete time, equivalently, as $n \to \infty$ the chain is geometrically ergodic with decay parameter δ if as $n \to \infty, p_{ij}(n) - \pi_j = \mathrm{O}(\delta^n)$ for some $0 < \delta < 1$. [J. F. C. Kingman, The exponential decay of Markov transition probabilities, *Proc. Lond. Math. Soc.*, 13(1), 337–358, 1963]

decision rule (alias: **strategy; decision function**) Arises in decision theory and game theory inter alia. Broadly, a function (or map) from a suitable state space to a set of feasible actions. In statistical decision theory the state space comprises possible 'states of nature', commonly represented by a parameter, and possibly augmented by observations that are usually given as random variables.

In game theory, the state space comprises possible configurations of the system, that may be called decision nodes, where an action (or 'move') is required from one or more players (or agents). If choices are made from a set of feasible actions, randomly at the behest of the decision maker(s), one speaks of a **randomized decision rule** (or strategy). [D. Blackwell & M. Girshick, *Theory of Games and Statistical Decisions.* Wiley, 1954; reprinted by Dover, 1979]

decision theory (alias: **decision analysis**) Broadly, this encompasses all those analytical problems arising when it is required to choose one of a set of possible actions, in the knowledge that each possible action leads to one of a set of possible consequences. The purpose of the analysis is, in general, the quest for a decision rule (or strategy), that is optimal according to some suitable criterion. If the consequences of actions are described in the framework of probability theory, then one speaks of decisions under risk or statistical decision theory; otherwise one may speak of decisions under uncertainty, or Knightian decision theory. [S. French, *Decision Theory.* Ellis Horwood, 1993]

decision tree Arises in decision problems in which choices can be (or must be) made sequentially and contingently; in the context of games, it may be called a **game tree**. Nodes represent moves or decision points, edges represent the actions or chance outcomes that may occur. [K. Binmore, *Fun and Games*, 1992, Heath]

decoding rule Arises in the context of signal-transmission through noisy channels. In the simplest standard framework the input to the channel is a random vector **C** of length n, drawn from a set S of M code words; the channel is a family of probability distributions parameterized by the possible code words, and the output is a random n-vector **Y** whose conditional distribution is $p(\mathbf{y} \mid \mathbf{C} = \mathbf{c}), \mathbf{y} \in T$. A decoding rule partitions the state-space T of **Y** into $M + 1$ sets corresponding to the M codewords together with error, or decoding failure, F, yielding a decoding function from T to $S \cup \{F\}$.
Examples of such rules include:

(a) The **ideal observer rule** (also called the minimum error rule, maximum a posteriori (or MAP), rule, or optimal decoder). This decodes an observed output **y** as that **c** maximizing $p(\mathbf{c} \mid \mathbf{y})$, (or F if no such **c** exists).
(b) The **maximum likelihood rule**, which decodes **y** as that **c** maximizing $p(\mathbf{y} \mid \mathbf{c})$, (or F, if no such **c** exists).
(c) For binary channels, in which **c** and **y** are binary strings of length n, the **Hamming rule** decodes **y** as that **c** minimizing the Hamming distance between **y** and **c**, (or F if no such unique **c** exists), (where the Hamming distance is the number of locations at which **c** and **y** differ). In the special case of a binary channel with **C** uniformly distributed on S, these three rules agree, in general.

[D. MacKay, *Information Theory, Inference, and Learning Algorithms.* Cambridge University Press, 2003]

decomposable (i) A characteristic function $\phi(t)$ (or a random variable X having $\phi(t)$ as its chf), is said to be decomposable if it can be written as $\phi(t) = \phi_1(t)\phi_2(t)$, where ϕ_1, and ϕ_2 are the characteristic functions of non-degenerate (and independent) random variables Y and Z respectively. If no such factorization exists, then $\phi(t)$ (and X), is said to be indecomposable. The theory of such decompositions is known as the **arithmetic of distributions**, though such factorization need not be unique and it is possible that $\phi(t) = \phi_1(t)\phi_2(t) = \phi_2(t)\phi_2(t)$ for $\phi_1(t) \neq \phi_2(t)$, called the Khinchin phenomenon.

(ii) Much more narrowly, a r.v. X is **strongly decomposable** if $X = Y + Z$, where Y and Z are independent, and $Y = d(X)$ for some Borel function $d(\cdot)$ called the decomposition function. [E. Lukacs, *Characteristic Functions*, 2nd edn. Griffin, 1970; J. Hoffman-Jørgensen *et al.*, Strong decomposition of random variables, *J. Theor. Probab.*, 20, 211–220, 2007]

decomposition theorem of Markov chains The classic theorem for Markov chains on a countable state space S, which asserts that S may be partitioned uniquely into a set T of transient (or inessential) states, and a countable collection $C_1, C_2, \ldots$ of closed irreducible sets of persistent (or recurrent) states. Each C_r is either aperiodic, or may be further divided into a finite number of cyclically moving subsets. (For other types of decomposition theorem, *see* Doeblin; Doob–Meyer; Hoeffding; Lebesgue; Lévy; martingale; spectral; Wald; etc.) [A. Shiryayev, *Probability*, 2nd edn. Springer, 1996]

default contagion model Arises in financial mathematics, where it is intended to provide a framework for valuing assets and instruments when the hazard rates for their default are correlated. [H. Haworth *et al.*, Modelling Bonds and credit default swaps with contagion, *Quantitative Finan.*, 8(7), 669–680, 2008]

defective distribution (alias: **dishonest distribution; improper distribution; pseudo-distribution; sub-probability distribution**) Arises in considering a random variable X such that $\mathbb{P}(|X| < \infty) < 1$; which is to say that X is not almost surely finite, but may take values $+\infty$ or $-\infty$ in the extended real line. The function $F(x) = \mathbb{P}(X \leq x)$ is not now a proper distribution function, as $\lim_{x\to-\infty}F(x) = \mathbb{P}(X = -\infty) \geq 0$, $\lim_{x\to\infty}F(x) = 1 - \mathbb{P}(X = +\infty) \leq 1$. Both $F(x)$ and X may be termed defective or dishonest; the term **improper** is also used, but should not be confused with its different meaning in 'improper prior'. [W. Feller, *Introduction to Probability Theory and its Applications*, Vol. 2, 2nd edn. Wiley, 1971]

de Finetti's theorem Supplies a connection between the concepts of exchangeability and independence for an infinite sequence of random variables. Formally, let $\mathbf{X} = (X_1, X_2, \ldots)$ be an infinite sequence of random variables such that $(X_1, X_2, \ldots, X_r)$ is exchangeable for each $r \geq 2$. Then there is a sigma-field conditional on which the components of $(X_1, \ldots, X_r)$, for any r are independent and identically distributed. Equivalently, there is a measure on the space of distribution functions such that the joint distribution of $(X_1, \ldots, X_n)$ is a mixture by that measure (or convex combination) of the joint distributions of a sequence of independent random variables so distributed. The original result of de Finetti, stated for an exchangeable sequence of Bernoulli random variables, may be given in this form as

$$\mathbb{P}(X_1 = 1, X_2 = 1, \ldots, X_r = 1, X_{r+1} = 0, \ldots, X_n = 0) = \int_0^1 \theta^r(1-\theta)^{n-r}dF(\theta),$$

for some F, which may be called the de Finetti (or mixing) measure. [J. F. C. Kingman, Uses of exchangeability, *Ann. Probab.*, 6(2), 183–197, 1978]

degenerate distribution If the distribution function $F(x)$ of a r.v. X is such that $F(c) = 1$, and $\lim_{x\uparrow c} F(x) = 0$, for some finite c, then $F(x)$ is said to be degenerate; the r.v. X, which is a.s. constant (i.e. $\mathbb{P}(X = c) = 1$, is a degenerate r.v. Any constant r.v. (i.e. $X(\omega) = c, \omega \in \Omega$) is therefore degenerate. (The term improper is sometimes used, but is better avoided in this context.) Likewise, a measure is degenerate if concentrated at point. By extension, when the distribution of a random vector in n dimensions is concentrated on a subspace of $r < n$ dimensions, then it is said to be degenerate (or singular). [L. Breiman, *Probability*. Addison-Wesley, 1968; reprinted by SIAM, 1992]

degenerate normal distribution (alias: **singular multinormal distribution**) Arises when a random vector $\mathbf{X} = (X_1, \ldots, X_n)$ has a multivariate normal distribution (which is to say that all linear combinations $Y = \sum_{i=1}^n a_i X_i$ are normally distributed) but there is also a non-zero vector $(b_1, \ldots, b_n)$ and constant c, such that $\mathbb{P}(\sum_{r=1}^n b_r X_r = c) = 1$. More formally, if $\mathbf{X}$ is a zero-mean multinormal r.vr with $n \times n$ covariance matrix V, and the $m \times n$ matrix M has rank ρ less than m, then $\mathbf{Y} = M\mathbf{X}$ has the degenerate (multivariate) normal distribution denoted by $N(0, MVM^{\mathrm{T}})$. The distribution of $\mathbf{Y}$ is concentrated on a linear ρ-dimensional subspace of $\mathbb{R}^m$. [W. Feller, *Introduction to Probability Theory and its Applications*, Vol. 2, 2nd edn. Wiley, 1971]

degraded channel Arises in communication theory in the context of broadcast channels, and other related channel networks. In the simplest case, the input X yields two outputs, Y and Z; if the sequence (X, Y, Z) comprises a Markov chain then the channel is said to be physically degraded. If there is a distribution $q(z|y) = \mathbb{P}(Z = z|Y = y)$ such that $\mathbb{P}(Z = z|X = x) = \sum_y q(z|y)\mathbb{P}(Y = y|X = x)$, then the channel is said to be stochastically degraded. [T. Cover & J. Thomas, *Elements of Information Theory*, 2nd edn. Wiley, 2006]

degree of belief (alias: **credence**) Arises in the interpretation of probability that is sometimes called epistemological, or, in the words of A. de Morgan: 'Probability is a feeling of the mind.' In this framework, to say that $\mathbb{P}(A) = \frac{1}{2}$ is to say that your opinion is equally divided as to whether A will occur or not. There is a further subdivision of degrees of belief, as follows: if your opinion is a personal and private assessment of the chances of some event A occurring then one speaks of personalist (or subjective) degrees of belief; such probabilities may be measured (in principle) by considering hypothetical bets. Alternatively, your opinion may be claimed as the assessment of the chances of A occurring that would be arrived at by anyone sharing suitable agreed rules of logic and inference. In this case one speaks of logical (or objective) degrees of belief; such probabilities are intended to be arrived at by some calculus of logical support. [D. Gillies, *Philosophical Theories of Probability*. Routledge, 2000]

degree of randomness Arises in the context of a cryptodeterministic process having random initial conditions, with a non-random mechanism for evolution over time. Let $X(t)$ be such a process, and suppose that for some n it is the case that conditional on the values of $X(t_1), \ldots, X(t_n)$, for some $t_1 < t_2 < \cdots < t_n$, the process $X(t), t > t_n$, is (with probability 1) a non-random function of $X(t_1), \ldots, X(t_n)$. If n is the smallest such n, then $X(t)$ is said to have n degrees of randomness; otherwise, if this is impossible for any finite set of values $t_1, \ldots, t_n$, then $X(t)$ is said to be **completely random**. [J. Moyal, Stochastic processes and statistical physics, *J. Roy. Stat. Soc. B*, 11(2), 150–211, 1949]

Delaporte distribution Arises in insurance mathematics as a model for the distribution of claims. It can be represented either as the sum of two independent random variables having respectively a Poisson distn and a negative binomial distn, or as a mixed Poisson distn in which the mixing density (of the random parameter of the Poisson distn) is that of a shifted gamma random variable taking values in $[c, \infty), c > 0$. [G. Willmot & B. Sundt, On evaluation of the Delaporte distribution, *Scand. Actuarial J.*, 89, 101–113, 1989]

de la Vallée Poussin theorem In the context of probability this asserts that a collection of finite mean r.vs $X_n, n \geq 1$, is uniformly integrable iff there is some convex function g increasing on $[0, \infty]$, with $\lim_{x \to \infty} x^{-1} g(x) = \infty$, such that $\sup_n \mathbb{E}g(|X_n|) < \infty$. [M. Rao & R. Swift, *Probability Theory with Applications*, 2nd edn. Springer, 2006]

delayed renewal process (alias: **general (or modified) renewal process**) Essentially, this is an ordinary (or pure) *renewal process* started at a random time $X_0 \geq 0$. That is, a sequence $\{T_r, r \geq 0\}$, of random points in $\mathbb{R}^+$, such that $T_0 = X_0$, and $T_n = X_0 + Y_1 + \cdots + Y_n, n \geq 1$, where the $Y_r \geq 0$ are i.i.d. non-negative random variables. If $\mathbb{E}Y_1 = \int_0^\infty 1 - F_Y(y)\,\mathrm{d}y < \infty$, and X_0 has distribution function $F(x) = (\mathbb{E}\,Y_1)^{-1} \int_0^x 1 - F_Y(y)\mathrm{d}y$, then this delayed renewal process is a strictly stationary point process, and is therefore called a **stationary renewal process** (or an **equilibrium renewal process**). [S. Resnick, *Adventures in Stochastic Processes*. Birkhäuser, 1992]

delta (alias: **delta coefficient**) Arises in financial mathematics in the context of a portfolio, or a derivative, whose value (or pricing) function at time t, based on a single underlying asset with price process $S(t)$, is $V(t, S(t))$. Then the associated delta (coefficient) is $\Delta = \frac{\partial V}{\partial s}(t,s)|_{s=S(t)} = \frac{\partial V}{\partial S}$. It is a measure of the so-called sensitivity of V to the asset S. In the classic market (Black–Scholes) model, when one seeks to replicate (or hedge) a derivative (or option) with pricing function $V(t, s)$, then the delta $\Delta = \frac{\partial V}{\partial S}$ of V equals the amount of underlying asset $S(t)$ needed in a self-financing replicating portfolio. (*See* greeks.) [T. Björk, *Arbitrage Theory in Continuous Time*, 3rd edn. Oxford University Press, 2009]

Dembo–Kagan–Shepp theorem Asserts that the maximal correlation coefficient between the partial sums $S_m = \sum_{r=1}^m X_r$ and $S_n = \sum_{r=1}^n X_r$ of a sequence $X_r, r \geq 1$, of i.i.d. finite-variance non-degenerate r.vs, is $\left(\frac{m}{n}\right)^{\frac{1}{2}} \equiv \rho(S_m, S_n), m \leq n$, where ρ is the ordinary correlation coefficient. [A. Dembo *et al.*, On the maximal correlation coefficient, *Bernoulli*, 7(2), 343–350, 2001]

de Méré's problem One of a number of problems posed to B. Pascal in 1654 by Antoine Gombaud, Chevalier de Méré. These included the well-known problem of the points; the problem usually known as de Méré's is to show that the chance of obtaining at least one 6 in four rolls of a die is greater than the chance of obtaining at least one double 6 in 24 rolls of two dice. [A. Hald, *History of Probability and Statistics and their Applications Before 1750*. Wiley, 1990]

demimartingale Let $X_n, n \geq 1$, be a sequence of finite-mean r.vs such that for any componentwise non-decreasing function f, and all $n \geq 1$, $\mathbb{E}\{(X_{n+1} - X_n)f(X_1, \ldots, X_n)\} \geq 0$, whenever the expectation exists; then the sequence $X_n, n \geq 1$, is a demimartingale. If this condition is required to hold for non-negative componentwise non-decreasing f, then the sequence is called a demisubmartingale. Such objects arise naturally in the context of a sequence $Y_n, n \geq 1$, of associated r.vs, with zero mean; for in this case the partial sums $X_n = \sum_{r=1}^n Y_r$ comprise a demimartingale. [C. Newman & A. Wright, Associated random variables and martingale inequalities, *Z. Wahrscheinlichkeits Theorie verw. Gebiete*, 59, 361–371, 1982]

De Moivre's theorem (alias: **binomial central limit theorem; de Moivre–Laplace theorem**) Establishes an approximation to the binomial $\mathrm{B}(n, p)$ distribution in terms of the normal distribution, when the number n of trials is sufficiently large. Explicitly, if $b(x) = \binom{n}{x} p^x q^{n-x}, 0 \leq x \leq n; q = 1 - p$ then for $x = np + d$, where $d = O(\sqrt{n})$, $b(x) \sim (2\pi npq)^{-\frac{1}{2}} \exp\left\{-\frac{(x-np)^2}{2npq}\right\}$. It follows that for a r.v. X with distribution $b(x)$, one can set $d = y(npq)^{1/2}$ and write

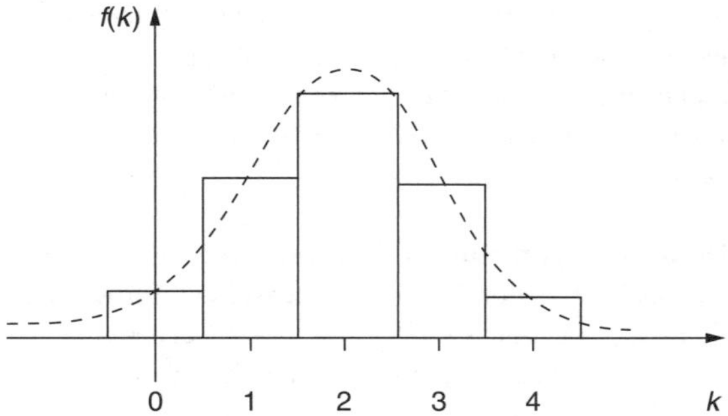

Binomial B (4,½) versus its normal approximation.

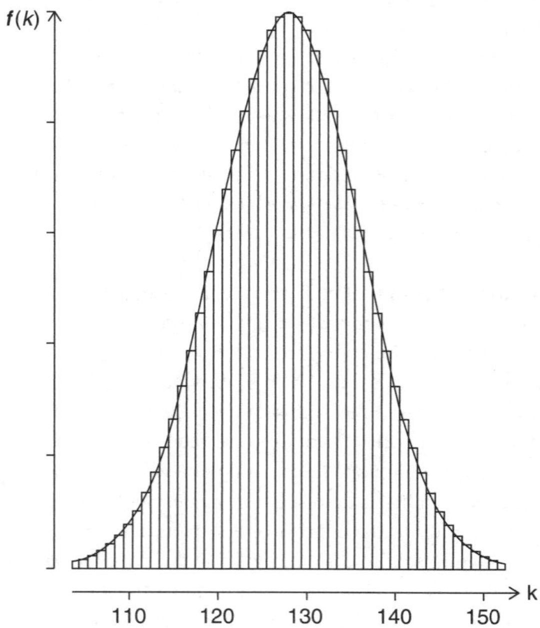

Binomial B (256,½) versus its normal approximation.

$$\mathbb{P}(|X - np| \le d) = \sum_{|x-np|\le d} b(x)$$

$$\simeq \int_{-y}^{y} (2\pi)^{-\frac{1}{2}} \exp(-\tfrac{1}{2}x^2)\mathrm{d}x.$$

[A. de Moivre, Approximatio ad Summam Terminorum Binomii $(a+b)^n$ in *Seriem Expansi*, 1733; English translation in *The Doctrine of Chances*, 1738, printed by Woodfall, London; W. Feller, *Introduction to Probability Theory and its Applications*, Vol. 1, 3rd edn. Wiley, 1968]

Dempster–Shafer theory (alias: **theory of belief functions**) Broadly, any of a number of mathematical structures modelling the degree of support for a proposition or hypothesis H, given incomplete information, and in the light of various evidential statements, which may be true with appropriate probabilities. A notable feature of the theory is the construction of a belief function $b(H)$, defined on the propositions (or hypotheses) (sometimes called a support function), and a plausibility function $C(H)$, where $0 \le b(H) \le C(H) \le 1$. The theory can be seen as a generalization of some types of epistemological probabilities (also called subjective, or Bayesian, in which probability is a feeling of the mind); in this interpretation $[b(H), C(H)]$ is seen as an interval representation of the degree of probability that we may attach to H. [G. Shafer, *A Mathematical Theory of Evidence*. Princeton University Press, 1976]

Denisov's path decomposition theorem Arises in the context of the Wiener process $W(t), t \ge 0$ (i.e. Brownian motion). Let Y be the time in $[0,1]$ at which $W(t)$ achieves its minimum for $0 \le t \le 1$. Then the two processes

$$M_1 = Y^{-1/2}[W((1-t)Y) - W(Y)],\ 0 \le t \le 1;$$
$$M_2 = (1-Y)^{-1/2}[W(Y(1-t)+t) - W(Y)],\ 0 \le t \le 1;$$

are independent Brownian meanders that are also independent of Y. [I. Denisov, A random walk and a Wiener process near a maximum, *Theor. Probab. Appl.*, 713–716, 1985]

density (alias: **probability density function; frequency function; density of a probability distribution**) Arises in its simplest form wherever the distribution function $F(x)$ of a r.v. X has the form $F(x) = \int_{-\infty}^{x} f(u)\,\mathrm{d}u$, for all real x, where the non-negative function $f(x)$ is called the density of X, and $f(x) = F'(x)$ is the derivative of $F(x)$. (This occurs if and only if $F(x)$ is absolutely continuous.) More generally, for a random vector $\mathbf{X}$ in $\mathbb{R}^d$ with absolutely continuous distn $F(\mathbf{x})$, the joint density $f(\mathbf{x})$ is such that

$$F(\mathbf{x}) = \int_{-\infty}^{x_1} \cdots \int_{-\infty}^{x_d} f(\boldsymbol{u})\, du_1 \cdots du_d, \mathbf{x} \in \mathbb{R}^d.$$

In the context of measure theory, let $\mathbb{P}$ and $\mathbb{Q}$ be probability measures on a suitable space such that $\mathbb{P}$ is absolutely continuous wrt (i.e. dominated by) $\mathbb{Q}$; which is to say that $\mathbb{Q}(A) = 0 \Rightarrow \mathbb{P}(A) = 0$. Then there is a function f such that, for all suitable A, $\mathbb{P}(A) = \int_A f \mathrm{d}\mathbb{Q}$; and f may

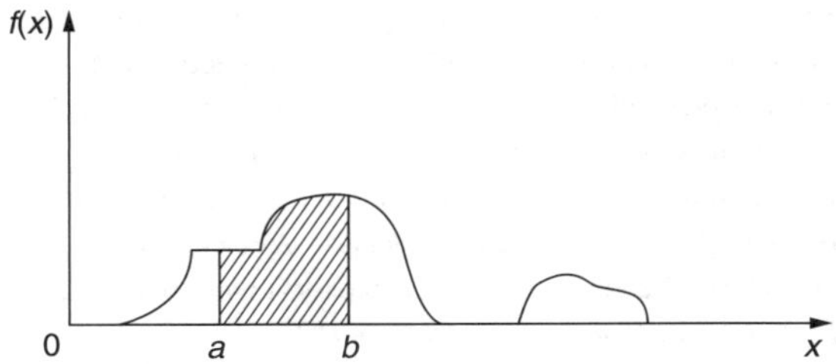

For X with density $f(x)$, the probability $\mathbb{P}(a < X \leq b)$ is the shaded area.

be called the density of $\mathbb{P}$ relative to $\mathbb{Q}$. (It may also be called the Radon–Nikodym derivative, in which case it is often denoted by $f = \mathrm{d}\mathbb{P}/\mathrm{d}\mathbb{Q}$.) In this framework, the simple density above is seen as that relative to the Lebesgue measure. [D. Pollard, *A User's Guide to Measure Theoretic Probability*. Cambridge University Press, 2002]

density-quantile function Arises in considering the quantile function $F^{-1}(x)$ of a distribution function $F(x)$, $x \in \mathbb{R}$; formally $F^{-1}(x)$ is the left-continuous inverse of $F(x)$, so that for continuous $F, F^{-1}(y) = \inf[x : F(x) = y]$. When F is absolutely continuous with density $f(x)$, the density-quantile function is defined as $f(F^{-1}(x))$, and the **quantile-density fn** as $1/f(F^{-1}(x))$. [M. Csörgö, *Quantile Processes with Statistical Applications*. SIAM, 1983]

departure process (alias: **output process**) Arises in queues and network processes, where customers (or objects, or packets) arrive at a service point (or node, or queue) as a point process in time, and emerge at times comprising a point process called the departure (or output) process. Then a natural task is to characterize the departure process as a function of the arrival process; a classic result of this type is Burke's theorem. [R. Disney & P. Kiessler, *Traffic Processes in Queuing Networks*. Johns Hopkins University Press, 1987]

dependence (alias: **statistical dependence; association**) Broadly random variables X and Y are (statistically) dependent if they are not independent (i.e. their joint distn does not factorize appropriately). Most narrowly, X and Y may be dependent in the usual functional sense, which is to say that one of them is determined as a non-random function of the other; it may be said to be completely dependent. In intermediate cases it is natural to formalize concepts of dependence and seek measures of dependence; there are a great many such. They divide roughly into those concepts and measures that are symmetric in X and Y, and those that are aymmetrical; the latter type arising naturally when one of the r.vs is seen as being influenced by, or arising as a response to, the other (but not only in that case). Examples include:

(a) X and Y are positively (lower) quadrant dependent if $\mathbb{P}(X \leq x, Y \leq y) \geq \mathbb{P}(X \leq x)\mathbb{P}(Y \leq y)$ for all x and y; this is a symmetrical concept.
(b) If $\mathbb{P}(X \leq x \mid Y = y)$ is non-increasing in y, then X is positively regression dependent on Y; and this is an asymmetrical concept of dependence.

A measure of dependence between X and Y may be supplied by any distance (or divergence) function $d(X, Y) \geq 0$, of X and Y (or of their distributions); where $d(X, Y) = d(Y, X)$ in the symmetric case. Such measures are often conventionally normalized, so that $d(X, Y) = 0$ when X and Y are independent, while $d(X, Y) = 1$ when one r.v. is completely dependent on the other. A. Rényi listed a set of desirable properties that any such measure should possess; the resulting axioms are satisfied only by the maximal correlation coefficient. Weakening Rényi's constraints yields a wider class of measures, including the **Schweizer–Wolff measure** given by

$$\sigma(X, Y) = 12\int_0^1 \int_0^1 |C(u, v) - u\,v|\,\mathrm{d}u\,\mathrm{d}v,$$

where $C(x, y)$ is the copula corresponding to the joint distribution $F(x, y)$ of X and Y.

Other measures of dependence include (in random order), Pearson's (or 'the') correlation coefficient; Spearman's ρ coefficient; Kendall's τ; Pearson's divergence coefficient, i.e. for a population: $\phi^2 = \mathbb{E}\left[\frac{f(X, Y)}{f_X(X)\, f_Y(Y)}\right] - 1$, where f and f_X, f_Y are the joint and marginal densities of X and Y; Blomquist's q; the correlation ratio for X or Y, i.e. $\mathrm{var}\,[\mathbb{E}(X|Y)]/\mathrm{var}\,X$; the mean square contingency; Goodman–Kruskal's τ; Theil's entropy (or uncertainty) coefficient; mutual information; copula; and many more. [E. Lehmann, Some concepts of dependence, *Ann. Math. Stat.*, 37(5), 1137–1153, 1966; B. Schweizer & E. Wolff, Non-parametric measures of dependence for random variables, *Ann. Stat.*, 9(4), 879–885, 1981]

dependence function *see* copula

De Pril transform Let $p(x), x \geq 0$, be a probability distribution on the non-negative integers with $p(0) > 0$. Then the De Pril transform of p, denoted by $D_p(y)$, $y \in [1, 2, \ldots]$, is the function defined recursively by

$$D_p(y) = p(0)^{-1}\left(yp(y) - \sum_{r-1}^{y-1} D_p(r)\,p(y-r)\right),\ y \geq 1.$$

Solving for $p(y)$ we have also that

$$p(y) = \frac{1}{y}\sum_{r=1}^{y} D_p(r)p(y-r),\ y \geq 1.$$

If $p*q$ is the correlation of two such distributions p and q, then it can be shown that $D_{p*q} = D_p + D_q$. The function $D_p(y)$ finds particular applications in insurance mathematics. [B. Sundt, On some properties of De Pril transforms, *ASTIN Bull.*, 25, 19–31, 1995]

depth function Arises in seeking to extend the concepts of median and tail region, from their natural origins in distributions on $\mathbb{R}$, to distributions on $\mathbb{R}^d$, $d \geq 2$. The depth $d(\mathbf{x})$ of a point $\mathbf{x}$ with respect to a distribution $F(\mathbf{x}), \mathbf{x} \in \mathbb{R}^d$, is seen as a measure of how central the point $\mathbf{x}$ is in $F(\mathbf{x})$; The points of maximum depth define 'centre(s)' that are generalizations of the median(s) m in $\mathbb{R}$ of any $F(x)$; and the function $d(\mathbf{x})$ supplies an ordering outward from the centre(s), with higher values indicating centrality, and lower values indicating 'outlyingness' that generalizes the idea of the tails $F(x), x < m$, and $1 - F(x), x > m$, of any $F(x)$ on $\mathbb{R}$. A classic example is the **Tukey half-space depth** $d_T(\mathbf{x}, F)$, which is defined for each $\mathbf{x}$ as the smallest probability that the distn $F(\mathbf{x})$ assigns to any of the closed half spaces including the point $\mathbf{x}$. The set of points at which d_T is maximal comprises the Tukey (or half-space, or generalized) medians. In one dimension $d_T(x, F) = \min\{F(x), 1 - F(x)\}$, and the Tukey median is the usual median. There are several other depth functions, e.g.:

(a) **Simplicial depth**, d_s, which is the probability that $\mathbf{x}$ is contained in a random $d+1$ simplex whose $d+1$ vertices are i.i.d.r.vs having the distribution $F(\mathbf{x})$.

(b) **Mahalanobis depth**, d_M, which is defined for a distn $F(\mathbf{x})$ having covariance matrix $\mathbf{V}$ and mean vector $\boldsymbol{m}$ as $d_M = \{1 + (\boldsymbol{x} - \boldsymbol{m})^{\mathrm{T}}\mathbf{V}^{-1}(\boldsymbol{x} - \boldsymbol{m})\}^{-1}$.

(c) **Majority depth**, d_V, which is defined as $\mathbb{P}(\mathbf{x} \in H(X_1, \ldots, X_d))$, where for any $(y_1, \ldots, y_d)$ determining a hyperplane in $\mathbb{R}^d, H(y_1, \ldots, y_d)$ is the more likely half space under the distribution $F(\mathbf{x})$.

[Y. Zuo & R. Serfling, General notions of statistical depth function, *Ann. Stat.*, 28(2), 461–482, 2000]

derandomization Arises in the context of computational complexity and probabilistic computation. On one hand it may refer directly to the process of replacing a randomized (or probabilistic) algorithm for a problem of given complexity by a non-random algorithm; on the other hand it may refer to the conjecture that the class of problems solvable in polynomial time is identical to the class solvable in bounded-error probabilistic polynomial time. (Which may be denoted by P ≡ BPP.) [R. Impagliazzo, Using hard problems to derandomize algorithms: 'Randomization and approximation techniques', *Lecture Notes Comp. Sci.*, 1269, 165–173, 1997]

derangement problem (alias: **Montmort's problem; matching problem; problem des rencontres; hat-check problem; Treize; etc.**) Arises when an ordered pack of cards is shuffled and taken in random order. Any reordering of n distinguishable objects is called a derangement if no object is in its original position. The problem is to determine the probability that a reordering picked uniformly from the $n!$ permutations is a derangement. More generally, one asks for the probability p_r that exactly r objects are in their original position and the rest deranged; the answer is

$$p_r = \frac{1}{r!}\left\{\frac{1}{2!} - \frac{1}{3!} + \cdots + \frac{(-1)^{n-r}}{(n-r)!}\right\} \to \frac{1}{r!}\mathrm{e}^{-1} \text{ as } n \to \infty.$$

Setting $r = 0$ recovers the answer to the derangement problem. [P. de Montmort, *Essay d'Analyse sur les Jeux de Hazard*. Paris, 1708; 2nd edn, 1713, reprinted by Chelsea, 1980; G. Blom *et al.*, *Problems and Snapshots from the World of Probability*. Springer, 1994]

derivative security *see* contingent claim

derived Markov process (alias: **subordinated Markov process**) Arises when a Markov process $X(t)$, $t \geq 0$, is sampled (i.e. evaluated) at a random operational time $T(u), u \geq 0$, that is independent of $X(t)$ (where we assume that $X(t)$ has countable state space S). Thus, for each fixed $t \geq 0$, let $F(s,t), 0 \leq s < \infty$, be the distribution function of a random variable $T(t)$, where the random process $T(t), t \geq 0$, has stationary independent non-negative increments. Denoting the transition probabilities of $X(t)$ by $p_{ij}(t)$, the random process $Y(t) = X(T(t))$ is Markov with transition probabilities

$$q_{ij}(t) = \int_0^\infty p_{ij}(s)\mathrm{d}_s F(s,t).$$

The chain $Y(t)$ is said to be derived from $X(t)$ [or subordinated to $X(t)$] with deriving function $F(s,t)$ (or subordinator $T(t)$). When $t = 0$, $T(0)$ is assumed to be degenerate at 0. [A. Stam, Derived stochastic processes, *Compositio Math.*, 17, 102–140, 1965–1966]

Derman's theorem A ratio limit theorem of particular application to a null-recurrent irreducible Markov chain $X_n, n \geq 0$, taking values in $\{0, 1, 2, \ldots\}$. Let the irreducible recurrent chain X_n have n-step transition probabilities $p_{ij}(n)$, where $p_{ij}(1) = p_{ij}$, and $i, j \geq 0$. Define the ratio

$$R(n,k) = \sum_{t=0}^{n} p_{kk}(t) \Big/ \sum_{t=0}^{n} p_{00}(t).$$

Then the theorem asserts that there is one and only one set of numbers $\lambda_k, k \geq 0$, such that $\lambda_k = \lim_{n\to\infty} R(n,k)$ and $\lambda_k = \sum_{r=0}^{\infty} \lambda_r p_{rk}$ (which is to say that λ_k is an invariant measure for the chain); and the chain is null-recurrent iff $\sum_{k=0}^{\infty} \lambda_k = \infty$. This is called Derman's construction of invariant measure. [C. Derman, A solution to a set of fundamental equations in Markov chains, *Proc. Amer. Math. Soc.*, 5(2), 332–334, 1954]

Derrida's random energy model Arises in statistical mechanics as an idealized disordered model for ferromagnetism. The state space S is $\{0,1\}^M$, and the probability of a configuration $\mathbf{x}$ is given by the distribution $p(\mathbf{x}) = Z^{-1}\exp(-\beta E(\mathbf{x}))$, where Z is the partition function. In Derrida's REM the $E(\mathbf{x})$ are independent zero-mean r.vs having a normal $N(0, M\sigma^2)$ distribution, $\sigma^2 > 0$. In its generalized form, GREM, the $E(\mathbf{x})$ have a covariance function that depends on the lexicographic distance between points of S. [B. Derrida, Random energy model, *Phys. Rev. B*, 24, 2613–2626, 1981]

descriptive complexity *see* Kolmogorov complexity

determinacy problem *see* moment problem

determinate distribution (alias **M-determinate distn (or measure)**) Arises in the context of the moment problem. Let M be the set of probability distributions on $\mathbb{R}$ having all moments $m_n(F) = \int x^n \mathrm{d}F(x)$ finite, $n \geq 0$. Distributions F and G are equivalent (or **moment-equivalent**) if $m_n(F) = m_n(G)$ for all n, and this yields a collection of equivalence classes of distributions. If the equivalence class of F has more than one member, then F is said to be indeterminate (or M-indeterminate); if F is the only member, then F is said to be determinate (or M-determinate). A classic example of an indeterminate distribution is the lognormal (Heyde's lemma); **Berg's thm** shows that for a standard normal r.v. X, the distribution of X^{2n+1} is indeterminate for $n \geq 1$, and the distribution of $|X|^a$ is indeterminate for $a > 4$, but determinate for $0 < a \leq 4$.

If the support of F is $\mathbb{R}^+$ (or an unbounded subset of $\mathbb{R}^+$), then F is said to be **Stieltjes-(in)determinate**, if the support of F is $\mathbb{R}$, then F is **Hamburger-(in) determinate**. It is a corollary of **Chihara's thm** that a Stieltjes-determinate distn that is Hamburger-indeterminate must be of a discrete unbounded r.v. in a subset of $\mathbb{R}^+$. [J. Stoyanov, Stieltjes classes for moment indeterminate distns, *J. Appl. Probab.*, 41, 281–294, 2004; T. Chihara, Indeterminate Hamburger moment problems, *Pacific J. Math.*, 22(3), 475–484, 1968]

deviation Arises in considering the extent to which a r.v. X (or a random process) departs from its expectation m (or its median, or other measure of centrality); thus $X - m$ is the (signed) deviation of X from m. On the one hand, we may consider the tails of the distribution of the deviation, and bounds for these; in this context a bound for the deviation $u(\mathbf{X}) - \mathbb{E}u(\mathbf{X})$ of a function $u(\mathbf{X})$ of a r.vr $\mathbf{X} = (X_1, \ldots, X_n)$, $\mathbb{P}(u(\mathbf{X}) - \mathbb{E}u(\mathbf{X}) > a) < r(a,n,u), a > 0$, is called a deviation inequality with rate function r; and asymptotic bounds as $a \to \infty$ in a suitable manner are called large deviation inequalities. Bounds of this type for $\mathbb{P}(|u - \mathbb{E}u| > a)$ are often called concentration inequalities, when they hold for all $a > 0$, rather than asymptotically. On the other hand, we can seek one or another function (or functional) of the distribution of the deviation as a measure of departure from the mean. Elementary examples include the mean absolute deviation $\mathbb{E}|X - m|$ and the standard deviation $\sigma = (\mathrm{var}\, X)^{1/2}$. Generalizing the standard deviation, a deviation measure $d(X)$ of any r.v. X with finite variance is a real-valued function such that:

(a) $d(X + c) = d(X)$ for any constant c;
(b) $d(X + Y) \leq d(X) + d(Y)$ for all X and Y;
(c) $d(aX) = ad(X)$ for $a \geq 0$;
(d) $d(X) > 0$ for non-degenerate X, while $d(c) = 0$ for any constant c.

Other examples of such measures include the **semideviations** $[\mathbb{E}\{\max(0, X - m)\}^2]^{1/2}$ and $[\mathbb{E}\{\min(0, X - m)\}^2]^{1/2}$. [R. Rockafellar *et al.*, Generalized deviations in risk analysis, *Finan. Stoch.*, 10(1), 51–74, 2006]

Devroye's inequality Related to the Efron–Stein, and Steele, inequalities. This gives a bound for the variance of a real-valued function $f(X_1, \ldots, X_n)$ of i.i.d.r.vs, such that for $1 \leq r \leq n$, $\sup_{x_1,\ldots,x_n} \sup_{x_r \neq y_r}$ $|f(x_1, \ldots, x_n) - f(x_1, \ldots, x_{r-1}, y_r, x_{r+1}, \ldots, x_n)| \leq c_r$.

Then $\mathrm{var} f(X_1, \ldots, X_n) \leq \frac{1}{4}\sum_{r=1}^{n} c_r^2$. [L. Devroye, Exponential inequalities in nonparametric estimation, in

Nonparametric Functional Estimation, G. Roussel (ed.). Kluwer, 1991]

De Vylder approximation Arises in the ruin process of insurance mathematics. It supplies an approximation to the ruin probability by replacing the actual risk reserve process by a process having exponentially distributed claims, whose parameters are determined by requiring that the first (lowest order) three moments of the two processes agree. [J. Grandell, *Aspects of Risk Theory*. Springer, 1991]

Dickman distribution Arises in the theory of prime numbers, where the Dickman function $d(u)$, is continuous for $u \geq 0$, and is defined by:
(a) $d(u) = 1, 0 \leq u \leq 1$, and
(b) $ud'(u) + d(u-1) = 0, u > 1$.
Then $f(x) = \mathrm{e}^{-\gamma} d(x), x \geq 0$, is a probability density when γ is Euler's constant, and may be called the Dickman distribution. (Other related densities that have been given the same name are $g(x) = d\{(1-x)/x\}/x, 0 < x < 1$, and $h(x) = d(x-1)/x, 1 < x < \infty$.) A random variable X with the Dickman density $f(x)$ is also characterized by the property that X has the same distribution as $u(\hat{X}+1)$, where U is uniform on $(0, 1]$, and independent of $\hat{X}$, which has the same density as X. [M. Penrose & A. Wade, Random minimal directed spanning trees and Dickman-type distributions, *Adv. Appl. Probab.*, 36, 691–714, 2004]

die A die (plural dice) is a classic randomizer, most commonly taking the form of a cube (or other symmetrical object), and constructed with the intention that each face has the same chance to be distinguished (as uppermost, or lowermost) when the die comes to rest after being thrown. The mathematical abstraction of a die simply asserts that each one of a given number n, where $n = 6$ most often, of events is equally likely to occur. If the distribution is not uniform, the die is said to be biased, and when $n = 2$ the randomizer is often called a **coin**. [R. Epstein, *Theory of Gambling and Statistical Logic*, 2nd edn. Elsevier, 2009]

difference time Arises in considering a random walk, or a Markov process, on a state space S; for any two states j and k of S denote the hitting time of k from j (i.e. the interval, or number of steps, to reach k from j), by $h(j,k)$. Then the difference time is $h(j,k) - h(k,j)$. *See also* commute time. [J. Jonasson & O. Schramm, On the cover time of planar graphs, *Electr. Commun. Probab.*, 5, 85–90, 2000]

differential entropy Arises in extending the definition of entropy from r.vs taking values in a countable set to continuous r.vs taking values in $\mathbb{R}^n$. For example, the Shannon differential entropy of the r.v. $X \in \mathbb{R}$, with density $f(x)$ is defined as $h(X) = -\int_{\mathbb{R}} f(x) \log f(x) \mathrm{d}x = -\mathbb{E} \log f(X)$, by analogy with the definition $H(X) = -\mathbb{E} \log p(X)$, for the entropy of a simple r.v. X with distribution $p(\cdot)$. The entropy $h(X)$ is not scale invariant, in fact $h(aX+b) = h(X) + \log|a|$, $a \neq 0$. To avoid this undesirable property **Jaynes differential entropy** is defined as

$$h_J(X,m) = -\int_{\mathbb{R}} f(x) \log\left(\frac{f(x)}{m(x)}\right) \mathrm{d}x = -\mathbb{E}\log\{f(X)/m(X)\},$$

where $m(x)$ is a measure on $\mathbb{R}$, called the reference measure. The relative differential entropy between two densities $f(x)$ and $g(x)$, defined as $d(f,g) = \int_{\mathbb{R}} f(x) \log\left(\frac{f(x)}{g(x)}\right) \mathrm{d}x$, is scale invariant; and $d(f,g) = -h_J(X,g)$. [E. Jaynes, Information theory and statistical mechanics, in *Brandeis Lectures* (1963), reprinted in E. T. Jaynes, *Papers on Probability, Statistics, and Statistical Physics*, R. Rosenkrantz (ed.). Kluwer, 1989]

diffusion limited aggregation (alias: **Witten–Sander model**) Arises as a model for aggregation (or coagulation, and related physical processes). The process may be formalized and implemented on a lattice, with an initial particle present at the origin. Subsequent particles are launched from random points on some distant boundary, and execute independent random walks until reaching a vertex adjacent to a particle of the cluster, at which point the walk stops and the particle joins the cluster. More generally curve segments, or other objects, may likewise aggregate suitably. [T. Witten & L. Sander, Diffusion-limited aggregation, *Phys. Rev. Lett.*, 47(19), 1400–1403, 1981]

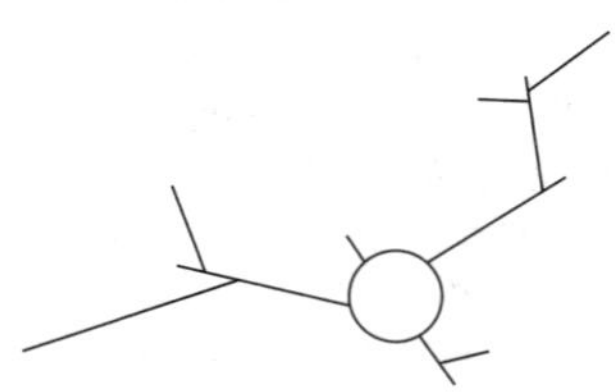

Aggregation of line segments on the perimeter of a disc, in its early stages.

diffusion process Arises as a type of mathematical model for a great many physical processes, which range from the classic (and fundamental) Brownian motion of a small particle in a fluid, to the basic model for the movement of asset prices. Formally, a continuous parameter strong Markov process $X(t), a \leq t \leq b$, is said to be a diffusion process if, with probability 1, its sample paths are continuous. Such diffusions may be required to obey extra regularity conditions. In particular, a diffusion $X(t)$ is a **regular diffusion** in the open interval I (say) if for any $x, y \in I, \mathbb{P}\big(T_y < \infty \,|\, X(0) = x\big) > 0$, where T_y is the first-passage time of $X(t)$ to y. Diffusion processes may be obtained (or constructed, or characterized) in several different ways. The first was N. Wiener's definition of the Wiener process (also called Brownian motion) as the continuous limit of scaled random walk S_n. In this context we note that if $\lim_{n\to\infty} \frac{1}{n} \mathbb{E} S_n{}^2$ exists then it is called the **diffusion constant** of S_n. The second construction of Brownian motion was that of Lévy, via the limit of a suitable sum of orthonormal functions. The general diffusion has been characterized as:

(a) A Markov process whose increments obey a small displacement condition, together with conditions on the first and second moments of the instantaneous increments. (The names of A. Kolmogorov and W. Feller are strongly associated with this.)
(b) A Markov process whose generator is expressed as a suitable differential operator (Feller and others); and the variant of this expressed in terms of the characteristic operator of the process (E. Dynkin and others).
(c) In terms of suitable continuous martingales, e.g. P. Lévy's characterization of the Wiener process, and the Stroock–Varadhan characterization of the general diffusion.
(d) One-dimensional diffusions may be characterized by the total positivity of their transition densities $p(t, x, y), (x, y) \in \{x_r, y_r\}, 1 \leq r \leq n$.
(e) As the strong continuous solution of a stochastic differential equation $\mathrm{d}X(t) = a(X)\mathrm{d}W(t) + b(X)\mathrm{d}t$, where $W(t)$ is the Wiener process. In this case one often speaks of an Itō diffusion.
(f) The Wiener process may be characterized as a Gaussian process; and then any other diffusion (on its natural scale) is a time-changed version of the Wiener process.

[S. Karlin & H. Taylor, *A Second Course in Stochastic Processes*. Academic Press, 1981]

dimension Arises naturally in numerically quantifying the content of sets that are uncountably infinite, but for which the classical measures of size (such as area, length, Lebesgue measure, etc.) do not provide a useful description. Classic examples of such sets include fractals (which may be random), and a number of sets generated by the Wiener process. There are many ways of defining dimension, which will not necessarily yield the same numerical content for a given set. The earliest, and perhaps most important, is Hausdorff dimension, which is derived from Hausdorff measure. Another often seen is the box-counting (or box) dimension, which is also called the entropy dimension, Kolmogorov (entropy) dimension, metric dimension, information dimension, Minkowski dimension, log-density dimension, capacity dimension, etc. If $S \neq \emptyset$ is a bounded subset of $\mathbb{R}^d$, and $N(\delta)$ is the smallest number of sets, of diameter not greater than δ, needed to cover S, then (when it exists) $d(S) = \lim_{\delta\to 0}\left[\frac{\log N(\delta)}{-\log \delta}\right]$ is the box dimension of S.

For another type, the **Fourier dimension** of $A \in \mathbb{R}$ is the supremum of $b \in [0, 1]$ such that for some probability measure $\mathbb{P}$, supported on A, with chf $\phi(t), |\phi(t)| \leq C|t|^{-b/2}$, for all t and some finite constant c. Famous Hausdorff dimensions include:

(a) The zeros of the Wiener process $W(t), 0 \leq t \leq 1$, which has Hausdorff dimension $\frac{1}{2}$.
(b) The graph of the Wiener process has dimension 3/2.
(c) The path of a Wiener process in $\mathbb{R}^d, d \geq 2$, has dimension 2.

[K. Falconer, *Fractal Geometry*, 3rd edn. Wiley, 2014]

dimer model Arises in statistical mechanics as a model for physical processes including e.g. the two-dimensional Ising model. The state space of such a process comprises those sets of edges such that each vertex of an underlying graph is included (or covered) exactly once. This is also called a dimer covering, or perfect matching. In a related problem, if disjoint edges are selected at random, one may ask for the properties of the number X_n of unmatched vertices of a graph with n vertices. For example, if adjacent pairs are picked at random from the set $[0, 1, \ldots, n]$; then $\lim_{n\to\infty} \mathbb{E} X_n / n = \mathrm{e}^{-2}$.

***Left:* A dimer covering. *Right:* Unmatched vertices (monomers) exist.**

[R. Kenyon, *Lectures on Dimers, in Statistical Mechanics*, S. Sheffield & T. Spencer (eds). American Mathematical Society, 2007]

Dirac measure Arises naturally in probability theory as the distribution of a degenerate random variable. In general, let $\mathcal{F}$ be a σ-field of subsets of some non-empty set Ω; then the Dirac measure δ_x is defined as a function $\delta_x(\cdot)$ on the members A of $\mathcal{F}$ by

$$\delta_x(A) = \begin{cases} 1 \text{ if } x \in A, \\ 0 \text{ if } x \notin A. \end{cases}$$

Thus $(\Omega, \mathcal{F}, \delta_x)$ is a probability space in which the probability is concentrated at x (often described as a point-mass of unit probability). It is often seen e.g. in the so-called nth empirical measure m_n of the sequence $X(r), r \geq 1$, of i.i.d.r.vs given by

$$m_n = \frac{1}{n}\sum_{r=i}^{n} \delta_{X(r)}.$$

Similarly, a discrete random variable Y can be represented as a linear combination, $\sum \mathbb{P}(Y = y(r))\delta_{y(r)}$. [T. Tao, *Introduction to Measure Theory*. American Mathematical Society, 2011]

directed graphical model *see* Bayes network

directional distribution *see* spherical distribution

directly given process *see* canonical random process

Dirichlet distribution Arises naturally when considering a random vector $\mathbf{Z} = (Z_1, \ldots, Z_n)$ whose entries comprise a discrete probability distribution. A simple non-trivial example of a probability distribution for Z is the density $f(\mathbf{z})$, on the $(n-1)$-dimensional simplex $\sum_{r=1}^{n} z_r = 1$ with each $z_r \geq 0$, given by

$$f(z) = \frac{\Gamma(a_1 + a_2 + \cdots + a_n)}{\Gamma(a_1)\Gamma(a_2)\cdots\Gamma(a_n)} z_1{}^{a_1-1} z_2{}^{a_2-1} \cdots z_n{}^{a_n-1},$$

where $a_r > 0, 1 \leq r \leq n$; this is the Dirichlet density. It can be seen as a multivariate version of the beta density, to which it reduces when $n = 2$. A convenient representation of Z is as follows: Let $X_1, \ldots, X_n$ be independent gamma-distributed random variables, with respective parameters $(1, a_r), 1 \leq r \leq n$. Then the vector $\mathbf{Z} = (X_1, \ldots, X_n)/\sum_{r=1}^{n} X_r$ has the Dirichlet distn, and is independent of $\sum_{r=1}^{n} X_r$. Furthermore, the vector $(Y_1, \ldots, Y_n - 1)$ defined by $Y_r = X_r/X_n$ has the $(n-1)$ dimensional form of the so-called **inverted Dirichlet distn**, with density

$$f(y_1, \ldots, y_{n-1}) = \frac{\Gamma(a_1 + \cdots + a_n)}{\Gamma(a_1)\ldots\Gamma(a_n)} \frac{y_1^{a_1-1} y_2^{a_2-1} \cdots y_{n-1}^{a_{n-1}-1}}{(1 + y_1 + \cdots + y_{n-1})^a},$$

where $a = a_1 + \cdots + a_n$, and $0 < y_r < \infty, 1 \leq r \leq n-1$. [K. W. Ng, G.-L. Tian, & M.-L. Tang, *Dirichlet and Related Distributions*. Wiley, 2011]

Dirichlet–Ferguson process (alias: **Dirichlet process; Ferguson distribution**) A type of random process whose realizations (or sample paths) are probability measures. It can be seen as an extension of the Dirichlet random variable, for which any outcome is a discrete probability distribution. The formal definition involves two probability spaces: the first space is $(\Omega, \mathcal{F}, \mathbb{P})$, called the base space, where typically $\Omega \in \mathbb{R}^n$, and $\mathcal{F}$ comprises the usual Borel σ-field; the second space is $(\pi, \mathcal{G}, \mathbb{Q})$, called the distributional space, where π is the space of probability measures on $(\Omega, \mathcal{F})$, so that $\mathbb{P} \in \pi$. Given a base measure $m(\cdot)$ on $(\Omega, \mathcal{F})$, and a parameter a called the precision, the corresponding Dirichlet–Ferguson process is a probability measure $\mathbb{Q}$ such that for any finite measurable partition $A_1, \ldots, A_n$ of Ω, the vector $\mathbf{X} = (\mathbb{P}(A_1), \ldots, \mathbb{P}(A_n))$ has the Dirichlet distribution with parameters $a_r = am(A_r), 1 \leq r \leq n$. This process was defined by T. Ferguson, who named it the Dirichlet process. A related generalization is called the **Pitman–Yor process**. [T. Ferguson, Bayesian analysis of some nonparametric problems, *Ann. Stat.*, 1(2), 209–230, 1973]

Dirichlet form Arises in the stochastic analysis of Markov processes, particularly in connecting classical (analytic) potential theory and probabilistic potential theory, and in the solution of SDEs. A closable, non-negative definite, symmetric bilinear form on a Hilbert space is called a Dirichlet form if it satisfies an additional unit contraction property (that is sometimes called the Markovian property). On certain conditions, a Dirichlet form may be associated with the transformation semi-group p_t of a Markov process by the property that $p_t f = T_t f$ a.e., $t \geq 0$, where T_t is the generator associated with the Dirichlet form; if the semi-group is unique, one speaks of Markov uniqueness. [M. Fukushima *et al.*, *Dirichlet Forms and Symmetric Markov Processes*, 2nd edn. de Gruyter, 2011]

Dirichlet game *see* chip-firing game

Dirichlet problem (alias: **probabilistic Dirichlet problem**) Arises naturally from the connexion between analytical and probabilistic potential theory. The classical Dirichlet problem seeks a function f that satisfies Laplace's equation (i.e. is harmonic), on an open set D, and is equal to a given continuous function g on the boundary ∂D of D. For a suitably regular ∂D, the probabilistic solution yields $f(x), x \in D$, in terms of the Wiener process started at $x \in D$, denoted by $W_x(t), t \geq 0$, as $f(x) = \mathbb{E}(g(W_x(T)))$, where T is the hitting time of ∂D by $W_x(t)$; and $f(x)$ is thus unique.

An analogous representation holds for the so-called discrete Dirichlet problem, arising in the context of a non-negative bounded function h that is harmonic in the sense that $h = Ph$ in D, for a stochastic transition matrix P, where $h = g$ on ∂D. [J. Doob, *Classical Potential Theory and its Probabilistic Counterpart*. Springer, 1984]

discounting Arises in financial and insurance mathematics, economic sequential decision processes, dynamic programming, and optimal control, where the concept of discounting is used to quantify the difference between the value of a reward (or cost) incurred now, and the value of a reward (or cost) incurred in the future. There are many reasons for discounting; two basic examples are inflation and interest. If the value of money is falling against a constant value numeraire at a continuously compounded rate $I(t)$, then a future (time t) reward R is discounted by $e^{-\int_0^t I(u)\mathrm{d}u}$, i.e. its value now is $Re^{-\int_0^t I(u)\mathrm{d}u}$. Conversely, if a constant value riskless bond accrues continuously compounded interest at rate $B(t)$, then the present cost of ensuring a reward R at time t is $Re^{-\int_0^t B(u)\mathrm{d}u}$; that is, R multiplied by a discount factor. More generally, such discount processes are random, and the price at time $t < T$ of an instrument (reward) to be paid at time T may be given by $R(t) = \mathbb{E}(D_t R(T))$, where the expected value is with respect to some 'no arbitrage' equivalent martingale measure. In such a case, D_t is called the discount process.

Another classic example arises in sequential decision procedures, in which a selected action a yields a sequence $C_n, n \geq 1$, of future rewards; the expected discounted reward, with discount factor $d < 1$, is $\mathbb{E}\sum d^n C_n$. [E. Feinberg & A. Shwartz, *Handbook of Markov Decision Processes*. Kluwer, 2002]

discrepancy (alias: **discrepancy metric**) For probability distributions $\mathbb{Q}(A)$ and $\mathbb{P}(A)$, defined for all the connected intervals $A \in [0, 1]$, the discrepancy $d(\mathbb{P}, \mathbb{Q})$ is $\sup_A |\mathbb{P}(A) - \mathbb{Q}(A)|$. More generally, for any metric space Ω, with the usual Borel σ-field, supporting probability measures $\mathbb{P}$ and $\mathbb{Q}$,

$$d(\mathbb{P}, \mathbb{Q}) = \sup_B |\mathbb{P}(B) - \mathbb{Q}(B)|,$$

where the supremum is taken over all closed balls $B \in \Omega$. The discrepancy was first introduced as a measure of uniformity for the distribution of an infinite sequence $u_n \in [0, 1], n \geq 1$ (or, more generally, for any infinite sequence modulo 1); given the first n terms of the sequence $d(u, n) = \sup_{0 \leq a \leq b \leq 1} |n(b-a) - |\{u_1, ..., u_n\} \cap [a, b]||$, where the final term denotes the number of values of the sequence lying in $[a, b]$.

The word discrepancy is also used in statistical literature in its everyday sense to denote an inconsistent or surprising difference (*see also* total variation distance). [A. Gibbs & F. Su, Choosing and bounding probability metrics, *Int. Stat. Rev.*, 70(3), 419–435, 2002]

discrete distribution (alias: **discrete probability measure**) A basic concept of probability, arising when a sample space, or a r.v. defined on it, has a finite (or countable) number of points or values. Formally, given a sample space Ω, suppose that there is a countable subset $S = \{\omega_1, \omega_2, \ldots\}$ of points of Ω, and that for each ω_r there is a non-negative number called the probability of ω_r, and denoted by $p_r = \mathbb{P}\{\omega_r\}$, where $\sum_r \mathbb{P}\{\omega_r\} = 1$. Then (p_r) is discrete probability distribution on Ω with support set S. If $\mathcal{G}$ is the set of all subsets of S, and for any $A \in S, \mathbb{Q}(A) = \sum_{\omega_r \in A} p_r$, then $(S, \mathcal{G}, \mathbb{Q})$ is a discrete probability space. If $\mathcal{F}$ is any σ-field of subsets of Ω, and for $B \in \mathcal{F}$, we have $\mathbb{Q}(B) = \sum_{w_r \in B} p_r$, then $\mathbb{Q}(\cdot)$ may be called a discrete probability measure on the space $(\Omega, \mathcal{F})$.

A real-valued random variable X on a probability space $(\Omega, \mathcal{F}, \mathbb{P})$ has a distribution $m(\cdot)$ on $\mathbb{R}$ defined by $m(A) = \mathbb{P}(X \in A)$ for any Borel set A. The random variable X and its distribution function $F(x)$ are both said to be discrete if $m(\cdot)$ is a discrete probability measure on $\mathbb{R}$, which is to say that $m(\cdot)$ has countable support $S \subseteq \mathbb{R}$. A random variable on a discrete probability space is necessarily discrete. If the support S is finite, then the r.v. and the distn may be called **simple**. *See* probability mass function. [P. Billingsley, *Probability and Measure*, 3rd edn. Wiley, 1995]

discrete random process (alias: **countable state space process**) A random (i.e. stochastic) process $X(t)$, $t \in I$, where $X(t)$ takes values in a state space S that is at most countable. The term is not to be confused with that in discrete-parameter random process, in which the parameter set I is countable and discrete. [J. Doob, *Stochastic Processes*. Wiley, 1953]

discrete random variable *see* discrete distribution

discrete renewal theorem *see* Erdös–Feller–Pollard theorem

discrete skeleton (alias: ***h*-skeleton**) Arises in the analysis of random processes in continuous time, where it is designed that suitable properties of the process $X(t), t \geq 0$, coincide with those of the discrete-time process $S(n) = X(t_n), n \geq 1$. Thus the discrete skeleton at scale $h > 0$, or h-skeleton, of $X(t)$, is the discrete process $S(n) = X(nh), n \geq 0$. Under suitable conditions (continuity, or uniform continuity) the limit as $n \to \infty$ of a function $f(S)$ of $S(n)$ coincides with $\lim_{t\to\infty} f(X(t))$, as may other extremal properties (*see* Croft–Kingman lemma).

For a random process $Y(t)$ such that $Y(T_r)$ constitutes an imbedded Markov chain at the (possibly random) times $0 < T_1 < T_2 < \ldots$, the process $S(r) = Y(T_r), r \geq 1$, is sometimes called a **Markov skeleton process**. [J. F. C. Kingman, Ergodic properties of Markov processes and their discrete skeletons, *Proc. Lond. Math. Soc.*, 13(1), 593–604, 1963]

dishonest (alias: **explosive**) Broadly, a stochastic process $X(t), t \geq 0$, taking values in a state space (or domain) S, is said to be dishonest (or explosive) if for some finite t , $\mathbb{P}(X(t) \in S) < 1$, so that its distribution is defective (or improper, or substochastic). Otherwise, if $\mathbb{P}(X(t) \in S) = 1$ for all t, the process is honest (or non-explosive). Under suitable regularity conditions one defines $T = \inf_{t\geq 0}\{t : X(t) \notin S\}$, often called the lifetime (or explosion time) of $X(t)$: It is usual to augment S by a state Δ (the cemetery state), and define $X(t) = \Delta$ for $t \geq T$; the process is honest on $S \cup \Delta$; Δ is often the so-called point at infinity, in the one point compactification of S. A dishonest Markov process is sometimes said to be sub-Markov and for a Markov jump process the lifetime T may arise as a limit point of times of jumps, as in the dishonest birth process. (Defective r.vs are sometime called dishonest.) [D. R. Cox & H. Miller, *Theory of Stochastic Processes*. Methuen, 1965; reprinted by CRC Press, 2001]

dishonest distribution *see* defective distribution

disintegration (alias: **decomposition**) A measure theoretic concept that has applications in the theory of conditional probability measures. A probability measure is said to be disintegrable if it can be written as the expectation with respect to a so-called marginal (or mixing) probability measure of a family of conditional probability measures. The disintegration theorem supplies conditions for this to occur; and can be seen as an extension of Fubini's theorem or a type of converse to the construction of a product probability measure. Subject to suitable measurability conditions it yields a regular conditional probability (known as a disintegrating measure, or kernel in measure theory, with the marginal also being called a mixing measure). [O. Kallenberg, *Foundations of Modern Probability*, 2nd edn. Springer, 2002; J. Chang & D. Pollard, Conditioning as disintegration, *Statistica Neerlandica*, 51(3), 287–317, 1997]

disordered system Disorder, in its usual meaning, arises in many physical contexts. More narrowly, it is used when models are formed in terms of graphs and lattices, which may be random. Typically interest lies in random walks (which may be standard or self-avoiding), or random clusters, etc. In a standard base model, the parameters for the statistical properties of edges and nodes are constant, i.e. equal over the edge and node sets respectively. In a disordered system the parameters are indexed by the edges and/or nodes. A classic example is Freeman Dyson's 1953 model for glass, comprising a chain of random masses linked by springs with random spring constants. Equally classically, a random walk on a disordered lattice is effectively a random walk in a random environment. [J. Ziman, *Models of Disorder*. Cambridge University Press, 1979]

disorder problem *see* change-point problem

disparity *see* divergence

dispersion *see* variance; concentration; entropy

dispersion order (alias: **Bickel–Lehmann dispersive order**) A formal characterization of the concept of spread (or scatter) of random variables about their state space. Let F and G be the distribution functions of X and Y respectively, and denote their right-continuous inverses by F^{-1} and G^{-1}. Then X is said to be less dispersed than Y, or smaller than Y in dispersive (or dispersion) order, if $F^{-1}(v) - F^{-1}(u) \leq G^{-1}(v) - G^{-1}(u)$, for all $0 < u < v < 1$. [P. Bickel & E. Lehmann, Descriptive statistics for non-parametric models, IV: contributions to statistics, in *Mem. J. Hajek*, J. Jurečková (ed.). Reidel, 1979]

dissipative Markov chain A Markov chain taking values in a countable state space S may be called dissipative if all its states are transient or null-recurrent; if the chain almost surely ends up in the set of non-null recurrent states then it is said to be non-dissipative. Otherwise, it may be termed semi-dissipative. In the context of Markov chains on a more general state space S, a set $A \subseteq S$ is called dissipative if it is a countable union of transient sets; and then $A^c = S \backslash A$ may be called the conservative part of S. [R. Tweedie, Sufficient conditions for ergodicity and recurrence of Markov chains on a general state space, *Stoch. Process. Appl.*, 3(4), 385–403, 1975; F. Foster, Markoff chains with an enumerable member of states and a class of cascade processes, *Math. Proc. Camb. Phil Soc.*, 47(1), 77–85, 1951; E. Nummelin, *General Irreducible Markov Chains and Non-negative Operators*. Cambridge University Press, 1984]

dissociation A weak form of independence. Let $\{X(\mathbf{n})\}$ be a collection of random variables indexed by r-vectors of integers, $\mathbf{n} = \{i_1, \ldots, i_r\}$. Consider two subsets A and B of $\{X(n)\}$ such that the set of integers appearing in indices of A is disjoint from the set of integers appearing in indices of B. If every such pair of families A and B are independent of each other, then $\{X(\mathbf{n})\}$ is said to be dissociated. [W. McGinley & R. Sibson, Dissociated random variables, *Math. Proc. Camb. Phil. Soc.*, 77, 185–188, 1975]

distance (alias: **probability metric; divergence; dissimilarity**) Arises in quantifying the difference between probability distributions (or measures, or random objects), particularly in the context of determining convergence of a sequence of such objects. There are many such distances, even when considering only distributions on $\mathbb{R}$, and real-valued random variables; and several have various aliases. In the strict sense a distance $d(x,y) \in \mathbb{R}^+, x, y \in S$, is a metric on some space S, while dissimilarity measures that fail to be metrics in some way are called divergences; many authors refer to both types as distances. As usual $d(x,y)$ is a metric on S when:
(a) it is symmetric: $d(x,y) = d(y,x)$;
(b) $d(x,z) \le d(x,y) + d(y,z)$: the triangle inequality;
(c) $d(x,y) = 0$ iff $x = y$.
There are several types:

Compound metrics are defined on the set of jointly distributed r.vs X and Y on some space; in this case $X \equiv Y$ means that $X = Y$ a.s.

Simple metrics are defined on the set of r.vs on some space; here $X = Y$ means that X and Y have the same distribution.

Yet more weakly, for **primary metrics**, $X \equiv Y$ means that some properties of X and Y are the same, e.g. $\mathbb{E}X = \mathbb{E}Y$, or var $X =$ var Y together with $\mathbb{E}X = \mathbb{E}Y$.

If (a) fails to hold then d is a quasimetric (or skew metric); if (b) fails to hold then d is a semimetric; if (c) fails to hold then d is a pseudometric (or a semimetric, for some authors). And various other weakenings of (a) (b) (c), exist. In probability theory, such non-metric distances may be called by a variety of general names; divergence; dissimilarity (or similarity, viewed differently); separation; dissociation; deviation; statistical distance, etc., where use depends on context.

In the other direction, metrics that are in addition homogeneous and regular are called **ideal metrics**, where regularity entails $d(x+z, y+z) \le d(x,y)$, and homogenity entails $d(ax, ay) = |a|^h d(x,y)$ for some $h \in \mathbb{R}$ and any $a \in \mathbb{R}$. [A. Gibbs & F. Su, Choosing and bounding probability metrics, *Int. Stat. Rev.*, 70(3), 419–435, 2002; S.-H. Cha, Comprehensive survey on distance/similarity measures between probability density functions, *Int. J. Math. Models Method. Appl. Sci.*, 1(4), 300–307, 2007; S. Rachev, *Probability Metrics and the Stability of Statistical Models*. Wiley, 1991; S. Rachev *et al.*, *The Method of Distances in the Theory of Probability and Statistics*. Springer, 2013]

distance covariance A measure of dependence (i.e. correlation), between random vectors $X \in \mathbb{R}^p$ and $Y \in \mathbb{R}^q$, defined as follows: let the joint characteristic function of X and Y be $\phi_{X,Y}(s,t)$, with marginal chfs $\phi_X(s)$ and $\phi_Y(t)$, $s \in \mathbb{R}^p, t \in \mathbb{R}^q$. Define $c(p) = \pi^{\frac{1}{2}(1+p)}/\Gamma(\frac{1}{2}(1+p))$, where Γ is the gamma fn. Then the distance covariance $C(X,Y) > 0$ is given by

$$C^2 = \{c(p)c(q)\}^{-1} \int_{\mathbb{R}^{p+q}} |\phi_{X,Y}(s,t) - \phi_X(s)\phi_Y(t)| \|s\|^{-1-p} \|t\|^{-1-q} \mathrm{d}s\mathrm{d}t,$$

where $\|s\|$ and $\|t\|$ are the usual Euclidean norm of s and t. Usefully, $C = 0$ iff X and Y are independent. $C(X,X)$ is the distance variance, and $R(X,Y) = C(X,Y)/\{C(X,X)C(Y,Y)\}^{1/2}$ is the distance correlation. [G. Szekely, & M. Rizzo, Brownian distance covariance, *Ann. Appl. Stat.*, 3(4), 1236–1265, 2009]

distortion *see* distance; rate distortion theory

distribution (alias: **probability distribution, or function, or law, or measure**) Broadly, the function, often denoted by $\mathbb{P}(\cdot) \ge 0$, that assigns a non-negative real number, representing probability, to suitable subsets of a set (or space) representing the possible values of a random object, or outcomes of a random experiment. More narrowly, when referring to the underlying probability space $(\Omega, \mathcal{F}, \mathbb{P})$ many authors call $\mathbb{P}(\cdot)$, defined on the subsets of the σ-field $\mathcal{F}$, a probability measure. The term distribution is then reserved for those distributions of probability that are induced on another measurable space $(S, \mathcal{A})$ by a measurable mapping $X(\omega)$ from Ω to S. Then, for $B \in \mathcal{A}, \mathbb{P}(X \in B) = \mathbb{P}(\omega : X(\omega) \in B)$ is called the distribution (or **law**) of X. In particular, when X takes values in $\mathbb{R}^n$, the probability $\mathbb{P}(X_1 \leq x_1, \ldots X_n \leq x_n)$ is called the distribution function of X, and is denoted by $F(x_1, \ldots, x_n)$. [O. Kallenberg, *Foundations of Modern Probability*, 2nd edn. Springer, 2002]

distribution function (alias: **cumulative distribution function; cdf**) For a real-valued random variable X, having probability distribution $\mathbb{P}_X$, its distribution function is defined as $F(x) = \mathbb{P}_X(X \leq x)$ (also denoted by $F_X(x)$.) If X is almost surely finite (i.e. $\mathbb{P}_X(|X| < \infty) = 1$), then $F(x)$ is non-decreasing, right-continuous, and obeys $\lim_{x \to -\infty} F(x) = 0$, and $\lim_{x \to \infty} F(x) = 1$. Either of $\mathbb{P}_X$ or $F(x)$ may be called the **law** of X. More formally, for X defined on a probability space $(\Omega, \mathcal{F}, \mathbb{P})$, it can be shown that for any Borel set B in $\mathbb{R}, \mathbb{P}(X \in B) = \mathbb{P}(\omega : X(\omega) \in B) = \mathbb{P}_X(B) = \int_B \mathrm{d}F(x)$. It is therefore usual to speak (loosely) of the distribution $\mathbb{P}$ of X. A distribution function may be discrete, absolutely continuous (with a density), or singular, or a mixture of these. The function $1 - F(x) = \mathbb{P}(X > x)$ is called the **complementary distn fn**, or **right tail** or **survival fn**, or **exceedance**. Some writers define a left-continuous cdf as $F = \mathbb{P}(X < x)$.

Likewise, for a random vector $\mathbf{X}$, defined on a probability space $(\Omega, \mathcal{F}, \mathbb{P})$, with an induced joint probability distribution on $\mathbb{R}^n$, the joint distribution function is defined by $F(x_1, \ldots, x_n) = \mathbb{P}(X_1 \leq x_1, \ldots, X_n \leq x_n)$, having properties analogous to those of $F(x)$. [O. Kallenberg, *Foundations of Modern Probability*, 2nd edn. Springer, 2002; J. F. C. Kingman & S. Taylor, *Measure & Probability*. Cambridge University Press, 1966]

divergence *see* distance

dixie cup problem *see* collector problem

Dobrushin's central limit theorem Supplies a general CLT for discrete-time Markov chains that are not necessarily homogeneous. One simple special case is this: let $X_r^{(n)}, 1 \leq r \leq n < \infty$ be an array of non-homogeneous Markov chains; the rth transition function being $P_r^{(n)}$, with Dobrushin ergodic coefficient being $a_r^{(n)} = a\left(P_r^{(n)}\right)$. If $|X_r^{(n)}| \leq C < \infty$, and $\text{var}|X_r^{(n)}| \geq c > 0$, and $\min_{1 \leq r \leq n-1} a_r^{(n)} n^{1/3} \to \infty$ as $n \to \infty$, then the array obeys a central limit theorem (i.e. $S^{(n)} = \sum_{r=1}^{n} X_r^{(n)}$ is asymptotically normally distributed when suitably scaled and shifted). [R. Dobrushin, Central limit theorems for non-stationary Markov chains, I, II, *Theor. Probab. Appl.*, 1(1), 65–80, 329–383, 1956]

Dobrushin's ergodicity coefficient A type of coefficient of ergodicity (specifically, the type corresponding to the L_1 norm: $\|\mathbf{x}\| = \sum_{r=1}^{n} |x_r|, \mathbf{x} \in \mathbb{R}^n$). In its simplest form for a stochastic matrix $P = (p_{ij})$, the ergodicity (or contraction) coefficient is

$$d(P) = \sup \frac{1}{2} \sum_k |p_{ik} - p_{jk}|,$$

where the upper bound is taken over all i and j. Equivalently, $a(P) = 1 - d(P)$ is sometimes taken as the ergodicity coefficient, where $d(P)$ may be called the delta (or contraction) coefficient. More generally, let $P(x, y)$ be a stochastic kernel (or transition function), which is to say that for fixed x, $P(x, y)$ is a probability measure; and for fixed y, $P(x, y)$ is a measurable function of x. Then Dobrushin's ergodicity coefficient is $d(P) = \sup|P(x, A) - P(z, A)|$, or, equivalently $a(P) = 1 - d(P)$, where the upper bound is taken over all x, z, and suitable sets A, and $P(x, A)$ is the probability of transition from x to the set A.

For finite stochastic matrices P, Q, and $R = PQ, d(PQ) = d(R) \leq d(P)d(Q)$, which may be called **Dobrushin's inequality**. [R. Dobrushin, Central limit theorems for non-stationary Markov chains, I, II, *Theor. Probab. Appl.*, 1(1), 65–80, 329–383, 1956]

Dobrushin's lemma Arises in the theory of stationary point processes, where it supplies sufficient conditions for a process to be orderly; it can thus be seen as a converse of **Korolyuk's thm**; *see* Kinchin's theorem. Formally, let the point process $N(s, t) \equiv$ the number of points in the interval (s, t) be (crudely) stationary (which is to say that the distribution of $N(s_1 + h, t_1 + h), \ldots, N(s_r + h, t_r + h)$ is independent of h for fixed r and any $(s_i, t_i), 1 \leq i \leq r$). Then if points never coincide (a.s.) and $\mathbb{E}N(0, 1)$ is finite, we have $\mathbb{P}(N(0, h) > 1)/h \to 0$ as $h \to 0$, which is to say that N is orderly. [M. Leadbetter, On three basic results of stationary point processes, *Proc. Amer. Math. Soc.*, 19(1), 115–117, 1968]

Doeblin condition Arises when considering the ergodic behaviour of a Markov chain, where the term is used to denote several different conditions on the transition matrix (or kernel) of the chain. One simple strong form for a chain with n-step transition matrix $p_{ij}(n)$ on a countable state space S, is that there exists some finite n and state k such that $\inf_j p_{jk}(n) = a > 0$; this ensures that an ergodic chain is geometrically ergodic. One form of the condition (called Hypothesis (D) by J. Doob) for a chain with kernel $p(x, B)$ is that there exists a finite strictly positive measure $m(B)$ on the Borel sets B of the state space, and some $a > 0$, and some finite n, such that the nth iterate $p^{(n)}(x, B)$ of the transition kernel $p(x, B)$ satisfies $p^{(n)}(x, B) \leq 1 - a$ whenever $m(B) \leq a$, for every $x \in S$. Then the chain has a stationary distribution, and S can be partitioned into a finite number of ergodic sets. More strongly, Doeblin's minorization condition (Hypothesis (D′)) requires a probability measure $m(\cdot)$ on S, some $a > 0$, and finite n, such that $p^{(n)}(x, B) \geq am(B)$; this ensures geometric ergodicity. [S. Meyn & R. Tweedie, *Markov Chains and Stochastic Stability*, 2nd edn. Cambridge University Press, 2009; J. Doob, *Stochastic Processes*. Wiley, 1953]

Doeblin coupling (alias: **classic coupling**) Arises in the context of stochastic processes, especially in the evolution of Markov chains. Thus, let $X_n, n \geq 0$ be a Markov chain; a Doeblin coupling may be achieved by defining Y_n to be a Markov chain independent of X_n, but having the same transition probabilities. Let the bivariate process (X_n, Y_n) evolve until the first time T at which $X_T = Y_T$, and require that $X_n = Y_n$ for $n \geq T$. Such a process is a Doeblin coupling; more generally one may require only that X_n and Y_n (prior to coupling) are conditionally independent given (X_0, Y_0). For a chain with transition matrix $P = (p_{ij})$ the transition probabilities $p_{ij,kl}$ of the coupled process are given by $p_{ij,kl} = p_{ik}p_{jl}, i \neq j$, while $p_{ii,kk} = p_{ik}$.

This is sometimes called, but should not be confused with, basic coupling. [T. Lindvall, *Lectures on the Coupling Method*. Wiley, 1992; reprinted by Dover, 2002]

Doeblin's decomposition Arises in extending the elementary decomposition theorem of countable state space, discrete-time Markov chains to Markov chains on a general measure space $\{\Omega, \mathcal{F}\}$. Under Doeblin's condition (or, more generally, Doob's condition that the process has a stationary measure), Ω may be decomposed into a transient set (or state) and a collection of ergodic in decomposable (or irreducible) sets. Furthermore, the restriction of the transition function to each irreducible ergodic set has a unique stationary measure, and any stationary measure for the chain is a linear combination of these. This may be called **Doeblin's ergodic thm**. [S. Orey, *Limit theorems for Markov chain transition probabilities*, 1971, van Nostrand]

Doeblin–Gnedenko conditions (or theorem) (alias: **Gnedenko–Doeblin theorem**) Arise in characterizing the domain of attraction of the non-Gaussian stable laws. Thus, let $S_n = \sum_{r=1}^{n} X_r$, where X_r are i.i.d.r.vs with distn $F(x)$. Then there exist sequences a_n and $b_n, n \geq 1$, and a stable distn $G(x)$, index $\alpha \in (0, 2)$, such that $(S_n - a_n)/b_n$ converges weakly to $G(x)$ as $n \to \infty$, iff as $x \to \infty$,

$\mathbb{P}(X < -x)/\mathbb{P}(|X| > x) \to l$,

$\mathbb{P}(X > x)/\mathbb{P}(|X| > x) \to r$, where $l + r = 1$, and for all $\lambda > 0$,

$\mathbb{P}(|X| > \lambda x)/\mathbb{P}(|X| > x) \to \lambda^{\alpha}$.

Note that the term Doeblin–Gnedenko thm is also sometimes used to denote a characterization of the convergence (to infinitely divisible distributions) of null arrays.

[O. Kallenberg, *Foundations of Modern Probability*, 2nd edn. Springer, 2002; D. Neuenschwander, Probabilities on simply connected nilpotent Lie groups: on the Doeblin–Gnedenko Laws for the domain of attraction of stable laws, *Int. J. Pure Appl. Math.*, 55(2), 187–199, 2009]

Doeblin–Lévy–Kolmogorov–Rogozin inequality *see* Kolmogorov–Rogozin inequality

Doeblin's ratio limit theorem Arises in the context of discrete-time homogeneous irreducible, recurrent Markov chains on a countable state space S. Let X_n, $n \geq 1$, be such a chain with n-step transition probabilities $p_n(j, k); j, k \in S$. Then there is a set of finite positive numbers $r_j, j \in S$, such that

$$\lim_{m\to\infty} \frac{\sum_{n=1}^{m} p_n(i,j)}{\sum_{n=1}^{m} p_n(h,k)} = \frac{r_j}{r_k}; i, j, h, k, \in S.$$

If the chain is non-null then $\pi_j = r_j / \sum_{k \in S} r_k$ is the stationary distribution; if null then r_j is a stationary measure for $p_1(j, k)$, *see* Derman's theorem. [K. L. Chung, *Markov Chains*. Springer, 1967]

Doeblin's universal laws The characteristic function $\phi(t) = \mathbb{E}e^{itX}$ of X belongs to the domain of partial attraction of the (non-degenerate) characteristic function $\theta(t)$ iff there exist sequences a_r, b_r, and $n_r, r \geq 1$, where

a_r and b_r are constants, and the n_r are integers, such that $n_r \to \infty$ and then

$$\left\{\phi\left(\frac{t}{a_r}\right)\mathrm{e}^{-\mathrm{i}tb_r}\right\}^{n_r} \to \theta(t).$$

The limit $\theta(t)$ is necessarily infinitely divisible, and W. Doeblin showed that there are functions $\phi(t)$ that belong to the domain of partial attraction of every possible infinitely divisible $\theta(t)$; the corresponding distributions are called universal laws. [W. Feller, *Introduction to Probability Theory and its Applications*, Vol. 2, 2nd edn. Wiley, 1971]

dog-flea model *see* Ehrenfest urn

Doléans-Dade exponential (alias: **stochastic (or semi-martingale, or Doléans) exponential (after C. Doléans-Dade, née Doléans)**) Arises as the a.s. unique solution of the SDE $\mathrm{d}Z_t = Z_t - \mathrm{d}X_t$, where X_t is a semi-martingale with $X_0 = 1$ (and Z can be seen as analogous to the ordinary exponential arising as the solution of $\mathrm{d}z = z\mathrm{d}x$). It is given by

$$Z_t = \exp\left(X_t - \frac{1}{2}[X]_t^{\mathrm{c}}\right)\prod_{s\le t}(1+\Delta X_s)\exp(-\Delta X_s),$$

where ΔX_s denotes jumps of X, and $[X]^{\mathrm{c}}$ is the quadratic variation of the continuous part of X. If X is a continuous semi-martingale, the product on the right is 1. [O. Kallenberg, *Foundations of Modern Probability*, 2nd edn. Springer, 2002]

Doléans-Dade measure (alias: **Doléans measure**) Arises in a construction of the Itô stochastic integral as a linear isometry. For a continuous finite-variance martingale X_t, on $(\Omega, \mathcal{F}, \mathbb{P})$, with quadratic variation process $[X]_t$, one representation of the Doléans measure $m(B)$ of a set B in the Boolean algebra generated by the so-called predictable rectangles $(s,t] \times A, 0 \le s < t < \infty$, $A \in \mathcal{F}$, is given by $M(B) = \mathbb{E}\int_0^\infty I_B(s)\mathrm{d}[X]_s$, where $I_B(\cdot)$ is the usual indicator function. [K. L. Chung & R. Williams, *Introduction to Stochastic Integration*, 2nd edn. Birkhäuser, 1990]

domain of attraction Arises in the context of a suitable sequence $S_n = S_n(X_1, \ldots, X_n), n \ge 1$, of functions of a sequence of independent r.vs $X_r, r \ge 1$, having a common distribution function $F(x) = \mathbb{P}(X_r \le x)$. If there are norming (i.e. scaling and shifting) constants a_n, and $b_n > 0, n \ge 1$, such that the distribution of $(S_n - a_n)/b_n$ converges, as $n \to \infty$, to some distn $G(x)$, then $F(x)$ is said to be in the domain of attraction of $G(x)$. Classic examples include $S_n = \sum_{r\ge1}^n X_r$, in which case $G(x)$ is necessarily stable, and $S_n = \max\{X_1, \ldots, X_n\}$, in which case $G(x)$ is necessarily max-stable (i.e. an extreme value distribution). [B. Fristedt & L. Gray, *A Modern Approach to Probability*. Birkhäuser, 1997]

dominated convergence theorem (alias: **Lebesgue's dominated convergence theorem; DCT**) Arises in the theory of integration, where it supplies a condition for the integral of the limit of a suitably convergent sequence of functions to equal the limit of their sequence of integrals. In the context of a sequence of random variables $X_n, n \ge 1$, it takes this form: If, as $n \to \infty, X_n$ converges almost surely (or in probability) to X, and $|X_n| \le Z$ for some r.v. Z with $\mathbb{E}Z < \infty$, then $\lim_{n\to\infty}\mathbb{E}X_n = \mathbb{E}X$. There are a number of useful extensions of this, e.g. **Pratt's lemma**: suppose that $X_n \le Y_n \le Z_n$, where $X_n \to^P X$, $Y_n \to^P Y$, and $Z_n \to^P Z$. If $\mathbb{E}X_n \to \mathbb{E}X$ and $\mathbb{E}Z_n \to \mathbb{E}Z$, then $\mathbb{E}Y_n \to \mathbb{E}Y$. [A. Gut, *Probability: A Graduate Course*, 2nd edn. Springer, 2013]

dominated ergodic theorem Let $X_n, n \le 1$, be a stationary sequence of random variables, and write $M = \sup_{n\ge1}\frac{1}{n}\sum_{r=1}^n X_r$. Broadly, a dominated ergodic theorem supplies a bound on the moments of M in terms of those of X_1; and one form is this: there is a constant $C < \infty$ such that:

(a) for given $r > 1, \mathbb{E}|M|^r \le C\mathbb{E}|X_1|^r$;

(b) for given $r \ge 0$,

$$\mathbb{E}\{|M|(\log_+|M|)^r\} \le C\{1 + \mathbb{E}|X_1|\{\log_+|X_1|\}^{r+1}\}.$$

In this case, and in the more general context of measure-preserving transformations, one speaks of Wiener's theorem, or the Wiener–Yosida–Kakutani theorem; when the X_n are i.i.d., it may be called the Marcinkiewicz–Zygmund theorem. [O. Kallenberg, *Foundations of Modern Probability*, 2nd edn. Springer, 2002]

Donsker's theorem (alias: **Donsker's invariance principle; functional CLT; Donsker–Kolmogorov–Skorokhod theorem.**) Broadly, this can be seen as an extension of the classical central limit theorem for sequences of random variables to suitable sequences of random functions, which are shown to converge (weakly, in a suitable space of functions) to the Wiener process, or a functional thereof. One form of the theorem asserts that for i.i.d.r.vs $X_n, n \ge 1$, each with zero mean and unit variance, the process $X(t,n) = n^{-1/2} \sum_{r\le nt} X_r, t \in [0,1], n \ge 1$, converges (in a suitable sense) to the Wiener process $W(t)$ on $[0,1]$ (and, more strongly, $f(X(t_n))$ converges to $f(W(t))$ for a continuous functional of paths of $W(t)$). The invariance of this result under changes in the actual distribution of X_n leads to the term invariance principle. M. Donsker's first theorem of this type addressed the convergence of a sequence of empirical distribution functions to the Brownian bridge. [R. Dudley, *Uniform Central Limit Theorems*, 2nd edn. Cambridge University Press, 2014]

Donsker–Varadhan theory Describes the large deviations of the sojourn time (or occupation measure) of a class of Markov processes. [M. Donsker & S. Varadhan, Asymptotic evaluation of certain Markov process expectations for large time, I–IV, *Commun. Pure Appl. Math.*, 28(1), 1–47, 28(2), 279–301, 1975; 29(4), 389–461, 1976; 36(2), 183–212, 1983]

Doob decomposition Asserts that a discrete-parameter (or time) **submartingale** denoted by $X_n, n \geq 0$, can be expressed in the form $X_n = M_n + S_n$, where M_n is a martingale (with respect to the same filtration $\mathcal{F}_n$), and S_n is an a.s. unique increasing process that is predictable wrt $\mathcal{F}_n$ (i.e. S_n is $\mathcal{F}_{n-1}$-measurable for all $n \geq 1$), and may be called the compensator of X_n. [J. Doob, *Stochastic Processes*. Wiley, 1953]

Doob–Dynkin lemma A measure-theoretic description of functional relationships between random variables. Let $X(\omega)$ be a measurable function from $(\Omega, \mathcal{F})$ to $(S, \mathcal{G})$, and denote by $\sigma(X)$ the sub sigma-field of elements A of $\mathcal{F}$ given by $A = \{\omega : X(\omega) \in B\}$ for $B \in \mathcal{G}$. Then the random variable $Y(\omega), \omega \in \Omega$, is $\sigma(X)$-measurable iff there is a real-valued measurable function $f(s), s \in S$, such that $Y = f(X)$. The case when $(S, \mathcal{G}) = (\mathbb{R}^d, \mathcal{B})$ is Doob's lemma. [M. M. Rao, *Measure Theory and Integration*, 2nd edn. Marcel Dekker, 2004]

Doob's *h*-transform (alias: **Doob's, or Doob's harmonic transform; *h*-transform**) Arises naturally in considering a Markov process $X(t) \in S$ whose future is required to satisfy some suitable condition, e.g. $X(t)$ never leaves $A \subseteq S$; or $X(T) \in A$ for some fixed $T > 0$; or $X(t)$ visits A before $B \subseteq S$; etc. Forming the h-transform of the process consists in replacing the original transition function of $X(t)$ with the new conditional laws arising from the constraint. Thus, suppose $X(t)$ is sure to visit $A \cup B \subseteq S$, and let $h(x)$ be the probability that A is visited first, starting from $X(0) = x$, so that for $x \notin A$, $h(x) = \sum_y p_{xy} h(y)$, where p_{xy} is the transition matrix of the discrete chain X. Then the associated h-transform, for the chain ${}_hX$ that does visit A first, comprises the transition matrix ${}_hp_{xy} = p_{xy} h(y)/h(x)$.

In general, a Doob h-transform replaces the transition function $p(t; x, \mathrm{d}y)$ by ${}_hp = p(t; x, \mathrm{d}y)h(y)/h(x), 0 < h(x) < \infty$ (or by 0 otherwise), where h is a harmonic (or, more generally, superharmonic) function for p. The idea extends to diffusions, where e.g. the h-transform of the Wiener process for the harmonic function $h(x) = x, x > 0$, corresponds to the Wiener process conditional on its never revisiting the origin; and this process is essentially the same as the distance of the three-dimensional Wiener process from the origin. [J. Doob, *Classical Potential Theory and its Probabilistic Counterpart*. Springer, 1984]

Doob–Lévy martingale *see* closed martingale

Doob L log L maximal inequality *see* Doob L_p inequality

Doob L_p inequality (alias: **Doob maximal inequality**) Arises in martingale theory, where it bounds the moments of the running maximum of a suitable process. Let $X_r, 0 \leq r \leq n$, be a martingale (or a positive submartingale), then for $p > 1$,

$$\mathbb{E}\Big(\max_{1\leq r\leq n} |X_r|^p\Big) \leq \Big(\frac{p}{p-1}\Big)^p \mathbb{E}(|X_n|^p);$$

(or, in terms of the L_p norm $\|\cdot\|$, $\|\max|X_r|\| \leq \frac{p}{p-1}\|X_n\|$). For $p = 1$, we have Doob's L log L maximal inequality

$$\mathbb{E}\Big(\max_{1\leq r\leq n} |X_r|\Big) \leq \frac{\mathrm{e}}{\mathrm{e}-1}\{1 + \mathbb{E}(|X_n|\log|X_n|\}.$$

[J. Doob, *Stochastic Processes*. Wiley, 1953]

Doob's maximal inequality (alias: **maximal inequality; Doob–Kolmogorov inequality; Doob–Lévy inequality**) Arises in martingale theory, where it supplies a bound for the tail of the distribution of the running maximum of a submartingale $X_r, 1 \leq r \leq n$. Thus

$$\mathbb{P}\Big(\max_{1\leq r\leq n} X_r \geq x\Big) \leq x^{-1}\mathbb{E}\Big(X_n^+ I\Big(\max_{1\leq r\leq n} X_r \geq x\Big)\Big),$$

for $x > 0$, where $I(\cdot)$ is the usual indicator function. It follows that for a martingale (or for a positive submartingale) X_n and $p > 1$, $x^p \mathbb{P}(\max_{1\leq r\leq n}|X_r| \geq x) \leq \mathbb{E}(|X_n|^p)$, which may perhaps be the result most often called Doob's inequality. The special case arising when M_n is a martingale, so that M_n^2 is a submartingale, giving $\mathbb{P}(\max_{1<r\leq n}|M_n|>t)\leq t^{-2}\mathbb{E}M_n^2, t>0$, is often called the Doob–Kolmogorov inequality. [J. Doob, *Stochastic Processes*. Wiley, 1953]

Doob–Meyer decomposition Supplies conditions such that an rcll submartingale $X(t), t\geq 0$, can be written uniquely (a.s.) as the sum of a martingale $M(t)$ and some non-decreasing predictable process $A(t)$ (called the compensator); it thus extends the Doob decomposition for discrete-parameter martingales to continuous time. Formally, the theorem holds when $X(t)=0$, and $X(T)$ is uniformly integrable over the collection of all finite stopping times T; such a submartingale is said to be of class D. [P. Meyer, A decomposition theorem for supermartingales, *Illinois J. Math.*, 6(2), 193–205, 1962; P. Meyer, Decomposition of supermartingales: the uniqueness theorem, *Illinois J. Math.*, 7(1), 1–17, 1963]

Doob's optional stopping theorem (alias: **optional stopping (or sampling, or systems) theorem; OST**) Arises in martingale theory, where it may take a number of forms. Informally, because martingales arise naturally as models for fair games (and sub- or super-martingales for unfair games), elementary forms of the OST can be interpreted as asserting this basic truth: no reasonable system of betting can turn a fair game into an advantageous game, or an unfair game into a fair game. (An unreasonable betting system is one that relies on knowledge of the future.) More formally, let, $X_n, n\geq 1$, be a martingale, S and T being stopping times for the martingale (i.e. wrt the same filtration $\mathcal{F}_n$). Then we have:

(i) **Optional stopping**. Define

$$Z_n=\begin{cases}X_n \text{ for } n\leq T,\\ X_T \text{ for } n>T.\end{cases}$$

Then $Z_n, n\geq 1$, is itself a martingale (wrt $\mathcal{F}_n$), and may be called X_n stopped at T, or a stopped martingale. The stopping time T is allowed to take the value $+\infty$ with non-zero probability.

(ii) **Optional sampling**. For $S\leq T$, an optional sampling thm supplies sufficient conditions for the martingale condition to hold in the form $\mathbb{E}(X_T\mid\mathcal{F}_S)=X_S$, with $\mathbb{E}X_T=\mathbb{E}X_S=\mathbb{E}X_1$. An easy sufficient condition requires that T is almost surely bounded. More generally, it is sufficient that $\mathbb{P}(T<\infty)=1$, with any one of the following additional conditions:
   (a) $|X_n|\leq c<\infty$ a.s., for some constant c;
   (b) $X_n, n\geq 1$ is a uniformly integrable sequence;
   (c) $\mathbb{E}T<\infty$, and $\mathbb{E}(|X_{n+1}-X_n||\mathcal{F}_n)\leq c<\infty$, a.s., for $n<T$; or
   (d) $\mathbb{E}|X_T|<\infty$, and $\mathbb{E}(X_nI(T>n))\to 0$ as $n\to\infty$.

(iii) **Optional systems**. The martingale $X_n, n\geq 0$, transformed by the system sequence $S_n\geq 1$ yields the so-called martingale transform sequence

$$Z_n=Z_{n-1}+S_n(X_n-X_{n-1}), n\geq 1,$$

where S_n is required to be $\mathcal{F}_{n-1}$-measurable (and is interpreted as the nth stake in betting applications). An optional systems theorem asserts that $Z_n, n\geq 1$, is a martingale when $\mathbb{E}|S_n|<\infty$. The special case when $S_n\in\{0,1\}$ is called **optional skipping** and this is also called **Halmos' thm**.

Similar theorems hold for sub- and super-martingales, with equality replaced appropriately by inequality. [J. Doob, *Stochastic Processes*. Wiley, 1953]

Doob's theorem Any one of numerous results, but perhaps most commonly a martingale convergence thm such as this: let $X_n, n\geq 1$ be a martingale. Then it converges in mean iff there exists Z such that $X_n\to Z$ a.s. as $n\to\infty$, and $X_n=\mathbb{E}(Z\mid\mathcal{F}_n), n\geq 1$, where $\mathcal{F}_n$ is the associated filtration. This, in turn, holds iff $X_n, n\geq 1$, is a uniformly integrable sequence. It is usual to write $Z=X_\infty$, and to call the martingale closed, regular, closable, or a Doob–Lévy martingale. The r.v. Z may be called the **last element**. The submartingale convergence theorem (*see* martingale converence theorem) may also be called Doob's theorem, as may a number of martingale inequalities.

Outside martingale theory, the term Doob's theorem probably most often refers to the proof that a stationary, Markov, Gaussian process with continuous sample paths is the Ornstein–Uhlenbeck process with a covariance function of the form $c(s,t)=\sigma^2\exp\{-\lambda|s-t|\}$, for some $\lambda>0$, together with the corresponding spectral density, and specification of the transition densities. [J. Doob, Brownian movement and stochastic equations, *Ann. Math.*, 43(2), 351–369, 1942, reprinted in *Selected Papers on Noise and Stochastic Processes*, N. Wax (ed.). Dover, 1954; J. Doob, *Stochastic Processes*. Wiley, 1953]

double exponential distribution (alias: **Laplace distribution; two-sided or bilateral exponential**) Arises in a standard form (centred at 0) as the distribution of the difference between independent $\text{Exp}(\lambda)$ and $\text{Exp}(\mu)$ random variables, with density

$$\frac{(\lambda+\mu)}{\lambda\mu}f(x) = \begin{cases} e^{-\lambda x} & \text{for } x \geq 0 \\ e^{\mu x} & \text{for } x \leq 0. \end{cases}$$

The symmetrical form, when $\lambda = \mu$, may be called the Laplace distn, having been introduced by P. Laplace in 1774 as a possible distn of errors; in 1781 he proposed the alternative

$$f(x) = \frac{1}{2\lambda}\log\frac{\lambda}{|x|}, \quad |x| < \lambda,$$

called the two-sided (or double) **logarithmic density**. The double exponential distn is not to be confused with the doubly exponential distn of extreme value distributions theory. [S. Kotz *et al.*, *The Laplace Distribution and Generalizations*. Birkhäuser, 2001]

double Poisson distribution A term used to denote several things. Efron's double Poisson distn arises in the context of a generalized exponential family of distns, and has probability function

$$f(x) = c\frac{e^{-x}x^x}{x!}\left(\frac{e\mu}{x}\right)^{\theta x}, \quad x = 0, 1, 2, \ldots,$$

where μ and θ are non-negative parameters, and c is a normalizing constant. Thomas's double (or doubly) Poisson distn arises when a r.v. Λ is Poisson with parameter μ, and conditional on $\Lambda = \lambda$, X is Poisson with parameter λ. Then unconditionally X has a double Poisson distn, with pgf $\exp\{\mu(e^{s-1} - 1)\} = G_X(s)$.

The term is also loosely used to denote a bivariate Poisson distn; and sometimes a mixture of two Poisson distns. [B. Efron, Double exponential families, *J. Amer. Stat. Ass.*, 81(395), 709–721, 1986; M. Thomas, A generalization of Poisson's binomial limit, *Biometrika*, 36(1–2), 18–25, 1949]

doubling rate *see* Kelly betting

doubly compounded Poisson process *see* Cox process; compounding

doubly stationary A term with several meanings:

(a) Given a stationary sequence $X_n, n \geq 0$, an associated sequence $Y_n, n \in \mathbb{Z}$, that is stationary and such that Y_n and X_n have the same distributions for $n \geq 0$, the pair may be called doubly stationary, especially when $Y_0 = X_0$, and $Y_{-n}, n \geq 1$, is seen the 'past' of $X_n, n \geq 1$.

(b) If $X(n,m), n, m \in \mathbb{Z}$ is an array of r.vs such that the covariance $\text{cov}(X(n+s, m+t), X(n,m)) = \rho(s,t)$, is a function of s and t, then $X(n,m)$ may be called doubly stationary.

(c) Let $X(t), t \in \mathbb{R}$ (or $t \in \mathbb{Z}$), be a symmetric α-stable random process having the same distribution as a representation in the form of a stochastic integral $\int f(t,u)dM(u)$, where $M(\cdot)$ is a symmetric stable random measure, and $f(t,u)$ is a random process of L_α-functions. If $f(t,u)$ is stationary, then $X(t)$ is said to be doubly stationary.

[S. Cambanis *et al.*, Ergodic properties of stationary stable processes, *Stoch. Process. Appl.*, 24, 1–18, 1987]

doubly stochastic matrix A square matrix with non-negative entries in which each row sum and each column sum is 1. **Birkhoff's thm** asserts that such a matrix is a convex combination of permutation matrices; and **Horn's thm** is that $d_1, \ldots, d_n$ are the diagonal elements of such a matrix iff $\sum_{r=1}^{n}(1 - d_r) \geq 2 - \min_{1\leq r\leq n} d_r$. Similarly, a **doubly stochastic measure** is a distn on the unit square whose two marginals are uniform U[0, 1]. [A. Horn, Doubly stochastic matrices, *Amer. J. Math.*, 76(3), 620–630, 1954]

drift A term arising naturally in the context of random processes (with discrete or continuous time-parameter) whose increments have non-zero mean μ. Thus e.g. a random walk, $S_n = \sum_{r=1}^{n} X_r$, is said to have positive (resp. negative) drift if $\mathbb{E}X_1 > 0$ (resp. < 0). If $\mathbb{E}X_1 > 0$ (resp. < 0) then $\lim_{n\to\infty} S_n = \infty, a.s.$, (resp. $-\infty$), and the walk is said to drift to $+\infty$ (resp. $-\infty$). Otherwise the walk takes arbitrarily large values, both positive and negative, infinitely often, and is said to oscillate. Likewise, for a diffusion X defined by the SDE $dX = \mu dt + \sigma dW(t), t \geq 0$, μ is called the drift (or drift coefficient); if X converges to some part of the boundary of its state space, it may be said to drift there (e.g. to $+\infty$ or $-\infty$). In higher dimensions one may speak of a drift vector $\boldsymbol{\mu}$. [W. Feller, *Introduction to Probability Theory and its Applications*, Vol. 1, 3rd edn. Wiley, 1968; Vol. 2, 2nd edn. Wiley, 1971]

drift condition (alias: **Foster–Lyapunov drift condition**) Denotes any of several sufficient (and sometimes necessary) conditions for a Markov chain on a suitable state space to have an invariant distribution, and then for the ergodicity (or, more strongly, geometric ergodicity) of the chain (or, further, strong ergodicity). They appear as natural extensions of Foster's condition for the ergodicity of a Markov chain (irreducible and aperiodic) on $\mathbb{Z}^+$, with transition matrix p_{ij}, i.e. there exists a non-negative function V_i, some $\varepsilon > 0$, and some finite integer m, such that $\sum_k p_{jk}V_k \leq V_j - \varepsilon$, for $j \geq m$, while $\sum_k p_{jk}V_k < \infty$ for $j < m$.

If the RHS of the first condition is $(1-\varepsilon)V_j$, then the chain is geometrically ergodic; while if the function V_i is bounded, then the chain is strongly ergodic. More generally, for a suitable Markov chain on a general state space the equivalent condition for ergodicity is the existence of a small set C, a function $v(\cdot) \geq 1$ (that is bounded on C), and $b > 0$ such that $\mathbb{E}\{v(X_n+1) - v(X_n) \mid X_n\} \leq bI(X_n \in C) - 1$. For geometric ergodicity the condition takes the form $\mathbb{E}[v(X_n+1) - av(X_n) \mid X_n)] \leq bI(X_n \in C)$, for some $0 < a < 1$. [R. Tweedie, Drift conditions and invariant measure for Markov chains, *Stoch. Process. Appl.*, 92(2), 345–354, 2001; S. Meyn & R. Tweedie, *Markov Chains and Stochastic Stability*, 2nd edn. Cambridge University Press, 2009]

duality Arises in the usual mathematical sense whenever two stochastic objects are interchangeable in the sense that to solve a problem for one (primal problem) automatically solves a problem for the other (dual problem). In particular, the duality principle for a random walk $S_n = \sum_{r=1}^{n} X_r, n \geq 1$, is the observation that $(S_1, \ldots, S_n)$ has the same distribution as $(S_n - S_{n-1}, S_n - S_{n-2}, \ldots, S_n)$, which follows from the fact that $(X_1, \ldots, X_n)$ and $(X_n, \ldots, X_1)$ have the same distribution. From this follows the duality lemma: for any interval $I \in \mathbb{R}^+$ we have

$$\mathbb{P}(\{S_r < S_n; 0 \leq r \leq n-1\} \cap \{S_n \in I\}) \\ = \mathbb{P}(\{S_r > 0, 0 \leq r \leq n\} \cap \{S_n \in I\}.$$

The above duality can be seen as taking the steps of the first walk in reverse order; another form of duality is seen as reflection, thus: if Q is a standard single server queue with service time distribution F_S and interarrival time distn F_X, then its dual process Q_d is the queue with service time distn F_X, and interarrival time distn F_S. [W. Feller, *Introduction to Probability Theory and its Applications*, Vol. 2, 2nd edn. Wiley, 1971]

duality lemma *see* ladder process

Dubins embedding A proof of the Skorokhod embedding theorem by construction in terms of the limit of a sequence of binary splitting martingales, i.e. a zero-mean martingale X_n such that, conditional on a non-null event of the form $X_0 = x_0, \ldots, X_n = x_n$, the r.v. X_{n+1} takes at most two values, a.s. [L. Dubins, On a theorem of Skorokhod, *Ann. Math. Stat.*, 39, 2094–2097, 1968]

Dubins graph An infinite non-homogeneous random graph on the vertex set $[1, 2, 3, 4, \ldots]$ generated thus: each edge jk, $1 \leq j < k$ is present independently with probability λ/k, where $0 \leq \lambda \leq 2$. Shepp's theorem asserts that Dubins graph is a.s. disconnected for $\lambda \leq \frac{1}{4}$, but a.s. connected for $\lambda > \frac{1}{4}$. [L. Shepp, Connectedness of certain random graphs, *Israel J. Math.*, 67(1), 23–33, 1989]

Dubins inequality Arises in the context of a non-negative supermartingale $Y_n, n \geq 0$, where it supplies a bound on the tail of the distribution of the number U of upcrossings of the interval $[a, b]$ by the supermartingale Y_n. That is, for $r \geq 0$, $\mathbb{P}(U \geq r) \leq \left(\frac{a}{b}\right)^r \mathbb{E}(\min\{1, Y_0/a\})$. Hence $\mathbb{E}U \leq \frac{a}{b-a}\mathbb{E}(\min\{1, Y_0/a\})$. [L. Dubins, Rises and upcrossings of non-negative martingales, *Ill. J. Math.*, 6(2), 226–241, 1964]

Dubins–Savage casino (alias: **subfair primitive casino**) A classic model of probability theory, in which a gambler can bet any proportion of his/her current wealth, losing the stake with probability $1 - w$, or winning r times the stake with probability w, where $0 < w < 1, r > 0$, and $w(1 + r) - 1 < 0$. [L. Dubins & L. Savage, *Inequalities for Stochastic Processes*, McGraw-Hill, 1965; reprinted by Dover, 1976]

Dubins–Savage inequality Arises as a bound for the distribution of the partial sums $S_n = \sum_{r=1}^{n} X_r$ of a sequence of random variables $X_n, n \geq 1$, where $m_n = \mathbb{E}(X_n \mid X_1, \ldots, X_n - 1) < \infty$, and $V_n = \text{var}(X_n \mid X_1, \ldots, X_n - 1)$. Let a and b be any positive constants. Then the probability that there is $n < \infty$ such that $S_n \geq b + \sum_{r=1}^{n} m_r + a\sum_{r=1}^{n} V_r$ is less than $(1 + ab)^{-1}$. [L. Dubins & L. Savage, A Tchebycheff-like inequality for stochastic processes, *Proc. Natl. Acad. Sci. USA*, 53(2), 274–275, 1965]

Dubins–Savage theorem Any of a large number of results determining optimal strategies in certain idealized casinos. [L. Dubins & L. Savage, *Inequalities for Stochastic Processes*, McGraw-Hill, 1965; reprinted by Dover, 1976]

Dubins–Schwarz theorem Asserts that any continuous martingale started from 0 is the time-change of a suitably stopped Brownian motion (i.e. Wiener process). The extension of this to suitable continuous local martingales is the Dambis–Dubins–Schwarz theorem (or representation). The multidimensional version is called **Knight's thm.** [D. Revuz, & M. Yor, *Continuous Martingales and Brownian Motion*, 3rd edn. Springer, 1999]

Dudley's condition (alias: **Dudley's entropy bound; Dudley's theorem**) Supplies a sufficient condition for a real-valued Gaussian process to be a.s. continuous and bounded. If the process has stationary increments, then the Fernique–Talagrand theorem gives a necessary and sufficient condition for a.s. continuity. (Also called the Dudley–Fernique–Talagrand thm.) [R. Dudley, The sizes of compact subsets of Hilbert spaces and continuity of Gaussian processes, *J. Funct. Anal.*, 1, 290–330, 1967]

Dudley's theorem (alias: **Dudley's representation**) Broadly, this asserts that suitable functionals of the Wiener process $W(t), t \geq 0$ (i.e. Brownian motion), can be expressed as a stochastic integral wrt $W(t)$. Formally, let $\mathcal{F}_t$ be the natural filtration for $W(t)$, satisfying the usual conditions, and let X be an $\mathcal{F}_T$-measurable r.v. that is almost surely finite, $0 < T < \infty$. Then there is a progressively measurable (i.e. non-anticipating) process $Y(t)$ such that, with probability 1, we have $\int_0^T Y^2 \mathrm{d}t < \infty$, and $X = \int_0^T Y \mathrm{d}W(t)$. [R. Dudley, Wiener functionals as Itô integrals, *Ann. Probab.*, 5(1), 140–141, 1977]

Duffie–Kan model In financial mathematics, this is an arbitrage-free interest rate model for the instantaneous short rate $r(t)$ and the yields on a number of bonds with maturity dates $t + T_i$, denoted by $Y_i(t), 1 \leq i \leq m$. It postulates an affine set of suitable SDEs for $\mathbf{X} = (r, Y_1, \ldots, Y_m)$ (i.e. the drift and covariance coefficients are linear in the entries of $\mathbf{X}$). [D. Duffie, & R. Kan, A yield-factor model of interest rates, *Math. Finan.*, 6(4), 379–406, 1996]

Dufresne's relation (alias: **Dufresne's identity (or perpetuity, or formula)**) Arises in mathematical finance and other applications of diffusion process. Let $D(t, \mu) = \int_0^t \exp\{2W(s) + 2\mu s\}\mathrm{d}s$, $t > 0$, where $W(t)$ is the standard Wiener process, and $\mu \in \mathbb{R}^+$. Then the distribution of $\{D(t, -\mu)\}^{-1}$ is the same as that of $\{D(t,\mu)\}^{-1} + 2X$, where X is independent of $D(t, \mu)$ and has a gamma density with parameter μ. In particular, $D(\infty, -\mu)$ is called Dufresne's perpetuity, arising as the continuous limit for the value of a discrete time perpetuity in the presence of a random discount factor. Then $D(\infty, -\mu)$ has the distribution of $(2X)^{-1}$; and, furthermore, $D(\infty, -\mu)$ has the distribution of the first hitting time of 0 by a Bessel process of dimension $2(1 - \mu)$, started at the point 1. [H. Matsumoto & M. Yor, On Dufresne's relation, *Adv. Appl. Probab.*, 35(1), 184–206, 2003; P. Salminen & M. Yor, Perpetual integral functional as hitting and occupation times, *Electr. J. Probab.*, 10, 11, 371–419, 2005]

Duncan–Mortensen–Zakai equation Arises in the context of the optimal filtering problem for diffusion processes. In its most basic form, one has a diffusion $X(t), t \geq 0$ (called the state process) and an observation process $Y(t), t \geq 0$, such that $\mathrm{d}Y = \mu(X, Y)\mathrm{d}t + \sigma(Y)\mathrm{d}W(t)$, where $W(t)$ is a standard Wiener process independent of $X(t)$. The filtering problem is to determine the conditional distribution of the state $X(t)$, conditional on $Y(s), 0 \leq s \leq t$. The required solution satisfies a stochastic PDE, which is called the Duncan–Mortensen–Zakai eqn. [M. Zakai, On the optimal filtering of diffusion processes, *Probab. Theor. Rel. Fields*, 11(3), 230–243, 1969]

Dunford's theorem Any of a number of results, often this one: let $X_n, n \geq l$, be a uniformly integrable sequence of r.vs. Then there exists a subsequence that is weakly convergent (in L_1). There is also a Dunford mean ergodic thm, of the von Neumann type. [O. Kallenberg, *Foundations of Modern Probability*, 2nd edn. Springer, 2002]

duration of play problem *see* gambler's ruin problem

Dutch book A combination of bets that must yield a loss to the gamester, on the basis of arithmetic and logic (i.e. without involving any laws of probability). Such books may be synchronic (simultaneous bets) or diachronic (sequential bets). They have been used in attempts to justify applying the mathematical (Kolmogorov) axioms of probability to subjective probabilities, a step which is convenient but requires justification. The so-called Dutch book argument runs as follows: it is assumed that anyone who has a set of subjective (personal, or Bayesian) probabilities is prepared to bet accordingly, at fair odds. Then the Dutch book thm shows that such a gambler is safe against a Dutch book iff their probabilities obey the axioms of mathematical probability. Hence (the argument asserts) these axioms are perforce required by rationality. This reasoning is not universally regarded as valid. [M. Eaton & D. Freedman, Dutch book against some 'objective' priors, *Bernoulli*, 10(5), 861–872, 2004]

Dvoretzky–Erdos–Kakutani inequality Supplies a bound for the rate of convergence of an empirical distn fn $F_n(x)$ to the parent distribution $F(x)$ of n i.i.d.r.vs (cf Glivenko–Cantelli theorem). It asserts that, for $t > 0$,

$$\mathbb{P}\left(\sqrt{n} \sup_x (F_n(x) - F(x)) > t\right) \leq C \exp(-2t^2),$$

for some constant C; P. Massart has shown that C can be taken as 1, when $\exp(-2t^2) \leq \frac{1}{2}$. [P. Massart, The tight constant in the D-K-W inequality, *Ann. Probab.*, 18(3), 1269–1283, 1990]

Dvoretzky–Erdös–Kakutani theorem Determines the question of points of highest multiplicity for paths of the Wiener process in d-dimensions. (For a function $f(t) \in \mathbb{R}^d, t > 0$, a point $x \in \mathbb{R}^d$ has multiplicity m if $|f^{-1}(\{x\})| \geq m$.) Then the D–E–K thm asserts that:
(a) for $d \geq 4$, the path has only simple points $(m = 1)$;
(b) for $d = 3$, the path has double points, but not triples $(m = 2)$;
(c) for $d \leq 2$; the path has points of multiplicity C, where C is the cardinality of the continuum.
[A. Dvoretzky *et al.*, Points of multiplicity C of plane Brownian paths, *Bull. Res. Council Israel*, 7(F)(4), 175–180, 1958]

Dvoretzky's problem *see* Shepp's covering theorem

Dvoretzky–Robbins theorem Any of several results on Rényi's parking problem (also called the one-dimensional random packing problem); most commonly the CLT proved in that context. [A. Dvoretzky & H. Robbins, On the parking problem, *Publ. Math. Inst. Hung. Acad. Sci.*, 9, 209–224, 1964]

Dvoretzky's stochastic approximation theorem Addresses the convergence of a sequence of transformations that are subject to noise (random errors) at each stage; it extends the earlier results of Robbins & Monro and Kiefer & Wolfowitz. One form is this: let the sequences $X_n, T_n, Y_n, n \geq 1$, satisfy $X_{n+1} = T_n(X_1, \ldots, X_n) + Y_n(X_1, \ldots, X_n)$, where $\mathbb{E}(Y_n \mid X_1, \ldots, X_n) = 0$ and $\sum \mathbb{E}{Y_n}^2 < \infty$. If $|T_n| \leq \max\{a_n, (1 + b_n)|X_n| - c_n\}$ for sequences of real numbers a_n, b_n, and c_n such that $a_n \to \infty, \sum b_n < \infty$, and $\sum c_n = \infty$, then X_n converges to 0 a.s. as $n \to \infty$. Another form addresses mean-square convergence. [C. Derman & J. Sacks, On Dvoretzky's stochastic approximation theorem, *Ann. Math. Stat.*, 30(2), 601–606, 1959]

Dvoretzky–Wald–Wolfowitz theorem Arises in considering minimax strategies for games and decisions. Essentially it asserts that if the number of possible decisions is finite, and the number of unknown states (or parameters) is finite, with each having a non-atomic distribution, then non-randomized decision functions will do. [A. Dvoretzky *et al.*, Elimination of randomization in certain statistical procedures and zero-sum-two-person games, *Ann. Math. Stat.*, 22(1), 1–21, 1951]

Dwass's random walk Arises in the context of an ordinary Galton–Watson branching process, started from a single individual, where the family sizes are $X_1, X_2, \ldots$, taken in the usual lexicographical order of birth. Define the left-continuous random walk $S_n = \sum_{r=1}^{n}(X_r - 1)$, with $S_0 = 0$. Then the first arrival of this walk at -1 corresponds to extinction of the branching process, and the hitting time of -1 by the walk equals the size of the total population. Furthermore, the branching process can be reconstructed from the walk. See the related concepts of height process, contour process, and hitting-time theorem. [M. Dwass, The total progeny in a branching process and a related random walk, *J. Appl. Probab.*, 6, 682–686, 1969]

Dyer–Frieze–McDiarmid inequality Arises in linear programming with random costs, where it supplies a general bound on the expected value of the objective function. [M. Dyer *et al.*, On linear programs with random costs, *Math. Program.*, 35, 3–16, 1986]

dynamical function Arises in the theory of stationary processes, where it supplies a dependence measure generalizing the autocovariance (see codifference). Let X_n be a stationary process, not necessarily of finite variance. Then the dynamical function $d(s,t,k)$ is defined as $d(s,t,k) = \mathbb{E}\ \exp\{i(sX_n + k + tX_n)\}$. [A. Janicki & A Weron, *Simulation and Chaotic Behavior of Alpha-stable Stochastic Processes*. Dekker, 1994]

dynamically complete Arises in financial mathematics, where a market model (i.e. a price process for a collection of tradable assets), is said to be dynamically complete if it is possible to replicate every suitable contingent claim (i.e. an option that is integrable, and measurable for the underlying filtration), by using a self-financing trading strategy. Such continuous implementation of a portfolio strategy is called **dynamic hedging** (of the option). [L. T. Nielsen, *Pricing and Hedging of Derivative Securities*. Oxford University Press, 1999]

dynamic reversibility Arises in reversing Markov chains that are not themselves reversible in equilibrium, but would be if some states were interchanged. Formally, suppose that to each state $s \in S$ there is associated a conjugate state s^+, such that $(s^+)^+ = s$. Then the Markov chain $X(t) \in S$ is said to be dynamically reversible if the conjugate chain $X^+(t)$ reversed (i.e. $(X^+(T-t)$ for any T), has the same joint distributions as $X(t), 0 \leq t \leq T$. [F. Kelly, *Reversibility and Stochastic Networks*, Wiley, 1979; reprinted by Cambridge University Press, 2011]

dynamic storage allocation A classic process arising in operations research, with applications in computing. One form is this: there is an unbounded sequence of storage areas, each of unit size, represented by $\{1, 2, \ldots\}$. Objects of random integer sizes arrive at random to be stored; an object of size x requires a consecutive sequence of storage areas of size at least x, the actual allocation being according to a rule, e.g. nearest to the origin. Each object remains for a random time, and then leaves the space free for subsequent objects. [O. Aven, E. Coffman, & Y. Kogan, *Stochastic Analysis of Computer Storage*. Reidel/Kluwer, 1987]

Dynkin's criterion (or condition) Supplies a sufficient condition for a function $Y(t) = f(X(t))$ of a homogeneous Markov process $X(t), t \geq 0$, also to be a Markov process. There are several other such conditions, whose authors are not (yet) eponymous. [L. Rogers & J. Pitman, Markov functions, *Ann. Probab.*, 9(4), 573–582, 1981]

Dynkin's formula Arises in the context of a suitable continuous-time Markov process $X(t), t \geq 0$, having (infinitesimal) generator A, e.g. a Feller process or Itô diffusion. It asserts that the expected value of a suitable function $f(X(T))$, of $X(t)$ evaluated at a finite-mean stopping time T, is given by $\mathbb{E}f(X(T)) = f(X(0)) + \mathbb{E}\int_0^T Af(X(t))\mathrm{d}t$. [S. Karlin & H. Taylor, *A Second Course in Stochastic Processes*. Academic Press, 1981]

Dynkin's game (alias: **Dynkin's stopping problem**) Is a generalization of elementary optional stopping problems to the context of game theory. There are three pay-off (or reward) processes, $X(t), Y(t)$, and $Z(t), t \geq 0$, together with two stopping times S and T; these are seen as separately decidable by two players A and B in the game interpretation, i.e. A chooses S and B chooses T. Then the outcome of the game is

$$R(S,T) = \begin{cases} X(S) & \text{if } S < T, \\ Y(T) & \text{if } T < S, \\ Z(T) & \text{if } S = T. \end{cases}$$

The game problem is to determine conditions under which the game has a value V, i.e. given by

$$V = \inf_T \sup_S \mathbb{E}R(S,T) = \sup_S \inf_T \mathbb{E}R(S,T).$$

[J. Neveu, *Discrete-parameter Martingales*. North-Holland, 1975]

Dynkin–Lamperti theorem (alias: **D-L arcsine law(s)**) Arises in renewal theory for the partial sums $S_n = \sum_{r=1}^n X_r$ of i.i.d.r.vs X_r with distn $F(x) 0 \leq x < \infty$, where $N(t) = \max\{n : S_n \leq t\}$ is the renewal function. Define $Y(t) = t - S_{N(t)}$ (the age), and $Z(t) = S_{N(t)+1} - t$ (the current life), and $u(t) = \mathbb{E}(N(t) + 1)$ (the renewal function). Then one form of the D-L theorem asserts that if $1 - F(x) = x^{-a}K(x), 0 < a < 1$, where $K(tx)/K(t) \to 1$ as $x \to \infty$, then the limiting distributions as $t \to \infty$ of $Y(t)/t$ and $Z(t)/t$ have, respectively, the densities $f(y) = \frac{1}{\pi}\sin(a\pi)y^{-a}(1-y)^{a-1}, 0 < y < 1$ and $g(z) = \frac{1}{\pi}\sin(a\pi)z^{-a}(1+z)^{-1}, z > 0$. [N. Bingham, C. Goldie & J. Teugels, *Regular Variation*. Cambridge University Press, 1987]

Dynkin's pi-lambda theorem (alias: **Dynkin's π-system lemma**) Arises in considering the extension of a probability measure on a field of events to the sigma-field generated by the field, particularly its uniqueness. This finds applications in constructing probability spaces for random processes. Formally, given a set Ω, a π-system $\mathcal{A}$ is a collection of subsets such that for $A \in \mathcal{A}$ and $B \in \mathcal{A}$, we have $A \cap B \in \mathcal{A}$. Furthermore, a collection of sets L is called a λ-system (or d-system, or delta class, or Dynkin system, etc.) if $\Omega \in L$, and also:

(i) $A \in L$ implies that $A^c \in L$.

(ii) If $A_1, A_2, \ldots \in L$ and $A_i \cap A_j = \emptyset$ for $i \neq j$, then $\bigcup_r A_r \in L$.

Dynkin's lemma asserts that if $\mathcal{A} \in L$, then the σ-field generated by $\mathcal{A}$ is also included in L. *See* monotone class theorem. [P. Billingsley, *Probability and Measure*, 3rd edn. Wiley, 1995]

Dyson's theorem Arises in the context of a particular $n \times n$ Hermitian matrix-valued stochastic process, defined as follows: let $B_{jk}(t)$ and $W_{jk}(t), 1 \leq j, k \leq n$ be a family (having $2n^2$ members) of independent standard Wiener processes (i.e. Brownian motion). Define the Hermitian matrix-valued process $X(t)$ such that $X_{kk}(t) = B_{kk}(t)$ and $X_{jk}(t) = \frac{1}{\sqrt{2}}\left(B_{jk}(t) + \mathrm{i}W_{jk}(t)\right)$ for $j < k$, and $X_{kj}(t) = \overline{X_{jk}}(t)$, where $\overline{X}$ is the complex conjugate of X. Then Dyson's thm asserts that the n eigenvalues of $X(t)$, denoted by $\lambda_1(t), \ldots, \lambda_n(t)$, evolve as independent Wiener processes conditioned never to meet. They satisfy the SDEs

$$\mathrm{d}\lambda_r = \sum_{j \neq k} \frac{\mathrm{d}t}{\lambda_j - \lambda_k} + \mathrm{d}W_r(t),$$

where $W_r(t), 1 \leq r \leq n$ are independent Wiener processes. This is called **Dyson's Brownian motion.** [F. Dyson, A Brownian motion model for the eigenvalues of a random matrix, *J. Math. Phys.*, 3, 1191–1198, 1962]

E

earth mover's distance Arises as a measure of dissimilarity between two distributions, where 'distribution' may be interpreted quite generally; the concept was introduced to compare images, and also to compare textured objects. The two distributions (often called signatures) are $(x_r, p_r), 1 \leq r \leq m$, and (y_r, q_r), $1 \leq r \leq n$, where x_r and y_r are points in a space equipped with a distance $d(x_i, y_j) = |x_i - y_j|$, and p_r and q_r are weights. These are informally envisaged as piles of earth of size p_r at points x_r that are to fill holes of size q_r at points y_r. The earth mover's distance (EMD), is defined as the minimum amount of work required to do so, i.e. as the value of the associated linear program. In probability applications x_r and y_r are values of r.vs X and Y with distributions p_r and q_r; in this case the EMD is equivalent to the Mallows, or Wasserstein, or minimum L_1 distance, being a special case of the Kantorovich metric. [E. Levina & P. Bickel, The earthmover's distance is the Mallows distance, *IEEE ICCV, Proc.*, 2, 251–257, 2001]

east process An interacting particle system that can be described as an asymmetric one-dimensional constrained Ising model. The sites of the system are integer points on the real line, or a subset thereof, and each site is either empty or carries one particle, denoted by the state space $\{0, 1\}$. The state of any site $n+1$ can change only if site n is occupied by a particle. Sites that can change flip from $0 \longrightarrow 1$ at rate p or from $1 \longrightarrow 0$ at rate $1-p, 0 < p < 1$, independently as a Markov process. [D. Aldous & P. Diaconis, The asymmetric one dimensional constrained Ising model, *J. Stat. Phys.*, 107(5–6), 945–975, 2002]

Eddington's controversy Arises from the following question: If A, B, C, D each speak the truth once in three times (independently), and A affirms that B denies that C declares that D is a liar, what is the probability that D is speaking the truth?

The astrophysicist A. Eddington heard of this problem from the astronomer A. Crommelin, who felt the answer might be useful in evaluating reports from the 1919 expeditions sent to verify Einstein's predictions on the observations of the eclipse; Crommelin attributes the problem to an article in *Granta* satirizing Cambridge Tripos questions. The controversy follows from the different answers depending on how one interprets hearsay denials. Eddington favoured 25/71 as the solution, but 13/41 is more popular. [H. Wallis Chapman, Eddington's probability problem, *Math. Gazette*, 20(241), 298–308, 1936]

Eden–Richardson model (alias: **last-passage percolation**) A classic random growth process on the square lattice, that is a Markov process denoted by $X(t)$, $t \geq 0$, where $X(t)$ is a random subset of $\mathbb{Z}^2$ generated as follows. At $t = 0, X(0) = \{0\}$, where the origin is seen as occupied by a particle (or infected). For any $t > 0$, any site in $\mathbb{Z}^2 \backslash X(t)$ is occupied by a particle (or infected) with probability rate proportional to the member $r(t)$ of infected neighbours, $0 \leq r \leq 4$, independently of other infections. As $t \longrightarrow \infty, X(t)/t$ converges to a convex set that is not a circle, a result that may be called **Richardson's shape thm**. The connexion with last-passage percolation is seen by considering infection of vertices as flow along edges, with the origin as source. [D. Richardson, Random growth in a tessellation, *Proc. Camb. Phil. Soc.*, 74, 515–525, 1973]

edge process *see* bond process

Edgeworth expansion (or series) Broadly, this is an asymptotic series representation of a suitable distribution function $F(x)$ (or of its density $f(x)$ if it has one) in terms of the cumulants κ_r of $F(x)$ (or equivalently the central moments μ_r of $F(x)$), and a family of orthogonal polynomials, and $\Phi(x)$ and $\phi(x)$ (the standard normal distribution function and its density). The expansion finds natural application to the central limit theorem, so the usual form of the Edgeworth expansion is this: let $F_n(x)$ be the distn fn of the normalized sum $Z_n = \sum_{r=1}^{n} X_r$ of zero-mean, unit variance, non-lattice i.i.d.r.vs $X_r, r \geq 1$, with cumulants κ_s. Then as $n \longrightarrow \infty$,

$$F_n(x) - \Phi(x) - \phi(x) \sum_{r=3}^{m} n^{-\frac{1}{2}r+1} E_r(x) = o\left(n^{-\frac{1}{2}m+1}\right),$$

where $E_r(x)$ is a polynomial of degree r depending only on cumulants of order not exceeding m; the explicit form

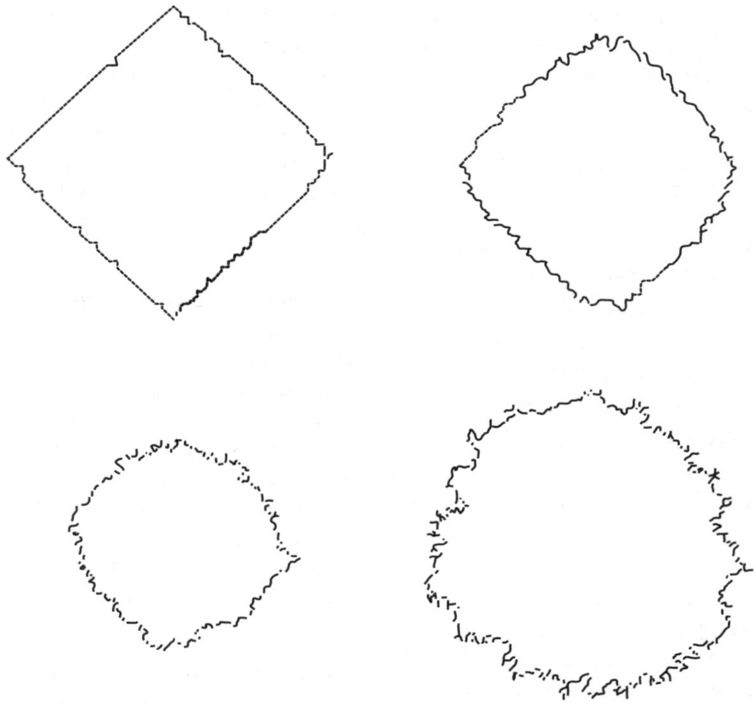

Eden–Richardson models for various rates of infection and times.

of $E_r(x)$ is given in terms of Hermite (or Chebyshov–Hermite) polynomials. Similar expressions may be given for the density $f_n(x)$ if it exists, and for the distribution of sums of arithmetic r.vs. These series may also be considered in various rearranged forms, which may be termed Gram–Charlier expansions, or Gauss–Hermite series, of various types. [W. Feller, *Introduction to Probability Theory and its Applications*, Vol. 2, 2nd edn. Wiley, 1971; S. Blinnikov & R. Moessner, Expansions for nearly Gaussian distributions, *Astron. Astrophys. Suppl. Ser.*, 130, 193–205, 1998; P. Hall, *The Bootstrap and Edgeworth Expansion*. Springer, 1995]

Edgeworth–Sargan density Denotes a probability density of the form

$$f(x) = \phi(x)\left\{1 + \sum_{r=1}^{m} a_r H_r(x)\right\}, \quad x \in \mathbb{R},$$

for suitable constants $a_1, \ldots, a_m$, where $\phi(x)$ is the standard normal density, and $H_r(x)$ is the rth-order Hermite polynomial defined by the property that

$$\frac{\mathrm{d}^r}{\mathrm{d}x^r}\phi(x) = (-1)^r \phi(x) H_r(x).$$

[I. Mauleon & J. Perote, Testing densities with financial data: empirical comparison of the Edgeworth-Sargan density to students *t*, *Euro. J. Finan.*, 6(2), 225–239, 2000]

Edwards–Anderson model *see* spin-glass model

Efron's dice An example of non-transitivity in a probabilistic context, thus: four cubical dice bear the numbers $(4,4,4,0,0,0)$; $(3,3,3,3,3,3)$; $(6,6,2,2,2,2)$; $(5,5,5,1,1,1)$; and when rolled show the scores X_A, X_B, X_C, X_D respectively. Then

$$\mathbb{P}(X_A > X_B) = \mathbb{P}(X_B > X_C) = \mathbb{P}(X_C > X_D) = \mathbb{P}(X_D > X_A) = \frac{2}{3}.$$

A similar example is known as **Gale's roulette**. Three spinners are marked to give equal probability 1/3 to the elements of the three triples $(2,4,9)$; $(1,6,8)$; $(3,5,7)$, and yield *X*, *Y*, *Z* as outcomes. Then $\mathbb{P}(X > Y) = \mathbb{P}(Y > Z) = \mathbb{P}(Z > X) = 5/9$. *See* circular triad. [H. Tijms, *Understanding Probability*, 3rd edn. Cambridge University Press, 2012]

Efron–Stein inequality Supplies a concentration inequality for functions of independent random variables

$X_1, \ldots, X_n$. In its first form the function $g(X_1, \ldots, X_{n-1})$ is assumed symmetric in its $n-1$ variables, and the X_r are assumed to be i.i.d. Then, writing $g_i = g(X_1, \ldots, X_{i-1}, X_{i+1}, \ldots, X_n)$, we have

$$\operatorname{var} g(X_1, \ldots, X_{n-1}) \leq \mathbb{E}\sum_{r=1}^{n}\left(g_r - \sum_{K=1}^{n} g_K\right)^2.$$

More strongly, let $\mathbf{X} = (X_1, \ldots, X_n)$ be independent but not necessarily identically distributed, with $\hat{X} = (\hat{X}_1, \ldots, \hat{X}_n)$ being an independent copy of $\mathbf{X}$. Further, suppose $g(X_1, \ldots, X_n)$ is any, not necessarily symmetric function of $X_1, \ldots, X_n$, and write $g_i = g(X_1, \ldots, X_{i-1}, \hat{X}_i, X_{i+1}, \ldots, X_n)$. Then the **Efron–Stein–Steele inequality** is $\operatorname{var} g \leq \frac{1}{2}\mathbb{E}\sum_{r=1}^{n}(g - g_r)^2$; with equality iff g is linear in its arguments. [J. Steele, An Efron–Stein inequality for non symmetric statistics, *Ann. Stat.*, 14(2), 753–758, 1986]

Efron's theorem Arises in stochastic ordering; in the case when $\mathbf{Y}$ dominates $\mathbf{X}$ (i.e. $\mathbf{X}$ is stochastically smaller than $\mathbf{Y}$ if $\mathbb{E}g(\mathbf{X}) \leq \mathbb{E}g(\mathbf{Y})$ for any function $g(\mathbf{x}) : \mathbb{R}^n \to \mathbb{R}$ that is increasing in each of its arguments, and such that the expectations are finite). Let $\mathbf{X} = (X_1, \ldots, X_n)$, and denote any r.vr with the distribution of $\mathbf{X}$ conditional on $\sum_{r=1}^{n} X_r = u$ by $\mathbf{X}(u)$. Then Efron's theorem asserts that for $u < v, \mathbf{X}(v)$ dominates $\mathbf{X}(u)$. [B. Efron, Increasing properties of Pólya frequency functions, *Ann. Math. Stat.*, 36, 272–279, 1965]

Egorov's (or Egoroff's) theorem Arises in measure theory, where it asserts the almost uniform convergence of a pointwise convergent sequence of functions. In the probabilistic context, it is this: let $X, X_n, n \geq 1$, be r.vs defined on $(\Omega, \mathcal{F}, \mathbb{P})$ such that, for all $\omega \in A \in \mathcal{F}, \lim_{n\to\infty} X_n(\omega) = X(\omega)$. Then for $\varepsilon > 0$ there is event B such that $\mathbb{P}(B) < \varepsilon$, and $\lim_{n\to\infty} \sup_{\omega\in A\setminus B} |X(\omega) - X_n(\omega)| = 0$. [S. Resnick, *A Probability Path*. Birkhäuser, 1999]

Ehrenfest urn (alias: **dog-flea model; Ehrenfests' urn (note, not Ehrenfest's urn)**) Introduced as a model for various phenomena in statistical mechanics, such as heat flow or the diffusion of gases. In one formulation, an urn is supposed to contain n particles that may be of two types A and B. At each stage a particle is picked uniformly random and its type changed (from A to B, or B to A). The number of particles of type A at time $t \in \mathbb{Z}^+$ constitutes a Markov chain whose transition matrix is $n^{-1}K$, where K is the Sylvester–Kac matrix (or Kac matrix).

There are many modifications of this model; for example: at each stage if the particle is of type A it is changed to B with probability α, or not with probability $1-\alpha$; if of type B it is changed to A with probability β, or not with probability $1-\beta$. [S. Karlin & J. McGregor, Ehrenfest urn models, *J. Appl. Probab.*, 2(2), 352–376, 1965]

Ehrenfest wind-tree model (alias: **stochastic pin-ball**) Arises in statistical mechanics, where it can be seen as a simpler version of the Lorentz gas model. In the wind-tree model a particle (the 'wind') moving in the plane is directed at a collection (the 'forest') of randomly placed scattering objects; (the 'trees') which are squares in the basic model, with which it collides, and is reflected according to the rules of optics. *See* Lorentz gas model. [P. Ehrenfest and T. Ehrenfest, *The Conceptual Foundations of the Statistical Approach in Mechanics*. Dover, 2002 (first published in 1912 in German)]

eigentime identity Arises in the context of a finite state-space (size s) irreducible reversible Markov chain $X_n, n \geq 0$ (or, in continuous time, $X(t), t \geq 0$), where it relates expected hitting times of the chain to the eigenvalues of the transition matrix $P = (p_{ij})$ (denoted by $1 = \lambda_1 > \lambda_2 \geq \cdots \geq \lambda_s \geq -1$ in the discrete time case, and $0 = \lambda_1 \leq \lambda_2 \leq \cdots \leq \lambda_s$ in the continuous-time case). Denoting the stationary distribution by $(\pi_1, \ldots, \pi_s)$ and the expected hitting time of j from i by $\mathbb{E}_i \mathrm{T}_j$, the eigentime identity is

$$\sum_j \pi_j \mathbb{E}_i T_j = \begin{cases} \sum_{r\geq 2} \lambda_r^{-1}, & \text{in continuous time,} \\ \sum_{r\geq 2} (1-\lambda_r)^{-1}, & \text{in discrete time.} \end{cases}$$

[D. Aldous & J. Fill, *Reversible Markov Chains and Random Walks on Graphs*. Cambridge University Press, to appear; see www.stat.berkeley.edu/~aldous/RWG/book.html]

elastic boundary (alias: **radiation boundary condition**) A type of regular, i.e. non-singular, boundary, of a diffusion process (which is to say the process reaches the boundary in finite time with non-zero probability). On reaching such a boundary the process spends no time there, but is either reflected with probability r, or killed with probability $1-r$. If $r = 0$, the boundary is called killing. [S. Karlin & H. Taylor, *A Second Course in Stochastic Processes*. Academic Press, 1981]

elementary renewal theorem Arises in considering the renewal function $m(t) = \mathbb{E}N(t), t \geq 0$, of the renewal process $N(t) = \max\{n : T_n \leq t\}$, where $T_0 = 0$, $T_n = \sum_{r=1}^{n} X_r, n > 1$, and $X_1, X_2, \ldots$ is a sequence of i.i.d.r.vs such that $0 < \mathbb{E}X_1 < \infty$. Then the theorem asserts that $\lim_{t\to\infty} m(t)/t = 1/\mathbb{E}X_2$. If X_1 has a different distribution, yielding a delayed renewal process, the elementary renewal theorem still holds as stated. [S. Resnick, *Adventures in Stochastic Processes*. Birkhäuser, 1992]

Elfving's problem *see* embedding

ellipse (or ellipsoid) of concentration (alias: **concentration ellipse (or ellipsoid)**) Arises in considering how an n-dimensional random vector $\mathbf{X}$, with mean vector $\mathbf{m}$ and covariance matrix V, is concentrated about its mean. The ellipsoid of concentration E is one measure, defined by the equation $(\mathbf{x} - \mathbf{m})^T V^{-1}(\mathbf{x} - \mathbf{m}) = n + 2$. It has the property that a random vector $\mathbf{Y}$, that is uniform on the interior of E, has the same mean and covariance matrix as $\mathbf{X}$. When $n = 2$, E reduces to an ellipse of concentration. [H. Cramér, *Mathematical Methods of Statistics*. Princeton University Press, 1946]

elliptical distribution A type of symmetrical distribution arising as a generalization of the multivariate normal, among others. The random n-vector $\mathbf{X}$ has an elliptical distn if, for some column-vector $\mathbf{m}$ and positive definite matrix V, its characteristic function takes the form $\phi(\mathbf{t}) = \exp(\mathrm{i}\mathbf{t}^T\mathbf{m})\psi(\mathbf{t}^T V\mathbf{t})$, where $\psi(\cdot)$ is a function called the characteristic generator such that $\psi(t_1^2 + \cdots + t_n^2)$ is the chf of some random vector. The distribution, and its density if it has one, is said to be **elliptically symmetric**. In two dimensions the standard elliptical r.vr (U, V) has a representation as $U = X, V = \rho X + \sqrt{1-\rho^2}Y$, for $|\rho| < 1$, where (X, Y) is uniformly distributed on the unit circle. [K.-T. Fang, S. Kotz, & K. Ng, *Symmetric Multivariate and Related Distributions*. Chapman & Hall, 1987]

Ellsberg paradox Arises in making decisions with uncertain outcomes in which uncertainty is both probabilistic (there is a distribution on the states of nature) and there is ignorance about the true states of nature (*see* Knight uncertainty). The paradox arises from a class of decision problems (choice situations) in which many people make choices that are consistent neither with rational decision making in the sense of maximising expected utility (Savage's theory) nor with various other systems of rational choice based on risk aversion. It can be explained by some suitable hypothesis expressing aversion to uncertainty (or ambiguity), i.e. avoiding choices with unknown odds. [D. Ellsberg, Risk, ambiguity, and the Savage axioms, *Q. J. Econ.*, 75(4), 643–669, 1961]

embedding (alias: **imbedding**) Broadly, a random process $X(t), t \in \mathbb{R}(\text{or } t \in \mathbb{Z})$, is said to be embedded in a random process $Y(t)$ if any realization of $Y(t)$ yields a realization of $X(t)$; thus the fdds of $X(t)$ are determined (in principle) by those of $Y(t)$. Given some process $Y(t)$, various processes $X(t)$ may be found to be embedded in it; conversely, given some process $X(t)$, the probability space may be extended to support $Y(t)$ in which $X(t)$ is embedded. For example:

1. The successive ladder heights and ladder indices of a random walk comprise (possibly terminating) imbedded renewal processes.
2. The successive visiting times (or epochs, or instants) of a Markov chain to some state comprise an imbedded renewal process.
3. The jumps of a birth–death process in continuous time comprise a random walk in discrete time.
4. The Skorokhod imbedding thm asserts that any finite-variance zero mean r.v. X is embeddable in the Wiener process $W(t)$ in the sense that $W(T)$ has the distribution of X for some stopping time T.
5. The discrete h-skeleton X_n of a Markov process $X(t)$ is embedded in $X(t)$, as is also the jump chain of $X(t)$.

Elfving's problem, also called the **embedding problem for a Markov chain** with $n \times n$ transition matrix P, asks whether it can be expressed as $P = P(1)$, where $P(t), t \geq 0$, are transition probabilities of a continuous-time Markov chain. Thus P is said to be embeddable if for some Markov generator (i.e. Q-matrix), $P = \mathrm{e}^Q$, where e^{Qt} is a Markov stochastic matrix for all $t \geq 0$. [E. B. Davies, Embeddable Markov matrices, *Electr. J. Probab.*, 15(47), 1474–1486, 2010; J. F. C. Kingman, The imbedding problem for finite Markov chains, *Probab. Theor. Rel. Fields*, 1(1), 14–24, 1962; K. Athreya & S. Karlin, Embedding of urn schemes into continuous time Markov branching processes, *Ann. Math. Stat.*, 39(6), 1801–1817, 1968]

Emery's structure equation Arises in seeking to construct a class C of normal (local) martingales (i.e. for $X \in C, X(0) = 0$ and the compensator of X is t), that have the representation property. A necessary condition on the law of X and its quadratic variation $[X, X]_t$ is $[X, X]_t = t + \int_0^t H(s)X(s)$, for some suitable predictable (previsible) process $H(s)$. Examples having a solution include: $H = 0$. the Wiener process; $H = \text{constant} \neq 0$, a Poisson process; and $H = bX(t-)$, the so-called Azéma martingales for $-2 < b < 0$. [P. Protter, *Stochastic Integration and Differential Equations*, 2nd edn. Springer, 2005]

empirical distribution (alias: **empirical cdf; empiric distribution**) Arises naturally in the context of a random sample, that is a collection of i.i.d.r.vs $X_1, \ldots, X_n$ having cdf $F(x)$. The empirical cdf for $F(x)$, denoted by $F_n(x)$ for $x \in \mathbb{R}$, is the right-continuous step-function having jumps of size $\frac{1}{n}$ at the points $X_1, \ldots, X_n$, where $F_n(-\infty) = 0$ and $F_n(\infty) = 1$. Formally, $F_n(x) = \frac{1}{n}\sum_{r=1}^{n} I(X_r \leq x)$, where $I(\cdot)$ is the usual indicator. The Glivenko–Cantelli theorem asserts that $\lim_{n\to\infty}\sup_x|F_n(x) - F(x)| = 0$ a.s. [P. Billingsley, *Probability and Measure*, 3rd edn. Wiley, 1995]

empirical measure (alias: **empirical probability measure**) Generalises the concept of empirical distribution to a collection of i.i.d. random elements of a set Ω, denoted by $S_1, \ldots, S_n$, and defined on $(\Omega, \mathcal{F}, \mathbb{P})$; the measure m_n places a probability mass n^{-1} at the points $S_1, \ldots, S_n$, in particular, for $A \in \mathcal{F}$, the empirical probability of A is $\mathbb{P}_n(A) = \frac{1}{n}\sum_{r=1}^{n} I(S_r \in A)$. The Vapnik–Chervonenkis theory supplies conditions for the uniform convergence of $\mathbb{P}_n(A)$ to $\mathbb{P}(A)$. [R. Dudley, *Real Analysis and Probability*. Cambridge University Press, 2002]

empirical process Arises as a suitably standardized form of an empirical distribution $F_n(x)$, or measure m_n, and is used in proving central limit theorems and invariance principles. The basic form is $\pi_n(x) = \sqrt{n}(F_n(x) - F(x))$. When the index is randomized as $N(n)$, with mean $\mu_n \to \infty$, as $n \to \infty$, one obtains so-called Kac empirical processes such as $K_n(x) = \sqrt{n}(F_{N(n)}(x) - F(x))$; in this case $N(n)$ is often Poisson with parameter λn, and one speaks of **Poissonized** (or **modified**) **empirical processes** [S. van de Geer, *Empirical Processes in M-estimation*. Cambridge University Press, 2000]

Engelbert–Schmidt zero-one law Arises in considering continuous additive non-decreasing functionals of the Wiener process $W(t), t \geq 0$. Let $g(x) \geq 0, x \in \mathbb{R}$ be Borel measurable. Then the following are equivalent:

(a) For any compact set $C \subseteq \mathbb{R}$, $\int_C g(x)\mathrm{d}x < \infty$.

(b) $\mathbb{P}\left(\int_0^t g(W(s))\mathrm{d}s < \infty \quad \text{for all } t \geq 0\right) > 0$.

(c) $\mathbb{P}\left(\int_0^t g(W(s))\,\mathrm{d}s < \infty \quad \text{for all } t \geq 0\right) = 1$.

[I. Karatzas & S. Shreve, *Brownian Motion and Stochastic Calculus*, 2nd edn. Springer, 1998]

Engset model Arises in queueing theory as a model for a telephone exchange in which there are M customers generating calls (independently, with an exponential waiting time until a customer makes a call), and $K < M$ lines available. Calls arriving when all lines are busy are lost, and the fraction of calls lost in equilibrium is given by the Engset loss formula. As $M \to \infty$, one recovers the **Erlang model**. Engset's model is essentially the same as **Palm's machine repair problem**, in which M machines are serviced by K engineers. [H. Tijms, *A First Course in Stochastic Models*. Wiley, 2003]

enhanced diffusion *see* anomalous diffusion

enlargement *see* filtration

ensemble (alias: **statistical ensemble**) Broadly, a collection of entities that are similar in some way, where the nature of the similarity varies according to context. The term is particularly used in physics to denote collections of configurations of particles that have approximately the same energy, or pressure, etc.; such configurations are usually assumed to be equiprobable. The term is also used to denote a collection $X_t(\omega), \omega \in \Omega$, of realizations of a random process X_t, or a sample $X_t(\omega_r), 1 \leq r \leq n$, of these. Then one speaks of the **ensemble average** when the expectation is taken over ω, in contrast to the **time average** $\frac{1}{t}\int_0^t X_u\,\mathrm{d}u$; if equal, the process may be called ergodic. By extension the term arises in other random processes of interest to physicists; random matrices are often called ensembles, and a collection of points i.i.d. uniformly distributed in some space may be called a Poisson ensemble.

entrance boundary Arises naturally in considering a diffusion $X(t), t \geq 0$, on an interval $(a, b) \in \mathbb{R}$, where the end point b (say), is inaccessible in that it cannot be reached in finite time for any starting point $X(0) \in (a, b)$. However, if the process can be started at $X(0) = b$, which is to say that for some $t > 0$, and any $x \in (a, b)$, the probability p_x of hitting $x \in (a, b)$, given $X(0) = b$, is positive, then b is said to be an entrance boundary. The process so started runs on the completed state space $(a, b]$. Otherwise, if $p_x = 0$ for all $t > 0$, the boundary is said to be natural. A classic example is the radial part of an n-dimensional Wiener process for $n \geq 3$ (i.e. a Bessel process), for which the origin is an entrance boundary and $+\infty$ is a natural boundary. The idea extends to Markov processes on other state spaces, in which the entrance boundary is supplied by adjoining points at the frontier of the state space, given in terms of a suitable metric (completion). *See* Martin boundary and exit boundary. [S. Karlin & H. Taylor, *A Second Course in Stochastic Processes*. Academic Press, 1981; J. Kemeny *et al.*, *Denumerable Markov Chains*, 2nd edn. Springer, 1978]

entropy ergodic theorem *see* asymptotic equipartition property

entropy function Any of several functions defined to measure disorder, or dispersion, or uncertainty, or indeterminacy, or diversity, or mixing, or complexity, etc., depending on the context; originating in thermodynamics the concept finds applications in communication, quantum mechanics, statistics, biology, computational theory, etc. Entropy functions are commonly defined either from a set of plausible axioms, or as functions defined by theorems (or 'operationally' in the terminology of the natural sciences); thus the Shannon entropy of a simple r.v. with probabilities $p_1, \ldots, p_n$ is axiomatically the unique function $H(p_1, \ldots, p_n)$ that is continuous and symmetric in its arguments, monotone increasing in n, and satisfies a suitable branching property, with $H(\frac{1}{2}, \frac{1}{2}) = 1$; while H also appears operationally as the lower bound in the noiseless coding (data compression) theorem.

Broadly, a function $h(p_1, \ldots, p_n)$ on the set of such distributions $(p_1, \ldots, p_n)$ is a plausible entropy function if it is suitably (i.e. Shur) concave; for example, an important class of entropies take the form $h(\mathrm{p}) = \sum_{r=1}^{n} g(p_r)$, where $g(x) \in \mathbb{R}$ is concave on $[0, 1]$:

(a) the Shannon entropy $-\sum p_r \log p_r$;
(b) the quadratic entropy $\sum p_r(1 - p_r)$;
(c) the Havrda–Charvát entropy $(1 - 2^{1-a})^{-1}(1 - \sum p_r^a), a \neq 1, a > 0$, which yields the Shannon entropy as $a \to 1$. Often seen in the physics literature is the related so-called q-entropy (or Tsallis entropy, or non-extensive entropy), $(q - 1)^{-1}(1 - \sum p_r^q), 0 < q \neq 1$; and the term Cressie–Read entropy is seen in statistics literature;
(d) the Aczél–Daróczy entropy $-\sum p_r^a \log p_r / \sum p_r^a$, $a > 0$; being related to the I. J. Good's index of heterogeneity $\sum p_r^a(-\log p_r)^b$, for suitable a and $b > 0$;
(e) the Arimoto entropy $(2^{a-1} - 1)^{-1}\{(\sum p_r^{1/a})^a - 1\}$, $0 < a \neq 1$; (also sometimes called the a-norm entropy);
(f) the Burg entropy $\sum \log p_r$.

Another class of entropy functions is epitomized by the Rényi entropy $(1 - a)^{-1}\log(\sum p_r^a), 0 < a \neq 1$, which also yields the Shannon entropy as $a \to 1$. The case $a = 2$ may be called the collision entropy in physics literature, while the limit $a \to \infty$ yields the entropy function $h(\boldsymbol{p}) = -\log\{\max_r p_r\}$, called the min-entropy. Many such generalized entropies have been suggested for simple (and discrete) distributions. When seeking the entropy of a r.v. with density $f(x)$, it is customary to speak of differential entropy, so that Shannon's differential entropy is $-\int_{\mathbb{R}} f(x) \log f(x) \mathrm{d}x$; generalized differential entropies may be often defined analogously with integrals replacing sums. In either case, continuous or discrete, if the distribution is a joint distribution of several r.vs, one may refer to **joint entropy**; and given a sequence X_r, $r \geq 1$, of r.vs, with joint entropy of $X_1, \ldots, X_n$ denoted by H_n, the limit of $\frac{1}{n}H_n$ as $n \to \infty$ (if it exists) may be called the **asymptotic entropy**.

In the same manner, if the distribution is the conditional distribution of X given Y, then one speaks of conditional entropy, denoted by $H(X \mid Y)$; in this context the **erasure entropy** of a random vector $X_1, \ldots, X_n$ is $\sum_{r=1}^{n} H(X_r \mid X_1, \ldots, X_{r-1}, \ldots, X_{r+1}, \ldots, X_n)$ (with the natural convention for $r = 1$ and $r = n$).

Other contexts and applications lead to yet more entropy functions, such as:

(i) the Hartley entropy of a set A of size $|A|$ is $\log |A|$;
(ii) in quantum theory, the von Neumann entropy is $-$trace $(\rho \log \rho)$, for density operators ρ;
(iii) free entropy is a function analogous to entropy that is defined in the context of free probability; but note that the term has another meaning in physics;
(iv) Kolmogorov–Sinai entropy arises when considering dynamical systems.

See also Linnik information; differential entropy; Kullback–Leibler divergence; relative entropy; entropy rate. R. Gray, *Entropy and Information Theory*, 2nd edn. Springer, 2011; J. Aczél & Z. Daróczy, *Measures of Information and their Generalizations*. Academic Press, 1975; A. Rényi, On measures of entropy and information, *Proc 4th Berk. Symp. Math. Stat. Probab.*, 1, 547–561, 1961; J. Havrda & F. Charvát, Quantification method of classification processes: concept of structural a-entropy, *Kybernetika*, 3(1), 30–35, 1967; N. Martin & J. England, *Mathematical Theory of Entropy*. Addison–Wesley, 1981; reprinted by Cambridge University Press, 1984]

entropy gap *see* maximum entropy

entropy power inequality (alias: **Shannon power inequality; Shannon–Stam inequality**) Arises in the context of the differential entropy $H(\mathbf{X})$ of a jointly continuously distributed r.vr $\mathbf{X} = (X_1, \ldots, X_n)$. The entropy power is $N(\mathbf{X}) = \frac{1}{2\pi e}\exp\left(\frac{2}{n}H(\mathbf{X})\right)$, when it exists. Then the Shannon entropy power inequality asserts that for an independent r.vr $\mathbf{Y} = (Y_1, \ldots, Y_n)$, $N(\mathbf{X}+\mathbf{Y}) \geq N(\mathbf{X}) + N(\mathbf{Y})$, with equality if (and also only if, by the Shannon–Stam thm) $\mathbf{X}$ and $\mathbf{Y}$ are multivariate normal r.vs, whose covariance matrices are suitably proportional. [A. Dembo *et al.*, Information theoretic inequalities, *IEEE Trans. Inform. Theor.*, 37(6), 1501–1518, 1991]

entropy rate (alias: **asymptotic entropy**) Arises in the context of a suitable random process $X_r, r \geq 1$, having average entropy per symbol denoted by $H_n = \frac{1}{n}H(X_1, \ldots, X_n)$, $n \geq 1$. Then the entropy rate is $\lim_{n\to\infty}H_n$, when the limit exists; more generally it may be defined as lim sup H_n, as $n \to \infty$. When the sequence X_r is stationary, we have $H = \lim_{n\to\infty}H_n = \lim_{n\to\infty}H(X_n \mid X_1, \ldots, X_{n-1})$. And, in particular, if $X_n, n \geq 1$, is a stationary Markov chain with transition matrix p_{ij}, then $H = \lim_{n\to\infty}H(X_n \mid X_{n-1}) = -\sum_i \pi_i \sum_j p_{ij} \log p_{ij}$, where π_i is the stationary distribution. [T. Cover & J. Thomas, *Information Theory*, 2nd edn. Wiley, 2006]

entry time *see* hitting time

envelope method (alias: **squeezing (or squeeze) method**) A general method for simulating from non-standard distributions that is intended to be computationally inexpensive. If $d(x)$ is the density from which it is desired to simulate a sample, and there exist functions $l(x)$ and $u(x)$ such that $l(x) \leq d(x) \leq u(x)$, one form of the method is this. Suppose that U is a uniform random variable, independent of a random variable X having density $f(x)$, where $d(x) \leq f(x)$. Given X, do this:

if $Uf(X) > u(X)$, reject X,
or: if $Uf(X) < l(X)$, accept X,
or: if $Uf(X) \leq d(X)$, accept X,
or: reject X.

Conditional on acceptance, X has density $d(x)$. The upper and lower functions $u(x)$ and $l(x)$ are said to be envelopes of $d(x)$. Note that these are not envelopes in the classical geometric sense of the term, so they may be called **squeezing fns** to avoid the clash of nomenclature. [G. Fishman, *Discrete-Event Simulation*, 2001, Springer; D. Kroese *et al.*, *Handbook of Monte Carlo methods*, 2011, Wiley]

epidemic model Any of a number of mathematical models formulated to describe the progress of an infectious disease in some population. The classic example is the stochastic S-I-R model (i.e. susceptible-infectious-removed), in which the population $n = S(t) + I(t) + R(t)$, where (S, I, R) is a Markov process. Transitions $(S \to S-1, I \to I+1)$ occur at rate λIR, and transitions $(I \to I-1, R \to R+1)$ occur at rate μI, for suitable

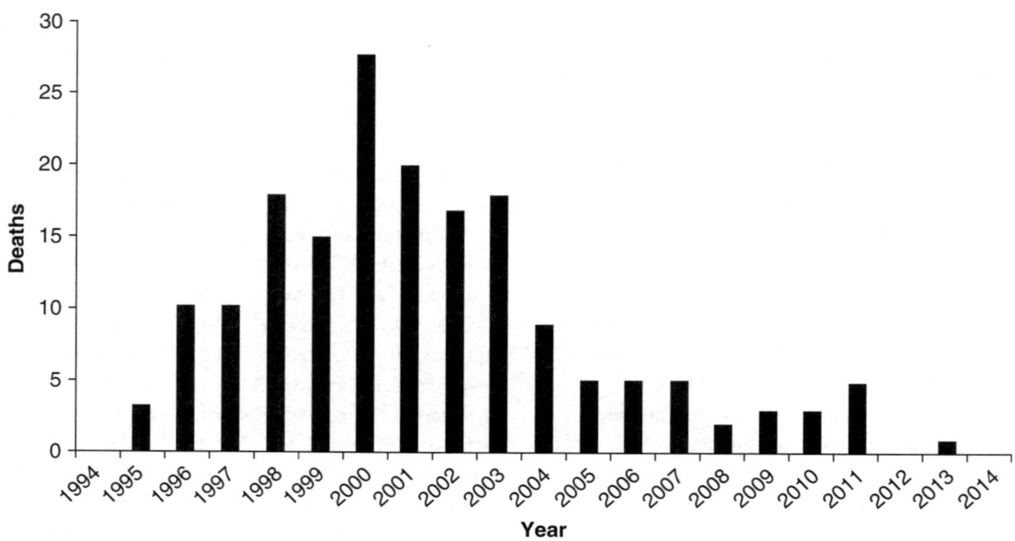

Realization of vCJD epidemic in Britain.

constants λ and μ. The state $R = n$ is absorbing, and the epidemic stops when $I = 0$. [D. Daley & J. Gani, *Epidemic modelling*, 1999, Cambridge University Press]

epistemic probability *see* logical probability; subjective probability; probability

equalizer rule Arises in decision theory, where given a random variable X, a loss function $L(d(X), \theta)$ depends on a decision $d(X)$ and an unknown state of nature θ. A decision rule $d(\cdot)$ such that the risk $R(d, \theta) = \mathbb{E}L(d(X), \theta)$ is a constant function of θ is called an equalizer rule. [J. Berger, *Statistical Decision Theory and Bayesian Analysis*, 2nd edn. Springer, 1985]

equatorial distribution *see* girdle distribution

equi-integrable *see* uniformly integrable

equilibrium (alias: **stationary; invariant; steady state; statistical equilibrium**) Let $X(t)$ be a random process taking values in a state space S, with parameter $t \in T \subseteq \mathbb{R}$ seen as time. If the distribution $\mathbb{P}_t$ of $X(t)$ does not depend on t, then $X(t)$ is said to be in (statistical) equilibrium. Likewise if $\lim_{t\to\infty} \mathbb{P}_t = \mathbb{P}$ for some distribution $\mathbb{P}$, then $X(t)$ is said to converge to equilibrium, and the set of states in S on which $\mathbb{P}$ is not zero may be called equilibrium states. In particular, the stationary distribution of an ergodic Markov chain is often called the equilibrium distribution. In the context of a renewal process on the line with inter-event distribution $F(x)$ (that is not arithmetic), having mean $\mathbb{E}X = \int_0^\infty 1 - F(x)\mathrm{d}x < \infty$ the distribution $G(x) = \frac{1}{\mathbb{E}X}\int_0^x 1 - F(y)\mathrm{d}y$ is called the equilibrium distribution of $F(x)$, because the corresponding renewal process delayed by a time having distn $G(x)$ is stationary; it is also called the **first-moment distn** (and sometimes, an integrated tail distn); $G(x)$ is also called the residual distribution, because it is the limiting distribution of the residual (or current, or excess) life, for the above renewal process. *See* integrated tail distribution. [Y. Suhov & M. Kelbert, *Probability and statistics by example*, Vol. 2, 2008, Cambridge University Press]

equitable class Arises in the context of a homogeneous Markov chain $X_n \in \mathbb{Z}$, with last-exit taboo probabilities ${}_iP_{ij}(n) = \mathbb{P}(\{X_n = j\} \cap \{X_k \neq i, 0 < k < n\} \mid X_0 = i), n \geq 1$. Defining $e_{ij} = \sum_{n=1}^{\infty} {}_iP_{ij}(n)$, a subset C of $\mathbb{Z}$ is an equitable class if $e_{ij} = 1$ for every i and j in C. Any such class is recurrent, but a recurrent class is equitable iff for all $i \in C, \sum_{j\in C} \mathbb{P}(X_n = i \mid X_0 = j) = 1$, i.e. the transition matrix restricted to C is doubly stochastic. [K. L. Chung, Remarks on taboo probabilities, *Illinois J. Math.*, 1961, 5(3), 431–435]

equivalence theorem (alias: **Blackwell's comparison of experiments theorem**) Arises in the context of decision theory, where it asserts that a method for comparing (ordering) experiments based on the idea of sufficiency is equivalent to a method of ordering (comparing) experiments based on the idea of value. [J. Crémer, A simple proof of Blackwell's comparison of experiments theorem, *J. Economic Theor.*, 1982, 27, 439–443].

equivalent Random variables (or more generally, random objects taking values in the same space) that are defined on the same probability space $(\Omega, \mathcal{F}, \mathbb{P})$ are strongly equivalent, or a.s. equal, or **modifications** of each other if $\mathbb{P}(X = Y) = 1$. If not necessarily defined on the same probability space, random variables X and Y are weakly (or stochastically) equivalent, or equal in distribution, or versions of each other, if $\mathbb{P}(X \leq x) = \mathbb{P}(Y \leq x)$ for all x. Strong equivalence implies weak equivalence; and more strongly, if $X(\omega) = Y(\omega)$ for all $\omega \in \Omega$, then X and Y are identical.

Similarly, random processes $X(t)$ and $Y(t), t \geq 0$, defined on the same $(\Omega, \mathcal{F}, \mathbb{P})$ are modifications of each other, or strongly equivalent, if $\mathbb{P}(X(t) = Y(t)) = 1$ for all t. When not necessarily defined on the same space they are versions of each other, *or* weakly (or stochastically) equivalent if they have the same finite dimensional distributions. More strongly, $X(t)$ and $Y(t)$ are **indistinguishable** if $\mathbb{P}(X(t) = Y(t) \text{ for all } t) = 1$; when $X(t)$ and $Y(t)$ are modifications of each other and continuous (or right-continuous) then they are indistinguishable. Note that some authors interchange the meanings of 'modification' and 'version' so that 'version' is stronger. [S. Ethier & T. Kurtz, *Markov Processes*, 1986, Wiley; D. Revuz, & M. Yor, *Continuous Martingales and Brownian Motion*, 3rd edn. Springer, 1999]

equivalent martingale measure *see* risk-neutral measure

equivalent measures *see* absolute continuity

equivalent sequences (alias: **Khinchin equivalence**) *see* tail equivalence

equivocation A term introduced by C. Shannon in his theories of communication and secrecy, referring to a number of important conditional entropies namely:

(a) For a noisy channel with input X and output Y, the equivocation is the conditional entropy $H(X \mid Y)$, interpreted as the ambiguity (about X) given the received signal Y.

(b) For a secrecy (or crypto-) system (M, K, C), where M is the set of messages, K the set of keys, and C the set of encryptions, we have
  (i) the key equivocation $H(K \mid C)$, and
  (ii) the message equivocation $H(M \mid C)$.

[C. Shannon, Mathematical theory of communication, *Bell Syst. Tech. J.*, 1948, 27, 379–423, 623–656; & loc cit 1949, 28, 656–715]

erasure channel Arises in communication theory as a type of noisy channel. In one definition an erasure channel is obtained by subjecting the output Y of a channel to a random binary sequence, $X_r \in \{0, 1\}, r \geq 1$. The rth symbol Y_r of Y is erased or left unchanged according as $X_r = 0$ or $X_r = 1$.

In the memoryless case, $X_r, r \geq 1$, comprises a sequence of i.i.d. Bernoulli r.vs; more generally it may be a renewal process, in which case one speaks of **burst erasure**.

Note that in every case it is known by the receiver whether or not erasure occurred; this is not to be confused with the **unreliable channel** where erasures may occur unknown to the receiver. [T. Cover & J. Thomas, *Information Theory*, 2nd edn. Wiley, 2006]

Erdős–Feller–Kolmogorov–Petrovski test Arises in the context of the law of the iterated logarithm, describing large deviations of the partial sums $S_n = \sum_{r=1}^{n} X_r$ of a sequence $X_r, r \geq 1$, of i.i.d.r.vs having distribution function $F(x)$ with zero mean and unit variance. One form is this: let $f(n), n \geq 1$, be a monotonically increasing sequence of positive numbers, and define the test function $\sum_{n=1}^{\infty} n^{-1} f(n) \exp\{-\frac{1}{2}(f(n))^2\} = \tau$. If, as $x \to \infty$, $\int_{|x|>y} x^2 \mathrm{d}F(x) = O\{(\log\log y)^{-1}\}$ then $\mathbb{P}(S_n > n^{\frac{1}{2}} f(n)$ infinitely often) is either 0 or 1 according as τ is finite or infinite. *See also* Kolmogorov–Erdös test. [W. Feller, The law of the iterated logarithm, *Ann. Math.*, 47(4), 631–638, 1946]

Erdős–Feller–Pollard theorem (alias: **discrete renewal theorem**) A theorem about power series arising in the discrete renewal problem, which supplies an approach to the classic limit theorem for discrete-time Markov chains on a countable state space. The E-F-P theorem considers the function $P(x) = \sum_{r=0}^{\infty} p_r x^r$, where $p_r \geq 0$, $\sum_{r=0}^{\infty} p_r = 1$, $m = \sum_r r p_r \leq \infty$, and $P(x)$ is not a power series in x^d for any $d > 1$ (which excludes periodicity). Then it asserts that $1 - P(z)$ has no zeros in $|z| < 1$, and the series defined by $U(x) = (1 - P(x))^{-1} = \sum_{r=0}^{\infty} u_r x^r$, is such that $\lim_{r\to\infty} u_r = m^{-1}$, where the limit is 0 if $m = \infty$. *See* Blackwell's renewal theorem. [P. Erdös *et al.*, A property of power series with positive coefficients, *Bull. Amer. Math. Soc.*, 55, 201–204, 1949]

Erdős–Kac theorem Arises as a type of central limit theorem in probabilistic number theory. An important special case of the theorem applies to the prime divisor function $d(m)$, which equals the number of distinct prime factors of the integer m. Let K_n be the number of integers in $[1, \ldots, n]$ such that, for arbitrary $x \in \mathbb{R}, d(m) < \log\log n + x(2\log\log n)^{\frac{1}{2}}$. Then $\lim_{n\to\infty} \frac{1}{n} K_n = \pi^{-\frac{1}{2}} \int_{-\infty}^{x} \exp(-y^2)\mathrm{d}y$.

Note that there are other results entitled to the label Erdös–Kac theorem, such as: let $X_r, r \geq 1$, be a sequence of i.i.d.r.vs, with zero mean and unit variance, and let $S_n = \sum_{r=1}^{n} X_r$. Then for $a > 0$,

$$\lim_{n\to\infty} \mathbb{P}\left(\max_{1\leq r\leq n} S_r < a n^{1/2}\right) = \left(\frac{2}{\pi}\right)^{1/2} \int_0^a \exp\left(-\frac{1}{2}x^2\right)\mathrm{d}x,$$

and

$$\lim_{n\to\infty} \mathbb{P}\left(\max_{1\leq r\leq n} |S_r| < a n^{1/2}\right) = \frac{4}{\pi}\sum_{r=0}^{\infty} \frac{(-1)^r}{2r+1} \exp\left\{-\frac{(2r+1)^2\pi^2}{8a^2}\right\},$$

together with similar results for the limit as $n \to \infty$ of $\mathbb{P}\left(\sum_{r=1}^{n} S_r^2 < a n^2\right)$, and $\mathbb{P}\left(\sum_{r=1}^{n} |S_r| < a n^{3/2}\right)$. [P. Erdös & M. Kac, The Gaussian law of errors in the theory of additive number theoretic functions, *Amer. J. Math.*, 62(1), 738–742, 1940; P. Erdös & M. Kac, On certain limit theorems of the theory of probability, *Bull. Amer. Math. Soc.*, 52(4), 292–302, 1946]

Erdős–Littlewood–Offord theorem Arises in the context of the so-called Littlewood–Offord problem, which in one form is this: Let $a_1, \ldots, a_n$ be real numbers such that $|a_r| \geq 1$, and let $X_1, \ldots, X_n$ be i.i.d.r. vs such that $\mathbb{P}(X_r = 1) = \mathbb{P}(X_r = -1) = \frac{1}{2}$, and set $S_n = \sum_{r=1}^{n} a_r X_r$. The E-L-O theorem asserts that for any interval I of length 1, $\mathbb{P}(S_n \in I) = \mathrm{O}(n^{-1/2})$, as $n \to \infty$.

This is called the concentration function (or probability), and the result may be called an anti-concentration theorem. Such estimates have applications in combinatorics and random matrices. More strongly, if $|a_r - a_s| \geq 1$ for $r \neq s$, then $\mathbb{P}(S_n \in I) = \mathrm{O}(n^{-3/2})$, a result due to Sárkozy & Szemerédi, following Erdös & Moser. A more general d-dimensional form is called the **Halász inequality**. [T. Tao & V. Vu, A sharp inverse Littlewood–Offord theorem, *Random Struct. Algor.*, 37(4), 525–539, 2010]

Erdős–Rényi law of large numbers The best known (from 1970) of a class of results called Erdös–Rényi–Shepp laws, the first of which is **Shepp's law** (1964). Let $X_1, X_2, \ldots$ be i.i.d. non-trivial random variables with moment generating function $M(t)$ such that $M(t) < \infty$ for $0 < t < T$. Define $m(a) = \min_{t\in(0,T)} \mathrm{e}^{-at}M(t)$; for any a such that $m(a) < 1$. Setting $S_n = \sum_{i=1}^{n} X_i$, with $S_0 = 0$, and $R(n,k) = \max_{0\leq r\leq n-k} \frac{1}{k}\sum_{i=r+1}^{r+k} X_i, \quad 1 \leq k \leq n$, the assertion of the theorem is that, as

$$n \to \infty, R\left(n, \left[-\frac{\log n}{\log m(a)}\right]\right) \to a,$$

with probability 1, where $[x]$ is the integer part of x. *See* law of the single logarithm. [P. Deheuvels & L. Devroye, Limit laws of Erdös–Rényi–Shepp type, *Ann. Probab.*, 15(4), 1363–1386, 1987]

Erdős–Rényi random graph *see* random graph

Erdős–Révész theorem Arises in considering the length L_n of the longest consecutive run of heads in n flips of a fair coin. It asserts that with probability 1, for any $\varepsilon \geq 0$, there is $m < \infty$ such that for all $n > m, L_n \geq [\log_2 n - \log_2 \log_2 \log_2 n + \log_2 \log_2 \mathrm{e} - 2 - \varepsilon]$. [P. Révész, Regularities and irregularities in a random 0, 1 sequence, *Stat. Papers*, 31(1), 95–101, 1990]

Erdős–Taylor theorem Any one of a large number of results on the nature and structure of the sample paths of the simple symmetric random walk S_n on $\mathbb{Z}^d$ (i.e. the lattice of integer points in Euclidean space of d dimensions), in which steps are independently made with probability $\frac{1}{2d}$ to any of the $2d$ neighbours. Particularly noteworthy are the laws of the iterated logarithm for $\|S_n\|$ and $T_n = \inf_{k\geq n}\|S_k\|$, the so-called infimum (or infima) process.

In this context, the Erdös–Taylor conjecture was that the number of visits to the most frequently visited site in the first n steps of such a walk in $\mathbb{Z}^2$ is asymptotic to $(\log n)^2/\pi$. This survived for 40 years, until proved to be true by Dembo, Peres, Rosen, & Zeitouni. [P. Erdös & S. Taylor, Some intersection properties of random walk paths, *Acta Math. Hungar.*, 11(3–4), 231–248, 1960; P. Erdös & S. Taylor, Some problems considering the structure of random walk paths, *Acta Math. Hungar.*, 11(1–2), 137–162, 1960; A. Dembo *et al.*, Thick points for planar Brownian motion and the Erdös–Taylor conjecture, *Acta Math.*, 186, 239–270, 2001]

Erdős–Turán theorem (or law) Supplies a normal approximation for the order (as a group element) X_n of a randomly (uniformly from $n!$ possibilities) chosen permutation of n objects. Formally, as $n \to \infty$,

$$\mathbb{P}\left(\left\{\frac{1}{3}(\log n)^3\right\}^{-\frac{1}{2}}\left\{\log X_n - \frac{1}{2}(\log n)^2\right\} < x\right) \sim \Phi(x),$$

where $\Phi(x)$ is the standard normal distn.

Another result that is termed the Erdös–Turán theorem arises in considering the zeros of a random polynomial $\sum_{r=0}^{n} X_r z^r = g(z)$. If $N(a,b)$ denotes the number of roots of $g(z) = 0$ lying in $0 \leq a < \arg z < b \leq 2\pi$, then

$$|N(a,b) - (b-a)n/(2\pi)| < 16\left\{n \log\left(\sum_1^n |X_r|/\sqrt{X_0 X_n}\right)\right\}^{1/2}.$$

Weak conditions on the sequence $X_r, r \geq 0$, ensure that the empirical distribution of the arguments of the zeros is asymptotically uniform as $n \to \infty$. [P. Erdös & P. Turán, Some problems of a statistical group theory III, *Acta Math. Acad. Sci. Hungar.*, 18, 309–320, 1967; P. Erdös & P. Turán, On the distribution of roots of polynomials, *Ann. Math.*, 51(1), 105–119, 1950]

ergodic A term used with several subtly different meanings in the context of stationary random processes (or suitable processes having a stationary distribution); and more generally, dynamical systems (called flows in continuous time) generated by a measure-preserving transformation. Given such a random process $X_n(\omega), n \geq 1, \omega \in \Omega$, ergodic theorems (yielding ergodic properties) investigate the convergence (a.s., or in mean) of so-called sample- (or time-) averages of the form $\langle X \rangle_n = \frac{1}{n}\sum_{r=1}^{n} X_r$. In general, X has the ergodic property (or is ergodic) for $g(\cdot) \in \mathbb{R}$ if $\langle g(X) \rangle_n$ converges a.s. as $n \to \infty$. (If the convergence is in mean-square, one speaks of second-order ergodic properties or theorems.) A classic example is Birkhoff's ergodic theorem for a suitable sequence X_n of integrable random variables, that asserts $\frac{1}{n}\sum_{r=1}^{n} X_r \rightarrow^{\text{a.s.}} \mathbb{E}(X \mid \mathcal{I})$, where $\mathcal{I}$ is the invariant σ-field (the σ-field of events that are invariant under shifts; note that $\mathcal{I} \subseteq \mathcal{F}_\infty$, where $\mathcal{F}_\infty$ is the tail σ-field of the sequence). This may be called the ergodic decomposition (according to $\mathcal{I}$).

The process (or measure-preserving transformation) is said to be ergodic (or to have the property of ergodicity) if the invariant σ-field $\mathcal{I}$ is trivial for the stationary distribution, i.e. all events in $\mathcal{I}$ have probability 0 or 1. In this case convergence is a.s.to $\mathbb{E}X$ = constant. This is what is usually meant by ergodicity in the context of stationary processes; broadly speaking, a process (or flow, or dynamical system) is ergodic if the long-run empirical distribution of values of a sample path (or trajectory) can be taken as good approximation to the stationary distribution of the process. More strongly, a discrete-time irreducible Markov chain X_n taking values in a countable state space is said to be ergodic if it is aperiodic, as well as recurrent and non-null.

The origins of the word ergodic lie in statistical mechanics, in the case when all states of some system were assumed accessible and equally likely, and therefore visited approximately equally often by any trajectory or sample path; this is the ergodic hypothesis. The stationary distribution may be called the ensemble, and its expected value the ensemble average. [R. Gray, *Probability, Random Processes, and Ergodic Properties*, 2nd edn. Springer, 2009]

ergodic Markov chain A non-explosive (i.e. regular), continuous-time, irreducible Markov chain $X(t), t \geq 0$, taking values in a countable state-space S, with transition probabilities $p_{jk}(t), t \geq 0$, is said to be ergodic (or recurrent non-null) if there is a probability distribution $\pi_r, r \in S$, such that as $t \to \infty$, for all $j, k \in S, p_{jk}(t) \to \pi_k$; or equivalently such that $\sum_k |p_{jk}(t) - \pi_k| \to 0$. Such a process satisfies the so-called ergodic theorem for Markov processes, which in one form asserts that for a bounded function $g(\cdot) \in \mathbb{R}$,

$$\mathbb{P}\left(\lim_{t\to\infty} \frac{1}{t}\int_0^t g\Big(X(s)\Big)\mathrm{d}s \to \sum_r \pi_r g(r)\right) = 1.$$

This may also be called the weak ergodic theorem, while $p_{jk}(t) \to \pi_k$ is the (strong) ergodic theorem.

For a Markov chain $X_n, n \geq 1$, in discrete time, with n-step transition matrices $p_{jk}(n)$, there is the complication that such chains may be periodic, and there are therefore (broadly) two schools. One (following W. Feller) defines ergodicity by analogy with the continuous-time case; thus an irreducible Markov chain (and each of its states) is said to be ergodic if each state is aperiodic and non-null recurrent, with $\lim_{n\to\infty} p_{jk}(n) = \pi_k > 0$. (Some authors call this case strong ergodicity, and speak of a weakly ergodic state (and chain) if $\pi_k = 0$.)

Another school (following J. Doob) defines ergodicity by analogy with the more general concept of ergodic, which concerns itself with Cesàro averages, rather than pointwise convergence of distributions. In this case, any non-transient irreducible chain is said to be ergodic. More narrowly $X(t)$ with a transition function $p_{jk}(t)$ is said to be strongly ergodic if as $t \to \infty$, $\sup_j \sum_k |p_{jk}(t) - \pi_k| \to 0$; and X_n with n-step transition probabilities $p_{jk}(n)$ is said to be **geometrically ergodic** if for each j and k there exists $C_{jk} < \infty$ and $r_{jk} \in (0, 1)$ such that $|p_{jk}(n) - \pi_k| \leq C_{jk} r_{jk}^n$, $n \geq 0$.

If $C_{jk} \leq C < \infty$ for some C, then it may be called **uniformly ergodic**. Note finally that for an ergodic Markov chain with stationary distn π_i and transition matrix p_{ij}, $\pi_i p_{ij}$ is called the **ergodic flow** from i to j. [W. Anderson, *Continuous-time Markov Chains*. Springer, 1991; J. Doob, *Stochastic Processes*. Wiley, 1953; W. Feller, *Introduction to Probability Theory and its Applications*, Vol. 1, 3rd edn. Wiley, 1968]

Erlang distribution *see* gamma distribution

Erlang loss system (alias: **Erlang loss formula (or model); Erlang B-formula (or 1-formula)**) Arises in a queueing model for a telephone exchange in which there are K lines to deal with calls which are assumed to arrive at the instants of a Poisson process of rate λ. After getting a line, calls are assumed to last for intervals that are i.i.d. exponentially Exp(μ) distributed r.vs (or, in an alternative model due to Erlang, are of the same constant duration). If callers queue for a line to become available when all K are busy, one speaks of Erlang's delay (or C-) model. If queueing callers abandon their wait, one has Erlang's A-system. If calls arriving when all K lines are busy are rejected and lost, then one speaks of Erlang's loss (or B-) system. In equilibrium, the probability that an arriving call is lost because lines are busy is

$$E = \frac{(\lambda/\mu)^K / K!}{\sum_{r=0}^{K} (\lambda/\mu)^r r!},$$

which is Erlang's loss formula; the system is said to be congested. More generally, when calls have an arbitrarily distributed duration (possibly degenerate), the corresponding formula is called the

Crommelin–Pollaczek formula. [A. Janssen & J. Van Leeuwaarden, Back to the roots of the M/D/s queue, and the work of Erlang, Crommelin and Pollaczek, *Stat. Neerlandica*, 62(3), 299–313, 2008]

Erlang model *see* Engset model

escape time *see* holding time

Esscher transform (alias: **exponential tilting; associated distribution; conjugate distribution; exponential twisting**) Arising in a method for approximation of the distribution of insurance claims in the theory of collective risk, this change of measure has since been used in random walks, renewal theory, option pricing, and the theory of large deviations. For a distribution $F(x)$ with moment generating function $M(t) = \int_{\mathbb{R}} e^{tx} dF(x)$, finite for $a \leq t \leq b$, the Esscher transform family (or exponentially tilted associate distributions) have distribution functions $F(t,x) = M(t)^{-1} \int_{-\infty}^{x} e^{ty} dF(y)$, $t \in [a,b]$, with mgf $\int_{\mathbb{R}} e^{sx} F(t, dx) = M(s+t)/M(t)$. It may also be called the ***t*-tilted** (or ***t*-twisted**) distn. [F. Esscher, The probability function in the collective theory of risk, *Scand. Actuarial J.*, 3, 175–195, 1932]

Esseen's bound *see* smoothing inequality

Esseen's lemma Any one of a number of bounds for the concentration function $Q(X,\theta)$ of a r.v. X, i.e. $Q(X,\theta) = \sup_x \mathbb{P}(x \leq X \leq x+\theta)$. Denoting the chf of X by $\phi(t)$, one such lemma asserts that for any $b > 0$, and $0 < a < \pi/\theta$, there are constants c and d such that

$$c\frac{\theta}{1+b\theta}\int_{-\frac{1}{2}b}^{\frac{1}{2}b} |\phi(t)|^2 dt \leq Q(X,\theta) \leq \frac{d}{a}\int_{-a}^{a} |\phi(t)| dt.$$

[C. Esseen, On the concentration function of a sum of independent random variables, *Probab. Theor. Rel. Fields*, 9, 290–308, 1968]

essentially complete *see* complete class

essentially convergent Arises in considering a series $S_n = \sum_{r=1}^{n} X_r$ of independent r.vs $X_r, r \geq 1$. If there exist real numbers $c_r, r \geq 1$ (called centering constants) such that $\sum_{r=1}^{n}(X_r - c_r)$ converges a.s. as $n \to \infty$, then the series is said to be essentially convergent. By the essential convergence lemma this occurs if the symmetrized series $\sum_{r=1}^{n}(X_r - Y_r)$ converges a.s., where $Y_r, r \geq 1$ is an independent copy of the sequence $X_r, r \geq 1$. Otherwise, by the zero-one law, S_n is said to be essentially divergent as $\sum_1^n (X_r - b_r)$ diverges a.s. for any sequence $b_r, r \geq 1$. [M. Lóeve, *Probability Theory*, 4th edn. Springer, 1977]

essential state A classification arising in the context of a Markov chain X_n, taking values in a countable state space Ω, when the chain is not irreducible. If it is possible to get from $j \in \Omega$ to $k \in \Omega$ in a finite number of steps, then we say k is accessible from j, and write $j \to k$. A state j is essential if for all k such that $j \to k$, we also have $k \to j$; otherwise j is said to be inessential (or unessential) and it occurs finitely often with probability 1 (or, what is the same, with positive probability it is possible to escape from it without ever returning).

Any equivalence class C of communicating states (under the equivalence relation '$j \to k$ and $k \to j$'), has states that are either all essential, or all inessential. The decomposition theorem extends to assert that Ω can be partitioned into disjoint sets that are one of:

(a) an essential non-null recurrent closed set C (where C is closed iff $\sum_{k \in C} p_{jk} = 1$ for $j \in C$);
(b) an inessential transient set;
(c) an essential transient closed set;
(d) an essential recurrent class that is not closed.

The concept extends to Markov chains on a suitable measure space $(\Omega, \mathcal{F})$, where a set A is inessential if, when started from $x \in A$, the chain revisits A finitely often with probability 1; otherwise it is essential. *See*: evanescent [D. Levin *et al.*, *Markov Chains and Mixing Times*. American Mathematical Society, 2009]

Etemadi's inequality *see* Lévy–Ottaviani inequality

Etemadi's theorem A form of the strong law of large numbers for a sequence of identically distributed random variables $X_r, r \geq 1$. Formally, if $\mathbb{E}|X_1| < \infty$ and the sequence is pairwise independent (weaker than the usual assumption of independence) then as $n \to \infty$, $\frac{1}{n}\sum_{r=1}^{n} X_r \to \mathbb{E}X_r$ with probability 1. [N. Etemadi, An elementary proof of the strong law of large numbers, *Probab. Theor. Rel. Fields*, 55, 119–122, 1981]

European option Arising in financial mathematics this denotes a derivative contract (contingent claim), of whatever structure, that can be exercised only at the fixed expiry date (also called maturity date).

evanescent A term used in several distinct ways in the theory of stochastic processes. First, given a probability space $(\Omega, \mathcal{F}, \mathbb{P})$, and a set T (seen as the parameter set of a random process), a random set A is a subset of $\Omega \times T$ such that the indicator $\mathrm{I}_A(t, \omega), t \in T$ is a random variable for all t, as a function of $\omega \in \Omega$. Then a random set A is evanescent (for $\mathbb{P}$) if there is a null event $B \in \Omega$(i.e. $\mathbb{P}(B) = 0$), such that $A \subseteq B \times T$. Equivalently, $I_A(t, \omega)$ is indistinguishable from 0 (as a function of t).

Second, if the state space S of a Markov chain takes the form $A \cup C$, where A is absorbing and C is an irreducible class of transient states, then C is evanescent if the chain remains in C finitely often with probability 1; or if, more strongly, the expected time of the final visit to C is finite for any starting point. (*See* essential state.)

Third, in a Wold (or Wold-like) decomposition of a stationary random field in two or more dimensions, the non-random (i.e. deterministic) non-harmonic component may be called evanescent; it has Lebesgue measure zero in the corresponding spectral decomposition. [J. Doob, *Classical Potential Theory and its Probabilistic Counterpart*. Springer, 1984]

evanescent process *see* $\mathbb{P}$-null process

Evans's immortal particle representation A method of representing a conditional superprocess, where the conditioning event is non-extinction. The representation can be visualized as the evolution of a particle that generates new copies of the underlying superprocess while never becoming extinct itself. [S. Evans, Two representations of a conditioned superprocess, *Proc. Roy. Soc. Edin.*, 123A, 959–971, 1993]

EVD *see* extreme value distributions

event Informally, an event is an assertion about some phenomenon of interest, e.g. 'the flipped coin showed a head', or 'the random variable is positive.' All such statements arise from the outcomes of what are called experiments, where the universal set Ω (or sample space), of all possible outcomes of an experiment is called the certain event, with probability $\mathbb{P}(\Omega) = 1$. Any other event is a subset A of Ω, having a well-defined probability $\mathbb{P}(A)$. In particular, the empty set $\emptyset$, called the impossible event, has probability $\mathbb{P}(\emptyset) = 0$. Other particular events include:

(a) an elementary (or simple) event is a single point (or sample point) $\omega \in \Omega$, with probability $\mathbb{P}(\omega)$;
(b) the complementary event of A is $\Omega \backslash A = A^c$, with probability $\mathbb{P}(A^c) = 1 - \mathbb{P}(A)$;
(c) a compound event is a union of simple events, or formed by union or intersection of other events;
(d) disjoint (or incompatible, or mutually exclusive) events A and B are such that $A \cap B = \emptyset$;
(e) an induced event is generated by random objects in terms of the appropriate inverse function, e.g. for a random variable $X(\omega)$ defined on $\Omega, A_x = \{\omega : X(\omega) \leq x\}$;
(f) a **null event** A has zero probability (but is not necessarily impossible).

Formally, given a probability space $(\Omega, \mathcal{F}, \mathbb{P})$, an event is any member of the σ-field $\mathcal{F}$ of subsets of Ω, on which $\mathbb{P}$ is defined. [W. Feller, *Introduction to Probability Theory and its Applications*, Vol. 1, 3rd edn. Wiley, 1969; Vol. 2, 2nd edn. Wiley, 1971]

Everett–Ulam–Harris theorem *see* Kesten–Stigum theorem

Ewens sampling formula *see* Pitman's sampling formula

exceedance Arises naturally in the theory of extreme values of a random process $X_n, n \in \mathbb{Z}^+$(or $X_t, t \in \mathbb{R}^+$), where for a given sequence of positive numbers $\tau_n, n \geq 1$ the exceedance process for $X_r, 1 \leq r \leq n$, at time n is $A_r = X_r I(X_r \geq \tau_n), 1 \leq r \leq n$; and $I(X_r \geq \tau_n)$ is the usual indicator function. The **excess process** is $B_r = (X_r - \tau_n)^+$, $1 \leq r \leq n$. For any n, the indices of indicators taking the value 1 comprise the **exceedance point process**. [M. Falk *et al.*, *Laws of Small Numbers*, 2nd edn. Birkhäuser, 2001]

exceedance distribution (alias: **distribution of exceedances**) Arises in comparing one random sample $X_1, \ldots, X_n$ with another $Y_1, \ldots, Y_m$, where all these $m + n$ r.vs are i.i.d. with continuous distribution function $F(x)$. Let X_r be the rth largest of the Xs (i.e. $r - 1$ of $X_1, \ldots, X_n$ are larger than X_r). Then the probability that exactly j of $Y_1, \ldots, Y_m$ exceed X_r is

$$e_j = r\binom{n}{r}\binom{m}{j} \Big/ \left\{(n+m)\binom{n+m-1}{r+j-1}\right\}, \quad 0 \leq j \leq m.$$

[E. Gumbel & H. von Schelling, The distribution of the number of exceedances, *Ann. Math. Stat.*, 21, 247–262, 1950]

excess *see* kurtosis

excessive function Arises in G. Hunt's generalization of classical potential theory, in which analogues of several classical concepts are defied in terms of the transition kernel $p(t,x,y)$ of a Markov chain in continuous time on a suitable measurable state space. The associated transformation (operator) P_t is defined for suitable functions $f(y)$ by $P_t f(x) = \int p(t,x,y)f(y)\mathrm{d}y$, and the role of positive superharmonic functions in the classic theory is played by excessive functions $f(\cdot)$, where $f \geq 0$ is excessive if $P_t f(x) \leq f(x)$, and $\lim_{t\to 0} P_t f = f$.

In particular, f is called invariant excessive if equality holds for all $t > 0$. Similarly, f is said to be λ-excessive if $e^{-\lambda t}P_t f \leq f$ and $\lim_{t\to 0} e^{-\lambda t}P_t f = f$; and invariant λ-excessive if equality holds for $t > 0$. When $p(t,x,y)$ is the transition function of a (d-dimensional) Wiener process, Hunt's theory effectively coincides with the classical case, so that excessive functions are superharmonic, and invariant excessive functions are harmonic. In discrete time, excessive functions are called super regular. [J. Doob, *Classical Potential Theory and its Probabilistic Counterpart*. Springer, 1984; G. Hunt, Markoff processes and potentials I, II, III, *Illinois J. Math.*, II, 44–93 & 316–369, 1957, and 2, 151–213, 1958].

excess life (alias: **residual waiting time; remaining (or residual) life; forward recurrence time; overshoot; etc.**) Arises in a renewal process, which is a (simple) point process on $\mathbb{R}^+$ (seen as time, with the process initiated at 0), such that the intervals (or lifetimes) between points (or renewals) are i.i.d. non-negative random variables. For fixed $t > 0$, the interval to the next point (renewal) is the excess life. denoted by $E(t)$, with distribution $R(t,x) = \mathbb{P}(E(t) \leq x)$. Writing $F(x)$ for the distribution of intervals between successive points, and $m(t)$ for the **renewal function**, we have

$$R(t,x) = F(t+x) - F(t) + \int_0^t R(t-s,x)\mathrm{d}F(s)$$
$$= F(t+x) - F(t) + \int_0^t F(t+x-s)$$
$$- F(t-s)\mathrm{d}m(x).$$

If $F(x)$ is not arithmetic, and has finite mean μ, then $\lim_{t\to\infty} R(t,x) = \mu^{-1}\int_0^x (1-F(x))\mathrm{d}x$, the equilibrium distn of $F(x)$. *See also* Lorden's inequality, size-biased distribution, and inspection paradox. [W. Feller, *Introduction to Probability Theory and its Applications*, Vol. 2, 2nd edn. Wiley, 1971]

exchangeable (alias: **interchangeable; symmetrically dependent; permutable**) Random variables $X_1, \ldots, X_n$ are exchangeable if all their $n!$ permutations, $X_{\pi(1)}, \ldots, X_{\pi(n)}$, have the same joint distribution. An i.i.d. finite collection of r.vs is exchangeable, but not conversely. The set $X_1, \ldots, X_n$ is **spreadable** (or contractable or has the selection property) if for all $1 \leq k \leq n$, the subset $X_1, \ldots, X_k$ has the same distribution as $X_{s(1)}, \ldots, X_{s(k)}$, for any $1 \leq s(1) < \ldots < s(k) \leq n$. Exchangeable finite sequences are spreadable, but not conversely. An infinite sequence of r.v. $X_r, r \geq 1$, is exchangeable (resp. spreadable) if for all n, $X_1, \ldots, X_n$ is exchangeable (resp. spreadable).

De Finetti's theorem asserts that $X_r, r \geq 1$ are exchangeable iff they are conditionally i.i.d. given some σ-field, and the **Ryll–Nardzewski thm** asserts the same for spreadable r.vs $X_r, r \geq 1$. Exchangeability, spreadability, and conditionally i.i.d. are thus equivalent for infinite sequences. [O. Kallenberg, *Probabilistic Symmetries and Invariance Principles*. Springer, 2005]

exchange problem (or paradox) *see* two-envelopes problem

excited random process *see* self-interacting random walk; Hawkes process

exclusion process A type of interacting particle system, in which each site of a set of vertices (such as the integer points of $\mathbb{Z}^d$, or the nodes of a tree) is occupied by at most one particle. The dynamics of the process let particles either move from one site to another, or appear (or disappear), subject to the constraint that two particles can never be at one site. [T. Liggett, *Stochastic Interacting Systems*. Springer, 1999]

excursion Arises naturally in the context of Markov processes X, where (broadly speaking) for any fixed point a in the state space, an excursion from a is a part of the path (or segment) of the process between first leaving a after a visit, and the first return to a thereafter. The possible excursions comprise the excursion space, and the concatenated excursions comprise the excursions process. Thus, for example, let $X_n, n \geq 0$, be an irreducible Markov chain on a countable state space S with transition matrix $p_{j,k}$. Then an excursion from any recurrent state a is a finite sequence $ax_1x_2\cdots x_{t-1}, x_r \neq a, 1 \leq r \leq t-1$, which terminates at the first-passage time t of the chain to a, to be followed by the next excursion $a\ldots$. The sequence $a, ax_1, ax_1x_2, \ldots, ax_1\cdots x_{t-1}, a\cdots$ is the excursions process. Likewise in continuous time an excursion is that part of the process $X(t)$ lying between the conclusion S_n of a holding time at a, and the subsequent nth first-passage time T_n to a, where $T_0 = 0, S_n = \inf\{t > T_{n-1} : X(t) \neq a], n \geq 1$, and

$T_n = \inf\{t > S_n : X(t) = a\}, n \geq 1$. For fixed s such that $X(s) \neq a$, and suitable $N, s \in (S_N, T_N)$, and this gives the excursion straddling s. For a continuous Markov process, such as the Wiener process $W(t)$ the excursions process comprises the sequence of path segments on the intervals during which $W(t) \neq a$; these are intervals of constancy of the local time function at a. The detailed analysis of this process, yielding a representation as a Poisson point process on a suitable space, is called Itô excursion theory. [L. Rogers & D. Williams, *Diffusions, Markov Processes and Martingales*, Vol. 2, 2nd edn. Wiley, 1994; reprinted by Cambridge University Press, 2000]

excursion law *see* Itô excursion theory

exercise date, or time, or boundary *see* option

exit boundary Arises naturally in considering a diffusion $X(t), t \geq 0$, in an interval $[a, b] \in \mathbb{R}$, where a (say) is a point that is effectively absorbing, in that sample paths never return to (a, b) after visiting a. More formally, if T_b is the first-passage time to b from $X(0) = x_0$, then

$$\lim_{b \to a} \lim_{x_0 \to a} \mathbb{P}(T_b < t \,|\, X(0) = x_0) = 0, \quad t > 0.$$

Exit boundaries also arise in the context of Markov chains in continuous time on a countably infinite state space S, where the jumps of the chain occur at times $T_n, n \geq 1$. If $T = \lim_{n \to \infty} T_n < \infty$, then T is the time of the first explosion, and the behaviour of the jump chain at this limit is described in terms of a suitable compactification of S called the exit boundary (or fictitious states). *See* Martin boundary and entrance boundary. [K. L. Chung, On the boundary theory for Markov chains, *Acta Math.*, 110(1), 19–77, 1963]

exit measure (alias: **exit distribution**) Arises in the context of random processes evolving in an open subset D of some suitable space-time domain. In its simplest form, one envisages a branching diffusion process $Z(t), t \geq 0$, where the initial population $Z(0)$ lies in an open subset D of $\mathbb{R}^d$. For any particle A that ever finds itself outside D, there is a point on the boundary ∂D of D first hit by the first ancestor of A to exit D. Assigning unit mass to each such point yields the exit measure of $Z(t)$. The concept finds applications in superprocesses (measure valued) arising as limits of branching diffusions, and in the (path valued) Brownian Snake, etc. One also speaks of the exit measure for a random process $X(t)$ that is induced by $X(\infty) = \lim_{t \to \infty} X(t)$, on a suitable space. [E. Dynkin, Superprocesses and partial differential equations, *Ann. Probab.*, 21(3), 1185–1262, 1993]

exotic option Arises in mathematical finance where it refers to any option that is not a standard (or so-called **'vanilla'**) option. There are many, which fall into half-a-dozen broad types:

(a) Packages of vanilla options, called package options.
(b) Options with discontinuous or non-linear pay-offs.
(c) Options that depend in some way on the sample path of the underlying asset up to the final exercise date, e.g. its extremes, visits to barrier values of intermediate times, average, etc. These are often called **look back options**, e.g. the Russian option which can look back to any previous time (*see* Shepp–Shiryaev problem).
(d) Options involving more than one currency or numeraire.
(e) Those involving two or more assets, called **rainbow options** in general, of which the commonest are spread options whose value depends on the difference in price between two assets (or the same asset at different times).
(f) **Compound options** in which the assets underlying the options are themselves options.

[F. De Weert, *Exotic Option Trading*. Wiley, 2008]

expectation (alias: **mathematical expectation; expected value; mean; first moment; average**) For a suitable random variable X, the expected value, or expectation, of X is denoted by $\mathbb{E}X$, and supplies a useful gauge of centrality. In its simplest form, when X is a discrete random variable, $\mathbb{E}X = \sum_x x\mathbb{P}(X = x)$, provided that the sum is absolutely convergent (i.e. $\sum_x |x|\mathbb{P}(X = x) < \infty$). If X is continuous with density $f(x)$, then $\mathbb{E}X = \int_{\mathbb{R}} xf(x)\mathrm{d}x$, provided that $\mathbb{E}|X| < \infty$. In general, for a r.v. X defined on $(\Omega, \mathcal{F}, \mathbb{P})$ with distn fn $F(x), x \in \mathbb{R}$, $\mathbb{E}X = \int_\Omega X(\omega)\mathrm{d}\mathbb{P}(\omega) = \int_{\mathbb{R}} x\mathrm{d}F(x)$, provided that the integrals exist. An important special case arises when X is the indicator of an event A, and then $\mathbb{E}X = \mathbb{P}(A)$.

More generally, the extended expectation $\mathbb{E}X$ of X is given, whenever the limit exists, by $\frac{1}{n}\sum_{r=1}^n X_r \to \mathbb{E}X$ a.s., as $n \to \infty$, where $X, X_1, X_2, \ldots$ are i.i.d.r.vs. The concept of expectation arises naturally from the empirically observed stability of statistical ratios (alias sampling stability), thus: Let $\mathbf{r} = (r_1, \ldots, r_N)$ be the realizations (actual outcomes) of N repetitions of some experiment with a discrete state space $S = (x_1, x_2 \ldots)$, where x_r has probability p_r. The stability of statistical ratios (supporting the frequency interpretation of probability) entails that x_j will appear in **r** on about Np_j occasions. Thus the average of the observations is

$$\frac{1}{N}\sum_{k=1}^{N} r_k \simeq \frac{1}{N}\sum_{x_j\in S} Np_jx_j = \sum_{x_j\in S} p_jx_j = \mathbb{E}X.$$

This observation led to the first definition of the expectation $\sum p_r x_r$ as the fair price to pay for the reward from a set of possible pay-offs $x_1, \ldots, x_M$ with respective probabilities $p_1, \ldots, p_M$. The role of $\mathbb{E}X$ as a gauge of centrality is reinforced by the observation that $s(a) = \mathbb{E}(X-a)^2$ (when finite), is minimized by $a = \mathbb{E}X$, yielding an analogy with the centre of mass in physics. [P. Whittle, *Probability via Expectation*, 4th edn. Springer, 2000]

expected discrimination information *see* Kullback–Leibler divergence

expected log-likelihood ratio *see* Kullback–Leibler divergence

expected posterior loss (alias: **posterior risk; posterior expected loss**) Arises in decision theory when using Bayes' rule to choose among decision functions and actions, on the basis of an observation x, being the realization of a random variable X. Thus, let $L(d, \theta)$ be the loss entailed by decision d when the true state of nature is θ, and let $p(\theta\,|\,x)$ be the posterior distribution of θ given x. Then the expected posterior loss, or posterior risk, to be minimized by choice of d is $R(x) = \mathbb{E}(L(d,\theta)\,|\,x) = \int_\Theta L(d,\theta)p(\theta\,|\,x)\mathrm{d}\theta$. [G. Parmigiani & L. Inoue, *Decision Theory*. Wiley, 2009]

expected quadratic loss *see* mean-square error

expected utility Arises in decision theory and its many applications in game theory, economics, statistics, etc., where naive criteria fail. In decision making under risk (i.e. with random outcomes), one naive rule is to seek to maximize expected monetary value, or to minimize risk (= expected loss).

The Saint Petersburg paradox invalidates the universal correctness of this, and Daniel Bernoulli proposed the more general principle of maximizing expected utility, for some suitable utility function. The **Von Neumann–Morgenstern thm** (*see* utility theory) shows that a decision maker under risk acts rationally (in some sense) iff preferences are represented by maximizing expected utility; though Allais's paradox suggests that real individuals may not be rational in the required way. This theory, which postulates a known distribution on the outcomes, was generalized in L. J. Savage's subjective expected utility theory. Here it is assumed only that the decision maker has beliefs about the states of the world, and is prepared to make a rational collection of bets on these states; and, further, has rational preferences for actions in the light of their consequences. Savage's theory shows that, given suitably rational betting and action preferences, there is a representation in terms of the maximization of the expected value of a utility fn.

Ellsberg's paradox (and others) suggest that real individuals fail to emulate these demands of rationality, and the theory has been prolifically extended as a result. *See also* moral expectation. [S. Barbera, P. Hammond, & C. Seidl, *Handbook of Utility Theory*. Springer, Vol. I, 1999 & Vol. II, 2004]

exploration process *see* contour process

exploration time *see* cover time

explosion Arises in the context of random processes in continuous time, and is of particular importance for Markov processes. Broadly, let $X(t)$ be a random process in a discrete state-space; where $X(t)$ is a pure jump process, so that the time H_r spent in the rth state is positive. Formally, we define the jump times $J_r, r \geq 0$, by $J_0 = 0$, and $J_n = \{\inf t > J_{n-1} : X(t) \neq X(J_n)\}$ (with inf $\emptyset = \infty$). Thus $J_n. = H_1 + \cdots + H_n$. If $Z = \lim_{n\to\infty} J_n < \infty$, then the process has made infinitely many jumps in finite time, and this is called an explosion. Non-explosive processes are often called regular; explosive processes may be called dishonest. Continuous processes $X(t), t \in \mathbb{R}$, such as a diffusion in d dimensions, may also explode. Let $T_n = \inf\{t \geq 0 : \|X(t)\| > n\}$, where $\|X\|$ is distance from the origin, then $X(t)$ is said to be explosive (or dishonest) if $\lim_{n\to\infty} T_n = \tau < \infty$. [R. Bhattacharya & E. Waymire, *Stochastic Processes with Applications*. Wiley, 1990; reprinted by SIAM, 2009]

explosive *see* dishonest

exponential distribution A probability distribution on $[0,\infty)$ having density $f(x) = \lambda \mathrm{e}^{-\lambda x}$, $x \geq 0$, where $\lambda = (\mathbb{E}X)^{-1} > 0$. It arises naturally as a continuous limit of a sequence of geometric distributions, and also as the maximum entropy distribution on $[0,\infty)$ with given finite mean. It is a special case of the gamma distn, having chf $\frac{\lambda}{\lambda - \mathrm{i}t}$. It is notable for having the lack-of-memory property. The density of $-X$ on $(-\infty, 0]$ is sometimes called the reverse exponential distn; and the density of $X + c$ may be called the shifted exponential. The function $g(x) = \exp(-(ax)^b)$ is called the **stretched exponential fn** (or Kohlrausch fn by physicists), and some writers therefore refer to the r.v. X having density $f(x)\,\alpha\, g(x), x > 0; a, b > 0$, as a stretched exponential; a few others call the Weibull distn a stretched exponential. [W. Feller, *Introduction to Probability Theory and its Applications*, Vol. 2, 2nd edn. Wiley, 1971]

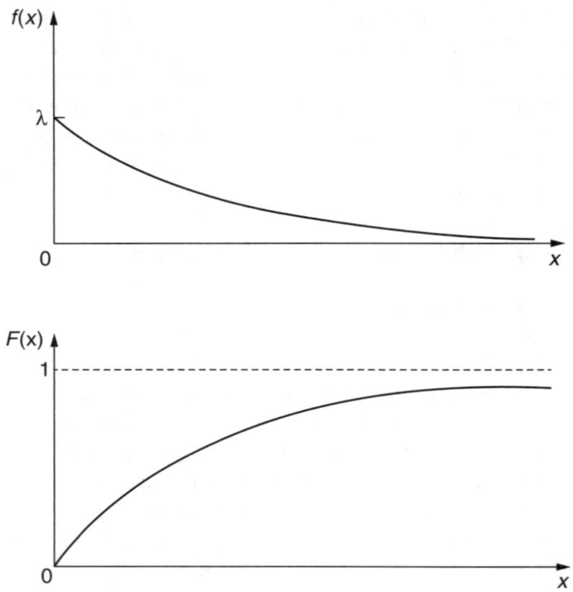

Top: **Density of an exponential random variable.**
Bottom: **Distribution of an exponential random variable.**

exponential family (alias: **Koopman–Darmois (–Pitman) (or v.v.) family**) A collection of probability distributions that includes many of the more important types, and arises in many applications. Its simplest form is the so-called natural exponential family in the univariate case, where $f(x, \theta)$ is a density (or probability mass fn in the discrete case) for each value of θ in some parameter set Θ, and can be written as $f(x, \theta) = b(x) \exp\{\theta x - a(\theta)\}, x \in S \subseteq \mathbb{R}$, for suitable functions $a(\theta)$ and $b(x)$. More generally, the exponential dispersion family is a two-parameter set in which, for $\theta \in \Theta$ and $\lambda \in \Lambda, f(x, \theta, \lambda) = b(x, \lambda) \exp\{\lambda(\theta x - a(\theta))\}$ is a density (or pmf). The concept generalizes to vectors of variables $\mathbf{x}$ and parameters $\boldsymbol{\theta}$. The subset of distributions for which $\operatorname{var} X = (\mathbb{E}X)^q, q \in \mathbb{R}\backslash(0, 1)$, has been termed the **Tweedie distns** by B. Jørgensen. Those families are also called generalized linear models, and dispersion models, respectively. [B. Jørgensen, *Theory of Dispersion Models*. Chapman & Hall, 1997]

exponential formula for reliability *see* ageing

exponential growth rate *see* Malthusian parameter

exponential inequality In probability theory this usually refers to any of a class of bounds for the probability in the tail of a distribution where, very often, the distribution is that of the partial sums $S_n = \sum_{r=1}^n X_r$ of a suitable sequence of r.vs. In this group one has the inequalities of Bennett, Bernstein, Hoeffding, Kolmogorov, Prokhorov, etc., of the general form $\mathbb{P}(S_n > t) \leq \exp\{-a(n, t)\}$, for $t > 0$. One may speak of a quadratic or linear exponential inequality according as $a \sim t^2$ or $a \sim t$; other functions $a(n, t)$ are often seen. Such exponential inequalities may hold for two quantities jointly, as in Freedman's inequality. They may also hold in continuous time, as in e.g. Lundberg's inequality for the ruin probability, and the maximal inequalities for continuous time martingales, such as e.g. for the Wiener process $W(t)$, $\mathbb{P}\left(\sup_{0 \leq s \leq t} W(s) \geq \theta t\right) \leq \exp\{-\theta^2 t/2\}$, $\theta > 0$. [M. Ledoux & M. Talagrand, *Probability in Banach Spaces*. Springer, 1991]

exponential martingale Arises as the suitably scaled exponential function of another martingale; such martingales have important applications in mathematical finance, among other areas. The most elementary arises from a random walk $S_n = \sum_{r=1}^n X_r$, where the X_r are i.i.d.r.vs, with $K(u) = \log\{\mathbb{E}\exp(uX_1)\} < \infty$ in an interval including 0. Then for u such that $K(u) < \infty, M_n = \exp(uS_n - nK(u)), n \geq 0$, with $M_0 = 1$, is a martingale, often called an exponential martingale. In continuous time, an example arises in the context of the Wiener process: the process $W(t), t \geq 0$, is a Wiener process iff for each $\lambda \in \mathbb{R}$, and $t \geq 0$, $M(t) = \exp\left(\lambda W(t) - \frac{1}{2}\lambda^2 t\right)$ is a martingale (wrt the same filtration). Indeed, any continuous $Z(t), t \geq 0$, is a local martingale iff, for some continuous local martingale $X(t)$,

$$Z(t) = \exp\left\{X(t) - \frac{1}{2}[X]_t\right\},\ t \geq 0,$$

where $[X]_t$ is the quadratic variation. Such general exponential martingales arise as the solution of the SDE $dZ(t) = Z(t)dX(t)$; *see* Doléans–Dade exponential. [K. L. Chung & R. Williams, *Introduction to Stochastic Integration*, 2nd edn. Birkhäuser, 1990]

exponential tightness Arises in large deviation theory, where it plays a role analogous to that of tightness in the theory of weak convergence of probability measures. Formally, a family $\mathbb{P}_n(\cdot)$ of probability measures on the Borel σ-field of subsets of a measurable topological space Ω is exponentially tight if for $0 < a < \infty$ there exists a compact set C_a such that $\lim_{n\to\infty}\sup\frac{1}{n}\log\mathbb{P}(C_a^c) \leq -a$. [A. Dembo & O. Zeitouni, *Large Deviations*, 2nd edn. Springer, 1998]

exponential tilting *see* Esscher transform

exponential twisting *see* Esscher transform

exponential weighting (or smoothing) *see* moving average

exponential Wiener process *see* geometric Brownian motion, in Brownian motion

extended random variable A random variable X taking values in the extended real line $\overline{\mathbb{R}} = [-\infty, \infty]$. Formally, $X(\omega)$ is defined on a measurable space $(\Omega, \mathcal{F})$, taking values in $\overline{\mathbb{R}}$, and such that $\{\omega : X(\omega) \in B\} \in \mathcal{F}$, for every set B in the Borel σ–field of subsets of $\overline{\mathbb{R}}$. [A. Shiryayev, *Probability*, 2nd edn. Springer, 1996]

extension theorem Broadly, given a function f from some domain $D \subseteq S$, an extension theorem investigates the existence and properties (e.g. uniqueness) of any function $\hat{f}$ defined on S, such that $\hat{f}(x) = f(x)$ for all x in D. The extension $\hat{f}$ is said to be determined if it is unique, and f is the restriction of $\hat{f}$ to D. Such theorems arise in probability because probability spaces are very often constructed by assigning probabilities to a limited class of events, and then extending the distribution to a larger field of events. Important extension theorems include those of Caratheodory, Daniell–Kolmogorov, and Ionescu Tulcea. Where the extension is not unique, it is usual to seek to restrict the extension to a version having desirable properties (e.g. suitably continuous, when extending a random process). Another expression of the concept arises in the extension theorem for chfs, which asserts that if a chf $\phi(t)$ is analytic (holomorphic) in a region including the origin, then it is determined uniquely for all t. [W. Feller, *Introduction to Probability Theory and its Applications*, Vol. 2, 2nd edn. Wiley, 1971]

extinction probability *see* branching process

extremal index Arises in the context of a strictly stationary sequence of r.vs $Y_r, r \in \mathbb{Z}^+$, having common marginal distribution function $F(y)$, with running maximum $M_n = \max_{1\leq r\leq n} Y_r$. Suppose that there exists a sequence of functions $C_n(t)$ such that for each $t > 0, \lim_{n\to\infty} n(1 - F(C_n(t))) = t$, and that for some $\theta \in [0, 1], \lim_{n\to\infty}\mathbb{P}(M_n \leq C_n(t)) = e^{-\theta t},\ t > 0$. Then θ is called the extremal index of the sequence. Now define a sequence X_r; $r \in \mathbb{Z}$, of i.i.d.r.vs with the common distn fn $F(x)$ of the Y_r, and running maximum $R_n = \max_{1\leq r\leq n} X_r$. Then R_n has an asymptotic distn of the form $G(x) = \lim_{n\to\infty}\mathbb{P}((R_n - b_n)/a_n \leq x)$ iff M_n has an asymptotic distn of the form $(G(x))^\theta = \lim_{n\to\infty}\mathbb{P}((M_n - b_n)/a_n \leq x)$, for the same $G(x)$ and constants $b_n, a_n > 0$, for $n \geq 1$. [M. Leadbetter & H. Rootzén, Extremal theory for stochastic processes, *Ann. Probab.*, 16(2), 431–478, 1988]

extreme value distributions (alias: **max-stable distribution; EVD; GEV**) A class comprising all those distributions that can arise as the limiting distribution, as $n \to \infty$, of the suitably scaled running maximum $M_n = \max_{1\leq r\leq n} X_r$ of a sequence $X_r, r \geq 1$, of i.i.d.r.vs; it coincides with the class of max-stable distns, where $F(x)$ is a max stable distribution fn if for a sequence of constants $a_r \in \mathbb{R}, b_r > 0, r \geq , 1$, and all $m, \{F(x)\}^m = F((x - a_m)/b_m)$. Formally, the **Fisher–Tippett–Gnedenko thm** asserts this: if, for some non-degenerate $G(x)$, $\lim_{m\to\infty}\mathbb{P}((M_m - a_m)/b_m \leq x) = G(x)$, (at continuity points of G), then one can choose sequences $a_m \in \mathbb{R}$ and $b_m > 0$ such that the limiting distribution has the form

$$G(x) = \exp(-(1 + cx)^{-1/c}), \quad \text{for } c \in \mathbb{R} \text{ and } 1 + cx > 0.$$

This may be called the **Jenkinson–von Mises** (or vice versa, or either name alone) **representation** or parameterization of the generalized extreme value distribution (or GEV), and the shape parameter c may be called the extreme value index; for $c = 0$ one takes

$$G(x) = \exp(-\exp(-x)), x \in \mathbb{R},$$

which has been called the extreme value, or Fisher–Tippett or **Fisher–Tippett–Gumbel distn**, it is perhaps most often now called the **Gumbel (or type I) extreme value distn**.

For $c > 0$ one obtains a **Fréchet, or type II, extreme value distn**, which is given in standard form as $F(x) = \exp(-x^{-a}), a > 0, x > 0$. For $c < 0$ one obtains a **Weibull, or type III extreme value distn**, which is given in standard form as $W(x) = \exp(-(-x)^a), x < 0, a > 0$. The Gumbel (Type I) distribution has mean γ, where γ is Euler's constant, and chf $\phi(t) = \Gamma(1 - \mathrm{i}t)$, where Γ is the gamma function. [P. Embrechts *et al.*, *Modelling Extremal Events*. Springer, 1997, corrected 4th printing 2003; R Fisher & L. Tippett, Limiting forms of the frequency distribution of the largest, or smallest member of a sample, *Math. Proc. Camb. Phil. Soc.*, 24(2), 180–190, 1928; S. Kotz & S. Nadarajah, *Extreme Value Distributions: Theory and Applications*. Imperial College Press, 2000]

extreme value theory *see* extreme value distribution

F

factor graph Arises naturally as a means of representing the structure of dependence among a large number of jointly distributed random variables; applications arise in coding theory, Markov random fields, and network theory. Formally, suppose that the joint distribution has a factorization as $f(x_1, \ldots, x_n) = \prod_{i \in I} g_i(S_i)$, where $S_i = \{x_{i(1)}, \ldots, x_{i(n_i)}\}$ is a subset of $\{x_1, \ldots, x_n\}$, and I is an index set. The associated factor graph is bipartite, with one set of nodes corresponding to the argument x_r, and another set corresponding to the functions g_s. An edge joins x_r and g_s if $x_r \in S_s$. [M. Mézard & A. Montanari, *Information, Physics, and Computation*. Oxford University Press, 2009]

factorial cumulant Let X be a discrete random variable with pgf $G(t) = \mathbb{E}t^X$, and hence factorial moment generating function $G(1+z)$. Then, by analogy with the ordinary cumulants, the function $C(z) = \log G(1+z) = \sum_{r=0}^{\infty} \frac{z^r}{r!} C_{(r)}$ is called the factorial cumulant generating function, and the coefficient $C_{(r)}$ is the rth factorial cumulant. For example, if X is Poisson (λ), $C(z) = \lambda z = zC_{(1)}$. [J. Douglas, *Analysis With Standard Contagious Distributions*. International Cooperative Publishing House, 1980]

factorial distribution Broadly, the term may be used to describe any distribution expressed in terms of factorials, which may be rising $a^{\overline{n}} = a(a+1)\ldots(a+n-1)$, or falling $a^{\underline{n}} = a(a-1)\ldots(a-n+1)$. In particular, it refers to the distribution

$$\mathbb{P}(X = x) = \frac{(n-1)(m-1)^{\underline{n-1}}}{(m+x)^{\underline{n}}}, \quad x = 0, 1, 2, \ldots,$$

where $m > n - 1 \geq 1$. It is a special case of a hypergeometric distribution. Note that some authors use the term factorial distribution to mean a factorizable distn, i.e. one such that $\mathbb{P}(X = x, Y = y) = \mathbb{P}(X = x)\mathbb{P}(Y = y)$, being the distribution of independent r.vs. [W. Marlow, Factorial distributions, *Ann. Math. Stat.*, 36(3), 1066–1068, 1965]

factorial moment For a non-negative integer valued r.v. X, the rth factorial moment is $\mathbb{E}X^{\underline{r}} = \mathbb{E}(X(X-1)\ldots(X-r+1))$, (and is 1 for r = 0). The definition may be extended to more general lattice (arithmetic) distns in the natural way. Bivariate factorial moments are of the form $\mathbb{E}(X^{\underline{r}} Y^{\underline{s}})$. The factorial moment generating function is defined as

$$\sum_{r=0}^{\infty} \mathbb{E}X^{\underline{r}} \frac{z^r}{r!} = \mathbb{E}\sum_{r=0}^{\infty} X^{\underline{r}} \frac{z^r}{r!}, \text{ given convergence,}$$

$= \mathbb{E}(1+z)^X = G(1+z)$, where $G(t) = \mathbb{E}t^X$ is the pgf of X. [P. Moran, *Introduction to Probability Theory*. Oxford University Press, 1968]

factorial moment bound Asserts that for a non-negative integer-valued r.v. X, and integer t, $\mathbb{P}(X \geq t) \leq \inf_{0 \leq n \leq t-1} \mathbb{E}\left(\frac{X^{\underline{n}}}{t^{\underline{n}}}\right)$, where $t^{\underline{n}}$ denotes the falling factorial. [P. Naveau, Comparison between the Chernoff and factorial moment bounds, *Amer. Stat.*, 51(1), 40–41, 1997]

Faddeev postulates (alias: **Faddeev axioms**) Arise in characterizing the Shannon entropy $H(p_1, \ldots, p_n) = -\sum_{r=1}^{n} p_r \log p_r$, of a probability distribution $p_1, \ldots, p_n$, as a measure of uncertainty (or information, etc.). Of several suggestions, D. Faddeev's are perhaps the simplest:

(a) $H(p, 1-p)$ is continuous in p for $0 \leq p \leq 1$, and $H(\frac{1}{2}, \frac{1}{2}) = 1$.
(b) $H(p_1, \ldots, p_n)$ is symmetric in its arguments for $n \geq 2$.
(c) For $0 \leq t \leq 1$, and any $p_1, \ldots, p_n$

$$H(tp_1, (1-t)p_1, p_2, \ldots, p_n) = p_1 H(t, 1-t) + H(p_1, \ldots, p_n).$$

[A. Rényi, On measures of entropy and information, *Proc. 4th Berkeley Symp. Math. Stat. Probab.*, 1, 547–561, 1961]

failure-rate function *see* ageing

fairness In the context of probability, a lottery, game, gamble, or investment is said to be fair if the net random return R (i.e. the reward or prize net of any non-returnable fee or stake) has zero expectation. This idea is fundamental to the initiation of probability in the hands of Pascal, Fermat, and Huygens. By extension, a coin or die is called fair if it is unbiased. More generally, consider a collection of simple betting odds $b_1, \ldots, b_n$, where b_r is the return (net of stake) on a unit bet on one of the corresponding n outcomes. These are called fair odds if no set of bets can ensure a gain. By the arbitrage theorem the odds are fair iff $\sum_{r=1}^{m}(1+b_r)^{-1} = 1$. Yet more generally a sequence of games is called fair (or absolutely fair) if the rewards $X_1, X_2, \ldots$ satisfy $\mathbb{E}X_1 = 0$, and $\mathbb{E}(X_{n+1} \mid X_1, \ldots, X_n) = 0$. (Then $Y_n = \sum_{r=1}^{n} X_r$ comprises a martingale, and the X_r are martingale differences.) The sequence may be termed subfair or superfair if equality is replaced by the appropriate inequality. The finiteness of the expectation is necessary for this fairness, as the St Petersburg paradox demonstrates. Furthermore, fairness is not necessarily preserved in an unbounded sequence of fair games, and even if it is, such a sequence may be unfavourable to the player in some sense other than unfairness. [W. Feller, *Introduction to Probability Theory and its Applications*, Vol. 1, 3rd edn. Wiley, 1968; Vol. 2, 2nd edn. Wiley, 1971]

fallacy of the transposed conditional *see* prosecutor's fallacy

false positive paradox *see* base-rate fallacy test

family-history space *see* Ulam–Harris framework

Fano's inequality Arises in communication theory, where it may be used to prove a lower bound for the probability of error in transmission by a channel. Formally, let $\eta = \mathbb{P}(X \neq Y)$, and $h = -\eta\log_2 \eta - (1-\eta)\log_2(1-\eta)$. Then Fano's inequality is $H(X \mid Y) \leq h + \eta\log_2(|A|-1)$, where $H(X \mid Y)$ is the conditional entropy. [R. Fano, *Transmission of Information*. M.I.T., 1961]

Farlie-Gumbel–Morgenstern distribution (alias: **FGM distribution; Morgenstern distribution**) A bivariate distribution of a continuous random vector (X, Y), having distribution function of the form $F(x,y) = F_X(x)F_Y(y)(1 + ag(F_X(x))\,h(G_Y(y)))$, where $g(z) \to 0$ and $h(z) \to 0$ as $z \to 1$, and a, g, and h satisfy constraints to ensure that $F(x,y)$ is indeed a proper distn fn with absolutely continuous marginal distributions.

The term **Morgenstern distn** commonly denotes the special case when $g(z) = h(z) = 1 - z$, with $|a| \leq 1$. [I. Bairamov, S. Kotz, & M. Bekçi, New generalized FGM distributions, *J. Appl. Stat.*, 28(5), 521–536, 2001]

Fatou's lemma Arises in considering the interchange of taking expectations and taking limits for a sequence of random variables $X_1, X_2 \ldots$. One form is this: assume that $X_n \geq Y$, for some r.v. Y with finite expectation. Then

$$\mathbb{E} \lim_{n\to\infty} \inf_{m\geq n} X_m \leq \lim_{n\to\infty} \inf_{m\geq n} \mathbb{E}X_m.$$

The dominated convergence theorem is a corollary. [T. Tao, *Introduction to Measure Theory*. American Mathematical Society, 2011]

Favard length *see* Buffon's needle

favourite site Arises in simple, usually symmetric, random walks S_n, started from $S_0 = 0$, on the integers $\mathbb{Z}$ (or, more generally, on $\mathbb{Z}^d$). Let $V(n, x)$ be the number of occasions on which the walk visits the point x during its first n steps, and set $L(n) = \sup_y V(n,y)$, the maximum of this local time. Then the favourite sites are $A(n) = \{x \in \mathbb{Z} : V(n,x) = L(n)\}$, of which the largest is given by $M(n) = \max_{x\in A} x$. The Erdös–Révész thm asserts that

$$\limsup_{n\to\infty} \frac{M(n)}{(2n\log\log n)^{1/2}} = 1,$$

while **Kesten's thm** is

$$\limsup_{n\to\infty} \frac{L(n)}{(2n\log\log n)^{1/2}} = 1.$$

More generally, the **Csáki–Révész–Shi thm** defines the joint limit set of these two sequences. [E. Csáki, P. Révész, & Z. Shi, Favourite sites, favourite values, and jump sizes for random walk and Brownian motion, *Bernoulli*, 6(6), 951–975, 2000]

Fazekas–Klesov strong law Establishes a general type of strong law of large numbers, given a suitable maximal inequality. Formally, let $S_n = \sum_{r=1}^{n} X_r$ be the partial sums of a sequence $X_1, X_2, \ldots$ of r.vs, and assume that for some $p > 0$, and all n, $\mathbb{E}(\max_{1\leq k\leq n}|S_k|^p) \leq \sum_{r=1}^{n} a_r$, where $a_r \geq 0$, $r \geq 1$. If $0 < b_1 \leq b_2 \leq \cdots$ is an unbounded sequence such that $\sum_{r=1}^{\infty} a_r/b_r^p < \infty$, then $\lim_{n\to\infty} S_n/b_n = 0$, with probability 1. [I. Fazekas & O. Klesov, A general approach to the SLLN, *Theory. Probab. Appl.*, 45(3), 436–439, 2002]

F distribution (alias: **Fisher's F distribution; Snedecor's F distribution; Fisher–Snedecor distribution (or v.v.); variance ratio distribution**) The distribution of the ratio R of two independent chi-squared r.vs U and V, each divided by its number of degrees of freedom r and s respectively. Thus $R = \frac{U/r}{V/s}$. It was so named, and denoted $F(r, s)$, by G. Snedecor, after R. Fisher, who considered the distn of the related $Z = \frac{1}{2}\log R$ – usually called **Fisher's Z distn** – in connexion with the variance ratio test. The r.v. R has density

$$f(x) = \frac{r^{r/2}\, s^{s/2}\, x^{r/2-1}\, \Gamma\left(\frac{1}{2}(r+s)\right)}{(s+rx)^{(r+s)/2}\, \Gamma\left(\frac{1}{2}r\right)\Gamma\left(\frac{1}{2}s\right)}, \quad x \geq 0.$$

If U in the numerator is replaced by W having a non-central chi-squared distribution, with r degrees of freedom and non-centrality parameter θ, and being independent of V, then $N = \frac{W/r}{V/s}$ has the non-central F distribution denoted by $F(r, s; \theta)$. If, further V has a non-central chi-squared distn then N has the doubly non-central F distribution. The usual multivariate generalization of F has what is called **Wilks's lambda distn.** [N. Johnson *et al.*, *Continuous Univariate Distributions*, Vol. 2. Wiley, 1995]

f-divergence (alias: **ϕ-divergence, Ali–Silvey divergence, Csiszár divergence, Csiszár–Morimoto divergence**) A class of functions, denoted by $d_f(\mathbb{P}, \mathbb{Q})$, supplying a measure of the divergence (or separation, or discrimination, or (loosely) distance, etc.) between two probability distributions on the same space S, where $\mathbb{P}$ is dominated by $\mathbb{Q}$ (or absolutely continuous wrt $\mathbb{Q}$).

It is defined in terms of a continuous convex function $f(x), x \geq 0$, together with (optionally) an increasing function $u(x), x \in \mathbb{R}$. Then in the simplest basic form, for discrete probability distributions p_r and $q_r, r \geq 1$, $d_f(\mathbb{P}, \mathbb{Q}) = d_f(p, q) = \sum_r q_r f\left(\frac{p_r}{q_r}\right)$; or, more generally, $d_{f,u}(p,q) = u\left(d_f(p,q)\right)$. It is usual to take $f(1) = 0$, so that the minimum value of d_f is zero when $\mathbb{P} = \mathbb{Q}$. In general

$$d_f(\mathbb{P}, \mathbb{Q}) = \int_S f\left(\frac{\mathrm{d}\mathbb{P}}{\mathrm{d}\mathbb{Q}}\right)\mathrm{d}\mathbb{Q},$$

where $\mathrm{d}\mathbb{P}/\mathrm{d}\mathbb{Q}$ is the Radon–Nikodym derivative. Particular choices of convex function $f(x)$ yield many classic measures of divergence (some of which have several aliases); perhaps most commonly seen are these:

1. $f(x) = x\log x$, and $f(x) = -\log x$, yield the Kullback–Leibler divergence $d_{\mathrm{KL}}(\mathbb{P}, \mathbb{Q})$ and $d_{\mathrm{KL}}(\mathbb{Q}, \mathbb{P})$. Then $(x-1)\log(x)$ yields Jeffrey's divergence, $d_{\mathrm{J}} = d_{\mathrm{KL}}(\mathbb{P}, \mathbb{Q}) + d_{\mathrm{KL}}(\mathbb{Q}, \mathbb{P})$.
2. $f(x) = \frac{1}{2}|x-1|$, or in another convention $f(x) = |x-1|$, yields the total variation distance (also called Kolmogorov variation distance).
3. $f(x) = \frac{1}{a-1}(x^a - 1)$, $0 < a \neq 1$, yields the Hellinger divergence $d_{\mathrm{H}}(\mathbb{P}, \mathbb{Q})$; this is often seen in the case when $a = \frac{1}{2}$, when it may also be called Bhattacharya's (or Kailath's) distance. The function $-\log d_H, 0 < a < 1$, is a Chernoff divergence (*see also* Rényi divergence).
4. $f(x) = (x-1)^2$, or $f(x) = x^2 - 1$, yields the χ^2 or Pearson divergence, while $f(x) = |x-1|^a, a \geq 1$, yields the so-called χ^a-divergence.
5. $f(x) = \frac{1}{2}(\sqrt{x} - 1)^2$ yields Kolmogorov's divergence d_{K}, while $(2d_{\mathrm{K}})^{1/2}$ is Matusita's divergence.
6. $f(x) = \frac{4}{1-\alpha^2}\left(1 - x^{(1+\alpha)/2}\right)$, $\alpha \neq \pm 1$, yields Csiszár's I-divergence, also called Havrda–Charvát or Tsallis, or Amari divergence, which is essentially (after a change of parameter) the same as the Cressie–Read power divergence; it is perhaps most often referred to simply as the α-divergence in the context of *f*-divergences.

[S. Ali & S. Silvey, A general class of coefficients of divergence of one class from another, *J. Roy. Stat. Soc. B*, 28(1), 131–142, 1966; M. Basseville, Divergence measures for statistical data processing; an annotated bibliography, *Signal Processing*, 93(4), 621–633, 2013; F. Liese & I. Vajda, Divergences and informations in statistics and information theory, *IEEE Trans. Inf. Theor.*, 52(10), 4394–4412, 2006]

Fechner distribution A family of distributions used to describe skewness by using different scale parameters on either side of a suitable centre. In standard form, centred at the origin, the family of densities is

$$f(x) = c\exp\left\{-\frac{1}{2}B_a(x)\right\}, \quad \text{where } B_a(x) = \begin{cases} x^a, & x \geq 0 \\ (-bx)^a, & x \leq 0 \end{cases},$$

and $c^{-1} = (1 + b^{-1})2^{1/a}\,\Gamma(a^{-1})/a$. The case $a = 2$ yields a type of skew normal density. [G. Barnard, Sophisticated theory and practice in quality improvement, *Phil. Trans. Roy. Soc. Lond. A*, 327(1596), 581–589, 1989]

Fefferman's inequality In its martingale form, in discrete time this can be stated as follows: Let $X(t)$ and $Y(t)$ be martingales with respect to the same filtration $\mathcal{F}_t$, on the same space, $t \geq 0$.

Assume that $\sup_t \mathbb{E}|X(t)| < \infty$, and that $Y(t)$ is a BMO_2 martingale, which is to say that $\inf\{C :$ for any stopping time T, $\mathbb{E}\{(Y(\infty) - Y(T))^2 \mid \mathcal{F}_T\} < C^2\} = \|Y\|_{\mathrm{BMO}_2} < \infty$. Then

$$\mathbb{E}\int_0^\infty |\mathrm{d}[X,Y]_t| \le 2\mathbb{E}\left([X,X]_\infty^{\frac{1}{2}}\right)\|Y\|_{\mathrm{BMO}_2},$$

where $[X, Y]_t$ is the quadratic covariation process. A generalization to $\sup_t \mathbb{E}|X(t)|^p < \infty$ is called the Fefferman-Garsia inequality. [D. Revuz, & M. Yor, *Continuous Martingales and Brownian Motion*, 3rd edn. Springer, 1999]

Feinstein's theorem An extension of the noisy coding theorem to ergodic channels that have finite memory. [A. Feinstein, On the coding theorem and its converse for finite-memory channels, *Inf. Control*, 2, 25–44, 1959]

Feldman–Hájek theorem (alias: **Hájek–Feldman theorem**) Asserts that Gaussian probability measures on the same space are either singular or equivalent, and supplies a number of necessary and sufficient conditions determining which of these holds. *See also* Kakutani's dichotomy. [A. Shiryayev, *Probability*, 2nd edn. Springer, 1996]

Feller–Arley process *see* birth–death process

Feller boundary classification Arises in considering Markov processes in continuous time, including both diffusion processes and jump processes. In W. Feller's scheme, a boundary of the state space falls into one of the following four types:

(a) Natural: from which the process cannot be started, nor can the process reach it in finite expected time.
(b) Entrance: which the process cannot reach, but from which it may be started.
(c) Exit: which the process can reach, but never return from.
(d) Regular: which the process can both arrive at and depart from; boundary conditions of various types can be imposed, called Feller's boundary conditions which Feller called the generalized classical boundary conditions.

Exit and regular boundaries are called accessible; natural and entrance boundaries are inaccessible. [W. Feller, The parabolic differential equations and the associated semi-groups of transformations, *Ann. Math.*, 55(3), 468–519, 1952; W. Feller, The birth and death processes as diffusion process, *J. Math. Pure Appl.*, 38(3), 301–345, 1959]

Feller class (alias: **Feller distribution**) The random variable X (or its distribution F), is said to be in the Feller class if

$$\liminf_{t\to\infty} \frac{\mathbb{E}\{X^2 I(|X| \le t)\}}{t^2\mathbb{P}(|X| > t)} > 0,$$

where $I(|X| \le t)$ is the usual indicator function. The Feller class includes the domain of attraction of all stable distributions. [W. Feller, On regular variation and local limit theorems, *Proc. 5th Berkeley Symp. Math. Stat. Probab.*, 2, 373–388, 1967]

Feller condition *see* Feller boundary classification; Feller property; Lindeberg–Feller theorem

Feller's construction (alias: **Feller decomposition**) Commonly refers to the construction of a realization of a continuous-time Markov chain $X(t), t \ge 0$, with a suitable (conservative and stable) Q-matrix, through its holding times and jump matrix. The method is sketched thus: starting from state j, the chain remains there for a holding time that is exponentially distributed with parameter $q_i = -q_{ii}$, and then jumps to state k with probability q_{jk}/q_j, being the (j, k)th entry in the stochastic transition matrix of the so-called jump chain. This generates the so-called Feller minimal process, in that it stops if there is an explosion, i.e. if the sum of the successive holding times is finite. W. Feller interpreted the matrix of the jump chain as a road system, with the probabilities of selecting given routes at jumps, while the Q-matrix determines the speed of travel along the roads. These ideas were extended by J. Doob, and rediscovered by subsequent writers, so that the implementation of the method is known under numerous aliases, including: dynamic (or kinetic) Monte Carlo; residence-time algorithm; n-fold way; Bortz–Kalos–Lebowitz method; Gillespie, or Doob–Gillespie method; etc.

The term Feller's construction may also refer to the representation of suitable non-singular diffusions on the line as a time and space transform of the Wiener process; also called Feller's canonical form. [W. Feller, *Introduction to Probability Theory and its Applications*, Vol. 2, 2nd edn. Wiley, 1971; W. Feller, On the boundaries and lateral conditions for the Kolmogorov differential equations, *Ann. Math.*, 65(3), 527–570, 1957]

Feller continuity *see* Feller property

Feller coupling Denotes a method for constructing a random permutation of $\{1, \ldots, n\}$ that supplies a coupling between the cycle-counts $C_r(n)$ (i.e. the number of cycles of length r in the permutation), as $n \to \infty$, with a collection of independent Poisson r.vs $X_r, r \ge 1$, with

$\mathbb{E}X_r = r^{-1}$. [R. Arratia *et al.*, *Logarithmic Combinatorial Structures: A Probabilistic Approach.* EMS, 2003]

Feller's decomposition *see* Feller's construction

Feller diffusion A diffusion whose **Fokker–Planck** (or forward) eqn for the transition density $f(t,x)$ of the process has the form

$$\frac{\partial f}{\partial t} = \frac{\partial^2}{\partial x^2}(axf) - \frac{\partial}{\partial x}\{(bx+c)f\},$$

for suitable values of a, b, and c, with corresponding SDE $\mathrm{d}X = +(bX+c)\mathrm{d}t + \sqrt{aX}\mathrm{d}W(t)$.

The case $c = 0$ was derived by W. Feller as a continuous limit of a suitably scaled and shifted ordinary branching process. The case $b = 0$, with $a = 4$, is the square of a c-dimensional Bessel process. The general case is sometimes called Feller mean-reverting square root diffusion. [W. Feller, Two singular diffusion problems, *Ann. Math.*, 51(1), 173–182, 1951; W. Feller, Diffusion processes in genetics, *Proc. 2nd Berkeley Symp.*, 227–246, 1951]

Feller diffusions Arise as that subset of the family of Feller processes whose sample paths are a.s. continuous. Choosing the underlying probability space to be the set of continuous paths yields a so-called canonical Feller diffusion. [O. Kallenberg, *Foundations of Modern Probability*, 2nd edn. Springer, 2002]

Feller–Dynkin process *see* Feller property

Feller equations (alias: **Feller's integro-differential equations; Kolmogorov–Feller (or v. v.) equations**) Arise in considering continuous time but purely discontinuous Markov processes $X(t), t \geq 0$ (i.e. those whose only transitions are jumps), on a general state space S (equipped with a suitable Borel field including the points of S and S itself). The conditional probability that $X(t) \in A$, given $X(s) = x, s < t$, is denoted by $p(s,\ x;\ t,\ A)$ and the Feller equations are a pair of integrodifferential equations (being backward and forward equations) for p. Feller's theorem establishes the existence of a unique, possibly substochastic, solution to these. [W. Feller, On the integro-differential equations of purely discontinuous Markoff process, *Trans. Amer. Math. Soc.*, 48(3), 488–515, 1940]

Feller–McKean chain A classic example of a Markov chain in continuous time all of whose states are instantaneous, i.e. given the stationary transition function $p_{jk}(t), t \geq 0$, of the chain, we have $\lim_{t\to 0}\frac{1}{t}(1-p_{jj}(t)) = \infty$, for all j. Other early examples of such chains are those of R. Dobrushin and D. Blackwell. [D. Blackwell, Another countable Markov process with only instantaneous states, *Ann. Math. Stat.*, 29(1), 313–316, 1958]

Feller minimal process (alias: **Feller minimal solution**) Arises in considering continuous-time Markov chains $X(t), t \geq 0$, on a countable state space S, with transition probability matrix $P(t) = p_{jk}(t)$, and Q-matrix $P'(0) = Q$ (assumed to be stable), for which there are corresponding sets of (Kolmogorov) backward, and forward equations. The uniqueness problem asks if the solution P is unique given Q. In answer, Feller's minimal solution is a transition probability matrix $\hat{P}(t) = \hat{p}_{ij}(t)$ (possibly substochastic, i.e. dishonest) satisfying both the backward and forward equations, with $\hat{P}'(0) = Q$, which is minimal in that $\hat{p}_{ij}(t) \leq p_{ij}(t)$ for any other solution $P(t)$. Further, if $\hat{P}(t)$ is stochastic (i.e. honest) then it is the unique such solution of the backward and forward equations. Feller's construction uses Q to generate realizations of the minimal process through holding times in states and jumps between them; a limit point of jumps is interpreted as arrival at an (abstract) exit boundary, and non-minimal processes are constructed by allowing the process to restart according to some return distribution on S. Feller minimal solutions are similarly defined for semi-Markov processes. [W. Feller, On the integro-differential equations of purely discontinuous Markoff process, *Trans. Amer. Math. Soc.*, 48(3), 488–515, 1940; W. Feller, On semi-Markov processes, *Proc. Natl. Acad. Sci.*, 51(4), 653–659, 1964]

Feller–Orey theorem Generalizes the classical renewal theorem for sums $S_n = \sum_{r=1}^{n} X_r$ of i.i.d. non-negative r.vs to the case when X_r are not concentrated on a half-line. [W. Feller & S. Orey, A renewal theorem, *J. Math. Mech.*, 10, 619–624, 1961]

Feller–Pareto distribution The distribution of the r.v. $X = a + b(U/V)^c$, where U and V are independent r.vs each having a gamma distribution, and a, b, c, are real constants. The distribution of the r.v. $Y = a + b(W^{-1} - 1)^c$, where W is a r.v. having a beta distribution, is also termed a Feller–Pareto distn. [C. Kleiber & S. Kotz, *Statistical Size Distributions in Economics and Actuarial Sciences*. Wiley, 2003]

Feller process *see* Feller property

Feller property (alias: **Feller condition; Feller continuity**) A regularity property that may be required of (i.e. a condition on) the transition measure of a Markov process in continuous time; or, equivalently, a condition on its semi-group (or generator, or evolution operator). If such a condition is satisfied, then one speaks of a Feller process, or of a Feller–Dynkin process, or a Feller semi-group, etc.; it is important because Feller processes satisfy the strong Markov property, and have a version with rcll paths. Broadly speaking, the condition requires the development of the process to be suitably (weakly) continuous in its initial state; the property is variously formulated depending on the specific context. In the basic case, when the state space is $S = \mathbb{R}^n$, and $p_t, t \geq 0$, is a (stochastic) transition semi-group acting on $f \in C_0$ (the space of real continuous functions on S vanishing at infinity), then p_t is a **Feller semi-group** if $\lim_{t\to 0} p_t f = f$, for all $f \in C_0$, and then the corresponding Markov process $X(t), X(0) = x_0$ (or the collection, for $x_0 \in S$), is said to be a Feller process. In terms of $X(t, x_0) \in \mathbb{R}^n$, one form of the Feller property makes the standard assumption that, as $t \downarrow 0, X(t, x_0)$ converges in probability to x_0, and requires also that for all t, as $y \to x_0, X(t, y)$ converges in distribution to $X(t, x_0)$. Other forms of the latter condition require that for an appropriate family of functions A, for all $g \in A, \mathbb{E}g(X(t, x_0))$ is continuous in x_0; the strength of the condition depending on the nature of A. [O. Kallenberg, *Foundations of Modern Probability*, 2nd edn. Springer, 2002]

Feller–Reuter–Riley property Refers to the particular form of the Feller property arising in a Feller process, in the special case of a countable state space; for a non-negative matrix p_{ij} with bounded row sums the property is equivalent to $\lim_{i\to\infty} p_{ij} = 0$, for all j. [G. Reuter & P. Riley, The Feller property for Markov semi-groups on a countable state space, *J. Lond. Math. Soc.*, 2(5), 267–275, 1977]

Feller–Reuter theorem (alias: **Reuter's criterion**) Arises when considering a continuous-time Markov chain $X(t), t \geq 0$, on a countable state space S, with transition matrix $P(t)$, having Q-matrix (q_{ij}). It asserts that $P(t)$ is unique for given (q_{ij}) iff for some $\lambda > 0$, $(\lambda - q_{ii})x_i = \sum_{j\neq i} q_{ij}x_j,\ 0 \leq x_i \leq 1,\ i \in S$, has only the solution $x_i = 0$. [G. Reuter, Denumerable Markov processes and the associated contraction semigroups, *Acta Math.*, 97(1–4), 1–46, 1957; G. Reuter, Denumerable Markov processes II, *J. Lond. Math. Soc.*, 34, 81–91, 1959]

Feller semi-group *see* Feller property

Feller's strong law A law of large numbers for a sequence $X_r, r \geq 1$, of i.i.d.r.vs such that $\mathbb{E}|X_1|^{1+\delta} = \infty$ for some $0 \leq \delta < 1$, i.e. a criterion determining whether the event $L = \{|\sum_{r=1}^n X_r| > a_n \text{ for infinitely many } n\}$ has probability zero, or probability one, for some monotone sequence $a_r, r \geq 1$.

In particular, if $\mathbb{E}|X_1| = \infty$, and a_n/n is increasing in n, L has probability 1 or 0 according as $\sum_{r=1}^{\infty} \mathbb{P}(|X_r| \geq a_r)$ diverges or converges; and when $a_n = n, \mathbb{P}(L) = 1$. [W. Feller, A limit theorem for random variables with infinite moments, *Amer. J. Math.*, 68(2), 257–262, 1946]

Feller's test for explosion Supplies a criterion for deciding whether or not a homogeneous diffusion $X(t)$ on $\mathbb{R}$ may explode to infinity in a finite time or not. Let $X(t)$ satisfy the SDE $\mathrm{d}X = \mu(X)\mathrm{d}t + \sigma(X)\mathrm{d}W$, where $\mu(x)$ and $\sigma(x) > 0$ are bounded and continuous on finite intervals, and define the scale $s(x) = \int_c^x \exp(-2\int_c^u \mu(y)/\sigma^2(y)\mathrm{d}y)\mathrm{d}u,\ c \in \mathbb{R}$.

Then $X(t)$ does not explode iff

$$\int_c^{\infty} [s(\infty) - s(x)]m(x)\mathrm{d}u = \int_{-\infty}^{c} [s(x) - s(-\infty)]m(x)\mathrm{d}x = \infty,$$

where $\frac{1}{m(x)} = \frac{1}{2}\sigma^2(x)\,\frac{\mathrm{d}s}{\mathrm{d}x}$. More generally, for a diffusion on $[a, b]$, with $c \in [a, b]$, the test requires the divergence of $\int_c^b (s(b) - s(x))m\mathrm{d}x$, and $\int_c^a (s(a) - s(x))m\mathrm{d}x$. The extension to a multivariate diffusion process is called **Khasminskii's test**. [L. Rogers, & D. Williams, *Diffusions, Markov Processes and Martingales*, Vol. 2, 2nd edn. Wiley, 1994; reprinted by Cambridge University Press, 2000]

Feller's theorem Several theorems might be called Feller's theorem, so the term is rarely used unqualified. If it is, it perhaps most often denotes the converse part of the Lindeberg–Feller theorem. [W. Feller, *Introduction to Probability Theory and its Applications*, Vol. 1, 3rd edn. Wiley, 1968; Vol. 2, 2nd edn. Wiley, 1971]

Feller's weak law of large numbers (alias: **Kolmogorov–Feller WLLN**) Arises for a sequence $X_1, X_2, \ldots$ of i.i.d.r.vs that do not necessarily have an expected value. It asserts that as $n \to \infty$, $\frac{1}{n}\{\sum_{r=1}^n X_r - n\mathbb{E}(XI\{|X| \leq n\})\}$ converges in probability to 0, iff $\lim_{x\to\infty} x\mathbb{P}(|X| > x) = 0$. [A. Gut, An extension of the Kolmogorov–Feller WLLN, with an application to the St. Petersburg game, *J. Theor. Probab.*, 17(3), 767–779, 2004]

Ferguson distribution *see* Dirichlet–Ferguson process

Ferguson–Klass representation Arises as a description of independent increments processes that are not Gaussian, having randomly placed jump discontinuities of random heights. Broadly, it can be seen as a Poisson approximation to such a process, expressed as a countable sum that takes the jumps in decreasing order of size. It also can be thought of as analogous to the Wiener representation of the Wiener process as a normal approximation expressed as a sum of normal r.vs. The generalization to Banach-space-valued random vectors is called the Ferguson–Klass–LePage representation. [T. Ferguson & M. Klass, A representation of independent increment processes without Gaussian components, *Ann. Math. Stat.*, 43(5), 1634–1643, 1972]

Fernique–Talagrand theorem Arises in the context of Gaussian processes, where it supplies a characterization of the properties of continuity and boundedness of sample paths. The relevant necessary and sufficient condition is often called Fernique's condition, originally introduced as sufficient and then shown to be necessary. [M. Talagrand, *The Generic Chaining*. Springer, 2005]

Feynman–Kac formula (alias: **F–K stochastic representation**) Arises in connexion with diffusion processes and their associated differential equations; it finds extensive applications in financial mathematics. In a popular formulation, the Feynman–Kac thm represents the solution $y(x,t)$ (assumed to exist), of the pde

$$\frac{\partial y}{\partial t} + a(x,t)\frac{\partial y}{\partial x} + \frac{1}{2}b^2(x,t)\frac{\partial^2 y}{\partial x^2} = c(x,t)y, \quad 0 \le t \le T,$$

with $y(x,T) = g(x)$, as a functional of the associated diffusion process $X(t)$ generated by the SDE $\mathrm{d}X = a\mathrm{d}t + b\mathrm{d}W(t)$, where $X(0) = x$, and W is the Wiener process, i.e.

$$y(x,t) = \mathbb{E}\Big\{\exp\Big(-\int_t^T c(X(s),s)\mathrm{d}s\Big)g(X(T)) \,|\, X(t) = x\Big\}, \quad 0 \le t \le T,$$

for suitably well-behaved functions a, b, c, and g. [P. Mörters & Y. Peres, *Brownian Motion*. Cambridge University Press, 2010]

fictitious state An element of the state space of a Markov chain that is not intrinsic in the natural construction of the chain, but is augmented to it. For example, if a Markov chain on a countable state space can explode, its state space may be augmented by a suitable fictitious state denoted by $\{\infty\}$, or Δ. In the context of the method of phases, the state space of a non-(or semi-) Markov process may be augmented by fictitious states to yield one having the Markov property. A state k may also be called fictitious if the transition probability $p_t(j,k) = 0$ for all j, and all $t > 0$. [K. L. Chung, *Markov Chains With Stationary Transition Probabilities*, 2nd edn. Springer, 1967]

fidelity criterion Arises in communication theory, when considering the problem of encoding a discrete (or discretized) source. It may be that the encoding need not be perfect; in fact some level of distortion of the input vector $\mathbf{X}$ may be permitted, as measured by some distance function $d(\mathbf{X},\mathbf{Y})$ between $\mathbf{X}$ and its encoding $\mathbf{Y}$. The permitted level of distortion Δ is called a fidelity criterion, and one may speak of an ε-fidelity criterion if, for some $\varepsilon > 0, \mathbb{P}(d(\mathbf{X},\mathbf{Y}) \le \Delta) \ge 1 - \varepsilon$.

Note that this should not be confused with the fidelity similarity, alias Bhattacharyya coefficient, or the Hellinger affinity. [C. Shannon, Coding theorems for a discrete source with a fidelity criterion, in *Collected Papers*. IEEE, 1993]

Fill's algorithm (alias: **Fill's exact (or perfect) sampling**) Arises in the context of Markov chain Monte Carlo (McMC) methods for generating a sample from a probability distribution $\boldsymbol{\pi}$. The algorithm uses acceptance-rejection sampling to produce an exact sample, in a bounded random time, provided that the underlying chain (whose stationary distribution is $\boldsymbol{\pi}$) is stochastically monotone on a finite state space. The extension to continuous state spaces is called the **FMMR algorithm**. [J. Fill, M. Machida, D. Murdoch, & J. Rosenthal, Extension of Fill's perfect rejection sampling algorithm to general chains, *Random Struct. Algor.*, 17, 290–316, 2000]

filtering (alias: **prediction problem**) A concept arising in filtering problems which, broadly, encompass the tasks of reconstructing or predicting a partially observed sequence (as e.g. when a signal is corrupted by noise). Formally, there is a sequence $(X(t), Z(t))$ of jointly distributed pairs of r.vs (or pair of r.vrs), where t is seen as time, and either $t \in \mathbb{Z}^+$ or $t \in \mathbb{R}^+$ (say). The sequence $Z(t), t \ge 0$, is observable while $X(t), t \ge 0$, is not observable, and the problem is to supply an estimate $\hat{X}(t)$ of $X(t)$, that is in some sense optimal, given the observations $Z(s), 0 \le s \le t$. If one seeks minimum mean-square error, then the projection theorem yields $\hat{X}(t) = \mathbb{E}(X(t) \,|\, (Z(s), 0 \le s \le t)$, and this is called a (non-linear) filter. If $\hat{X}$ is linear in Z, which may obtain in the multinormal case, one has a linear filter, Being a function of the past (of Z) at time t, it may be called a causal, or physically realizable filter (as filters looking into the future are not). The name arises by analogy with physical filters that remove unwanted quantities, such as particles or wavelengths.

A **predictive filter** likewise seeks an estimate of $X(t+u), u > 0$, at time t. More generally, one may seek an estimate of some function $h(X(t))$. Useful filters are recursive, which is to say, e.g. in discrete time, that $\hat{X}(t+1) = r(\hat{X}(t), Z(t+1), t)$, for some function r; classic examples include the Kalman and Wiener–Kolmogorov filters. In continuous time, a standard presentation of the problem postulates a non-observed process with SDE $\mathrm{d}X = a(X)\mathrm{d}t + b(X)\mathrm{d}W_1$, $t \geq 0$, and an observed process with SDE $\mathrm{d}Y = \alpha(X)\mathrm{d}t + \beta\mathrm{d}W_2$ and one seeks a representation for the estimate $\mathbb{E}(X(t) \mid Y(s)$, $0 \leq s \leq t)$. The problem may be generalized, and also formalized in terms of the filtration generated by the independent Wiener processes W_1, and W_2, and of $Y(t)$. The linear case $a(X) = aX$ and $\alpha(X) = \alpha X$ corresponds to the Kalman–Bucy filter. Under suitable conditions, the estimate $\hat{X}(t)$ satisfies an SDE called the optimal non-linear filtering equation. One such is called the Kushner–Stratonovich eqn, and others include the Duncan–Mortensen–Zakai eqn, and Fujisaki–Kallianpur–Kunita eqn. [B. Øksendal, *Stochastic Differential Equations*, 6th edn. Springer, 2003]

filtration Arises naturally in the context of random processes in seeking to formalize the idea of 'what is known about the process by time t'. In general, given a measurable space $(\Omega, \mathcal{F})$, such as appears in a probability space $(\Omega, \mathcal{F}, \mathbb{P})$, a filtration is any non-decreasing sequence $(\mathcal{F}_t)$ of sub-sigma-fields of $\mathcal{F}$, i.e. $\mathcal{F}_s \subseteq \mathcal{F}_t \subseteq \mathcal{F}$ for $s < t < \infty$, where t takes values in, e.g. $[0, T]$, or $\mathbb{R}^+$, or $\{0, 1, 2, \ldots\}$. In the context of a probability space, the quadruple $[\Omega, \mathcal{F}, \mathbb{P}, (\mathcal{F}_t)]$ is called a **filtered probability space**, or a **stochastic base** (or basis). And then a sequence of random variables $X(t)$, such that $X(t)$ is $\mathcal{F}_t$ measurable, is called a stochastic sequence on that base; or an adapted sequence. If $\mathcal{F}_t$ is the sequence generated by $X(s), 0 \leq s \leq t$, i.e. $\mathcal{F}_t$ is the smallest σ-field for which $X(s), 0 \leq s \leq t$, is measurable for all t, then one speaks of the natural filtration for $X(t), t \geq 0$. It is convenient if $\mathcal{F}_t$ is right-continuous in t (i.e. $\mathcal{F}_{t+} = \mathcal{F}_t$ for all $t \geq 0$), and it is also convenient if $\mathcal{F}_t$ is complete (i.e. $\mathcal{F}_0$, and hence $\mathcal{F}_t$ for all $t \geq 0$, contains all subsets of sets A such that $\mathbb{P}(A) = 0$; the so-called $\mathbb{P}$-null sets); if both hold then $\mathcal{F}_t$ is said to satisfy the usual conditions. If a non-complete filtration is enlarged to satisfy the usual conditions; one speaks of the usual augmentation. [I. Karatzas & S. Shreve, *Brownian Motion and Stochastic Calculus*, 2nd edn. Springer, 1998]

Fine's theorem *see* qualitative probability

finite-dimensional distributions (alias: **fdds; fidis**) Arise in the context of a random process $X(t)$, $t \in T$, i.e. a collection of r.vs defined on a probability space $(\Omega, \mathcal{F}, \mathbb{P})$, where T is a suitable index (or parameter) set, often called the time, e.g. $\mathbb{R}, \mathbb{N}, [a, b]$, etc. For the collection C of all finite subsets $t_1, \ldots, t_n$ of distinct members of T, the joint distributions of the random vectors $X(t_1), \ldots, X(t_n)$ are called the finite-dimensional distributions of $X(t)$. Conversely, if a given collection C of such joint distributions on a set T satisfy Kolmogorov's consistency conditions, then there is a probability space $(\Omega, \mathcal{F}, \mathbb{P})$, and a random process $X(t, \omega), t \in T$, defined on it having C as its fdds.

The fdds define a probability measure $\mathbb{Q}$ on $(\mathbb{R}^T, \mathcal{G})$ (where $\mathcal{G}$ is the Borel field $\mathcal{B}^T$, completed as usual), and then $\mathbb{Q}$ supplies the probability of properties of $X(t, \omega)$, $t \in T$, provided such properties are in $\mathcal{G}$. (For random processes in continuous time, properties of interest may not be events with a probability, unless $X(t, \omega)$ is constrained in some way, or $(\Omega, \mathcal{F}, \mathbb{P})$ chosen suitably.) [P. Billingsley, *Probability and Measure*, 3rd edn. Wiley, 1995]

finite horizon *see* decision theory

first-come first-served *see* queueing theory

first digit law *see* Benford's distribution

first fundamental theorem of finance *see* arbitrage theorem

first moment *see* expectation

first moment distribution *see* equilibrium distribution

first-order A term applied with a variety of meanings in the context of a stationary random process $X_t, t \in T$. Thus:

(a) A random process $X(t), t \in T$, is called first-order stationary if it has constant mean, $\mathbb{E}X(s) = \mathbb{E}X(t)$ for all s and t in T.
(b) A random measure (or point-process) $\mu(A), A \in \mathbb{R}^d$ is first-order stationary if $\mu(\cdot)$ is invariant under shifts (and hence is a multiple of Lebesgue measure).

(c) A stationary random process $X_n, n \in \mathbb{Z}$, is first-order autoregressive (resp. moving average) if for a sequence of orthogonal r.vs $Z_n, n \in \mathbb{Z}$ (with zero mean and variance σ^2), X_n has the representation $X_n = \alpha X_{n-1} + Z_n, \alpha \neq 1$ (resp. $X_n = Z_n + \beta Z_{n-1}$).
[D. Cox & H. Miller, *Theory of Stochastic Processes*. Methuen, 1965; reprinted by Chapman & Hall, 2001]

first-passage decomposition (alias: **first-entrance decomposition**) Arises naturally in the context of Markov processes, where it expresses transition probabilities (or probability functions) in terms of first-passage probabilities and recurrence probabilities. Thus, e.g., let $X_n, n \geq 0$, be a Markov chain with stationary transition probabilities $p(j,k)$, and n-step probabilities $p_n(j,k), n \geq 2$, where j and k take values in a countable state space S. Define $T_k = \inf\{n \geq 1 : X_n = k\}$. Then

$$p_n(j,k) = \sum_{m=1}^{n} \mathbb{P}(T_k = m) p_{n-m}(k,k).$$

There is a related, analogous last-exit decomposition, written in terms of the taboo probabilities:

$$t_n(j,k) = \mathbb{P}(\{X_n = k\} \cap \{X_r \neq j,\ 1 \leq r \leq n-1\} \,|\, X_0 = j), \quad \text{for } j \neq k.$$

[R. Syski, *Passage Times for Markov Chains*. IOS Press, 1992]

first-passage distribution *see* Lévy distribution

first-passage percolation Arises as a model for the flow of fluid through a porous medium, and can be seen as a graph-theoretic generalization of renewal theory. In the simplest case, the graph is the cubic lattice $\mathbb{Z}^d$; that is, the points of $\mathbb{Z}^d$ with edges added between all points $\mathbf{x} = (x_1, \ldots, x_d)$ and $y = (y_1, \ldots, y_d)$, such that $\sum_1^d |x_r - y_r| = 1$. To each edge e is attached a random variable $T(e)$, the collection of such r.vs being i.i.d., and $T(e)$ is thought of as being the time taken for fluid to flow along e. At time $t = 0$, fluid is supplied at the origin, and first-passage percolation investigates the behaviour of the set $W(t)$ of wet points of $\mathbb{R}^d$ at time t, i.e. those vertices reached by fluid along edges up to time t. [G. Grimmett, *Percolation*, 2nd edn. Springer, 1999]

first-passage time (alias: **first-entrance (or entry) time; reaching time; hitting time**) Arises in the context of a random process $X(t)$, taking values in a state space S, with time-parameter t. Broadly, the first-passage time T of $X(t)$ to a state $a \in S$, or suitable set of states $A \subseteq S$, is the instant at which it first takes the value a, or a value in A; the concept is important for Markov processes in which the Markov property may hold at T. Thus, let $X_n, n \geq 0$, be a Markov chain; we define two types of first-passage time:

$$T_A = \min\{n \geq 0 : X_n \in A\} \equiv \text{non-negative hitting time};$$
$$T_A^+ = \min\{n \geq 1 : X_n \in A\} \equiv \text{positive hitting time}.$$

If $X_0 \notin A$, then the two are equal; whereas if $X_0 \in A$, then $T_A = 0$ and T_A^+ may be called the first return (or recurrence) time. Some authors distinguish between the two by calling T_A the hitting time of A, but this usage is not universal. For $X(0) \in A$, the first-passage time to A^c may be called the first-leaving time, or first-exit time. When the set $A(t)$ (or state $a(t)$), depends on time, one speaks of a moving boundary.

In continuous time, for a Markov chain $X(t)$ that spends a time J in its initial state before jumping, we define:

$$T_A = \min\,[t \geq 0 : X(t) \in A];$$
$$T_A^+ = \inf\,[t > J : X(t) \in A].$$

As above, $T_A = T_A^+$ unless $X(0) \in A$, and then T_A^+ is the first return time. For diffusions, with continuous sample-paths, the analogously defined T_A and then T_A^+ always coincide. [J. Norris, *Markov Chains*. Cambridge University Press, 1997]

first-step analysis (alias: **first-step decomposition**) Arises naturally in the context of Markov chains, where it supplies a method for evaluating a class of functionals of homogeneous chains. The basic technique is to condition on the first step of the chain, and invoke the Markov property. It can be seen as generating a type of backward equation. *See* first-passage decomposition. [P. Brémaud, *Markov Chains*. Springer, 1999]

first variation process (alias: **tangent process**) Arises in the context of Malliavin calculus, with consequent applications in mathematical finance. Let $X(x,t)$ be an Itô diffusion given by the SDE $\mathrm{d}X = a(X)\mathrm{d}t + b(X)\mathrm{d}W(t), t \geq 0$, where $X(x,0) = x$, for suitably smooth functions $a(\cdot)$ and $b(\cdot)$, with $b(x) \neq 0$. The first variation process is $V(t) = \frac{\partial}{\partial x}X(x,t), t \geq 0$, and is the solution of the SDE $\mathrm{d}V = \{a'(X)\mathrm{d}t + b'(X)\mathrm{d}W\}V,\ t \geq 0$; with $V(0) = 1$. The **Malliavin derivative** $D_t X_T$ may then be represented as a linear transformation of V. More generally, for a d-dimensional diffusion $X_1, \ldots, X_d$, with initial value $x_1, \ldots, x_d$, the first variation is the Jacobian $\dfrac{\partial(X_1, \ldots, X_d)}{\partial(x_1, \ldots, x_d)}$. Higher-order variations arise from higher-order derivatives. [G. Di Nunno, B. Øksendal, & F. Proske, *Malliavin Calculus for Lévy Processes*. Springer, 2009]

Fishburn–Shepp inequality *see* XYZ theorem

Fisher–Bingham distribution *see* Bingham distribution

Fisher's distribution Several probability distributions have been associated with the name of R. Fisher, including:

1. Fisher's B distribution, which is the density of the square root of a non-central chi-squared r.v.
2. Fisher's F distribution, which is properly the F distribution, named by G. Snedecor for Fisher.
3. Fisher's logarithmic series distribution, more often called the logarithmic distribution.
4. **Fisher–von Mises–Langevin** (or FVML) distribution, which is a spherical distribution having density (wrt Lebesgue measure on the sphere) given by $f \propto \exp(\kappa \boldsymbol{\mu}^T \mathbf{x}), \mathbf{x} \in S^{k-1}$, for some location (vector) parameter $\boldsymbol{\mu}$ and $\kappa > 0$. As it is the only spherical density for which the maximum likelihood estimator of μ is the sample mean, a property that characterizes the normal density on $\mathbb{R}$, it is also called the **spherical normal distn**. The circular (two-dimensional) case is often called **von Mises distn**; the three-dimensional case is often called **Fisher's distn**, but would perhaps be better called **Arnold's distn** after its originator (predating Fisher by 12 years). In higher dimensions one often speaks of the **Langevin distn**, or **Watson–Williams distn.** It arises as the conditional joint distn of independent normal r.vs $X_1, \ldots, X_k$, with mean vector $\boldsymbol{\mu}$, having $\operatorname{var} X_r = \kappa^{-1}$, given that $\sum_{r=1}^{k} X_r^2 = 1$.
5. Fisher–Tippett distn, which is a class of distributions arising in extreme value theory, with which are also associated the names of M. Fréchet, E. Gumbel, and W. Weibull.
6. Fisher's Z distribution; arises as the distribution of the random variable $Z = \frac{1}{2}\log R$, where R has the F distribution.
7. Fisher's non-central (or extended) hypergeometric distn, which arises as the conditional distributions of independent binomial random variables X and Y, given their sum $Z = X + Y$, where X is $\mathrm{B}(n, \alpha)$ and Y is $\mathrm{B}(m, \beta), 0 < \alpha, \beta < 1$. It has pmf of the form

$$\mathbb{P}(X = x \,|\, Z = z) \propto \binom{m}{x}\binom{n}{z-x}\left\{\frac{\alpha(1-\beta)}{\beta(1-\alpha)}\right\}^x.$$

[[G. Watson, *Statistics on Spheres*. Wiley, 1983; J. Bennett (ed.), *Collected Papers of R. A. Fisher*, Vols 1–5. Adelaide University Press, 1971–1974]

Fisher's equation (alias: **Fisher–Kolmogorov–Petrovskii–Piscounov equation**) Introduced by R. Fisher to describe the advance of advantageous genes, the equation arises in a number of situations in which particles both react and diffuse in space (being therefore a special case of so-called reaction diffusion models). In one spatial dimension the equation takes the form $\frac{\partial f}{\partial t} = D\frac{\partial^2 f}{\partial x^2} + R(f)$, where $t \geq 0$ is seen as time, $R(f)$ is the reaction term, and $D\frac{\partial^2 f}{\partial x^2}$ arises from diffusion. In Fisher's case $R(f) = af(1-f)$; in the case of branching diffusion $R(f) = a(G(f) - f)$, where $G(s)$ is the pgf of a non-negative integer valued r.v. [R. Fisher, The wave of advance of advantageous genes, *Ann. Eugenics*, 7, 355–369, 1937]

Fisher–Geary theorem *see* Geary's theorem

Fisher–Girshick–Hsu–Roy distribution (alias: **F–G–H–R (and variations)**) This is the joint distribution of the eigenvalues of a random matrix arising as the sample covariance matrix of a normal sample. Specifically, let X be an $m \times n$ matrix whose entries are i.i.d. zero-mean normal r.vs. Then the joint density of the eigenvalues $\lambda_1 \geq \lambda_2 \geq \cdots \geq \lambda_m$ of the so-called Wishart matrix $M = XX^T$, for $n \geq m$, is

$$f(\lambda) \propto \prod_{r=1}^{m} \lambda_r^{(n-m-1)/2} \prod_{r<s} (\lambda_s - \lambda_r) \exp\left(-\frac{1}{2}\sum_{r=1}^{m} \lambda_r\right).$$

[R. A. Fisher, The sampling distribution of some statistics obtained from non-linear equations, *Ann. Eugenics*, 9, 238–249, 1939; M. A. Girshick, On the sampling theory of roots of determinantal equations, *Ann. Math. Stat.*, 10, 203–204, 1939; P. L. Hsu, On the distribution of roots of certain determinantal equations, *Ann. Eugenics*, 9, 250–258, 1939; S. N. Roy, p-statistics or some generalizations in the analysis of variance appropriate to multivariate problems, *Sankhya*, 4, 381–396, 1939]

Fisher's information In its simplest form, this arises for a random variable X having a density $f(x, \theta)$ depending on a parameter θ, for which f is differentiable. Then Fisher's information is

$$J(f) = \mathbb{E}\left(\left\{\frac{\partial}{\partial\theta}\log f(X, \theta)^2\right\}\right) = -\mathbb{E}\left(\frac{\partial^2}{\partial\theta^2}\log f(X, \theta)\right).$$

Fisher named it **the intrinsic accuracy**, a choice supported by the Cramér–Rao theorem that (in regular cases) the variance var $\hat{\theta}$ of any unbiased estimator $\hat{\theta}$ of θ satisfies var $\hat{\theta} \geq J(f)^{-1}$. When θ is a location parameter, i.e. $f(x,\theta) = p(x-\theta)$, we have (for differentiable p)

$$J(f) = \int_{\mathbb{R}} \frac{1}{p(x)} \{p'(x)\}^2 \mathrm{d}x = 4\int_{\mathbb{R}} \left(\frac{\mathrm{d}}{\mathrm{d}x}\sqrt{p(x)}\right)^2 \mathrm{d}x,$$

which may be called Linnik's information (or functional), or Fisher's information with respect to location parameter, or simply (and confusingly) Fisher's information; or the Linnik–Fisher or Fisher–Linnik-(information) functional; and is so defined for every differentiable p, irrespective of parameterization.

Thus, for a differentiable probability density f on $\mathbb{R}^d$, the Fisher–Linnik information is

$$J(f) = \int_{\mathbb{R}^d} \frac{1}{f(x)} |\nabla f|^2 \mathrm{d}x = 4\int_{\mathbb{R}^d} |\nabla \sqrt{f}|^2 \mathrm{d}x,$$

where ∇ is the gradient operator. For $X \in \mathbb{R}$, with density f and finite variance σ^2, the standardized Fisher–Linnik information is

$$J_S(X) = \sigma^2 \mathbb{E}\left\{\left(\frac{f'(X)}{f(X)}\right)^2\right\} - 1 = \sigma^2 \mathbb{E}\left\{\left(\frac{f'(X)}{f(X)} + \frac{X-\mu}{\sigma^2}\right)^2\right\}.$$

As with other forms of differential entropy, one may define the Fisher–Linnik relative information (or information distance) between densities f and g (of r.vs X and Y) by

$$\begin{aligned} J(f;g) &= \int f(x)\left(f'(x)/f(x) - g'(x)/g(x)\right)^2 \mathrm{d}x \\ &= 4\int g(x)\left(\frac{\mathrm{d}}{\mathrm{d}x}\sqrt{\frac{f}{g}}\right)^2 \mathrm{d}x, \end{aligned}$$

and then $J_S(X) = \sigma^2 J(f;h)$, where h is the density of an $N(\mu,\sigma^2)$ r.v. [O. Johnson, *Information Theory and the Central Limit Theorem*. Imperial College Press, 2004]

Fisher–Snedecor distribution *see* F distribution

Fisher–Tippett–Gnedenko theorem *see* extreme value distributions

Fisher–Tippett–Gumbel distribution *see* extreme value distributions

Fisher–von Mises distribution *see* Fisher's distribution

Fisher–Wright model *see* Wright–Fisher model

Fisher–Yates shuffle *see* shuffle

fishing problem A term used to encompass a type of optimal stopping problem arising in sequential search problems, including bug-hunting, both figuratively (in software testing) and literally (in ecology). In one formulation there is a random number N of objects (that we call fish, for definiteness) having (say) i.i.d. sizes $X_r, 1 \leq r \leq N$. The fisherman catches fish at (say) the instants T_r of a renewal process from time $t = 0$, with an increasing cost function $C(t), t \geq 0$. The problem is to determine the time T to stop fishing that maximizes the reward (amount of fish) less the cost of catching them. More generally, one can fish in separate pools, in which case the problem is to determine an optimal order in which to fish the pools, together with the optimal change-times; this may be called the **foraging problem**. [N. Starr, Optimal and adaptive stopping based on capture times, *J. Appl. Probab.*, 11, 294–301, 1974]

Fisk distribution *see* log-logistic distribution

Fisk–Stratonovich integral (alias: **Stratonovich integral; symmetrized stochastic integral**)
A stochastic integral related to the Itô integral, but having different properties; most notably these two:

(a) The F-S integral obeys the classical chain rule formula under change of variables, unlike the Itô integral.
(b) The F-S integral is not (in general) a martingale, unlike the Itô integral.

Let $Y(t)$ be a continuous martingale, and $X(t)$ a process adopted to the filtration generated by $Y(t), t \geq 0$ (or more generally, in Meyer's extension, let $X(t)$ and $Y(t)$ be continuous semi-martingales adapted to the same filtration), then the Fisk–Stratonovich integral of X against Y is directly defined as the mean square limit $\left(\text{as } \max_r(t_{r+1}^n - t_r^n) \to 0\right)$ of $\frac{1}{2}\sum_{r=0}^{n-1}\left(X(t_{r+1}^n) + X(t_r^n)\right)\left(Y(t_{r+1}^n) - Y(t_r^n)\right)$, for $0 = t_0^n < \cdots < t_n^n = t$. It may be denoted by $\int_0^t X \circ \mathrm{d}Y$, or $\int_0^t X \partial Y$, among other notations.

For semi-martingales having the decompositions $X(t) - X(0) = M(t) + A(t), Y(t) - Y(0) = N(t) + B(t)$, for continuous local martingales M and N, we have

$$\int_0^t X \circ dY = \int_0^t XdN + \int_0^t XdB + \frac{1}{2} < M, N>_t,$$

where the first term in the RHS is an Itô integral, and the final term is the cross-variation (or covariation). [I. Karatzas & S. Shreve, *Brownian Motion and Stochastic Calculus*, 2nd edn. Springer, 1998]

Fisz's theorem Supplies a characterization of the exponential distn, and a converse to the Rényi–Epstein–Sobel thm (see Sukhatme's theorem). In its simplest form, let X_1 and X_2 be absolutely continuous i.i.d.r.vs such that max $\{X_1, X_2\}-$ min$\{X_1, X_2\}$ is independent of min$\{X_1, X_2\}$. Then X_1 and X_2 are exponentially distributed. More generally, for absolutely continuous i.i.d.r.vs $X_1, \ldots, X_n$ with order statistics $X_{(1)}, \ldots, X_{(n)}$, then $X_{(r+1)} - X_{(r)}$ is independent of $(X_{(1)}, X_{(2)} - X_{(1)}, \ldots, X_{(r)} - X_{(r-1)}), 1 \leq r \leq n-1$, iff X_1 has an exponential density. [J. Galambos, Characterization of certain populations by independence of order statistics, *J. Appl. Probab.*, 9, 224–230, 1972]

FKG inequality (alias: **Harris–FKG inequality; Fortuin–Kasteleyn–Ginibre inequality; (positive) correlation inequality**) Arises in the study of random graphs and related problems, particularly in statistical physics, often grouped under the title probabilistic combinatorics. Broadly, it asserts that for suitable probability measures on graphs, increasing events (or r.vs) are positively correlated. It supplies a generalization of Harris's lemma and Harris's inequality. One form is this: let $\Omega = \{0,1\}^S$, where S is some finite set (typically the edge set of a graph or lattice), with its resulting natural partial order. If $\mathbb{P}(\omega)$ is a strictly positive probability measure on Ω having the so-called FKG (or lattice) condition for any $\omega_1, \omega_2 \in \Omega$:

$$\mathbb{P}(\min\{\omega_1, \omega_2\})\, \mathbb{P}(\max\{\omega_1, \omega_2\}) \geq \mathbb{P}(\omega_1)\, \mathbb{P}(\omega_2),$$

then for increasing real-valued functions, $X(\omega)$ and $Y(\omega)$ on $\Omega, \mathbb{E}(XY) \geq \mathbb{E}X\mathbb{E}Y$. *See* Holley's inequality. [G. Grimmett, *Probability on Graphs.* Cambridge University Press, 2010]

FKK equation (alias: **Kushner–Stratonovich equation**) *see* filtering

flash of flashes A pathological example arising in the theory of Markov processes, demonstrating that for a suitable conservative Q-matrix Q, there may be a continuum of (i.e. uncountably many) transition probability functions $p_{ij}(t)$, each satisfying the backward and forward eqns, such that their corresponding honest Markov processes have the given Q as their Q-matrix. Its construction is based on the Lévy flash, being a Markov process on $\mathbb{Z}$ that has mean holding time $\frac{1}{\lambda_r}$ in state r, before jumping to $r+1$ with probability 1, where $\sum_{r=-\infty}^{\infty} \lambda_r^{-1} < \infty$. When a flash arrives at $+\infty$, it instantaneously starts a new flash at $-\infty$. [D. Kendall, Some further pathological examples in the theory of denumerable Markov processes, *Q. J. Math. Oxford*, 7, 39–56, 1956]

Fleming–Viot process (alias: **F-V superprocess**) A measure-valued continuous Markov process (and thus a diffusion) arising from the theory of population genetics. The evolving measure in question corresponds to the distribution of gametic types (or alleles) of a population of individuals (assumed constant) over the set of possible types, which is assumed to be a compact metric space (usually comprising a subset of $\mathbb{R}^d$). It may be obtained as a suitable continuous limit of a rescaled discrete model; and (very roughly speaking) can be seen as the Dawson–Watanabe process conditioned to have constant total measure. [W. Fleming & M. Viot, Some measure-valued Markov processes in population genetics theory, *Indiana Uni. Math. J.*, 28(5), 817–843, 1979]

Flesaker–Hughston model Arises in financial mathematics as a model for interest rates (term structure model) in which the rates are determined by postulating a model for the bond (zero-coupon) price $B(t, T)$ of the form

$$B(t,T) = \int_T^\infty f(y) M(t,y) \mathrm{d}y \Big/ \int_t^\infty f(y) M(t,y) \mathrm{d}y, \; 0 \leq t \leq T < \infty,$$

for a positive non-random $f(y)$, and a family of positive martingales $M(t, T)$ wrt a filtration $\mathcal{F}_t$. This is intimately connected to the so-called **potential approach** in which $B(t, T) = \mathbb{E}(S(T) \mid \mathcal{F}_t)/S(t)$, where $S(t)$ is a positive supermartingale converging to zero (and hence a potential). Particularizing $M(t, y)$ or $S(t)$, specifies the model. [N. Bingham & R. Kiesel, *Risk-neutral Valuation*, 2nd edn. Springer, 2004]

flip-flop chain (or process) *see* two-state process

Flood's fiancé(e) problem *see* secretary problem

flow-equivalent method (alias: **Chandy–Herzog–Woo theorem; Norton's theorem**) A technique for analysing suitable networks of queues by choosing some server S (or some subset of servers), and replacing all other servers by a notional node whose behaviour seems the same to S as the original servers. The method is analogous to E. Norton's short-circuit approach to electrical networks. [K. Chandy *et al.*, Parametric analysis of queueing networks, *IBM J. Res. Dev.*, 19(1), 36–42, 1975]

flow parameter Arises in bounds for the rate of convergence of Markov chains, in a similar manner to the related conductance. Formally, let p_{ij} be the transition matrix of an irreducible ergodic reversible Markov chain, with stationary distn π_i, and state space S. If L denotes a proper subset of S, the flow parameter is

$$\phi = \sup_L \left\{ \left(\sum_{i\in L} \pi_i \sum_{j\in L^c} \pi_j \right) \Big/ \left(\sum_{i\in L} \sum_{j\in L^c} \pi_i p_{ij} \right) \right\}.$$

[D. Aldous & J. Fill, *Reversible Markov Chains and Random Walks on Graphs*. Cambridge University Press, to appear; see www.stat.berkeley.edu/~aldous/RWG/book.html]

fluctuation theory Arises in the context of the sequence of partial sums $S_n = \sum_{r=1}^n X_r$, of r.vs $X_1, X_2, \ldots$ that are usually i.i.d. Fluctuation theory addresses itself to the behaviour of a number of functions of the set $\sum = [S_1, \ldots, S_n]$ that are of natural interest. Examples include the maximum and minimum elements of $\sum$; the number that are positive; the ladder heights and times of the sequence, and so on. Key results include those of Baxter, Pollaczek–Spitzer, and Sparre Andersen. [S. Karlin & H. Taylor, *A Second Course in Stochastic Processes*. Academic Press, 1981]

fluid Markov model (alias: **Markovian fluid model**) A piecewise linear, or more generally piecewise deterministic, Markov process that has continuous sample paths; a classic example is the dam process (fluid queue). [L. Breuer, *From Markov Jump Processes to Spatial Queues*. Kluwer, 2003; L. Rogers, Fluid models in queueing theory and Wiener–Hoff factorization of Markov chains, *Ann. Appl. Probab.*, 4(2), 390–413, 1994]

fluid queue *see* dam process

$\mathcal{F}$-measurable A term arising in the definition of a random variable $X(\omega)$ as a real-valued function on a space Ω equipped with a sigma-field $\mathcal{F}$ of subsets (so that $(\Omega, \mathcal{F})$ is a measurable space). Such a function is a r.v. if $\{\omega \in \Omega : X(\omega) \le x\} \in \mathcal{F}$, for each $x \in \mathbb{R}$, and $X(\cdot)$ is said to be $\mathcal{F}$-measurable. More generally, when $X(\omega)$ takes values in a measurable space $(S, \mathcal{G})$, it is said to be an S-valued r.v. if, for all $A \in \mathcal{G}, X^{-1}(A) = \{\omega : X(\omega) \in A\} \in \mathcal{F}$. [O. Kallenberg, *Foundations of Modern Probability*, 2nd edn. Springer, 2002]

Fokker–Planck equation A term commonly used by physicists to denote the forward equations for the transition density of a diffusion. It arises from the work of A. Fokker (1914) and M. Planck (1917), similar equations having been derived by A. Einstein (1905), and M. Smoluchowski (1906), all in the context of Brownian motion. R. Fisher (1930) developed such equations independently in the context of genetics, and A. Kolmogorov (1931) developed a rigorous theory in greater generality. The corresponding term for fractional diffusions is **Fokker–Planck–Zaslavsky eqn**. [C. Gardiner, *Stochastic Methods: A Handbook*, 4th edn. Springer, 2010]

folded distribution (alias: **half-distribution**) For a random variable X having distribution function $F(x)$, the corresponding folded distn $G(x)$ is that of $|X|$, so that

$$G(x) = \begin{cases} F(x) - \lim\limits_{y\uparrow -x} F(y), & x \ge 0, \\ 0, & x < 0. \end{cases}$$

When X is a zero-mean normal r.v., the distn of $|X|$ is sometimes called half-normal. The term d-folded distn has been used to refer to that of the r.v. $Y = X$ modulo d, perhaps more often called a wrapped distribution.

Földes–Révész identity Arises in the context of the Wiener process (Brownian motion) $W(t), t \ge 0$. For $0 < a < b$, let $L(s, T_b)$ be the local time of $W(t)$, at level s, up to the time $T_b = \inf\{t : L(0,t) > b\}$; and let $T_{\sqrt{a}}(R)$ be the first-passage time to $\sqrt{a}$ by a two-dimensional Bessel process R, started at 0. Then the Földes–Révész identity asserts that $T_{\sqrt{a}}(R)$ has the same distribution as

$$\int_0^\infty I(0 < L(s, T_b) < a) \mathrm{d}s,$$

where $I(\cdot)$ is the usual indicator fn. Both r.vs have Laplace transform $I_0(\sqrt{2a\theta})$, where I_0 is the usual modified Bessel fn. [A. Földes & P. Révész, On hardly visited points of Brownian motion, *Probab. Theor. Rel. Fields*, 91, 71–80, 1992]

Föllmer–Schweizer decomposition Arises as a generalization of the Kunita–Watanabe decomposition, and has applications in financial mathematics when the price process $S(t)$ is a semi-martingale (rather than a martingale) with decomposition $S(t) - S(0) = M(t) + A(t)$, where $M(t)$ is a martingale. Then the F-S decomposition of a suitable random variable X (typically a claim), with respect to $S(t)$, takes the form

$$X = a + \int_0^t Y(u)\mathrm{d}S(u) + B(t),$$

where $B(t)$ is a martingale orthogonal to $M(t)$. The process $Y(t)$ in this determines a minimal quadratic risk hedging strategy for the claim. [M. Musiela & M. Rutkowski, *Martingale Methods in Financial Modelling*, 2nd edn, 3rd printing. Springer, 2009]

foretold time *see* stopping time

forgetfulness *see* lack of memory

forget time *see* access time

Forsythe's method *see* von Neumann's comparison method

Fortet–Mourier distance For probability measures $\mathbb{P}(\omega)$ and $\mathbb{Q}(\omega)$ defined on some metric space Ω, with metric $\|\cdot\|$, the Fortet–Mourier distance of order r is given by $d_r(\mathbb{P}, \mathbb{Q}) = \sup_{g \in S} |\int g(\omega)(\mathbb{P}(\mathrm{d}\omega) - \mathbb{Q}(\mathrm{d}\omega))|$, provided that both $\mathbb{P}$ and $\mathbb{Q}$ have finite rth moment, and g satisfies $|g(\omega) - g(\lambda)| \leq \|\omega - \lambda\| \max\{1, \|\omega\|^{r-1}, \|\lambda\|^{r-1}\}$. It can be seen as generalizing the earth movers distance (alias Wasserstein or Kantorovich metric), to which it reduces when $r = 1$. [S. Rachev, *Probability Metrics and the Stability of Stochastic Models*. Wiley, 1991]

Fortuin–Kasteleyn–Ginibre inequality *see* FKG inequality

forward-backward stochastic differential equation *see* stochastic integral

forward equation *see* backward and forward equation

Foster's criteria (or conditions, or theorems) (alias: **Foster–Lyapunov theorems (or drift conditions); test function methods**) Supply conditions for the transience, or null (and non-null) recurrence of an irreducible Markov chain on a countable state space S, with transition probabilities p_{ij}. One form is this:

(a) Let $s \in S$ be some fixed state. Then the chain is transient iff there is a bounded real-valued function $g(i), i \in S$, such that for all $j \neq s$, $g(j) = \sum_{k \neq s} p_{jk} g(k)$.

(b) Let $F \subseteq S$ be a finite subset of states.
  (i) Then the chain is recurrent if there is function $g(i)$ such that (as $s \to \infty$) $g(s) \to \infty$, and $g(j) \geq \sum_{k \in S} p_{jk} g(k)$, for $j \notin F$.
  (ii) The chain is non-null recurrent if there is some $a > 0$, and a function $g(i)$, such that $\inf_{s \in S} g(i) > -\infty$, and
$$\sum_{k \in S} p_{jk} g(k) \leq g(j) - a, \quad j \notin F, \text{and}$$
$$\sum_{k \in S} p_{jk} g(k) < \infty, \quad j \in F.$$

More general forms are called **drift conditions**. [S. Asmussen, *Applied Probability and Queues*, 2nd edn. Springer, 2003. Note that this is F. G. Foster]

Foster–Kendall theorem Arises in the context of a Markov chain $X_n, n \geq 0$, with stationary transition probabilities p_{ij}, where $i, j \in \{0, 1, 2, \ldots\}$. It asserts that if there is a non-negative, finite-valued function $g(i)$ on the states of X_n, such that $\lim_{i \to \infty} g(i) = \infty$ and for all i, $g(i) \geq \sum_{j=0}^{\infty} p_{ij} g(j)$, then:

(i) the chain has no null recurrent states;
(ii) any non-null recurrent class is finite;
(iii) for all i, $\sum_{j=0}^{\infty} \lim_{n \to \infty} \frac{1}{n} \sum_{r=0}^{n} p_{ij}(r) = 1$,

(where $p_{ij}(r)$ is the r-step transition probability), which is to say that the chain is non-dissipative. [D. Kendall, On non-dissipative Markov chains with an enumerable number of states, *Proc. Camb. Phil. Soc.*, 47, 633–634, 1951]

Foster–Pakes process A modification of the ordinary Galton–Watson branching process in which immigration is permitted only when the size of the population is zero. [A. Pakes, Some results for non-supercritical Galton–Watson processes with immigration, *Math. Bio.*, 24(1–2), 71–92, 1975. Note that this is J. H. Foster]

four functions theorem *see* Ahlswede–Daykin inequality

Fourier transform *see* characteristic function

fractal Broadly (but not exclusively) a set (or object) having some property of self-similarity under change of scale. The Wiener process $W(t)$ (Brownian motion in one dimension), has such a property in that for any $a > 0$, the processes $W(t)$ and $a^{-\frac{1}{2}} W(at)$ have the same distn. The Hausdorff (fractal) dimension of $W(t)$ is 3/2 a.s., as is also its Kolmogorov entropy, or box dimension. In two or more dimensions, the sample paths of $W(t)$ have Hausdorff (and box) dimension 2, a.s. while the fractal dimension of the frontier of a two-dimensional Wiener process is 4/3.

Yet more generally, a multifractal $X(t)$ is seen as a random process such that, for $a > 0, X(at)$ has the same distribution as $M(a)X(t)$ for independent random functions $M(a)$ and $X(t)$; more precisely, the process $X(t)$ having stationary increments is a multifractal process if $\mathbb{E}(|X(t)|^r) = c(r)t^{b(r)+1}$, for suitable (non-random) functions b and c of r, defined on some interval (assumed to include [0,1]). [K. Falconer, *Fractal Geometry*, 3rd edn. Wiley, 2014]

fractional Brownian motion (alias: **fractal B.M.; Wiener helix (or spiral); fBm (or fBM); fractional Wiener process**) Arises as a continuous, zero-mean Gaussian process $X(t), t \geq 0$, having the following self-similarity property: for any $a > 0$, the process $X(at)$ and $a^{-H}X(t)$ have the same distribution, for some constant $H \in (0, 1)$. It may also be defined as a continuous, zero-mean Gaussian process having covariance function $c(s, t) = \mathbb{E}(X(s)X(t)) = \frac{1}{2}(t^{2H} + s^{2H} - |s - t|^{2H})$, where H is called the Hurst index (or self-similarity parameter). In particular, $\mathbb{E}(X(t)^2) = t^{2H}$. The process has stationary (i.e. homogeneous) increments, which is to say that $X(t + s) - X(t)$ has the distn of $X(s)$, for $s, t \geq 0$; and its sample paths have Hausdorff (and box) dimension $2 - H$. The increments are positively or negatively correlated according as $H > 1/2$ or $H < 1/2$; and are independent when $H = 1/2$, when the process reduces to ordinary Brownian motion. The increment process may be called fractional Gaussian noise (fGn). There are several integral representations of $X(t)$, either in terms of the Wiener process $W(t)$, or another f BM of index $K \neq H$; they differ according as the process is defined for $t \geq 0$ or for $t \in \mathbb{R}$. More generally, the multifractional Brownian motion replaces H by a suitable function of t. Also more generally, for a constant vector $\mathbf{H} \in (0, 1)^d$, the continuous zero-mean Gaussian random field $X(t)$ defined on $[0, 1]^d$ is a **fractional Brownian sheet** if

$$\mathbb{E}(X(s)X(t)) = \frac{1}{2^d}\prod_{r=1}^{d}(t_r^{2H_r} + s_r^{2H_r} - |s_r - t_r|^{2H_r}).$$

[F. Biagini *et al.*, *Stochastic Calculus for Fractional Brownian Motion*. Springer, 2008]

fractional diffusion Arises as a model for anomalous diffusion in which $X(0) = 0$ and var $X(t) \propto t^\alpha, \alpha \neq 1$. In one formulation of the theory, the time or space derivatives in conventional diffusion equations may be replaced by suitable fractional derivatives. [Y. Mishura, *Stochastic Calculus for Fractional Brownian Motion*. Springer, 2008]

fractional Gaussian noise *see* fractional Brownian motion

fractional Lévy process (or motion) Broadly, a process defined analogously to fractional Brownian motion, in any one of several distinct ways. For example, let $W(t), t \geq 0$, be the Wiener process and $T(t)$ an independent subordinator (i.e. an a.s. non-decreasing Lévy process), yielding a symmetric Lévy process $L(t) = W(T(t))$. Then the process $L_H(t) = X(T(t))$ obtained by replacing $W(\cdot)$ by a fractional Brownian motion $X(\cdot)$ may be called a fractional Lévy motion.

Alternatively, one may consider any one of a number of integral representations of fractional Brownian motion in terms of the Wiener process $W(t)$ (where the integral may be stochastic Weyl, or fractional Riemann–Liouville, etc.). Then the process arising when $W(t)$ is replaced by a suitable Lévy process in the integral may be called a fractional Lévy process. Some authors use the term linear fractional stable motion for such a process, to distinguish it from the first definition above. Thus, if $L(t)$ is a two-sided Lévy process, $t \in \mathbb{R}$, then

$$X(t) = \frac{1}{\Gamma(1 + \frac{1}{2}H)}\int_{-\infty}^{\infty}(t - s)_+^{2H} - (-s)_+^{2H}L(\mathrm{d}s)$$

may be called a fractional Lévy process (or stable motion) for $0 < H < 1$ (where $x_+ = \max\{x, 0\}$). [T. Marquardt, Fractional Lévy processes with an application to long memory moving average processes, *Bernoulli*, 12(6), 1099–1126, 2006]

fractionally integrated autoregressive moving-average process (alias: **fractional ARIMA**) A time series $X(n)$ that is stationary and satisfies $Q(B)(1 - B)^d X_n = R(B)Z_n$, for suitable polynomials $Q(\cdot)$ and $R(\cdot)$ of the backward shift operator B, where $d \in \left(-\frac{1}{2}, \frac{1}{2}\right)$. [P. Brockwell & R. Davis, *Time Series*, 2nd edn. Springer, 1991]

fractional Ornstein–Uhlenbeck process A process $Y(t)$ defined in terms of fractional Brownian motion $X(t)$ by analogy with a representation of the ordinary Ornstein–Uhlenbeck process in terms of the standard Brownian motion $W(t)$. Since there are several such representations, a number of processes may be called fractional Ornstein–Uhlenbeck. For example, one may define $Y(t)$ by the Lamperti transformation of $X(t)$, i.e. $Y(t) = e^{-tH}X(e^t)$, where H is the Hurst parameter of $X(t)$. Alternatively one may seek $Y(t)$ as the stationary solution $Z(t)$ of $dZ(t) = -\alpha Z(t)dt + dX(t), \alpha > 0$, which is the Langevin equation driven by fractional Brownian motion.

The processes $Y(t)$ and $Z(t)$ are both stationary Gaussian processes, but not the same. The second, $Z(t)$, is usually termed the fractional O-U process. [P. Cheridito *et al.*, Fractional Ornstein–Uhlenbeck processes, *Electr. J. Probab.*, 8(3), 1–14, 2003]

fractional white noise The derivative of fractional Brownian motion, as defined in the context of a suitable fractional version of Hida white noise theory (*see* white noise). [F. Biagini *et al.*, *Stochastic Calculus for Fractional Brownian Motion*. Springer, 2008]

fragmentation process Broadly, any process seeking to describe the evolution (in discrete or continuous time) of a collection of sets, called blocks (modeling objects), that repeatedly split at random into two (or more) sub-blocks according to some (usually) homogeneous rule. Real-world examples include polymer or DNA chains, and rocks. Classic elementary examples are stick-breaking, and Kolmogorov's rock-crushing (or mass fragmentation) problem. In the basic formulation, a unit block (measured by mass, diameter, length, etc.) is divided at random; each sub-block is then independently subject to division according to the same distribution, when suitably scaled. In basic models, other properties of blocks (such as their shape, site, location, or age) do not affect this division; in more general models they may. Partition-valued fragmentation processes have a genealogical structure, and more complicated processes may both fragment and coagulate (*see* coalescence). [J. Bertoin, *Random Fragmentation and Coagulation Processes*. Cambridge University Press, 2006]

frailty Arises in the context of ageing and reliability models, where it seeks to account for random variation (due to the environment, say) among otherwise similar components (or individuals). At its simplest, it is assumed that there is a random so-called frailty factor A such that, conditional on $A = a$ the lifetime T of a component has the form $\mathbb{P}(T > t | A = a) = (1 - F(t))^a$, for some distribution function $F(t)$. [M. Finkelstein, *Failure Rate Modelling for Reliability and Risk*. Springer, 2008]

Frank distribution A bivariate distribution from a one-parameter family that has the joint distribution on $[0,1]^2 : F(x,y) = \log_a\{1 + (a^x - 1)(a^y - 1)/(a - 1)\}$, $a \neq 1$, where the base of the logarithm is the parameter a. This yields a family of bivariate distributions having prescribed marginals $F_X(x)$ and $F_Y(y)$ in the form $F = \log_a\{1 + (a^{F_X(x)} - 1)(a^{F_Y(y)} - 1)/(a - 1)\}$, $a \neq 1$. [C. Genest, Frank's family of bivariate distributions, *Biometrika*, 74(3), 549–555, 1987]

Fréchet distribution *see* extreme value distributions

Fréchet expectation Let X be a random element in a space S, equipped with a metric (or pseudometric) d, such that $\mathbb{E}\{d(X,y)^2\} < \infty$ for at least one $y \in S$. Then the Fréchet expectation of X is the set $\phi = \{y : \mathbb{E}\{d(X,y)^2\} = \inf_x \mathbb{E}\{d(X,x)^2\}\}$. If S is a Hilbert space, then ϕ is a unique element of S. [M. Fréchet, Les éléments aléatoires de nature quelconque dans un espace distancié, *Ann. Inst. H. Poincaré*, 10, 215–310, 1948]

Fréchet–Hoeffding bounds (alias: **Hoeffding–Fréchet bounds**) Arise in the context of multivariate joint distributions and copulas. For the basic form, let $F(x,y)$ be a bivariate joint distribution function with marginal distribution functions $F_X(x)$ and $F_Y(y)$. Then the Fréchet–Hoeffding bounds are $\max\{F_X(x) + F_Y(y) - 1, 0\} \leq F(x,y) \leq \min\{F_X(x), F_Y(y)\}$.

A related inequality, often called the Hoeffding–Fréchet bounds, asserts that for a r.v. U uniformly distributed on (0,1), $\mathbb{E}\{F_X^{-1}(1 - U)F_Y^{-1}(U)\} \leq \mathbb{E}(XY) \leq \mathbb{E}\{F_X^{-1}(U)F_Y^{-1}(U)\}$, when the expectations exist, and where $F_X^{-1}(\cdot)$ and $F_Y^{-1}(\cdot)$ are the right-continuous inverse functions of F_X and F_Y.

In terms of a copula $C(x,y)$, being a joint distn with uniform marginals, we have $\max\{x + y - 1, 0\} \leq C(x,y) \leq \min\{x,y\}$; and in n dimensions

$$\max\left\{\sum_{r=1}^{n} x_r + n - 1, 0\right\} \leq C(x_1, \ldots, x_n) \leq \min\{x_1, \ldots, x_n\}.$$

[R. Nelsen, *Introduction to Copulas*, 2nd edn. Springer, 2006; C. Burgert & L. Rüschendorf, On the optimal risk allocation problem, *Stat. Decisions*, 24(1), 153–171, 2006]

Fréchet–Shohat theorem (alias: **method of moments; moment convergence theorem**) Arises in the context of the so-called second-limit theorem, which seeks to deduce the convergence of a sequence of distributions from the convergence of the moments of the sequence of distributions. The following result, being a special case of their more general second-limit theorem, is usually what is meant by the Fréchet–Shohat theorem. Let $F_n(x), n = 1, 2, \ldots$ be a sequence of distributions with moments $m_n^{(r)} = \int_{-\infty}^{\infty} x^r \mathrm{d}F_n(x), \quad r = 0, 1, \ldots$. If $\lim_{n\to\infty} m_n^{(r)} = m^{(r)}$ exists for all $r = 0, 1, \ldots$, then there is at least one distribution $F(x)$ having $m^{(r)}$ as its moments, and a subsequence of $(F_n(x))$ converges to $F(x)$ at all continuity points of $F(x)$. If $F(x)$ is uniquely determined by $m^{(r)}$, then $\lim_{n\to\infty} F_n(x) = F(x)$, at all continuity points of $F(x)$. [M. Fréchet & J. Shohat, A proof of the generalized second-limit theorem, *Trans. Amer. Math. Soc.*, 33(2), 533–543, 1931]

Freedman's inequality Arises in the context of a martingale M_n whose differences almost surely satisfy $|\Delta M_k| = |D_k| \le c > 0, 1 \le k \le n$. Then, for $x, y > 0$,

$$\mathbb{P}\Big(M_n \ge x, \sum_{k=1}^{n} \mathbb{E}(D_k^2 \,|\, M_1, \ldots, M_{k-1}) \le y\Big)$$
$$\le \exp\Big(-\frac{1}{2}x^2/(y + cx)\Big).$$

The level crossing times of such martingales are hence shown to have approximately the same distribution as those of Brownian motion. [D. Freedman, On tail probabilities for martingales, *Ann. Probab.*, 3(1), 100–118, 1975]

Freedman's theorem Arises in considering an irreducible recurrent Markov chain $X(t)$ on a countable state space, S. It asserts that for an increasing sequence of finite subsets S_n of S, whose union is S, the sequence of chains $X_n(t)$, obtained by watching $X(t)$ while it is in S_n for each n, converges to $X(t)$ as $S_n \to S$, as do the sequences of Q-matrices and transition functions of X_n, to those of $X(t)$. [D. Freedman, *Approximately Countable Markov Chains*. Holden Day, 1972]

free entropy Arises in free probability as the analogue of Shannon's entropy in mathematical probability. It is additive for free r.vs that are freely independent, and for a single free r.v. X with distribution $\mathbb{P}(\cdot)$ it is given by

$$H(X) = \int\int \log|x - y| \mathbb{P}(\mathrm{d}x)\mathbb{P}(\mathrm{d}y) + \frac{1}{2}\log 2\pi + \frac{3}{4}.$$

[D. Voiculescu, Free entropy, *Bull. Lond. Math. Soc.*, 34(3), 257–278, 2002]

free Markov field (alias: **Nelson's free (Markov) field**) A classic example of a Gaussian Markov random field with applications in quantum field theory. [E. Dynkin, Markov processes and random fields, *Bull. Amer. Math. Soc.*, 3, 975–999, 1980]

free lunch A concept arising in the fundamental theorems of asset pricing, which include results establishing a necessary and sufficient condition for the existence of an equivalent martingale measure, in terms of a feasible stock portfolio strategy. If $S(t)$ is the stock price (a real-valued bounded semi-martingale) and $H(t)$ is a suitable function that represents an admissible strategy for trading in the stock (and the relevant bond or numeraire), then the final pay-off is the stochastic integral $p = \int_0^\infty H(t)\mathrm{d}S(t)$. The no free lunch condition with vanishing risk (or NFLVR) requires that there must be no sequence of strategies $H_n, n \ge 1$, such that as $n \to \infty$, $p_n = \int_0^\infty H_n \mathrm{d}S \to p_\infty \in [0, \infty]$, with $\mathbb{P}(p_\infty > 0) > 0$; while also $p_n \ge -a_n$ for all n, where $a_n > 0$ converges uniformly to 0 as $n \to \infty$. The theorem asserts that S obeys NFLVR iff there is an equivalent martingale measure for S. [F. Delbaen & W. Schachermeyer, A general version of the fundamental theorem of asset pricing, *Math. Ann.*, 300, 463–520, 1994]

free probability A type of probability theory that deals not with real-valued random variables, but with random objects comprising a possibly non-commutative algebra A including a unit element u. This algebra is equipped with a linear expectation functional $\mathbb{E}$ such that $\mathbb{E}u = u$; and a property of free independence, defined by $\mathbb{E}$, for subalgebras of A. These properties arise naturally in the context of suitable $n \times n$ random matrices, as $n \to \infty$; and in quantum mechanics. *See* quantum probability. [D. Voiculescu *et al.*, *Free Random Variables*. American Mathematical Society, 1992]

Freidlin–Wentzell theorem (alias: **F–Venttsel theorem; F–Ventcel theorem; F–Ventzell theorem**) Proves a large deviations principle for a suitable Itô diffusion process $X(t)$, satisfying (for $\varepsilon > 0$, and suitable $a(\cdot)$), $\mathrm{d}X(t) = a(X(t))\mathrm{d}t + \sqrt{\varepsilon}\mathrm{d}W(t), \ t \ge 0$, which is seen as a random perturbation of the so-called dynamical system $\dot{x} = a(x(t))$. *See* Schilder's theorem. [M. Freidlin & A. Wentzell, *Random Perturbations of Dynamical Systems*, 2nd edn. Springer, 1998]

Freivalds's technique (alias: **Freivalds's matrix multiplication checker**) A randomized method, which is an example of a so-called fingerprinting technique, that decides whether or not $AB = C$, where A, B and C are given $n \times n$ matrices. The checker takes a random $\{0,1\}^n$-valued vector $\mathbf{x}$, uniformly distributed on the 2^n possibilities, and computes $f = (AB - C)\mathbf{x}$. If $f = 0$, the checker asserts that $AB = C$, otherwise it asserts that $AB \neq 0$. If $AB = C$, the assertion is correct with probability 1, if $AB \neq 0$ it is correct with probability $p \geq 1/2$. [R. Motwani & P. Raghavan, *Randomized Algorithms*. Cambridge University Press, 1995]

frequency function *see* density

frequency probability (alias: **frequentist theory of probability; statistical probability**) An interpretation of probability that derives it from proportion or relative frequency; that view arises naturally from the empirical observation of the stability of statistical ratios. Consequently, the relative frequency of an event in a sufficiently long sequence of trials is taken as an approximation to the probability of the event occurring at any given future trial. It is tempting to seek to define probability as a limit of such relative frequencies, but this steps beyond the bounds of empiricism. The frequency interpretation competes with subjective probability (which sees probability as a feeling of the mind). An early proponent was John Venn. [J. Venn, *The Logic of Chance*, 3rd edn. Macmillan, 1858; J. von Plato, *Creating Modern Probability*. Cambridge University Press, 1994]

Freund's distribution A bivariate exponential distribution that arises constructively in the context of reliability, thus: two components A and B have lifetimes that are exponentially distributed, with respective parameters λ and μ, and independent except that if A is first to fail the residual lifetime of B is $\text{Exp}(b)$, while if B fails first the residual lifetime of A is $\text{Exp}(a)$. Then the joint mgf of the total lifetimes is

$$M(s,t) = \frac{1}{\lambda + \mu - s - t}\left\{\frac{a\mu}{a-s} + \frac{\lambda b}{b-t}\right\}.$$

[J. Freund, A bivariate extension of the exponential distribution, *J. Amer. Stat. Ass.*, 56, 971–977, 1961]

Friedman's urn (alias: **generalized Pólya urn**) A classic urn model that has been much generalized in the context of adaptive sequential decisions. In its original form the urn contains R_n red balls and B_n blue balls at time n. At each integer time $n \geq 0$ one ball is drawn at random and replaced in the urn together with $a > 0$ balls of the same colour as the drawn ball, and $b > 0$ balls of the other colour. The asymptotic behaviour of R_n as $n \to \infty$ is strikingly different, depending on whether $\frac{a-b}{a+b}$ is > 0, < 0, or $= 0$. [D. Freedman, Bernard Friedman's urn, *Ann. Math. Stat.*, 36(3), 956–970, 1965]

frog model A percolation model arising in statistical mechanics. Informally, at any time $n \geq 0$ each vertex of a graph is occupied by random number of active and inactive frogs. Active frogs perform a random walk on the graph, waking inactive frogs (who become active) on arrival at any vertex. Frogs may be assumed to have random lifetimes, at the conclusion of which they are removed. *See also* activated random walk. [L. Fortes *et al.*, The critical probability for the frog model, *J. Appl Probab.* 41, 292–298, 2004]

Frostman's lemma Arises in the context of random sets, particularly fractals. Let A be a closed subset of $\mathbb{R}^d$, and let $H_a(A) > 0$ be its a-dimensional Hausdorff measure. Then there is a probability measure $\mathbb{P}(\cdot)$, and a positive constant c, such that for all Borel sets B, $\mathbb{P}(B) \leq c|B|^a$, where $|B|$ is the diameter of B. [P. Mörters & Y. Peres, *Brownian Motion*. Cambridge University Press, 2010]

Fubini's theorem A classic theorem relating repeated integrals to a corresponding multiple integral, which has the following representation in probabilistic terms: let $Z = Z(\omega, \lambda)$ be either suitably measurable and non-negative, or integrable with respect to the product measure $\pi = \mathbb{P} \times \mathbb{Q}$, on the usual product space of the probability spaces $(\Omega, \mathcal{F}, \mathbb{P})$ and $(\Lambda, \mathcal{G}, \mathbb{Q})$. Then

$$\begin{aligned}\mathbb{E}Z &= \int_{\mathbb{P}\times\mathbb{Q}} Z(\omega,\lambda)\pi(\mathrm{d}\omega,\mathrm{d}\lambda) \\ &= \int_\Omega \mathbb{P}(\mathrm{d}\omega)\int_\Lambda Z(\omega,\lambda)\mathbb{Q}(\mathrm{d}\lambda) = \int_\Lambda \mathbb{Q}(\mathrm{d}\lambda)\int_\Omega Z(\omega,\lambda)\mathbb{P}(\mathrm{d}\omega).\end{aligned}$$

The version of the theorem assuming only non-negativity may be called **Tonelli's thm** or the Fubini–Tonelli thm. The equivalent result for the product of countably many measures and probability spaces is called the **Fubini–Jessen thm.**

fugacity (alias: **activity**) A type of parameter appearing in a class of models arising in statistical mechanics. For example, let the vertices of a graph V be occupied or not, denoted by $\{0,1\}$, and let the probability $\pi(\boldsymbol{\omega})$ of a feasible configuration $\boldsymbol{\omega} = (\omega_1, \ldots, \omega_{|V|}) \in \Omega = \{0,1\}^V$ take the form $\pi(\boldsymbol{\omega}) \propto \prod_{r\in V}\lambda_r^{\omega_r}$, for numbers $\lambda_r > 0$. Then λ_r is called the fugacity (or activity) at the

rth vertex. In the special case, when $\lambda_r = \lambda, \pi(\boldsymbol{\omega}) \propto \lambda^{n(\boldsymbol{\omega})}$, where n is the number of occupied vertices in the feasible configuration $\boldsymbol{\omega}$. [J. van der Berg & J. Steif, Percolation and the hardcore lattice gas model, *Stoch. Proc. Appl.*, 49, 179–197, 1994]

Fuk–Nagaev inequalities Supply bounds for the probability in the tail of the distribution of sum $S_n = \sum_{r=1}^{n} X_r$ of independent, but not necessarily identically distributed, r.vs $X_1, X_2, \ldots$, in the form

$$\mathbb{P}(S_n \geq x) \leq \sum_{r=1}^{n} \mathbb{P}(X_r \geq y_r) + A_t(x, y_1, \ldots, y_n),$$

where $x > 0, y_r > 0, 1 \leq r \leq n$, and A_t is a function of the truncated moments of order t of $X_1, \ldots, X_n$, taking different forms according as $0 < t \leq 1, 1 \leq t \leq 2$ or $t \geq 2$. [D. Fuk & S. Nagaev, Probability inequalities for sums of independent random variables, *Theor. Probab. Appl.*, 16(4), 643–660, 1971]

Fukushima's decomposition A decomposition arising in the context of suitable functions of Markov processes associated with some Dirichlet forms, which represents the function uniquely as the sum of two additive functionals, one of which is a martingale. It can be seen as analogous to the Doob–Meyer decomposition for semi-martingales. [M. Fukushima *et al.*, *Dirichlet Forms and Symmetric Markov Processes*, 2nd edn. Walter de Gruyter, 2011]

full information best-choice problem *see* Gilbert–Mosteller problem

functional Broadly, a function $f(\cdot)$ whose domain Φ is a set of functions. Such functions may be real-(or complex-)valued, or take values in some set of functions other than Φ, and are typically required to satisfy a measurability condition. Usually Φ is some function space, which is to say that it is a vector or topological space, or both, such spaces typically comprise functions required to satisfy some attractive constraints such as: the nature of their support, continuity, smoothness, differentiability, limiting behaviour, and so on. The number of function spaces is unbounded, but popular examples include:

Besov space – which generalizes Sobolev space;
BMO space – with functions of bounded mean oscillation;
Hölder space – whose functions have suitably continuous derivatives;
L_p space – having finite moments of order p;
Orlicz space – generalizing L_p by replacing the functions x^p by Young functions;
Schwartz space – having rapidly decreasing derivatives;
Skorokhod space – of rcll functions;
Sobolev space – whose derivatives have suitably finite moments;
VMO space – with functions of vanishing mean oscillation;
Zygmund space – with a Zygmund norm for suitably continuous fns.

Functionals arise naturally in the applications of probability to analysis; and, importantly in the theory of random processes, where the sample paths of the process comprise Φ. Examples of such functionals include: hitting probabilities and densities, stochastic integrals, and other integrals, such as that in the characteristic functional of a random process $X(t)$; $\phi(\theta(\cdot)) = \mathbb{E}\exp\{\mathrm{i}\int_a^b X(t)\theta(t)\mathrm{d}t\}$, defined for a suitable set of $\theta(t)$ such that the integral exists. [A. Bobrowski, *Functional Analysis for Probability and Stochastic Processes*. Cambridge University Press, 2005]

functional CLT *see* Donsker's theorem

fundamental martingale Arises in the context of a fractional Brownian motion $Z(t)$, with Hurst (or self-similarity) parameter $H \in (0, 1)$, where the process $X(t)$ given by $X(t) = \int_0^t k(t,s)\mathrm{d}Z(s)$, where $k(t,s) = \{2H\beta(\frac{3}{2} - H, H + \frac{1}{2})\}s^{\frac{1}{2}-H}(t-s)^{\frac{1}{2}-H}$ (and $\beta(\cdot,\cdot)$ is the beta-function), is an independent increments process called the fundamental martingale. [I. Norros *et al.*, An elementary approach to a Girsanov formula, and fractional Brownian motions, *Bernoulli*, 5(4), 571–587, 1999]

fundamental matrices (alias: **(discrete) Green functions**) Two distinct matrices that are useful in the theory of Markov chains:

(a) Let P be the transition matrix of a Markov chain with countable state space S that has at least one absorbing or recurrent state, and at least one transient state (such a chain is called absorbing by some authors, or transient by others). The matrix $G = \sum_{n=0}^{\infty} P^n$ is called the Green (or potential), matrix of the chain; and then the restriction of G to the set of transient states is called the fundamental matrix N. Writing P as $P = \begin{pmatrix} A & O \\ B & Q \end{pmatrix}$, where Q denotes transitions among transient states, this means $N = \sum_{n=0}^{\infty} Q^n$.

(b) Let P be the transition matrix of a non-null irreducible recurrent (i.e. ergodic) chain, having stationary distribution π and let Π be a matrix all of whose rows are π. Then the fundamental matrix is $Z = \sum_{n=0}^{\infty}(P - \Pi)^n$, so that $z_{ij} = \sum_{r=1}^{\infty}(p_{ij}(n) - \pi_j)$. (Note the convention that $M^0 = I$, the identity matrix, for a square matrix M.)

[C. Grinstead & J. Snell, *Introduction to Probability*, 2nd edn. American Mathematical Society, 1988]

fundamental probability set (alias: **reference set; FPS**) The smallest (i.e. minimal) algebra $\mathcal{A}$ of events (in some given sample space Ω) that is sufficient to contain all the events whose probability it may be required to know. Sometimes, by extension, the FPS is deemed to include the actual probability measure $\mathbb{P}(\cdot)$ on the algebra $\mathcal{A}$. The term FPS is sometimes used as a synonym for sample space, but this seems pointless. [J. Neyman, *Lectures and Conferences on Mathematical Statistics and Probability*, 2nd edn. USDA, 1952]

fundamental theorem of stochastic integration *see* Bichteler–Dellacherie theorem

fundamental theorems of asset pricing Arise in the classic presentation of mathematical finance addressed to asset pricing, where the term denotes a theorem that relates real world constraints and attributes of markets, such as no-arbitrage (or no free lunch) and completeness, to technical conditions in mathematical models, such as the existence and uniqueness of an equivalent martingale measure for the price process. Broadly, under suitable conditions, the first fundamental theorem is that which asserts that the market is arbitrage free iff there exists at least one equivalent martingale measure, and the second fundamental theorem asserts that (in the absence of such arbitrage) the market is complete iff the equivalent martingale measure is unique. Forms of these theorems are known under the title Harrison–Kreps–Pliska thm, Dalang–Morton–Willinger thm, and Delbaen–Schachermeyer thm; forms of the second theorem may be known under the title Harrison–Pliska thm, among others. [M. Musiela & M. Rutkowski, *Martingale Methods in Financial Modelling*, 2nd edn. Springer, 2005; F. Delbaen & W. Schachermeyer, The fundamental theorem of asset pricing for unbounded stochastic process, *Math. Ann.*, 312(2), 215–250, 1998]

Furry process *see* birth process

Furstenberg–Kesten theorem Arises in the context of a stationary sequence $X_1, X_2, \ldots$, taking values in the set of $r \times r$ matrices. It establishes that the sequence $n^{-1}\log\|Y_n\|$, where $\|Y_n\| = \|X_n, X_{n-1}, \ldots, X_1\|$ converges (under very general conditions, for the matrix norm $\|A\| = \max_i \sum_j |a_{ij}|$), both a.s. and in mean.

[H. Furstenberg & H. Kesten, Products of random matrices, *Ann. Math. Stat.*, 31(2), 457–469, 1960]

Fussell–Vesely importance *see* importance measure

future-independent *see* causal

fuzzy measure *see* Choquet capacity

G

Gale's roulette *see* Efron's dice; non-transitivity

Gál–Koksma theorem (alias: **Gaal–Koksma theorem**) A strong law of large numbers for a sequence $X_1, X_2, \ldots$, of weakly dependent random variables. One form is this: suppose that for $1 < p < \infty$, and some real-valued function $f(\cdot)$, we have for any $m \geq 1$, $\mathbb{E}[|\sum_{r=m}^{n+m} X_r|^p] \leq f(n)$; where $f(n)n^{-1-p}$ is monotone decreasing in n, and $\sum_{n=0}^{\infty} f(n)n^{-1-p} < \infty$. Then $\frac{1}{n}\sum_{r=1}^{n} X_r \overset{\text{a.s.}}{\to} 0$, as $n \to \infty$. [S. Chobanyan, S. Levental, & H. Salchi, Strong law of large numbers under a general moment condition, *Electr. Commun. Probab.*, 10, 218–222, 2005]

Gallager's constant Arises in communication theory, where it appears in an upper bound for the redundancy in both a compact binary prefix code for a uniformly distributed source, and in a compact binary Huffman coding of a source with arbitrary distribution. It is $g = 1 + \log_2\log_2 e - \log_2 e \simeq 0.086\,071$. [R. Gallager, Variations on a theme by Huffman, *IEEE Trans. Inf. Theor.*, 24(6), 668–674, 1978]

Gallot–Kounias inequalities *see* Kounias inequality

Galmarino's test Arises in the context of stopping (Markov) times for a continuous random process $X(t, \omega), t \in \mathbb{R}^+, \omega \in \Omega$, with natural filtration $\mathcal{F}_t$. Then $T(\omega), \Omega \to [0, \infty]$, is an $\mathcal{F}_t$ stopping time iff, for any $\lambda, \omega \in \Omega, T(\omega) = t$ together with $X(s, \omega) = X(s, \lambda)$ for all $s \leq t$ implies that $T(\lambda) = t$. Roughly speaking, T is an $\mathcal{F}_t$ stopping time iff it depends only on the present and the past. An extension asserts that T is an $\mathcal{F}_{t+}$ stopping time iff it may also depend on the 'infinitesimally near future'. [K. Itô & H. McKean, *Diffusion Processes and their Sample Paths.* Springer, 1965]

Galton's paradox An example of the errors that can arise when attempting short cuts in the solution of problems of chance. The problem in question is the chance of three flipped fair coins all showing the same (all heads or all tails). An erroneous argument asserts that because two coins must show the same, and it is an even chance if the third is heads or tails, the required probability is 1/2; the correct chance being 1/4. [F. Galton, A plausible paradox in chances. *Nature*, 49(1268), 365–366, 1894]

Galton–Watson process *see* branching process

gambler's fallacy (alias: **Monte Carlo fallacy**) Any of a number of incorrect assessments of probabilities or expectations, made in the special context of gaming. (There are obvious equivalents in economics and finance.) The error most often given the name 'Gambler's fallacy' is the belief that, in a sequence of demonstrably independent similar trials, the relative non-occurrence of an event over a period tends to increase its probability of occurrence in the future. The fallacious statement often includes the words, 'by the law of averages'. Another common fallacy is the belief that a feasible system of gambling can transform a disadvantageous game into an advantageous game; see fairness and martingale transform. [E. Poulton, *Behavioural Decision Theory.* Cambridge University Press, 1994]

gambler's ruin problem (alias: **duration of play problem**) A classic problem in probability, first proposed and solved by B. Pascal in 1656. It was then solved by P. Fermat in the same year, and then also by C. Huygens, who included it as the fifth of his problems in his book *De Ratiociniis in Ludo Aleae*. In its standard form, two gamblers A and B make a series of wagers in i. i.d. trials. At any trial A (resp. B) wins a unit from B (resp. A) with probability p (resp. $1 - p = q$), or loses a unit with probability $1 - p$ (resp. p); $0 < p < 1$. Initially A has a units and B has b units; if either player is reduced to 0 units, that player is said to be ruined. Hence the gambler's ruin problem asks for the probability of ruin for each player. More precisely, it asks for the probability that A (resp. B) is ruined at the nth trial, and for the expected time until ruin occurs (either unconditionally, or conditional on the event that A (say) is ruined). This may be called the problem of the duration of play. [A. Hald, *History of Probability and Statistics, and their Applications before 1750.* Wiley, 1990]

gambler's rule Arises when you play a sequence of independent games with probability n^{-1} of winning each, and ask how many games m it is necessary to play until the chance of enjoying at least one win exceeds 1/2. The gambler's rule asserts that m is approximately $\frac{2}{3}n$. In fact more precisely; for $n \leq 27$, $m = \lceil \frac{2}{3}n \rceil$ = the integer ceiling of $\frac{2}{3}n$; and as $n \to \infty, m \sim n \log 2 \simeq 0.693n$. [J. Pitman, *Probability*. Springer, 1993]

game theory Broadly, game theory can be seen as an extension (or type) of decision theory, in which there are two or more decision-makers (called players, or agents), capable of formulating strategies for taking a number (or sequence) of actions, whose results (or pay-offs) may affect any of those in the game. There is a very extensive taxonomy of games, based on, for example: whether the players cooperate or not; the number of players; whether the game is zero-sum (competitive) or not; whether it is symmetrical between the players or not; whether it has a random or chance input or/and outcomes (or not); is one-off, or sequential; and on how much each player knows about the other players and outcomes of actions. An important link between game theory and probability arises from the fact that optimal strategies are quite often necessarily randomized strategies. The subject is extremely fertile of paradoxes, e.g. the Battle of the sexes; the Prisoners' dilemma; the Tragedy of the commons; and the Volunteer's dilemma; among many more. [J. Webb, *Game Theory*. Springer, 2007]

game tree *see* decision tree; influence diagram

gamma distribution (alias: **Pearson's type III distribution**) A distribution on the positive real axis, having a density that in standard form is $f(x) = \frac{1}{\Gamma(\tau)} x^{\tau-1} e^{-x}$, $x > 0$, where $\tau > 0$ is called the shape parameter, and $\Gamma(\tau)$ is the gamma function. It can be seen as the continuous counterpart of the negative binomial distn, and arises in a number of statistical and other applications. It is infinitely divisible. More generally, with scale parameter λ, it has chf $\phi(t) = \left(\frac{\lambda}{\lambda - it}\right)^{\tau}$, with mean τ/λ and variance τ/λ^2. Special cases include:

(a) τ is an integer n, in which case it arises as a sum of n i.i.d. Exp(λ) r.vs, and is also called the Erlang distribution;

(b) $\tau = n/2$ and $\lambda = \frac{1}{2}$, in which case it arises as a sum of squares of i.i.d. $N(0, 1)$ r.vs, and is also called the chi-squared distn.

A density that may be called a generalized gamma is, in a standard form, $f(x) = bx^{\tau-1}e^{-x^b}/\Gamma(\tau/b)$, $x > 0, b > 0, \tau > 0$.

A multivariate gamma density may be specified very generally as any whose marginal distributions all have a gamma density; or, more narrowly, as a density whose multivariate characteristic function is of the form $\phi(t_1, \ldots, t_d) = \phi(\mathbf{t}) = \{1 + A(\mathbf{t})\}^{-\tau}$, where $\frac{\partial^2 A}{\partial t_r^2} = 0$ for all r (which is to say that $A(\mathbf{t})$ is an affine polynomial in $t_1, \ldots, t_d$), for suitable values of the coefficients in A. For example, $(1 - at_1 - bt_2 + ct_1t_2)^{-\tau}$ is the mgf of such a probability density if $a > 0$, $b > 0$, and $ab > c > 0$. [P. Bernardoff, Which multivariate gamma distns are infinitely divisible?, *Bernoulli*, 12(1), 169–189, 2006]

gamma martingale Arises as a generalization of the idea of continuous local martingale, as a stochastic process taking values in a suitable Riemannian manifold. [M. Emery, *Stochastic Calculus in Manifolds* (with an appendix by P. Meyer). Springer, 1989]

gamma process *see* Lévy process; Moran gamma process

gap diffusion *see* quasi-diffusion process

Garsia-Neveu inequality (alias: **norm inequality**) Extends the Burkholder–Davis–Gundy inequality as follows: let $X(t)$ be a continuous (or, more generally either right- or left-continuous) increasing predictable process (for a filtration $\mathcal{F}_t$); and let there exist $Z \geq 0$ such that with probability 1, $\mathbb{E}[X(\infty) - X(t)|\mathcal{F}_t] \leq \mathbb{E}[Z|\mathcal{F}_t]$, $t \geq 0$. Then $\{\mathbb{E}|X(\infty)|^r\}^{1/r} \leq r\{\mathbb{E}|Z|^r\}^{1/r}$, $r \geq 1$. [O. Kallenberg, *Foundations of Modern Probability*, 2nd edn. Springer, 2002]

Garsia–Rodemich–Rumsey inequality An inequality for certain functions of a real variable that has applications to random processes, in particular, to Gaussian processes. Let $g(u)$ and $p(u)$ be non-negative and even on $\mathbb{R}$ and $[-1, 1]$ respectively; $p(u)$ is continuous and non-decreasing for $u \geq 0$, with $p(0) = 0$; $g(u)$ is non-decreasing for $u \geq 0$, with $g(\infty) = \infty$. Then for any given continuous function $f(x), x \in [0, 1]$, such that

$$\int_0^1 \int_0^1 g\left[\frac{f(x) - f(y)}{p(x - y)}\right] dx dy \leq c < \infty,$$

we have

$$|f(s)-f(t)| \le 8\int_0^{|s-t|} g^{-1}(4cu^{-2})\mathrm{d}p(u).$$

[A. Garsia *et al.*, A real variable lemma and the continuity of paths of some Gaussian processes, *Indiana Univ. Math. J.*, 1971, 20(6), 565–578]

Gärtner–Ellis theorem Arises in the theory of large deviations, where it establishes a large-deviations principle for a suitable sequence of r.vs $X_n, n \ge 1$, without making explicit assumptions about its dependency structure; the theorem rests on the existence and sufficiently good behaviour of $\phi(\theta) = \lim_{n\to\infty} a_n^{-1} \log(\mathbb{E}\exp(\theta X_n))$, where $a_n \uparrow \infty$ is an appropriate sequence of constants (e.g. $a_n = n$). [A. Dembo & O. Zeitouni, *Large deviations techniques and applications*, 2nd edn. Springer, 2009]

Gauss–Camp–Meidell inequality *see* Vysochanskij–Petunin inequality

Gauss' characterization of the normal distribution Asserts that (under suitable conditions), the maximum likelihood estimator of a location parameter for a density f on $\mathbb{R}$ coincides with the mean of a random sample from f iff f has a normal distribution. [A. Azzalini & M. Genton, On Gauss' characterization of the normal distribution, *Bernoulli*, 13(1), 169–174, 2007]

Gauss' distribution (or law) *see* normal distribution

Gaussian channel *see* Shannon–Hartley formula

Gaussian coefficient *see* Brownian coefficient

Gaussian ensemble *see* random matrix

Gaussian free field Arises in statistical mechanics as a model for random surfaces, such as the interface between two phases. Broadly speaking, it is a Gaussian process whose covariance structure is determined by the Green function generated by a suitable Markov transition function, e.g. by the step probabilities of a random walk on $\mathbb{Z}^2$, (say). [S. Sheffield, Gaussian free fields for mathematicians, *Probab. Theor. Rel. Fields*, 139(3–4), 521–541, 2007]

Gaussian isoperimetric inequality Let $\phi_n(\boldsymbol{x}) = \prod_1^n \phi(x_r)$ be the standard normal product measure on $\mathbb{R}^n$, where ϕ is the standard normal density with cdf $\Phi(x)$, and define the t-fattening (or blow up) $A(t) = \{x \in \mathbb{R}^n : x$ is within distance t of some point of $A\}$ for any $A \subseteq \mathbb{R}^n$. Then $\int_{A(t)} \phi_n(d\mathbf{x}) \ge \Phi\big(t + \Phi^{-1}\big(\int_A \phi_n(d\mathbf{x})\big)\big)$, with equality if A is a half-space. It may be called the **Borell–Sudakov–Tsirelson inequality**. [S. Boucheron *et al.*, *Concentration Inequalities*, Oxford University Press, 2013]

Gaussian orthogonal ensemble *see* random matrix

Gaussian (or normal) process A random process $X(t), t \in T$, is said to be Gaussian if $\sum_{r=1}^n a_r X(t_r)$ is normally distributed for any choice of $n \in \mathbb{Z}^+, t_r \in T$, and $a_r \in \mathbb{R}$; or, equivalently, if the joint distn of any finite collection $X(t_r), 1 \le r \le n$, is multivariate normal. More generally, the process may be vector-valued; several processes may be jointly Gaussian, and $X(t)$ having two or more parameters may be called a Gaussian random field. More particularly, let $c(s, t)$ be a (possibly complex-valued) function such that $c(t_j, t_k), 1 \le j, k \le n$ is non-negative definite, and $c(s,t) = \overline{c(t,s)}$; then there is a zero-mean (possibly complex-valued) Gaussian process having $c(s, t)$ as its covariance function. Classic examples of Gaussian processes include Brownian motion (Wiener process) and the Ornstein–Uhlenbeck process, whose quadratic variation exists. Other Gaussian processes with this property include the **Baxter processes**, whose mean function has bounded first derivative, and whose covariance function is continuous in s and t, and has uniformly bounded second derivatives for $s \ne t$. A yet more general class of Gaussian processes with this property are the **Gladyshev processes**. [J. Doob, *Stochastic Processes*. Wiley, 1953; T. Hida & M. Hitsuda, *Gaussian Processes*. American Mathematical Society, 1993; M. Lifshits, *Lectures on Gaussian Processes*. Springer, 2012]

Gaussian scale mixture (alias: **normal scale mixture**) A class of random variables such that any member X can be represented as Z/V, where Z has a standard normal distribution and is independent of the random variable $V > 0$. Equivalently, the r. v. X has a compound (or mixed) distribution of the form $\mathbb{P}(X \le x) = \int_0^\infty \Phi(vx)\mathrm{d}F_V(v), \ -\infty < x < \infty$, where F_V is the mixing distribution. For example, if $F_V(v)$ is the folded normal distn on $[0,\infty)$, then X is Cauchy; and if $\frac{1}{2}V^2$ has an exponential density then X has the two-sided (or double-) exponential (or Laplace) density. The idea extends naturally to the multivariate Gaussian scale mixture. [D. Andrews & C. Mallows, Scale mixtures of normal distributions, *J. Roy. Stat. Soc.*, 36(1), 99–102, 1974]

Gaussian unitary ensemble *see* random matrix

Gauss inequality Let X be a continuous random variable with a density having a single mode at m. Then for $t > 0$,

$$\mathbb{P}(|X - m| \geq t) \leq \frac{4}{9t^2}\mathbb{E}\left\{(X - m)^2\right\}$$
$$= \frac{4}{9t^2}\left\{\operatorname{var} X + (m - \mathbb{E}X)^2\right\}.$$

For $t^2 < \frac{4}{3}\mathbb{E}\{(X - m^2)\}$ this is improved by $\mathbb{P}(|X - m| \geq t) \leq 1 - t/\{3\mathbb{E}\{(X - m)^2\}\}^{\frac{1}{2}}$. The case $m = \mathbb{E}X$ is often seen, in which case $\mathbb{P}(|X - m| \geq t) \leq \frac{4}{9t^2}\operatorname{var} X$, for $t > 0$. *See* Vysochanskij–Petunin inequality. [F. Pukelsheim, The three sigma rule, *Amer. Stat.*, 48(2), 88–91, 1994]

Gauss–Kuzmin distribution Arises in the context of the continued-fraction expansion of a r.v. U distributed uniformly in $(0, 1)$, so that $U = 1/(X_1 + 1)/(X_2 + \cdots)$ where $X_r, r \geq 1$ are random positive integers. Writing $V_n = 1/(X_{n+1} + 1)/(X_{n+2} + \cdots)$, it was shown by C. Gauss. R. Kuzmin, and P. Lévy that $\lim_{n\to\infty}\mathbb{P}(V_n \leq x) = \log_2(1 + x)$, $0 \leq x \leq 1$, and hence that

$$\lim_{n\to\infty}\mathbb{P}(X_n = k) = \log_2\left(1 + \frac{1}{k(k+2)}\right)$$
$$= -\log_2\left(1 - \frac{1}{(k+1)^2}\right), \quad k \geq 1.$$

Either may be called the Gauss–Kuzmin distn. The Gauss–Kuzmin–Wirsing thm describes the rate of convergence of $\mathbb{P}(V_n \leq x)$ for each x. [E. Wirsing, On the theorem of Gauss–Kusmin–Lévy, *Acta Arithmetica*, 24, 507–528, 1974]

Gauss–Laplace distribution *see* normal distribution

Gauss–Markov process (alias: **Gaussian Markov process**) One that is both a Gaussian process, and possessed of the Markov property. *See* Doob's theorem.

Gauss–Poisson *see* Hermite distribution

Gauss–Poisson process A type of Poisson cluster process in which clusters have either one or two points, so that it comprises a process of correlated pairs of points. The name arises because, by analogy with a Gaussian process, it is characterized by its second moment density. Formally, the G-P process has probability generating functional $G(f) = \exp\{\int (f(x) - 1)\mathrm{d}\mathbb{P}(x) + \iint (f(x) - 1)(f(y) - 1)\mathrm{d}\mathbb{Q}(x, y)\}$, for suitable measures $\mathbb{P}$ and $\mathbb{Q}$ defined on spaces S and $S \times S$ respectively. [D. Daley & D. Vere-Jones, *Introduction to the Theory of Point Processes*, 2nd edn. Springer, Vol. II, 2008]

Gauss–Winckler inequality An inequality for the moments of a non-negative continuous random variable X having a density $f(x)$, that is monotone non-increasing for $x > 0$. One form asserts that for $-1 < r < s$, $\{(1 + r)\mathbb{E}(X^r)\}^{1/r} \leq \{(1 + s)\mathbb{E}(X^s)\}^{1/s}$. The case $r = 2$, $s = 4$, was stated without proof by C. Gauss, and a more general case (without complete proof) by A. Winckler. **Faber's inequality** asserts that, for $0 \leq r < s < t$,

$$((1 + s)\mathbb{E}X^s)^{t-r} \leq ((1 + r)\mathbb{E}X^r)^{t-s}\left((1 + t)\mathbb{E}X^t\right)^{s-r}.$$

The elementary extensions to densities with a single mode at the origin, and resulting bounds for $\mathbb{P}(|X| > t)$, have also been called Gauss–Winckler inequalities. *See* Camp–Meidell–Narumi inequality. [F. Avkhadiev, A simple proof of the Gauss–Winckler inequality, *Amer. Math. Mon.*, 112(5), 459–462, 2005]

Geary's theorem (alias: **Fisher–Geary theorem**) Supplies a characterization of the normal distribution, in terms of a random sample $X_1, \ldots, X_n$. Let $F(x)$ be the distribution function of the i.i.d.r.vs X_r, $1 \leq r \leq n$, and define

$$\overline{X} = \frac{1}{n}\sum_{r=1}^{n} X_r, \quad \text{and} \quad S^2 = \frac{1}{n}\sum_{r=1}^{n}(X_r - \overline{X})^2.$$

R. Fisher showed that if $F(x)$ has a normal distn, then $\overline{X}$ and S^2 are independent. R. Geary showed that if $F(x)$ has all its moments finite, then the independence of $\overline{X}$ and S^2 implies that $F(x)$ is normal. The assumption of finite moments was later much weakened by E. Lukacs and others. The joint distn of $\overline{X}$ and S^2 is called **Helmert's distn**. [Y. Wang, On characterization of certain probability distributions, *Proc. Camb. Phil. Soc.*, 71, 347–352, 1972]

Gebelein–Lancaster theorem Arises in the context of maximal correlation. Let X and Y be jointly normally distributed and define the random variables $U = u(X)$ and $V = v(X)$ for arbitrary (but suitable) functions $u(\cdot)$ and $v(\cdot)$. Then the correlation coefficients satisfy the so-called Gebelein inequality $\rho(U, V) \leq |\rho(X, Y)|$, in which equality holds (attained by the maximal correlation coefficient of X and Y) iff $u(X) = aX$ and $v(Y) = bY$ for constants a and b, such that $ab \neq 0$. [C. Klaassen & J. Wellner, Efficient estimation in the bivariate normal copula model, *Bernoulli*, 3(1), 55–77, 1997]

gelation model *see* Marcus–Lushnikov process

Gelenbe network (alias: **G-network; generalized queueing network**) A generalization of the standard network of queues (or Jackson network) which extends the field of applications of such models to neural networks and data-processing networks. The principal differences from the classical model lie in the introduction of the concepts of 'negative customers', and externally driven processes affecting the queue-discipline (i.e. the movement of customers between queues), called triggers. [E. Gelenbe, G-networks with triggered customer movement, *J. Appl. Probab.*, 30, 742–748, 1993]

Gelfand–Pinsker problem Arises in communication theory when one seeks the capacity of a channel with side information, i.e. a state-dependent channel. [S. Gelfand & M. Pinsker, Coding for a channel with random parameters, *Problem. Control Inf. Theor.*, 9(1), 19–31, 1980]

genealogical process Broadly, a class of processes arising in population genetics, and the mathematical models describing population genetics, that describe the genealogy (or line of descent, or ancestry) of any 'individual' (in the broadest sense of that term), or a random sample of individuals, in the population at time $t \geq 0$. Such processes effectively take values in the path space of the underlying evolutionary process. The term 'historical process' is often used in the context of general measure-valued superprocesses; while the term 'genealogical process' has been used for the Fleming–Viot probability-measure-valued process. [N. Bingham & C. Goldie (eds), *Probability and Mathematical Genetics* (selected chapters). Cambridge University Press, 2010]

generalized central limit theorem *see* central asymptotic theorem

generalized diffusion *see* quasi-diffusion process

generalized mean *see* Kolmogorov–Nagumo mean

generalized Pólya urn *see* Friedman's urn

generalized X *see* X (for any given value of X)

generating function (alias: **transform; generating series**) A device, appearing in several forms, that is often used in probability theory. Broadly, if a (possibly infinite) sequence of numbers $a_0, a_1, a_2, \ldots$ appears as the coefficients of a power series in an auxiliary symbol s, thus: $a(s) = a_0 + a_1 s + a_2 s^2 + \cdots = \sum_{r=0}^{\infty} a_r s^r$, then $a(s)$ is the (ordinary) generating function of (a_n). In some (e.g. combinatorial or number theoretic) applications, $a(s)$ is a formal power series, which is to say that it is an element in the usual ring of such series.

For applications in probability, s is generally assumed to be a complex number, so that $a(s)$ exists inside the radius of convergence R such that $R^{-1} = \limsup_{n\to\infty} |a_n|^{1/n} \in [0, \infty]$. In probability theory, commonly met generating functions (g.fs) include: the probability g.f. (or pgf); the moment g.f. (or mgf), the cumulant g.f. (or cgf); and the characteristic fn (or chf). Note that these may also be seen as transforms of the distribution function of a suitably defined random variable, e.g. the mgf as a Laplace transform, and chf as a Fourier transform.

More generally, if a_n, $-\infty < n < \infty$, is a doubly infinite sequence, its g.fn is a Laurent series for $s \in \mathbb{C}$, that may converge in an annulus. Furthermore, if a_{mn}, $-\infty < m, n < \infty$, is an array, the bivariate g.f. is $a(s, t) = \sum_{m,n} a_{mn} s^m t^n$, with a natural extension to multivariate g.fns. Yet more generally, if $W_n, n \geq 0$, is another sequence then $W(s) = \sum_0^{\infty} W_n a_n s^n$ is the weighted g.f. of (a_n) with weights (W_n). In particular, when $W_n = (n!)^{-1}$, $W(s)$ is called the exponential g.f. of (a_n) (and is also termed the Borel transform of $\sum_0^{\infty} a_n s^n$). Several other types of generating series (or transforms) exist, including e.g. the Lambert g.f. $L(s) = \sum_{r=1}^{\infty} \frac{a_r s^r}{1-s^r}$, and the Dirichlet g.f. $d(s) = \sum_{r=1}^{\infty} a_r r^{-s}$. *See also* universal generating function. [W. Feller, *Introduction to Probability Theory and its Applications*, Vol. 1, 3rd edn. Wiley, 1968; Vol. 2, 2nd edn. Wiley, 1971; P. Flajolet & R. Sedgewick, *Analytic Combinatorics*. Cambridge University Press, 2009]

generating functional Arises in the context of random processes in time and space, particularly point processes, as a function $G(f)$ of a sufficiently rich set of auxiliary (or test) functions f, where it plays a role equivalent to that of generating functions for random variables. Thus, for a stochastic process $X(t), t \in \mathbb{R}$, and a suitable function $f(\cdot)$, the characteristic functional is $\Phi(f) = \mathbb{E}\big(\exp\{\mathrm{i}\int f(t)X(t)\mathrm{d}t\}\big)$; or, for a point process $N(t)$, $\Phi(f) = \mathbb{E}\big(\exp\{\mathrm{i}\int f(t)\mathrm{d}N(t)\}\big)$. The Laplace functional is defined similarly, e.g. for a point process X, seen as a random countable subset S, of (say) $\mathbb{R}^d$, the

Laplace functional is $L(f) = \mathbb{E}\big(\exp\{-\sum_{X\in S} f(X)\}\big)$, for suitable real $f(\cdot)$. Likewise, the probability generating functional may be defined for $0 < f \leq 1$ as the Laplace functional of $-\log f$. Thus, e.g. for a point process $N(x)$ on $\mathbb{R}^+$,

$$\begin{aligned} G(f) &= \mathbb{E}\Big\{\exp\Big(\int_0^\infty \log f(t)\mathrm{d}N(t)\Big)\Big\} \\ &= \mathbb{E}\Big(\exp\sum_r \log f(X_r)\Big), \text{ where } X_r \text{ are the points of } N \\ &= \mathbb{E}\Big\{\prod_r f(X_r)\Big\}, \text{ when the expectation exists.} \end{aligned}$$

[D. Daley & D. Vere-Jones, *Introduction to the Theory of Point Processes*, 2nd edn. Springer, Vol. II, 2008]

generator (alias: **infinitesimal generator**) Arises in the analysis of Markov processes in continuous time, where it effectively takes over the role of the transition distribution (or kernel) in discrete time, in seeking to describe what happens 'next' in the evolution of the process when, with continuous parameter, there is no implicit unit of time. It takes the form of an operator G on a suitable set of functions $f(\cdot)$ on the state space S of the process. Informally, the operator G has the property that, as $h \downarrow 0$,

$$\mathbb{E}f(X(t+h)|X(t)=x) = f(x) + Gf(x)h + o(h).$$

For example, if $P(t)$ is the matrix of transition probabilities $p_{ij}(t)$ of a (right-continuous) Markov chain with finite state space, having Q-matrix Q, then $P(t)$ is a solution of the forward equations $\frac{dP}{dt} = P(t)Q$, with $P(0) = I$ (the identity matrix), and is given by $P(t) = \sum_{n=0}^{\infty} \frac{t^n}{n!} Q^n = \exp(tQ)$. Thus Q is the generator in this case. Under appropriate conditions on P and Q, this also holds on a countable state space.

More generally, the transition kernels of a Markov process define a collection of operators $U(t)$, on functions on the state space, that form a semi-group obeying $U(t+s) = U(s)\,U(t)$, for $s, t > 0$, because of the Markov property. For suitably well-behaved processes, there exists a generator of this semi-group that supplies the required description of the Markov process. For example, if $X(t)$ is continuous, with continuous state space, then G is an elliptic differential operator A; and in the classic case of Brownian motion $A = \frac{1}{2}\frac{\partial^2}{\partial x^2}$. In general, the generator is an unbounded linear operator, and the properties of the semi-group are established using the resolvent (also called λ-potential). [S. Karlin & H. Taylor, *A Second Course in Stochastic Processes*. Academic Press, 1981; R. Bass, *Stochastic Processes*. Cambridge University Press, 2011]

geometrical probability (alias: **geometric probability; stochastic geometry**) Broadly, the branch of probability in which the events of interest are most naturally described in terms of the properties and configurations of geometrical objects. Classically the underlying geometry is Euclidean, and the events are defined by random points, lines, circles, polygons, convex sets, etc., with their higher-dimensional generalizations. Famous problems include Buffon's needle, Sylvester's problem, and Crofton's theorem.

Both in the classical framework, and in other geometries, a key problem is to construct a suitable measure to describe a geometrical object picked at random. One natural procedure is to seek those measures that are invariant under the defining group of transformations; this is part of the subject of **integral geometry**. [W. Kendall & I. Molchanov, *New Perspectives in Stochastic Geometry*. Oxford University Press, 2010; L. Santalo, *Integral Geometry and Geometric Probability*, 2nd edn. Cambridge University Press, 2004]

geometric Brownian motion *see* Brownian motion

geometric distribution Broadly, a distribution whose terms comprise a geometric series. In particular, the distribution on the non-negative integers given by $\mathbb{P}(X=x) = p_X(x) = q^x p, x = 0, 1, 2, \ldots,$ $0 < p = 1 - q < 1$. It arises naturally as the number of flips of a biased coin up to but not including the first head H, where $\mathbb{P}(H) = p$. For any integer k, and geometric r.v. X, the r.v. $X + k$ has a geometric distn; in particular, for $Y = X + 1$,

$$\mathbb{P}(Y=y) = p_Y(y) = q^{y-1}p, y = 1, 2, \ldots,$$

which arises naturally as the number of flips of the coin up to and including the first head. Both $p_X(x)$ and $p_Y(y)$ have been called 'the geometric distribution'. Geometric distributions have the lack of memory property, and the maximum entropy property among discrete distributions on $\mathbb{Z}^+$ with given mean; it is also characterized by the property that for i.i.d. geometric r.vs X and Y, $\min\{X, Y\}$ and $\max\{X, Y\} - \min\{X, Y\}$ are independent; this is **Ferguson's lemma**.

The sum of i.i.d. geometric r.vs has a negative binomial distn, of which the geometric is thus a special case. The distribution of the number of flips required to first obtain k consecutive heads has been called the **order-k geometric distn**.

[T. Ferguson, A characterization of the geometric distn, *Amer. Math. Mon.*, 72, 256–260, 1965]

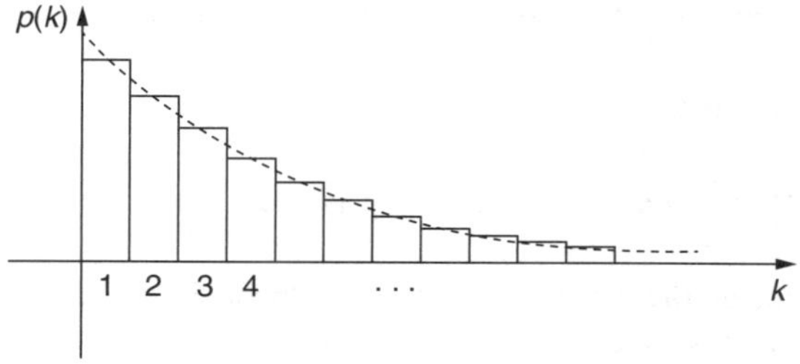

Histogram of a geometric distribution together with its continuous approximation. (*dashed line*).

geometric drift condition *see* drift condition

geometric ergodicity Arises for an irreducible Markov chain, with a stationary distribution π, when the distribution of the chain converges to π geometrically (or exponentially) fast. In the simplest case of a chain on a countable state space with transition probabilities $p_{ij}(n)$, i.e. for some $0 < \rho < 1$, and constants c_{ij}, $|p_{ij}(n) - \pi_j| \leq c_{ij}\rho^n$, for all n. The property may be characterized either in terms of a drift condition or in terms of Dobrushin's ergodicity coefficient. [D. Isaacson, A characterization of geometric ergodicity, *Probab. Theor. Rel. Fields*, 49, 267–273, 1979]

geometric expectation A term arising in the literature of financial mathematics where, for any random variable X such that $\mathbb{E}(\log X)$ exists, the quantity $\exp\{\mathbb{E}(\log X)\}$ may be called the geometric expectation of X. (Note: this is to be confused neither with the expectation of a geometric r.v. nor with concepts of expectation arising in geometrical probability.) [M. Bean, *Probability: The Science of Uncertainty*. American Mathematical Society, 2001]

geometric stable distribution The distribution F of a r.v. X is called geometric stable if, for a sequence of i.i.d.r.vs $X_r, r \geq 1$, with common distn F, there exist a geometric r.v. Y with parameter q, and $a(q) > 0$ and $b(q)$, such that as $q \to 0$ the distribution of $Z = a(q)\sum_{r=1}^{Y}(X_r + b(q))$ converges to F. If $b(q) = 0$, F may be called strictly geometric stable.

Examples include the exponential and Laplace distns; the chf of such a distn is of the form $\phi = \left(1 - \log \psi(t)\right)^{-1}$, where ψ is the chf of an α-stable distn. [S. Kotz *et al.*, *The Laplace Distribution and Generalizations*. Birkhäuser, 2001]

geometric Wiener process *see* Brownian Motion

germ-grain process (or model) *see* Boolean model

GEV *see* extreme value distributions

Gibbs distribution (alias: **Boltzmann–Gibbs (or Gibbs–Boltzmann) or Boltzmann distribution (or measure, or state, or ensemble)**) Arises in statistical mechanics in describing the thermodynamics of gases (Boltzmann's distribution), and more generally in statistical physics to describe energetic systems in equilibrium. J. W. Gibbs's underlying argument is that the probability of any configuration of the system should be proportional to a negative exponential function of the energy of that configuration. Thus, very broadly, a distribution of the form

$$f(x) = Z^{-1}\exp\{-E(x)/T\}, \quad x \in \Omega,$$

may be called a Gibbs distn, where Ω is the set of configurations, $E(x) \geq 0$ is called the energy function, T is called the temperature, and $Z = \sum_{x\in\Omega}\exp\{-E(x)/T\}$ is called the partition function. (Gibbs himself called the exponent $-E(x)/T$ the **index of probability**.)

More narrowly, the configuration space Ω is of the form V^S, where V is a set of sites, seen as comprising a graph or lattice, at each of which there is a r.v. taking values in S. A subset c of V is called a clique if any two sites of c are neighbours in V, and in many physical (and other) applications the energy function takes the form $\mathbb{E}(x) = \sum_c \phi_c(x)$, where the sum is over cliques of V and $\phi_c(x)$ is called a (clique) potential, and $E(x)$ is now called a Gibbs potential. Some authors reserve the name Gibbs distn to this case; then the Hammersley–Clifford theorem asserts that the r.vs taking values in V^S comprise a Markov random field. Extending this idea to the case when V (and possibly S) is infinite may be accomplished in various ways; one approach is to specify local conditional probabilities and seek to verify that they generate a Gibbs distn on the entire configuration space. This is the DLR (or Dobrushin–Lanford–Ruelle) problem; the existence of distinct solutions may be called a **phase transition**. [P. Brémaud, *Markov Chains*. Springer, 1999]

Gibbs inequality (alias: **Shannon's inequality; information inequality**) Asserts that the relative entropy (i.e. the Kullback–Leibler divergence) is non-negative. Thus for distributions $p_1, \ldots, p_n$ and $q_1, \ldots, q_n$, $\sum_r p_r \log \frac{p_r}{q_r} \geq 0$, with equality iff $p_r = q_r, 1 \leq r \leq n$; and, likewise, for densities $p(x)$ and

$q(x)$, $\int p(x)\log\{p(x)/q(x)\}\mathrm{d}x \geq 0$, with equality iff $p(x) = q(x)$ almost everywhere. In either case the relative entropy is $+\infty$ if the support of p is not included in the support of q. [T. Cover & J. Thomas, *Elements of Information Theory*, 2nd edn. Wiley, 2006]

Gibbs sampling (alias: **heat bath; Glauber dynamics**) Arises as a form of Metropolis–Hastings Markov chain Monte Carlo simulation, designed to sample from the distribution π of a random field, i.e. a collection of r.vs $X_V \in S$ indexed by the vertices v of a graph V (often a lattice) taking values in S^V. The basic feature of the method is that only one site x (or edge, in the dual problem) is updated at each step of the process. It is said to be a 'single flip rule', or to proceed by 'local moves', in contrast to 'cluster flip' rules such as Wolff dynamics or the Swendsen–Wang method. The site x is updated according to the conditional distribution of x, given the current values of the rest of the field, under the distribution π. There are many variants of this basic idea, with a slightly unsettled nomenclature. [D. Levin *et al.*, *Markov Chains and Mixing Times*. American Mathematical Society, 2009]

Gilbert–Elliott channel Arises in communication theory as a basic model for a channel where errors occur in bursts. It is a binary symmetric channel in which the error (or crossover) probability is given by the current value of a Markov chain having two states. [E. Gilbert, Capacity of a burst noise channel, *Bell Syst. Tech. J.*, 39, 1253–1265, 1960; M. Mushkin & I. Bar-David, Capacity and coding for the Gilbert-Elliott channels, *IEEE Trans. Inf. Theor.*, 1989, 35(6), 1277–1290]

Gilbert's graph A random graph defined on a random point process in $\mathbb{R}^2$ (e.g. a Poisson process) by joining every pair of points (x, y), such that $|x - y| \leq 1$, by an undirected edge. When the underlying process is Poison of rate λ, there exists a critical $\lambda_0 \in (0, \infty)$ such that there is or is not an infinite connected component (a.s.) according as $\lambda > \lambda_0$ or $\lambda < \lambda_0$. [R. Meester & R. Roy, *Continuous Percolation*. Cambridge University Press, 1996]

Gilbert–Mosteller problem (alias: **full information best-choice problem**) Given a sequence of i.i.d.r.vs uniform on [0, 1], which are revealed in order (with no recall), the problem is to determine which stopping rule yields $T \in \{1, \ldots, n\}$ to maximize the probability $\mathbb{P}(X_T = \max\{X_1, \ldots, X_n\})$. [J. Gilbert & F. Mosteller, Recognizing the maximum of a sequence, *J. Amer. Stat. Ass.*, 61(313), 35–73, 1966]

Gilbert–Shannon–Reed shuffle *see* shuffle

Gini index *see* mean difference; Lorenz curve

Gini–Simpson index *see* quadratic entropy

girdle distribution An axial distribution on a sphere that has its mode (or median, or mean) on (or, more loosely, is concentrated around), a latitude. If it has rotational symmetry about the axis, it is a symmetric girdle distn; if the latitude is a great circle, it may be called equatorial. [N. Fisher *et al.*, *Statistical Analysis of Spherical Data*. Cambridge University Press, 1987]

Girko's law *see* circular law; random matrix

Girsanov's theorem *see* Cameron–Martin–Girsanov theorem

Gittins index *see* bandit process

Gladyshev process *see* Gaussian process

Glauber dynamics *see* Gibbs sampling

Glivenko–Cantelli theorem This asserts the uniform convergence, with probability 1, of the empirical distribution function. Let $X_1, \ldots, X_n$, be i.i.d. with distn fn $F(x)$, and define the empirical distribution function

$$F_n(x) = \frac{1}{n}\sum_{r=1}^{n} I(X_r \leq x),$$

where $I(\cdot)$ is the usual indicator function. Then

$$M = \sup_{x\in\mathbb{R}} |F_n(x) - F(x)| \xrightarrow{\text{a.s.}} 0, \quad \text{as } n \to \infty.$$

More precisely, the Dvoretzky–Kiefer–Wolfowitz–Massart inequality asserts that for any integer n and $t \in \mathbb{R}$,

$$\mathbb{P}\Big(\sqrt{n}\sup_{x\in\mathbb{R}} |F_n(x) - F(x)| > t\Big) \leq 2\exp(-2t^2).$$

This idea can be extended to suitable sequences of empirical measures, in which case the equivalent inequalities may be called Vapnik–Chervonenkis (type) bounds. [A. Van der Vaart & J. Wellner, *Weak convergence and Empirical Processes*. Springer, 1996; G. Shorack & J. Wellner, *Empirical Processes with Applications to Statistics*. SIAM, 2009]

Gnedenko's central limit theorem A local limit theorem for a sequence $X_n, n \geq 1$, of zero-mean and unit variance i.i.d.r.vs taking values in the lattice $a, a \pm h, a \pm 2k, \ldots$, where k is the largest positive number such that this holds (i.e. it is the maximal span). Then, as $n \to \infty$,

$$\frac{\sqrt{n}}{h}\mathbb{P}\left(\frac{1}{\sqrt{n}}\sum_{r=1}^{n} X_r = \frac{na+kh}{\sqrt{n}}\right) - \frac{1}{\sqrt{2\pi}}\exp\left\{-\frac{1}{2}\frac{(na+kh)^2}{n}\right\} \to 0,$$

uniformly for $k \in \mathbb{Z}$. [B. Gnedenko & A. Kolmogorov, *Limit Distributions for Sums of Independent Random Variables*, 2nd edn, Addison-Wesley, 1968. Note that several results in this book have been termed as a **Gnedenko–Kolmogorov thm**, by subsequent authors, particularly central limit theorems and the characterization of stable laws]

Gnedenko-Doeblin theorem *see* Doeblin–Gnedenko theorem

Gnedenko–Khinchin theorem *see* convergence to types

Gnedenko–Korolyuk theorem (or distribution) Arises in comparing the empirical distributions of independent random samples X_r and $Y_r, 1 \leq r \leq n$, having the same continuous distribution function $F(x)$, where it gives the distribution

$$\mathbb{P}\left(\sup_{x\in\mathbb{R}}\left|\sum_{r=1}^{n} I(X_r \leq x) - \sum_{r=1}^{n} I(Y_r \leq x)\right| \geq \frac{t}{n}\right)$$

$$= 2\binom{2n}{n}^{-1}\sum_{r=1}^{[n/t]}(-1)^{r+1}\binom{2n}{n-rt}, \quad \text{for integer } t \geq 0.$$

[B. Gnedenko & V. Korolyuk, On the maximum discrepancy between two empirical distributions, *Dokl. Akad. Nauk. SSSR*, 80, 525–528, 1951]

Gnedenko's theorem *see* extreme value distributions

Goldie–Steutel–Bondesson class A class of infinitely divisible distributions developed as the extension to $\mathbb{R}^d, d > 1$, of the Bondesson class $B(\mathbb{R}^+)$, where $B(\mathbb{R}^+)$ comprises the set of distributions including all mixtures of exponential r.vs, and is closed under convolutions and convergence in distn. [O. Barndorff-Nielsen *et al.*, Some classes of multivariate infinitely divisible distributions admitting stochastic integral representations, *Bernoulli*, 12(1), 1–33, 2006]

Goldstein–Kac process Arises as a diffusion model in which a particle has finite speed c. For a particle moving in one dimension, whose changes of direction occur at the instants of a Poisson process $N(t)$ of rate λ, the resulting position $X(t)$ at time t (starting from the origin) is given by integrating the telegraph process $T(t)$, thus:

$$X(t) = T(0)\int_0^t (-1)^{N(s)}\mathrm{d}s, \quad t \geq 0,$$

where $\mathbb{P}(T(0) = +c) = \mathbb{P}(T(0) = -c) = \frac{1}{2}$. Then the probability density $f(x,t)$ of the position x of the particle at time t, $-ct \leq x \leq ct$, satisfies the telegraph equation

$$\frac{\partial^2 f}{\partial t^2} + 2\lambda\frac{\partial f}{\partial t} = c^2\frac{\partial^2 f}{\partial x^2}.$$

[A. Kolesnik & E. Orsingher, A planar random motion with an infinite number of directions controlled by the damped wave equation, *J. Appl. Probab.*, 42, 1168–1182, 2005]

Gompertz–Makeham law Arises in actuarial mathematics as a classic model for the so-called force of mortality in compiling life tables. Its basic assumption is that the hazard rate (or function) of the lifetime of an individual takes the form $h(x) = a + be^{cx}$, at age x, for some positive a, b, and c. [D. Daley & D. Vere-Jones, *Introduction to the Theory of Point Processes*, 2nd edn. Springer, Vol. I, 2003; A. Marshall & I. Olkin, *Life distributions*, 2007, Springer]

Goncharov's theorem A central limit theorem arising in the context of a random permutation $\pi_n = (e(1), \ldots, e(n))$ that is picked uniformly at random from the $n!$ possibilities. A cycle of length r is a sequence $i_1, \ldots, i_r$ such that $e(i_1) = i_2, \ldots, e(i_{r-1}) = i_r, e(i_r) = i_1$; let there be C_r such cycles in $\pi_n, 1 \leq r \leq n$. Then the theorem asserts that $\mathbb{P}\left(\frac{S_n - \log n}{(\log n)^{1/2}} \leq x\right) \to \Phi(x)$, as $n \to \infty$, where $\Phi(x)$ is the $N(0,1)$ distn fn, and $S_n = \sum_{r=1}^{n} C_r$.

By Foata's theorem, there is a one–one correspondence between permutations with C_r cycles and permutations with C_r record values, so the same CLT holds for the number of record values in π_n (where e_1 is a record, and there is a record in the rth place if $e_r > e_i$ for all $i < r$).

Another theorem of A. Goncharov asserts that the joint distribution of $(C_1, \ldots, C_n)$ converges as $n \to \infty$ to that of $(X_1, X_2, \ldots)$, where X_r are independent Poisson r.vs having parameters $r^{-1}, r \geq 1$. [R. Arratia & S. Tavaré, The cycle structure of random permutations, *Ann. Probab.*, 20(3), 1567–1591, 1992]

Good's index A measure of diversity (or entropy) that for a probability distribution $p_1, \ldots, p_n$ takes the form $g(p) = \sum_{r=1}^{n} p_r{}^a(-\log p_r)^b$, for suitable $a, b > 0$. [I. Good, The population frequencies of species, and the estimation of population parameters, *Biometrika*, 40(3/4), 237–264, 1953]

good lambda inequality (alias: **Burkholder–Gundy good lambda inequality**) Arises in martingale theory (and related areas, e.g. harmonic functions), in this basic form: a pair (X, Y) of random variables is said to satisfy the good lambda inequality if for every $\lambda > 0$ and $0 < a < \infty$, $\mathbb{P}(X \geq b\lambda, Y < a\lambda) \leq f(a, b)\ \mathbb{P}(X \geq \lambda)$ for some b, and suitable $f(a, b)$. [D. Burkholder, Distribution function inequalities for martingales, *Ann. Probab.*, 1(1), 19–42, 1973]

good rate function *see* large deviations

googol game (or problem) A version of the secretary problem reformulated as a two-person zero-sum game. [A. Gnedin, A solution to the game of googol, *Ann. Probab.*, 22(3), 1588–1595, 1994]

Gordon–Newell network (alias: **closed Jackson network**) A network of queues such that the vector of queue lengths is an ergodic Markov process, and the number of customers circulating among the queues is constant (either by requiring no entrances and exits, or by requiring that an exiting customer is replaced instantly by a replacement with the same properties). In a simple formulation the rth queue is of M/M/s_r type, where each of the s_r servers has service time that is exponential with parameter μ_r, and departing customers instantly join another (or the same) queue, with first-come-first-served discipline throughout. Then the **Gordon–Newall thm** shows the equilibrium distribution to have a product form (over nodes of the network), though the processes in question are dependent. [F. Kelly, *Reversibility and Stochastic Networks*, Wiley, 1979; reprinted by Cambridge University Press, 2011]

grade correlation A measure of dependence between r.vs X and Y, having distns F_X and F_Y, given by $\rho(F_X(X), F_Y(Y))$, where ρ is the usual (Pearson) correlation. It is a population form of **Spearman's rho** (or rank correlation) from statistical theory.

Gram–Charlier expansion *see* Edgeworth expansion

Granger causality (alias: **Granger–Wiener (or v.v.) causality**) Arises in forecasting a time series $(X_r, r \geq 1)$. At time n the future X_{n+1} is first predicted by $\hat{X} = f(X_1, \ldots, X_n, Y_1, \ldots, Y_n)$ in terms of the past $(X_r, r \leq n)$ of X, and also in terms of the past $(Y_r, r \leq n)$ of some appropriate sequence $(Y_r, r \geq 1)$ of possible explanatory variables (each Y_r may be a vector). Now suppose that there is another time series $(G_r, r \geq 1)$ and X_{n+1} is also predicted by $X* = g(X_1, \ldots, X_n, Y_1, \ldots, Y_n, G_1, \ldots, G_n)$ including the past $(G_r, r \leq n)$ of G. If X^* is a better predictor of X_{n+1} than $\hat{X}$, according to some suitable criterion of merit (such as cost) then (G_r) is said to Granger-cause (X_r), and the effect is called G-causality. *See* Wiener causality. [G. Kirchgässner & J. Wolters, *Introduction to Modern Time Series Analysis*. Springer, 2008]

greedy rule *see* one-stage look-ahead policy

greeks Functions arising in financial mathematics, in the context of a portfolio of options (or derivatives) based on a single underlying asset with price function $S(t)$ at time t. Essentially, for the price $P(S(t), t)$ of the portfolio, the greeks supply a measure of the local dependence on $S(t)$ and the parameters of the model, in terms of the first differential coefficient (and sometimes the second) of $P(S(t), t)$ with respect to S, and the parameters of interest. Those most commonly seen are:

$$\text{delta: } \Delta = \frac{\partial P}{\partial S}(S, t)$$

$$\text{gamma: } \Gamma = \frac{\partial^2 P}{\partial S^2}$$

$$\text{rho: } \rho = \frac{\partial P}{\partial r}, \text{ where } r \text{ is the interest rate}$$

$$\text{theta: } \Theta = \frac{\partial P}{\partial t}, \text{ where } t \text{ is time}$$

$$\text{vega: } \nu = \frac{\partial P}{\partial \sigma}, \text{ where } \sigma \text{ is the volatility.}$$

If any first derivative is zero, then the portfolio is said to be neutral with respect to that parameter. [T. Bjőrk, *Arbitrage Theory in Continuous Time*, 3rd edn. Oxford University Press, 2009]

Greenberg–Hastings model A classic cellular automation (of the type suggested by S. Ulam and J. von Neumann) arising as a model for an excitable medium. Each site of a lattice (or graph) can be excited, weary, or rested; at each time-step rested sites become excited if they have at least one excited neighbour, excited sites become weary, and weary sites become rested. The initial configuration is random, and interest centres on the distribution of excited states in the long run. [R. Durrett & J. Steif, Rigorous results for the Greenberg–Hastings model, *J. Theor. Probab.*, 4(4), 669–690, 1991]

Green's function (alias: **the Green function**) Broadly, this arises in the context of differential equations (and related problems) when the solution may be written in terms of an integral with the so-called Green function as a kernel in the integrand; the nomenclature varies among users and applications. In its original, classic formulation, one requires the solution of Laplace's (or Poisson's) eqn in a region in three dimensions, with given values v on its surface, S (the Dirichlet problem). Then Green's fn is of the form $G(x, y) = |x - y|^{-1} + H(x, y)$, where H is harmonic in the region, and the required solution is of the form $\int_S v \frac{\partial G}{\partial n} \mathrm{d}S$. Note that some authors call H the Green function; while others refer to G as the Green potential and $\frac{\partial G}{\partial n}$ as the Green function. These terms, and allied concepts arise in probability because various functionals of Brownian motion satisfy Laplace's, Poisson's, and related differential eqns.

Thus, let $p(t; x, y)$ be the transition density of Brownian motion $X(t)$ in a suitable domain D of $\mathbb{R}^d$, where the boundary of D is absorbing. Then D is said to be Greenian if $d \geq 3$; or, if $d \leq 2, X(t)$ is absorbed (i.e. exits D) in finite time a.s. The function $g_D(x, y) = \int_0^\infty p(t; x, y)\mathrm{d}t, \quad x, y \in D$, is called the Green function, and for suitable functions $f(x)$, the Green operator (or Green potential, or Green potential operator) G_D is defined by $G_D f = \int g_D(x, y) f(y) dy$. If, more strongly, the expected exit-time D is finite, the D may be called Green-bounded.

By extension, similar terms are often applied to similar objects in the context of Markov chains on a discrete state space with transition probabilities $p_n(x, y)$; where the Green function $G(x, y) = \sum_n p_n(x, y)$ is also called the **fundamental matrix**. [K. L. Chung, *Green, Brown, and Probability & Brownian Motion on the Line*. World Scientific, 2002]

Greenwood–Yule distribution (alias: **Greenwood–Yule–Eggenberger–Pólya distribution; generalized negative binomial distribution; gamma Poisson distribution**) A distribution on the non-negative integers obtained by allowing the parameter of a Poisson distribution to have a gamma density with parameters $\lambda > 0$ and $\gamma > 0$ (a compounding). Then the distn has pgf $\mathbb{E}s^X = G_X(s) = \left\{\frac{\lambda}{\lambda-(s-1)}\right\}^\gamma$, which is the standard negative binomial distn for integer γ. [W. Feller, On a general class of contagious distns, *Ann. Math. Stat.*, 14(4), 389–400, 1943]

Griffeath's theorem *see* coupling inequality

Griffith's association measure Let $\phi(s, t)$ be the joint characteristic function of non-constant random variables X and Y, having respective characteristic functions $\phi_X(s) = \phi(s, 0)$, and $\phi_Y(t) = \phi(0, t)$. Then Griffith's association measure is

$$\gamma = \limsup_{\substack{s \to 0 \\ t \to 0}} \frac{|\phi(-s, t) - \phi_X(-s)\phi_Y(t)|}{\left(1 - |\phi_X(-s)|^2\right)^{\frac{1}{2}} \left(1 - |\phi_Y(t)|^2\right)^{1/2}}.$$

When X and Y have finite variance γ equals their correlation. For all bivariate stable laws, whose exponent is not unity, $\gamma = 0$ iff X and Y are independent. [R. Griffiths, Linear independence in bivariate distributions, *Austral. J. Stat.*, 14, 182–187, 1972]

Griffiths–Engen–McCloskey law (alias: **GEM distribution with parameter θ**) Arising in mathematical genetics, this is the distn of a r.vr $(X_1, \ldots, X_n)$ generated in terms of i.i.d. Beta $(1, \theta)$ r.vs $Z_1, \ldots, Z_n$ by the rules $X_1 = Z_1$, and $X_k = Z_k \prod_{r=1}^{k-1}(1 - Z_r)$, $2 \leq k \leq n$. It is an example of a stick-breaking distribution, or residual allocation model. The Poisson–Dirichlet law with parameter θ is that of the decreasing order statistics of $X_1, \ldots, X_n$. [W. Ewens, *Mathematical Population Genetics*, 2nd edn. Springer, 2004]

Grigelionis's theorem Supplies a necessary and sufficient condition for the convergence of the superposition of n independent point processes on $[0, \infty)$ to converge to a (possibly non-homogeneous) Poisson process as $n \to \infty$. The term is also used for the generalization by J. Goldman and P. Jagers. [B. Grigelionis, On the convergence of the sums of random step processes to a Poisson process, *Theor. Probab. Appl.*, 8(2), 177–182, 1963]

Grimmett's formula *see* Hammersley–Clifford theorem

Grimmett's lemma *see* Hammersley–Clifford theorem

Gross's logarithmic Sobolev inequality *see* log Sobolev inequality

grouping lemma Asserts that grouped subsets of families of independent events or random variables are independent. One general form is this: let $\{\mathcal{F}_t; t \in T\}$ be an independent collection of σ-fields, and let $A_i; i \in I$ be a non-trivial collection of disjoint subsets of T. Then $\sigma_i = \sigma(\mathcal{F}_s; s \in A_i)$, where σ_i is the minimal σ-field generated by $\{\mathcal{F}_s; s \in A_i\}$, is an independent family. [S. Resnick, *A Probability Path*. Birkhäuser, 1999]

grouping property (or axiom) Arises in characterizing Shannon's entropy function $H(p)$ of a probability distribution $\mathbf{p}$. It has several forms, e.g. for any m, and $n > m$, and $\mathbf{p} = (p_1, \ldots, p_n)$, we have: $H(\mathbf{p}) = H(a,b) + aH(p_1/a, \ldots, p_m/a) + bH(p_{m+1}/b, \ldots, p_n/b)$, where $a = \sum_{r=1}^{m} p_r$ and $b = \sum_{r=m+1}^{n} p_r$. The weaker axiom of additivity suffices to characterize Rényi entropy (*see* entropy). [A. Rényi, On measures of entropy and information, *Proc. 4th Berkeley Symp. Math. Stat. Probab.*, 1, 547–561, 1961]

Gruss inequality Asserts that for random variables X and Y such that $a \leq X \leq A$ and $b \leq Y \leq B$ a.s., $\text{cov}(X,Y) \leq \frac{1}{4}(A-a)(B-b)$. [J. Steele, *The Cauchy–Schwarz Master Class: An Introduction to the Art of Mathematical Inequalities*. Cambridge University Press, 2004]

GUE *see* random matrix

Gumbel distribution *see* extreme value distributions

H

Hájek–Feldman dichotomy *see* Feldman–Hájek theorem

Hájek–Rényi–Chow inequality (alias: **Chow's inequality**) An extension of the Hájek–Rényi inequality to submartingales (alias semi-martingales in the original terminology of Y. Chow, following J. Doob). Let $Y_n, n \geq 1$ be a submartingale (for a filtration $\mathcal{F}_n$), and $c_n > 0, n \geq 1$ a non-increasing sequence; then, for $t > 0$,

$$\mathbb{P}\left(\max_{1\leq r\leq n} c_r Y_r \geq t\right) \leq t^{-1}\left\{c_1 \mathbb{E} Y_1^+ + \sum_2^n c_r \mathbb{E}(Y_r^+ - Y_{r-1}^+)\right\},$$

where $Y_r^+ = \max\{0, Y_r\}$.

[Y. Chow, A martingale inequality and the law of large numbers, *Proc. Amer. Math. Soc.*, 11(1), 107–111, 1960]

Hájek–Rényi inequality (alias: **Kolmogorov–Hájek–Rényi inequality**) A generalization of Kolmogorov's maximal inequality, with applications to laws of large numbers. Let $X_n, n \geq 1$, be zero-mean finite-variance independent r.vs, and $c_n > 0, n \geq 1$, a non-increasing sequence; then for $t > 0$, and $m \leq n$,

$$\mathbb{P}\left(\max_{m\leq r\leq n} c_r |X_1 + \cdots + X_r| > t\right)$$

$$\leq t^{-2}\left\{c_m^2 \sum_{r=1}^m \mathbb{E}(X_r^2) + \sum_{r=m+1}^n {c_r}^2 \mathbb{E}({X_r}^2)\right\}.$$

[J. Hájek & A. Rényi, Generalization of an inequality of Kolmogorov, *Acta. Math. Acad. Sci. Hungar.*, 6(3–4), 281–283, 1955; Z. Wang, *The Hájek–Rényi Inequality*, MS thesis, University of Regina, 2007]

Halász's Poisson approximation Arises in probabilistic number theory where it asserts that for any subset H of the primes, the number of prime factors in H of a random integer chosen uniformly from $[1, \ldots, n]$ is asymptotically Poisson with parameter $\lambda_n = \sum \frac{1}{p}$, as $\lambda_n \to \infty$, where the sum is over primes in H that are less than n. [H. K. Hwang, Asymptotics of Poisson approximation to random discrete distributions, *Adv. Appl. Probab.*, 31, 448–491, 1999]

half distribution *see* folded distribution

half-invariants *see* cumulants

Halfin–Whitt regime *see* heavy traffic

Hall's lemma A theorem about hitting times of Brownian motion $X(t)$ in the unit disk D in the plane. Let C be a subset of D, and let $a(C)$ be its radial projection from the origin into the unit circle, i.e. $a(C) = \{e^{i\theta} : re^{i\theta} \in C\}$. For $X(0) = 0$, let T, T_C, and T_a be the hitting times by $X(t)$ of the unit circle, C, and $a(C)$ respectively. Then the lemma asserts the existence of a constant k, which does not depend on C, such that $\mathbb{P}(T_a < T) \leq k\,\mathbb{P}(T_C < T)$. A similar result holds for subsets of the unit ball in $\mathbb{R}^d, d > 2$. [R. Bass, *Probabilistic Techniques in Analysis*. Springer, 1995]

Halphen distribution *see* inverse Gaussian distribution

halting state Arises in the context of a stopping time T for a Markov chain $X_n, n \geq 0$. A state s is a halting state for T if $X_n = s$ entails that $T \leq n$. [D. Levin et al., *Markov Chains and Mixing Times*. American Mathematical Society, 2009]

Hamburger moment problem *see* moment problem

Hamilton–Jacobi–Bellman equation The continuous-time analogue of the Bellman equation, which arises when the principle of optimality (or dynamic programming principle) is invoked in (possibly stochastic) optimal control theory with continuous-time parameter, yielding a development and applications of the classical Hamilton–Jacobi extremal theory in stochastic optimal control. [J. Yong & X. Zhou, *Stochastic Controls: Hamiltonian Systems and HJB Equations*. Springer, 1999]

Hammersley–Clifford theorem Asserts that a random field on a graph G (e.g. a lattice) has the Markov property with a positive distribution (i.e. is a Markov random field) iff the random field has a distribution $\mathbb{P}(\cdot)$ that can be factorized over the cliques (i.e. complete subgraphs) of G; that is, there exists a so-called potential function $\phi_K(x)$, defined on the cliques K of G, such that for any configuration x of the field

$$\mathbb{P}(x) \propto \exp\Big(\sum_K \phi_K(x)\Big) = \prod_K \exp\, \phi_K(x).$$

Such a distribution is called a Gibbs state (or ensemble), so the theorem establishes the equivalence of Gibbs fields and Markov fields. **Grimmett's lemma** supplies a natural formula for the potential function in terms of the Markov field, using the Möbius inversion theorem. *See also* Brook's factorization theorem. [G. Grimmett, A theorem about random fields, *Bull. Lond. Math. Soc.*, 5(1), 81–84, 1973]

Hammersley's law (or theorem) Let $X_m, m \geq 1$, be a sequence of i.i.d.r.vs, each uniformly distributed in $[0, 1]$, and let H_n be the number of terms in the longest monotone-increasing subsequence of $X_1 \ldots, X_n$. The original theorem asserts that as $n \to \infty$, $n^{-\frac{1}{2}} H_n \to c$, where c is a constant, and the convergence is in probability. It has been shown, more strongly, that the convergence is with probability 1, and $c = 2$. This is related to **Hammersley's particle process** on $[0, 1]$: at time 0 there are no particles, and at each time step a new particle appears independently and uniformly in $[0, 1]$ while the particle next to its right (if any) is deleted. The number of particles at the nth step is distributed as H_n. The two-dimensional version is **Hammersley's path process**. [J. Hammersley, A few seedlings of research, *Proc. 6th Berkeley Symp. Math. Stat. Probab.*, I, 345–394, 1972]

Hammersley's random walk theorem *see* self-interacting random walk

Hammersley's theorem Arises in the context of a subadditive process, $X_{mn}, 0 \leq m < n$, for $m, n \in \mathbb{Z}^+$, where it is additionally supposed that the process is superconvolutive, so that for all $x \in \mathbb{R}$, the distn fns satisfy $F_{X_{0,m+n}}(x) \geq \int_{-\infty}^{\infty} F_{X_{0,m}}(x-y)\mathrm{d}F_{X_{0,n}}(y)$. Then the theorem establishes a large-deviations principle, i.e. the existence and properties of the limit $\lim_{n\to\infty} n^{-1}\mathbb{P}(X_{0n} \leq nx)$. *See* Gärtner–Ellis theorem and Kingman's subadditive ergodic theorem. [G. Grimmett, Large deviations in subadditive processes, and first-passage percolation; in *Particle Systems, Random Media and Large Deviations*, R. Durrett (ed.). American Mathematical Society, 1985]

Han's inequalities Arise in information theory as bounds for the entropy and conditional entropy of subsets of a set of random variables $\{X_1, \ldots, X_n\} = S$. Denote the Shannon entropy by $H(\cdot)$, let

$$\left\{S_i, 1 \leq i \leq \binom{n}{r}\right\}$$

be the set of distinct subsets of S of size r, and define

$$A(r) = \binom{n}{r}^{-1} \sum_i \frac{1}{r} H(S_i), \quad 1 \leq r \leq n.$$

Then $A(1) \geq A(2) \geq \cdots \geq A(n)$. A similar set of inequalities for the conditional entropy, denoted by $H(\cdot|\cdot)$, involves

$$C(r) = \binom{n}{r}^{-1} \sum_i \frac{1}{r} H(S_i|S_i^c), \quad 1 \leq r \leq n,$$

for which $C(1) \leq C(2) \leq \cdots \leq C(n)$. [A. Dembo et al., Information theoretic inequalities, *IEEE Trans. Inf. Theor.*, 37(6), 1501–1518, 1991]

Han–Kobayashi region Arises in network information theory in the context of the discrete memoryless interference channel (and the Gaussian interference channel), where it supplies an achievable rate region. [T. Han & K. Kobayashi, A new achievable rate region for the interference channel, *IEEE Trans. Inf. Theor.*, 27(1), 49–60, 1981]

hard-sphere model (alias: **Matérn hard-sphere (or hard-core) model**) A type of point process that arises as a model for the spatial distribution of particles that repel each other, or are solid spheres that cannot overlap or intersect (the name soft-sphere or soft-core is given to those that can overlap or intersect). It thus has the property that no two points are closer to each other than some minimum distance s, called the separation. It may be realized in $\mathbb{R}^d$ in several ways. For example, a classic Matérn model takes a spatial Poisson process of rate λ, and deletes all those points that are closer than s to each other. The result is a non-Poisson renewal process with mean rate $\lambda \exp[\mathrm{e}^{-\lambda V(s)}]$, where $V(s)$ is the volume of the s-ball in $\mathbb{R}^d$. Alternatively, the sequential allocation process is realized by dropping spheres of radius s randomly (in some subset of $\mathbb{R}^d$) and rejecting those that infringe on a previously allocated sphere.

Analogously, similar models may be defined on other structures that include a concept of distance or proximity.

For example, such a process may be defined on the independent sets of the graph G, i.e. a set I of vertices in G such that no two vertices of I are adjacent. [D. Stoyan et al., *Stochastic Geometry and its Applications*, 2nd edn. Wiley, 1995]

Hardy–Littlewood transform Arising in the theory of functions, this has applications in financial and insurance mathematics. Let X have density $f(x)$ and distribution function $F(x)$, with inverse (or quantile) function $F^{-1}(y)$. Then the distribution function F_H defined by

$$F_H{}^{-1}(y) = \begin{cases} \dfrac{1}{1-y}\displaystyle\int_y^1 F^{-1}(u)\mathrm{d}u, & y < 1 \\ F^{-1}(1), & y = 1, \end{cases}$$

is the Hardy–Littlewood transform of F, and a random variable having this distn fn may be called the Hardy–Littlewood transform $H(X)$ of X. Alternatively, defining the barycentre function

$$b(x) = \frac{1}{1-F(x)}\int_x^\infty y\,\mathrm{d}F(y),$$

then $F_H(x)$ is the distribution function of $b(X)$. [L. Carraro et al., On Azéma-Yor processes, their optimal properties, and the Bachelier drawdown equation, *Ann. Probab.*, 40(1), 372–400, 2012]

harmonic A term with several applications in mathematics; it arises in probability through the concept of harmonic function, i.e. a solution $h(\cdot)$ of Laplace's equation. Theorems of G. Green and C. Gauss entail that h is such a solution iff it satisfies a local averaging property: the value $h(x)$ at the centre x of a ball B in the domain of interest is the average of $h(\cdot)$ over the surface of B. But suitable functionals of Brownian motion $W(t)$ also have this property (because it is a strong Markov process), and are therefore said to be harmonic (wrt $W(t)$). More strongly, an averaging property holds over any suitable surface S, provided that the integrand is weighted by the hitting density (first-passage density) of S by $W(t)$, which may be called the harmonic measure on S. (It is classically equivalent to the electrostatic charge distribution induced on an earthed conductor at S by a point charge at $W(0)$.)

This motivates a more general definition for a suitable continuous strong Markov process $X(t)$ having a hitting distribution on S: the function $h(\cdot)$ is harmonic wrt X if $h(X)$ has such a weighted averaging property for all suitable S. By extension, similar uses of the word harmonic are applied to a homogeneous Markov chain X with countable state space having transition matrix P. Any bounded solution h of $h = Ph$ is said to be harmonic (or regular, or invariant). In particular, if P corresponds to symmetric simple random walk (on $\mathbb{Z}^d$, say) then $I - P$ is called the discrete Laplacian. Furthermore, if $h \le Ph$, then h is said to be subharmonic (or subregular); if $h \ge Ph$ then it is superharmonic (or excessive, or super-regular). Because such processes generate natural martingales, the term harmonic arises in martingale theory likewise. [T. Liggett, *Continuous Time Markov Processes: An Introduction*. American Mathematical Society, 2010]

harmonic explorer A self-avoiding random walk in a simply connected subset of a planar lattice, in which the distribution of any step depends on a harmonic measure (i.e. a hitting probability) that is a function of the path of the walk up to that point. It arises in the context of percolation problems, and in a suitably framed scaling limit the path converges to a **Schramm–Loewner evolution**.

For definiteness, consider a subset D of the hexagonal lattice whose boundary comprises two connected sets of hexagons, each set coloured red or blue. The walk begins at a red–blue interface and turns right (or left) at red (or blue) hexagons until it meets an uncoloured hexagon H. This is coloured by the colour of the boundary point first encountered by a symmetric random walk started at H, and the walk proceeds. [O. Schramm & S. Sheffield, Harmonic explorer and its convergence to SLE4, *Ann. Probab.*, 33(6), 2127–2148, 2005]

harmonic renewal measure *see* renewal theory

harmonizable process (alias: **harmonized process**) Arises in the context of finite-variance (i.e. second-order) random processes that are not necessarily stationary. If $X(t)$ is such a process with covariance $c(s,t)$, then $X(t)$ is said to be harmonizable if there exists a second-order process $S(t)$, whose covariance is of bounded variation on $\mathbb{R}^2$, such that $X(t) = \int \mathrm{e}^{\mathrm{i}ty}\mathrm{d}S(y)$. Similarly, the covariance $c(s,t)$ is harmonizable if there exists a bounded-variation covariance $\sigma(s,t)$ such that

$$c(s,t) = \iint e^{ity-isz}\, d\sigma(y,z).$$

The sufficient condition $\int\int_{-\infty}^{\infty} |d\sigma(y,z)| < \infty$ is called **Loéve's condition**. If the condition of bounded variation is relaxed, entailing a more general form of the integral, some authors speak of weakly harmonizable processes, and reserve strongly harmonizable for the above. [M. Loéve, *Probability Theory*, 4th edn. Springer, 1978]

harness (alias: **affine process; serial harness**) Arose initially as a type of discrete-parameter stochastic cellular automaton proposed by J. Hammersley to describe long-range misorientation in the crystalline structure of metals. More generally, and in continuous time, the finite-mean process $X(t)$ is a harness if, for all $s < t < u < v$,

$$\mathbb{E}\left\{\frac{X(u)-X(t)}{u-t} | \mathcal{F}(s,v)\right\} = \frac{X(v)-X(s)}{v-s},$$

where $\mathcal{F}(s,v)$ is the σ-field generated by $\{(X(y),X(z));\ y \le s, z \ge t\}$. The only continuous harnesses with finite variance are drifting Wiener processes, and all centred finite-variance harnesses have a covariance function of the form $C(s,t) = \alpha + \beta s + \gamma t + \delta st$, for $s \le t$, and constants α, β, γ, and δ. [W. Bryc et al., Quadratic harnesses, q-commutations, and orthogonal martingale polynomials, *Trans. Amer. Math. Soc.*, 359(11), 5449–5483, 2007]

Harris's branching-process theorem Any one of a number of theorems on ordinary and multitype branching processes, but perhaps most often this basic result: Let $Z_n, n \ge 0$, be an ordinary Galton–Watson branching process with $0 < Z_0 < \infty$, and family size distribution $\{p_1, p_2, \ldots\}$ having finite variance, mean $m > 1$, and such that $p_1 < 1$. Defining $W_n = Z_n m^{-n}$, the theorem asserts the existence of an absolutely continuous r.v. $W \in (0,\infty)$ such that W_n converges in mean square to W as $n \to \infty$, and $\mathbb{E}(W|Z_0 = 1) = 1$. *See* Kesten–Stigum theorem. [T. Harris, Some mathematical models for branching processes, *Proc. 2nd Berkeley Symp. Math. Stat. Probab.*, 305–328, 1951]

Harris chain *see* Harris recurrence

Harris correspondence A term referring to the one-one correspondence between finite rooted trees (as generated by a Galton–Watson branching process that is extinguished with probability 1), and the paths of finite simple random walks on the non-negative integers (as generated by a simple random walk that returns to its origin with probability 1, and stops on the first occasion when it does so). The bijection is demonstrated by exploring the tree in an orderly fashion from the root, yielding a so-called contour process.

For this reason, a non-homogeneous simple r.w. with step probabilities p_r and q_r, of positive and negative steps respectively, is called a Harris random walk if it is killed on revisiting its origin. It corresponds to a Galton–Watson branching process in which the family size of an individual in the rth generation is geometric with distribution $f(x) = p_{r+1}^x\, q_{r+1}$, for $p_r + q_r = 1, x \ge 0, r \ge 0$.

The path of the walk may then be imbedded in Brownian motion $W(t)$ by identifying successive level crossings of $W(t)$ with the step probabilities of the walk; this has been called the **Harris imbedding**. *See also* contour process. [T. Harris, First passage and recurrence distributions, *Trans. Amer. Math. Soc.*, 73(3), 471–486, 1952; J. Pitman, Combinatorial stochastic processes, in *Ecole d'Eté de Probabilités de Saint-Flour*, J. Picard (ed.). Springer, 2002, p. xxxii]

Harris's epidemic model *see* contact process

Harris–FKG inequality *see* Harris inequality; FKG inequality

Harris–Keane coin-tossing problem Arises as a simple model for distinguishing between possible sceneries, in a random walk through the scenery, in the presence of a type of noise. [D. Levin et al., A phase transition in random coin tossing, *Ann. Probab.*, 29, 1637–1669, 2001]

Harris inequality (alias: **Harris's lemma**) A correlation inequality arising in T. Harris's half of the Harris–Kesten thm on percolation, in which bonds on a lattice are independently open or closed with probability p. Its basic form is this: For $1 \le i \le m$ and $1 \le j \le n$ let a_i and b_j be finite non-empty sets of links (or bonds), where no assumption is made about mutual exclusiveness. Then set $A = \cup_1^m A_i$ and $B = \cup_1^n B_j$ where A_i (resp. B_j) is the event that all bonds in A_i (resp. B_j) are open. Then $\mathbb{P}(A \cap B) \ge \mathbb{P}(A)\,\mathbb{P}(B)$. *See* FKG inequality, Holley's inequality, and Ahlswede–Daykin inequality. [T. Harris, A lower bound for the critical probability in a certain percolation process, *Proc. Camb. Philos. Soc.*, 56, 13–20, 1960]

Harris–Kesten theorem Asserts that for bond percolation on the planar square lattice $\mathbb{Z}^2$, the critical probability is 1/2. That is to say, when edges are present independently with probability p, percolation does not occur for $p \leq 1/2$, (there is a.s. no infinite cluster), but percolation does occur for $p > 1/2$, (there is an infinite cluster with positive probability). [B. Bollobás & O. Riordan, A note on the Harris-Kesten theorem, *Euro. J. Combin.*, 28(6), 1720–1723, 2007]

Harrison–Kreps doubling strategy Arises in financial mathematics in the context of the Black–Scholes model as a self-financing portfolio that achieves a positive net return (arbitrage) in finite time with probability 1. It is analogous to the classic martingale betting system and is inadmissible for similar reasons, e.g. a limit on borrowing. [L. Tyge Nielsen, *Pricing and Hedging of Derivative Securities*. Oxford University Press, 1999]

Harrison–Kreps–Pliska theorems *see* fundamental theorems of asset pricing

Harris random walk *see* Harris correspondence

Harris recurrence (alias: **φ-(or phi-)recurrence**) Arises in the context of a phi-irreducible Markov process (X_n or X_t), on a measurable space $(S, \mathcal{F})$, where it supplies a condition (the Harris recurrence condition) for the chain to have a stationary measure. The requirement (Harris's condition C) is that there exists a countably additive sigma-finite measure $m(A)$ on the sets of the sigma-field $\mathcal{F}$ such that for any A with $m(A) > 0$, for all $s \in S, \mathbb{P}(X_n \in A$ infinitely often $|X_0 = s) = 1$.

In continuous time this may be written in the form $\int_0^\infty I(X(t) \in A)\mathrm{d}t = \infty$, with probability 1, for a suitable Markov process $X(t)$. The measure $m(A)$ is often denoted as $\varphi(A)$, and one speaks of φ-irreducibility, while the property of Harris-(or $\varphi-$) recurrence is equivalent to being suitably regenerative. It follows that X_n has an invariant measure (which may be called Harris measure) that dominates $m(\cdot)$, and is unique if the chain is non-null (i.e. positive); this may be called Harris's theorem. [T. Harris, The existence of stationary measures for certain Markov processes, *Proc. 3rd Berkeley Symp. Math. Stat. Probab.*, 2, 113–124, 1956; K. Athreya & P. Ney, A new approach to the limit theory of recurrent Markov chains, *Trans. Amer. Math. Soc.*, 245, 493–501, 1978]

Hartley's information (alias: **Hartley's formula (or law, or measure, or entropy)**) Arises in communication theory as the measure (or amount) of information necessary to characterize (i.e. uniquely distinguish) the elements of a set A of cardinality $|A|$ (i.e. A contains $|A|$, elements); these may be called symbol sequences, or messages, or transmissions. The Hartley information of such a set is $H(A) = \log |A|$, where the choice of logarithmic base determines the unit of information. This measure coincides with Shannon's information in a r.v. X, when X is uniformly distributed on $[1, \ldots, |A|]$. The terms **self-information** and **surprise** have also been used, and extended to an event E of probability $\mathbb{P}(E)$ as $H(E) = -\log \mathbb{P}(E)$. [R. Hartley, Transmission of information, *Bell Syst. Tech. J.*, 7(3), 535–563, 1928]

Hartman-Watson distribution Any member of three families of distributions arising originally in the context of spherical distributions in $\mathbb{R}^n, n > 1$, where they supply a link between the 'normal' distn as generated by a Wiener process on the sphere, and the 'normal' distn favoured in directional statistics (see **Fisher–Arnold–Langevin–Von Mises distn**, in Fisher distribution); in fact the latter is expressed as the distn of a suitably stopped Wiener process, and some writers therefore speak of the H-W mixing distn. Explicitly, the principal H-W distn function $F_s(x)$ *on* $\mathbb{R}^+$ has Laplace transform $\mathrm{L}(\theta) = \int_0^\infty \mathrm{e}^{-\theta x}\mathrm{d}F_s(x) = I_{\sqrt{2\theta}}(s)/I_0(s)$, for any fixed positive value of the parameter s, and where $I_n(s)$ is the modified Bessel function of the first kind. Other Hartman–Winter distribution functions are of the form $I_t(x)/I_s(x)$ and $K_s(x)/K_t(x),\ \ x \geq 0$, for $0 \leq s < t$, where $K_m(x)$ is the modified Bessel function of the second kind. [P. Hartman & G. Watson, 'Normal' distribution functions on spheres, and the modified Bessel functions, *Ann. Probab.*, 2(4), 593–607, 1974]

Hartman-Winter LIL *see* law of the iterated logarithm

Hastings' approximation A very widely used (via computer programs) method for numerically approximating the inverse $x = \Phi^{-1}(p)$ of the standard normal distribution $\Phi(x)$ for a given probability p, by means of Chebyshov-type polynomial approximation. [B. Bailey, Alternatives to Hastings' approximation to the inverse of the normal cumulative distribution function, *Appl. Stat.*, 30(3), 275–276, 1981]

Hastings–Levitov model (alias: **planar aggregation (or clustering) model**) This is a continuum planar growth model for stochastic aggregation. The model is formulated in terms of a sequence of iterated conformal maps. The sequence is indexed by a parameter a, where the resulting transformation of the complex plane is called the **bump fn**. At each iteration, the effect of the random map is to add a fresh 'bump' randomly to the existing cluster. Note that for a suitable choice of the parameter a, the addition is typically bump-like, in that the cluster is the union of overlapping circles. For other choices of the parameter, the 'bump' may take the form of a randomly added line-segment, or arc, called a 'strike'. The resulting random clusters, or agglomerations have fractal properties. [F. Johansson Viklund et al., Scaling limits of anisotropic Hastings-Levitov clusters, *Ann. Inst. H. Poincare Probab. Stat.* 2012, 48(1), 235–257. Note that this is M. B. Hastings]

Hastings McMC *see* Metropolis–Hastings method

hat-check problem *see* derangement problem

Hausdorff dimension *see* dimension

Hausdorff measure *see* measure

Hausdorff moment theorem Arises from the (Hausdorff) moment problem. It asserts that a sequence $(\mu_r; r \geq 0)$ comprises the moments of a probability distribution fn $F(x)$ on $[0, 1]$ (i.e. $\mu_r = \int_0^1 x^r \mathrm{d}F(x)$ for all r), iff $\mu_0 = 1$ and the sequence is completely monotone, which is to say that $(-1)^n \Delta^n \mu_r \geq 0, n \geq 0$, where $\Delta\mu_r = \mu_{r+1} - \mu_r$ is the usual difference operator. Furthermore, $F(x)$ is given uniquely at its points of continuity by

$$F(x) = \lim_{n\to\infty} \sum_{k \leq nx} \sum_{j=k}^{n} \binom{n}{j} \binom{j}{k} (-1)^{j-k} \mu_j, \quad 0 \leq x \leq 1, n \geq 0.$$

[R. Mnatsakanov & A. Hakobyan, Recovery of distributions via moments, *IMS Lecture Notes*, 57, 252–265, 2009]

Havrda–Charvát entropy *see* entropy

Hawkes process (alias: **linear self-(or mutually) exciting process; infectivity model**) Broadly, a point process on the line (with counting function $N(t), t \in \mathbb{R}$), whose intensity function $m(t)$ is a linear function of the history of the process prior to t, i.e. $\{N(s), s \leq t\}$. Formally, $m(t) = \mu + \int_{-\infty}^{t} w(t-s)\mathrm{d}N(s)$, where $\mu > 0$, and $w > 0$ is a weighting function such that $\int_0^\infty w(t)\mathrm{d}t < 1$. The process may be alternatively characterized as a type of cluster process, where the cluster centres comprise a Poisson process and (conditional on that process), each centre independently initiates a type of branching process thus: an individual i in the nth generation of any cluster, at the point t_i initiates a Poisson process of rate $w_i(t - t_i), t > t_i$, contributing to the $(n+1)$th generation. The union of the underlying process and all the descendant branching clusters comprises the Hawkes process. This generalises in a natural way to $\mathbb{R}^d, d > 1$. [J. Møller & J. Rasmussen, Perfect simulation of Hawkes processes, *Adv. Appl. Probab.*, 37, 629–646, 2005]

Hawkins–Ulam theorem Arises in the context of a super critical Galton–Watson branching process Z_n, with expected family size $m > 1$, and finite non-zero variance, where it asserts that $W_n = Z_n m^{-n}$ converges in distribution to a r.v. W such that $M(t) = \mathbb{E}e^{-tW}$ satisfies $M(mt) = G(M(t)), \mathrm{Re}(t) \geq 0$; where $G(s)$ is the family-size pgf of the process Z_n. [T. Harris, *The Theory of Branching Processes*. Springer, 1963; reprinted by Dover, 2002]

hazard (rate) function *see* ageing

heaps process (alias: **Tsetlin library; move-to-front process**) Denotes a type of self-organizing filing system of objects $B_j, j \geq 1$, in which objects are removed independently one at a time, at random from a heap (or shelf), with probability p_j that B_j is selected; the selected object is returned to the top of the heap (or leftmost point of the shelf). [P. Donnelly, The heaps process, libraries, and size-biased permutations, *J. Appl. Probab.*, 28(2), 321–335, 1991]

heat bath *see* Gibbs sampling

Heath–Jarrow–Morton framework (alias: **HJM forward interest rate model**) Arises in financial mathematics as a model for the behaviour of the instantaneous forward interest rate curve (or function) $f(t, T), 0 \leq t \leq T$. It assumes that for any given $T > 0$, $f(t, T)$ satisfies an SDE of the form (in the one-factor case)

$$\mathrm{d}f(t, T) = a(t, T)\mathrm{d}t + b(t, T)\mathrm{d}W(t),$$

where W is a Wiener process, and a and b are suitable adapted processes that must satisfy a so-called drift condition to ensure that the associated bond prices are arbitrage-free. [T. Björk, *Arbitrage Theory in Continuous Time*, 3rd edn. Oxford University Press, 2009]

heavy tail *see* tail

heavy traffic A concept arising naturally when considering congestion in a queue, or in a network of queues. In its simplest form, the traffic intensity τ of a single server queue (e.g. GI/G/1 with service and inter-arrival times having finite variance) may be defined as the mean number of customers that arrive during a service period. For $\tau > 1$ the queue is unstable (which in the Markov case is to say that the process is not positive recurrent). For $\tau < 1$, the queue is stable, but as $\tau \uparrow 1$, the queue approaches instability; this is called the heavy traffic regime. When suitably scaled, various quantities may have non-degenerate limits as $\tau \uparrow 1$, yielding so-called heavy-traffic theorems. Examples include: (i) **Kingman's thm** for the equilibrium waiting time W, asserting that as $\tau \uparrow 1$, $(1-\tau)W$ converges in distn to an exponential r.v.; and (ii) the realizations of such a queue, under a suitable (diffusive) scaling, converge to a type of reflected Brownian motion.

In queues with many servers other types of heavy-traffic limit may occur, e.g. when the number of servers and customers increase without bound, and $\tau \to 1$, while the probability that all servers are busy is held constant; this may be called the **Halfin–Whitt regime**. For networks of queues, instability may occur even while all traffic intensities are strictly less than 1. [W. Whitt, *Stochastic-process Limits*. Springer, 2002; M. Bramson, Stability and heavy traffic limits for queueing networks, in *Perplexing Problems in Probability*, M. Bramson & R. Durrett (eds). Birkhäuser, 1999]

hedging Arises in financial mathematics, especially in the context of option pricing. Broadly, a hedge is a self-financing portfolio of assets (or investment position) that removes (ideally) or reduces (in practice) the risk of loss in some other asset or function of assets (such as an option). Formally, and more narrowly, a portfolio with value function $V(t), 0 \le t \le T$, is said to hedge (or replicate, or duplicate, or attain, or reach, etc.) an option X with exercise date T if $V(T) = X$ with probability 1; such a hedge is risk-free (or riskless). If every option (or claim) is so hedgeable, then the market is complete, and at time t, $V(t)$ is the no-arbitrage price of the option. More widely, in incomplete markets, one may have risk-neutral (non-unique) hedging portfolios such that $\mathbb{E}V(T) = \mathbb{E}X$, and the hedging strategy must be chosen according to some empirical principle. For example, a strategy minimizing $\mathbb{E}(V(T) - X)^2$ may be called a variance optimal (or dynamically optimal) hedge. [T. Björk, *Arbitrage Theory in Continuous Time*, 3rd edn. Oxford University Press, 2009]

height process *see* contour process

Heisenberg uncertainty *see* uncertainty

Hellinger distance A measure of distance between probability densities $h(x)$ and $g(x)$ on $\mathbb{R}$ (or, more generally, between measures having densities h and g wrt some dominating measure). It is given by

$$d(h,g) = \frac{1}{\sqrt{2}}[\int_{\mathbb{R}} (\sqrt{h} - \sqrt{g})^2 \mathrm{d}x]^{\frac{1}{2}} == 1 - \int_{\mathbb{R}} \sqrt{hg}\mathrm{d}x,$$

where $\int_{\mathbb{R}} (hg)^{1/2} \mathrm{d}x$ is called the Hellinger affinity. (Some authors adopt a different scale by omitting the factor $1/\sqrt{2}$ in the second eqn above.) It is a type of *f*-divergence, with $f(x) = 1 - \sqrt{x}$. [D. Pollard, *A User's Guide to Measure Theoretic Probability*. Cambridge University Press, 2002]

Helly–Bray theorem (or lemma) Any of a number of results linking the convergence of a sequence of distribution functions (or probability measures) to the convergence of a corresponding sequence of integrals against a suitable continuous function. Those most often seen are these:

H-B 1: Let $F_n(x)$ converge vaguely (also called weakly) to $F(x)$ on an interval $[a, b]$, where a and b are continuity points of $F(x)$. Then for any continuous real-valued $g(x), x \in \mathbb{R}$, $\lim_{n\to\infty} \int_a^b g(x)\mathrm{d}F_n(x) = \int_a^b g(x)\mathrm{d}F(x)$. This is extended to integrals over the whole real line for continuous functions $g(x)$ such that $g(+\infty) = g(-\infty) = 0$.

H-B 2: Let $F_n(x)$ converge completely (also called weakly by some authors) to $F(x)$ on $\mathbb{R}$, then for any continuous and bounded fn $g(x)$ on $\mathbb{R}$, $\lim_{n\to\infty} \int_{-\infty}^{\infty} g(x)\mathrm{d}F_n(x) = \int_{-\infty}^{\infty} g(x)\mathrm{d}F(x)$.

[M. Loève, *Probability Theory*, 4th edn. Springer, 1978]

Helly's theorem (alias: **Helly's selection (or extraction, or weak compactness) theorem**) Arises in probability theory in the context of any sequence $F_n(x), n \ge 1, x \in \mathbb{R}$, of distn functions, where it asserts that there exists a right-continuous non-decreasing function $F(x)$ (i.e. a possibly defective distn fn), and a subsequence $F_{r_k}(x), k \ge 1$ such that $F_{r_k}(x)$ converges weakly (or vaguely, according to some writers) to $F(x)$ as $r_k \to \infty$, $0 \le F(x) \le 1$. More strongly, there is a subsequence $F_{r_k}(x)$ converging completely (or weakly, for some authors) to a proper distribution $F(x)$ iff for any $\varepsilon > 0$ there is a finite interval $(a(\varepsilon), b(\varepsilon)]$ such that $F_n(b) - F_n(a) > 1 - \varepsilon$ for all n; that is, the sequence is tight (or equicontinuous at ∞, or mass-preserving, or stochastically bounded). *See* Prokhorov's theorem. [W. Feller, *Introduction to Probability Theory and its Applications*, Vol. 2, 2nd edn. Wiley, 1971]

Helmert's distribution *see* Geary's theorem

hereditary process *see* Papangelou conditional intensity

Herfindahl index *see* quadratic entropy

Hermite distribution (alias: **Gauss–Poisson (or v. v.) distribution**) Arises as a special case of a generalized Poisson distribution, having pgf of the form $G(s) = \exp\{a(s-1) + b(s^2-1)\}$, or more generally $G(s) = \exp\{a(s-1) + b(s^m-1)\}, m \geq 2$. The probabilities and factorial moments can be simply written in terms of suitable Hermite polynomials (generalized in the case when $m > 2$). [C. Kemp & A. Kemp, Some properties of the Hermite distribution, *Biometrika*, 52(3 & 4), 381–394, 1965; C. Kemp & A. Kemp, An alternative derivation of the Hermite distribution, *Biometrika*, 53(3–4), 62–628, 1966]

Hermite polynomials Arise in several areas of probability, and are closely linked with the standard normal density $\phi(x)$ by the fact that $\phi(x)H_n(x) = (-1)^n \frac{\mathrm{d}^n}{\mathrm{d}x^n}\phi(x)$ where $H_n(x)$ is the nth Hermite polynomial having exponential generating function

$$E(x,t) = \sum_{n=0}^{\infty} H_n(x)\frac{t^n}{n!} = \exp\left(xt - \frac{1}{2}t^2\right).$$

These are not to be confused with the standard Hermite polynomials ${}_sH_n(x)$ having exponential gen. fn. $S(x,t) = \exp(2xt - t^2)$, which are generally in use outside probability theory. The polynomials $H_n(x)$ are orthogonal on $\mathbb{R}$ with weight function $\mathrm{e}^{-\frac{1}{2}t^2}$; while ${}_sH_n(x)$ are orthogonal on $\mathbb{R}$ with weight function e^{-t^2}; $H_n(x)$ may be called the rescaled Hermite polynomial. [W. Schoutens, *Stochastic Processes and Orthogonal Polynomials*. Springer, 2000]

Heston's model Arises in financial mathematics where it seeks to price options on an asset with stochastic volatility by assuming that the asset price $S(t), t \geq 0$, is a diffusion obeying the SDE $\mathrm{d}S = \mu S\mathrm{d}t + \sqrt{V(t)}S\mathrm{d}W_1(t)$, where the volatility $\sqrt{V(t)}$ is an Ornstein–Uhlenbeck process obeying $\mathrm{d}\sqrt{V} = -\beta\sqrt{V}\mathrm{d}t + \alpha\mathrm{d}W_2(t)$, and (W_1, W_2) is a bivariate Wiener process whose correlation at any time t is ρ, for suitable μ, α, and β. [S. Heston, A closed-form solution for options with stochastic volatility with applications to bond and currency options, *Rev. Finan. Stud.*, 6(2), 327–343, 1993]

Hewitt–Savage zero-one law Extends Kolmogorov'szero-onelaw (for the tail events of an i.i.d. sequence of r.vs) to symmetric (or exchangeable) events of an i.i.d. sequence $X_1, X_2, \ldots$. (The class of symmetric events includes the tail events.) Let A be such a symmetric event, which is to say that its indicator I_A is unchanged by any permutation of any finite subset of the sequence: $I_A(X_1, \ldots X_n, X_{n+1}, \ldots) = I_A(X_{\pi_1}, \ldots, X_{\pi_n}, X_{n+1}, \ldots)$, where $(\pi_1, \ldots, \pi_n)$ is a permutation of $(1, \ldots, n)$. Then either $\mathbb{P}(A) = 0$ or $\mathbb{P}(A) = 1$. [E. Hewitt & L. Savage, Symmetric measures on Cartesian products, *Trans. Amer. Math. Soc.*, 80(1), 470–501, 1955]

Heyde's example Arises in the moment problem, where it shows that the lognormal distribution, possessing moments of all orders, is not determined by those moments. [C. Heyde, On a property of the lognormal distn, *J. Roy. Stat. Soc. B*, 25(2), 392–393, 1963]

Heyde's theorem Arises in the context of the Hsu–Robbins–Erdös thm on the complete convergence of the partial sums $S_n = \sum_{r=1}^{n} X_r$ of a sequence of i.i.d.r.vs $X_1, X_2, \ldots$. It asserts that if $\mathbb{E}X_1 = 0$ and $\mathbb{E}X_1^2 < \infty$, then as $\varepsilon \downarrow 0$

$$\varepsilon^2 \sum_{r=1}^{\infty} \mathbb{P}(|S_r| > \varepsilon r) = \mathbb{E}X_1^2.$$

For another theorem of C. Heyde, *see* Skitovich–Darmois theorem. [C. Heyde, A supplement to the strong law of large numbers, *J. Appl. Probab.*, 12, 173–175, 1975]

Hida–Cramér canonical representation (alias: **Lévy–Hida–Cramér representation; moving average representation**) Very broadly, this arises in the context of the general problem of deciding when and how a Gaussian process $X(t), t \in \mathbb{R}$, can be represented as an instantaneous function of a Markov process. In particular, in its simplest form (and subject to a number of technical constraints) if there exists a Gaussian process $Z(t)$ having independent increments, and a suitable kernel $f(t,u)$ such that $X(t) = \int_{-\infty}^{t} f(t,u)\mathrm{d}\,Z(u)$, then this may be called a Hida–Cramér representation, or forward canonical representation (if the integral is taken over $[t, \infty)$, it is a backward representation). Such canonical representations can be proved to exist when $X(t)$ is a second-order stationary mean-square continuous Gaussian process. A classic example, due to Lévy, is the representation of a stationary Ornstein–Uhlenbeck process $X(t)$ as $X(t) = \int_{-\infty}^{t} e^{a(s-t)} dW(s)$, $t \geq 0$, where $W(t)$ is the standard Wiener process, and $a > 0$. [T. Hida, Canonical representations of Gaussian processes and their applications, *Mem. Coll Sci. Univ. Kyoto*, 33, 109–155, 1960]

Hida's theorem (or calculus) *see* white noise

hidden Markov process (or model) (alias: **latent Markov model; partially observed Markov sequence**) Broadly, in its simplest form, a bivariate discrete-time Markov chain $(H_n, Y_n), n \geq 0$, taking values in a discrete state space, such that the r.vs $Y_n, n \geq 0$, are conditionally independent given the sequence $H_n, n \geq 0$, and (still conditionally) for each k, Y_k is function only of H_k (and possibly k). The sequence is partially observed, where the Y_n comprise the observations, while the H_n are unobserved and may be called the hidden variables (or the regime, or the state). Interest centres on describing the H_n, given the Y_n, and this can be seen as a special case of filtering. A classic example arises in coding theory, when the input H_n to a discrete memoryless channel is a Markov chain, yielding the output Y_n. (In the statistical literature some of the resulting recursive filters and predictors are called **Baum–Welch** (or **forward–backward**) **algorithms**). [A. Fraser, *Hidden Markov Models and Dynamical Systems*. SIAM, 2008]

Hill's theorem *see* Benford's distribution

Hirschman's uncertainty inequality *see* uncertainty

historical process *see* genealogical process

history-dependent process *see* reinforced processes

hit-and-run algorithm (alias: **random directions algorithm**) A Markov chain Monte Carlo method for generating points that are uniformly distributed on (or uniformly sampled from) a bounded convex region S. The method uses a type of reversible random walk (as its Markov chain), generated thus: if the walk is currently at x, generate a random line L through x whose direction is uniformly distributed over the unit sphere, and then pick a point y uniformly at random on the line segment $L \cap S$; the next step of the walk is to y, and the procedure is repeated. [R. Smith, Efficient Monte Carlo procedures for generating points uniformly distributed over bounded regions, *Oper. Res.*, 32, 1296–1308, 1984]

hit-or-miss *see* Monte Carlo method

hitting time *see* first-passage time

hitting-time theorem (alias: **Kemperman's formula; Otter–Dwass formula**) Asserts that for a left-continuous random walk on the integers, beginning at a positive integer i, the conditional probability of first taking the value 0 at time n, given that the walk is at 0 at time n, is $\frac{i}{n}$. Formally, let $S_n = i + \sum_{r=1}^{n} X_r$ be the random walk, where the steps (X_r) are i.i.d.r.v. taking values in $\{-1, 0, 1, \ldots\}$. Let $T = \inf\{n \colon S_n = 0\}$, where $T = \infty$ if no such n exists, and $T = 0$ when $i = 0$. Then the hitting-time theorem is that for $i > 0$, $\mathbb{P}(T = n) = \frac{i}{n}\mathbb{P}(S_n = 0)$.

The Otter–Dwass formula replaces T by the total number of descendants (progeny) in an ordinary Galton–Watson branching process (on the event of extinction), with i initial members, whose family sizes have the distribution of $X_1 + 1$. *See* Dwass's random walk. [W. Kager, The hitting time theorem revisited, *Amer. Math. Mon.*, 118(8), 735–737, 2001]

Hitsuda integral *see* Skorokhod integral

Hoeffding's combinatorial CLT *see* combinatorial central limit theorem

Hoeffding's decomposition Arises in the context of a so-called U-statistic of a random sample $X_1, \ldots, X_n$, given by $U = \sum h(X_{\pi_1}, \ldots, X_{\pi_m})/\{n(n-1)\cdots(n-m+1)\}$, where the sum is over all permutations of m different integers $1 \leq \pi_r \leq n$, and $h(x_1, \ldots, x_m) \in \mathbb{R}$. The decomposition represents U as a sum of suitable uncorrelated symmetric functions of the Xs. [W. Hoeffding, A class of statistics with asymptotically normal distribution, *Ann. Math. Stat.*, 19, 293–325, 1948]

Hoeffding–Fréchet *see* Fréchet–Hoeffding

Hoeffding's identity Let $F_X(x)$ and $F_Y(y)$ be the distn fns of r.vs X and Y with joint distn $F(x, y)$, and finite covariance $\mathrm{cov}(X, Y)$, then

$$\mathrm{cov}(X, Y) = \int_{-\infty}^{\infty}\int_{-\infty}^{\infty}\{F(x, y) - F_X(x)F_Y(y)\}\mathrm{d}x\mathrm{d}y.$$

More generally, for differentiable non-decreasing functions $g(x)$ and $h(y)$, $\mathrm{cov}(g(X), h(Y)h) = \int\int g'(x)h'(y)\{F(x, y) - F_X(x)F_Y(y)\}\mathrm{d}x\mathrm{d}y$, when the LHS exists. [E. Lehmann, Some concepts of dependence, *Ann. Math. Stat.*, 37(5), 1137–1153, 1966]

Hoeffding's inequality Any one of several inequalities that supply upper bounds for the tail of the distribution of a sum of bounded r.vs, where the bounds depend only on the endpoints of the ranges of the summands X_r and the mean (and possibly the variance) of each X_r. The bounds arise from applying Bernstein's method (see Chernoff's bounds). The most commonly seen Hoeffding tail inequality is for independent r.vs $X_1, \ldots, X_n$, such that $a_r \leq X_r \leq b_r$ with probability 1. Then for $t > 0$, and $S_n = \sum_{r=1}^{n} X_r$,

$$\mathbb{P}(S_n - \mathbb{E}S_n \geq t) \leq \exp\left\{-2t^2 / \sum_{r=1}^{n}(b_r - a_r)^2\right\}$$

with similar inequalities for $\mathbb{P}(S_n - \mathbb{E}S_n \leq -t)$ and $\mathbb{P}(|S_n - \mathbb{E}S_n| \geq t)$. The case when S_n is Binomial $\mathrm{B}(n, p)$, corresponding to Bernoulli r.vs $X_1, \ldots, X_n$, may be called Chernoff's inequality.

The inequalities also hold when $X_1, \ldots, X_n$ is a bounded martingale difference sequence, where $a_r \leq X_r \leq b_r$, so that S_n is a martingale. In this case the terms Hoeffding's bounded martingale difference inequ., Hoeffding–Azuma inequ., Azuma–Hoeffding inequ. or **Azuma's inequality** are all used. [W. Hoeffding, Probability inequalities for sums of bounded random variables, *J. Amer. Stat. Ass.*, 58(301), 13–30, 1963]

Hoffmann–Jørgensen/Klass–Nowicki inequality In its basic form this supplies a bound for the tail of the distribution of the running maximum

$$S_n = \max_{1 \leq r \leq n}\left\|\sum_{k=1}^{r} X_k\right\|,$$

where $X_k, k \geq 1$, is a sequence of independent symmetric random objects taking values in a Banach space. It generalizes Hoffman–Jørgensen's inequality, also known as the KHJ inequality as it extends an inequality of J.-P. Kahane. [A. Gut, *Probability: A Graduate Course*, 2nd edn. Springer, 2013; M. Klass & K. Nowicki, An improvement of Hoffmann-Jørgensen's inequality, *Ann. Probab.*, 28(2), 851–862, 2000; P. Hitczenko & S. Montgomery-Smith, Measuring the magnitude of sums of independent random variables, *Ann. Probab.*, 29(1), 447–466, 2001]

Hoggar's theorem *see* log-concave distribution

Hölder's inequality (alias: **Hölder–Rogers inequality**) An inequality for integrals of suitable measurable functions on a set S. In the context of probability theory it asserts that for random variables X and Y, and positive p, q, r such that $\frac{1}{p} + \frac{1}{q} = \frac{1}{r}$,

$$(\mathbb{E}(|XY|^r))^{1/r} \leq (\mathbb{E}(|X|^p))^{1/p}(\mathbb{E}(|Y|^q))^{1/q},$$

with equality iff $\mathbb{P}(a|X|^p = b|Y|^q) = 1$ for some constants a and b, not both zero. [J. Steele, *The Cauchy–Schwarz Master Class: An Introduction to the Art of Mathematical Inequalities*. Cambridge University Press, 2004]

holding time (alias: **sojourn time; escape time; inter-jump (or event, or arrival) time; waiting time; residence time**) Arises in the context of Markov processes in discrete time, and Markov (or semi-Markov) processes in continuous time whose sample paths are (pure) jump processes, i.e. a.s. right-continuous and constant except for isolated jumps. For any non-absorbing state k of such a process, the holding time is $H = \inf\{t > 0 : X(t+s) \neq k | X(s) = k\}$. For homogeneous continuous-time Markov chains, such a holding time is exponentially distributed; the state k may be called the holding point of H. (Note that 'sojourn time' is also used to denote occupation time, and 'waiting time' also has other uses.) [J. Norris, *Markov Chains*. Cambridge University Press, 1997]

Ho–Lee model Arises in financial mathematics as a model for the short-rate of interest $r(t)$, where it entails that in continuous time $r(t)$ obeys the SDE $\mathrm{d}r = a(t)r\mathrm{d}t + \sigma\mathrm{d}W$, and in the usual (no-arbitrage) framework supplies the term structure of interest rates (i. e. bond price processes), when $a(t)$ is suitably calibrated. [T. Björk, *Arbitrage Theory in Continuous Time*, 3rd edn. Oxford University Press, 2009]

Holevo bound (or theorem) (alias: **Kholevo bound; Gordon–Levitin–Holevo bound**) Arises in the context of the transmission of classical (Shannon) information via a noiseless quantum channel, where it supplies an upper bound for the information accessible to the recipient. Formally, the sender encodes a r.v. X having distribution $p_r, 1 \leq r \leq n$, by means of a set of states $\rho_r, 1 \leq r \leq n$ of a quantum system, yielding an expected density operator $\rho = \sum_{r=1}^{n} p_r \rho_r$. The recipient performs some measurement whose outcome supplies a r.v. Y, with distribution $\boldsymbol{q}$. Then the information J accessible to the recipient satisfies

$$J = \max I(X;Y) \leq S(\rho) - \sum_r p_r S(\rho_r) = H,$$

where I is the mutual information, and $S(\rho)$ is the Von Neumann entropy given by

$$S(\rho) = -\mathrm{Trace}(\rho \log \rho).$$

The function H may be called the Holevo information, and $H \leq \log n$. If the receiver's freedom of measurement is limited then there is a similar but tighter bound called the (Holevo–) Schumacher–Westmoreland–Wootters bound. [A. Holevo, Bounds for the quantity of information transmitted by a quantum information channel, *Prob. Inf. Transmission*, 9, 177–183, 1973]

Holley's inequality Arises as a generalization of the FKG inequality; given two probability distributions μ_1 and μ_2 on a finite distributive lattice, it supplies a condition for one to dominate (i.e. be stochastically larger than) the other. In a common application, the relevant state space Ω for the measure is that of a random graph, $\Omega = \{0,1\}^E$, partially ordered by inclusion, where it asserts that if $\mu_1(A \cup B)\mu_2(A \cap B) \geq \mu_1(A)\mu_2(B)$ for any $A, B \in \Omega$, then μ_1 is stochastically larger than (i.e. dominates) μ_2. (Which is to say that for any non-decreasing function f on $\Omega, \mathbb{E}_1 f(\omega) \geq \mathbb{E}_2 f(\omega)$.) [R. Holley, Remarks on the FKG inequalities, *Commun. Math. Phys.*, 36, 227–231, 1974]

holospoudic array *see* null array

Holtsmark distribution In one dimension, and in standard form, this has a density that is symmetric about the origin, and is a stable (i.e. Lévy-stable) distribution with exponent $3/2$, having characteristic function $\phi(t) = \exp(-|t|^{3/2})$. It arises naturally as the distribution of the random gravitational field at the origin, due to a Poisson process of suitably random astronomical objects (e.g. stars and galaxies). [S. Chandrasekhar, Stochastic problems in physics and astronomy, *Rev. Modern Phys.*, 15(1), 1–89, 1943]

homogenous process Broadly, a random process (or field) is said to be homogeneous if its joint distns, or conditional distns, or covariance functions, are invariant under some type of transformation. In the case of a rotation, the process may be called isotropic, e.g. if $X(\mathbf{r})$ lies on the unit sphere $|\mathbf{r}| = 1$ then X is strongly (or weakly) homogenous according as $X(\mathbf{r}) - X(\mathbf{s})$ (or $\mathrm{cov}(X(\mathbf{r}), X(\mathbf{s}))$ depends only on the angle between the unit vectors $\mathbf{r}$ and $\mathbf{s}$. Similarly, in the case of translation the process may be called strongly or weakly stationary. In particular, a random process $X(t), t \in T$, with $\mathbb{P}(X(t) = y | X(s) = x) = p(x, y; s, t)$ is called time or space-homogenous according as

$$p(x, y; s, t) = f(x, y; t - s) \quad \text{or} \quad p(x, y; s, t) = g(y - x; s, t).$$

A Markov process that is time-homogeneous is often said to have **stationary transition probabilities** (see stationarity).

honest *see* dishonest

honest time Arises in the context of a filtration $\mathcal{F}_t, t \geq 0$, on a complete probability space $[\Omega, \mathcal{F}, \mathbb{P}]$, that satisfies the usual conditions, i.e. the filtration is increasing, right-continuous, and includes the $\mathbb{P}$-null sets of $\mathcal{F}$. A random time T (i.e. a non-negative $\mathcal{F}$-measurable r.v.), is said to be honest if, for every t, T is equal to some $\mathcal{F}_t$-measurable r.v. R_t on the event $\{T < t\}$. All stopping times T are honest, as is seen by setting $R_t = T \wedge t$. [P. Meyer et al., Birth and death of Markov processes, *Proc. 6th Berkeley Symp. Math. Stat. Probab.*, 3, 293–305, 1972]

Hoppe–Pólya urn (alias: **Hoppe's urn**) Arises in the context of mathematical genetics as a model for mutation in a population. An urn contains one ball (sometimes called the mutator) of mass μ, and balls of various colours (that are not the colour of the mutator), each of mass 1. At any time n a ball is selected at random in proportion to its mass and returned to the urn with another ball of the same colour, if the selected ball was not the mutator; if the mutator is selected it is returned together with a ball of a new colour, different from all those present. [F. Hoppe, Pólya-like urns and the Ewens sampling formula, *J. Math. Biol.*, 20, 91–94, 1984]

hopping process *see* interacting particle process

horizon Arises in sequential decision theory and related fields, e.g. optimal stopping, Markov decision processes, and stochastic control. Broadly, the horizon of such a decision process is that time τ after which the values of the process are independent of the optimal policy at any time prior to τ. Such a horizon is called finite or limited, and certainly obtains if the process is observed for only a finite time, or the cost function depends only on the process over a finite interval. Otherwise the horizon is infinite, or unlimited. Policies may be devised in which the horizon is a function of the current time-point of the process, leading to shrinking, receding, or rolling horizons. [N. Bäuerle & U. Rieder, *Markov Decision Processes with Applications to Finance*. Springer, 2011]

Hörmander's condition (alias: **H's hypothesis; H's theorem**) Originally arose as a sufficient (and sometimes necessary) condition for a suitable second-order differential operator to yield a smooth fundamental solution of the Cauchy (i.e. initial value) problem. This theorem then also supplies a condition for the solution of a stochastic differential equation to have a smooth density, because such a density satisfies a second order partial differential equation. An alternative, essentially probabilistic, proof of this result from the SDE viewpoint is supplied by the Malliavin calculus (which was developed initially to do just that). [D. Nualart, *The Malliavin Calculus and Related Topics*. Springer, 2006]

Hotelling analysis *see* Karhunen–Loève expansion

house-selling problem (alias: **job search problem; asset-selling problem**) A classic problem of optimal stopping, which is related to the secretary problem. In a broad formulation an asset (possibly a vector) is to be disposed of by an agent who receives a sequence of random offers to buy it; there are holding costs, and possibly a horizon or deadline (either of which may be random). There are many variants. *See* Cayley–Moser problem. [F. T. Bruss, & T. Ferguson, Multiple buying or selling with vector offers, *J. Appl. Probab.*, 34, 959–973, 1997]

Howard policy improvement (alias: **policy improvement method**) Arises in the context of Markov decision processes, where it supplies an iterative procedure that generates a policy yielding a solution to the Bellman (or dynamic programming) equation for the optimal policy and its corresponding value V. The method begins with an arbitrary policy π_0 having corresponding value V_0 when the policy π_0 is used; it then finds the optimal policy π_1 using the value function V_0, yielding a new value V_1 corresponding to π_1. This procedure is iterated until successive policies are identical, and hence optimal (as solutions of Bellman's optimality equation). **MacQueen extrapolation** is an improved version yielding convergent upper and lower bounds for the value V. [R. Howard, *Dynamic Programming and Markov Processes*, 1960, MIT Press; J. MacQueen, A modified dynamic programming method for Markovian decision problems, *J. Math. Anal. Appl.*, 14, 38–43, 1966]

h-skeleton *see* discrete skeleton

Hsu–Robbins–Erdös theorem (or SLLN) Arises in the context of the complete convergence of a sequence S_n, $n \geq 1$, of r.vs. In the case when $S_n = \frac{1}{n}\sum_{r=1}^{n} X_r$ comprises the sequence of running averages of i.i.d.r.vs X_n, $n \geq 1$, the theorem (which is an SLLN) asserts that S_n converges completely to $\mathbb{E}X_1$ iff var $X_1 < \infty$. [A. Gut, Complete convergence for arrays, *Periodica Mathem. Hungarica*, 25(1), 51–75, 1992]

***h*-transform** *see* Doob's harmonic transform

Hu correspondence Arises in information theory when considering linear relationships satisfied by the Shannon entropy $H(X)$ of a simple r.v. X, and Shannon's mutual information $I(X;Y)$ between simple r.vs X and Y. If sets $A_1, A_2, \ldots$ and an additive function $m(A_r), r \geq 1$, on those sets are put into a certain correspondence with H and I (and conditional forms of these), then any linear relationship (including inequality) between these entropies and mutual informations is true iff the corresponding linear relationship for the set function $m(\cdot)$ is true. Thus the first Hu correspondences are

$$H(X) \leftrightarrow m(A)$$
$$I(X;Y) \leftrightarrow m(A \cap B)$$
$$H(X,Y) \leftrightarrow m(A \cup B)$$
$$H(X|Y) \leftrightarrow m(A \backslash B)$$
$$I(X;Y|Z) \leftrightarrow m((A \cap B) \backslash C)$$

etc., and it is seen, for example, that

$$I(X;Y) = H(Y) - H(Y|X) \leftrightarrow m(A \cap B) = m(B) - m(B \backslash A).$$

[K. Hu, On the amount of information, *Theor. Probab. Appl.*, 7, 439–447, 1962]

Hull–White model Arises in financial mathematics, where it denotes any one of several interest-rate models used in pricing derivatives based on interest rates. The simplest of the Hull–White models has one factor and postulates an SDE for the short-term interest rate $r(t)$ of the form $\mathrm{d}r = \{\theta(t) + a(t)(b - r)\}\mathrm{d}t + \sigma(t)r^{\beta}\mathrm{d}W$, (for suitable $\beta, b, a(t), \theta(t), \theta(t)$, and $\sigma(t)$), where $W(t)$ is the Wiener process. The case $\beta = 0$ forms an extended Vasicek model, and $\beta = 1/2$ is an extended

Cox–Ingersoll–Ross model. [J. Hull & A. White, Pricing Interest-Rate-Derivative securities, *Rev. Finan. Stud.*, 3(4), 573–592, 1990]

Hu–Meyer formula Arises in stochastic analysis (or calculus). In its original classical formulation it states a relationship between multiple stochastic multiple integrals of Itô (or Wiener–Itô) type, and multiple Stratonovich stochastic integrals (also considered by Wiener). More generally, it refers to similar identities between other types of stochastic multiple integral. [M. Farré et al., Multiple Stratonovich integral and Hu-Meyer formula for Lévy processes, *Ann. Probab.*, 38(6), 2136–2169, 2010]

Hungarian construction *see* Komlós–Major–Tusnády construction

Hunt's lemma (alias: **Hunt's dominated convergence theorem; Blackwell–Dubins theorem**) A useful extension of the dominated convergence theorem that can be seen as a generalized martingale convergence theorem. Let $X_n, n \geq 1$, be a sequence of r.vs that converges a.s. to X. If for some r.v. Y, and for all n, $|X_n| < Y$, a.s., where $\mathbb{E}Y < \infty$, then for any monotone increasing (or decreasing) sequence of sigma-fields $\mathcal{F}_n$, with limit $\mathcal{F}_\infty$, we have $\lim_{n\to\infty}\mathbb{E}(X_n|F_n) = \mathbb{E}(X|\mathcal{F}_\infty)$, both a.s. and in mean. [D. Blackwell & L. Dubins, Merging of opinions with increasing information, *Ann. Math. Stat.*, 33(3), 882–886, 1962]

Hunt process A Markov process having a number of desirable properties as follows: let $X(t), t \geq 0$, be a homogeneous, right-continuous process (possibly substochastic) whose suitably completed (i.e. augmented) natural filtration is $\mathcal{F}_t$. Then $X(t)$ is a Hunt process if:

1. it has the strong Markov property wrt $\mathcal{F}_t$, and
2. it is quasi-left-continuous, which is to say that for any sequence $T, T_1, T_2, \ldots$ of $\mathcal{F}_t$–adapted stopping times such that $T_n \uparrow T \, as \, n \to \infty$,

$$\mathbb{P}(\{\lim_{n\to\infty} X(T_n) = X(T)\} \cap \{T < \infty\}) = \mathbb{P}(T < \infty).$$

Feller processes are also Hunt processes. [K. L. Chung & J. Walsh, *Markov Processes, Brownian Motion, and Time Symmetry*, 2nd edn. Springer, 2005]

Hunt's representation formula Arises in the context of Lévy processes in a Lie group, where it supplies a type of Lévy–Khinchin formula for the infinitesimal generator of the process. [M. Liao, *Lévy Processes in Lie Groups*. Cambridge University Press, 2004]

Hunt's stopping time theorem Extends Doob's optional stopping time theorem as follows: let $X(t)$, adapted to the filtration $\mathcal{F}_t$, be either a martingale or a non-negative integrable submartingale, with S and T being bounded stopping times such that $S \leq T$. Then $\mathbb{E}(X(T)|\mathcal{F}_S) \geq X(S)$, with equality if $X(t)$ is a martingale. The result holds for arbitrary stopping times $S \leq T$ whenever $X(t), t \geq 0$, is uniformly integrable and $\lim_{t\to\infty} X(t) = X(\infty)$. [G. Hunt, *Martingales et Processus de Markov*. Dunod, 1966]

Hunt's theorem Any one of several results in the probabilistic potential theory developed by G. Hunt; this is a generalization of classical potential theory that is obtained by defining the corresponding concepts in terms of the stationary transition probabilities (or kernel) of a Markov process on a suitable space. Conversely, potential theory concepts may be interpreted in probabilistic terms, with applications to Markov processes and martingales. Perhaps that most often referred to as Hunt's theorem relates the hitting probabilities of a set to a suitably normalized equilibrium potential function. Another result often called Hunt's theorem is the strong Markov property of the Wiener process. [G. Hunt, Some theorems concerning Brownian motion, *Trans. Amer. Math. Soc.*, 81, 294–319, 1956; G. Hunt, Markov processes and potentials, I, II, & III, IV, *J. Math.*, 1, 44–93 & 316–369, 1957 & 2, 151–213, 1958]

Hurst exponent (alias: **Hurst parameter; Hurst coefficient; Hurst index; self-similarity exponent**) Arises in the context of random processes, $X(t), t \geq 0$, that exhibit self-similar behaviour (or self-affinity, or fractal paths) in the sense that for any $a \geq 0$, and some $H \in (0, 1)$, the processes $X(at)$ and $a^H X(t)$ have the same finite-dimensional distributions. In such a case, H is called the Hurst exponent of the process; where values of H in $(\frac{1}{2}, 1)$ correspond to processes having long-range dependence (called persistence or trend reinforcing), and values of H in $(0, \frac{1}{2})$ correspond to chaotic and highly fluctuating processes (also called anti-persistence, or mean reverting). A classic example of such a process is fractional Brownian motion. [P. Embrechts & M. Maejima, *Self-similar Processes*. Princeton University Press, 2002]

Huygens' problems Appear in C. Huygens' short treatise on probability, which is arguably the first book on probability: *De Ratiociniis in Ludo Aleae* (*Calculations in Games of Chance*), 1657; a Dutch version was issued in 1660, and an English translation by J. Arbuthnot in

1692. The book arose from Huygens' visit to Paris in 1655, where he learned of the problem of the points. The first and third problems were posed by P. Fermat, and the fifth (gambler's ruin problem) by B. Pascal.

1. Two players A and B roll two dice in the order ABBAABBA. . ., where A wins with a total of 6, and B wins with a total of 7. Then $\mathbb{P}$ (A wins) : $\mathbb{P}$ (B wins) = 10 355:12 276.
2. Three players A, B, and C, take a ball at random from an urn with four white and eight black balls, with replacement, in the order ABCABC . . .. The first white ball wins. Then $\mathbb{P}$ (A wins) : $\mathbb{P}$(B wins) : $\mathbb{P}$(C wins) = 9:6:4.
3. A player takes four cards at random without replacement from a pack of 40 comprising ten each of four different colours. Then the probability that the cards are of all different colours is $\frac{1000}{9139}$.
4. With the urn of problem 2, if seven balls are taken without replacement, the probability that exactly three are white is $\frac{35}{99}$.
5. Two players A and B start with 12 counters, and roll three dice repeatedly. At each roll, A gains a counter from B if the total score is 14, or gives a counter to B if the total is 11, until one or other has all the counters. Then $\mathbb{P}$ (A wins) : $\mathbb{P}$ (B wins) $= 244\,140\ 625 : 282\,429\,536\,481 = 5^{12} : 9^{12}$.

[A. Hald, *History of Probability and Statistics and their Applications Before 1750*. Wiley, 1990]

hyperbolic Brownian motion *see* Brownian motion

hyperbolic distribution (alias: **Barndorff–Nielsen distribution**) A distribution with density given (in standard form) by

$$f(x) = \frac{(a^{-1}+b^{-1})^{-1}}{c\sqrt{ab}K_1(c\sqrt{ab})} \exp\left\{-\frac{1}{2}(a+b)(c^2+x^2)^{\frac{1}{2}} + \frac{1}{2}(a-b)x\right\},$$

where $x \in \mathbb{R}$, and a, b, and c are positive constants, and $K_1(\cdot)$ is a modified Bessel function of the second kind. The name arises from the fact that the logarithm of f defines a hyperbola, and the density is generalized to higher dimensions by making it a hyperboloid. It may be further extended, yielding generalized hyperbolic distns that include the inverse Gaussian distn and variance-gamma distn. *See also* inverse Gaussian distribution. [O. Barndorff-Nielsen, Exponentially decreasing distributions for the logarithm of particle size, *Proc. R. Soc. Lond. A*, 353, 401–419, 1977]

hyperbolic secant distribution (alias: **sech-squared distribution; logistic distribution**) Has the density (in standard form) given by

$$f(x) = \frac{1}{2}\operatorname{sech}\frac{\pi x}{2}, \quad x \in \mathbb{R}$$

with distn fn $F(x) = \frac{2}{\pi}\tan^{-1}\left(\exp\frac{\pi x}{2}\right)$, and characteristic function $\phi(t) = \operatorname{sech} t, t \in \mathbb{R}$. [D. Vaughan, The generalized secant hyperbolic distn and its properties, *Commun. Stat. Theor. Meth.*, 31(2), 219–228, 2002]

hypercontractive inequality Broadly, hypercontractivity arises in the context of an operator T (typically linear) mapping an L_p space boundedly into an L_q space for $1 \le p < q$. Then, denoting the L_p-norm by $\|\cdot\|_p$, an inequality of the form $\|Tf\|_q \le \|f\|_p$, that holds for all suitable f, is called hypercontractive. For example, in the context of (pseudo-)Boolean functions, let f be a real-valued function on the probability space $\{-1,1\}^n$ equipped with the uniform probability distribution. This can be written uniquely as a multilinear polynomial in the form $f = \sum_S f_S \prod_{r\in S} x_r$, $S \subseteq \{1, 2, \ldots, n\}$, for suitable coefficients f_S, where $x_r \in \{-1, 1\}$. Defining (for $0 < a < 1$) $T_a(f) = \sum f_S a^{|S|} \prod_{r\in S} x_r$, the Bonami–Becker hypercontractive inequ. asserts that $\|T_a f\|_2 \le \|f\|_{1+a^2}$, where $\|X\|_p = (\mathbb{E}|X|^p)^{1/p}$. By duality, for $1 \le d \le m$ and for a suitable function g_d, this is equivalent to $\|g\|_r \le (r-1)^{-\frac{d}{2}}\|g\|_2$; inequalities of this form are also called hypercontractive. If $q = \infty$, then one speaks of an ultracontractive inequality. [E. Friedgut & V. Rödl, Proof of a hypercontractive estimate via entropy, *Israel J. Math.*, 125, 369–380, 2001]

hyperexponential distribution Has a density of the form $f(x) = \sum_r p_r \lambda_r e^{-\lambda_r x}$, $x > 0$, $\lambda_r > 0$, $p_r > 0$, where $\sum_r p_r = 1$. It is thus a mixture of a collection of exponential densities. More generally, $p_r \in \mathbb{R}$, requiring conditions ensuring that $f(x) \ge 0$ for all x; this is not a mixture as above, and may be called **generalized hyperexponential**. [C. Harris et al., A note on generalized hyperexponential distributions, *Commun. Stat. Stoch. Models*, 8(1), 179–191, 1992]

hypergeometric distribution Arises in sampling without replacement from a population of two types (note: sampling with replacement yields the binomial distn). Thus if r balls are drawn at random from an urn containing m mauve balls and w white balls, with no replacement, the probability of obtaining k mauve and $r-k$ white balls is

$$p(k) = \binom{m}{k}\binom{w}{r-k}\bigg/\binom{m+w}{r}$$
$$= \left\{\binom{w}{r}\bigg/\binom{m+w}{r}\right\}\left\{\binom{r}{k}\binom{m}{k}\bigg/\binom{w-r+k}{k}\right\}.$$

The second form explains the name, via the clear connection with the hypergeometric fn H:

$$H = \sum_{k=0}^{\infty} x^k \binom{r}{k}\binom{m}{k}\bigg/\binom{w-r+k}{k}.$$

The distribution has mean $rm/(m+w)$ and variance $r\left(\frac{m+w-r}{m+w-1}\right)\frac{mw}{(m+w)^2}$.

It also arises naturally as the conditional distribution of X, given $X+Y=r$, when X and Y are independent binomial r.vs with parameters (m, p) and (w, p) respectively. More generally, if X and Y are multinomial with parameters $(m, p_1, \ldots, p_t)$ and $(w, p_1, \ldots, p_t)$ respectively, the conditional distn of X given $X+Y=(n_1, \ldots, n_t)$ is

$$f(x_1, \ldots, x_t) = \binom{n_1}{x_1}\cdots\binom{n_t}{x_t}\bigg/\binom{\sum_1^t n_i}{\sum_1^t x_i},$$

the multivariate hypergeometric distn arising when sampling from a population of t types without replacement. More generally, still, if X has parameters $(m, p_1, \ldots, p_t)$ and Y has parameters $(w, q_1, \ldots, q_t)$ then X given $X+Y$ has the non-central multivariate hypergeometric distn. Finally, a wider class of **hypergeometric-factorial** distributions may be based on the more general hypergeometric series, yielding e.g.

$$\mathbb{P}(X=x) = \frac{\Gamma(c-b)\Gamma(c-a)}{\Gamma(c)\Gamma(c-a-b)}\frac{a^{\bar{x}}b^{\bar{x}}}{x!c^{\bar{x}}}, \quad x = 0, 1, 2, \ldots,$$

where the rising factorial $a^{\bar{x}} = a(a+1)\cdots(a+x-1)$ is 1 if $x=0$. There are also two **non-central hypergeometric distns**, named after R. Fisher and K. Wallenius respectively. [A. Kemp, A wide class of discrete distributions and the associated differential equations, *Sankhyā, Indian J. Stat.*, 30(4), 401–410, 1968; A. Fog, Calculation methods for Wallenius' noncentral hypergeometric distribution, *Commun. Stat. Sim. Comp.*, 37(2), 258–273, 2008]

hyperspherical distribution see spherical distribution

hypoexponential distribution Arises as the distribution of the sum of $n \geq 2$ independent exponential r.vs with respective parameters $\lambda_1, \lambda_2, \ldots \lambda_n$, where $\lambda_i \neq \lambda_j, i \neq j$. [S. Ross, *Introduction to Probability Models*, 11th edn. Academic Press, 2014]

I

Ibragimov's theorem Any of a number of central limit theorems (or functional CLTs) for a sequence of r.vs obeying various mixing conditions, including the uniformly strong (or Ibragimov) mixing condition. In particular, the **Billingsley–Ibragimov CLT** holds for a suitable sequence of martingale differences. [P. Billingsley, *Convergence of Probability Measures*, 2nd edn. Wiley, 1999]

ideal probability metric A probability distance satisfying two extra conditions; thus a metric $d(X,Y)$, defined for pairs of r.vs X and Y, is called ideal of order r if:

(i) $d(aX,aY) = |a|^r d(X,Y)$, for any real a, any X, any Y, and some real r. This is the homogeneity property;

(ii) $d(X+Z, Y+Z) \leq d(X,Y)$, for any X and Y, with Z independent of (X,Y). This is the (weak) regularity property; the (strong) regularity property requires the inequality to hold for all X, Y, and Z.

[S. Stoyanov *et al.*, Probability metrics with applications in finance, *J. Stat. Theor. Practice*, 2(2), 253–277, 2008]

identically distributed Random variables X and Y are identically distributed if their respective distribution functions are the same; or, if they have densities, their respective densities are the same. This is equivalent to the condition that $\mathbb{P}(X \in B) = \mathbb{P}(Y \in B)$ for any Borel set of $\mathbb{R}$. This definition holds likewise for random vectors, with $\mathbb{R}^n$ replacing $\mathbb{R}$; and similarly for random elements. [B. Liu, *Uncertainty Theory*, 2nd edn. Springer, 2007]

identification entropy *see* quadratic entropy

I-divergence *see* Kullback–Leibler divergence

IFRA *see* ageing

Ignatov's theorem Let $X_1, X_2, \ldots$ be a sequence of i.i.d.r.vs having common distribution $F(x)$. The term X_n is a record value (or 1-record value) if $X_n \geq X_i$ for $1 \leq i \leq n$. (Thus X_1 is the first record value.) Further, the term X_n is an r-record value if $X_i \geq X_n$ for exactly r values of i in $\{1, \ldots, n\}$. That is to say X_n is an r-record if it has rank r in the collection $\{X_1, \ldots, X_n\}$. Let L_r denote the vector of r-record values in order of their occurrence; thus L_1 is simply the sequence of ascending ladder values of $X_1, X_2, \ldots$. Then Ignatov's theorem is that the vectors $L_1, L_2, L_3, \ldots$ are independent and identically distributed, each being a non-homogeneous Poisson process. [R. Engelen *et al.*, Ignatov's theorem: a new and short proof, *J. Appl. Probab.*, 25, 229–236, 1988]

images *see* reflection principle

imbed *see* embedding

imbedding *see* embedding

immigration process Arises naturally in the context of population models, which in their simplest form entail isolated (closed) groups whose size evolves according to essentially internal dynamics. Increments to the population arising independently, and generated by exterior dynamics, constitute immigration. The most elegant expression of the concept is perhaps the **Foster–Williamson thm**: let Z_n be a subcritical ($\mathbb{E}Z_1 \leq 1$) ordinary Galton–Watson branching process with $Z_0 = 1$, and $\mathrm{E}s^{Z_1} = G(s)$. Assume stationary immigration, such that the nth generation is augmented by Y_n, where $(Y_n, n \geq 1)$ are i.i.d. with $\mathbb{E}s^{Y_1} = H(s)$, independently of $Z_1, \ldots, Z_{n-1}$. Then, as $n \to \infty$, Z_n converges to a proper random variable iff

$$\int_0^1 \frac{1 - H(s)}{G(s) - s}\, \mathrm{d}s < \infty.$$

[K. Athreya & P. Ney, *Branching Processes*. Springer, 1972; reprinted by Dover, 2004]

importance measure (alias: **reliability importance; influence function**) Any of a large number of functions (often called indices) indicating the relative importance (or influence) of a component in the functioning of a system. Examples arise in the theories of reliability, voting, games, statistical physics, and control; very commonly the models have Boolean structure functions, and may be static or time dependent.

For example, consider a system S with independent components where $p_r = \mathbb{P}(X_r = 1) = 1 - \mathbb{P}(X_r = 0)$ is the probability that the rth component is working (not failed) and X_r is the indicator of that event. (And $p_r(t) = \mathbb{P}(T_r > t)$ is the probability that the rth component has not failed by t, in the time-dependent case.) Let the Boolean structure function of S (assumed coherent), be $\phi(X_1, \ldots, X_n) \in \{0, 1\}$, so that ϕ is the indicator of the event that S is working, and write $\mathbb{P}(\phi = 1) = h(p_1, \ldots, p_n)$. Then the **Birnbaum importance** of the rth component C_r is the probability that S fails if C_r fails (which is to say that C_r is a pivotal or critical component of S), namely

$$\begin{aligned} I_r(\mathbf{p}) &= \mathbb{P}(\phi(X_1, \ldots, 1, X_{r+1}, \ldots, X_n) \\ &\quad - \phi(X_1, \ldots, 0, X_{r+1}, \ldots, X_n) = 1) \\ &= h(p_1, \ldots, 1, p_{r+1}, \ldots p_n) - h(p_1, \ldots, 0, p_{r+1}, \ldots p_n) \\ &= \frac{\partial h}{\partial p_r}, \end{aligned}$$

by Russo's lemma for monotone Boolean functions. When $p_r = p$ for all r, $I_S = \int_0^1 I_r(p)\,\mathrm{d}p$ has been called the **Shapley–Shubik index** of C_r; and when $p_r = 1/2$ for all r, $I_r(\frac{1}{2}, \ldots, \frac{1}{2})$ has been called the **Penrose (or P.–Banzhaf or P.–B.–Coleman)** index.

Note that Birnbaum's importance for C_r depends only on the system structure and the failure probabilities of the other components, which can be seen as a potential drawback. In the time-dependent case, $\mathbf{p} = \mathbf{p}(t)$, where T_r has density $f_r(t)$, the **Barlow–Proschan importance** is $I_b = \int_0^\infty I_r(\mathbf{p}) f_r(t)\mathrm{d}t$. Other importance functions include:

- **Criticality importance**: the probability that C_r is both critical and not working, given the system fails.
- **Fussell–Vesely** (or **Vesely–Fussell**) **importance**: the conditional probability given S fails that at least one set of components (including C_r), that is minimally sufficient to cause S to fail, has failed.
- **Improvement potential:** the difference between the reliabilities of the system in which C_r always works, and the actual system.
- **Natvig measure:** like the Barlow–Proschan measure, this is a weighted time-integral of the Birnbaum importance, with weight function proportional to $-\mathbb{P}(T_r > t) \log\{\mathbb{P}(T_r > t)\}$.

[B. Natvig, *Multistate Systems Reliability Theory with Applications*. Wiley, 2011; G. Kalai & S. Safra, *Threshold Phenomena and Influence, Computational Complexity and Statistical Physics*, A. Percus *et al.* (eds). Oxford University Press 2006]

importance sampling Arises in Monte Carlo methods (for calculating an integral or an expectation) as a technique for variance reduction. In the basic case, if $g(x)$ is to be integrated, then the optimal density $f(x)$ of the points simulated in Monte Carlo integration is proportional to $g(x)$; in practice, one often uses a conveniently tractable density q, (of a r.v. Y) that is sufficiently 'close' to g, where $q > 0$ is called the importance density. Then, for an expectation: $\mathbb{E}g(X) = \int g(x)p_X(x)\mathrm{d}x = \int (g(x)/q(x))p_X(x)q\mathrm{d}x = \mathbb{E}(g(Y)p_X(Y)/q(Y))$, with likelihood ratio p_X/q.

[R. Rubinstein & D. Kroese, *Simulation and the Monte Carlo Method*, 2nd edn. Wiley, 2008]

importance sampling identity (alias: **change of measure identity**) Let $f(x)$ and $g(x)$ be the probability densities (or mass functions) of r.vs X and Y respectively, and $h(x)$, $x \in \mathbb{R}$, a function such that $g(x) = 0 \Rightarrow f(x)h(x) = 0$. Then

$$\mathbb{E}h(X) = \mathbb{E}\left\{\frac{h(Y)f(Y)}{g(Y)}\right\}.$$

[S. M. Ross, Using the importance sampling identity to bound tail probabilities, *Prob. Eng. Inf. Sci.*, 12(4), 445–452, 1998]

impossible event *see* event

improper *see* defective distribution

improper distribution *see* defective distribution

IMRL *see* ageing

inaccessible boundary *see* Feller boundary classification

inclusion-exclusion bounds Arise from the basic elementary counting principle that in a set S, whose members may possess any of n possible properties, the number having at least one such property equals the number having exactly one property, less the number having exactly two, plus the number having exactly three, and so on. Applications in probability follow by letting S be the sample space, and letting a 'property' be membership of an event A_r, $1 \le r \le n$. Thus, define $s_r = \sum_I \mathbb{P}(\cap_{i \in I} A_i)$, where the summation is over all subsets I of $\{1, \ldots, n\}$ such that $|I| = r$. Set $a_r = \mathbb{P}$ (at least r of the n events occur), and $e_r = \mathbb{P}$ (exactly r of the n events occur), and then Jordan's identities (Charles (Karoly), not Camille) assert that

$$a_r = s_r - \binom{r}{r-1}s_{r+1} + \binom{r+1}{r-1}s_{r+2} - \cdots + (-1)^{n-r}\binom{n-1}{r-1}s_n,$$

$$e_r = s_r - \binom{r+1}{r}s_{r+1} + \binom{r+2}{r}s_{r+2} - \cdots + (-1)^{n-r}\binom{n}{r}s_n.$$

The inclusion-exclusion bounds follow from the fact that for either series terminated at the kth term the resulting error is smaller in absolute value that the first omitted term, and has the same sign. For $r = 1$, the first such truncations are called Boole's inequalities (or Boole–Fréchet), and the remainder are called the Bonferroni inequalities. For $r \geq 2$, one speaks of inclusion-exclusion or Bonferroni, or Fréchet, or (occasionally) Jordan inequalities. [J. Galambos & I. Simonelli, *Bonferroni-type Inequalities with Applications*. Springer, 1996]

increasing convex ordering *see* stop-loss ordering

increasing failure rate *see* ageing

independence (alias: **stochastic independence; statistical independence; mutual independence; indifference**) A fundamental concept in probability theory, where it formalizes the popular intuition about events A and B such that knowing A (resp. B) occurs (or does not occur) does not change the probability that B (resp. A) occurs. Formally: A and B are independent if $\mathbb{P}(A \cap B) = \mathbb{P}(A)\mathbb{P}(B)$; or, equivalently in terms of the conditional probabilities, $\mathbb{P}(A \mid B) = \mathbb{P}(A)$, or $\mathbb{P}(B \mid A) = \mathbb{P}(B)$.

Likewise, random variables X and Y defined on the same space $(\Omega, \mathcal{F}, \mathbb{P})$ are independent if, for all x and y, the events $A = \{\omega : X(\omega) \leq x\}$ and $B = \{\omega : Y(\omega) \leq y\}$ are independent, i.e. $\mathbb{P}(\{X \leq x\} \cap \{Y \leq y\}) = \mathbb{P}(X \leq x)\mathbb{P}(Y \leq y)$, or equivalently in terms of distribution functions:

$$F(x, y) = \mathbb{P}(\{X \leq x\} \cap \{Y \leq y\}) = F_X(x)F_Y(y),$$

or, equivalently, for any Borel sets A and B in $\mathbb{R}$:

$$\mathbb{P}(\{X \in A\} \cap \{Y \in B\}) = \mathbb{P}(X \in A)\mathbb{P}(Y \in B);$$

or, for r.vs with a joint density, $f(x, y) = f_X(x)f_Y(y)$.

More generally, a family $A_n, n \in I$, of events is said to be (mutually, or jointly) independent if, for any subset $n_1, \ldots, n_k$ of the index set I,

$$\mathbb{P}(A_{n_1} \cap \cdots \cap A_{n_k}) = \prod_{r=1}^{k} \mathbb{P}(A_{n_r}),$$

with a similar formulation for the independence of a family of r.vs.

More weakly, the family $A_n, n \in I$, is **pairwise independent** if, for any $i \neq j$,

$$\mathbb{P}(A_i \cap A_j) = \mathbb{P}(A_i)\mathbb{P}(A_j),$$

with a similar definition for r.vs. This is implied by, but does not in general imply, independence. Families may also be triplewise (etc.) independent, defined analogously. Note that the n-wise independence of $A_1, \ldots, A_n$ does not imply their independence.

Events A and B are **conditionally independent** given C if $\mathbb{P}(A \cap B \mid C) = \mathbb{P}(A \mid C)\mathbb{P}(B \mid C)$, as may be families of events and r.vs, defined similarly. This is not implied by, nor does it imply, independence. If for all $n \in I$ we have

$$\begin{aligned}&\mathbb{P}(B \cap A_n \mid A_1, \ldots, A_{n-1}) \\ &\quad = \mathbb{P}(B \mid A_1, \ldots, A_{n-1})\, \mathbb{P}(A_n \mid A_1, \ldots, A_{n-1}),\end{aligned}$$

then this is called **chained independence** and is necessary and sufficient for the independence of B and the family A_n, $n \in I$.

More generally, let $(\Omega, \mathcal{F}, \mathbb{P})$ be a probability space, with two sub-sigma-fields $\mathcal{G}$ and $\mathcal{H}$ of $\mathcal{F}$. Then $\mathcal{G}$ and $\mathcal{H}$ are independent if, for every $A \in \mathcal{G}$ and $B \in \mathcal{H}$, $\mathbb{P}(A \cap B) = \mathbb{P}(A)\mathbb{P}(B)$.

The concept extends to non-commutative probability (arising e.g. in quantum mechanics), as exemplified by free probability, in various ways according to the context. This yields free independence, Boolean independence, monotone independence, tensor independence, Q-independence, and more. [W. Feller, *Introduction to Probability Theory and its Applications*, Vol. 1, 3rd edn. Wiley, 1968; Vol. 2, 2nd edn. Wiley, 1971; K. L. Chung, *Course in Probability Theory*, 2nd edn revised. Academic Press, 2001]

independence axiom Arises in the Von Neumann–Morgenstern theory of expected utility, where, for any three lotteries (i.e. random rewards), R_1, R_2, R_3, and any $0 < a < 1, R_2$ is preferred to R_1 iff $aR_2 + (1 - a)R_3$ is preferred to $aR_1 + (1 - a)R_3$. This is sometimes called the **independence of irrelevant alternatives**, R_3 being irrelevant here, and also the **cancellation condition**. [M. DeGroot, *Optimal Statistical Decisions*. Wiley, 1970]

independence sampler A special case of the Metropolis–Hastings Markov chain Monte Carlo method, in which at each stage the proposed value of the next state of the chain is simulated from a distribution independent of its current state. [K. Lange, *Applied Probability*, 2nd edn. Springer, 2010]

independent increments process *see* additive process

index of probability *see* Gibbs distribution

indicator metric A (compound) probability metric defined for r.vs X and Y (on the same probability space) by $d(X, Y) = \mathbb{P}(X \neq Y) = \mathbb{E}\{I(X \neq Y)\}$. [S. Rachev,

Probability Metrics and the Stability of Stochastic Models. Wiley, 1991]

indicator random variable A basic but important r.v. useful in counting problems. For any event A, the indicator of A is

$$I(A) = I_A(\omega) = \begin{cases} 1 & \text{for } \omega \in A, \\ 0 & \text{otherwise.} \end{cases}$$

Thus $\mathbb{E}I(A) = \mathbb{P}(A)$. It is often seen in the form $I(\{X \leq x\})$, in which case $\mathbb{E}I = F(x) = \mathbb{P}(X \leq x)$. It is related to the Iverson bracket [.], in which notation $I(A) = [\omega \in A]$ and $I(\{X \leq x\}) = [X \leq x]$.

indifference *see* independence

indifference curve *see* utility

indifference principle (alias: **principle of insufficient reason (or of indifference)**) Arises in assigning the values of the probabilities of the elementary (or point) events s in a sample space S. It is applied, in the case when S is finite, to assign equal probability $\mathbb{P}(s) = |S|^{-1}$ to all the points of S, when there is no reason (either intuitive, or experimental, or of any other type) to believe that any outcome is more or less likely than another. J. M. Keynes pointed out that it amounts to an assumption of symmetry (logical or physical) between the outcomes. In any context other than such special finite state spaces, the principle fails in this naive form; the maximum entropy principle has been suggested as more general. [J.M. Keynes, *A Treatise on Probability*. Macmillan, 1921; reprinted by Dover, 2004]

indistinguishable *see* equivalent

induced order statistic concomitant

induced probability (alias: **push forward probability measure (or law)**) Informally, this is seen as the image (or transferred version) on $\mathbb{R}$ of the underlying measure $\mathbb{P}$, (of some probability space $(\Omega, \mathcal{F}, \mathbb{P})$), by a random variable $X(\omega) \in \mathbb{R}, \omega \in \Omega$. Formally, let $\mathbb{Q}(B) = \mathbb{P}(X^{-1}(B)) = \mathbb{P}(\{\omega : X(\omega) \in B\})$, for any Borel set B in $\mathbb{R}$. Then $\mathbb{Q}(\cdot)$, is the probability measure on $\mathbb{R}$ induced by $X(\cdot)$. With a suitable completion, this yields an induced probability space $(\mathbb{R}, \mathcal{B}, \mathbb{Q})$, and the r.v. $\hat{X}$ on $(\mathbb{R}, \mathcal{B}, \mathbb{Q})$ defined by $\hat{X}(x) = x$ may be called X in its canonical form; *see* canonical random variable. [J. Kingman & S. Taylor, *Introduction to Measure and Probability*. Cambridge University Press, 1966]

inductive probability *see* logical probability

infectivity model *see* Hawkes process

infinite divisibility A random variable X is said to be n-divisible if its distribution $F(x)$ is the distribution of the sum of n i.i.d.rvs $X_{1n}, \ldots, X_{nn}$. Equivalently F is said to be n-divisible if it is the n-fold self-convolution of some distribution function $F_n(x)$; or equivalently, the characteristic function ϕ of X is n-divisible if $\phi = \phi_n^n$, for some chf $\phi_n(t)$.

If X (or F, or ϕ) is n-divisible for any $n \in \mathbb{N}$, then it is said to be infinitely divisible. If the r.vs $X_{1n}, \ldots, X_{nn}$ above are not identically distributed then X is said to be n-decomposable, and infinitely decomposable if n-decomposable for any n. This is clearly weaker, as seen by considering the uniform density on $[-1, 1]$, which is infinitely decomposable, but for no $n > 2$ is it n-divisible. The stable distributions comprise an important subset of the infinitely divisible distributions.

There are several canonical representations of the chf of the general infinitely divisible distn, e.g. Lévy–Khinchin, Kolmogorov's, and **Feller's thm** asserting that a distn on the non-negative integers is infinitely divisible iff it is a compound Poisson distn. [F. Steutel & K. Van Harn, *Infinite Divisibility of Probability Distributions on the Real Line*. Marcel Dekker, 2004]

infinitely often A concept arising e.g. in considering random processes with unbounded index set, where a natural interest lies in the long-term occupancy (or not) of some given set of states. Formally, if $A_n, n \geq 1$, is a sequence of events then any element ω of the event

$$L_s = \limsup_{n\to\infty} A_n = \bigcap_{n=1}^{\infty} \bigcup_{m=n}^{\infty} A_m$$

has the property that for any $n, \omega \in A_m$ for some $m > n$, which is to say that L_s comprises the elements ω that are in infinitely many A_n. One therefore writes:

$$L_s = \{A_n \text{ occurs infinitely often}\} = \{A_n \text{ i.o.}\}.$$

Similarly,

$$L_i = \bigcup_{n=1}^{\infty} \bigcap_{m=n}^{\infty} A_m = \liminf_{n\to\infty} A_n$$

comprises the elements that occur in all but finitely many A_n, which is to say ultimately always, and some writers denote this by $L_i = \{A_n \text{ ult.}\}$. [P. Billingsley, *Probability and Measure*, 3rd edn. Wiley, 1995]

infinitesimal array *see* null array

infinitesimal generator *see* generator

infinitesimal matrix *see* Q-matrix

infinitesimal parameter (or mean, or variance, etc.) (alias: **instantaneous parameter (or mean, or variance, etc.)**) Arise in a diffusion process $X(t) \in [a, b] \subseteq \mathbb{R}$, $t \geq 0$, where they characterize the diffusion by specifying its local behaviour for any t and value x of $X(t)$. Roughly speaking, they describe the behaviour of displacements over infinitesimal (i.e. arbitrarily small) time periods. Thus, as $h \downarrow 0$, define the parameters a and b by

$$\mathbb{E}(X(t+h) - X(t) \,|\, X(t) = x) = a(t,x)h + o(h);$$
$$\mathbb{E}(\{X(t+h) - X(t)\}^2 \,|\, X(t) = x) = b(t,x)h + o(h).$$

Then a is called the infinitesimal (or instantaneous) mean, or drift, or displacement; and b is called the infinitesimal (or instantaneous) variance, or diffusion, or volatility. They appear in the SDE associated with $X(t)$ thus: $\mathrm{d}X = a\mathrm{d}t + \sqrt{b}\mathrm{d}W$; and in the backward eqn for the transition density $f(t, y; s, x)$ thus

$$\frac{\partial f}{\partial s} = -a\frac{\partial f}{\partial x} - \frac{1}{2}b\frac{\partial^2 f}{\partial x^2},$$

and in the forward equation as

$$\frac{\partial f}{\partial t} = -\frac{\partial}{\partial y}(af) + \frac{1}{2}\frac{\partial^2}{\partial y^2}(bf).$$

More generally, for a diffusion with killing (*see* killed process), where the probability that the process is killed in $(t, t+h)$ given $X(t) = x$ is $c(x)h + o(h)$, the parameter c is called the infinitesimal killing rate, and appears as an additional term, $-cf$, in the backward and forward equations. [A. Borodin & P. Salminen, *Handbook of Brownian Motion*, 2nd edn. Birkhäuser, 2002]

inflated-parameter distribution *see* compounding

influence *see* measure

influence diagram Arises in decision, and game, theory as a graphical representation of a problem's (or game's) structure, in which nodes represent stages (or states of play), and directed edges from a node are decisions (or moves) which may be made by chance or choice (or a mixture of these). *See* decision tree. [K. Binmore, *Fun and Games*. Heath, 1992]

informant *see* score function

information Very broadly, 'information' means 'communicated knowledge'; or, from another viewpoint, it indicates the reduction either of uncertainty or of ignorance (partial or complete). The concept therefore naturally requires mathematical formalization in several probabilistic contexts, as well as in some areas of communication theory arising in computer science and signal processing, such as e.g. descriptive complexity. Instances include:

- The information gained by knowing that an event of probability $\mathbb{P}(A)$ has occurred. As defined by R. Hartley, this is $S(A) = -\log_2 \mathbb{P}(A)$, often called the self-information (or surprise) in A.
- The expected information gained by the observation (realization) of a simple r.v. X. In Shannon's theory, this is given by the entropy $H(X) = -\sum_x \mathbb{P}(X = x) \log_2\{\mathbb{P}(X = x)\}$.
- In a strict sense uncertainty (potential for gaining information) and information (reducing uncertainty) should be carefully distinguished, but usage and custom are somewhat lax in this. Some writers use the **negentropy** $= \log_2 n - H(X)$, where n is the number of points in the state space of X.
- The information gained by observing a continuous r.v. X with density $f(x)$ may be quantified by the (Shannon) differential entropy

  $$H(X) = -\int_{\mathbb{R}} f(x)\log f(x)\mathrm{d}x,$$

  or by the (Jaynes) differential entropy

  $$R(X) = -\int_{\mathbb{R}} f(x)\log\left(\frac{f(x)}{m(x)}\right)\mathrm{d}x,$$

  where $m(x)$ is a suitable reference measure on $\mathbb{R}$. In this case the negentropy is defined using that of a normal distn as the base value.
- The information obtained on receiving a string of symbols. In Kolmogorov's descriptive information theory (or complexity theory, or algorithmic entropy) this is specified in terms of the length of the shortest suitably defined program required to reproduce it (i.e. describe it).

More generally, one can ask for the information obtained about some quantity X (random or non-random), that is itself unknown, but is functionally related to some other observable variate Y, e.g.:

- when X is a r.v. jointly distributed with Y, in Shannon's theory the mutual information is $I(X; Y) = H(X) + H(Y) - H(X, Y)$;
- when $X = \theta$ is an unknown (but not random) parameter, in Fisher's theory the information is $I(\theta)$, defined as the expected value of the so-called observed information given by $\frac{\partial^2}{\partial\theta^2} \log f_Y(Y; \theta)$, where $f_Y(Y, \theta)$ is the density of the observed Y. *See* Fisher's information.

Many other information functions exist, notably in quantum theory*; see also* Lindley's information measure; Linnik information; and entropy. [C. Arndt, *Information Measures*. Springer, 2004]

information central limit theorem Any of several results proving a CLT by showing suitable convergence of relative entropy, e.g. **Barron's CLT**; and **Linnik's CLT**. [O. Johnson, *Information theary and the CLT*, 2004, Imperial]

information distance (or divergence) Any one of a number of distances (or divergences) defined in terms of an information function. That most commonly seen is the Kullback–Leibler divergence, defined in terms of Shannon information, but information distance may also be defined in terms of Kolmogorov complexity (descriptive information). Consequent theorems may be termed 'information theoretic', and some such divergences are also metrics. [M. Deza & E. Deza, *Encyclopedia of Distances*. Springer, 2009]

information geometry A term used to describe a type of differential geometry on the manifold of probability distributions, in which e.g. distance is based on the Kullback–Leibler (or information) divergence, and the definition of curvature is given in terms of Fisher's information. [S. Amari & H. Nagaoka, *Methods of Information Geometry* [English translation]. American Mathematical Society, 2000]

information inequality *see* Gibbs inequality

information rate *see* source

information source *see* source

information-theoretic limit theorem Broadly, any limit theorem for a sequence X_n of r.vs which establishes that the relative entropy (i.e. Kullback–Leibler divergence) between the distribution F_n of X_n and some distribution F tends to zero as $n \to \infty$; F_n is said to converge to F in relative entropy. The classic result is the information central limit theorem, and there is also an information theoretic law of small numbers (*see* rare events) establishing convergence in entropy to a Poisson distn. [O. Johnson, *Information Theory and the Central Limit Theorem*. Imperial College Press, 2004]

information theory *see* communication theory

innovation (alias: **(one-step) prediction error (or residual)**) Arises in the prediction of a time-series X_n, $n \in \mathbb{Z}$, and in the construction of such time-series. Informally, the innovation Z_n at time n is that 'part' of X_n which is uncorrelated with, $X_r, r \leq n-1$, the past of the sequence. Formally, let X_n have finite variance, so that the minimum mean-squared-error predictor of X_n given all X_r, $r \leq n-1$, is $\hat{X}_n = \mathbb{E}(X_n \mid X_{n-1}, X_{n-2}, \ldots)$. By the properties of conditional expectation, $Z_n = X_n - \hat{X}_n$ is uncorrelated with X_r, $r \leq n-1$, and is called the innovation at time n. By extension, if a finite-variance process Y_n has the representation $Y_n = \sum_{r=0}^{\infty} a_r Z_{n-r}$, for $a_0 = 1$, and $Z_n, n \in \mathbb{Z}$, is a sequence of uncorrelated r.vs, then Z_n is called the innovations process for Y_n. In the stationary case, Z_n is called white noise; classic examples include the Kalman filter and the Wold decomposition. [R. Shumway & D. Stoffer, *Time Series Analysis and its Applications*, 3rd edn. Springer, 2011]

insensitivity In the context of random processes, especially in connexion with the theory of queues, it sometimes happens that a quantity such as the stationary distribution (or some other performance measure) does not depend on the distribution of one or more components of the model (such as the service time of customers) but only on its mean. Such a property is then said to be insensitive to the component in question. Classic examples are supplied by the M/G/∞ queue, and Erlang's M/G/s loss system. The stationary distribution of M/G/1 is also insensitive to the service-time distribution, and hence so also is the probability of an arriving customer C (in equilibrium) not having to wait. However, C's waiting time is sensitive to the service time distribution. Symmetric queues also have the property of insensitivity. [S. Asmussen, *Applied Probability and Queues*, 2nd edn. Springer, 2003]

inspection paradox (alias: **waiting-time p.; audience (or congregation) p.; length-biased (or size-biased) sampling**) A 'paradox' of random selection (or sampling) arising from the fact that different methods of selection yield different results. It is most easily seen (and was first observed) in the context of audiences (or congregations). Suppose that the number of students attending n lectures comprise the r.vs $X_1, \ldots, X_n$. The university samples lectures uniformly at random, and reports the expected average class size as $a = \mathbb{E}\frac{1}{n}\sum_{r=1}^{n} X_r$. The student body pick a student uniformly at random and report the expected average class size seen by a random student as

$$\mathbb{E}\left(\sum_{r=1}^{n} X_r^2 \Big/ \sum_{r=1}^{n} X_r\right) \geq a.$$

The same effect is observed in renewal theory, where the lifetime of the component in service at time t is stochastically larger than the typical lifetime; it seems that its life has been prolonged by inspecting it, hence the name. [W. Feller, *Introduction to Probability Theory and its Applications*, Vol. 2, 2nd edn. Wiley, 1971; S. Ross, *Introduction to Probability Models*, 11th edn. Academic Press, 2014]

instantaneous parameters (or mean, or variance, etc.) *See* infinitesimal parameter (or mean, etc.)

instantaneous state (alias: **unstable state**) Arises in the context of a Markov process $X(t), t \geq 0$, evolving by jumps; informally, the process jumps out of such a state immediately following its arrival. Formally, let the process have (stationary) Markov transition function $p(t; x, A)$ on a suitable space S, where $\lim_{t \to 0} p(t; x, \{x\}) = 1$ for all x (called the regularity, or continuity, or normality, condition) and assume that paths of $X(t)$ are rcll. If $a = \mathbb{P}(\inf\{t > 0 : X(t) \neq X(0)\} = 0 \,|\, X(0) = x) = 1$, then x is instantaneous; otherwise $a = 0$, and the state x is said to be stable. Note that x is absorbing (or a trap) if $\mathbb{P}(\inf\{t > 0 : X(t) \neq X(0)\} = \infty \,|\, X(0) = x] = 1$. The property is related to the Q-matrix, or q-function given by

$$q(x) = \lim_{t \downarrow 0} \frac{1}{t}\Big(1 - p(t; x, \{x\})\Big).$$

If $q\ (x) = \infty$ then x is **instantaneous**; if $q(x) < \infty$ then x is **stable**; if $q(x) = 0$ then x is **absorbing**. [J. Doob, *Stochastic Processes*. Wiley, 1953]

insufficient reason *see* indifference principle

insurance mathematics (alias: **assurance mathematics**) Broadly, insurance is the process of buying and selling risk (or uncertainty); the theory of insurance has evolved in parallel with that of probability, stemming from the early work of J. De Witt, J. Hudde, E. Halley, A. De Moivre, etc., on annuities. The customer has the risk, and pays the insurer a fee (the premium) to remove, reduce, or cap it. By combining a suitably large number of sufficiently independent (or uncorrelated) risks, the company uses the law of large numbers (and the convergence theorems of probability) to reduce its own risk to any required level. [T. Rolski *et al.*, *Stochastic Processes for Insurance and Finance*. Wiley, 1999]

integral geometry *see* geometrical probability

integrated Brownian motion *see* Brownian motion

integrated tail distribution (alias: **second-tail distribution**) Given a distribution function $F(x), x \in \mathbb{R}$, the integrated tail distribution $S(x)$ is defined by $1 - S(x) = \min(1, \int_x^\infty (1 - F(y)\mathrm{d}y)$, which is non-trivial (i.e. proper) if the integral converges. *See* equilibrium distribution. [S. Foss *et al.*, *Introduction to Heavy-tailed and Subexponential Distributions*, 2nd edn. Springer, 2013]

intensity function (alias: **rate**) Arises in the context of a random point process X in $\mathbb{R}^d$ (or some more general space), where $N(A)$ is the counting function that gives the number of points of X in $A \subseteq \mathbb{R}^d$ (for suitable sets A), having mean measure $m(A) = \mathbb{E}N(A)$.

If there is a suitable positive function $\lambda(\mathbf{x})$ such that $m(A) = \int_A \lambda(\mathbf{x})\mathbf{dx}, \mathbf{x} \in \mathbb{R}^d$, then $\lambda(\mathbf{x})$ is called the intensity function, or field; and $\lambda(x)$ may be called the intensity rate when $d = 1$. (The function $m(A)$ may be called the intensity measure, especially when X takes values in spaces other than $\mathbb{R}^d$.) A classic example is given by the Poisson processes either homogeneous of rate (or intensity) λ, or non-homogeneous with rate or (intensity) $\lambda(\mathbf{x})$. One may refer to $\lambda(\mathbf{x})$ as the complete intensity function, to distinguish it from a **conditional intensity fn**, which arises from the mean measure $m(A|\{0\} \in X)$, i.e. the expected number of points in A given that the origin (for example) is a point of the process. [D. Daley & D. Vere-Jones, *Introduction to the Theory of Point Processes*, 2nd edn. Springer, Vol. I, 2003]

intensity matrix *see* Q-matrix

intensity process A term that is sometimes used in the context of a doubly-stochastic (or Cox) process when the intensity function is itself random. Otherwise, it is used to denote the hazard-rate function of r.vs in the context of a random process; thus, e.g. if $T \in \mathbb{R}^+$ is a stopping time for $\mathcal{F}_t$, $t \geq 0$, then $\lambda(t) = \lim_{h \downarrow 0} h^{-1}\mathbb{P}(t < T \leq t+h \mid \mathcal{F}_t)$ may be called the intensity process (or rate) of T.

interacting particle process (or system) (alias: **interactive particle system; random (or probabilistic) cellular automaton**) Any of a large and varied class of random processes designed to model the behaviour of objects that are distributed in some space, where their evolution among possible states depends on some form of mutual stochastic cooperation; such cooperation (or interaction) is often local in nature (e.g. between near or nearest neighbours). Applications arise in statistical physics, mathematical biology, economics, and chemistry, etc. The 'objects' may be variously seen as particles that may have one of several properties (e.g. a spin one of two or more directions, or polarity, or existence, etc); individuals that have one of several attributes; cells that may be in one of several states; and so on. In some applications the objects may be envisaged as moving between sites, with the possibilities of annihilation, coalescing, or creating new particles. Classic examples include the contact model, exclusion processes, the Ising model, and the voter model. Formally, the basic model is a random process $X(t)$, $t \geq 0$, taking values in a space $\{0, 1, \ldots, r\}^S$, where S is a countable set of sites (or vertices), each of which may be in one of the states $\{0, 1, \ldots, r\} = F$. In the simplest case only one site $v \in S$ can change at any time t; where the distribution of its next state is a function $k(v, X(t))$ of v and the state $X(t)$ of the system at time t. In many applications two states v and w may change simultaneously (e.g. by swapping) according to a distribution that is a function $k(v, w, X(t))$; more general interactions are possible. The function k is called the speed (or rate, or interaction) function, and under suitable conditions on k, the process $X(t)$ will have the Markov property.

Such processes are important because they may exhibit so-called emergent properties seen in real-world systems; these are variously called self-organization, long-range order, lock-in, phase changes, long-range correlation, magnetization, etc., depending on the context. [T. Liggett, *Interacting Particle Systems*. Springer, 2005]

interest rate model Arises in financial mathematics as a mathematical idealization in continuous time of empirical (or actuarial) interest rates, which are compounded in discrete time. Thus, a constant interest rate r in discrete time entails that the value $p(n, N)$ at time n (the present value), of a dollar received at time $N > n$ is given by $p(n, N) = (1 + r)^{-(N-n)}$, where $(1 + r)^{-1}$ is called the discount factor (*see* discounting). More generally, if interest rates comprise a (possibly random), sequence $r_1, r_2, \ldots$, then $p(n, N) = \prod_{i=n+1}^{N}(1 + r_i)^{-1}$. Therefore in continuous time the present value $p(t, T)$ of a dollar received at $T > t$ is given by $p(t, T) = \exp\{-\int_t^T \rho(s)\mathrm{d}s\}$, for some function $\rho(t)$, or more generally by $p(t, T) = \exp\{-\int_t^T \rho(t, s)\mathrm{d}s\}$.

Here $\rho(t, T)$ is called the instantaneous (continuously compounded) forward rate of interest, and because it is a function of both t and T, one speaks of the term structure of interest rates. Then $r(t) = \rho(t, t)$ is called the instantaneous short-term interest rate, or short (or spot) rate. Any interest rate model postulates a particular form for one of the random processes ρ, p, or r. Thus a forward rate model assumes that ρ satisfies some SDE such as e.g. in the Heath–Jarrow–Morton framework; whereas a short rate model assumes an SDE for r, e.g. the Black–Derman–Toy, or Cox–Ingersoll–Ross, or Vasicek models. [T. Björk, *Arbitrage Theory in Continuous Time*, 3rd edn. Oxford University Press, 2009]

inter-jump time *see* holding time

intermediate arcsine law *see* arcsine law

intermittent convergence A term introduced by P. Lévy in considering central limit theorems for the partial sums $S_n = \sum_{r=1}^{n} X_r$ of i.i.d.r.vs X_r, $r \geq 1$. For a class of distns $F(x)$ of X_r, it is the case that there exist subsequences n_i, $i \geq 1$, and m_j, $j \geq 1$, both tending to infinity, such that a suitable scaling of S_{n_i} has a normal distn in the limit as $n_i \to \infty$, while S_{m_j} has no normal limit as $m_j \to \infty$ for any scaling. [P. Moran, *Introduction to Probability Theory*. Oxford University Press, 1968]

interpretation of probability (alias: **concept (or meaning) of probability**) Arises in the necessary task of allocating a meaning, in terms of observable (real-world) phenomena, to the statements (or jargon) of any theory of probability (whether of the standard Kolmogorov type, or another). Such interpretations are unavoidable because first they suggest a suitable framework (i.e. axioms) for the theory; second they suggest what theorems may be useful (the directions in which the system may be developed and explored); third they tell us what the results 'really' mean outside the system; and fourth they may supply a check (where experiments are possible) on the validity of the theory; as well as supplying empirical (measured) values for the unspecified 'probabilities' of the axioms.

Probability has several interpretations, including:

1. Classical. Arises in elementary appearances of chance, where there is a finite number of outcomes, such that no one has any reason to be considered more likely than another (*see* indifference principle). This can be seen as an interpretation in terms either of symmetry, or lack of knowledge; it has been extended to more general cases under the name maximum entropy principle.
2. Frequentist. Arises in the context of experiments that can be repeated independently under essentially the same circumstances. Then the empirically observed stability of statistical ratios suggests that in the long run the probability $\mathbb{P}(A)$ of an event A is approximately the proportion of occasions an which A occurs. This interpretation seems unavoidable, but has the drawback that any finite set of observations is consistent with any assertion about long run behaviour. In particular, the Kolmogorov axioms of probability entail probabilities with no frequentist interpretation.
3. Logical probability (or normed belief). Arises as a generalization of deductive logic in which the probability of some statement being true, i.e. that event, is the degree to which it is logically confirmed (or supported) by the available evidence (or knowledge) E. (The usual deductive logic is included when all events have probability 0 or 1, i.e. all statements are true or false.)
4. Personal. Arises from the assumption that for any event A (or the truth of any proposition A), one can formulate a numerical degree of belief $\mathbb{P}(A)$ that A occurs (or the proposition A is true), where $0 \leq \mathbb{P}(A) \leq 1$. If the individual formulating this degree of belief does so in a reasonably rational way, as measured by betting preferences among a suitable set of offered wagers, then $\mathbb{P}(A)$ is indeed a probability function in the usual sense.

Various other interpretations of probability yield theories of probability that have different axioms e.g. Dempster–Shafer theory.

[C. Howson, Theories of probability, *British J. Phil. Sci.*, 46, 1–32, 1995]

intersection exponent Arises in considering independent paths of simple random walks, or independent paths of Brownian motions, and in other related contexts. Broadly, as the time t increases, the exponent gives the rate of decrease of the probability that paths (or sets of paths) do not intersect by time t. Thus, for a basic example, let $S_0(t), S_1(t), \ldots, S_r(t), t \geq 0$, be $r+1$ independent simple random walks on the planar or cubic lattice (i.e. on $\mathbb{Z}^d$ for d either 2 or 3). Denote the probability that S_0 meets none of the r walks $S_1, \ldots, S_r$ at any time $t \in \{1, \ldots, n^2\}$ by $a_d(n, r)$. If, for some $\eta_d, a_d(n,r) = n^{-\eta_d + o(1)}$, then η_d is the intersection exponent. An essentially similar definition holds for paths of Brownian motion. More generally, one defines a similar exponent $\eta_d(r, m)$ for m walks (or BMs) avoiding r other walks (or BMs). [G. Lawler *et al.*, Values of Brownian intersection exponents: I & II, *Acta Math*, 187, 237–273; 187, 275–308, 2001; & III *Ann. Inst. H. Poincaré B*, 38(1), 109–123, 2002]

interval-stationary *see* stationarity

interval-valued probability *see* lower and upper probability

invariance principle (alias: **Erdös–Kac invariance principle**) Arose initially in the context of the partial sums $S_n = \sum_{r=1}^{n} X_r$ of i.i.d.r.vs obeying the central limit theorem, where the limiting normal distn (as $n \to \infty$) does not depend on the distribution $F(x)$ of the X_s. Then considering suitable functionals of the sequence (e.g. $\mathbb{P}(\max\{S_1, \ldots, S_n\} < a\sqrt{n})$, and $\mathbb{P}(\sum_{r=1}^{n} S_r^2 < a_n^2)$, as $n \to \infty$), it is shown that the limit exists and is independent of $F(x)$. That is to say it is invariant under (or insensitive to) changes in $F(x)$, and the limit may be determined explicitly by a suitable choice of F. More strongly, a suitably scaled continuous path formed from $\{S_n, n \geq 1\}$ converges (in an appropriate metric space of continuous functions) to a path of Brownian motion, $W(t)$; and the invariance principle can be applied to various measurable and continuous functionals of S_n via the corresponding functional of $W(t)$. *See* Donsker's theorem and Strassen's invariance principle.

[O. Kallenberg, *Foundations of Modern Probability*, 2nd edn. Springer, 2002]

invariance relation (alias: **invariant relation; rate (or intensity) conservation principle (or law)**) Arises in queueing theory, where it denotes any of several relationships (between queueing quantities) that remain true under suitable types of changes in the queueing discipline, and/or arrivals process, service times, etc. Rate conservation laws relate customer averages to time averages; this being equivalent to invariant relations between Palm distributions and stationary distns in point processes (*see* Little's law).

The term also denotes invariance of various performance measures under changes of distn or discipline; *see* insensitivity. [P. Brémaud, An elementary proof of Sengupta's invariance relation and a remark on Miyazawa's conservation principle, *J. Appl. Probab.*, 28, 950–954, 1991]

invariant distribution (or measure) (alias: **stationary (or equilibrium, or steady-state) measure or distribution; invariant probability measure**) Arises in considering discrete-time Markov processes. Let $P = (p_{ij}), i, j \in S$, be the transition matrix of a Markov chain on a countable state space S, then a vector $\boldsymbol{\pi} = (\pi_i), i \in S$, having non-negative entries is an invariant distribution for P if $\sum_{i \in S} \pi_i = 1$ and $\pi_j = \sum_{i \in S} \pi_i p_{ij}, j \in S$. More generally, a vector $\lambda = (\lambda_i), i \in S$, of non-negative terms is an invariant measure if $\lambda_j = \sum_{i \in S} \lambda_i p_{ij}, j \in S$. If also, $\bar{\lambda} = \sum_{i \in S} \lambda_i < \infty$ then $\lambda / \bar{\lambda}$ is an invariant distribution. More generally, if $k(x, A)$ is the transition kernel of a time-homogenous Markov process, then $\lambda(\cdot) \geq 0$ is an invariant measure for k when $\lambda(A) = \int k(x, A) \lambda(dx)$, $A \in S$. It is an invariant distribution if also $\lambda(S) = 1$.

The term stationary is used because of this property: let the initial distribution of the chain $X_n, n \geq 0$, be $\boldsymbol{\nu} = (\nu_i)$, $i \in S$, then X is a stationary process if ν is an invariant distribution for P (or $\nu(\cdot)$ an invariant probability measure for k, respectively). The terms equilibrium (or steady-state) arise from this property: suppose that for some $i \in S$, and all j in S, the n-step transition probabilities $p_{ij}(n)$ obey $\lim_{n \to \infty} p_{ij}(n) = \pi_j$, where $\boldsymbol{\pi} \neq 0$. Then $\boldsymbol{\pi}$ is invariant for P.

The existence and nature of a stationary measure (or distribution) for a Markov process is of great importance to its evolution over time:

1. If an invariant distribution $\boldsymbol{\pi}$ exists, then any state $i \in S$ with $\pi_i > 0$ is recurrent. Furthermore, if the chain is irreducible and there is an invariant distribution then all states are recurrent.
2. If X_n is an aperiodic irreducible chain on S then either there is a unique invariant distribution $\boldsymbol{\pi}$, and the distribution of the chain converges to $\boldsymbol{\pi}$ as $n \to \infty$; or there is no invariant distribution, and $\lim_{n \to \infty} \mathbb{P}(X_n = i \mid X_0 = j) = 0$ for all $i, j \in S$.

In continuous time, similarly, for a Markov process with Q-matrix $Q = (q_{ij}), i, j \in S$, and transition probabilities $P(t) = p_{ij}(t), t > 0$, a measure $\boldsymbol{\lambda}$ is invariant for Q if $\lambda_j = \sum_i \lambda_i q_{ij}$, and a measure $\boldsymbol{\pi}$ is invariant for P if $\pi_j = \sum_i p_{ij}(t) \pi_i$, being an invariant distribution if $\sum_j \pi_j = 1$. For a process with transition semi-group T_t, π is invariant for T_t, *if* $\pi T_t = \pi$, for all t. In the context of transient Markov chains that may have quasi-stationary distributions, the concept of **subinvariance** arises. Thus e.g. in continuous time, for some $m \geq 0$, a vector of positive terms $\lambda = (\lambda_i), i \in S$, is called an m-subinvariant measure on S for P if

$$\sum_{i \in S} \lambda_i p_{ij}(t) \leq \mathrm{e}^{-mt} \lambda_j, \quad t \geq 0, j \in S;$$

and likewise an m-subinvariant measure S for Q if $\sum_{i \in S} \lambda_i q_{ij} \leq -m\lambda_j, j \in S$.

In the case of equality one speaks of m-invariance, which reduces to ordinary invariance when $m = 0$. [W. Anderson, *Continuous-time Markov Chains*. Springer, 1991; J. Norris, *Markov Chains*. Cambridge University Press, 1997; G. R. Grimmett, *Probability on Graphs*. Cambridge University Press, 2010; J. von Neumann, *Invariant Measures*. American Mathematical Society, 1999]

invariant event Arises in the context of stationary processes, and those that are ergodic. Let $X_n, n \in T = \{1, 2, 3, \ldots\}$, be a stationary (in the strict sense) random process, defined on the natural (canonical) probability space $(\mathbb{R}^T, \mathcal{B}^T, \mathbb{P})$ (where $\mathcal{B}^T$ is the product Borel sigma-field of Borel subsets of $\mathbb{R}$). Let τ be the shift operator on vectors x_r, $r \in T$, that maps $(x_1, x_2, x_3, \ldots)$ to $(x_2, x_3, \ldots)$; this is a measure-preserving transformation with stationary measure $\mathbb{P}$. An event A in $\mathcal{B}^T$ is said to be invariant if $A = \tau^{-1} A$, and the invariant sigma-field is the set of all such invariant events. [R. Durrett, *Probability: Theory and Examples*, 4th edn. Cambridge University Press, 2010]

inventory theory A classic branch of operations research that deals with the decisions and policies needed to maintain an optimal supply of objects for some purpose. Demand, costs, rewards, and supply may each be random. *See* single-period model. [E. Porteus, *Foundations of Stochastic Inventory Theory*. Stanford University Press, 2002]

inverse Gaussian distribution (alias: **Wald distribution; Gaussian (or Brownian) first-passage time distribution**) A probability distribution with density

$$f(x) = \frac{b}{\sqrt{2\pi}} x^{-3/2} \exp\left\{ -\frac{(ax - b)^2}{2x} \right\}, \quad x > 0,$$

where $a, b > 0$. It arises naturally in the context of the standard drifting Wiener process, $D(t) = at + W(t)$, with drift $a > 0$, where the density of the first-passage time T of $D(t)$ to $b > 0$, from $D(0) = 0$, is $f(x)$ above. (If $a < 0$ and $b > 0$ have opposite signs, then $f(x)$ is defective, reflecting the fact that $\mathbb{P}(T < \infty) < 1$ in this case.) The cumulant generating function of T is $K(\psi) = \log\{\mathbb{E}e^{-\psi T}\} = ab - b(a^2 + 2\psi)^{1/2}$, whose inverse function

$$\psi(K) = -\frac{a}{b}K + \frac{1}{2b^2}K^2,$$

is the cumulant generating function of a normal r.v., and these are inverse variates, hence the name. When $a = 0$, it is commonly called the first-passage, or Lévy distribution.

More generally, the generalized inverse Gaussian distn (or Sichel distn) has density

$$f(x) = \left(\frac{a}{b}\right)^{\frac{1}{2}\nu} \frac{1}{2K_\nu(\sqrt{ab})} x^{\nu-1} \exp\left\{-\frac{1}{2}(ax + b/x)\right\},\ x > 0,$$

where K_ν is a modified Bessel function with index ν. For $\nu \le 0$ it arises as the distn of a first-passage time of a suitable time-homogeneous diffusion. It has mgf

$$M(\theta) = \left(\frac{a}{a - 2\theta}\right)^{\frac{1}{2}\nu} K_\nu\left(\{b(a - 2\theta)\}^{\frac{1}{2}}\right) / K_\nu(\sqrt{ab}),$$

and is hence infinitely divisible. [B. Jørgensen, *Statistical Properties of the Generalized Inverse Gaussian Distribution.* Lecture notes in statistics, 9, Springer, 1982]

inverse local time *see* local time

inverse moment (alias: **negative (or reciprocal) moment**) For $r > 0$, the rth inverse moment of the r.v. X is $\mathbb{E}(X^{-r})$, when the expectation exists; likewise the rth inverse factorial moment is

$$\mathbb{E}\left\{\frac{1}{(X + r)\ldots(X + 1)}\right\},$$

when it exists. Inverse moments of strictly positive r.vs arise in several applications; when $M(t) = \mathbb{E}e^{tX}$ is the mgf of an a.s. non-negative r.v. X, then

$$\mathbb{E}\left\{\frac{1}{(1 + X)^r}\right\} = \frac{1}{(r - 1)!}\int_0^\infty t^{r-1}e^{-t}M(-t)dt.$$

[N. Garcia & J. Palacios, On inverse moments of non-negative random variables, *Stat. Probab. Lett.*, 53(3), 235–239, 2001]

inverse probability (alias: **posterior distribution**) A term coined by A. de Morgan to distinguish two types of probability in the work of previous authors, the other type being direct probability, that arise in the context of a random variable X whose distribution depends on a parameter (or state of nature), θ, and is denoted by the direct probability $p(x \mid \theta)$. In statistical applications, after X has been observed, $p(x \mid \theta)$ is called the likelihood function of the unknown parameter θ. It is natural in seeking to make inferences about θ to treat it as a random variable with distribution $\pi(\theta)$, as one does with other quantities that are unknown before experiments. Then, by conditional probability, one obtains the inverse probability, nowadays called the posterior probability,

$$p(\theta \mid x) = \pi(\theta)p(x \mid \theta) \Big/ \int \pi(\theta)p(x \mid \theta)d\theta.$$

This is the start of what is often called Bayesian inference. Some writers see it as a serious problem that $\pi(\theta)$ is not naturally open to interpretation in frequentist terms; it is evidently to be seen as a subjective probability. [A. Dale, *History of Inverse Probability from Thomas Bayes to Karl Pearson*, 2nd edn. Springer, 1999]

inverse transform method (alias: **inverse transformation technique; inversion sampling (or simulation)**) A fundamental method for sampling from a distn $F(x)$, i.e. simulating a realization of a r.v. X with distn fn $F(x)$. Let $F(x)$ be a continuous distribution function with inverse function $F^{-1}(y) = \inf\{x : F(x) \ge y,\ 0 \le y \le 1\}$, and let U be a r.v. uniform on [0, 1]. Then $F^{-1}(U)$ has distribution function $F(\cdot)$; and, for X with distn fn $F(x)$, $F(X)$ is uniform on [0,1]. If, more strongly, $F(x)$ is strictly increasing, then we have more simply that $F^{-1}(y) = [x : F(x) = y]$ as the inverse function. Hence any simulation of U (i.e. a sample from the uniform distn on [0,1]), yields a sample from $F(x)$ simulating a value of X. For a discrete distribution, analogously, we realize a value of X with distribution $\mathbb{P}(X = x_j) = p_j,\ j = 0, 1, 2, \ldots$, by setting $X = x_j$ if $\sum_{r=0}^{j-1} p_r \le U < \sum_{r=0}^{j} p_r$. *See* probability integral transform. [L. Devroye, *Non-uniform Random Number Generation.* Springer, 1986; L. Rüschendorf, On the distributional transform, Sklar's theorem, and the empirical copula process, *J. Stat. Planning Infer.*, 139(11), 3921–3927, 2009]

inverse variates (alias: **Tweedie variates**) A class of pairs of distributions (of respective r.vs X and Y), such that the cumulant generating function $K_X(t) = log(\mathbb{E}e^{-tX})$ of one is proportional to the inverse function of the cgf of the other, i.e. $K_X(t) \propto K^{-1}{}_Y(t)$. Examples include:

1. Exponential (or gamma) and Poisson distributions with parameter λ;$K_X(t) \propto \log\frac{\lambda}{\lambda + t}$ and $K_Y(t) = \lambda(e^{-t} - 1)$.

2. Normal and inverse Gaussian distributions.
3. Geometric (or negative binomial) and Bernoulli distributions such that $K_X(t) \propto \log\left(\frac{p}{e^t - q}\right)$ and $K_Y(t) = log(q + pe^{-t})$, where $q + p = 1$, $p > 0$.

[M. Tweedie, Inverse statistical variates, *Nature*, 155, p. 453, 1945]

inversion theorem Arises generally in the context of any map (or function) T that is one–one, with inverse function T^{-1}. An inversion theorem simply supplies a suitable so-called inversion formula for T^{-1}. In probability the term commonly refers to an inversion formula for one of the several functional transforms (or representations) of random variables and processes, e.g. characteristic fn; probability (or moment) generating fn; spectral representation; etc.

Ionescu Tulcea ergodic theorem An abstract ergodic theorem for suitable sequences taking values in a Banach space such that $\mathbb{E}\|X\|^p < \infty$ for some $p \in [1, \infty)$. By establishing a strong maximal inequality (i.e. a dominated ergodic theorem) it provides a unified proof of a.s. convergence for suitable martingales, and the pointwise ergodic theorem. [A. Ionescu Tulcea & C. Ionescu Tulcea, Abstract ergodic theorems, *Trans. Amer. Math. Soc.*, 107(1), 107–24, 1963]

Ionescu Tulcea extension theorem Arises in constructing a suitable infinite product measure, appropriate for a given random process. Unlike the Daniell–Kolmogorov theorem, in which the extension is from the family of finite dimensional distns, the Ionescu Tulcea extension is specified by a suitable sequence of regular conditional probability distns. [C. Ionescu Tulcea, *Mesures dans les espaces produits, Atti. Accad. Naz. Lincei. Rend.*, 7, 208–211, 1949–50]

irreducible (alias: **non-decomposable**) Arises naturally in the context of a Markov chain X_n, $n = 0, 1, 2, \ldots$ (or $X(t), t \geq 0$), taking values in a countable state space S. For i and j in S, the state i is said to lead to j (or j is said to be reachable, or accessible, from i), if the chain started in i takes the value j in finite time with positive probability. If every state of S leads to every other state of S, then the chain is called irreducible (and the states are said to intercommunicate, or communicate, with each other). For a Markov chain on an uncountable state space S (with a sigma-field $\mathcal{F}$ of measurable subsets) this elementary concept is of limited use because points of S will commonly be reached with probability 0 in finite time. In this case, if there is a non-zero measure $\psi(\cdot)$ on $(S, \mathcal{F})$ such that for all $x \in S$ and any $A \in \mathcal{F}$ with $\psi(A) > 0$ the chain started in x enters A in finite time with positive probability, then the chain is said to be $\psi-$irreducible for ψ. As in the case of countable S, for an aperiodic $\psi-$irreducible Markov chain X_n that has a stationary distn $\pi(\cdot)$, the distn of the chain converges to $\pi(\cdot)$ in total variation, as $n \to \infty$. [S. Meyn & R. Tweedie, *Markov Chains and Stochastic Stability*, 2nd edn. Cambridge University Press, 2009]

irrelevance A type of conditional independence arising in Bayesian networks and machine learning and related fields. Given a random vector $(X_0, X_1, \ldots, X_n, Y)$, where $X_0, \ldots, X_n$ are seen as variables that are potentially informative about Y, the variable X_0 is said to be irrelevant to Y if it is conditionally independent of Y given any subset of $\{X_1, \ldots, X_n\}$. If X_0 is not conditionally independent of Y given the vector $(X_1, \ldots, X_n)$ then it is strongly relevant to Y. If neither of these, then it is weakly relevant. [A. Webb & K. Copsey, *Statistical Pattern Recognition*, 3rd edn. Wiley, 2011]

Irwin–Hall distribution The distribution of the sum of n i.i.d.r.vs each uniformly distributed on [0, 1]; a result that has been independently derived many times, the first proof being that of P. Laplace. The distribution is

$$F(x) = \frac{1}{n!}\sum_{r=0}^{[x]}(-1)^r\binom{n}{r}(x-r)^n, \quad 0 \leq x \leq n.$$

[H. Seal, Spot the prior references, *J. Inst. Act. Stud. Soc.*, 10, 255–258, 1950–51]

Irwin–Skellam distribution *see* Skellam's distribution

Ising model (alias: **Lenz–Ising model**) A classic problem of statistical physics that arose as a simple (idealized) description of ferromagnetism. Its importance arises from its status as the most tractable and effective model for a range of examples of cooperative phenomena, which generate long-range order on the basis of interactions that are only local, e.g. phase transitions. Formally, the model postulates a collection X of random variables, each attached to a vertex $v \in V$ of a set V of vertices (called sites) comprising a (finite) subset of the plane rectangular lattice, where X takes values in $[-1, +1]^V$. This describes the physical reality of a collection of molecules that may, e.g.:

(a) have a positive or negative spin, i.e. up or down with respect to an external field;

(b) be of two different types, labelled +1 or -1;
(c) be present or absent, labelled + 1 or -1.
It is assumed that the distribution of the configuration ω of the molecules (i.e. the value ω of X), depends only on interactions between neighbours on the lattice (and possibly an external field); to be precise: $p(\omega) = Z^{-1} \exp(T^{-1}H(\omega)), \omega \in [-1, +1]^V$, where Z is called the partition function, T the temperature, and H the Hamiltonian, given by

$$H(\omega) = -a\sum \omega_u\omega_v - b\sum_{v\in V} \omega_v,$$

where ω_v is the value of ω at the vertex v, and the first sum is over neighbouring vertices u and v. The celebrated **Onsager solution** supplies an exact form of Z for a square lattice of $n \times m$ sites, in the zero-field case when $b = 0$. Generalizations are numerous, including the Ashkin–Teller model, in which there are four spin states, and the Potts model, which allows any finite number of states. [S. Brush, History of the Lenz-Ising model, *Rev. Mod. Phys.*, 39(4) 883–893, 1967]

island model *see* stepping stone model

island problem A classic application of probability to inference in forensic (legal) identification. Its basic statement is this: in a population of $n+1$ individuals it is known that some individual C has committed a crime (the group being isolated on an island); given a suspect S who shares a property E with C (e.g. stature, proximity to the crime scene, etc.), the problem is to determine the probability that S actually is C. [D. Balding & P. Donnelly, Inference in forensic identification, *J. Roy. Stat. Soc. A*, 158(1), 21–53, 1995]

isonormal process Any of a family that (in a sense) comprises all zero-mean Gaussian random functions. Let H be a real inner (or scalar) product space (a Hilbert space having a countable orthonormal basis); then an isonormal process on H, denoted by $X(h)$, indexed by the elements of H, is a zero-mean Gaussian process with covariance function $c(h,k) = \mathbb{E}(X(h)X(k)) = <h,k>$, h, $k \in H$, where $<.,.>$ is the inner product on H. [R. Dudley, *Real Analysis and Probability*. Cambridge University Press, 2002]

Isserlis's theorem (alias: **Wick's theorem**) Arises in the context of the multivariate normal distribution, where it expresses the central joint moments of any order in terms of the covariances of the variables. For simplicity, let $X = \{X_1, \ldots, X_n\}$ have a standard multinormal distribution (with zero means and unit variances), and write $\mathbb{E}(X_rX_s) = \rho_{rs}$. Then, for odd $m \le n, \mathbb{E}(X_1X_2\cdots X_m) = 0$, and for even $m \le n$, $\mathbb{E}(X_1X_2\cdots X_m) = \sum \rho_{ab}\rho_{cd}\cdots\rho_{ij}$, where the summation is over every distinct selection of $m/2$ pairs $ab, cd, \ldots, ij$, from $\{1, 2, \ldots, m\}$. Joint moments of the form $\mathbb{E}(X_1^{t_1}X_2^{t_2}\ldots X_r^{t_r})$ are expressed by first using the theorem on the product of $\sum_{i=1}^r t_i$ variables Y_j, and then identifying $Y_1, \ldots, Y_{t_1}$, with X_1, and so on. Thus, e.g. $\mathbb{E}(X_1X_2X_3X_4) = \rho_{12}\rho_{34} + \rho_{13}\rho_{24} + \rho_{14}\rho_{23}$, and $\mathbb{E}(X_1^2X_2X_3) = \rho_{23} + 2\rho_{12}\rho_{13}$.

A later theorem of G-C. Wick in quantum theory takes a similar form, and the result can be extended to distributions such as e.g. the q-normal. [L. Isserlis, On a formula for the product-moment coefficient of any order of a normal frequency distribution in any number of variables, *Biometrika*, 12(1–2), 134–139, 1918]

Itakura–Saito divergence (alias: **likelihood ratio divergence**) Arises as a measure of distortion in coding theory, as a type of Bregman divergence that, in the context of a pair of probability distributions $(p_1, \ldots, p_n)$ and $(q_1, \ldots, q_n)$ such that $p_rq_r \ne 0, 1 \le r \le n$, takes the form

$$D_B(p,q) = \sum_{r=1}^{n}\left(\frac{p_r}{q_r} - \log\frac{p_r}{q_r} - 1\right).$$

This is a not a distance, since $D_B(p,q)$ is not necessarily equal to $D_B(q,p)$; the symmetrized form may be called the **cosh-divergence**, and denoted by $\mathrm{COSH}(p,q) = \frac{1}{2}\{D_B(p,q) + D_B(q,p)\}$. The forms for densities are the natural analogues. [R. Gray *et al.*, Distortion measures for speech processing, *IEEE Trans. ASSP*, 28(4), 367–376, 1980]

iterated Brownian motion *see* Brownian motion

Itô calculus Arises naturally as an extension to continuous time of the discrete-time martingale transform, and hence (in particular) can be seen as supplying a continuous-time version of the martingale systems theorems (*see* martingale transform), e.g. no realistic system of gambling can turn a fair game into a game advantageous to the player. This, in part, accounts for its widespread application in mathematical finance. Formally, the calculus can be presented as a type of integral calculus based on the Itô integral, and having theorems analogous to those of classical calculus (e.g. fundamental theorem of calculus; change of variable; integration by parts, etc.), of which Itô's formula is the most important. [R. Bass, *Stochastic Processes*. Cambridge University Press, 2011]

Itô decomposition *see* Lévy–Itô decomposition

Itô excursion theory Arises in the context of a strong Markov process $X(t), t \geq 0$ (having paths that are rcll), and whose compact state space S has a regenerative point a (or regenerative compact subset A). The theory considers the successive excursions of the process away from a (or A) and shows that they form a Poisson point process, called the excursion process, on the space of possible excursion paths; the characteristic (or mean) measure of this Poisson process is called the **excursion law**. [K. Itô, Poisson point process attached to Markov processes, *Proc. 6th Berkeley Symp. Math. Stat. Probab.*, 3, 225–240, 1971]

Itô's formula (alias: **Itô's lemma; Itô substitution rule**) Arises in Itô calculus in considering a function $f(X(t))$ of an Itô process $X(t)$, where it supplies a representation of $f(X)$ as an Itô process. It can be interpreted as a type of stochastic chain rule, or as a stochastic fundamental theorem of calculus, among other analogies. In its simplest form, let $f(W)$ be a continuously twice-differentiable function of the Wiener process $W(t)$. Then

$$f(W(t)) = f(0) + \int_0^t f'(W)\mathrm{d}W(s) + \frac{1}{2}\int_0^t f''(W)\mathrm{d}s.$$

More generally, let $X(t)$ be an Itô process with SDE $\mathrm{d}X = \mu(t,X)\mathrm{d}t + \sigma(t,X)\mathrm{d}W$, and suppose that $f(t,x)$ has continuous partial derivatives f_t, f_x, and f_{xx}, in a natural notation. Then $Y = f(t, X(t))$ is an Itô process, with representation

$$\begin{aligned} Y(t) - Y(0) &= \int_0^t \left\{ f_x(t,X)\mu(t,X) + f_t(t,X) + \frac{1}{2} f_{xx}(t,X)\sigma^2(t,X) \right\} \mathrm{d}t \\ &\quad + \int_0^t f_x(t,X)\sigma(t,X)\mathrm{d}W(t). \end{aligned}$$

The formula can be extended to multivariate Itô diffusions; and also to suitable semi-martingales, in which case it may be called the **Itô–Kunita–Watanabe formula**. [O. Kallenberg, *Foundations of Modern Probability*, 2nd edn. Springer, 2002]

Itô integral (alias: **Itô stochastic integral**) A type of stochastic integral that is basic to Itô calculus. It arises in considering the existence of formal integrals of the form $I = \int_0^T f\mathrm{d}W(t)$, where $W(t)$ is the Wiener process.

As is often the case, it is most conveniently defined (for a suitably measurable and adapted and finite variance function f) as the limit of a sequence of well-defined integrals of the form $I_n = \int_0^T f_n\mathrm{d}W(t)$, where each f_n is a random step (or simple) function on $[0, T]$. Formally:

1. For $0 = a_0 < a_1 < \cdots < a_n = T$, let $A_0, \ldots, A_{n-1}$ be finite variance r.vs such that A_r is a function of $W(s)$ only for $s \leq a_r$ (or, more precisely, each A_r is $\mathcal{F}_{a_r}$-measurable, where $\mathcal{F}_t$ is the filtration generated by $W(t)$). Let,
$$f_n(t) = \sum_{r=0}^{n-1} A_r I \;(a_r, a_{r+1}) = \begin{cases} 0 & \text{if } t \notin (0,T) \\ A_r & \text{if } a_r < t \leq a_{r+1}, \end{cases}$$
and define the stochastic (Itô) integral I_n of f_n as
$$I_n = I(f_n) = \sum_{r=0}^{n-1} A_r(W(a_{r+1}) - W(a_r)).$$
2. Next, note that I_n and f_n can be shown to satisfy a relationship called **Itô's isometry**, that is:
$$\mathbb{E}(|I_n|^2) = \mathbb{E}\int_0^\infty |f_n(t)|^2 \mathrm{d}t.$$
3. Next, observe that for any suitable function f (e.g. continuous), there is a sequence of random step functions f_n (as defined above), such that, as $n \to \infty$, $\int_0^\infty (f_n - f)^2 \mathrm{d}t \to 0$. It follows, using Itô's isometry, that the corresponding sequence of integrals I_n converges in mean square to an a.s. unique r.v. $I(T)$, and then $I(T)$ is called the Itô integral of f on $[0, T]$.
4. Next, it can be shown that there is a version of the sequence of integrals $I(t)$, $0 \leq t \leq T$, that is a continuous martingale with respect to the filtration $\mathcal{F}_t, 0 \leq t \leq T$.
5. Finally, we remark that the condition that f has finite variance can be relaxed to the weaker $\mathbb{P}\left(\int_0^T f^2 \mathrm{d}t < \infty\right) = 1$.

Given another $\mathcal{F}_t$–adapted process g, a random process defined by

$$X(t) = \int_0^t g\mathrm{d}t + \int_0^t f\mathrm{d}W(t)$$

is called an **Itô process**, or **Itô diffusion**; the definition may be written as $\mathrm{d}X = g\mathrm{d}t + f\mathrm{d}W$. [F. Klebaner, *Introduction to Stochastic Calculus with Applications*, 3rd edn. Imperial College Press, 2012]

Itô isometry *see* Itô integral

Itô process *see* Itô integral

Itô representation *see* martingale representation

J

jackknife inequality *see* Efron–Stein inequality

Jackson network Arises in queueing theory, where it denotes a system of queues in which customers (or jobs) completing service at one queue (or node) may then join another queue for further service (if the network is closed, they must do this, if it is open they may leave the system altogether). It is assumed that external arrivals at each queue comprise independent Poisson processes, that service times are independently exponentially distributed, that routings between queues are independent, and that each queue is FIFO with infinite waiting room. Then (for suitable parameters) the queue lengths $X_1(t), \ldots, X_J(t)$ comprise an ergodic Markov process, and in equilibrium the stationary distribution π has a product form $\pi(n_1, \ldots, n_J) = \prod_{r=1}^{J}\pi(n_r)$.

Thus the queue lengths $X_1, \ldots, X_J$ are independent in equilibrium, though, of course, the queueing processes $X_1(t), \ldots, XJ(t)$ are not. This result by J. R. Jackson can be seen as a generalization of the Jackson product rule by R. R. P. Jackson. In the case of an open network **Melamed's thm** asserts that equilibrium flow along any link is a Poisson process iff no customer traverses it twice. *See also* BCMP theorem; Gordon–Newell network; and Burke's theorem. [J. R. Jackson, Jobshop-like queueing systems, *Manage. Sci.*, 10(1), 131–142, 1963; T. Brown, Approximate versions of Melamed's theorem, *J. Appl. Probab.*, 33(2), 472–489, 1996]

Jackson product rule Arises in the context of queues in tandem, or series, where customers leaving a queue then immediately join the next queue for further service. Assuming a Poisson process of arrivals, independent of the independent exponentially distributed service times, the equilibrium distribution of queue lengths in the ergodic case has a product form: $\pi(n_1, \ldots, n_J) = \prod_{r=1}^{J}\pi(n_r)$. [R. R. P. Jackson, Random queueing processes with phase-type service, *J. Roy. Stat. Soc. B*, 18(1), 129–132, 1956]

Jacobi diffusion *see* Wright–Fisher model

Jacod's theorem Arises in the context of semi-martingales where it supplies a sufficient criterion for the preservation of the semi-martingale property under an enlargement (or expansion) of the filtration. [P. Protter, *Stochastic Integration and Differential Equations*, 2nd edn. Springer, 2005]

Janossy density (or function) Arises in the theory of point processes (or, in applications, a process of indistinguishable particles), where the nth Janossy function J_n is the joint density (or measure) of the positions of the points of the process on the event that there are exactly n such points; it is then taken to be symmetrical in its n arguments. Thus, informally $\mathbb{P}$ (there is one point in each of $(x_r, x_r + \mathrm{d}x_r,), 1 \le r \le n) = J_n(x_1, \ldots, x_n)\mathrm{d}x, \ldots, \mathrm{d}x_n$. [D. Daley & D. Vere-Jones, *Introduction to the Theory of Point Processes*, 2nd edn. Springer, Vol. II, 2008]

Janossy G-equation Arises in cascade theory as an integro-differential equation for the generating function of the numbers of particles in different energy levels; it is obtained by using a branching-process model. [K. Athreya & P. Ney, *Branching Processes*. Springer, 1972; reprinted by Dover, 2004]

Janson's inequality (alias: **Suen–Janson (or v. v.) inequality**) Any of several inequalities for the distribution of a sum S of a family F of indicator random variables having a particular dependence structure (that arises e.g. in random graphs). The inequality most often tagged as Janson's is an exponential upper bound for $\mathbb{P}(S = 0)$ in this setting.

Let H be a finite set and select each element r of H independently with probability p_r, yielding a random subset G. For any set A in a family F of subsets of H, let I_A be the indicator of the event that $A \subseteq G$, and let $S = \sum_{A \in F} I_A$. Define $m = \mathbb{E}S$ and $d = \frac{1}{2}\sum\sum\mathbb{E}(I_A I_B)$, where the sum is over $\{A \neq B; A \cap B = \emptyset\}$. Then $\mathbb{P}(S = 0) \le \exp(-m + d)$, and for $2d > m$ one has the extended Janson inequality

$$\mathbb{P}(S=0) \le \exp\left(-\frac{m^2}{m+2d}\right).$$

[S. Janson, Poisson approximation for large deviations, *Random Struct. Alg.*, 1(2), 221–229, 1990; S. Janson *et al.*, *Random Graphs*. Wiley, 2000]

Jaynes's entropy *see* differential entropy

Jaynes's principle *see* maximum entropy principle

Jeffreys's divergence A measure of difference between probability distributions that is a particular type of f-divergence, taking the form of a symmetrized Kullback–Leibler divergence. When p_r and q_r, $1 \le r \le n$, are probability distributions (with $p_r q_r \ne 0$), it is given by

$$d_J(p,q) = \sum_r (p_r - q_r) \log \frac{p_r}{q_r};$$

the sum is replaced in the obvious way by an integral when $p(x)$ and $q(x)$ are densities. [H. Jeffreys, *Theory of Probability*, 3rd edn. Oxford University Press, 1961]

Jensen's inequality Arises in the context of convex functions (or, equivalently, of concave functions). Convexity may be defined in several ways; here we note that a real-valued function $u(x), x \in \mathbb{R}$, is convex if for every $a \in \mathbb{R}$ there exists $m(a)$ such that $u(x) \ge u(a) + m(a)(x-a)$ for all x. If the inequality is strict ($>$) for all $x \ne a$, then $u(x)$ is strictly convex. Then for a r.v. X, with finite expectation, $\mathbb{E}u(X) \ge u(\mathbb{E}X)$, which is Jensen's inequality. If $u(\cdot)$ is concave, all the above inequalities are reversed. If $u(\cdot)$ is strictly convex, and $\mathbb{E}u(X) = u(\mathbb{E}X)$, then $\mathbb{P}(X = \mathbb{E}X) = 1$. Jensen's inequality extends to convex functions in higher dimensions, and there is a conditional form: for a suitable r.v. Y, and convex $u(\cdot)$, $\mathbb{E}(u(X) \mid Y) \ge u(\mathbb{E}(X \mid Y))$. [J. Steele, *The Cauchy–Schwarz Master Class: An Introduction to the Art of Mathematical Inequalities*. Cambridge University Press, 2004]

Jensen's method *see* uniformization

Jensen–Rényi divergence *see* Jensen–Shannon divergence

Jensen–Shannon divergence In its basic form this supplies a symmetrical measure of the difference between two probability distributions **p** and **q**, on the same support, in terms of the Kullback–Leibler divergence $d_{\mathrm{KL}}(\mathbf{p},\mathbf{q})$. Thus $d_{\mathrm{JS}}(\mathbf{p},\mathbf{q}) = \frac{1}{2}\{d_{\mathrm{KL}}(\mathbf{p},\mathbf{a}) + d_{\mathrm{KL}}(\mathbf{q},\mathbf{a})\}$ where **a** is the probability distribution $\mathbf{a} = \frac{1}{2}(\mathbf{p}+\mathbf{q})$. Hence

$$\begin{aligned} d_{\mathrm{JS}}(\mathbf{p},\mathbf{q}) &= \frac{1}{2}\left\{\sum_r p_r \log \frac{p_r}{a_r} + \sum_r q_r \log \frac{q_r}{a_r}\right\} \\ &= H(\mathbf{a}) - \frac{1}{2}\{H(\mathbf{p}) + H(\mathbf{q})\} \ge 0, \end{aligned}$$

where $H(\mathbf{p})$ is the Shannon entropy of the distribution **p**, and the final inequality holds by Jensen's inequality, because $H(\mathbf{p})$ is concave in **p**. This form of the definition leads to the concept of disparity among a collection of distributions $\mathbf{p}_k$, $1 \le k \le T$, on the same support, with weights W_k, where $\sum_{k=1}^T W_k = 1$. The Jensen–Shannon disparity (or divergence) among $\mathbf{p}_1, \ldots, \mathbf{p}_T$ is $D_{\mathrm{JS}}(\mathbf{p}_1, \ldots, \mathbf{p}_T) = H(\sum_{k=1}^T W_k \mathbf{p}_k) - \sum_{k=1}^T W_k H(\mathbf{p}_k)$.

A similar definition of disparity (or divergence among a weighted set of distributions) holds for any concave entropy function; and has e.g. been made for Jensen–Rényi disparity (divergence), and Jensen–Havrda–Charvát–Tsallis disparity (often using a subset of these names). *See* Rajski's distance. [B. Fuglede & F. Topsoe, Jensen–Shannon divergence and Hilbert space embedding, in *Proc. 2004 IEEE Symp. Inf. Theor.*, (ISIT), p. 30]

Jiřina process A type of branching process taking values in a continuous state-space. Introduced by M. Jiřina in both discrete and continuous time, the term commonly now denotes only discrete-time processes; while continuous time, continuous state-space branching processes are commonly called Crump–Mode(–Jagers) processes. [M. Jiřina, Stochastic branching processes with continuous state space, *Czechoslovak Math. J.*, 1958, 8, 292–313]

jitter A term arising in communications networks and systems where, broadly, it denotes the random (and otherwise irregular) displacements of a sequence of points (or values) that is intended to be regular or deterministic. For example, if the timing of some supposedly synchronous digital processes are randomly aberrant, then one speaks of clock jitter. Similarly, if the phase, frequency, or amplitude of some regular pulsed (or periodic) signal is subject to random displacements, the process is said to be jittered.

job search problem *see* house-selling problem

job-shop problem (alias: **makespan problem**) A classic optimization problem of operations research. For example, a finite set of machines are available for a finite set of jobs, where a job comprises a sequence of operations using the machines, generally with precedence constraints. Each operation takes a possibly random time to complete, with penalties for interruptions. The problem is to minimize the expected makespan (or objective function), e.g. the expected total time to complete the jobs on all the machines. (Thus the makespan is the time taken by the last machine to finish its allocated jobs.) If there are no precedence constraints, one speaks of an open-shop problem; if an unlimited number of jobs arrive in (possibly random) sequence, one speaks of a flow-shop problem. [M. Pinedo, *Scheduling*, 4th edn. Springer, 2012]

John–Nirenberg inequality Arises in considering functions of bounded mean oscillation; there are various applications in probability theory, e.g. for a continuous random process $X(t), t \geq 0$, adapted to a filtration $\mathcal{F}_t$. If there exists $C \in (0, \infty)$ such that for all $t, \mathbb{E}[|X(\infty) - X(t)| \mid \mathcal{F}_t] \leq C$, with probability 1, then there exists $a(C)$, depending only on C, such that $\mathbb{E}\{\exp(a \sup_{t\geq 0}|X(t)|)\} < \infty$. [R. Bass, *Stochastic Processes*. Cambridge University Press, 2011]

Johnson–Mehl tessellation (alias: **J-M germ-growth model**) Arises as a model for crystallization with random nucleation. A process of points $N(t)$ appears randomly in some space (e.g. $\mathbb{R}^d$) as time $t \geq 0$ increases. From the instant of its appearance each point initiates a sphere that grows at constant speed from size zero, unless the point falls within an already existing growing sphere, in which case it is instantly deleted. When growing spheres touch, growth ceases on their common frontier.

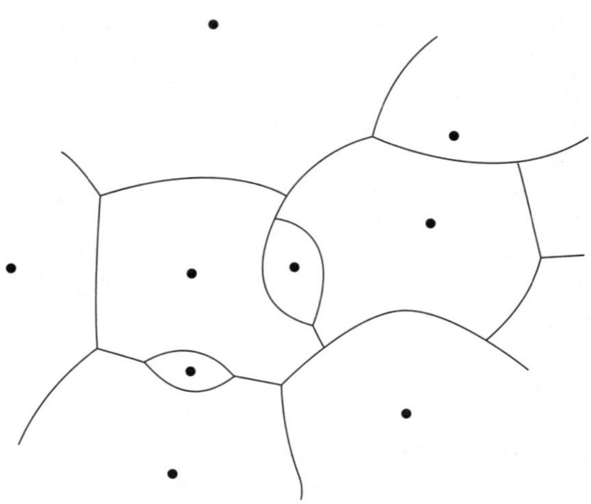

Part of a Johnson–Mehl tessellation.

(For a non-time-dependent Poisson point process, one obtains a Poisson–Voronoi mosaic.) [J. Møller, Random Johnson–Mehl tessellations, *Adv. Appl. Probab.*, 24, 814–844, 1992]

Johnson system (alias: **Johnson distributions**) A family of probability distributions $F_X(x)$ such that a relatively simple function of the corresponding random variable X is normally distributed. They are thus naturally defined by the relationship

$$Z = a + bf\left(\frac{X - c}{d}\right),$$

where Z is a standard normal r.v., f is a suitable function, and $a, b > 0, c$, and $d > 0$, are parameters. In Johnson's system f is monotonic in general, and in particular the cases $f_1(t) = \log t, f_2(t) = \log \frac{t}{1-t}$, and $f_3(t) = \log\{t + (1 + t^2)^{1/2}\}$, are prominent. The method extends to bivariate distributions. [N. Johnson, Systems of frequency curves generated by methods of translation, *Biometrika*, 36(1–2), 149–176, 1949; N. Johnson, Bivariate distributions based on simple translation systems, *Biometrika*, 36 (3–4), 297–304 1949]

joint (alias: **multivariate; multidimensional; jointly**) A term that arises in extending the concepts and attributes of one random variable to a larger number n of random variables $X_1, \ldots, X_n$ that are defined on some probability space $(\Omega, \mathcal{F}, \mathbb{P})$. As for a single r.v., the probability measure $\mathbb{P}$ induces a probability measure $\mathbb{Q}$ on $\mathbb{R}^n$ (with the appropriate σ-field), and this is the joint distribution of the r.vs. In particular, $F = \mathbb{P}(X_1 \leq x_1, \ldots, X_n < x_n)$ is their joint distribution function, with joint density $f(x_1, \ldots, x_n)$ if F is differentiable in all its arguments $x_1, \ldots, x_n$.

Their joint moments (if they exist) are $m(r_1, \ldots, r_n) = \mathbb{E}(X_1^{r_1} \ldots X_n^{r_n})$; their joint characteristic fn is $\phi(t_1, \ldots, t_n) = \mathbb{E}\exp(\mathrm{i}\sum_{r=1}^n t_r X_r)$ with a similar expression for the joint mgf (or pgf); they may have a joint entropy $H(X_1, \ldots, X_\mathrm{n})$; and so on. In particular, when $n = 2$, one speaks of a bivariate distn; less often, when $n = 3$, of a trivariate distn.

joint entropy *see* entropy function

Jones–Pewsey distribution *see* circular distribution

Jordan's identity *see* inclusion-exclusion bounds

Joseph effect *see* long-range dependence (and Hurst exponent)

J-shaped distribution Said of a univariate probability distribution, monotone on an interval $[a, b]$, that has its unique mode at b, and its minimum at a, thus resembling the letter J. Distributions resembling ᒐ are called reverse J-shaped. *See* U-shaped distribution and bell-shaped density.

jump (alias: **saltus; step**) Arises at a point $x \in \mathbb{R}$ at which a function $u(x) \in \mathbb{R}$ is discontinuous. In particular, the distribution function $F(x)$ of a r.v. X has a jump when $\mathbb{P}(X = x) > 0$. The set of such jumps of X is countable (and may be called the **spectrum**); *see* Lebesgue decomposition. If X is a discrete r.v. then $F(x)$ is said to be a (pure) jump function. [K. L. Chung, *Course in Probability Theory*, 2nd edn revised. Academic Press, 2001]

jump chain Arises in considering regular (conservative, stable, right-continuous) Markov chains $X(t)$ in continuous time on a countable state space, whose sample paths are pure jump processes. Such a process occupies a succession of states $X_1, X_2, \ldots$, which comprises the jump chain; if the sequence of (non-zero) holding times in these states is denoted by $H_1, H_2, \ldots$, then the successive jump times are $J_n = \sum_{r=1}^{n} H_r$. [J. Norris, *Markov Chains*. Cambridge University Press, 1997]

jump-diffusion process In its broadest sense this is simply a random process $X(t)$ with jumps at the instants of some point (or counting) process $N(t)$, whose paths between the jumps are diffusions. More narrowly, it denotes the (strong) solution of a general class of stochastic differential equations. [D. Applebaum, *Lévy Processes and Stochastic Calculus*, 2nd edn. Cambridge University Press, 2009]

jump process Arises in the context of random processes in continuous time $X(t)$ whose sample paths are a.s. piecewise constant (and usually taken to be right-continuous) except for jumps of size $X(t) - X(t-)$. If the jumps are isolated, so that there are only finitely many in any finite interval, one speaks of a pure jump process (or a step process). Limit points of jumps comprise explosions. Right-continuous pure-jump Markov processes have the strong Markov property. [O. Kallenberg, *Foundations of Modern Probability*, 2nd edn. Springer, 2002; L. Breiman, *Probability*. Addison-Wesley, 1968; reprinted by SIAM, 1992]

K

Kac empirical process *see* empirical process

Kac ergodic theorem (alias: **Kac's lemma**)

Arises in extending the ideas of Poincaré recurrence in dynamical systems to strongly stationary stochastic processes. In its basic form, for a strongly stationary sequence X of discrete r.vs $X_n, n \in \mathbb{Z}$, the theorem asserts that the mean recurrence time of a recurrent value a of X is proportional to the reciprocal of $\mathbb{P}(X_1 = a)$. More generally, for a measure-preserving ergodic transformation T of a probability space $(S, \mathcal{F}, \mathbb{P})$ onto itself, and any A such that $\mathbb{P}(A) > 0$, we define the first return time $H(x, A) = \inf\{n \geq 1 : T^n\, x \in A\}$, for each $x \in A$, and then the theorem asserts that $\int_A H(x, A)\mathbb{P}(dx) = 1$. [M. Kac, On the notion of recurrence in discrete stochastic processes, *Bull. Am. Math. Soc.*, 53(10), 1002–1010, 1947]

Kac formula

(1) Supplies an expression for the expected number Z_n of real distinct zeros of a random polynomial $K(x) = \sum_{r=0}^{n-1} A_r x^r$, where $A_0, A_1, \ldots$ is a sequence of i.i.d. standard normal r.vs. Explicitly

$$Z_n = \frac{4}{\pi}\int_0^1 \frac{1}{1-x^2}\left\{1 - n^2\left[\frac{x^{n-1}(1-x^2)}{1-x^{2n}}\right]^2\right\}^{\frac{1}{2}} \mathrm{d}x$$

$$\sim \frac{2}{\pi}\log n, \text{ as } n \to \infty.$$

More generally, the **Kac–Stevens thm** asserts that $Z_n \sim \frac{2}{\pi}\log n$ for a large class of zero-mean, unit-variance r.vs $A_0, A_1, \ldots$. By contrast, when $A_0, A_1, \ldots$ are, e.g. i.i.d. Cauchy r.vs, the **Logan–Shepp formula** for Z_n yields that, as $n \to \infty$, $Z_n \sim c \log n$, where

$$c \simeq 0.7413 > \frac{2}{\pi} \simeq 0.6366.$$

(2) *See* Feynman–Kac formula. [B. Logan & L. Shepp, Real zeros of random polynomials II, *Proc. Lond. Math. Soc.*, 18, 308–314, 1968]

Kac matrix (alias: **Clement matrix; Sylvester–Kac matrix**)

The tridiagonal $n+1$ by $n+1$ matrix of the form

$$K = \begin{pmatrix} 0 & n & & \cdots & & 0 \\ 1 & 0 & n-1 & & & \\ & 2 & 0 & n-2 & & \vdots \\ \vdots & & \ddots & \ddots & \ddots & \\ & & & n-1 & 0 & 1 \\ 0 & & \cdots & & n & 0 \end{pmatrix}.$$

It arises in the Ehrenfests' urn model, and also in a random walk on the hypercube. [A. Edelman & E. Kostlan, The road from Kac's matrix to Kac's random polynomials, in *Proc. 1994 SIAM Applied Linear Algebra Conference*, J. G. Lewis (ed.). SIAM, 1994]

Kac process *see* telegraph process

Kac random walk

Either of two random processes arising in connexion with a simple model proposed by M. Kac for a Boltzmann gas comprising n elastic particles whose velocities are randomly changed at random impacts in such a way that energy is conserved. One Kac random walk takes place on the $(n-1)$-dimensional sphere $S^{n-1} \subseteq \mathbb{R}^n$; the other is a random walk on the set of $n \times n$ rotation matrices. [R. Oliveira, On the convergence to equilibrium of Kac's random walk on matrices, *Ann. Appl. Probab.*, 19(3), 1200–1231, 2009]

Kac–Rice formula

A formula for the number of distinct zeros of a suitable random function $X(t), t \in \mathbb{R}$ (being a polynomial for Kac, or a random process for Rice). If $X(t)$ and its derivative $Y = X'(t)$ have joint density function, $f(t, x, y)$, then the expected number of zeros of $X(t)$ for $a < t < b$ is given by $Z = \int_a^b \int_{-\infty}^{\infty} f(t, 0, y)|y|\mathrm{d}y\mathrm{d}t$. [S. Rice, Mathematical theory of random noise, *Bell. System Tech J.*, 25, 46–156, 1945]

Kac's saying (or dictum)

Any of several maxims formulated by Mark Kac including: 'Primes play a game of chance', and 'where there is independence, there must be the normal law,' together with others attributed to him, e.g.: 'If you can't solve a problem exactly, then randomize.' [H. Kesten, The influence of Mark Kac on probability theory, *Ann. Probab.*, 14(4), 1103–1128, 1986]

Kac–Slepian paradox A paradox of conditional probability arising in considering the joint distribution of a random process $X(t)$ and its derivative $X'(t)$. It can be seen as a developed version of the Borel paradox. [M. Kac & D. Slepian, Large excursions of Gaussian processes, *Ann. Math. Stat.*, 30(4), 1215–1228, 1959]

Kac's theorem Any of several theorems, but most often perhaps this (also called the **Kac–Bernstein thm**), which supplies a characterization of the normal distn. For random variables X and Y define $U = X\cos\theta + Y\sin\theta$ and $V = -X\sin\theta + Y\cos\theta$: If U and V are independent for each θ then X and Y are zero-mean normal r.vs with the same variance. *See* Skitovich's theorem. [M. Kac, On a characterization of the normal distribution, *Amer. J. Math.*, 61, 726–728, 1939]

Kagan–Linnik–Rao theorem Supplies a characterization of the normal distn, one form is this: let $X_1, \ldots, X_n$ be i.i.d. zero-mean r.vs, and set $\overline{X} = \frac{1}{n}\sum_{r=1}^{n} X_r$. If $\mathbb{E}(\overline{X}\,|\,X_1 - \overline{X}, \ldots, X_n - \overline{X}) = 0$, then the Xs are normally distributed. [A. Kagan *et al.*, *Characterization Problems in Mathematical Statistics*. Wiley, 1973]

Kahane–Hoffmann–Jørgensen inequality *see* Hoffmann–Jørgensen inequality

Kahane–Khinchin inequality *see* Khinchin's inequality

Kahane–Peyrière theorem *see* Mandelbrot multiplicative cascade

Kahn–Kalai–Linial theorem (alias: **KKL theorem**) Any of a number of theorems arising in the context of Boolean functions, but typically this one: let $f(X_1, \ldots, X_n)$ be a Boolean (two-valued) function on the unit discrete cube $\{-1, +1\}^n$, so the sample space of $X_1, \ldots, X_n$ is the set of n-strings of +1 and −1. Let $\mathbb{P}(\cdot)$ be the uniform distn assigning probability 2^{-n} to each string, and assume that $\mathbb{P}(f = 1) = \mathbb{P}(f = -1) = 1/2$ (without essential loss of generality). The influence of the rth variable X_r is $I_r((f) = \mathbb{P}(f(X_1, \ldots, X_r, \ldots, X_n) \neq f(X_1, \ldots, -X_r, \ldots, X_n))$.

Then one KKL theorem asserts that there is some variable X_k that asserts a non-trivial influence on f, i.e. formally,

$$I_k(f) = \Omega\left(\text{var}\,(f(X))\frac{\log n}{n}\right);$$

where $a(n) = \Omega(b(n))$ if there exists $n_0 < \infty$ and $c > 0$ such that $a(n) \geq cb(n)$ for all $n \geq n_0$. A related theorem asserts (broadly speaking), that the values of such Boolean functions are determined (decisively influenced) by small sets of the variables $X_1, \ldots, X_n$, as $n \to \infty$. [J. Kahn *et al.*, The influence of variables in Boolean functions, in *Proc. 29th IEEE Symposium on the Foundations of Computer Science*, pp. 68–80, 1988]

Kailath–Segall formula (alias: **Kailath–Segall identity, or polynomials**) Arises in the context of a suitable semi-martingale, $X(t)$, $t \geq 0$ (with $X(0) = 0$), where the Kailath–Segall formula generates a sequence $P_n(t, X)$ of functionals of the martingale X that can be seen as generalizing the orthogonal polynomials used in series expansions of functionals of the Wiener process or Poisson process (being Wiener–Hermite or Charlier polynomials respectively). The formula is the recurrence $P_0(t, X) = 1, P_1(t, X) = X(t)$, and in general

$$P_n(t, X) = \int_0^t P_{n-1}(s-, X(s-))\mathrm{d}X(s).$$

When the process X has independent increments, the family $P_1, P_2, \ldots$ are orthogonal and complete, in that any suitable X can be written as a sum of the P_n. [A. Segall & T. Kailath, Orthogonal functionals of independent increment processes, *IEEE Trans. Inf. Theor.*, 22(3), 287–298, 1976]

Kakutani's dichotomy Arises in the context of infinite products of probability measures. Let $\mathbb{P}_n$ and $\mathbb{Q}_n$, $n \geq 1$, be probability measures on a sequence of measure spaces $(\Omega_n, \mathcal{F}^n)$, such that for all n, $\mathbb{Q}_n$ is absolutely continuous wrt $\mathbb{P}_n$. Then either $\prod_1^\infty \mathbb{Q}_n$ and $\prod_1^\infty \mathbb{P}_n$ are mutually singular, or $\prod_1^\infty \mathbb{Q}_n$ is absolutely continuous wrt $\prod_1^\infty \mathbb{P}_n$; and the latter occurs iff

$$\prod_n \int \left(\frac{\mathrm{d}\mathbb{Q}_n}{\mathrm{d}\mathbb{P}_n}\right)^{1/2} \mathrm{d}\mathbb{P}_n > 0.$$

[S. Kakutani, On equivalence of infinite product measures, *Ann. Math.*, 49, 214–224, 1948]

Kakutani's product martingale theorem Asserts that for the product $M_n = \prod_{r=1}^n X_r$, $r > 1$, of independent non-negative, unit-mean r.vs, the following are equivalent:

(a) $M_n, n \geq 1$ is a uniformly integrable martingale;
(b) $\prod_{r\geq 1}\mathbb{E}(\sqrt{X_r}) > 0$;

(c) $\mathbb{P}(M > 0) > 0$, where M is the finite a.s. limit of M_n as $n \to \infty$;
(d) $\mathbb{E}M = 1$.

[D. Williams, *Probability with Martingales*. Cambridge University Press, 1991]

Kakutani's theorem Any one of a number of theorems arising from the connexion between the Wiener process and harmonic functions; and hence also between the Wiener process and electrostatic quantities expressed in terms of solutions of Laplace's equation. The link was first established for the plane Wiener process by S. Kakutani. Examples include: (1) the fact that a compact set C in $\mathbb{R}^d$, with $0 \notin C$, is hit by the d-dimensional Wiener process with positive probability iff the (electrostatic) capacity of C is strictly positive (for a point charge at 0, i.e. with respect to the Green kernel $|x-y|^{2-d}$, $d \geq 3$, or $\log^+|x-y|^{-1}$, $d = 2$); and (2) the fact that the solution $u(x)$ of the Dirichlet problem for Laplace's eqn in D (for which $u = g$ on the boundary ∂D) is given by Kakutani's formula: $u(x) = \mathbb{E}[g(W(Tx, \partial D)))]$, where $W(Tx, \partial D))$ is the point at which the Wiener process started at x first hits ∂D (at time $T(x, \partial D)$) (subject to mild technical conditions on D and g). [S. Kakutani, Two-dimensional Brownian motion and Harmonic functions, *Proc. Imp. Acad.*, 20(10), 706–714, 1944]

Kallenberg's theorem Any of several results, but perhaps most often this: let $X_n(\cdot), n \geq 1$, and $X(\cdot)$ be point processes on $\mathbb{R}^d$, such that X is simple (having no multiple points). If, for any member A of a suitable family of sets, $\lim_{n\to\infty}\mathbb{P}(X_n(A) = 0) = \mathbb{P}(X(A) = 0)$, and $\lim_{n\to\infty}\mathbb{E}(X_n(A)) = \mathbb{E}X(A)$, then X_n converges in distribution to X (and one says that X_n converges weakly to X). [O. Kallenberg, *Foundations of Modern Probability*, 2nd edn. Springer, 2002]

Kallianpur–Robbins law Asserts that for an additive functional $V(t)$ of planar Brownian motion (such as the occupation time up to t of a Borel set in $\mathbb{R}^2$), $V(t)/\log t$ converges in distribution to an exponential r.v., as $t \to \infty$; for Brownian motion on the line the distribution of $V(t)/\sqrt{t}$ converges to that of the absolute value of a normal r.v. with zero mean. [P. Mörters, Almost sure Kallianpur–Robbins laws for Brownian motion in the plane, *Probab. Theor. Rel. Fields*, 118(1), 49–64, 2000]

Kallianpur–Striebel formula Arises in the non-linear filtering problem, where it supplies a Bayes-type formula for the family of conditional expectations representing the solution for the desired estimate of the unobserved variable. [A. Bain & D. Crisan, *Fundamentals of Stochastic Filtering*. Springer, 2008]

Kalman filter Arises in filtering, where it supplies an iterative solution for the optimal filtering problem in the case of a linear system subject to Gaussian white noise. The iteration is a simple first-order recursion, in that the optimal estimate of the $(n + 1)$-th state is a function only of the estimate of the previous state and the next observation. Formally, the model comprises the linear (vector) state process, $X_{n+1} = A_nX_n + W_n$, and the linear observation (or measurement) process, $Y_n = B_nX_n + Z_n$, where A_n and B_n are sequences of known matrices, and W_n and Z_n are mutually independent Gaussian white noise processes. The filter supplies a recursion for the minimum mean square estimator of $X_n, n \geq 1$, given the observations $Y_r, r \leq n$, in terms of the known moments of W_n and Z_n, and known matrices A_n and B_n.

In continuous time the equivalent problem and its recursive solution comprise the **Kalman–Bucy filter**. [R. Kalman & R. Bucy, New results in linear filtering and prediction theory, *Trans. ASME J. Basic Eng.*, 83, 95–108, 1961]

Kantorovich distance *see* earth movers distance; minimum L_p distance

Kantorovich inequality Let X be a positive simple random variable such that $0 < m \leq X \leq M < \infty$. Then

$$\mathbb{E}(X)\mathbb{E}\left(\frac{1}{X}\right) \leq 1 + \frac{(M-m)^2}{4mM} = \frac{(M+m)^2}{4mM}.$$

This Kantorovich inequality is implied by the earlier Schweitzer inequality, which asserts that for X uniformly distributed on $\{x_1, \ldots, x_n\}$ with $0 < m \leq x_r \leq M < \infty$,

$$\mathbb{E}(X)\mathbb{E}\left(\frac{1}{X}\right) \leq \frac{(M+m)^2}{4mM}.$$

[A. Clausing, Kantorovich-type inequalities, *Amer. Math. Mon.*, 89(5), 314 & 327–330, 1982]

Kantorovich–Rubinstein duality theorem *see* minimum L_p distance

Kaplan condition *see* Pakes's lemma

Kapur entropy *see* entropy

Karamata theory *see* regular variation

Karger's algorithm Any one of a number of randomized (i.e. Monte Carlo type) algorithms that seek a minimum cut (or all minimum cuts) of a connected graph with n vertices and m edges. One such method finds such a cut in $O(m \log^3 n)$ time, with high probability. [D. Karger, Minimum cuts in near-linear time, *J. ACM*, 47(1), 46–76, 2000]

Karhunen–Loève expansion (alias: **K-L decomposition (or transformation, or theorem); proper orthogonal decomposition; principal component analysis; Hotelling analysis**) Arises as the natural extension to random vectors and random processes of the classical representation of functions as a linear combination of orthogonal functions. Thus for a suitable random process $X(t)$, $a \le t \le b$ (or random vector $X_1, \ldots, X_n$), the Karhunen–Loève expansion takes the form

$$X(t) = \sum_{r=1}^{\infty} e_r(t) Z_r,$$

where $e_r(t)$ are the eigenfunctions generated by the covariance function $c(s,t) = \operatorname{cov}(X(s), X(t))$, and Z_r comprises an orthogonal sequence of r.vs whose variances are the corresponding eigenvalues λ_r. (The sum is finite in the case of a random vector.)

An important special case arises when $c(s,t) = s \wedge t$, $0 \le s, t \le 1$, and $e_r(t) = \sqrt{2} \sin\left\{\left(r - \frac{1}{2}\right)\pi t\right\}, r \ge 1$, and each Z_r is normal with var $Z_r = 4(2r-1)^{-1}\pi^{-2}$. Then $X(t) = \sum_{r=1}^{\infty} e_r(t) Z_r$ is the Karhunen–Loève expansion of a continuous version of the Wiener process. [R. Ash, *Information Theory (Appendix)*. Wiley, 1965; reprinted by Dover, 1990]

Karlin–McGregor theorem Any one of several theorems about the Markov birth–death process; but most often these:

(1) Karlin–McGregor coincidence theorem. In its simplest form this considers n independent realizations of a birth–death process $X(t)$ with transition matrix $p_{ij}(t)$, $t \ge 0$, $i, j \in \{0, 1, 2, \ldots\}$, started in states $i_1 < i_2 < \ldots < i_n$ at $t = 0$. Let A be the event that at $t > 0$ they are at $j_1 < j_2 < \cdots < j_n$ without any two ever having been simultaneously in the same state, i.e. coincident. Then $\mathbb{P}(A)$ is the modulus of the determinant

$$\begin{vmatrix} p_{i_1 j_1}(t) & \cdots & p_{i_1 j_n}(t) \\ \vdots & \ddots & \vdots \\ p_{i_n j_1}(t) & \cdots & p_{i_n j_n}(t) \end{vmatrix}.$$

(2) Karlin–McGregor representation theorem. Supplies an explicit representation for the transition probabilities of a (minimal) birth–death process in terms of a family of orthogonal polynomials and associated weights.

[S. Karlin & J. McGregor, The differential equations of birth and death processes and the Stieltjes moment problem, *Trans. Amer. Math. Soc.*, 85, 489–456, 1957; S. Karlin & J. McGregor, Coincidence properties of birth and death processes, *Pacific J. Math.*, 9(4), 1109–1140, 1959]

Karlin–Novikoff cut (or crossing) criterion (alias: **one-cut (or crossing) condition; Ohlin's lemma**) Arises in the context of stochastic ordering, in particular in considering increasing–convex (or stop-loss, or variability) orderings. In the simplest case, X is said to be smaller than Y in increasing convex order ($X \le_{icx} Y$) if $\mathbb{E}X$and $\mathbb{E}Y$ are finite and $\mathbb{E}g(X) \le \mathbb{E}g(Y)$ for all increasing convex functions $g(\cdot)$. Then the Karlin–Novikoff condition asserts that if $\mathbb{E}X \le \mathbb{E}Y$, and for some $c \in \mathbb{R}$ we have

$$\mathbb{P}(X \le x) \le \mathbb{P}(Y \le x), \quad x < c,$$
$$\mathbb{P}(X \le x) \ge \mathbb{P}(Y \le x), \quad x \ge c,$$

then $X \le_{icx} Y$. This is a special case of the more general Karlin–Novikoff k-cut criterion, applicable to k-increasing convex stochastic ordering. [S. Karlin & A. Novikoff, Generalized convex inequalities, *Pacific J. Math.*, 13(4), 1251–1279, 1963]

Karlin's solidarity theorem *see* solidarity theorem

Kazamaki's condition Arises in the context of the stochastic exponential, or Doléans-Dade exponential, of a continuous local martingale $M(t)$ with $M(0) = 0$, where it asserts (in one form) that if $\mathbb{E}\exp\left\{\frac{1}{2}M(T)\right\} \le c < \infty$ for all bounded stopping times T, and some constant c, then the Doléans-Dade exponential of $M(t)$ is a (uniformly integrable) martingale. It is implied by the Novikov condition that , $\mathbb{E}\exp\left\{\frac{1}{2}[M,M]_\infty\right\} < \infty$, $t \in [0, \infty]$, where $[M,M]_t$ is the quadratic variation process (*see* brackets process). [I. Karatzas & S. Shreve, *Brownian Motion and Stochastic Calculus*, 2nd edn. Springer, 1998]

Kellerer's theorem Arises in the context of a finite-mean random process $X(t), t \ge 0$, that is increasing in convex order, i.e. for any convex function $c(x) \in \mathbb{R}$, $x \in \mathbb{R}, \mathbb{E}c(X(t)), t \ge 0$, is increasing. Kellerer's theorem asserts that $X(t)$ has this property iff there exists a random process $Y(t)$ (possibly on another probability space) such that $Y(t)$ is a martingale, and $\mathbb{P}(Y(t) \le z) = \mathbb{P}(X(t) \le z)$

for all $t \geq 0$ and $z \in \mathbb{R}$. Such a process is called a **peacock**. [F. Hirsch *et al.*, *Peacocks and Associated Martingales, with Explicit Constructions*. Springer, 2011]

Kelly betting (or gambling) (alias: **proportional betting; Kelly strategy (or criterion, or rule, or system, or fraction, or formula, etc.); Kelly portfolio selection policy**) Arises in choosing an optimal strategy to adopt in betting on a non-terminating sequence of favourable (or, at least, fair) games. Broadly, a Kelly strategy (or criterion, etc.) seeks to maximize the asymptotic expected value of the logarithm of the gambler's fortune X_n at time n, namely

$$\Delta = \lim_{n\to\infty} \frac{1}{n} \mathbb{E} \log_2 \frac{X_n}{X_0};$$

such a strategy is said to be **log-optimal**, and Δ is called the (asymptotic) **doubling rate**. (Clearly $\Delta \geq 0$, since $\Delta = 0$ is attainable by not betting at all.) J. L. Kelly, Jr, considered two classic examples (which he also linked with Shannon's entropy and information rate):

1. A sequence of i.i.d. win-or-lose bets, with positive expected net gain, at each of which the gambler can bet any fraction ϕ of their current fortune. There is a value K of ϕ, called the **Kelly fraction**, that maximizes the doubling rate Δ.
2. A sequence of i.i.d. bets on a r.v. with several outcomes, called a horse-race for definiteness. The gambler can distribute their current fortune as bets-to-win on the several horses, and the odds are fair or superfair. Then the allocation of bets that maximizes the doubling rate backs each horse to win in proportion to their respective probabilities of winning. This is called Kelly proportional betting (the concept of proportional betting had been previously introduced and analyzed by W. Whitworth). The subfair case is more complicated, and the Kelly strategy involves a fraction, but not proportional betting. *See* Kelly–Breiman betting.

[J. Kelly, A new interpretation of information rate, *Bell Syst. Tech. J.*, 35, 917–926, 1956; G. Székely & D. Richards, Remain steadfast with the St. Petersburg paradox to quantify irrational exuberance, *Amer. Stat.*, 59(3), 235–239, 2005]

Kelly–Breiman betting An extension of Kelly betting in which bets may be made on a class of subsets of a finite set $S = \{1, \ldots, s\}$, and the game is favourable. Then Breiman shows that the Kelly (log-optimal) strategy is (asymptotically) optimal according to two further criteria:

1. Fix the number of games n and seek a strategy to maximize the fortune after these n plays.
2. Fix a level of required fortune n and seek a strategy to minimize the expected number of games needed to attain or exceed n. In both cases the log-optimal strategy is best asymptotically, as $n \to \infty$.

[L. Breiman, Optimal gambling systems for favourable games, *Proc. 4th Berkeley Symp. Math. Stat. Probab.*, 65–78, 1961]

Kelly's lemma Arises in considering the time reversibility of a (non-explosive) Markov chain on a discrete state space, S, having Q-matrix q_{ij}. If Q_{ij} is another non-explosive Q-matrix on S, and there is a distribution $\boldsymbol{\pi}$ on S such that $\pi_i q_{ij} = \pi_j Q_{ji}$ for all i and j in S then π is the stationary distribution of X, and of the chain $Y(t)$ obtained by reversing $X(t)$ in equilibrium, and Q_{ij} is the Q-matrix of $Y(t)$. [F, Kelly, *Reversibility and Stochastic Networks*. Wiley, 1979; reprinted by Cambridge University Press, 2011]

Kelvin's method of images *see* reflection principle

Kemeny's constant (alias: **seek time; random target lemma; right-averaging principle**) Arises in the context of an ergodic Markov chain $(X_n; n > 0)$ with transition matrix P, and stationary distribution π. Let $H = (h_{jk})$ be the matrix of expected hitting times of k from j, that is $h_{jk} = \mathbb{E}T_{jk} = \mathbb{E}\min\{n \geq 0 : X_n = k \,|\, X_0 = j\}$ (note that $h_{jj} = 0$). Then for all j, $\sum_k h_{jk}\pi_k = K$, a constant independent of j, where K is called Kemeny's constant. The statement can be regarded as asserting that the expected time for the chain to reach (make a first passage, or return to) a state chosen according to the equilibrium distribution does not depend on the distribution of the initial state X_0. [D. Levin *et al.*, *Markov Chains and Mixing Times*. American Mathematical Society, 2009; P. Doyle, The Kemeny constant of a Markov chain, arXiv: 0909.2636 vl, 2009]

Kemperman's formula *see* hitting-time theorem

Kendall–Cranston coupling A coupling of Brownian motions (or suitable diffusions) $X(t)$ and $Y(t)$ on a Riemannian manifold. Informally, it uses the idea of mirror coupling to envisage the increment $dY(t)$ as a suitable reflection of the increment $dX(t)$, where the 'reflecting surface' is determined by the minimal geodesic joining $X(t)$ and $Y(t)$. In Euclidean space, this reduces to mirror coupling. [E. Hsu, *Stochastic Analysis on Manifolds*. American Mathematical Society, 2002]

Kendall distribution function (alias: **bivariate probability integral transform**) Arises in considering the stochastic ordering of random vectors, and in relation to the copula of the bivariate distribution function $F(x,y)$ of a pair of r.vs X and Y. The Kendall distn fn $K(z)$ of (X,Y) is the distribution fn of the r.v. $Z = F(X,Y)$. The name was bestowed because the Kendall rank correlation coefficient τ has the property that $\tau = 4\mathbb{E}(Z) - 1$. [R. Nelsen, *Introduction to Copulas*, 2nd edn. Springer, 2006]

Kendall's formula (or identity) Arises in considering the first hitting time T_a, of the point $a > 0$, by a random process $X(t), t \geq 0$, started from $X(0) = 0$, where $X(t)$ has independent stationary increments and no positive jumps (i.e. it is a Lévy process that is upwardly skip-free). Then this formula holds: $\mathbb{P}(T_a \in (t, t + \mathrm{d}t))\mathrm{d}a = \frac{a}{t}\mathbb{P}(X(t) \in (a, a + \mathrm{d}a))\mathrm{d}t$, as $\mathrm{d}a, \mathrm{d}t \downarrow 0$; or, when $X(t)$ has a density $f_{X(t)}(a)$, so that T_a has a density $f_{T_a}(t)$, we have $tf_{T_a}(t) = af_{X(t)}(a)$. This can be seen as a continuous-time hitting-time theorem, or ballot theorem. [K. Borovkov & Z. Burq, Kendall's identity for the first crossing time revisited, *Electr. Commun. Probab.*, 6, 91–94, 2001]

Kendall's functional equation (alias: **Kendall–Takács equation**) Arises in considering the busy period B of the single server in a queue with Poisson process arrivals and a general distribution of service times, denoted by $M(\lambda)/G/1$ in Kendall's notation. If service times and B have respective Laplace transforms $S(\theta)$ and $B(\theta)$, then $B(\theta) = S(\theta + \lambda - \lambda B(\theta))$ is Kendall's functional equation. A dual form arises likewise for the wet period of a suitable dam. [D. Kendall, Some problems in the theory of queues, *J. Roy. Stat. Soc. B*, 13(2), 151–185, 1951; D. Kendall, Some problems in the theory of dams, *J. Roy. Stat. Soc. B*, 19(2), 207–233, 1957]

Kendall's generalized birth–death process A birth–death process, comprising a Markov chain in continuous time, in which the parameters $\lambda_n(t)$, $n \geq 0$, and $\mu_n(t)$, $n \geq 1$, take the form $\lambda_n = n\lambda(t)$ and $\mu_n = n\mu(t)$, being functions of the time parameter t. [D. Kendall, On the generalized 'birth-and-death' process, *Ann. Math. Stat.*, 19(1), 1–15, 1948]

Kendall's notation (or terminology) A system of abbreviations that denote various types of simple queueing systems. The first and simplest convention uses a triple in the from a/b/c, where a denotes the nature of the arrivals, b denotes the nature of the service times, and c denotes the number of servers.

For a:

M(λ): Poisson arrivals with parameter λ.

E(λ, r): a renewal process with gamma-distributed intervals.

D: deterministic (and usually at times of the form $dn, n \in \mathbb{Z}^+$).

GI: a renewal process.

G: not necessarily a renewal process, e.g. a Markov renewal process, or a Cox process. (Some authors use G instead of GI.)

For b (service times are almost always assumed independent):

M(λ): exponential with parameter λ.

E(λ, r): gamma, with parameters λ and r.

D: deterministic.

For c:

s: any integer $s \geq 1$ implying s servers are always available.

∞: implying that any arriving customer immediately begins service.

This may be augmented to:

$s + d$: where d denotes the size of the waiting room, or buffer, so that no more than $s + d$ customers can be in the system.

The notation is sometimes extended – one popular form uses a quadruple a/b/$s + d$/q, where the initial triple is as above, and q denotes the queueing discipline (assumed to be other than the usual first-come first-served (FCFS), or first-in first-out (FIFO); in those cases the triple notation suffices).

Possible values for q include:

LIFO: last in, first out

LCFS: last come, first served

SIRO: service in random order

BS: batch service

SS: shared service (by all present in the queue)

PS: priority service

PPS: preemptive priority service.

[D. Kendall, Stochastic processes occurring in the theory of queues and their analysis by the method of the imbedded Markov chain, *Ann. Math. Stat.*, 24(3), 338–354, 1953]

Kendall's rank correlation (alias: **Kendall's τ**) Arises in statistics as a measure of association (or dependence) between random samples from the distribution of two random variables X and Y (based on the idea of concordance function). For the r.vs themselves, Kendall's τ (which may be called the population τ coefficient) is defined thus: let$(\hat{X}, \hat{Y})$ be independent of (X, Y), with the same distribution. Then

$$\tau = \mathbb{P}\{(X - \hat{X})(Y - \hat{Y}) > 0\} - \mathbb{P}\{(X - \hat{X})(Y - \hat{Y}) < 0\}$$
$$= \rho\big(\mathrm{sign}(X - \hat{X}),\ \mathrm{sign}(Y - \hat{Y})\big).$$

See Kendall distribution and grade correlation. [G. Fredricks & R. Nelson, On the relationship between Spearman's rho and Kendall's tau for pairs of continuous random variables, *J. Stat. Plan. Infer.*, 137, 2143–2150, 2007]

Kendall's renewal theorem Arises in discrete renewal theory (or recurrent events) where it considers the geometric convergence of a renewal sequence $u_n, n \geq 0$, generated by the convolutional recurrence: $u_0 = 1$ and

$$u_n = u_0 p_n + u_1 p_{n-1} + \cdots + u_{n-1} p_1, \quad n \geq 1,$$

in which p_n, $n \geq 1$ is an aperiodic probability distribution with finite mean (so the renewal sequence is ergodic, and converges to u_∞ as $n \to \infty$). Then one form of Kendall's theorem asserts that the following are equivalent:

(a) For some $s > 1$, $\sum_1^\infty p_n z^n$ converges for all $|z| < s$.
(b) For some $t > 1$, $\sum_0^\infty |u_n - u_\infty| z^n$ converges for all $|z| < t$.
(c) For some $r > 1$, the holomorphic function $u(z)$ defined for $|z| < 1$ by $\sum_0^\infty u_n z^n = u(z)$ has an analytic extension to $|z| < r$, except for a simple pole at $z = 1$.

See Erdös–Feller–Pollard theorem. [S. Meyn & R. Tweedie, *Markov Chains and Stochastic Stability*, 2nd edn. Cambridge University Press, 2009]

Kendall's representation (alias: **Kendall's integral representation**) Arises in the context of Markov chains on a countable state space (in discrete or continuous time) where it supplies a representation for the transition probabilities $\big(p_{jk}(n)$ or $p_{jk}(t)\big)$ as Fourier (integral) transforms of suitable complex-valued measures. If the chain is time-reversible then the representation is a Laplace transform of real-valued functions. [D. Kendall, Integral representations for Markov transition probabilities, *Bull. Amer. Math. Soc.*, 64(6), 358–362, 1958]

Kendall's solidarity theorem *see* solidarity theorem

Kendall–Takács eqn *see* Kendall's functional equation

Kendall's taxicab problem A classic queue model in which it is assumed that taxis and customers arrive singly at a queueing point according to different (possibly non-homogeneous) independent Poisson processes. Taxis form a queue if customers are absent; customers form a queue in the absence of taxis. The queue length $X(t)$ at time t is the difference between two Poisson r.vs, having a Skellam distribution. [D. Kendall, Some problems in the theory of queues, *J. Roy. Stat. Soc. B*, 13, 151–185, 1951]

Kent distribution *see* Bingham distribution

kernel *see* transition function

Kerridge's inaccuracy Arises as a measure of how wrong it is to assert that a probability distribution is $\mathbf{q} = (q_1, q_2, \ldots)$ when in actual fact the true probability distribution is $\mathbf{p} = (p_1, p_2, \ldots)$. On a set of intuitively plausible assumptions (that extend the ideas of information theory) the 'inaccuracy' of this assertion is given by,

$$w(\mathbf{p}, \mathbf{q}) = -\sum_i p_i \log_2 q_i \geq -\sum p_i \log_2 p_i,$$

with equality iff $p_i = q_i$, for all i (by Gibbs inequality); where $w(\mathbf{p}, \mathbf{q}) = 0$ iff $p_i = q_i = 1$ for some i. It can be seen as adding doubt about the distribution to doubt about the outcome. It is sometimes called cross-entropy; *see* Kullback–Leibler divergence. [D. Kerridge, Inaccuracy and inference, *J. Roy. Stat. Soc. B*, 23(1), 184–194, 1961]

Kesten's hitting theorem (1) Any one of several results determining the strict positivity or otherwise of the probability that a right-continuous random process with stationary independent increments started from 0 ever hits a given point $r > 0$.

(2) Any one of several results concerning the asymptotic hitting probability of a connected set C in $\mathbb{Z}^d$ by a simple symmetric r.w. started from x, as $|x| \to \infty$. [H. Kesten, Hitting probabilities of single points of processes with stationary independent increments, *Memoirs Amer. Math. Soc.*, 93, 1969; H. Kesten, Hitting probabilities of random walks on $\mathbb{Z}^d$, *Stoch. Proc. Appl.*, 25(2), 165–184, 1987]

Kesten–Ney–Spitzer theorem Arises in the critical Galton–Watson branching process $Z_n, n \geq 0$ (with $Z_0 = 1, \mathbb{E}Z_1 = 1$, and $0 < \mathrm{var}\ Z_1 = \sigma^2 < \infty$), where it details the behaviour of the asymptotic properties of the transition probabilities $p_{ij}(n)$ of Z_n as $n \to \infty$. The principal result is that, for $a = \frac{1}{2}\sigma^2$,

$$\lim n^2 \exp\left(\frac{j}{an}\right) p_{ij}(n) = \frac{i}{a^2},$$

where the limit is taken as $j, n \to \infty$ with j/n bounded. [H. Kesten *et al.*, The Galton–Watson process with mean 1 and finite variance, *Theor. Probab. Appl.*, 11(4), 579–611, 1966]

Kesten's pattern theorem Arises in considering the number X_n of self-avoiding random walks of length n, and the number Y_n of self-avoiding polygons of length n, on the integer d-dimensional lattice $\mathbb{Z}^d$. In any such walk (or polygon) a pattern is a fixed finite sequence of steps whose form, suitably translated, may recur at various points in the path. Broadly, the theorem asserts that for an appropriately defined pattern p as $n \to \infty$, there is some $a > 0$ such that for all but geometrically few (i.e. their number decreases exponentially) the pattern p recurs at least an times. It is used to prove Kesten's ratio limit theorems for X_n, (e.g. for some constants b and c we have $|\frac{X_{n+2}}{X_n} - b^2| \le cn^{-1/3}$), and also in studying knotting probabilities for Y_n. [H. Kesten, On the number of self-avoiding walks, *J. Math. Phys.*, 4, 960–969 , 1963; & *J. Math. Phys.*, 5, 1128–1137, 1964]

Kesten's percolation theorem *see* Harris–Kesten theorem

Kesten–Spitzer random walk in random scenery In its basic form this comprises a simple r.w. $S_n = \sum_{r=1}^{n} x_r$ on $\mathbb{Z}$, where $S_0 = 0$, together with a further independent sequence of r.vs $\sigma(x), x \in \mathbb{Z}$ called the random scenery. Let $V(x, m)$ be the number of visits of the walk S_n to x up to time m. Then the cumulative scenery is

$$\begin{aligned} W_m &= \sum_{r=0}^{m} \sigma(S_r) \\ &= \sum_{x \in \mathbb{Z}} \sigma(x) V(x, m), \quad m \ge 0. \end{aligned}$$

Then a suitably scaled continuous version of W_n, with continuous parameter, converges to a self-similar process with stationary increments. [H. Kesten & F. Spitzer, A limit theorem related to a new class of self similar processes, *Probab. Theor. Rel. Fields*, 50, 5–25, 1979]

Kesten–Stigum theorem (alias: **Kesten–Stigum L log L theorem**) A limit theorem for multitype (with finitely many types) Galton–Watson branching processes that is essentially in its strongest form, in that the so-called L log L condition is necessary and sufficient. The simplest (one-type) form asserts this: let $Z_n, n \ge 0$, be a supercritical Galton–Watson branching process, with $Z_0 = 1$, and L a r.v. having the family-size distribution. Write $W_n = Z_n/(\mathbb{E}L)^n$, and let $W = \lim_{n\to\infty} W_n$ be the limit as $n \to \infty$ of this martingale. Then these are equivalent statements:

(a) $\mathbb{E}L \log L^+ < \infty$

(b) $\mathbb{P}(W = 0) = \lim_{n\to\infty} \mathbb{P}(Z_n = 0)$

(c) $\mathbb{E}W = 1$.

This extends a theorem of N. Levinson, and the multitype version generalizes a theorem of T. Harris, also called the **Everett–Ulam–Harris thm**.

[R. Lyons *et al.*, Conceptual proofs of L log L criteria for mean behaviour of branching processes, *Ann. Probab.*, 23(3), 1125–1138, 1995; J. Biggins & A. Kyprianou, Measure change in multitype branching, *Adv. Appl. Probab.*, 36(2), 544–581, 2004]

Kesten's theorem (alias: **Kesten's trichotomy; trichotomy theorem**) Arises in the context of the strong law of large numbers for a sequence $S_n = \sum_{r=1}^{n} X_r$, where the i.i.d.r.vs X_r have undefined mean, i.e. $\mathbb{E}X_1^+ = \mathbb{E}X_1^- = \infty$. Then the trichotomy theorem asserts that just one of these holds:

(a) $\mathbb{P}\left(\lim_{n\to\infty} \frac{1}{n} S_n = +\infty\right) = 1$

(b) $\mathbb{P}\left(\lim_{n\to\infty} \frac{1}{n} S_n = -\infty\right) = 1$

(c) $\mathbb{P}\left(\limsup_{n\to\infty} \frac{1}{n} S_n = +\infty \quad \text{and} \quad \liminf_{n\to\infty} \frac{1}{n} S_n = -\infty\right) = 1$, and the walk is said to be oscillatory.

[K. Erickson, The strong law of large numbers when the mean is undefined, *Trans. Amer. Math. Soc.*, 185, 371–381, 1973]

key renewal theorem *see* renewal theory

Khasminskii's test *see* Feller's test for explosion

Khatri–Šidak inequality Says that for a zero-mean multivariate normal r.vr $(X_1, \ldots X_n)$, $\mathbb{P}(\max_{1\le r\le n} |X_r| \le x) \ge \mathbb{P}(|X_1| \le x)\mathbb{P}(\max_{2\le r\le n}|X_r| \le x)$. *See* Šidak's inequality. [W. Li & O.-M. Shao, Gaussian processes, in *Handbook of Statistics*, Vol. 19. Elsevier, 2001, pp. 533–598; M. Lifshits, *Lectures on Gaussian Processes*. Springer, 2012]

Khinchin Note that this is also transliterated from the Cyrillic as Khintchine or Hinčin.

Khinchin's constant Arises in the theory of continued fractions as follows. Let X be randomly selected from $[0, 1]\backslash\mathbb{Q}$ according to a probability measure having the same null sets as Lebesgue measure, (where $\mathbb{Q}$ denotes the rationals) and let its continued fraction expansion be,

$$X = \frac{1}{Y_1 + \frac{1}{Y_2 + \frac{1}{\cdots}}}$$

where the Y_i are integers, and $Y_r \geq 0$ for $r \geq 1$. Then with probability 1, as $n \to \infty$, $(Y_1 Y_2 \cdots Y_n)^{1/n} \to K \simeq 2.685\,452\,001\ldots$, which is called Khinchin's constant. [A. Khintchine, *Continued Fractions*. Chicago University Press, 1964]

Khinchin's criterion Any one of several sufficient (or necessary and sufficient) conditions due to Khinchin; but most commonly one of these:

(a) A criterion for a stationary random process $X(t)$ to be ergodic, i.e. $X(t)$ is ergodic if its covariance function satisfies $C(s,t) = C(|t-s|) \to \operatorname{var} X(0)$, as $|t-s| \to \infty$.

(b) A criterion for a complex-valued function $\phi(t)$ of a real parameter $t \in \mathbb{R}$ to be the characteristic fn of some random variable, i.e. $\phi(t)$ is a chf iff there is a function $f(t), t \in \mathbb{R}$ such that

$$\phi(t) = \int_{-\infty}^{\infty} f(t+x)\overline{f(x)}\,\mathrm{d}x.$$

(c) A criterion (also called the **Khinchin–Shepp unimodality criterion**) for the distribution of r.v. $X \in \mathbb{R}$ to be unimodal (i.e. for its distribution function $F(x)$ to be convex for $x < m$ and concave for $x > m$ for some $m \in \mathbb{R}$, not necessarily unique), i.e. X has a unimodal distribution iff its distribution is the same as that of UY for some r.v. Y independent of the r.v. U which has a uniform distribution on (0,1). Formally, $\phi_X(t) = \int_0^1 \phi_Y(tu)\mathrm{d}u$, where ϕ_X and ϕ_Y are the chfs of X and Y respectively. This may be called **Khinchin's representation**.

[A. Khinchin, *Mathematical Foundations of Statistical Mechanics*. Dover, 1949; E. Lukacs, *Characteristic Functions*, 2nd edn. Griffin, 1970; W. Feller, *Introduction to Probability Theory and its Applications*, Vol. 2, 2nd edn. Wiley, 1971]

Khinchin equivalence *see* tail equivalence

Khinchin's factorization theorem Arises in considering the factorization of the characteristic function ϕ of a r.v. X as a product of two or more chfs, a process called the **arithmetic of distributions**. In terms of the distribution of X this corresponds to factorization in the semi-group of convolutions of distributions. The theorem asserts that ϕ can be factorized as $\phi_i \phi_n$, where ϕ_i is the chf of an infinitely divisible distn and ϕ_n is a countable product of chfs of (possible degenerate) distributions that are not decomposable. The factorization is not in general unique.

In this context, note that there exist distinct chfs $\phi \neq \Psi$ such that $\phi^2 = \phi\Psi$. Thus cancellation is invalid in the arithmetic of distributions, a fact called **Khinchin's phenomenon**. [E. Lukacs, *Characteristic Functions*, 2nd edn. Griffin, 1970]

Khinchin's inequality Supplies a bound for the magnitude of the sums of a type of infinite series. Its basic form is this: Let $X_r, r \geq 0$ be a sequence of i.i.d.r.vs with

$$\mathbb{P}(X_r = -1) = \frac{1}{2} = \mathbb{P}(X_r = +1).$$

Then for $0 < p < \infty$, and any constants c_r (possibly complex-valued), there exist positive constants $a(p)$ and $b(p)$ (depending on p only), such that

$$a\left(\sum_{r=0}^{\infty} |c_r|^2\right)^{1/2} \leq \left(\mathbb{E}\left\{\left|\sum_{r=0}^{\infty} c_r X_r\right|^p\right\}\right)^{1/p} \leq b\left(\sum_{r=0}^{\infty} |c_r|^2\right)^{\frac{1}{2}}.$$

When generalized to sequences taking values in a Banach space, this is often called the **Kahane–Khinchin inequality**. [J.-P. Kahane, *Some Random Series of Functions*, 2nd edn. Cambridge University Press, 1985]

Khinchin–Kolmogorov theorem; *see* Wiener–Khinchin theorem

Khinchin's law of the iterated logarithm Arises in considering the size of the fluctuations of $S_n = \sum_{r=1}^{n} X_r$, where the X_r are i.i.d.r.vs with $\mathbb{P}(X_1 = -1) = \mathbb{P}(X_1 = +1) = \frac{1}{2}$. Then with probability 1,

$$S_n > a(2n \log\log n)^{1/2} \begin{cases} \text{finitely often for } a > 1 \\ \text{infinitely often for } a < 1. \end{cases}$$

For the Wiener process Khinchin's LIL says that a.s.

$$\limsup_{t\to\infty} \frac{W(t)}{(2t \log\log t)^{\frac{1}{2}}} = 1 = \limsup_{t\to 0} \frac{W(t)}{\left(2t \log\log\left(\frac{1}{t}\right)\right)^{\frac{1}{2}}}.$$

[R. Dudley, *Real Analysis and Probability*. Cambridge University Press, 2002]

Khinchin–Lévy characterization *see* Lévy–Khinchin characterization

Khinchin's theorem Arises in considering a crudely stationary point process $N(\cdot)$, (i.e. for any compact intervals I_1, I_2 of equal length the r.vs $N(I_1)$, and $N(I_2)$ have the same distribution). Then the theorem asserts that

$$\lim_{t\downarrow 0}\frac{1}{t}\sum_{r=1}^{\infty}\mathbb{P}\Big(N([0,t])\geq r\Big)=\lambda$$

exists but may be infinite. If, as $t\downarrow 0, \mathbb{P}(N([0,t])>1]=\mathrm{o}(t)$, then the process $N(\cdot)$ is called orderly; this excludes multiple points in the sense that points a.s. never coincide. In this case **Korolyuk's thm** asserts that $\lambda=\mathbb{E}N(0,1)\leq\infty$. Conversely, **Dobrushin's lemma** asserts that is $\mathbb{E}N(0,1)<\infty$, and there are no multiple points, then the process is orderly. [M. Leadbetter, On three basic results in the theory of stationary point processes, *Proc. Amer. Math. Soc.*, 19, 115–117, 1968]

Khinchin's weak law of large numbers A form of the WLLN which asserts that for a sequence of pairwise independent i.i.d.rvs with finite mean μ, $\frac{1}{n}\sum_{r=1}^{n}X_r$ converges in probability to μ as $n\to\infty$. [K. L. Chung, *Course in Probability Theory*, 3rd edn. Academic Press, 2004]

Kholevo bound *see* Holevo bound

Kiefer process (alias: **Kiefer–Müller process**) Arises as a limiting approximation to a class of empirical processes; it is a centred Gaussian process on $[0,1]\times\mathbb{R}^+$ with covariance function

$$c(s_1,t_1;s_2,t_2)=(s_1\wedge s_2-s_1s_2)(t_1\wedge t_2).$$

It can be seen as a Brownian bridge in $0\leq s\leq 1$, for fixed t; and with independent increments on $t\geq 0$ for fixed s; and it may be constructed from a Brownian sheet $W(s,t)$, as

$$K(s,t)=W(s,t)-sW(1,t),\ 0\leq s\leq 1,\ t\geq 0.$$

[J. Kiefer, Skorokhod embedding of multivariate rvs, and the sample df, *Probab. Theor. Rel. Fields*, 24(1), 1–35, 1972]

Kiefer–Wolfowitz algorithm Arises in stochastic optimization, where it supplies a method for finding an extreme value of a suitable function $m(x)$ that is itself unknown, but is given for any x as the expected value of an observable random variable $Y(x)$, i.e. $m(x)=\mathbb{E}Y(x)$. [J. Kiefer & J. Wolfowitz, Stochastic estimation of the maximum of a regression function, *Ann. Math. Stat.*, 23(3), 462–466, 1952]

Kiefer–Wolfowitz recursion *see* Lindley's equation

killed Brownian motion *see* Brownian motion

killed process A concept with several applications in the theory of Markov processes; given any random process $X(t), t\geq 0$ (say), where X takes values in a state space S, one may kill X as follows. Introduce a r.v. $T\geq 0$, and augment the state space S with an extra state $\Delta\notin S$, and define

$$\hat{X}(t)=\begin{cases}X(t), & t<T\\ \Delta, & t\geq T.\end{cases}$$

Then $\hat{X}(t)$ is called $X(t)$ killed at T, with **coffin (or cemetery) state** Δ. In applications it is often useful to extend functions $f(X)$, defined on S, to take the value $f(\Delta)=0$. In the most elementary case, which is often seen, the process $X(t)$ is killed at a time T that is independent of $X(t), t\geq 0$, with an exponential distribution parameter K; in this case K is called the **killing rate**, and one speaks of **exponential killing**. Other killing times may depend on the process, e.g. it is often natural to kill a Markov process at its first-passage time to a set $A\subseteq S$, or at the time of its first explosion. A process may also be killed by a suitable functional of its path, e.g. if $A(X(t),t)$ is a continuous increasing additive functional, then $X(t)$ may be killed at $T=\inf\{t: A(X(t),t)\geq a\}$.

Killed Markov processes are most useful for killing times such that the killed process also has the Markov property for $t\leq T$; as, for example, in killing at a first-passage time. By contrast, this will not be true for a process killed at a second-passage time. [S. Karlin & H. Taylor, *A Second Course in Stochastic Processes*. Academic Press, 1981]

Kingman's coalescent Arises in population genetics as a finite state space Markov chain that describes the genealogy (i.e. the antecedent family-tree structure stemming from the most recent common ancestor), of n individuals randomly selected from a sufficiently large haploid population evolving in time. The key idea is that the coalescent is a random process that runs backwards in real (intergenerational) time, in contrast to classic models that attempt the harder task of keeping track of the whole population evolving in the conventional direction of time (forwards). [J. F. C. Kingman, Origins of the coalescent: 1974–1982, *Genetics*, 156(4), 1461–1463, 2000]

Kingman's formula (alias: **Kingman's equation**) Any one of several results due to J.F.C. Kingman, perhaps most often these:

(a) In queueing theory: in a G1/G/1 queue (in Kendall's notation) in equilibrium under conditions of heavy traffic, this formula supplies an approximation to the expected waiting time (before service) of an arriving customer as

$$\mathbb{E}W \simeq \frac{1}{2}\left(\frac{\sigma_A^2}{\mu_A^2}+\frac{\sigma_S^2}{\mu_S^2}\right)\frac{\mu_S}{\mu_A-\mu_S}\,\mu_S,$$

where σ_A^2, σ_S^2, and μ_A, μ_S are the variances and means of the inter arrival and service times respectively. It is also called the **VUT formula**, since the three factors can be seen as representing variability, utilization, and time in service respectively.

(b) In regenerative phenomena

(c) In Markov processes: for a suitable time-homogenous Markov process $X(t), t \geq 0$, Kingman's formula (or identity) supplies an expression (in terms of a triple Laplace transform) containing the joint distribution of the backward and forward recurrence times B and F, of any state x of X, at time t; i.e,

$$B = \inf\{s \geq 0 : X(t-s) = x\} \text{ and}$$
$$F = \inf\{s \geq 0 : X(t+s) = x\}.$$

(d) In genetics: *see* Kingman's coalescent.

[J. F. C. Kingman, On the algebra of queues, *J. Appl. Probab.*, 3(2), 285–326, 1966; J. F. C. Kingman, Homecomings of Markov processes, *Adv. Appl. Probab.*, 5(1), 66–102, 1973; N. Bingham & C. Goldie (eds), *Probability and Mathematical Genetics* (selected chapters). Cambridge University Press, 2010]

Kingman's heavy traffic theorem *see* heavy traffic

Kingman's paint box theorem *see* Kingman's representation theorem

Kingman's partition structure Arises in the context of population genetics (inter alia). A partition of an integer $n \geq 0$ is an unordered set of positive integers whose sum is n; or equivalently an unordered set of disjoint non-empty subsets S_i of $\{1, 2, \ldots, n\} = N_n$, with $\sum_i |S_i| = n$. A random partition is a r.v. A_n taking values in the set of all such partitions. Then a partition structure is a sequence of probability distributions P_n for A_n, $n \geq 1$, such that for all $m < n$ the random partition given by the non empty subsets of $\{S_i \cap N_m\}$ has distribution P_m. (The constraint arises from the real-world requirement that an m sub-sample from a sample of size n should have the same distribution as a direct m-sample.)

The structure is called exchangeable if it is invariant under permutations of N_n. *See* Kingman's representation theorem. [J. F. C. Kingman, Random partitions in population genetics, *Proc. Roy. Soc. Lond. A*, 361(1704), 1–20, 1978]

Kingman's p-function *see* regenerative phenomenon

Kingman's regenerative process *see* regenerative phenomenon

Kingman's representation (or paint box) theorem Arises in the context of Kingman's partition structure, where it can be seen as an analogue of de Finetti's theorem for exchangeable random variables, in that it represents any partition structure $P_n, n \geq 1$ as a mixture of conditional sampling distributions. Formally, let ∇_0 be the space of sequences $\mathbf{p} = (p_0, p_1, \ldots)$ such that $p_1 \geq p_2 \geq \cdots > 0 \leq p_0$, and $\sum_{r=0}^{\infty} p_r = 1$. For fixed $\mathbf{p}$ define independent r.vs $X_r, r \geq 1$, such that $\mathbb{P}(X_r = j) = p_j$, $j \geq 1$, and $\mathbb{P}(X_r = -r) = p_0$. Then given $\mathbf{p}$ let $\mathbb{P}_n(a_1, \ldots, a_n \,|\, \mathbf{p})$ be the probability that in $X_1, \ldots, X_n$ exactly a_j integers appear j times, $1 \leq j \leq n$. The representation theorem asserts that there is a unique measure $\mu(\cdot)$ on ∇_0 such that, for $n \geq 1$,

$$\mathbb{P}_n(a_1, \ldots, a_n) = \int \mathbb{P}_n(a_1, \ldots, a_n|\mathbf{p})\mu(d\mathbf{p}).$$

The alias arises from an interpretation of $\mathbf{p}$ as a paintbox of colours, the amount of the jth colour being p_j, and p_0 the amount (and hence the probability of using) an exceptional colour that has the property of looking different each time it is used. The measure $\mu(\cdot)$ may be called the directing measure (or canonical measure). [J. F. C. Kingman, The representation of partition structures, *J. Lond. Math. Soc.*, 18, 374–380, 1978]

Kingman's solidarity theorem *see* solidarity theorem

Kingman's subadditive ergodic theorem A limit theorem for subadditive processes that generalizes Birkhoff's ergodic theorem. Formally, let X_{st} be a set of r.vs indexed by the non-negative integers $0 \leq s < t$, and such that:

(a) $X_{su} \leq X_{st} + X_{tu}$ for $0 \leq s < t < u$;

(b) $\mathbb{E}|X_{0t}| < \infty$ for all t, and for some constant c and all t we have $\mathbb{E}X_{0t} \geq -ct$;

(c) for all $0 \leq s \leq t$ the joint distributions of X_{st} are the same as that of $X_{s+1,t+1}$.

Then the theorem asserts the existence of X and γ such that, as $t \to \infty$, $\frac{1}{t}X_{0t}$ converges to X both in mean and a.s.; and $\gamma = \inf_t \frac{1}{t}\mathbb{E}X_{0t} = \lim_{t\to\infty} \frac{1}{t}\mathbb{E}X_{0t}$; and $\gamma = \mathbb{E}X$. The Kingman-Liggett thm relaxes the assumptions of stationarity and subadditivity.
[J. F. C. Kingman, Subadditive ergodic theory, *Ann. Probab.*, 1(6), 883–909, 1973]

Kingman's theorem Any of several theorems, including those listed immediately above, but in this abbreviated appellation perhaps most commonly those theorems concerning Markov transition probabilities found in the references. [J. F. C. Kingman, Regenerative phenomena and the characterization of Markov transition probabilities, *Proc. 6th Berkeley Symp. Math. Stat. Probab.*, 3, 241–262, 1972; J. F. C. Kingman, *Regenerative Phenomena*. Wiley, 1972]

Kingman's upper bound Any one of a number of inequalities for the mean, variance, or tail of the stationary waiting time distribution in the general single server queue (GI/G/1 in Kendall's notation), as well as others. [J. F. C. Kingman, Inequalities in the theory of queues, *J. Roy. Stat. Soc.*, 32(1), 102–110, 1970]

Klass–Nowicki *see* Hoffman–Jørgensen inequality.

Klimov's rule (or problem) Arises in the context of a single server queue, at which C different types of customer arrive (in independent Poisson processes) having independent service times that also depend on their type, and such that a customer of type r incurs a cost C_r for each time-unit spent in the system. On completing any service the server must choose the next from those waiting. The rule supplies an optimal (minimum cost) strategy based on indices attached to each customer that are recursively computable. [H. Chen & D. Yao, *Fundamentals of Queuing Networks*. Springer, 2001]

knapsack problem *see* bin-packing

Knight's theorem (or lemma) Arises in the context of local times (or sojourns, or occupation times) of simple symmetric r.w. and of Brownian motion. The version for Brownian motion is usually called the Ray–Knight theorem. For the s.s.r.w $S_n, n \geq 0$, F. B. Knight considered local times at k via the **upstep process** $S(m,k)$ at k, where $S(m,k)$ is the number of steps of the walk S_n from k to $k+1$, counted up to the $m+1$th step from 0 to 1.

Then the analysis of the local time process follows from Knight's theorem, which asserts that for any $m \geq 0$, $S(m,k)$, $k \geq 0$, is a Markov chain on $\mathbb{Z}^+$, with $S(m,0) = m$, and transition probabilities

$$p_{ij} = 2^{\frac{1}{i+j}}\binom{i+j-1}{j}.$$

For Knight's representation theorem, *see* Dubins–Schwarz theorem. [F. B. Knight, Random walks and a sojourn process of Brownian motion, *Trans. Amer. Math. Soc.*, 109(1), 56–86, 1963]

Knight uncertainty Arises in decision theory as applied to economics and finance, where its precise meaning is uncertain. It seeks to characterize situations where the classical interpretations of probability (both objective and subjective) are inapplicable; according to its originator (F. H. Knight) a symptom of its presence in a collection of possible outcomes Ω is the absence (or impossibility) of an insurance market on Ω. Examples mostly arise when agents have asymmetric information, leading to market failure. An elementary instance of Knight uncertainty is given to a player of the shell game, also known as thimblerig. [S. Le Roy & L. Singell, Knight on risk and uncertainty, *J. Political Econ.*, 95(2), 394–406, 1987]

Knudsen process (alias: **stochastic (or random) billiards**) Arises in considering gas molecules whose only effective interactions are with the walls of some container. A particle travels linearly at unit speed in $\mathbb{R}^d$, $d \geq 2$, in the presence of obstacles, one of which may be a boundary enclosing all the other obstacles, or the particle may move in an unbounded region of obstacles. The boundaries are assumed to be nice (e.g. piecewise continuously differentiable) and on striking any boundary the particle is reflected in a random direction; at unit speed. Popular choices for the direction of reflection include:

(a) uniformly at random from the hemisphere (or semicircle for $d = 2$) of directions available;
(b) at random from the hemisphere according to the cosine (or Lambert's) distribution in which the direction is in an element $d\omega$ of solid angle with probability $\cos\theta\, d\omega$, where θ is the angle made by $d\omega$ with the normal to the boundary;
(c) a mixture of these and specular (i.e. mirror) reflection.

The path is Knudsen's stochastic billiards, and the points of reflection on the boundary comprise **Knudsen's random walk**. [M. Menshikov *et al.*, Asymptotic behavior of randomly reflecting billiards in unbounded tubular domains, *J. Stat. Phys.*, 132(6), 1097–1133, 2008]

Kochen–Stone lemma A theorem that generalizes the Borel–Cantelli lemmas, in which context it has this corollary: let $A_r, r \geq 1$, be a set of events; if $\sum_{r=1}^{\infty} \mathbb{P}(A_r) = \infty$, then

$$\mathbb{P}(\text{infinitely many } A_r \text{ occur})$$

$$= \mathbb{P}(\limsup_{n\to\infty} A_n) \geq \limsup_{n\to\infty} \frac{\left\{\sum_{r=1}^{n} \mathbb{P}(A_r)\right\}^2}{\sum_{1\leq r,\, s\leq n} \mathbb{P}(A_r \cap A_s)}.$$

In particular, if $\mathbb{P}(A_r \cap A_s) \leq C\,\mathbb{P}(A_r)\mathbb{P}(A_s)$ for all r, s, and some finite constant C, then the RHS $\geq C^{-1}$. If the A_r are independent, $C = 1$, yielding the Borel–Cantelli lemma. [S. Kochen & C. Stone, A note on the Borel–Cantelli lemma, *Illinois J. Math.*, 8(2), 248–251, 1964]

Kolmogorov's axioms The basic rules that underlie most of modern mathematical probability, which were elegantly and concisely summarized by A. N. Kolmogorov in 1933 from the work of many previous authors. They assert that probability (seen as a function) is a non-negative, normalized, countably additive measure over some sigma-algebra of events (i.e. suitable subsets of a universal set Ω).

Formally, as originally given, the axioms postulate:

(a) A **sigma-algebra** (i.e. σ-field) $\mathcal{F}$ of subsets of Ω, called events.

(b) $\Omega \in \mathcal{F}$, and is called the certain event.

(c) There is a function $\mathbb{P}(A) \geq 0$, defined for any $A \in \mathcal{F}$, called the probability of A.

(d) $\mathbb{P}(\Omega) = 1$.

(e) $\mathbb{P}(A \cup B) = \mathbb{P}(A) + \mathbb{P}(B)$ whenever $A \cap B = \emptyset$.

(f) If A_r, $r \geq 1$, is a decreasing sequence of events (i.e. $A_{n+1} \subseteq A_n$ for all n), whose limit is the empty set $\emptyset$, then $\lim_{n\to\infty} \mathbb{P}(A_n) = 0$.

Further concepts in probability such as independence, random variables, expectation, and conditioning are defined in terms of Ω, $\mathcal{F}$, and $\mathbb{P}$. In particular, conditional probability $\mathbb{P}(A|.)$ inherits the essentially same properties as $\mathbb{P}(\cdot)$. [G. Shafer & V. Vovk, The sources of Kolmogorov's Grundbegriffe, *Stat. Sci.*, 21(1), 70–98, 2006]

Kolmogorov's binomial inequality Let X_r, $r \geq 1$, be i.i.d. Bernoulli r.vs with mean $p \in [0, 1]$. Then the inequality asserts that for $\varepsilon \geq 0$

$$\mathbb{P}\left[\sup_{r\geq n} \left|\frac{1}{r}\sum_{i=1}^{r} X_i - p\right| \geq \varepsilon\right] \leq 2\exp\{-2n\varepsilon^2(1-\varepsilon)\}.$$

[D. Young *et al.*, A note on a Kolmogorov inequality, *Stat. Probab. Lett.*, 5, 217–218, 1987]

Kolmogorov's canonical form (or representation or formula) Asserts that $\phi(t)$ is the characteristic function of a finite-variance infinitely divisible r.v. iff it can be (uniquely) represented by,

$$\log \phi(t) = \mathrm{i}ct + \int_{-\infty}^{\infty} (\mathrm{e}^{\mathrm{i}tx} - 1 - \mathrm{i}cx)x^{-2}\mathrm{d}K(x)$$

for some $c \in \mathbb{R}$ and non-decreasing bounded function $K(x)$, such that $K(-\infty) = 0$. [E. Lukacs, *Characteristic Functions*, 2nd edn. Griffin, 1970]

Kolmogorov–Chentsov continuity theorem *see* Kolmogorov continuity theorem

Kolmogorov complexity (alias: **K-complexity; algorithmic complexity; program size complexity; descriptive complexity; Kolmogorov–Chaitin (algorithmic) complexity**) A concept that arose independently in separate areas of mathematics: formal inductive inference (R. Solomonoff in 1960); information theory and the concept of randomness (A. Kolmogorov in 1965); and the length of computer programs needed to compute binary strings (G. Chaitin, 1966). Broadly speaking, Kolmogorov's version of K-complexity arises in seeking to define the information content of, and the intrinsic randomness in, a given finite sequence (this is in contrast to Shannon information, which measures information as the entropy of an ensemble of possible strings, and to probability which only sees randomness in a sequence before its realization). Thus, the Kolmogorov (descriptive) complexity of a given (ordered) string M of symbols is the length K of the shortest computer program (binary) that exactly describes (i.e. determines) M. As a corollary, one says that a sequence (or string) M of length n is Kolmogorov (or descriptively) random, or incompressible, if its Kolmogorov complexity $K(M) \geq n$.

However, $K(M)$ is incomputable in the sense that it cannot be the output of a computer program with input M. More narrowly, if programs are required to be prefix-free, one speaks of the prefix-free Kolmogorov complexity $K_p(M)$, and then $K(M)$ is called the plain Kolmogorov complexity. The corresponding prefix-free randomness implies plain Kolmogorov randomness, but not conversely. For an infinite string M, let M_n be its initial segment of length n. If $K_p(M_n) \geq n - c$ for all n, and some $c > 0$, then M is said to be Martin-Löf random, or Kolmogorov–Levin–Chaitin random. *See* algorithmic probability. [R. Downey & D. Hirschfeldt, *Algorithmic Randomness and Complexity*. Springer, 2010; M. Li & P. Vitányi, *Introduction to Kolmogorov Complexity and its Applications*, 3rd edn. Springer, 2008]

Kolmogorov's condition *see* Kolmogorov's strong law of large numbers

Kolmogorov consistency conditions *see* Daniell–Kolmogorov theorem

Kolmogorov continuity theorem (alias: **Kolmogorov–Chentsov condition (or continuity theorem)**) This supplies a condition on the moments of a random process $X(t)$, sufficient to ensure that there is a version (or modification) of $X(t)$ that is almost surely continuous. If $t \in \mathbb{R}$, then a continuous version of $X(t)$ exists if there are positive constants a, b, and c, such that for all $h > 0$

$$\mathbb{E}(|X(t+h) - X(t)|^a) \leq ch^{1+b}.$$

For example, Brownian motion $W(t)$ in one dimension is such that $W(t+h) - W(t)$ is normal $N(0, h)$; thus $\mathbb{E}(W(t+h) - W(t))^4 = 3h^2$, and one may take $a = 4, b = 1$, and $c = 3$, in the above. More generally, if X is defined on $\mathbb{R}^d$, the right-hand side of the above condition is $C|h|^{d+b}, h \in \mathbb{R}^d$. [O. Kallenberg, *Foundations of Modern Probability*, 2nd edn. Springer, 2002]

Kolmogorov's cycle condition *see* reversibility

Kolmogorov's distance (or metric) (alias: **uniform metric; Kolmogorov–Smirnov metric**) For random variables X and Y, having distribution functions $F(x)$ and $G(x)$ respectively, $x \in \mathbb{R}$, the Kolmogorov distance is $d(F, G) = \sup_x|F(x) - G(x)|$, $x \in \mathbb{R}$. [A. Gibbs & F. Su, Choosing and bounding probability metrics, *Int. Stat. Rev.*, 70(3), 419–435, 2002]

Kolmogorov's distribution (alias: **Kolmogorov–Smirnov distribution**) Arises in the context of the discrepancy $|F_n(x) - F(x)|$ between a distribution function $F(x)$ and the empirical distribution $F_n(x)$ arising from a random sample from $F(x)$. It is the limit as $n \to \infty$ of the distribution of $D_n = \sqrt{n}\sup_x|F_n(x) - F(x)|$, given by: $K(x) = \sum_{r=-\infty}^{\infty}(-1)^r \exp\{-2r^2x^2\}$, $x>0$; where, subject to mild conditions on $F(x)$,

$$K(x) = \frac{\sqrt{2\pi}}{x}\sum_{r=1}^{\infty}\exp\{-(2r-1)\pi^2/(8x^2)\}.$$

Similarly, the Kolmogorov–Smirnov (or Smirnov) theorem asserts that the limit distribution of the suitably scaled collection of r.vs $\sup_x|F_n(x) - G_m(x)|$, as m, $n \to \infty$ with m/n fixed, where $G_m(x)$ is the empirical distribution of a random sample of size m from $F(x)$, is also $K(x)$. [W. Feller, On the Kolmogorov–Smirnov limit theorems for empirical distributions, *Ann. Math. Stat.*, 19(2), 177–189, 1948]

Kolmogorov–Doob inequality *see* Doob's maximal inequality

Kolmogorov equations *see* backward and forward equations; Chapman–Kolmogorov equations; Feller equations

Kolmogorov–Erdös test (alias: **Kolmogorov–Dvoretsky–Erdös–Feller–Petrovski test (and subsets of these)**) Arises in the law of the iterated logarithm in the context of the Wiener process $W(t)$, thus: let $u(t)$ be a positive function increasing to ∞ as $t \to \infty$, and define

$$T(u) = \int_1^{\infty}\frac{1}{x}u(x)\exp\left\{-\frac{1}{2}u(x)^2\right\}\mathrm{d}x.$$

Then $W(t) < u(t)t^{\frac{1}{2}}$ a.s. for all large enough t (which is to say that $t^{\frac{1}{2}}u$ is in the upper class of W) if $T(u) < \infty$; while $W(t_n) > u(t_n)t_n^{1/2}$ for an unbounded sequence $t_n \uparrow \infty$, a.s. (which is to say that $t^{\frac{1}{2}}u$ is not in the upper class of W) if $T(u) = \infty$. [N. Bingham, Variants on the law of the iterated logarithm, *Bull. Lond. Math. Soc.*, 18(5), 433–467, 1986]

Kolmogorov extension theorem *see* Daniell–Kolmogorov theorem

Kolmogorov–Feller equations *see* Feller equations

Kolmogorov–Feller weak law of large numbers *see* Feller's weak law of large numbers

Kolmogorov–Filippov problem *see* Kolmogorov's rock-crushing process

Kolmogorov–Fisher equation *see* Fisher's equation

Kolmogorov's formula (alias: **Kolmogorov–Szegö formula (or v.v.)**) Arises in considering a zero-mean, unit variance, covariance stationary random process

$X_n, n \in Z$, having spectral density $f(\lambda) > 0$, where it asserts that for any linear predictor $\hat{X}_n$ of X_n, given the past of the process, the variance of the prediction error (or innovation) $\hat{X}_n - X_n$ is bounded below by $\exp\{\frac{1}{2\pi}\int_{-\pi}^{\pi} \log\{2\pi f(\lambda)\}\mathrm{d}\lambda\}$. [P. Brockwell & R. Davis, *Time Series*, 2nd edn. Springer, 1991]

Kolmogorov's inequality (or maximal inequality) A type of maximal inequality. Let $X_1, X_2, \ldots$ be independent zero-mean r.vs having finite variances, and set $S_n = \sum_{r=1}^{n} X_r$. Then for $x > 0$.

$$\mathbb{P}(\max\{|S_1|, \ldots, |S_n|\} > x) \leq \frac{1}{x^2}\mathbb{E}S_n{}^2.$$

P. Lévy extended this to a (finite-variance) martingale , for which the above inequality remains true; and J. Doob extended it to suitable submartingale S_n (*see* Doob's maximal inequality). The one-sided version is called Marshall's inequality. Another Kolmogorov inequality arises when the r.vs X_r above obey $|X_r| \leq a$, a.s., in which case one has Kolmogorov's reverse maximal inequality:

$$\mathbb{P}(\max\{|S_1|, \ldots, |S_n|\} \leq x) \leq \frac{(x+a)^2}{\mathbb{E}S_n{}^2}.$$

This also holds for martingales with bounded differences. *See also* Hájek–Rényi–Chow inequality. [W. Feller, *Introduction to Probability Theory and its Applications*, Vol. 2, 2nd edn. Wiley, 1971]

Kolmogorov's law of the iterated logarithm Extends Khinchin's LIL (for i.i.d. Bernoulli r.vs) to a sequence of independent suitably bounded r.vs. Formally, let $X_1, X_2, \ldots$ be a sequence of zero mean independent r.vs with respective variances $\sigma_1^2, \sigma_2^2, \ldots$, and such that $s_n{}^2 = \sum_{r=1}^{n} \sigma_r{}^2 \to \infty$ as $n \to \infty$. If there is a sequence of positive constants $a_n \to 0$ as $n \to \infty$ such that $|X_n| \leq a_n s_n (\log\log s_n{}^2)^{\frac{1}{2}}$, a.s., then $\limsup_{n\to\infty} \sum_{r=1}^{n} X_r / \{2s_n{}^2 \log\log s_n{}^2\}^{\frac{1}{2}} = 1$, a.s. *See* law of the iterated logarithm. [N. Bingham, Variants on the law of the iterated logarithm, *Bull. Lond. Math. Soc.*, 18(5), 433–467, 1986]

Kolmogorov–Nagumo mean (or average) (alias: **generalized mean; quasilinear mean; quasi-arithmetic mean; de Finetti–K-N mean**) Generalizes the idea of mean value (or expectation) by including a suitable arbitrary function, arrived at through a characterization theorem. Thus, let $f(x), x \in \mathbb{R}$, be a continuous and strictly monotonic (increasing, or decreasing) function, with inverse $f^{-1}(\cdot)$. The Kolmogorov–Nagumo mean of a simple r.v. X, with $\mathbb{P}(X = x_r) = p_r, 1 \leq r \leq n$, is $\mathbb{E}_f X = f^{-1}\left(\sum_{r=1}^{n} p_r f(x_r)\right)$. It arises from this characterization (or Kolmogorov–Nagumo theorem): let a function $m(x)$ of $\mathbf{x} = (x_1, \ldots, x_n)$ (called the 'mean'), have the following properties (which are entailed by the intuitive concept of 'mean'):

(a) m is continuous and strictly monotone in each of $x_1, \ldots, x_n$;
(b) $m(x, x, \ldots, x) = x$, for all x;
(c) $m(\mathbf{x})$ is unchanged by any permutation of $x_1, \ldots, x_n$;
(d) $m(\mathbf{x})$ is unchanged if any subset $(x_1, \ldots, x_k)$ of $\mathbf{x}$ is replaced by the k-set $(s, s, \ldots, s)$, where $s = m(x_1, \ldots, x_k)$.

Then for some continuous monotone $f(\cdot)$,

$$m(x) = f^{-1}\left(\frac{1}{n}\sum_{r=1}^{n} f(x_r)\right).$$

The above extension to a nonuniform distribution over $(x_1, \ldots, x_n)$ was made by B. de Finetti, yielding what may be called the de Finetti–K.-N. representation. Two functions f and g are **KN-equivalent** if $\mathbb{E}_f X = \mathbb{E}_g X$ for any simple r.v. X, and this entails that $f = ag + b$, for constants $a \neq 0$ and b. Further, $\mathbb{E}_f(X + c) = \mathbb{E}_f(X) + c$ iff f is either a linear or an exponential function. [S. Ostasiewicz & W. Ostasiewicz, Means and their applications, *Ann. Oper. Res.*, 97, 337–355, 2000; G. Hardy *et al.*, *Inequalities*, 2nd edn. Cambridge University Press, 1952]

Kolmogorov's one-series theorem Asserts that the sum $S_n = \sum_{r=1}^{n} X_r$ of zero-mean, finite-variance, independent r.vs $X_1, X_2, X_3, \ldots$, converges in mean-square, and with probability 1, as $n \to \infty$, if $\sum_{r=1}^{\infty} \operatorname{var} X_r < \infty$. More generally, if the sequence of r.vs has non-zero means $m_r, r \geq 1$, then S_n converges a.s. and m.s. if $\sum_{r=1}^{\infty} m_r < \infty$ and $\sum_{r=1}^{\infty} \operatorname{var} X_r < \infty$; this may be called **Kolmogorov's two-series thm**. *See also* three-series theorem. [S. Varadhan, *Probability Theory*. American Mathematical Society, 2001]

Kolmogorov–Ornstein theorem *see* Ornstein's theorem

Kolmogorov–Pearson diffusion *see* Pearson system

Kolmogorov representation *see* Kolmogorov's canonical form

Kolmogorov's reverse maximal inequality *see* Kolmogorov's inequality

Kolmogorov's rock-crushing (or fragmentation) process A type of cascade (or branching) process. In its most elementary form a rock of mass 1 is divided according to some probability distribution, into two fragments, and at subsequent times all existing fragments are divided independently into two fragments according to the same distribution. Then the distribution of the suitably scaled mass of a randomly selected fragment in the nth generation converges to a lognormal density as $n \to \infty$. More generally, the **Kolmogorov–Filippov model** runs as a Markov process in continuous time in a similar fashion, except that the rate of division of any particle is proportional to its mass. [A. Filippov, On the distribution of the sizes of particles which undergo splitting, *Theor. Probab. Appl.*, 6(3), 275–294, 1961]

Kolmogorov–Rogozin inequality (alias: **Doeblin–Lévy–Kolmogorov–Rogozin inequality**) Arises in the context of the concentration function of a sum $S_n = \sum_1^n X_r$ of independent r.vs. Informally speaking, it supplies a limit to how 'spread out' the distribution of S_n is. Formally, denoting the concentration fn by $Q(X_r, \lambda) = \sup_x \mathbb{P}(x \leq X_r \leq x + \lambda)$, $\lambda \geq 0$, , it asserts that there is a universal constant c such that for any independent r.vs $X_r, r \geq 1$, and $0 < \lambda_1, \ldots, \lambda_n \leq 2\lambda$,

$$Q(S_n, \lambda) \leq c\lambda \left\{ \sum_{r=1}^{n} \lambda_r{}^2 [1 - Q(X_r, \lambda_r)] \right\}^{-\frac{1}{2}}.$$

[C. Esseen, On the Kolmogorov–Rogozin inequality for the concentration function, *Probab. Theor. Rel. Fields*, 5(3), 210–216, 1966]

Kolmogorov–Rozanov theorem Arises in considering strong mixing (i.e. alpha-mixing) of a stationary finite variance Gaussian process X. The theorem shows first that the (strong) mixing coefficient $\alpha(\tau)$ goes to zero as $\tau \to \infty$ iff the natural maximal correlation coefficient $\rho(\tau)$ goes to 0 as $\tau \to \infty$; second it identifies properties of the spectral distribution function F that entail strong mixing. [A. Kolmogorov & Y. Rozanov, Strong mixing conditions for stationary Gaussian processes, *Theor. Probab. Appl.*, 5(2), 204–208, 1960]

Kolmogorov–Sinai entropy (alias: **metric entropy; Kolmogorov entropy; KS-entropy; measure theoretic entropy**) Arises in the theory of dynamical systems (i.e. a process evolving according to a one–one invertible measure-preserving transformation on a probability space) and their ergodic properties. Formally, let T be such a transformation defined on $(\Omega, \mathcal{F}, \mathbb{P})$, $T : \Omega \to \Omega$, and $A = A_1, \ldots, A_n$ a finite partition of Ω, whose entropy (Shannon entropy) is $H(A) = -\sum_1^n \mathbb{P}(A_r) \log \mathbb{P}(A_r)$. Then the entropy of A relative to T is defined as $H(A, T) = \lim_{n \to \infty} n^{-1} H(TA, T^2A, \ldots, T^nA)$, where $H(., \ldots,.)$ is the Shannon joint entropy, and the limit exists by a form of Shannon–McMillan–Breiman theorem. Finally, the KS-entropy is the supremum of $H(A,T)$ over all partitions A. This leads to the so-called Kolmogorov–Sinai theory, in which the principal results include **Sinai's thm** (which asserts that there is a partition A such that the T^rA are independent), and Ornstein's theorem. (D. Ornstein, Some new results in the Kolmogorov–Sinai theory of entropy and ergodic theory, *Bull. Amer. Math. Soc.*, 77(6), 878–890, 1971]

Kolmogorov–Smirnov distribution *see* Kolmogorov's distribution

Kolmogorov's strong law of large numbers Refers to either of these assertions for a sequence of independent r.vs $X_1, X_2, \ldots, X_n$:

(a) If var $X_r = \sigma_r^2, r \geq 1$, then as $n \to \infty$, $\frac{1}{n}\sum_{r=1}^n (X_r - \mathbb{E}X_r) \to 0$ a.s., whenever $\sum_{r=1}^\infty r^{-2}\sigma_r^2 < \infty$, which may be called **Kolmogorov's condition** (sometimes called Kolmogorov's 1st SLLN).
(b) For i i.d.r.vs X_r, $\frac{1}{n}\sum_{r=1}^n X_r$ converges a.s., as $n \to \infty$, iff $\mathbb{E}|X_1| < \infty$, and then the limit is $\mathbb{E}X_1$, (sometimes called Kolmogorov's 2nd SLLN).

[O. Kallenberg, *Foundations of Modern Probability*, 2nd edn. Springer, 2002]

Kolmogorov–Szegö formula *see* Kolmogorov's formula

Kolmogorov's three-series theorem *see* three-series theorem

Kolmogorov's zero-one law Arises in considering an infinite collection of independent r.vs: $X_r, r \geq 1$. For many limiting properties of $X_1, X_2, \ldots, X_n$, as $n \to \infty$, the resulting events will not depend on any finite subset of the infinite sequence $X_1, X_2, \ldots$, and each is therefore called a tail event, and the collection of such events comprises a **sigma-algebra** called the tail σ algebra (or field) $\mathcal{T}$. The zero-one law asserts that all the events in $\mathcal{T}$ have probability 0 or 1; and that any $\mathcal{T}$-measurable r.v. is a.s. constant. [L. Breiman, *Probability*. Addison-Wesley,1968; reprinted by SIAM, 1992].

Komlós–Major–Tusnády approximation (alias: **K-M-T (or KMT) coupling (or embedding, or construction); Hungarian construction**) Arises in the context of the empirical process, and its convergence to a Brownian bridge, where it constructs a suitable coupling of the empirical distribution function (in standard form) to a Brownian bridge defined on the same probability space, permitting a good estimate of the rate of convergence. Formally, let $F_n(x), n \geq 1$, be the sequence of empirical distribution functions corresponding to the sequence $U_1, U_2, \ldots$, of i.i.d.r.vs uniform on [0,1], wlog. One form of the theorem then asserts the existence of a sequence of Brownian bridges $B_n(x)$, on the same probability space, such that almost surely we have

$$\sup_{o \leq s \leq 1} |n^{1/2}(F_n(s) - s) - B_n(s)| = \mathrm{O}\big(n^{-\frac{1}{2}}\log n\big).$$

One proof uses a coupling of binomial and normal distributions, whose tightness is bounded by **Tusnády's inequality**; one form of which constructs a coupling of X having the $B\big(n, \frac{1}{2}\big)$ distribution, and Y having the $N\big(\frac{1}{2}n, \frac{1}{4}n\big)$ distribution, such that,

$$|X - \frac{n}{2}| \leq |Y - \frac{n}{2}| + 1,$$

and $|X - Y| \leq \frac{1}{2n}(2Y - n)^2 + 1$. [A. Carter & D. Pollard, Tusnády's inequality revisited, *Ann. Stat.*, 32(6), 2731–2741, 2004; P. Massart, Tusnády's lemma 24 years later, *Ann. Inst. H. Poincaré B Probab. Stat.*, 38(6), 991–1007, 2002]

Komlós theorem Asserts that for a sequence of r.vs $X_n, n \geq 1$, such that $\sup_n \mathbb{E}|X_n| < \infty$, there is some r.v. Z with finite expected value, and a subsequence $S = X_{n_1}, X_{n_2}, \ldots$ such that as $r \to \infty$, $\frac{1}{r}\sum_{i=1}^{r} X_{n_i} \to Z$, a.s.; this remains true for all unbounded subsequences of S.

Another Komlós theorem asserts that for an $n \times n$ matrix M_n, whose n^2 entries are i.i.d.r.vs, $\mathbb{P}(M_n$ is of full rank$) \to 1$, as $n \to \infty$. [J. Komlós, A generalization of a problem of Steinhaus, *Acta Math. Acad Sci. Hung*., 18, 217–229, 1967]

Kondratiev space A space of stochastic test functions arising in white noise theory. [H. Holden *et al.*, *Stochastic Partial Differential Equations: A Modeling, White Noise Functional Approach*, 2nd edn. Springer, 2010]

Koopman–Darmois distribution *see* exponential family

Kornya approximation Arises in actuarial mathematics as a method for approximating the distribution F of aggregate claims in suitable insurance models. It is obtained by truncating an expansion of the characteristic function of F. [C. Hipp, Improved approximations for the aggregate claims distribution in the individual model, *ASTIN Bull.*, 16, 89–100, 1986]

Korolyuk's theorem *see* Khinchin's theorem; Gnedenko–Korolyuk theorem

Kounias inequality (alias: **Gallot–Kounias inequalities**) Arises in considering bounds for the probability $\mathbb{P}(\cup_1^n A_r)$ that any of a collection of events $A_1, \ldots, A_n$ occur. There are several; the basic (most often seen) upper bound (due to E. Kounias) is

$$\mathbb{P}\left(\bigcup_{r=1}^{n} A_r\right) \leq \min_k \left(\sum_{r=1}^{n} \mathbb{P}(A_r) - \sum_{r \neq k} \mathbb{P}(A_r \cap A_k)\right).$$

The basic lower bound (due to S. Gallot) is

$$\mathbb{P}\left(\bigcup_{r=1}^{n} A_r\right) \geq \mathbf{P}^T Q^{-1} \mathbf{P},$$

where $\mathbf{P}$ is the (column) vector with elements $\mathbb{P}(A_r), 1 \leq r \leq n$, and Q is the matrix with entries $\mathbb{P}(A_i \cap A_j) 1 \leq i, j \leq n$ and Q^{-1} is the (generalized) inverse of Q (and $\mathbf{P}^T$ is the transpose of $\mathbf{P}$). [C. Feng *et al.*, Some inequalities in functional analysis, combinatorics, and probability theory, *Elec. J. Combin.*, 17(1), paper 58, 2010]

***k*-out-of-*n* system** Arises in reliability theory, where it denotes a system with n components that functions iff at least k of the components are working correctly (or, at least, adequately). It is customary (though not universal) to assume that the performances of the components are independent random variables. In particular, 1-out-of-n is a parallel system, and n-out-of-n is a series system. [P. Boland & F. Proschan, The reliability of k out of n systems, *Ann. Probab.*, 11(3), 760–764, 1983]

Kramers equation *see* Langevin equation

Kramers–Moyal expansion Arises in the physics literature in the analysis of Markov processes. Given the appropriate Chapman–Kolmogorov equation, as an integral or integro-differential equation, the integrand is formally expanded as a Taylor series yielding a partial differential equation of infinite order in the space variable (and of order 1 in the time variable). Truncation at the term of order 2 yields a diffusion-type equation.

[C. Gardiner, *Handbook of Stochastic Methods*, 4th edn. Springer, 2010]

k-record *see* Ignatov's theorem

Krein condition (alias: **subsets of: Kolmogorov–Krein–Krylov–Szegö–Wiener condition**) Arises in the moment problem, where it gives a condition for a density $f(x)$ not to be determined by its moments (all of which are finite). Thus, if $f(x)$ is a density on $\mathbb{R}$ with finite moments, and $K_1 = \int_{-\infty}^{\infty} \frac{-\log f(x)}{1+x^2} \mathrm{d}x < \infty$, then $f(x)$ is not uniquely determined by its moments. If $f(x)$ is a density on $\mathbb{R}^+$ then the condition is $K_2 = \int_0^{\infty} -\frac{\log f(x^2)}{1+x^2} \mathrm{d}x < \infty$.

Conversely, if $K_1 = \infty$ and the so-called **Lin condition** holds, i.e. f is symmetric, differentiable, and for large positive x we have $-x\frac{f'(x)}{f(x)} \uparrow \infty$ as $x \to \infty$, then $f(x)$ is determined by its moments. This is **Lin's thm**. The same converse without the symmetry condition, holds on $\mathbb{R}^+$. [J. Stoyanov, Krein condition in probabilistic moment problems, *Bernoulli*, 6(5), 939–949, 2000]

Krickeberg decomposition Arises in martingale theory, e.g. in these forms for the case of discrete parameter:

(a) Let X_n be a submartingale wrt $\mathcal{F}_n, n \geq 0$, such that $\sup_n \mathbb{E}X_n^+ < \infty$. Then there exist a positive martingale M_n and a positive supermartingale S_n such that for all $n, X_n = M_n - S_n$, a.s.

(b) A martingale $X_n, n \geq 0$, is such that, $\sup_n \mathbb{E}|X_n| < \infty$ iff there exist positive L_1 bounded martingales Y_n and Z_n such that for all $n, X_n = Y_n - Z_n$ a.s.

[S. Resnick, *A Probability Path*. Birkhäuser, 1999]

Kullback–Leibler divergence (alias: **relative entropy; information (or I-, or directed) divergence; expected discrimination information; K-L criterion; expected log-likelihood ratio**) This gives a measure denoted by $d(\mathbf{p}, \mathbf{q})$, of the extent to which two probability distributions $\mathbf{p}$ and $\mathbf{q}$ differ from each other. It arises naturally in various contexts, each of which supplies a different interpretation. For example in information theory $d(\mathbf{p}, \mathbf{q})$ is seen as a measure of the cost (in efficient coding) of assuming the distribution of a source to be $\mathbf{q}$ when it is in fact $\mathbf{p}$; in statistics it is seen as the expected amount of information available to discriminate between the alternate hypotheses that the true distribution is $\mathbf{p}$ or is $\mathbf{q}$, on the prior assumption that it is $\mathbf{p}$.

In its simplest form $\mathbf{p} = (p_1, \ldots, p_n)$ and $\mathbf{q} = (q_1, \ldots, q_n)$ are simple distributions such that $p_r q_r \neq 0, 1 \leq r \leq n$, and then $d(p, q) = \sum_{r=1}^{n} p_r \log \frac{p_r}{q_r}$. Similarly, for densities $f(x)$ and $g(x)$ defined on a domain D where $f(x)g(x) \neq 0$, the relative differential entropy, or K-L divergence, is $d(f, g) = \int f(x) \log \frac{f(x)}{g(x)} \mathrm{d}x$, with a natural extension to measures $\mathbb{P}$ and $\mathbb{Q}$. The divergence $d(\mathbf{p}, \mathbf{q})$ is not symmetric; its natural symmetrical form appeared earlier, and is the Jeffreys's divergence, which is given by $J(\mathbf{p}, \mathbf{q}) = d(\mathbf{p}, \mathbf{q}) + d(\mathbf{q}, \mathbf{p})$.

In particular, when $\mathbf{q}$ is uniform, $d(\mathbf{p}, \mathbf{q}) = \log n - H(\mathbf{p})$, where $H(\mathbf{p})$ is the Shannon entropy of $\mathbf{p}$. (*See* negentropy.) Various other entropy functions have been used to give divergences in terms of relative entropy, e.g. for $a > 0$ and $a \neq 1$:

(i) The relative information of type a

$$d_a(\mathbf{p}, \mathbf{q}) = \frac{1}{(a-1)} \left\{ \sum_r p_r^a q_r^{1-a} - 1 \right\},$$

(ii) Renyi's information gain (or relative information) of order a.

$$d_R(\mathbf{p}, \mathbf{q}) = \frac{1}{a-1} \log \left\{ \sum_r p_r^a \, q_r^{1-a} \right\}.$$

These have natural differential forms and may be called generalized relative entropy, or K-L divergence. *See* information central limit theorem. [S. Kullback & R. Leibler, On information and sufficiency, *Ann. Math. Stat.*, 22(1), 79–86, 1951]

kumulant function *see* cumulants

Kunita–Watanabe decomposition *see* martingale representation theorem

Kunita–Watanabe inequality (or theorem) Either of two related inequalities arising in stochastic calculus, thus: Let $X(t)$ and $Y(t)$ be continuous local martingales, with $f(t)$ and $g(t)$ suitably measurable processes, $t \geq 0$. Then with probability 1,

$$I = \int_0^t |f||g||d < X, Y > | \leq \left(\int_0^t f^2 d\langle X, X \rangle \right)^{\frac{1}{2}} \left(\int_0^t g^2 d\langle Y, Y \rangle \right)^{\frac{1}{2}}$$
$$= I(f, X) I(g, Y),$$

(say), where the angle brackets denote the quadratic variation and covariation. Furthermore, for $p \geq 1$ and $p^{-1} + q^{-1} = 1$, $\mathbb{E}I \leq \|I(f, X)\|_p \|I(g, Y)\|_q$, where $\|.\|_p$ is the usual L_p norm. [D. Revuz & M. Yor, *Continuous Martingales and Brownian Motion*, 3rd edn. Springer, 1999]

Kunita–Watanabe representation (or decomposition) *see* martingale representation theorem

Kunita–Watanabe theorem Any of several results (see above) including this: let $M(t)$ be a square-integrable martingale, and $X_n(t),\ n \geq 1$, a sequence of suitably integrable and predictable functions such that each $Y_n = \int_0^t X_n \,\mathrm{d}M$ is an L_2-bounded martingale and Y_n is Cauchy convergent to Y (in the space of square-integrable martingales); then the limit exists as a square-integrable martingale of the form $\int_0^t Y \mathrm{d}M$.

Yor's thm generalizes this result to sequences Y_n that are Cauchy-convergent in L_p for $1 \leq p < \infty$. [F. Delbaen & W. Schachermayer, *The Mathematics of Arbitrage*. Springer, 2006]

kurtosis Broadly, for the probability distribution of a r.v. X the property of being strongly peaked at its mode. Usually, and more narrowly, the term refers to the scaled fourth central moment $k = \mathbb{E}(X-\mu)^4/(\operatorname{var} X)^2$ as a rough measure of this property; alternatively it may refer to a shifted version, $K = k - 3$, sometimes called excess kurtosis.

For a normal r.v. $X, K = 0$, and any r.v. for which $K = 0$ is said to be **mesokurtic**. If $K > 0$ the distribution is **leptokurtic**, while if $K < 0$ it is **platykurtic**. The concept extends to a random p-vector $\mathbf{X}$, with mean vector $\mathbf{m}$, and covariance matrix V, e.g. **Mardia's multivariate kurtosis** defined as $\mathbb{E}\{(X-m)^{\mathrm{T}}V^{-1}(X-m)\}^2$. [K. Mardia, Measures of multivariate skewness and kurtosis with applications, *Biometrika*, 57(3), 519–530, 1970]

Kushner–Stratonovich equation *see* filtering

Ky Fan distance A metric defined for real-valued r.vs X and Y by $k(X, Y) = \inf\{\varepsilon \geq 0 : \mathbb{P}(|X-Y| > \varepsilon) \leq \varepsilon\}$.

For r.vs taking values in a metric space (S, d), the generalized Ky Fan distance, or probability distance, is

$$g(X, Y) = \inf\{\varepsilon \geq 0 : \mathbb{P}(d(X, Y) > \varepsilon) \leq \varepsilon\}.$$

[R. Dudley, *Real Analysis and Probability*. Cambridge University Press, 2002]

L

Labouchere system A betting system, somewhat more cautious than the martingale, that operates thus: the gambler chooses a sequence $a_1, \ldots, a_n$, of positive numbers. The sum $s = a_1 + a_n$ of the first and last is wagered at evens (say). If the bet wins they are deleted; if the bet is lost then s is appended as an extra term $s = a_{n+1}$ in the series. This is continued until the sequence is empty (and note that if the sequence contains just one term, that is the bet). A variant, called the reverse Labouchere system, proceeds similarly except that the numbers comprising the current bet are deleted at a loss, or their sum s appended following a win. [S. Ethier, *The Doctrine of Chances*. Springer, 2010]

lace expansion Arises in considering a range of problems in probability on graphs, e.g. self-avoiding random walk, percolation, and random-cluster models, in a sufficiently high number of dimensions. A lace is a minimally connected graph that is disconnected by the removal of any edge (where the definition of 'connected' is not that of conventional graph theory). Configurations in the problem of interest are expressed in terms of laces that are easier to count. [G. Slade, *The Lace Expansion and its Applications*. Springer, 2006]

lack of anticipation Arises in the context of queues, networks, and related systems that are driven by a Poisson process $N(t)$ of arrivals, $t \geq 0$. Let $M(t)$ be some function of the system at time t, then the lack of anticipation property holds if $M(s)$ is independent of $N(t) - N(s)$ for all $0 \leq s < t < \infty$. Its importance lies in **Wolff's PASTA thm**, *see* arrivals theorem. [R. Wolff, Poisson arrivals see time averages, *Oper. Res.*, 30(2), 223–231, 1982]

lack of memory property (alias: **memoryless property; forgetfulness; loss of memory property**) Arises naturally in considering a system evolving in time whose future after time t depends only on its state at t and not on any previous state; *see* Markov property. In its simplest form, a positive continuous r.v. X is said to have the lack of memory property (or to be memoryless) if for all $s, t > 0$,

$$(*)\, \mathbb{P}(X > s + t \,|\, X > t) = \mathbb{P}(X > s),$$

and it follows that X must have an exponential density. Likewise, the r.v. taking values in the positive integers such that

$$\mathbb{P}(X > j + k \,|\, X > k) = \mathbb{P}(X > j), \quad j, k \geq 0,$$

has a geometric distribution.

More generally, if $(*)$ holds for some one t and all $s > 0, X$ is said to be **forgetful** at t. A theorem of G. Marsaglia and A. Tubilla asserts that if X is forgetful at t_1 and t_2 where t_1/t_2 is irrational, then X is exponentially distributed. If a r.v. is forgetful at a random time T, this may be called a strong lack of memory property; a theorem of B. Ramachandran asserts that for X and Y independent positive continuous r.vs, $\mathbb{P}(X > Y + s \,|\, X > Y) = \mathbb{P}(X > s)$ iff X has an exponential density. In higher dimensions, defining the survival fn $\overline{F}_X(\mathbf{x}) = \mathbb{P}(X_1 > x_1, \ldots, X_n > x_n)$, either $\overline{F}_X$, or equivalently $\mathbf{X}$, is said to be completely lacking in memory if $\overline{F}_X(\mathbf{x} + \mathbf{y}) = \overline{F}_X(\mathbf{x})\overline{F}_X(\mathbf{y})$ for all positive vectors $\mathbf{x}$ and $\mathbf{y}$, or said to be of **limited memory** if this is true only for $(\mathbf{x}, \mathbf{y})$ lying in a subset of $\mathbb{R}_+^{2n}$. [R. Shimizu, On a lack of memory property of the exponential distribution, *Ann. Inst. Stat. Math.*, 31(1), 309–313, 1979; A. Marshall & B. Olkin, Multivariate exponential and geometric distributions with limited memory, *J. Multivariate Anal.*, 53(1), 110–125, 1995]

ladder process Arising in the context of a random walk $S_n = \sum_{r=1}^{n} X_r$, $S_0 = 0$, where the X_r are i.i.d.r.vs, the concept was introduced by D. Blackwell in considering the renewal theorem. Broadly, the ascending (and descending) ladder processes are generated by the sequence of new record high (and record low) values of S_n as n increases. Their importance follows from the regenerative property apparent at the steps (or rungs) of the ladder. Formally, let L_1 be the first n such that $S_n > 0$: then L_1 is the first ladder time (or epoch, or index), and S_{L_1}, is the first ladder height (or rung).

The second ladder epoch L_2 is the smallest n such that $S_n > S_{L_1}$, yielding S_{L_2} as the second rung. The sequence $(L_r, S_{L_r}), r \geq 1$ is the ladder process. Subsequent visits of the walk to a ladder height S_{L_r} before the next ladder epoch L_{r+1} may be called weak ladder points (the first visit at (L_r, S_{L_r}) is then called a strong ladder point).

The descending ladder is defined likewise for $-S_n, n \geq 0$. The theory yields the so-called **duality lemma** relating the occupation measures of the ascending ladder heights and the values of the walk prior to becoming negative. The concept can be extended to Lévy processes, $X(t)$, with the complication that the epochs at which such processes attain their supremums $S(t)$ do not constitute a discrete subset of $\mathbb{R}$; this may be resolved by considering the local time of $S(t) - X(t)$.

Note that some authors use the term record point for ladder point. [D. Blackwell, Extension of a renewal theorem, *Pacific J. Math.*, 3, 315–320, 1953; S. Asmussen, *Applied Probability and Queues*, 2nd edn. Springer, 2003]

lag Broadly, the term denotes a time difference. Thus for a time series $X_n, n \in \mathbb{Z}$ the operator B such that $BX_{n+1} = X_n$ is the lag operator (or backshift). For a random process $X(t), t \in \mathbb{R}$ (or $\mathbb{Z}$), the process $L(t,\tau) = X(t-\tau)$ is X with lag τ, and where the covariance exists, denoted by $c(t,\tau) = \text{cov}(X(t), X(t-\tau))$, τ may be called the lag (or lead parameter). For a finite filter $f = \sum_{|\tau| \leq w} a(\tau)c(t,\tau)$, $a(\tau)$ may be called the lag window. [P. Brockwell & R. Davis, *Time Series*, 2nd edn. Springer, 1991]

Lagrange distribution (alias: **Lagrangian distribution**) A class of distributions whose general form is given by a Lagrange expansion. The simplest form arises naturally as the distribution of the total population T in a standard Galton–Walton branching process with family-size pgf $G(s)$. Then the pgf $G_T(s)$ of T is the solution $G_T = \lambda$ of the functional equation $\lambda = sG(\lambda)$, and Lagrange's expansion (or inversion formula) gives

$$G_T(s) = \sum_{n=1}^{\infty} \frac{1}{n!} s^n \left[\frac{\mathrm{d}^{n-1}}{\mathrm{d}u^{n-1}} G(u)^n\right]\Big|_{u=0},$$

the coefficients of s^n yielding the Lagrange distn of T. More generally, if the initial population is random with pgf $F(s)$, then the appropriate Lagrange expansion yields the distribution of T called a Lagrange distribution of the second kind.

This formal second expansion may still yield a probability distribution for suitable holomorphic (i.e. analytic) functions $f(z)$ and $g(z)$ (that are not necessarily pgfs, in place of $F(s)$ and $G(s)$), in which case it is called a general Lagrange distribution with pgf

$$G(s) = \sum_{n=1}^{\infty} \frac{s^n}{n!} \frac{\mathrm{d}^{n-1}}{\mathrm{d}u^{n-1}} \{g(u)^n f'(u)\}|_{u=0}.$$

Yet more generally, multivariate Lagrange distns arise in the context of multitype branching processes, and extended using the Lagrange–Good multivariate inversion formula, yielding **Lagrange–Good distns**. [I. Good, The Lagrange distributions and branching processes, *SIAM J. Appl. Math.*, 28(2), 270–275, 1975; P. Consul & F. Famoye, *Lagrangian Probability Distributions.* Birkhäuser, 2006]

Laguerre process (alias: **complex Wishart process**) Arises in random matrix theory. Let $C(t)$, $t \geq 0$, be an $n \times r$ matrix, whose entries are of the form $c_{jk}(t) = W_{jk}(t) + \mathrm{i}B_{jk}(t)$, where W_{jk} and B_{jk}, comprise independent collections of independent standard Wiener processes, and var $c_{jk}(t) = 2t$. Then $L(t) = C^{\mathrm{T}}C$, where C^T denotes transpose, is the Laguerre process of dimension n and size r. When the entries of C are real, $c_{jk} = W_{jk}(t)$, then L is the Wishart process. And when $r = 1$, the Laguerre process is a squared Bessel process. [N. Demni, The Laguerre process and generalized Hartman-Watson law, *Bernoulli*, 13(2), 556–580, 2007]

Laguerre tessellation *see* random mosaic

Lai's theorem Arises in considering the summability, by various classical methods, of a sequence $X_r, r \geq 1$, of i.i.d.r.vs. It relates Abel and Cesàro summability to the existence of $\mathbb{E}X_1$, by the assertion that the following are equivalent:

(a) $\mathbb{E}X_1 = \mu < \infty$;
(b) $\frac{1}{n}\sum_{r=1}^{n} X_r \to \mu$ with probability 1 as $n \to \infty$;
(c) $\lim_{x \uparrow 1}(1-x)\sum_{r=1}^{\infty} x^r X_r = \mu$ with probability 1.

Another theorem of T. Lai considers a type of law of the iterated logarithm for summability in the sense of Borel and Euler.
[T. Lai, Summability methods for i.i.d. random variables, *Proc. Amer. Math. Soc.*, 45(2), 253–261, 1974]

Lal's inequality *see* Berge's inequality

lambda distribution *see* Tukey's distribution, and Wilks's lambda distribution

lambda system *see* Dynkin's pi-lambda theorem

Lambert's distribution Arises in Knudsen's random walk and stochastic billiards as a **scattering (random) boundary condition**. In standard form it is a distribution over the hemisphere such that the probability of the element dA is proportional to cos θ dA, where θ is the angle made by the axis of symmetry with a radius through dA. (Therefore, some authors call it a cosine distn.)

In the oblique Lambert distn, the axis of symmetry is replaced by the direction of specular (mirror) reflection of an incident particle. [F. Comets *et al.*, Quenched invariance principle for the Knudsen stochastic billiard in a random tube, *Ann. Probab.*, 38(3), 1019–1061, 2010]

Lamperti distributions (or class, or laws) Arise in considering the occupation times of a class of stochastic processes, where the resulting distribution has density $f(x)$ for $0 < x < 1$ given by,

$$f(x) = a\frac{\sin c\,\pi}{\pi}\frac{x^c(1-x)^{c-1} + x^{c-1}(1-x)^c}{a^2x^{2c} + 2ax^c(1-x)^c\cos c\pi + (1-x)^{2c}},$$

when $a > 0$ and $0 < c < 1$. The corresponding Stieltjes transform of the distribution is

$$\int_0^1 \frac{\mathrm{d}F(x)}{x+\theta} = \frac{(1+\theta)^{c-1} + a\theta^{c-1}}{(1+\theta)^c + a\theta^c}.$$

A related distribution, on $\mathbb{R}^+$, also called Lamperti's distribution, arises as the density f of the ratio of independent one-sided stable distributions of exponent α, and Laplace transform $\mathbb{E}\mathrm{e}^{-\theta X} = \mathrm{e}^{-\theta^\alpha}$, $0 < \alpha < 1$. Then

$$f(x) = \frac{\sin\alpha\,\pi}{\pi}\frac{x^{\alpha-1}}{1 + 2x^\alpha\cos\alpha\pi + x^{2\alpha}}, \quad x > 0.$$

[J. Lamperti, An occupation time theorem for a class of stochastic processes, *Trans. Amer. Math. Soc.*, 88(2), 380–387, 1958; L. James, Lamperti-type laws, *Ann. Appl. Probab.*, 20(4), 1303–1340, 2010]

Lamperti's problem Broadly, resolving the question of recurrence or transience for random processes on $[0, \infty)$ with a (mean) drift that goes to zero as the values of the process go to ∞ (this may be called **Lamperti drift**). [M. Menshikov & A. Wade, Rate of escape and CLT for the supercritical Lamperti problem, *Stoch. Proc. Appl.*, 20(10), 2078–2099, 2010]

Lamperti process *see* self-similar process

Lamperti representation (alias: **Lamperti correspondence**) Arises in the context of self-similar Markov processes. In the case of a positive self-similar Markov process $X(t)$, J. Lamperti established a representation of X as the exponential of a suitably time-changed Lévy process (i.e. having independent increments); conversely there is a transformation of a suitable Lévy process that yields a positive self-similar Markov process. [J. Lamperti, Semi-stable Markov processes I, *Probab. Theor. Rel. Fields*, 22, 205–225, 1972; L. Chaumont *et al.*, The Lamperti representation of real-valued self-similar Markov processes, *Bernoulli*, 19(53), 2494–2493, 2013]

Lamperti's theorem Arises in answer to the question of what random processes can occur as the limit when a given random process is subject to a sequence of contractions of its space and time parameters. The theorem asserts that the limit is a self-similar process (also called semi-stable), being invariant under suitable changes of time and scale. [J. Lamperti, Semi-stable stochastic processes, *Trans. Amer. Math. Soc.*, 104(1), 62–78, 1962]

Lamperti's transformation Arises in the context of Lamperti's theorem on self-similar processes, where it asserts that for a (strictly) stationary process $X(t), t \in \mathbb{R}$, the process $S(t) = t^H X(\log t), t > 0, H > 0$, is self-similar (with Hurst parameter H). Conversely if $Y(t), t \geq 0$, is self-similar with parameter H, then $X(t) = \mathrm{e}^{-tH}Y(\mathrm{e}^t), t \in R$, is strictly stationary, and may be called the Lamperti transform of Y.

A classic example is supplied by the Wiener process $W(t)$ with Hurst parameter 1/2, where $X(t) = \mathrm{e}^{-\frac{1}{2}t}W(\mathrm{e}^t)$ is the stationary Ornstein–Uhlenbeck process. [K. Burnecki *et al.*, The Lamperti transformation for self-similar processes, *Yokohama Math. J.*, 44, 25–42, 1997]

lamplighter chain (or walk) A Markov chain taking values in $\{0, 1\}^V$, where V is the vertex set of a graph G; the values 0 and 1 are seen as representing the state (off or on) of lamps placed at each vertex. The chain is realized by a random walk on the vertices, at each step of which the lamp at the destination vertex is turned on (or off) according to some given rule. [D. Levin *et al.*, *Markov Chains and Mixing Times*. American Mathematical Society, 2009]

Lancaster probabilities (alias: **Lancaster class (or distribution)**) Arise in the expansion of bivariate distributions in terms of orthogonal polynomials determined by the marginal distributions. Thus, let the marginals $F(\cdot)$ and $G(\cdot)$ of a joint distribution $J(x, y)$ of (X, Y) be uniquely determined by their moments (all being finite). Then the Lancaster class of distributions (or probabilities) comprises all those distributions J such that for all $n \geq 1$

$$\mathbb{E}(X^n \mid Y = y) = \sum_{r=0}^{n} a_r y^r \quad \text{and} \quad \mathbb{E}(Y^n \mid X = x) = \sum_{r=0}^{n} b_r x^r;$$

that is, the conditional moments of order n are polynomials of degree n in the given variable.

There is an important characterization of the Lancaster class, as follows: let $p_n(x)$ and $q_n(x), n \geq 1$, be the L_2-complete families of orthogonal polynomials with respect to the distributions F and G respectively (i.e. $\int p_n(x)p_m(x)\mathrm{d}F(x) = 1$ for $m = n$ and 0 for $m \neq n$; and likewise for q_n's with $G(x)$). Then J is in the Lancaster class iff there is a so-called **Lancaster sequence** $r_n, n \geq 0$, such that $\mathbb{E}(p_n(X)q_m(Y)) = r_n$ for $m = n$ and 0 for $m \neq n$. In this case the Lancaster sequence is unique. The Lancaster problem is to characterize those sequences $r_n, n \geq 0$ that are a Lancaster sequence for some distribution. The set of Lancaster sequences is convex and compact; if $\sum_{n=0}^{\infty} r_n^2 < \infty$, the distribution has a density given by

$$J_{xy} = f(x,y) = F'(x)G'(y)\sum_{n=0}^{\infty} r_n p_n(x) q_n(y).$$

[H. Lancaster, The structure of bivariate distributions, *Ann. Math. Stat.*, 29, 719–736, 1958]

Lancaster's theorem *see* Gebelein–Lancaster theorem

Landau–Odlyzko theorem Arises in considering a random walk on a connected undirected graph with n vertices and diameter d. Let the vertex degree of the ith vertex be v_i, and the walk such that a step from the ith vertex any edge is chosen with probability v_i^{-1}. Then the eigenvalue λ having second largest modulus of the $n \times n$ transition matrix P of the walk satisfies $1 - |\lambda| \geq \frac{1}{n(d+1)m}$, where $m = \max_r v_r$. [H. Landau & A. Odlyzko, Bounds for eigenvalues of certain stochastic matrices, *Linear Alg. Appl.*, 38, 5–15, 1981]

Landau–Shepp theorem (alias: **Fernique–Landau–Shepp theorem**) Arises in considering the supremum $S = \sup_n |X_n|$, where $X_1, X_2, \ldots$, is a sequence of Gaussian r.vs with any means and covariances. One form is this: let $m = \sup_n \operatorname{var} X_n$. If $\mathbb{P}(S < \infty) > 0$, then $\mathbb{P}(S < \infty) = 1$, and $m < \infty$, and for all $0 < a < \frac{1}{2m}$ and all large enough t, $\mathbb{P}(S > t) < \exp(-at^2)$. [M. Marcus & L. Shepp, Sample behaviour of Gaussian processes, in *Proc. 6th Berkeley Symp. Math. Stat. Prob. II*. UCP and Cambridge University Press, 1972]

Langevin distribution *see* Fisher's distribution

Langevin equation The basic form arises as the (Newtonian) equation of motion for a particle at $X(t), t \geq 0$, with velocity $V(t)$, executing Brownian motion subject to a drag force $-aV(t)$ (attributable to viscosity) in addition to the driving force of random collisions.

It was written formally as $\frac{\mathrm{d}V}{\mathrm{d}t} = -a\,V + bL(t)$, where $L(t)$ is Gaussian white noise, and a formal integration yields an Ornstein–Uhlenbeck process. A rigorous analysis can be carried through in the framework of (Hida) white noise, using generalized functions. Perhaps more often, the Langevin equation is written as an SDE in the form $\mathrm{d}V = -aV\mathrm{d}t + b\mathrm{d}W(t)$, where $W(t)$ is the Wiener process, and integrated using the Itô calculus. [W. Coffey *et al.*, *The Langevin Equation*, 2nd edn. World Scientific, 2004]

Laplace–Borel transform (alias: **Borel–Laplace transform**) Denotes the relationship

$$G(z) = \int_0^{\infty} B(tz)\mathrm{e}^{-t}\mathrm{d}t,$$

where $G(z) = \sum_{n=0}^{\infty} p_n z^n$, is the ordinary generating function of $p_n, n \geq 0$, while $B(s)$ is the exponential, or Borel generating function $B(s) = \sum_{n=0}^{\infty} p_n s^n/n!$.

Laplace central limit theorem *see* De Moivre's theorem

Laplace distribution (or density, or error curve) *see* double exponential distribution

Laplace exponent (alias: **cumulant function**) Arises in the context of an infinitely divisible distribution F on a subset of $[0, \infty)$. A function $L(\theta)$ is the Laplace transform of an infinitely divisible r.v. X (having the distribution function F) iff $L(\theta) = \exp(-\Psi(\theta))$, where $\Psi(0) = 0$ and Ψ has completely monotone derivative. Then Ψ is the Laplace exponent of X (or F). Similarly, for a non-decreasing (a.s.) Lévy process $X(t) \geq 0$, we have $\mathbb{E}\mathrm{e}^{-\theta X(t)} = \exp(-t\Psi(\theta)), \theta \geq 0$, where $\Psi \geq 0$ is called the Laplace exponent of $X(t)$. (Some authors call it the

Bernstein function.) [J. Bertoin, *Lévy Processes*. Cambridge University Press, 1996]

Laplace functional *see* generating functional

Laplace motion *see* variance-gamma distribution

Laplace principle *see* large deviations theory

Laplace principle of insufficient reason *see* indifference principle

Laplace's rule of succession Arises from an application of conditional probability in a sequence of trials where the indifference principle is assumed to hold. Laplace's corollary (as it should properly be called; it was misnamed 'Rule' by John Venn), asserts that in such a case if an event occurs in n successive trials, the probability of its occurrence on a subsequent similar trial may be taken as $(n+1)/(n+2)$. Laplace notes that where the indifference principle fails, e.g. as it would for an astronomer observing 5000 years of sunrises, the chance of the sun rising tomorrow will be incomparably greater than the rule would suggest. [P. Laplace, *Philosophical Essay on Probabilities*, 5th edn, 1825; trans. by A. Dale, Springer, 1995]

Laplace's theorem *see* De Moivre's theorem

Laplace transform Is defined for a suitable real-valued function $f(x), x \geq 0$, and denoted by $M(t)$ where $M(t) = \int_0^\infty e^{-tx} f(x) dx$, for all values $t \in \mathbb{C}$ such that the integral exists. Introduced by P. Laplace in the context of probability theory, its importance arises from the case when $f(x)$ is a density, and then the transform is its moment generating function. More generally, a density on $\mathbb{R}$ may have a two-sided Laplace transform $L(t) = \int_{-\infty}^\infty e^{-tx} f(x) dx$. For a distribution function $F(x), x \geq 0$, that has no density, one may define the Laplace–Stieltjes transform $M(t) = \int_0^\infty e^{-tx} dF(x)$. The related Mellin transform is $\mathbb{E}X^t = \int_0^\infty x^t dF(x)$. [D. Widder, *The Laplace Transform*. Princeton, 1941; reprinted by Dover, 2011]

Laplace–Varadhan lemma *see* large deviations theory

Laplacian random walk A type of self-avoiding random walk that is defined on a suitable graph, G, commonly the (possibly infinite) integer lattice (i.e. $\mathbb{Z}^d$ or a subset of it). It is defined inductively as follows: let B denote the boundary of G, and let S_n denote the path of the first n steps of the walk in the interior of G with end vertex s_n. For any vertex x of G, let $h_n(x)$ be the probability that a random walk on the vertices of G, started from x, hits B before S_n. Then the next step of the walk is to any of the points x that are neighbours of s_n, with the probability of each such x being proportional to $h_n(x)$. More generally, if the probability of a step to x is proportional to $h_n(x)^b$, the walk is called Laplacian with parameter b, or Laplacian-b walk. [G. Lawler, The Laplacian-b walk and the Schramm-Loewner evolution, *Illinois J. Math.*, 50(3), 701–746, 2006]

Laplacian stochastic growth model Arises as an extension of the classical deterministic Laplacian growth in which the boundary ∂D of a growing object D grows at any point at a rate proportional to $\nabla\phi$, where ϕ is a solution of Laplace's equation outside D (given suitable boundary conditions). That is to say, it grows according to a harmonic measure. In the stochastic case, the object D grows by the random attachment of new objects, where the probability of accretion is proportional to $|\nabla\phi|$, or more generally $|\nabla\phi|^a$; and the size of the objects may be constant, or more generally proportional to $|\nabla\phi|^b$; and the shape of the object may vary between different models. Examples include diffusion limited aggregation, the Hastings–Levitov model, and the Laplacian random walk. [T. Vicsek, *Fractal Growth Phenomena*, 2nd edn. World Scientific, 1992]

large deviations theory Broadly, the theory of large deviations seeks to supply the exact rate of convergence of a sequence of probabilities that converges exponentially; or, more generally, to supply exponential bounds for a sequence of probabilities. The theory also provides similar results for associated sequences of expectations of a class of suitable functions. Its basic form arises in the context of the law of large numbers, where we have **Cramér's thm**: Let $X_r, r \geq 1$ be i.i.d.r.vs with mean $\mu < \infty$. Then for $x \geq \mu$,

$$\lim_{n\to\infty} \frac{1}{n} \log \mathbb{P}\left(\frac{1}{n} \sum_{r=1}^{n} X_r \geq x \right) = -\sup_{y\in\mathbb{R}} \{xy - \log \mathbb{E}e^{yX_1}\}.$$

More generally, a sequence $\mathbb{P}_n, n \geq 1$ of probability distributions on a complete separable metric space is said to satisfy a large-deviations principle, or LDP, with rate function $I(x)$ and scaling (or rate) constants

$a_n(\uparrow \infty \text{ as } n \to \infty)$ if $I(x) \geq 0$, and all sets $\{x : I(x) \leq c\}$ are compact, and for any closed set C,

$$\limsup_{n\to\infty} \frac{1}{a_n} \log \mathbb{P}_n(C) \leq -\inf_{y\in C} I(y),$$

and for any open set D

$$\liminf_{n\to\infty} \frac{1}{a_n} \log \mathbb{P}_n(D) \geq -\inf_{z\in D} I(z).$$

Examples of LDPs include:
(a) **Varadhan's LDP**, which generalizes Cramér's theorem to $\mathbb{R}^d, d \geq 2$;
(b) **Mogulskii's LDP**, which establishes a rate of convergence in the space of continuous functions of [0, 1] for the scaled continuized partial sum process for i.i.d.r.vs;
(c) **Sanov's LDP** for empirical distributions;
(d) **Schilder's LDP** for Brownian motion.

If $\mathbb{P}_n, n \geq 1$, satisfies the LDP above, and $f(\cdot)$ is a real-valued bounded continuous function on the space, then,

$$\lim_{n\to\infty} \frac{1}{a_n} \log \int \exp(a_n f(x))\mathbb{P}_n(\mathrm{d}x) = \sup_x\{f(x) - I(x)\},$$

which may variously be called a Laplace principle, a functional LDP, Varadhan's integral lemma, or the Laplace–Varadhan integral lemma.
[A. Dembo & O. Zeitouni, *Large Deviations and Applications*, 2nd edn, corrected. Springer, 2009]

last exit time (alias: **quitting time**) Arises in the context of a Markov process $X(t), t \in T$, as a natural and useful counterpart to the hitting (and first entrance) time. Formally, the quitting time L_A of a suitable set A in the state space of $X(t), 0 \leq t \leq T$, is $L_A = \sup\{0 \leq t < T : X(t) \in A\}$, with the convention that $\sup \emptyset = 0$. It finds applications through a natural dual of the first-passage decomposition, called the last exit decomposition. For example, given a Markov chain X_n, $n \geq 0$, with n-step transition matrix $p_{ij}(n)$, the last exit formula is

$$p_{ij}(n) = \sum_{r=0}^{n-1} p_{ii}(r)t_{ij}(n-r),$$

where $t_{ij}(m)$ is the probability of moving from i to j in m steps with no intervening visit to i. Applications are also found in the connection between potential theory and the Wiener process, e.g. this theorem: let $W(t)$ be a transient Wiener process in $\mathbb{R}^d$ with associated Green function $g(x, y)$, and L_A the last exit time from a compact set A. Then there is a so-called equilibrium measure $m(\cdot)$ on ∂A such that for any suitable $B \subseteq \partial A$

$$\mathbb{P}(W(L_A) \in B \mid W(0) = x) = \int_B g(x, y)m(\mathrm{d}y).$$

[K. L. Chung, Probabilistic approach in potential theory to the equilibrium problem, *Annales de l'Institut Fourier*, 23(3), 313–322, 1973]

last-passage percolation *see* Eden–Richardson model

Las Vegas algorithm A class of randomized algorithms, with a random run-time, whose distinguishing feature is that any answer returned is correct with arbitrarily small probability (the procedure is allowed to return no answer, denoted by '?'). The name was proposed to distinguish such algorithms from Monte Carlo algorithms, with (usually) fixed (i.e. deterministic) running time, whose probability of error is less e.g. than 1/3. [H. Hoos & T. Stützle, *Stochastic Local Search: Foundations and Applications*. Elsevier, 2005]

latent Markov model *see* hidden Markov process

lattice avalanche model *see* Abelian sandpile

lattice distribution A r.v. X has a lattice distribution if there exist reals a and α such that X takes values only in (i.e. its distribution is concentrated on) the set $\{a + \alpha n, n = 0, \pm 1, \pm 2, \ldots\}$. Then α is called a step of the distribution, and the largest α such that X is lattice is called the span. In particular, if $a = 0$, the distribution is called arithmetic. Such distributions have periodic characteristic functions; and $\phi(t)$ is the chf of a lattice r.v. iff $\phi(\theta) = 1$ for some $\theta \neq 0$. Otherwise X is said to be non-lattice; if, further, $\limsup_{|t|\to\infty}|\phi(t)| < 1$, (which is **Cramér's condition**) then X is said to be strongly non-lattice. More generally, a r.vr is lattice if each component is lattice, not necessarily with the same span.
[S. Asmussen, *Applied Probability and Queues*, 2nd edn. Springer, 2003]

Lau–Rao theorem Arises in characterizing the exponential and geometric distributions through their lack of memory property and other conditional independence properties. One form is this: let $f(x), x \geq 0$, be a non-zero, non-negative, integrable, Borel–measurable function. If, for some sigma-finite measure $m(x)$ on $[0, \infty)$, with $m(\{0\}) < 1$, we have

$f(x) = \int_0^\infty f(x+y)\mathrm{d}m(y)$, then one of these alternatives holds:

(a) $m(x)$ is not lattice and $f(x) \propto \mathrm{e}^{-\lambda x}$ a.e. in $[0,\infty)$, with $\int_0^\infty \mathrm{e}^{-\lambda x}\mathrm{d}m(x) = 1$;

(b) $m(x)$ is lattice with span h, and $m(x+nh) = f(x)\, q^n$ a.e. in $[0,\infty)$, with $\sum_{n=0}^{\infty} q^n m(\{nh\}) = 1$.

[C. Rao & D. Shanbhag, *Choquet-Deny Type Functional Equations with Applications to Stochastic Models*. Wiley, 1994]

Lauritzen–Spiegelhalter–Pearl procedure *see* belief propagation

law A term arising in the context of random variables and processes. Simply put, the law of a random variable X is its probability distribution on $\mathbb{R}$, as defined by the probability $\mathbb{P}(X \in B)$ for any suitable (i.e. Borel) subset B of $\mathbb{R}$. More formally, for X defined on $(\Omega, \mathcal{F}, \mathbb{P})$ the law of X is given by $\mathbb{P}(X^{-1}(B) = \mathbb{P}(\omega : X(\omega) \in B)$, and may be denoted by $\mathbb{P}_X(B)$. (*See* canonical random variable and induced probability.) The idea extends to a random process $X(t), t \in T$, taking values in $\mathbb{R}^T$, defined on a probability space $(\Omega, \mathcal{F}, \mathbb{P})$ so that $x(\omega) = X(t, \omega)$, i.e. $X(t)$ for given $\omega \in \Omega$. Then the law of X is given (for a suitable finite-dimensional subset B of $\mathbb{R}^T$), as above by $\mathbb{P}_X(B) = \mathbb{P}(\omega : X \in B)$.

Less formally, this amounts to the statement that for any $t_1, \ldots, t_n \in T$, and Borel subsets $B_1, \ldots, B_n$ of $\mathbb{R}$, we set $\mathbb{P}_X(B_1, \ldots, B_n) = \mathbb{P}(X(t_1) \in B_1, \ldots, X(t_n) \in B_n)$, so the law of $X(t)$ corresponds to its family of finite-dimensional distributions. [D. Revuz, & M. Yor, *Continuous Martingales and Brownian Motion*, 3rd edn. Springer, 1999]

Lawler's self-avoiding walk *see* loop-erased random walk

law of anomalous numbers *see* Benford's distribution

law of large numbers *see* strong law of large numbers; weak law of large numbers

law of rare events (alias: **law of small numbers; Poisson approximation**) Arises in considering the distribution of the number N that occur of a large number of unlikely events. Informally, if they are uniformly unlikely then N has approximately a Poisson distribution, if the events are sufficiently numerous. Formally, let I_{nr}, $1 \le r \le n, n \ge 1$ be indicator random variables such that:

(a) $\mathbb{P}(I_{nr} = 1) = \mathbb{E}I_{nr} = p_{nr}$;
(b) $I_{nr}, 1 \le r \le n$, are independent for each fixed n;
(c) $p_n = \sum_{r=1}^{n} p_{nr} \to \lambda$, as $n \to \infty$;
(d) $\max_{1 \le r \le n} p_{nr} \to 0$, as $n \to \infty$.

Then $N_n = \sum_{r=1}^{n} I_{nr}$ converges in distribution to a Poisson r.v. with parameter λ. *See* Poisson's theorem, Le Cam's inequality, and Stein's method.
[A. Barbour *et al.*, *Poisson Approximation*. Oxford University Press, 1992]

law of small numbers *see* law of rare events

law of the iterated logarithm (alias: **LIL**) Arises in the context of the partial sums $S_n = \sum_{r=1}^{n} X_r$ of a suitable sequence of i.i.d.r.vs $X_r, r \ge 1$, where, by the CLT (assumed to hold), we know that in the long-run (as $n \to \infty$) the magnitude $|S_n|$ of the sequence is of order of magnitude $n^{\frac{1}{2}}$. Nevertheless, the sequence $S_n/n^{\frac{1}{2}}$ enjoys some large but rare fluctuations. It is natural to seek to quantify these, in the sense of seeking the most slowly decreasing function $u(n)$ such that $u(n)S_n \to 0$ with probability 1 as $n \to \infty$. In this case the answer is given in terms of the sequence

$$L_n = \frac{S_n}{(2n \log\log n)^{1/2}},$$

giving the title 'law of the iterated logarithm' to the various theorems that result. Early LILs include:

(a) **Hartman–Winter LIL**. When S_n above comprises i. i.d.r.vs probability 1 as $\mathbb{E}X_1 = 0, \operatorname{var} X_1 = 1$, then $\limsup_{n\to\infty} L_n = 1 = -\liminf_{n\to\infty} L_n$, a.s. A classic earlier special case is **Khinchin's LIL** for Rademacher r.vs.
(b) **Kolmogorov's LIL** allows the summands to be differently distributed.
(c) V. Strassen showed that the finite-variance condition is necessary for the Hartman–Winter LIL, yielding the H-W-S LIL, often called 'the LIL'.
(d) **Strassen's LIL** denotes a functional law of the iterated logarithm, often called Strassen's invariance principle.
(e) **Khinchin's LIL** also denotes the assertion that for the Wiener process (in a continuous version) $\limsup_{t\to\infty} W(t)/(2t \log\log t)^{\frac{1}{2}} = 1$.

[N. Bingham, Variants on the law of the iterated logarithm, *Bull. Lond. Math. Soc.*, 18(5), 433–467, 1986]

law of the single logarithm Arises in the context of limit theorems for delayed partial sums, where it parallels the law of the iterated logarithm. A basic form is: let X_r,

$r \geq 1$, be i.i.d. zero-mean r.vs, with non-zero variance, and let $0 < a < 1$. Then these are equivalent:

$$\mathbb{E}X_1 = 0;\ \operatorname{var} X_1 = \sigma^2 < \infty;\ \text{ and }\ \mathbb{E}\left\{\frac{|X_1|^{2/a}}{(\log^+|X_1|+1)^{1/a}}\right\} < \infty, \tag{1}$$

$$\limsup_{n\to\infty}(X_n + \cdots + X_{n+[n^a]})/\{2(1-a)n^a \log n\}^{\frac{1}{2}} = (\operatorname{var} X_1)^{1/2} = -\liminf_{n\to\infty}(X_n + \cdots + X_{n+[n^a]})/\{2(1-a)n^a \log n\}^{\frac{1}{2}}. \tag{2}$$

See Erdös–Rényi law. [T. Lai, Limit theorems for delayed sums, *Ann. Probab.*, 2(3), 432–440, 1974]

law of thin numbers Arises in considering Rényi thinning of the sum $S_n = \sum_{r=1}^{n} X_r$ of i.i.d. non-negative integer-valued r.vs X_r, $r \geq 1$. The α-thinning (or **Rényi thinning**) of an integer-valued r.v. $X \geq 0$ is $\sum_{r=1}^{X} I_r$, where for each value of X, the I_r are i.i.d. Bernoulli (indicator) r.vs such that $\mathbb{E}I_1 = \alpha = \mathbb{P}(I_1 = 1)$. Then the law of thin numbers asserts that for $\mathbb{E}X_1 = \lambda < \infty$, as $n \to \infty$, the $\frac{1}{n}$-thinning of S_n converges in distribution (and in relative entropy, or information distance) to the Poisson distribution with parameter λ. [P. Harremoës *et al.*, Thinning, entropy, and the law of thin numbers, *IEEE Trans. Inf. Theor.*, 56(9), 4228–4244, 2010]

law of total probability *see* partition theorem

lazy random walk A random walk on a graph G is said to be lazy if at any step the probability of remaining at the current vertex (i.e. not moving) is at least 1/2. Likewise, a lazy Markov chain with transition matrix $P = (p_{ij})$ is lazy if $p_{ii} \geq \frac{1}{2}$ for all states i. Such lazy walks and chains are aperiodic, and a reversible lazy Markov chain has that property that all the eigenvalues of P are real and positive. Any Markov chain with matrix P can be made lazy by considering the chain with transition matrix $L = \frac{1}{2}(P + I)$, where I is the identity matrix. [G. Lawler & V. Limic, *Random Walk: A Modern Introduction*. Cambridge University Press, 2010]

lcrl Denotes the property of being left-continuous with right limits (sometimes denoted as càglàd by francophones). *See* rcll process.

LDP *see* large deviations theory

Leadbetter's theorem Arises in considering extreme values, in particular the maximum, of a strictly stationary sequence of r.vs $X_n, n \geq 1$. Broadly, it asserts that (in general) the classical results of extreme value theorem for i.i.d. sequences remain true in this stationary case provided only that the stationary sequence X_n satisfies the so-called distributional mixing condition D (also called **Leadbetter's condition D**). This condition is framed in terms of $M_L = \max\{X_{i_1}, \ldots, X_{i_r}\}$ and $M_R = \max\{X_{j_1}, \ldots, X_{j_m}\}$, where $0 < i_1 < i_2 < \cdots < i_r < j_1 < \cdots < j_m \leq n$, and $j_1 - i_r > l_n$. Then X_n satisfies $D(u_n)$ if there is a sequence of constants u_n and a sequence l_n, with $\lim_{n\to\infty} l_n/n = 0$, such that $|\mathbb{P}(\{M_L \leq u_n\} \cap \{M_R \leq u_n\}) - \mathbb{P}(M_L \leq u_n)\mathbb{P}(M_R \leq u_n)| \to 0$ as $n \to \infty$. *See also* extremal index. [M. Leadbetter, Extremes and local dependence in stationary sequences, *Probab. Theor. Rel. Fields*, 65, 291–306, 1983]

Lebesgue decomposition Arises in measure theory, and appears in two forms in probability theory:

(a) Decomposition of measure. Let $P(\cdot)$ and $Q(\cdot)$ be sigma-finite measures (e.g. probability measures) on a suitable measure space $(\Omega, \mathcal{F})$. Then P can be written uniquely as $P = P_c + P_s$, the sum of two measures, where P_c is absolutely continuous as with respect to Q (denoted by $P_c \ll Q$) and P_s and Q are mutually singular (i.e. are concentrated on disjoint subsets of Ω). Moreover (*see*: Radon–Nikodym) $P_c(A) = \int_A f dQ$ for all $A \in \mathcal{F}$, and some $\mathcal{F}$-measurable function f.

(b) Decomposition of distribution functions. Let $F(x)$ be a (possibly improper) distribution function on $\mathbb{R}$ (and thus of bounded variation and right-continuous); then F can be written uniquely as the sum of three (possibly improper) distribution functions $F = F_c + F_j + F_s$,, where F_c is absolutely continuous for Lebesgue measure, and $F_j + F_s$ is singular for Lebesgue measure; moreover, F_j is a discrete distribution (concentrated on a countable number of points of $\mathbb{R}$), and F_s, called the singular part of F, is continuous with $F'_s(x) = 0$ almost everywhere (though not constant).

See also Lévy–Itô decomposition.

[O. Kallenberg, *Foundations of Modern Probability*, 2nd edn. Springer, 2002]

Lebesgue's dominated convergence theorem *see* dominated convergence theorem

Lebesgue's function *see* Cantor distribution

Lebesgue measure Arises on the real line $\mathbb{R}$ (or more generally on $\mathbb{R}^n$) as a generalization of the concept of the length *b-a* of an interval $(a, b]$ (or the volume of a box in $\mathbb{R}^n$) to a larger family of sets $\mathcal{F}$. It comprises a countably additive measure $\lambda(S)$ on $S \in \mathcal{F}$, where $\mathcal{F}$ includes all the Borel sets $\mathcal{B}$, and also all sets $A \subseteq C \in \mathcal{B}$ where $\lambda(C) = 0$, together with all unions of such A and Borel sets $\mathcal{B}$, so that $\mathcal{F}$ is a sigma-algebra. (There are other constructions of $\lambda(\cdot)$.) Then Lebesgue measure provides a natural representation for the expected values of absolutely continuous random variables X having a density $f_X(x)$ with respect to Lebesgue measure, for then (either directly, or by the Radon–Nikodym theorem)

$$\mathbb{E}g(X) = \int f_X(x) g(x) \mathrm{d}x,$$

whenever $g(x)$ is integrable for the probability measure $\mathbb{P}_X(x)$ of X. For a random variable X whose distribution $F(x)$ has a singular part (*see* Lebesgue decomposition) one introduces the so-called Lebesgue–Stieltjes measure. This is defined similarly, except that the primary measure of an interval $I = (a, b]$ is given by $\lambda(I) = F(b) - F(a)$. Then for integrable $g(\cdot)$ we write $\mathbb{E}g(X) = \int g(x) \mathrm{d}F(x)$. An extension of Lebesgue measure to fractional dimensions is called Hausdorff measure. [M. Capiński & E. Kopp, *Measure, Integral and Probability*, 2nd edn. Springer, 2004]

Lebesgue–Radon–Nikodym theorem *see* Radon–Nikodym theorem

Le Cam contiguity *see* contiguity

Le Cam distance (alias: **Vincze–Le Cam distance**) A type of f-divergence between probability distributions $\mathbb{P}$ and $\mathbb{Q}$. When their densities f and g exist (e.g. for Lebesgue measure on $\mathbb{R}$), it takes the form

$$d(\mathbb{P}, \mathbb{Q}) = d(f, g) \propto \left\{ \int \frac{(f-g)^2}{(f+g)} \mathrm{d}x \right\}^{\frac{1}{2}}.$$

[I. Vajda, On metric divergences of probability measures, *Kybernetika*, 45(6), 885–900, 2009]

Le Cam's inequality Any one of a number of inequalities arising in Poisson approximation; most often this one: Let $S_n = \sum_{r=1}^{n} I_r$ be the sum of independent indicator (Bernoulli) r.vs, such that $\mathbb{E}I_r = \mathbb{P}(I_r = 1) = p_r$, and $\sum_{r=1}^{n} p_r = \lambda$. Then

$$\frac{1}{2}\sum_{r=0}^{\infty} |\mathbb{P}(S_n = r) - \mathrm{e}^{-\lambda}\lambda^r/r!| < \sum_{r=1}^{n} p_r^2.$$

[J. Steele, Le Cam's inequality and Poisson approximations, *Amer. Math. Mon.*, 1994, 101(1), 48–54].

Ledermann–Reuter representation (or **construction**) Arises in the context of Markov birth-and-death processes (and more generally Markov chains) in continuous time, where it supplies an integral representation for the transition matrix of the process through a truncation of the state space, followed by taking limits. *See* Karlin–McGregor theorem, and Kendall's representation. [W. Lederman & G. Reuter, Spectral theory for the dasferential equations of simple birth and death processes, *Philos. Trans. Roy. Soc. Lond. A*, 246, 321–369, 1954]

Legendre–Fenchel transform *see* Cramér's transform

Lehmer's generator A type of linear congruential generator of integer-valued pseudo-random numbers. It takes the form $X_{r+1} = mX_r$ modulo $s, r \geq 0$, where s is a suitable large prime number (or power of a prime) that is coprime with X_0; and $\min\{r : m^r = 1 \text{ modulo } s\}$ is sufficiently large. Division by s yields a pseudo-random sequence in $[0, 1]$. [S. Park & K. Miller, Random number generators: good ones are hard to find, *Ass. Comp. Mach.*, 31 (10), 1192–2001, 1988]

le montant Belge *see* Belgian progression

Lenglart inequality Broadly, this is a maximal inequality for an rcll random process $X(t)$ that is suitably dominated by some increasing process $A(t), t \geq 0$; it finds applications in martingale theory. One form is this: let $X(0) = A(0) = 0$ a.s., while $X(t)$ and $A(t)$ are adapted to the same filtration $\mathcal{F}_t$. If, for every finite stopping time T, $\mathbb{E}|X(T)| \leq \mathbb{E}A(T)$ (which is to say that X is Lenglart-dominated by A), then for any stopping time T, and $a, b > 0$,

$$\mathbb{P}\left(\sup_{0\leq s\leq T} |X(s)| \geq a\right) \leq \frac{1}{a}\mathbb{E}\min\{b, A(T)\} + \mathbb{P}(A(T) \geq b).$$

[J. Jacod & P. Protter, *Discretization of Processes*. Springer, 2012]

Lenglart's theorem *see* Cameron–Martin–Girsanov theorem

length-biased distribution *see* size-biased distribution

Lerch distribution A probability distribution on the positive (or non-negative) integers that can be seen as a generalized zeta (or Zipf) distribution. The first type has pmf

$$p(x) = \frac{p^x}{a(b+x)^c}, \quad x = 1, 2, 3, \ldots,$$

for suitable values of p, a, b, c, where

$$a = \sum_{r=1}^{\infty} \frac{p^r}{(b+r)^c},$$

which is related to the so-called Lerch transcendent (or Lerch zeta function), usually defined by

$$L(p, b, c) = \sum_{r=0}^{\infty} \frac{p^r}{(b+r)^c}.$$

The Lerch distribution on the non-negative integers is similarly defined in the obvious way in terms of L. [P. Zörnig & G. Altmann, Unified representation of Zipf distributions, *Comp. Stat. Data Anal.*, 19, 461–473, 1995]

Levin–Schnorr theorem *see* Schnorr's theorem

Lévy arcsine law *see* arcsine law

Lévy's area process (alias: **Lévy stochastic area process**) Arises in planar (two-dimensional) Brownian motion $(W_1(t), W_2(t))$, where W_1 and W_2 are independent standard Wiener processes. Then

$$L(t) = \int_0^t W_1(t)\mathrm{d}W_2(t) - \int_0^t W_2(t)\mathrm{d}W_1(t)$$

is twice the area swept out by the chord joining (W_1, W_2) to the origin, over $[0, t]$. The density of L is $f(x) = \{2t\cosh(\frac{\pi x}{2t})\}^{-1}$, $x \in \mathbb{R}$, corresponding to the chf $\mathbb{E}e^{i\theta L} = \{\cosh(t\theta)\}^{-1}$. [P. Lévy, Wiener's random function, and other Laplacian random functions, in *Proc. 2nd Berkeley Symp. Math. Stat. Probab.* University of California Press, 1951, pp. 171–187]

Lévy basis An infinitely divisible, independently scattered random measure, i.e. in detail:

Let $X(B), B \in \mathcal{B}$, be a collection of r.vs defined on the bounded Borel subsets of some subset D of $\mathbb{R}^d$. These comprise a Lévy basis if:

(a) $X(B)$ is infinitely divisible for every B; and
(b) for every collection $B_r, r \geq 1$, with $\cup_r B_r \in \mathcal{B}$, $X(\cup_r B_r) = \sum_r X(B_r)$, and also $X(B_1), X(B_2), \ldots$, are independent r.vs.

[K. Sato, *Lévy Processes and Infinitely Divisible Distributions.* Cambridge University Press, 1999]

Lévy central limit theorem *see* central limit theorem

Lévy chaos *see* chaos expansion

Lévy's characterization of Brownian motion Arises in martingale theory, and one form is this: let the random process $X(t), t \geq 0$, with $X(0) = 0$, generate the filtration $\mathcal{F}_t$. Then the following are equivalent:

(a) $X(t)$ is Brownian motion, i.e. the standard Wiener process;
(b) $X(t)$ is a continuous martingale for $\mathcal{F}_t$, with quadratic variation t;
(c) $X(t)$ and $X(t)^2 - t$ are continuous martingales for $\mathcal{F}_t$.

[H. Kunita & S. Watanabe, On square integrable martingales, *Nagoya Math. J.*, 30, 209–245, 1967]

Lévy concentration fn *see* concentration function

Lévy conformal invariance theorem *see* conformal invariance of Brownian motion

Lévy's continuity theorem *see* continuity theorem

Lévy–Cramér theorem *see* Cramér–Lévy theorem, and also continuity theorem

Lévy's dichotomy Asserts that for any two states i and j of a Markov chain in continuous time, with transition matrix $p_{ij}(t)$ (where $\lim_{t\downarrow 0} p_{ij}(t) = I$, the identity matrix), then exactly one of these holds:

(a) $p_{ij}(t) = 0, t > 0$;
(b) $p_{ij}(t) > 0, t > 0$.

[K. L. Chung, *Markov Chains With Stationary Transition Probabilities*, 2nd edn. Springer, 1967]

Lévy distribution (alias: **first-passage distribution of B.M.**) Arises as the density of the first-passage time T of the standard Wiener process to a point $x \neq 0$; i.e.

$$f(t) = \frac{|x|}{(2\pi t^3)^{\frac{1}{2}}} \exp\left(-\frac{x^2}{2t}\right), \quad t \geq 0.$$

It is the stable distribution with exponent 1/2, having chf $\phi(t) = \exp\left\{-|xt|^{\frac{1}{2}}\{1 - \mathrm{i}\,\mathrm{sign}\,t\}\right\}$ (*see* inverse Gaussian).

[W. Feller, *Introduction to Probability Theory and its Applications*, Vol. 2, 2nd edn. Wiley, 1971]

Lévy dust *see* Lévy flight

Lévy's equivalence theorem Asserts that for the partial sums $S_n = \sum_{r=1}^{n} X_r$ of a sequence $X_r, r \geq 1$, of independent random variables the following are equivalent:
(a) S_n converges with probability 1;
(b) S_n converges in probability;
(c) S_n converges in distribution.
Otherwise S_n diverges with probability 1. When $X_r \geq 0$ for all r, the assumption of independence is not necessary. *See also* Lévy's identity.
[R. Dudley, *Real Analysis and Probability*. Cambridge University Press, 2002]

Lévy–Feller diffusion (alias: **space-fractional diffusion**) Arises in models for anomalous diffusion (yielding, for example, advection-dispersion equations). It may be obtained as a suitable limit of random walks, and its evolution is governed by a fractional diffusion equation in which the usual second-order differential operator in the space variable is replaced by a Riesz–Feller potential (or pseudo-differential) operator.
[R. Gorenflo, Fractional diffusion: probability distributions and random walk models, *Physica A*, 305, 106–112, 2002]

Lévy–Feller–Khinchin theorem *see* central limit theorem

Lévy flight Broadly, a type of random flight in $\mathbb{R}^d$ in which the variance of the magnitude of any step is infinite. It is constructed sequentially from the origin. For the first step, first choose a direction uniformly at random and then step a distance X_1 in that direction, where X_1 is independent of the direction and X_1 has a distribution whose tail follows, either exactly or asymptotically, a power law, i.e. as $x \to \infty$,

$$\mathbb{P}(X_1 > x) = \mathrm{O}(x^{-m}),\ 0 < m < 2.$$

Successive steps are chosen independently, according to the same distribution. The special case when X_1 has a Cauchy density may be called a Cauchy flight. The set of end-point of steps (or corners of the walk) are said to be a **Lévy dust**.

In several applications it is convenient to attach to the set of line-segments between corners a sequence of independent r.vs whose distribution depends on the length of the edge to which they are attached. These represent the time taken by a particle starting at the origin to traverse the edges in order. The path of such a particle is called a **Lévy drive**; in the special case of constant speed (degenerate r.vs equalling the edge length) it is a **Lévy walk**.

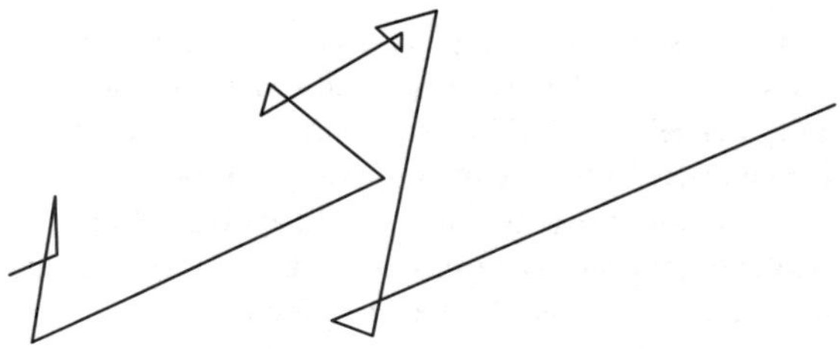

Lévy flight in the plane, illustrating the occasional big jumps typical of this type of random walk.

[M. Shlesinger *et al.*, *Lévy Flights and Related Topics in Physics*. Springer, 1995]

Lévy's fractional Brownian motion *see* fractional Brownian motion

Lévy–Hida–Cramér representation *see* Hida–Cramér canonical representation

Lévy's identity Arises in the context of Brownian motion i.e. the standard Wiener process $W(t), t \geq 0$. Let $M(t) = \sup_{0 \leq s \leq t} W(s)$ be the maximum of the process up to time t, and let $L(t)$ be the Lévy local time at zero of $W(t)$. Then the identity asserts that the vector random processes $(S(t) - W(t), S(t))$ and $(|W(t)|, L(t))$ have the same finite dimensional distributions (or law). [D. Revuz, & M. Yor, *Continuous Martingales and Brownian Motion*, 3rd edn. Springer, 1999 (Theorem 2.3, p. 240)]

Lévy–Itô decomposition Arises in the context of random processes with stationary independent increments (i.e. Lévy processes), where it asserts that the sample paths of such a process $X(t) \in \mathbb{R}^d, X(0) = 0$, can be written as the linear combination of a non-random drift, a standard Wiener process $W(t) \in \mathbb{R}^d$, and an independent Poisson process $N(t, x)$ on $\mathbb{R}^+ \times (\mathbb{R}^d - \{0\})$. [D. Applebaum, *Lévy Processes and Stochastic Calculus*, 2nd edn. Cambridge University Press, 2009]

Lévy–Khinchin formula (alias: **Lévy–Khinchin canonical form (or representation)**) Arises in characterizing infinitely divisible distributions, where it supplies an expression for the characteristic function $\phi(t)$ (or the Laplace transform, in suitable cases) of any such distribution in the form,

$$\log\phi(t) = \mathrm{i}at - \frac{1}{2}\sigma^2 t^2 + \int_{\mathbb{R}\setminus\{0\}} (\mathrm{e}^{\mathrm{i}tx} - 1 - \mathrm{i}txI\{|x| \le 1\})m(\mathrm{d}x),$$

for some $a \in \mathbb{R}$, $\sigma^2 \ge 0$, and measure $m(\cdot)$ such that $\int_{\mathbb{R}\setminus\{0\}} \min\{1, x^2\}m(\mathrm{d}x) < \infty$.

Conversely, any such ϕ is the chf of an infinitely divisible r.v. X. Here $m(x)$ is called a **Lévy measure** (or **Lévy kernel**). The RHS of the formula is called the **Lévy (or characteristic) exponent** or **Lévy symbol**, and the set $(a, \sigma^2, m(\cdot))$ may be called the **Lévy–Khinchin triplet** (or **characteristics**). When X is non-negative (and may be called a subordinator) the formula is commonly given in terms of the Laplace transform

$$\mathbb{E}\mathrm{e}^{-tX} = \exp\left\{-bt - \int_0^\infty (1 - \mathrm{e}^{-xt})m(\mathrm{d}x)\right\}.$$

[D. Applebaum, *Lévy Processes and Stochastic Calculus*, 2nd edn. Cambridge University Press, 2009]

Lévy's lemma (alias: **spherical isoperimetric inequality**) Arises in concentration of measure on the n-dimensional sphere S^n. Let $f(x) : S^n \to \mathbb{R}$ be a function with Lipschitz constant λ (i.e. $|f(x) - f(y)| \le \lambda||x - y||$ for all $x, y \in S^n$, where $||\cdot||$ denotes geodesic distance), and let X be uniformly distributed on S^n. Then

$$\mathbb{P}\left(|f(X) - \bar{f}| \ge \varepsilon\right) \le 2\exp\left\{\frac{-(n-1)\varepsilon^2}{2\lambda^2}\right\},$$

$\varepsilon > 0$, where $\bar{f}$ is any median of $f(X)$. [M. Ledoux, *The Concentration of Measure Phenomenon*. American Mathematical Society, 2001]

Lévy local time *see* local time

Lévy's maximal inequality Arises in considering the sum $S_n = \sum_{r=1}^n X_r$ of independent symmetric r.vs, where it asserts that for $t > 0$,

$$\mathbb{P}(\max_{1\le n\le m} |S_n| > t) \le 2\mathbb{P}(|S_m| > t).$$

Using the tail integral formula for expectation gives $\mathbb{E}\max_{1\le n\le m}|S_n| \le 2\mathbb{E}|S_m|$. For r.vs $X_r, r \ge 1$, that are independent but not necessarily symmetric, one has the more general form

$$\mathbb{P}\left(\max_{1\le n\le m} |S_n - \mu(S_n - S_m)| \ge t\right) \le 2\mathbb{P}(|S_m| \ge t),$$

where $\mu(Z)$ denotes a median of the r.v. Z. *See also* Doob's maximal inequality. [M. Loève, *Probability Theory*, 4th edn. Springer, 1978]

Lévy measure *see* Lévy–Khinchin formula

Lévy metric Defines a distance on the space of distribution functions of real-valued r.vs. Let $F(x)$ and $G(x), x \in \mathbb{R}$, be distribution functions; then the Lévy metric is

$$d(F, G) = \inf\{t > 0 : G(x - t) - t \le F(x) \le G(x + t) + t \text{ for all } x \in \mathbb{R}\}.$$

More generally, for probability measures $\mathbb{P}$ and $\mathbb{Q}$ on a metric space Ω, the **Prokhorov** (or **Lévy–Prokhorov**) **metric** is

$$d(\mathbb{P}, \mathbb{Q}) = \inf\{t > 0 : \mathbb{P}(B) \le \mathbb{Q}(B_t) + t \text{ for all Borel sets } B \text{ of } \Omega\},$$

where B_t is the t-extension (or t-neighbourhood, or t-fattening) of B, i.e. $B_t = \cup_{x\in B} B_t(x)$, the union of all open balls of radius t whose centres are in B. [P. Billingsley, *Convergence of Probability Measures*, 2nd edn. Wiley, 1999]

Lévy's modulus of continuity Arises in the context of Brownian motion (i.e. the standard Wiener process $W(t), t \ge 0$), where it supplies a non-random modulus of continuity for the a.s. continuous sample paths. Formally, let $\lambda(h) = \left(2t \log \frac{1}{t}\right)^{\frac{1}{2}}$; then for $s, t \in [0, 1]$, such that $|t - s| \le h$,

$$\lim_{h\downarrow 0} \sup_{s,t}\{|W(t) - W(s)|/\lambda(h)\} = 1,$$

with probability 1; thus $\lambda(h)$ is the Lévy (or exact) modulus of continuity. [P. Mörters & Y. Peres, *Brownian Motion*. Cambridge University Press, 2010]

Lévy multiparameter Brownian motion *see* Brownian motion

Lévy–Ottaviani inequality A maximal inequality for the partial sums $S_n = \sum_{r=1}^n X_r$ of independent r.vs $X_r, r \ge 1$, which asserts that for $t > 0$,

$$\mathbb{P}\left(\max_{1\le r\le n} |S_r| > 3t\right) \le 3\max_{1\le r\le n} \mathbb{P}(|S_r| > t).$$

The same inequality with 4 twice replacing 3 has been called **Etemadi's inequality**. [P. Hitchenko &

S. Montgomery-Smith, Measuring the magnitude of sums of independent random variables, *Ann. Probab.*, 29(1), 447–466, 2001]

Lévy–Pareto distribution *see* stable distribution

Lévy process (alias: **stationary independent increments process**) A natural continuous-time analogue of random walks that comprises an important class of random processes, including the Wiener and Poisson processes. Broadly $X(t) \in \mathbb{R}^d, t \geq 0$, is a Lévy process if:
(a) $X(0) = 0$;
(b) the random variables $X(t_1), X(t_2) - X(t_1), \ldots, X(t_n) - X(t_{n-1})$ are independent for any choice of $0 < t_1 < \cdots < t_n < \infty$;
(c) the distribution of $X(t+s) - X(t), s, t \geq 0$, does not depend on t.
If (c) does not hold, then $X(t)$ is simply an additive process. Some authors use a narrower definition requiring a Lévy process to satisfy also:
(a) $X(t)$ is stochastically continuous, i.e. for $a > 0$, $\lim_{h\to 0}\mathbb{P}(|X(t+h) - X(t)| > a) = 0$; and
(b) $X(t)$ is right-continuous, with left limits, i.e. rcll.
The only a.s. continuous Lévy process is drifting Brownian motion; all Lévy processes have a Lévy–Itô decomposition as a sum of this and a Poisson process and a singular continuous process of jumps. The chf of any Lévy process $X(t)$ is of the form $\phi(\theta) = \exp\{-t\psi(\theta)\}$, where ψ is the Lévy characteristic (or symbol) of X, having a Lévy–Khinchin representation. Lévy processes are semi-martingales, and an important subclass are subordinators. The generalization in which the Lévy characteristic is state dependent is called a Lévy-type process.

By analogy with fBM, a Lévy process may be used to define **fractional Lévy motion** in terms of a suitable stochastic (Riemann–Liouville fractional) integral. Such independent increments processes in higher dimensions may be called **Lévy sheets**. [D. Applebaum, *Lévy Processes and Stochastic Calculus*, 2nd edn. Cambridge University Press, 2009]

Lévy–Prokhorov metric *see* Lévy metric.

Lévy property A term used to denote a Lévy process, and the Lévy dichotomy; it also arises in considering sums $S_n = \sum_{r=1}^{n} X_r$ of independent random variables $X_r, r \geq 1$. Such a sequence is said to have the Lévy property with constants (a, b) if for finite sets $A \subseteq B \subseteq \mathcal{N}$, and all $t > 0$,

$$\mathbb{P}\left(\left|\sum_{r\in A} X_r\right| > at\right) \leq b\mathbb{P}\left(\left|\sum_{r\in B} X_r\right| > t\right).$$

[P. Hitchenko & S. Montgomery-Smith, Measuring the magnitude of sums of independent random variables, *Ann. Probab.*, 29(1), 447–466, 2001]

Lévy's representation *see* Lévy–Khinchin

Lévy's representation of excursions *see* Brownian excursion

Lévy sheet *see* Lévy process

Lévy stable distribution *see* stable distribution

Lévy symbol *see* Lévy process

Lévy's theorem Any of a large number of results, but perhaps this term most often refers to Lévy's characterization thm, Lévy's continuity theorem, Lévy's equivalence thm, the conformal invariance theorem, or Lévy's maximal inequality.

Lévy-type process *see* Lévy process

Lévy's upward theorem (alias: **Lévy's zero-or-one law**) Arises in the context of an increasing sequence of sigma-fields $\mathcal{F}_n, n \geq 1$. Defining the σ-field generated by the sequence as $F_\infty = \sigma(\cup_n \mathcal{F}_n)$, the theorem asserts that for an integrable r.v. X (defined on a suitable probability space), the sequence $\mathbb{E}(X \mid \mathcal{F}_n), n \geq 1$, is a uniformly integrable martingale (*see* Doob–Lévy martingale) that converges to $\mathbb{E}(X \mid \mathcal{F}_\infty)$ a.s. and in mean. The corresponding results for a decreasing sequence of sigma-fields is Lévy's downward theorem. The alias arises on considering any event $A \in \mathcal{F}_\infty$, which yields $\mathbb{E}(I(A) \mid \mathcal{F}_n) \to I_A$, a.s. [L. Rogers & D. Williams, *Diffusions, Markov Processes and Martingales*, Vol. 1, 2nd edn. Wiley, 1994; reprinted by Cambridge University Press, 2000; R. Durrett, *Probability, Theory and Examples*, 4th edn. Cambridge University Press, 2010]

Lévy walk *see* Lévy flight

Lévy's zero-or-one law *see* Lévy's upward theorem

Liapunov inequality *see* Lyapunov's inequality

life table (alias: **actuarial (or mortality) table**) Arises in actuarial mathematics, where it supplies the chances of survival at various ages (commonly integer years) for the members of a given population, and (usually) other related quantities. In its basic form it gives, for an individual age t, the empirical chance of survival to age $t+1$. From this a variety of other probabilities and expectations, e.g. the probability of survival to $t+k$ for any k, or the expected residual life at age t, may be computed. There are several standard forms of life table; see the early examples of such tables opposite. [M. Greenwood, The first life table, *Notes and Records Roy. Soc. Lond.*, 1(2), 70–72, 1938]

lifetime Arises in several branches of probability, where the term is inherited from some early or important application, and it commonly denotes the size of members of some class of random variables, e.g.:

- the times between the events of renewal process;
- the time from birth of an individual in a branching process until being replaced by a family;
- the time until a component fails in reliability and ageing;
- the duration of a random process until it is killed or explodes or reaches the (absorbing) boundary of its state space.

Lifshits–Shi theorem Asserts that the exit time T of planar Brownian motion from inside the parabolic domain $\{(x,y) : y > x^2\}$ satisfies $\lim_{t\to\infty} t^{-\frac{1}{3}} \log \mathbb{P}(T > t) = -3\pi^2/8$. [M. Lifshits & Z. Shi, First exit time of Brownian motion from a parabolic domain, *Bernoulli*, 8(2), 745–765, 2002]

lifted Markov chain Let X_n and $Z_n, n \geq 1$, be Markov chains on their respective state spaces S and U, and $g(\cdot)$ a surjective map from U to S, that induces a family of disjoint inverse sets, $L_s = g^{-1}(s) \subseteq U, s \in S$. The chain X_n is said to be lifted (or to admit a lifting, or to be liftable) to Z_n if the random process generated by transitions of Z_n among the subsets L_s is a Markov chain whose transition matrix is that of X_n. (*See* lumping.) [O. Chan & T. Lam, Lifting Markov chains to random walks on groups, *Combin. Probab. Comp.*, 14, 269–273, 2005]

likelihood Broadly, the term is sometimes used as a synonym for probability. More narrowly, it arises in decision theory and statistics when considering a family $f(x,\theta)$ of probability distributions depending on a parameter $\theta \in \Theta$, where Θ may be seen as states of nature, or competing hypotheses, etc., depending on the context. Then the corresponding likelihood function is defined by $l_x(\theta) \propto f(x,\theta)$, where the constant of proportionality is arbitrary, and $l_x(\theta)$ is a function of θ for given fixed x, where x is now seen as a parameter. The **likelihood ratio** $l_x(\theta_1)/l_x(\theta_2)$ is often useful. [A. W. F. Edwards, *Likelihood*. Cambridge University Press, 1972]

likelihood ratio dependence A measure of association (or dependence) between jointly continuous r.vs X and Y, defined in terms of the likelihood ratio of conditional densities:

$$\frac{f(y'|x)}{f(y|x)}, \quad y < y'.$$

If this is monotone non-decreasing as x increases, then X and Y (or their joint distribution), is said to be positively likelihood ratio dependent. For their joint density f this is equivalent to $f(x,y')f(x',y) \leq f(x,y)f(x',y')$ for $x < x', y < y'$. When the inequality is reversed, they are negatively likelihood ration dependent. [E. Lehmann, Some concepts of dependence, *Ann. Math. Stat.*, 37(5), 1137–1153, 1966]

likelihood ratio divergence *see* Itakura–Saito divergence; Kullback–Leibler divergence

likelihood ratio martingale Arises naturally in sequential decision theory, and also in the change of measure procedure for a suitable random process. In its simplest form, let $X_r, r \geq 1$, be i.i.d.r.vs with density $f(x)$ on $\mathbb{R}$, and let $g(x)$ be a density dominated by f (i.e. $g(x) = 0$ whenever $f(x) = 0$). Then the random process

$$R_n = \prod_{r=1}^{n} \frac{g(X_r)}{f(X_r)}, \quad n \geq 1, \quad \text{with } R_0 = 1,$$

is the likelihood ratio martingale. More generally, let $\mathbb{P}$ and $\mathbb{Q}$ be probability measures on a space $(\Omega, \mathcal{F})$, where $\mathbb{Q}$ is dominated by $\mathbb{P}$; and let $\mathcal{F}_n$ be an increasing sequence of sigma-fields in $\mathcal{F}$. Then the sequence R_n of Radon–Nikodym derivatives

$$R_n = \frac{d\mathbb{Q}}{d\mathbb{P}}\Big|_{\mathcal{F}_n}$$

is a martingale on $(\Omega, \mathcal{F}, \mathbb{P})$, adapted to $\mathcal{F}_n$. [T. Rolski *et al.*, *Stochastic Processes for Insurance and Finance*. Wiley, 1999]

likelihood ratio ordering Random variables X and Y with densities (or probability mass functions) $f_X(x)$ and $f_Y(x)$ are said to be likelihood ratio ordered if $f_X(x)/f_Y(x) \leq f_X(y)/f_Y(y)$ for all $x \leq y$; and then we say that X is larger than Y in likelihood ratio. It implies the

Two of several life tables given by A. De Moivre in *Annuities on Lives* (1752).

TABLE I, by Dr. *Halley*.

Age	Living	Age	Living	Age	Living	Age	Living	Age	Living	Age	Living
1	1000	16	622	31	523	46	387	61	232	76	78
2	855	17	616	32	515	47	377	62	222	77	68
3	798	18	610	33	507	48	367	63	212	78	58
4	760	19	604	34	499	49	357	64	202	79	* 49
5	732	20	598	35	*490	50	*346	65	192	80	41
6	710	21	592	36	481	51	335	66	182	81	34
7	692	22	586	37	472	52	324	67	172	82	28
8	680	23	580	38	463	53	313	68	162	83	23
9	670	24	574	39	454	54	302	69	152	84	19
10	661	25	*567	40	445	55	*292	70	142	*	*
11	653	26	560	41	436	56	282	71	*131		
12	646	27	553	42	427	57	272	72	120		
13	*640	28	546	43	*417	58	262	73	109		
14	634	29	539	44	407	59	252	74	98		
15	628	30	*531	45	397	60	242	75	*88		

TABLE II. by M. *Kersseboom*.

Age 0	Living 1400	Age	Living	Age	Living	Age	Living	Age	Living	Age	Living	Age	Living
1	1125	16	849	31	699	46	550	61	369	76	160	91	7
2	1075	17	842	32	687	47	540	62	356	77	145	92	5
3	1030	18	835	33	675	48	530	63	343	78	130	93	3
4	993	19	826	34	665	49	518	64	329	79	115	94	2
5	964	20	817	35	655	50	507	65	315	80	100	95	1
6	947	21	808	36	645	51	495	66	301	81	87	96	0.6
7	930	22	800	37	635	52	482	67	287	82	75	97	0.5
8	913	23	792	38	625	53	470	68	273	83	64	98	0.4
9	904	24	783	39	615	54	458	99	259	84	55	99	0.2
10	895	25	772	40	605	55	446	70	245	85	45	100	0.0
11	886	26	760	41	596	56	434	71	231	86	36	*	
12	878	27	747	42	587	57	421	72	217	87	28		
13	870	28	735	43	578	58	408	73	203	88	21		
14	863	29	723	44	569	59	395	74	189	89	15		
15	856	30	711	45	560	60	382	75	175	90	10		

usual stochastic order, but the converse does not hold. [M. Shaked & J. Shanthikumar, *Stochastic Orders and their Applications.* Academic Press, 1994]

lily-pond model Arises in geometrical probability as a germ-grain model (specifically a growth-maximal hardcore process) that comprises a random process of circles in $\mathbb{R}^2$ (or balls in $\mathbb{R}^d, d > 2$). It is constructed thus: given a spatial point process at time $t = 0$ (e.g. a Poisson process of points), each point P initiates the growth of a circle (or sphere) centred at P, each growing at constant rate (or, more generally, i.i.d. random rates). For any circle (or sphere) growth ceases at the first instant when it touches another, and the lily-pond process appears when all growth ceases. [G. Last & M. D. Penrose, Percolation and limit theory for the lily-pond model, *Random Struct. Alg.*, 42(2), 226–249, 2013]

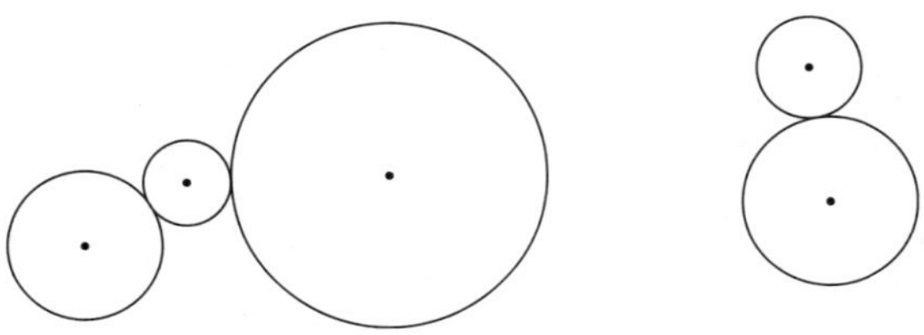

Lily-pond model.

Lin condition *see* Krein condition; moment problem

Lindeberg–Feller theorem (alias: **Lindeberg–Feller–Lévy theorem (or CLT)**) Arises in the field of central limit theorems, CLTs, where it gives necessary and sufficient conditions for a CLT to hold in the case of sums of independent, finite-variance r.vs, in terms of the so-called Lindeberg condition. Formally, let $X_k, k \geq 1$, be independent zero-mean (wlog) r.vs with distn fns $F_k(x)$, variances $\sigma_k^2, k \geq 1$, and set $s_n^2 = \sum_{k=1}^n \sigma_k^2 > 0$ for some finite n. The Lindeberg condition L holds if for each $a > 0$

$$L : \lim_{n\to\infty} \sum_{k=1}^{n} \int_{|x|>as_n} \left(\frac{x}{s_n}\right)^2 \mathrm{d}F_k(x) = 0.$$

In an alternative form of L, $|x| > as_k$ is the range of integration; in either case L entails that $\lim_{n\to\infty} s_n = \infty$. Then the Lindeberg CLT asserts that, under L, $|s_n|^{-1}\sum_{k=1}^n X_k$ converges in distribution to the standard normal distn. Conversely (proved independently by P. Lévy and W. Feller), if $|s_n|^{-1}\sum_{k=1}^n X_k$ converges in distn to $N(0,1)$, and $\lim_{n\to\infty} \frac{\sigma_n}{s_n} = 0$, then the condition L holds. Either form of L entails that

$$\hat{L} : \lim_{n\to\infty} \max_{1\leq k\leq n} \left(\frac{\sigma_k}{s_n}\right)^2 \to 0,$$

and this also has been called a Lindeberg condition. [L. Le Cam, The central limit theorem around 1935, *Stat. Sci.*, 1(1), 78–96, 1986]

Lindley's equation (alias: **Lindley process; Lindley recursion**) Arises naturally in the context of the classic single server queue (GI/G/1, in Kendall's notation), where it can be seen as describing the evolution of the (embedded) queue-length process, or as the process of waiting times of successive customers. Formally, it is a discrete-time random process W_n generated by the recursion

$$W_{n+1} = \max\{0, W_n + X_n\}, \quad n \geq 0,$$

with $W_0 = 0$, where $X_n, n \geq 1$, is a suitable sequence of i.i.d.r.vs. More generally, for the queue GI/G/s with several servers, the multivariate Lindley equation may be called a **Kiefer–Wolfowitz recursion.** [S. Asmussen, *Applied Probability and Queues*, 2nd edn. Springer, 2003]

Lindley's information measure Arises in Bayesian inference, where it applies Shannon's mutual information to supply a measure of the information provided about the value of a parameter by an experiment. Let the prior distribution of the parameter θ, taking values in Θ, be $\pi(\theta)$, and the posterior distribution be $p(\theta\,|\,x)$, given the experimental outcome x of a r.v. X with distribution $p(x)$. Then the Lindley measure of the information about the parameter to be expected on observing X is

$$\int_{\theta\in\Theta}\int_{x\in\mathbb{R}} p(x,\theta) \log \frac{p(x,\theta)}{p(x)\pi(\theta)} \mathrm{d}x\mathrm{d}\theta,$$

where $p(x,\theta) = p(x\,|\,\theta)\pi(\theta)$. [D. Lindley, On a measure of the information provided by an experiment, *Ann. Math. Stat.*, 27(4), 986–1005, 1956]

linear failure rate *see* ageing

linear self-exciting process *see* Hawkes process

line process *see* geometrical probability

Linnik condition Arises in large deviations theory, where it supplies a condition on the moments of a typical summand that is weaker than Cramér's condition. It is satisfied by X if $\mathbb{E}\exp(|X|^{1/(1+a)}) \le C(a) < \infty, a > 0$, for some constant $C(a)$. (Slight variants of this are also called Linnik's condition.) [Y. Prokhorov & V. Statulevičius (eds), *Limit Theorems of Probability Theory*. Springer, 2000]

Linnik distribution (alias: **Linnik law; α-Laplace distribution; symmetric geometric-stable distribution**) A distribution having characteristic function $\phi(t) = \{1 + \sigma^\alpha |t|^\alpha\}^{-1}, \sigma > 0, 0 < \alpha \le 2$. For $\alpha = 2$ this is the Laplace distn. It is a special case of the geometric-stable class of distributions, and there is a multivariate Linnik distn whose chf in standard form is

$$\phi(t_1, \ldots t_n) = \{1 + (t_1^2 + \cdots + t_n^2)^{\alpha/2}\}^{-1}, \quad 0 < \alpha \le 2.$$

The generalized Linnik law has standard chf

$$\phi(t) = (1 + |t|^\alpha)^b, \quad 0 < \alpha \le 2, b > 0.$$

[A. G. Pakes, Mixture representations for symmetric generalized Linnik Laws, *Stat. Probab. Lett.*, 37(3), 213–221, 1998]

Linnik information (alias: **Linnik information functional, or entropy**) Arises in Y. Linnik's proof of the information central limit theorem as a differential entropy function having the property of invariance under linear (homothetic) transformations. For a zero-mean finite-variance r.v. X with density $f(x)$, it is given by

$$L(X) = -\int_{\mathbb{R}} f \log f \, dx - \frac{1}{2} \log \int_{\mathbb{R}} x^2 f dx$$
$$= H(X) - \frac{1}{2} \log \operatorname{var} X$$

where $H(X)$ is the Shannon entropy of X. Then for $a \ne 0$, $L(aX + b) = L(X)$. For a random (column) vector $\mathbf{X}, L(\mathbf{X}) = H(\mathbf{X}) - ½\log\{\det\mathbb{E}(\mathbf{X}\mathbf{X}^T)\}$.

A few authors have (very confusingly) given the name 'Linnik's information' to the Fisher information with respect to location parameter, i.e. $J(f) = \int (f')^2 f dx$. [Y. Linnik, An information-theoretic proof of the central limit theorem with the Lindeberg condition, *Theor. Probab. Appl.*, 4(3), 288–299, 1959]

Linnik's theorem Any of a number of theorems about the family I_0 of infinitely divisible characteristic functions that do not admit indecomposable factors, for example the assertion that any convolution of a normal distribution with a Poisson distribution is in I_0. [Y. Linnik, *Decomposition of Probability Distributions*. Oliver and Boyd, 1964]

Lin's theorem *see* Krein condition

Liouville distribution A multivariate probability distribution (related to the Dirichlet distribution) whose general form of density on $\mathbb{R}_+^n$ is (for suitable functions $g(\cdot)$) given by,

$$f(x_1, \ldots, x_n) \propto g(x_1 + \cdots + x_n) \prod_{r=1}^{n} x_r^{a_r - 1}$$

where $x_r > 0, a_r > 0$, and $\int_0^\infty g(y) y^{\sum_1^n a_r - 1} dy < \infty$, and the constant of proportionality is determined by Liouville's integral. For the discrete Liouville distn, the x_r take non-negative integer values. [R. Gupta & D. Richards, The history of the Dirichlet and Liouville distributions, *Int. Stat. Rev.*, 69(3), 433–446, 2001]

Lipschitz percolation Arises in considering site percolation on the integer lattice $\mathbb{Z}^d$, where it considers random surfaces defined as follows: a function $F : \mathbb{Z}^{d-1} \to \mathbb{Z}^+$ is Lipschitz if for any x and y in $\mathbb{Z}^{d-1}$ with $||x - y|| = 1$, we have $|F(x) - F(y)| \le 1$. Then a Lipschitz surface is the subset $\{(x, F(x)) : x \in \mathbb{Z}^{d-1}\}$ of $\mathbb{Z}^d$. Interest centres on the probability of the event that for each $x \in \mathbb{Z}^{d-1}$ the sites of the surface are all open. [G. Grimmett & A. Holroyd, Geometry of Lipschitz percolation, *Ann. d L'Inst. H. Poincaré, Prob. Stat.*, 48(2), 309–326, 2012]

Little's law (alias: **Little's theorem, formula, or lemma**) Arises in the theory of queues and similar systems in which a stream of arriving items (or customers) join a queue for service by a system, and at the conclusion of their waiting time in the system join the stream of departures. Broadly, Little's law asserts that the mean number of items L in the system equals the mean waiting time W of an item in the system multiplied by the mean arrival rate λ, i.e. $L = \lambda W$. The word 'mean' may be interpreted as a time average or an expectation, depending on the framework in which the law is formulated; this framework may be deterministic (in which case means are time averages over given sample paths) or it may be stochastic (in which case means are mathematical expectations arising in stationary random processes). If the stationary process is ergodic, then suitable limiting time-averages coincide with the expected values. The law holds under a remarkable range of conditions.

More generally, if items have a weighting of their waiting time (which may be interpreted as profit, cost, or workload, etc.) there is a more general formula relating average costs in the system and average cost per item. This is called **Brumelle's law** (or formula), and denoted by $\lambda G = H$. [J. Little, Little's law on its 50th anniversary, *Oper. Res.*, 59(3), 536–549, 2011]

L log L condition (or criterion) Arises naturally in various areas of probability, e.g. the Kesten–Stigum L log L theorem, where it is the condition on a r.v. X that $\mathbb{E}X \log_+ X$ is finite ($\log_+ x = \log x$ for $x > 0$, or 0 if $x = 0$). Thus those martingales X_n such that $\sup_n \mathbb{E}\{|X_n| \log_+|X_n|\} < \infty$ are said to comprise the class $L \log L$.

The concept arises also in ergodic theory and functional analysis, where measurable functions $f(x)$ such that $|f| \log_+|f|$ is integrable comprise **Zygmund's *L* log *L* class**. *See* Doob L_p inequality. [J. Doob, *Stochastic Processes*. Wiley, 1953; D. Gilat, The best bound in the L log L inequality of Hardy and Littlewood and its martingale counterpart, *Proc. Amer. Math. Soc.*, 97(3), 429–436, 1986]

local independence The term 'locally independent' has several meanings, including:

(a) Given point processes $N(t)$ and $M(t)$, on the same probability space, with associated filtrations $\mathcal{F}_t^N$ and $\mathcal{F}_t^M$, N is locally independent if M if, for all t, the mean intensity of N at t is $\mathcal{F}_t^N$-measurable.
(b) In a suitable type of multivariate Markov process, a component X_i is independent of a component X_j if the transition probability of X_i is a constant function of X_j.

It has also been used as an alias for some types of asymptotic (or limiting) independence, and for conditional independence, among other uses. [V. Didelez, Graphical models for marked point processes based on local independence, *J. Roy. Stat. Soc. B*, 70, 245–264, 2008; T. Schweder, Comparable Markov processes, *J. Appl. Probab.*, 7 (2), 400–410, 1970; D. Allard *et al.*, Testing local independence between two point processes, *Biometrics*, 57(2), 508–517, 2001]

local lemma *see* Lovász local lemma

local limit theorem Arises in considering the pointwise convergence, to a probability density $f(x)$, of a sequence of densities $f_n(x), n \geq 1, x \in \mathbb{R}$, or of a suitable sequence of lattice distributions $p_n(x), n \geq 1$. That is, a local limit theorem asserts that as $n \to \infty, p_n(x) \to f(x)$, uniformly in appropriate x, while the usual so-called global or integral limit theorems assert only that $\lim_{n\to\infty} \int_{-\infty}^{x} f_n(u)\mathrm{d}u = F(x)$, (at continuity points of F), where $F(x)$ need not even have a density. Classic examples include the De Moivre–Laplace CLT and Gnedenko's CLT. *See also* Shepp's local limit theorem. [P. Billingsley, *Convergence of Probability Measures*, 2nd edn. Wiley, 1999]

local Markov property *see* causal Markov property

local martingale A generalization of the concept of a martingale in which the defining martingale property holds locally, in the following sense: the random process $X(t), t \geq 0$ (on the filtered probability space or stochastic basis $(\Omega, \mathcal{F}, \mathcal{F}_t, \mathbb{P})$), is a local martingale if there is a sequence of stopping times $T_k, k \geq 1$ (called a localizing or localization sequence), increasing to ∞ with probability 1 as $k \to \infty$, such that each stopped process $X(t \wedge T_n)$ is a uniformly integrable martingale (for $\mathcal{F}_t$). They arise naturally as martingale transforms in discrete time, and stochastic integrals in continuous time, and the property of being a local martingale is preserved under these operations (while martingales are not necessarily so preserved). A local martingale $X(t)$ is a true martingale iff for all $t > 0$, the family of r.vs $X(T)$, where T runs over all stopping times less than t, is uniformly integrable. [F. Klebaner, *Introduction to Stochastic Calculus with Applications*, 3rd edn. Imperial College Press, 2012]

local non-determinism Arises in the context of Gaussian processes, with applications to Gaussian random fields and more general processes. Informally, a Gaussian process $X(t)$ has this property if small increments of the process are relatively unpredictable given only a finite set of observations from the recent past. Formally, the condition for local non-determinism on a set T of a zero mean Gaussian process $X(t)$ is

$$\lim_{c\downarrow 0} \inf_{t_m - t_1 \leq c} \frac{\operatorname{var}\{X(t_m) - X(t_{m-1})|X(t_1), \ldots, X(t_{m-1})\}}{\operatorname{var}\{X(t_m) - X(t_{m-1})\}} > 0,$$

for arbitrary $t_1 < t_2 < \cdots < t_m, t_r \in T$; where it is assumed that $\mathbb{E}X(t)^2 > 0$ and $\mathbb{E}(X(t) - X(s))^2 > 0$ for all s and t in T. A theorem of S. Berman asserts that this is equivalent to a so-called property of approximately independent increments. [Y. Xiao, Properties of local nondeterminism of Gaussian and stable random fields, and their applications, *Ann. Fac. Sci. Toulouse Math.*, 15, 157–193, 2006]

local stability *see* Papangelou conditional intensity

local stationarity Arises in the context of a random process $X(t)$ that is not stationary, but such that the distribution of $X(t)$ (and hence its moments, if any) vary slowly as time passes. Of several formulations, the usual definition asserts that a zero mean process $X(t), t \in \mathbb{R}$ is locally stationary if its covariance function takes the form $c(s,t) = \phi(s-t)\psi(s+t)$, where $\phi(\cdot)$ is the covariance function of a stationary process, and ψ is a suitable non-negative function. [M. Genton & O. Perrin, On a time deformation reducing non stationary stochastic processes to local stationarity, *J. Appl. Probab.*, 41(1), 236–249, 2004]

local time (alias: **Lévy local time; vicinity measure; occupation density; sojourn time**) Arises in the context of the Wiener process and other diffusions on $\mathbb{R}$ (and related process taking values in a continuous state space), as a measure of the proportion of time the process spends at each point $a \in \mathbb{R}$ during any time interval $[0, t]$, denoted by $L(t, a)$. The concept may be formalized and defined in a number of ways; the first was **Lévy local time** for the Wiener process $W(t)$, given by

$$L(t,a) = \lim_{\varepsilon \downarrow 0} \frac{1}{2\varepsilon} \int_0^t I(\{a - \varepsilon < W(s) < a + \varepsilon\}) \mathrm{d}s,$$

where $I(A)$, is the usual indicator that the event A occurs. The function $L(t, a)$ can be informally interpreted as a random clock that progresses only when $W(t)$ is in an arbitrarily small neighbourhood (or vicinity) of a.

The function $S_a(T) = \inf\{t \geq 0 : L(t,a) > T\}$ is the **inverse local time**. Local time of $W(t)$ relates in a natural way to the occupation time of any set A by the process $W(t)$ through the identity $\int_{\mathbb{R}} g(x) L(t,x) \mathrm{d}x = \int_0^t g(W(s)) \mathrm{d}s$ a.s., for suitable functions $g(x)$, e.g. $g(x) = I(\{x \in A\})$.

Another formalization of the concept of local time $L(t, 0)$ of the Wiener process arises from the Doob–Meyer decomposition of the submartingale $|W(t)|, t \geq 0$, as the sum of a martingale M_t and an increasing process A_t. Since $W(t)$ is a martingale, A_t can increase only when $W(t) = 0$, and it can be shown that $|W(t)| = \int_0^t \mathrm{sign}[W(s)] \mathrm{d}W(s) + L(t, 0)$, which is Tanaka's formula. Note that the set Z of times s such that $W(s) = 0$ has Lebesgue measure zero; in fact it can be seen as a type of random Cantor set, so that (with probability 1) Z is a closed set of uncountably many points, none of which are isolated (i.e. every point of Z is a limit point of points of Z). *See also* Ray–Knight theorem. [I. Karatzas & S. Shreve, *Brownian Motion and Stochastic Calculus*, 2nd edn. Springer, 1998; J. Pitman & M. Yor, Hitting, occupation, and inverse local times of 1-dimensional diffusions, *Bernoulli*, 9(1), 1–24, 2003]

location parameter Arises in the context of linear transformations, $Y = a + bZ$, of a r.v. Z, where the collection for $b > 0, a \in \mathbb{R}$, is called the location-scale class (or family) for Z, with location parameter a, and **scale parameter**, b. If Z has mean 0 and variance 1, then a and b are the mean and standard deviation of Y, and Z may be called the canonical member.

More generally, if zero is a uniquely specified median of Z (where Z may have no moments), then Y is centred at its median. If $Q(p)$ is a pth-order quantile for Z, then $a + bQ(p)$ is a pth-order quantile for Y; $Q(p)$ has also been called a location parameter for Y, as quantifying the location of its probability measure in $[0, 1]$. [T. Ferguson, Location and scale parameters in exponential families, *Ann. Math. Stat.*, 33(3), 986–1001, 1962]

Loéve criterion Arises in considering the mean-square convergence of r.vs $X_1, X_2, \ldots, X_n$, having zero means and finite variances. Then the criterion asserts that X_n converges in mean square, as $n \to \infty$, iff $\mathbb{E}(X_n, X_m)$ converges to a finite constant as m and n go independently to infinity. [G. Lindgren, *Stationary Stochastic Processes*, CRC Press, 2013]

Loéve's condition *see* harmonizable process

Logan–Shepp formula *see* Kac lemma

logarithmic distribution (alias: **log-series distribution**) The probability distribution defined on the positive integers, for $0 < p < 1$, by

$\mathbb{P}(X = k) = \frac{1}{|\log(1-p)|} \frac{p^k}{k}, \quad k \geq 1.$

Its pgf is $G(s) = \frac{\log(1-sp)}{\log(1-s)}$.

The term has also been used to denote Benford's distribution; or more widely any distribution (discrete or continuous) of the form $a \log(x + b)$, on some subset of $\mathbb{R}$. [D. Kendall, On some models of population growth leading to RA Fisher's logarithmic series distribution, *Biometrika*, 35(1/2), 6–15, 1948]

log-concave distribution A property of probability distributions, densities, and measures that has widespread applications, particularly in ageing and reliability. It appears in various forms, thus:

(a) The non-negative integer-valued r.v. X, whose support S is a sequence of successive integers, is log-concave if its distribution $p(x) = \mathbb{P}(X = x)$ obeys $p(x)^2 \geq p(x+1)\, p(x-1)$, for all $x \in S$; e.g. the binomial, negative binomial, geometric, and Poisson distns are log-concave. More strongly, X is **ultralog-concave** if $xp(x)^2 \geq (x+1)p(x+1)\, p(x-1), x \in S$.
(b) The density $f(x)$ of a r.v. X is log-concave on an interval $S = [a, b] \in R$ if $\log f$ is concave on S. **Prékopa's thm** asserts that for a continuously differentiable log-concave density f, the distribution $F(x)$ of X is also log-concave. Examples include the normal and gamma distns.

(c) A probability measure $\mathbb{P}$ on $\mathbb{R}^n$ is log-concave if for all Borel sets A and $B\ (\neq \emptyset)$, and $0 < t < 1$, $\mathbb{P}((tA + (1-t)B) \geq \mathbb{P}(A)^t\, \mathbb{P}(B)^{1-t}$. The extended Prékopa theorem (which may be called the Prékopa–Leindler inequality), for suitable integrals of their joint distribution, then shows that the marginals of jointly distributed log-concave r.vrs are log-concave, as also are the distributions of sums of independent log-concave r.vs. In the discrete case, this is **Hoggar's thm.**

The property of log convexity is defined analogously. [O. Johnson & C. Goldschmidt, Preservation of log-concavity on summation, *ESAIM Probab. Stat.*, 10, 206–215, 2006]

log gamma distribution The distribution of $Y = \exp X$, when the distribution of X is a gamma density. (This definition is made by analogy with the lognormal density, etc.) The density of Y is $f_Y(y) = \frac{\lambda^r}{\Gamma(r)} \frac{(\log y)^{r-1}}{y^{\lambda+1}}$, $y > 1$.

Note that (rather confusingly) the distribution of $Z = e^{-X}$, with support $(0, 1)$, has also been called the log gamma distribution; and furthermore the distribution of $W = \log X$, with support $\mathbb{R}$ has been called log gamma.

logical entropy *see* quadratic entropy

logical probability (alias: **inductive probability; epistemic probability**) Broadly, any interpretation and consequent axiomatization of probability that sees a probability as a degree of logical support. ('Epistemic' is from the Greek word meaning knowledge.) It assigns probabilities to statements or propositions of a suitable type of logic, and seeks to determine how far one set of statements supports (or confirms, or implies) another. It can be seen as a generalization of deductive reasoning, and seeks to supply a rational method for inductive reasoning in the presence of uncertainty.

It may be contrasted with other types of subjective probability, which interpret probability as a degree of belief which need not be based on formal logic (though theories normally involve some kind of logical consistency, as expressed e.g. by betting preferences avoiding certain loss). [R. Carnap, *Logical Foundations of Probability*. University of Chicago Press, 1950]

logistic distribution *see* hyperbolic secant distribution

log-logistic distribution The distribution of $\exp\{(X-a)/b\}, b \neq 0$, when the r.v. X has a logistic distribution (*see* hyperbolic secant distribution). It is also called the **Fisk distn**.

lognormal distribution The distribution of e^X, when the r.v. X has a normal distribution. It arises naturally in Kolmogorov's rock-crushing process, and in the geometric Wiener process ($\equiv$ exponential Brownian motion). It has the property that it is not determined uniquely by its moments; *see* Heyde's example.

The density of the lognormal distn is

$$f(x) = \frac{1}{(x-a)\sqrt{2\pi\sigma^2}} \exp\left\{-\frac{1}{2}\left\{\frac{\log(x-a)-\mu}{\sigma}\right\}^2\right\}, \quad x > a,$$

where $\sigma > 0$ and $\mu \in \mathbb{R}$. [E. Crow & K. Shimuzu (eds), *Log Normal Distribution: Theory and Applications*. Dekker, 1988]

log Sobolev inequality (alias: **Gross's logarithmic Sobolev inequality; Gross–Sobolev logarithmic inequality**) Arising originally in the context of quantum field theory, this has since found applications in concentration of measure inequalities. In its basic form, a probability measure $\mathbb{P}$ on $\mathbb{R}^d$ satisfies a log Sobolev inequality if, for any suitable real-valued differentiable function $g(\cdot)$ on $\mathbb{R}^d$, and some constant a, we have $\int g^2 \log g^2\ \mathrm{d}\mathbb{P} \leq a\{\int |\mathrm{grad}\ g|^2\ \mathrm{d}\mathbb{P} + \int g^2 \log(\int g^2\ \mathrm{d}\mathbb{P})\mathrm{d}\mathbb{P})\}$, where grad g is the usual gradient vector, often denoted by ∇g. In the original formulation by L. Gross, $\mathbb{P}$ is a d-dimensional Gaussian (or normal) probability measure: $\mathrm{d}\mathbb{P}(x) = (2\pi)^{-d/2}\ \exp\{-|\mathbf{x}|^2/2\}d\mathbf{x}$; there is a substantially wider class of measures that have the same property. [L. Gross, Logarithmic Sobolev inequalities, *Amer. J. Math.*, 97(4), 1061–1083, 1975; M. Ledoux, Concentration of measure and logarithmic Sobolev inequalities, in *Séminaire de Probabilités*, XXXIII. Springer, 1999]

long-range dependence (or memory) (alias: **Joseph effect; long-range persistence**) Many empirical time series spend long periods entirely on one side of their mean value, with correspondingly rare crossings from one side to the other. Theoretical formalizations of this property, in which distantly separated terms exhibit marked dependence, vary according to the underlying model. Thus, a stationary time series $X_n, n \in \mathbb{Z}$, with autocorrelation $r(n)$, has long (or very long)-range dependence (or memory) if, as $n \to \infty, r(n) \sim n^{1-a}$ for some $a \in (0, 1)$, so that $\sum_{-\infty}^{\infty} |r(n)|$ diverges and the spectral distribution has a jump at zero; it has intermediate (or long) memory if $r(n) \sim n^{1-a}$ as $n \to \infty$, for $a < 0$. For an autoregressive moving-average process, $r(n) \sim b^n$ for some $b \in (0, 1)$, and one speaks of a short-memory process; while white noise has no memory at all. In another model, fBm (or other self-similar process) has long memory when its Hurst exponent H satisfies $H \in \left(\frac{1}{2}, 1\right)$. [P. Doukhan *et al.* (eds), *Long-range Dependence*. Birkhäuser, 2003; C. Heyde & Y. Yang, Defining long-range dependence, *J. Appl. Probab.*, 34 (4), 939–944, 1997; G. Samorodnitsky, Long range dependence, *Found. Trend. Stoch. Syst.*, 1(3), 163–251, 2006]

Longstaff–Schwartz model Arises in financial mathematics as a two-factor model for the short-term interest rate $r(t) = \mu X(t) + \theta Y(t)$, where $\mathrm{d}X = (a - bX)\mathrm{d}t + c\sqrt{X}\mathrm{d}W_1$, $\mathrm{d}Y = (d - eY)\mathrm{d}t + f\sqrt{Y}\mathrm{d}W_2$, for positive $a, b, c, d, e, f, \theta, \mu$, and uncorrelated Wiener processes W_1 and W_2. It may be reformulated as a pair of SDEs for $r(t)$ and its volatility $V(t)$. [F. Longstaff & E. Schwartz, Interest rate volatility and the term structure: a two-factor general equilibrium model, *J. Finan.*, 47(4), 1259–1282, 1992]

long tail *see* tail

look back option *see* exotic option

loop-erased random walk (alias: **Lawler's self-avoiding walk**) A random walk (on a graph G) whose paths are simple, i.e. no path visits any vertex twice. Such a walk is obtained from any ordinary random walk (X_n) (that is, one that steps to neighbouring vertices according to some prescribed distribution) as follows: the steps of the walk X_n are taken in order until the walk first revisits a vertex V. The loop in the path so formed (i.e. all steps between successive visits to V) is erased, and the walk then continues likewise with erasure of loops as they are formed. If the walk is simple random walk (i.e. steps have a uniform distribution on neighbours) then for any two vertices U and V, the path of the loop-erased walk started at u and stopped at v has the same distribution as the path from u to v obtained by selecting a spanning tree of G uniformly at random from all such spanning trees. This Lawler's walk may also be represented (i.e. the same distribution on the paths obtained) by a **Laplacian r.w.** [G. Lawler *et al.*, Conformal invariance of loop-erased random walks and uniform spanning trees, *Ann. Probab.*, 32(1B), 939–995, 2004]

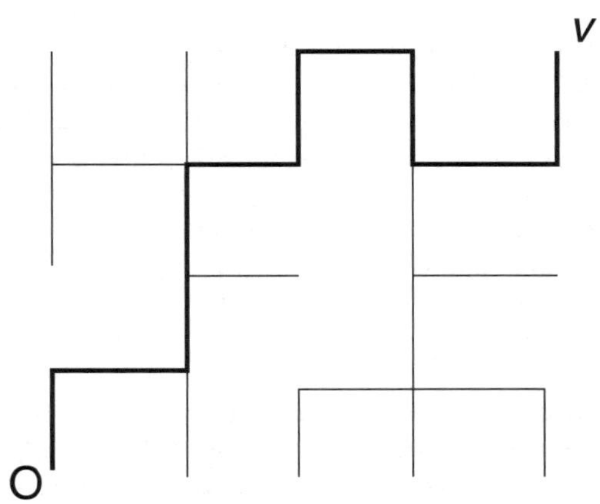

Uniform spanning tree on a small part of the square lattice. There is a unique (in bold) path from 0 to V, and such paths have the distribution of loop-erased random walk.

Lorden's inequality Any one of several inequalities arising in the context of a renewal process $N(t) = \max\{n : S_n \le t\}$, where $S_n = \sum_{r=1}^{n} X_r$ is the sum of i.i.d. positive r.vs with finite mean $\mu = \mathbb{E}X_1$. The excess (or residual) life, or overshoot, at t is $B(t) = S_{N(t)+1} - t$, and the inequalities supply bounds for the moments, and tail of the distribution, of $B(t)$. Thus, e. g. for $r \ge 2$,

$$\mathbb{E}B^r \le \frac{r+2}{r+1} \frac{\mathbb{E}X_1^{r+1}}{\mathbb{E}X_1};$$

and in particular $\mathbb{E}B \le \mathbb{E}(X_1^2)/\mathbb{E}X_1$. [G. Lorden, On excess over the boundary, *Ann. Math. Stat.*, 41, 520–527, 1970]

Lorentz distribution *see* Cauchy distribution

Lorentz gas model A type of random flight in a random environment that arises as a model for the motion of an electron through a field of much more massive particles. One basic form places strictly convex congruent obstacles (that never intersect) randomly at the vertices of a suitable lattice. A particle is then projected in this field with unit velocity, and reflected from any object it strikes according to the rules of optics. Simplified versions in two dimensions, where the objects may be two-sided mirrors, or squares, are sometimes called stochastic pin-ball; *see also* Ehrenfest wind-tree model. [G. Grimmett, Stochastic pin-ball, in *Random Walks and Discrete Potential Theory*, M. Picardello & W. Woess (eds), Symposia Mathematica, 39. Cambridge University Press, 1999, pp. 205–213]

Lorenz curve Arises naturally as an empirical measure of concentration when something (e.g. money), is distributed to n recipients (or sites). Formally, let $s_r = \sum_{i=1}^{r} x_r$ be the partial sums of an ordered set of real numbers $0 \le x_1 \le x_2 \le \cdots \le x_n < \infty$. Then define the function $L\left(\frac{r}{n}\right) = s_r/s_n$, with $L(0) = 0$; and the Lorenz curve $L(\theta)$ is the continuous linear interpolation of these

$n+1$ points, $0 \leq \theta \leq 1$. Broadly, it displays the proportion $L(\theta)$ of the total distribution that is held by the smallest θ-fraction of the recipients.

Very commonly, the numbers $x_1, \ldots, x_n$ are taken to be an ordered sample from a distribution function $F(x)$, having finite mean μ, and in this case the Lorenz curve is defined more generally by

$$L(\theta) = \mu^{-1} \int_0^\theta \pi(x) \mathrm{d}x,$$

where $\pi(x)$ is the appropriately defined inverse function $F^{-1}(x)$ of F.

The Lorenz curve may be equivalently defined implicitly by $L(F(x)) = \mu^{-1} \int_{-\infty}^{x} y \mathrm{d}F(y)$; and for $\mu > 0, L(\theta)$ is convex, continuous, increasing, and with a left derivative for $\theta \in [0, 1]$. Examples include:

(a) If $F(x)$ has zero variance, then $L(\theta) = \theta$.

(b) If $F(x) = 1 - \mathrm{e}^{-x/\mu}, x \geq 0$, then $L(\theta) = \theta + (1-\theta) \log(1-\theta)$.

(c) If $F(x) = 1 - x^{\frac{-\mu}{\mu-1}}; x > 1$, then $L(\theta) = 1 - (1-\theta)^{1/\mu}$.

From this is derived the **Gini index** (of inequality) given by $G = 2A$, where A is the area between $L(\theta)$ and $\theta, 0 \leq \theta \leq 1$. There are several representations for G, including

$$\begin{aligned} 2\mu G &= \int\int |x-y| \mathrm{d}F(x) \mathrm{d}F(y) = 2 \int F(x)(1-F(x)) \mathrm{d}x \\ &= 4 \int x \left[F(x) - \frac{1}{2} \right] \mathrm{d}F(x). \end{aligned}$$

[J. L. Gastwirth, The estimation of the Lorenz curve and Gini index, *Rev. Econ. Stat.*, 54(3), 306–316, 1972]

loss function A concept arising in the context of decision theory in its many forms, which include, e.g. game theory, prediction, control, estimation, and so on. Broadly there is an unknown quantity Y (which may be an unknown state, or unknown future), taking values in S; together with a set of actions (or decisions, or predictions) taking values in a set A; and also (possibly) a set of observations (or records) X, taking values in R, that are jointly distributed with Y. Very broadly speaking indeed, an action a taken when $Y = y$ and $X = x$, entails a penalty (or loss or cost, or negative reward or pay-off), that the agent bears in mind when choosing a. Expected loss is often called risk, and in the case of discrete state spaces, one speaks of a **loss matrix**. [J. Berger, *Statistical Decision Theory and Bayesian Analysis*, 2nd edn. Springer, 1985]

lossless A term arising in communication theory, where any encoding or compression of a source (of messages or data) is described as lossless if the original can always be recovered exactly from the encoding. Otherwise it is described as lossy coding (or lossy compression). Likewise, a channel is lossless if the input can always be inferred exactly from the output; otherwise it is described as noisy, or lossy. [K. Sayood, *Lossless Compression Handbook*. Academic Press, 2003]

loss network (alias: **loss system**) Broadly, any model for a communication system, or queueing network, in which there are capacity constraints but no waiting rooms (or buffer, or storage). Thus customers, or calls, or data packets are lost if they arrive at the system for service or routing, if any necessary part of the system would exceed capacity on acceptance into the network. The simplest, and classic, loss system is Erlang's telephone exchange model, *see* Erlang's loss system. [S. Zachary & I. Ziedins, Loss Networks, in *Queueing Networks: A Fundamental Approach*, R. Boucherie & N. van Dijk (eds). Springer 2011, pp. 721–728]

lost-games distribution A distribution on the non-negative integers $n, n+1, \ldots$ whose pgf is

$$G_{a,n}(z) = z^n p^a {}_2F_1\left(\frac{1}{2}a, \frac{1}{2}(a+1); a+1; 4pqz\right),$$

where a is an integer not less than n, and $p \geq q = 1-p \geq 0$, and ${}_2F_1$ is the generalized hypergeometric function. The name arises from a special case arising in the gambler's ruin problem. [A. Kemp & C. Kemp, Branching and clustering models associated with the lost-games distribution, *J. Appl. Prob.*, 6, 700–703, 1969]

lottery (alias: **casting (or drawing) lots; sortilege**) A classic randomizer, which in its original form entailed a random choice from a number of congruent objects, e.g. pieces of wood, pebbles, bones, shells, seeds, counters, tokens, straws, etc. More recently, the word has been used to denote any scheme for distributing rewards (or, occasionally, penalties) at random. [J. Haigh, *Taking Chances: Winning with Probability*, 2nd edn. Oxford University Press, 2003]

Lovász local lemma Given a family of events $A_1, \ldots, A_n$, the lemma supplies a sufficient condition for the probability that none of the events occur to be strictly positive. It has three main forms:

(a) Suppose that each A_i is independent of all but at most d of the other events, and $\mathbb{P}(A_i) \leq p < 1$. If $4pd \leq 1$, then with positive probability none of the A_i occur.

(b) With the events A_i having the same properties as in (a), the condition $ep(d+1) \leq 1$ is also sufficient for the probability that none occur to be positive.
(c) More generally, suppose that for each A_i, there is collection $C(A_i)$ of events from $A_1, \ldots, A_n$ such that A_i is independent of $\{A_1, \ldots, A_n\} \backslash \{A_i \cup C(A_i)\}$. If there are numbers $p_i \in (0, 1)$ such that for each A_i $\mathbb{P}(A_i) \leq p_i \prod_{A_j \in C(A_i)} (1 - p_j)$, then the probability that none of the A_i occur is not less than $\prod_i (1 - p_i) > 0$.

[R. Moser & G. Tardos, A constructive proof of the general Lovász local lemma, *J. ACM*, 57(2), 11:1–11:15, 2010]

Lovász–Winkler rule Any one of a number of stopping rules for a discrete time irreducible finite state-space Markov chain, such that the distribution of the state of the chain at the stopping time is either exactly or approximately the stationary distribution of the chain. [L. Lovász & P. Winkler, Efficient stopping rules for Markov chains, in *Proc. 27th Annual ACM Symposium on the Theory of Computing*. ACM, 1995, pp. 76–82]

lower and upper probability (alias: **interval-valued probability**) Arise in theories of probability that allow indeterminacy (or imprecision) in assigning probabilities to events. Roughly speaking, for any event A, the classical unique (additive) probability function $\mathbb{P}(A)$ is replaced by a pair of functions $\underline{\mathbb{P}}(A) \leq \overline{\mathbb{P}}(A)$ that define a probability interval $[\underline{\mathbb{P}}(A), \overline{\mathbb{P}}(A)]$ for A.

Formally, for a sample space Ω, and power-set sigma-field $\mathcal{F} = 2^{\Omega}$, the lower probability $\underline{\mathbb{P}}(\cdot)$ is a non-negative function defined on $\mathcal{F}$ such that $\underline{\mathbb{P}}(\Omega) = 1$, and $\underline{\mathbb{P}}(\emptyset) = 0$, and $\underline{\mathbb{P}}(\cdot)$ is super-additive, which is to say that whenever $A \cap B = \emptyset$,

$$\underline{\mathbb{P}}(A \cup B) \geq \underline{\mathbb{P}}(A) + \underline{\mathbb{P}}(B).$$

The conjugate upper probability $\overline{\mathbb{P}}(\cdot)$ of $\underline{\mathbb{P}}(\cdot)$ is given by $\overline{\mathbb{P}}(A) = 1 - \underline{\mathbb{P}}(A^c)$, and $\overline{\mathbb{P}}(\cdot)$ is subadditive. Such functions arise in a number of generalizations of the classical theory of probability, perhaps most notably in Dempster–Shafer theory, where they are interpreted as belief and plausibility. [P. Walley & T. Fine, Towards a frequentist theory of upper and lower probability, *Ann. Stat.*, 10(3), 741–761, 1982]

Loynes' theorem Any one of a large number of theorems about the stability of a queue, and systems of queues, with non-independent service and inter-arrival times. Their proofs follow from Loynes' lemma, which relates the limiting and stationary properties of a suitable type of stochastic recurrence. [R. Loynes, The stability of a queue with non-independent inter-arrival and service times, *Proc. Cambridge Philos. Soc.*, 58(3), 497–520, 1962]

L_p space (alias: **L_r space; L^p (or L^r) space**) Arises in probability as the space of all r.vs X such that $\mathbb{E}|X|^p < \infty; 1 \leq p \leq \infty$, with L_p norm $\|X\|_p = \{\mathbb{E}|X|^p\}^{1/p}$. In the case $p = \infty$, the norm is otherwise known as the Chebyshov, or essential supremum norm. L_p space is a Banach space, i.e. a complete normed space; it extends in a natural way to random vectors. For $0 < p < 1$, $\|X\|_p$ is not a norm for the space L_p. [M. Ledoux & M. Talagrand, *Probability in Banach Spaces*. Springer, 1991]

Lukacs' theorem Supplies a characterization of the gamma distribution, in this form: let X and Y be independent positive r.vs (that are not degenerate). If $X+Y$ and $\frac{X}{X+Y}$ are independent, then X and Y have gamma distributions, with possibly different shape parameter, but with the same scale parameter. Its multivariate extension is the **Lukacs–Olkin–Rubin thm**. [K. Bobecka & J. Wesołowski, Multivariate Lukacs theorem, *J. Multivariate Anal.*, 91(2), 143–160, 2004]

lumpability (alias: **aggregation; collapsibility; projection**) Arises naturally in considering an irreducible Markov chain $X_t, t \geq 0$, on a finite state space S that has been partitioned into disjoint subsets $S_i, \ldots, S_m$ whose union is S. The random process $Y_t \in \{1, \ldots, m\}$ that records only the subset occupied by X_t at time t is called a collapsed, or projected, or lumped, or aggregated process obtained from X_t. If Y_t is itself a Markov process for any initial distribution of X_t, then X_t is said to be (strongly) lumpable for the partition $\{S_1, \ldots, S_m\}$. If Y_t is Markov for at least one, but not all, initial distributions, then X_t is weakly lumpable. More formally, these properties may be expressed in terms of partial row and column sums (over the S_r) of the transition matrix P of X. [P. Buchholz, Exact and ordinary lumpability in finite Markov chains, *J. Appl. Probab.*, 31(1), 59–75, 1994]

Lundberg bounds, Lundberg exponent *see* Cramér–Lundberg model

Luria–Delbrück distribution A probability distribution on the non-negative integers that arises in mathematical genetics. If has pgf $G(s) = \exp\{\lambda(1-s)\{\log(1-s)\}/s\}, \lambda > 0$. [A. Kemp, Comments on the Luria-Delbrück distribution, *J. Appl. Probab.*, 31(3), 822–828, 1994]

Lusin space *see* Blackwell space

Lyapunov central limit theorem (alias: **Liapounov CLT; Lyapunov CLT**) A classical (proved in 1901) form of the CLT for sums of independent r.vs, thus: let $S_n = \sum_{r=1}^{n} X_r$ for independent non-degenerate r.vs $X_r, r \geq 1$, such that $a_r = \mathbb{E}(|X_r - \mathbb{E}X_r|^{2+\delta}) < \infty$ for some $\delta > 0$ and all r. If $\lim_{n\to\infty} \sum_{r=1}^{n} a_r / (\operatorname{var} S_n)^{1+\delta/2} = 0$ (which is called **Lyapunov's condition**), then as $n \to \infty$, $\mathbb{P}\left(\{S_n - \mathbb{E}S_n\}/(\operatorname{var} S_n)^{\frac{1}{2}} \leq x\right)$ converges (uniformly in x) to the standard normal distribution function. Lyapunov's condition implies the Lindeberg condition. [W. Adams, *The Life and Times of the Central Limit Theorem.* American Mathematical Society & LMS, 2009]

Lyapunov's inequality Asserts that for a r.v. X, and $0 \leq a < b < c$, $(\mathbb{E}|X|^b)^{c-a} \leq (\mathbb{E}|X|^a)^{c-b} (\mathbb{E}|X|^c)^{b-a}$. The case when $a = 0$ is most often seen. [H. Fischer, *A History of the Central Limit Theorem.* Springer, 2011]

M

machine interference problem (alias: **machine servicing (or repairman) problem**) A classic problem of queueing theory that arises in the operations research of many systems. In the basic formulation, m machines are working in the care of r repairmen (or servers, mechanics, etc.) whose services are required by breakdowns (or degradation) according to some stochastic process. If more than r machines require attention at any instant, then interference occurs, in the sense that some machines will be idle as their repair is delayed. Interest centres on the cost of idle times for both machines and their servers. *See* Erlang loss system. [L. Hague & M. Armstrong, A survey of the machine interference problem, *Euro. J. Oper. Res.*, 179(7), 469–482, 2007]

MacQueen extrapolation *see* Howard policy improvement

magic factor *see* Stein's method

majorization A concept arising in several branches of mathematics. In the context of random variables, its basic form is this: let X and Y be r.vs with finite means. Then Y majorizes X (written $X < Y$), if for all real-valued convex functions $g(x), x \in \mathbb{R}, \mathbb{E}g(X) \leq \mathbb{E}g(Y)$. An alternative characterization asserts that $X < Y$ iff, for all $a \in \mathbb{R}, \mathbb{E}(X-a)^+ \leq \mathbb{E}(Y-a)^+$, and $\mathbb{E}X = \mathbb{E}Y$.

See also stochastic ordering; and note that the term majorization has been used for other types, e.g. one may say that X is **tail-majorized** by Y if there is some finite $a > 0$ such that for all $x > a, \mathbb{P}(|X| > x) \leq \mathbb{P}(|Y| > x)$. [A. Marshall et al., *Inequalities: Theory of Majorization and its Applications*, 2nd edn. Springer, 2011]

makespan problem *see* job-shop problem

Malliavin calculus (alias: **stochastic calculus of variations**) Very broadly, in its basic form, the Malliavin calculus extends the idea of directional derivative to the infinite-dimensional Wiener (Gaussian) space. This yields an operator D_t, called the **Malliavin derivative**, such that $D_t \int_0^t g(s)\mathrm{d}W(s) = g(t)$, for suitable functions g, where W is the Wiener process. The formulation entails the existence of a formula for integration by parts, with other analogues of classical calculus, and was used by P. Malliavin to yield a probabilistic proof of Hörmander's theorem. The calculus has since been applied to derive regularity properties of Brownian functionals arising as strong solutions of suitable SDEs. [D. Nualart, *Malliavin Calculus and its Applications*. American Mathematical Society, 2009]

Mallows distance *see* earth mover's distance; minimum $\mathrm{L_p}$ distance

Malthusian parameter (alias: **exponential growth rate**) Arises in the context of branching processes and other growth models such that the expected population size at time t (or of the tth generation), is asymptotic to $\mathrm{e}^{\alpha t}$ for some α; in such a case α is called the Malthusian parameter. For example, in the simple Markov birth–death process $X(t) \in \mathbb{Z}^+$, with birth and death rates λ and $\mu, \mathbb{E}X(t) = X(0)\mathrm{e}^{(\lambda-\mu)t}$, and $\alpha = \lambda - \mu$. For an age-dependent (Bellman–Harris) branching process, with mean family size m and age distribution $G(x)$, the Malthusian parameter is the root (when it exists) α of $m\int_0^\infty \mathrm{e}^{-\alpha x}\mathrm{d}G(x) = 1$. More generally, for a multitype branching process with reproduction kernel $m(r, \mathrm{d}s \times \mathrm{d}t)$, a similar Malthusian parameter α arises under a number of conditions on $k(r, \mathrm{d}s, \alpha) = \int_0^\infty \mathrm{e}^{-\alpha t} m(r, \mathrm{d}s \times \mathrm{d}t)$. A process having such a parameter is called Malthusian. [P. Jagers, Stabilities and instabilities in population dynamics, *J. Appl. Probab.*, 29(4), 770–780, 1992]

Mandelbrot multiplicative cascade (alias: **multi-fractal**) Arises in the theory of fractals thus: let T be the rooted tree with q branches at each vertex (or node) directed away from the root (so $q = 2$ for a binary tree). To each vertex v of T attach a r.v. X_v, where the collection X_v are i.i.d.r.vs with distribution F (called the cascade distribution) of some r.v. X (called the cascaded variable). To any path $v_1 \cdots v_n$ of vertices (away from the root) there corresponds a product $X_{v_1} \cdots X_{v_n}$ yielding a random measure on the vertices at height n of T. The Mandelbrot–Kahane–Peyrière (or **MKP**) **structure fn** (a modified cumulant generating function) is $K(z, q) = \log_q \mathbb{E}X^z - (z-1)$, and the **Kahane–Peyrière thm** characterizes the limiting properties of the random measure at height n, as $n \to \infty$, in terms of $K(z, q)$. [B. Mandelbrot, *Fractals and Chaos*. Springer, 2004]

Mann–Wald theorem (alias: **continuous mapping theorem; mapping theorem**) Arises in considering a sequence X_n of r.vs that converges in distribution to X as $n \to \infty$ (or, equivalently, a sequence of distribution functions $F_n(x)$ converging to $F(x)$ at all the continuity points of F, as $n \to \infty$). In its basic form, it asserts that for any real-valued continuous function $h(x)$, the sequence $h(X_n)$ converges in distribution to $h(X)$ as $n \to \infty$.

More generally, let X_n converging in distribution to X take values in a separable metric space, and $h(\cdot)$ be a mapping from that on to another separable metric space, and let D be the set of discontinuity points of $h(\cdot)$. If $\mathbb{P}(D) = 0$, then $h(X_n)$ converges in distribution to $h(X)$. (Very commonly $\mathbf{X}_n$ and $\mathbf{X}$ take values in $\mathbb{R}^m$, and the joint distributions of $\mathbf{X}_n$ converge to those of $\mathbf{X}$, while $h(\cdot)$ takes values in $\mathbb{R}^n$.)

It follows that the convergence of X_n in probability to X, entails the convergence in probability of $h(X_n)$ to $h(X)$ as $n \to \infty$, for $h(\cdot)$ as above; and likewise for the almost sure convergence of $h(X_n)$ to $h(X)$ whenever $X_n \xrightarrow{\text{as}} X$. [P. Billingsley, *Convergence of Probability Measures*, 2nd edn. Wiley, 1999]

many-to-one theorem Arises in branching (and branching-diffusion, etc.) processes, where it supplies an expected value taken over all living particles in terms of the expected value of a quantity suitably defined for one particle (which is often called the spine). More generally, one has analogous many-to-few theorems. [R. Hardy & S. Harris, A conceptual approach to a path result for branching Brownian motion, *Stoch. Proc. Appl.*, 116, 1992–2013, 2006]

map-Airy distribution *see* Airy distribution

Marčenko–Pastur law (or distribution) Arises in the context of random matrix theory, where it supplies the limiting form of the empirical distribution of the eigenvalues of a class of $n \times n$ Wishart matrices as $n \to \infty$ (*see* Wishart distribution). Formally, let X_n be an $n \times m$ matrix whose entries are i.i.d. mean-zero r.vs (possibly complex) with finite variance, and let $W_n = \frac{1}{n} X_n X_n^{\mathrm{T}}$, where T denotes transpose. Suppose that $m, n \to \infty$ so that $\lim_{n\to\infty} \frac{m}{n} = c > 0$. Then the empirical distribution of the eigenvalues of W_n converges (weakly) to a distribution function $F(x)$ whose Stieltjes transform F_s satisfies

$$\theta F_s(\theta)^2 + (1 - c + \theta) F_s(\theta) + 1 = 0.$$

If $c > 1$, then F has the Marčenko–Pastur density (or law) given by

$$f(x) = \frac{1}{2\pi x}\left\{\left(x - (1-\sqrt{c})^2\right)\left((1+\sqrt{c})^2 - x\right)\right\}^{\frac{1}{2}},$$
$$(1-\sqrt{c})^2 < x < (1+\sqrt{c})^2.$$

If $c < 1$, then F is a mixture of this density with an atom of size $1 - c$ at zero. [V. Marchenko & L. Pastur, Distribution of eigenvalues for some sets of random matrices, *Math. USSR Sb*, 1(4), 457–483, 1967]

Marcinkiewicz theorem Asserts that if the characteristic function $\phi(t)$ of a r.v. X has the form $\exp(h(t))$, where $h(t)$ is a polynomial in t, then the order of $h(t)$ is not more than 2, so that X is either normally distributed or it is degenerate. [E. Lukacs, Some extensions of a theorem of Marcinkiewicz, *Pacific J. Math.*, 8(3), 487–501, 1958]

Marcinkiewicz–Zygmund inequalities Arise as a generalization of Khinchin's inequalities, in this form: let $1 \le p < \infty$, and let $X_r, r \ge 1$, be independent zero-mean r.vs such that $\mathbb{E}|X_r|^p < \infty$ for some $p \ge 1$ and all r. Then there exist positive constants a_p and b_p (depending on p only) such that for all $n \ge 1$

$$a_p \mathbb{E}\left\{\left(\sum_{k=1}^{n} X_k^2\right)^{p/2}\right\} \le \mathbb{E}\left(\left|\sum_{k=1}^{n} X_k\right|^p\right)$$
$$\le b_p \mathbb{E}\left\{\left(\sum_{k=1}^{n} X_k^2\right)^{p/2}\right\}.$$

[A. Gut, *Probability: A Graduate Course*, 2nd edn. Springer, 2013]

Marcinkiewicz–Zygmund theorem Any of several results, but in probability most often this strong law of large numbers, which strengthens Kolmogorov's classical strong law. Let $X_r, r \ge 1$, be i.i.d.r.vs, and let $0 < p < 2$. Then setting $S_n = \sum_{r=1}^{n} X_r$:

(i) If $\mathbb{E}|X_1|^p < \infty$, then $\lim_{n\to\infty} n^{-1/p}\{S_n - na\} = 0$, a.s., where $a = \mathbb{E}X_1$ for $1 \le p \le 2$, and a is an arbitrary real for $0 < p < 1$.

(ii) Conversely, if for some real a, with probability 1, $\lim_{n\to\infty} n^{-1/p}(S_n - na) = 0$, then $\mathbb{E}|X_1|^p < \infty$.

[A. Gut, *Probability: A Graduate Course*, 2nd edn. Springer, 2013]

Marcus–Lushnikov process (alias: **gelation (or coagulation, or coalescent) model**) A stochastic model for the aggregation of particles in which each particle is entirely described by its mass (position being irrelevant and ignored). It is essentially a random version of the deterministic Smoluchowski coagulation process. In its basic form it is a Markov process in which the state at any time comprises the masses of the separate particles, where each pair of particles with masses x and y coalesce into a single particle of mass $x + y$ at rate $K(x, y)$. [D. Aldous, Deterministic and stochastic models for coalescence (aggregation and coagulation): a review of the mean-field theory for probabilists, *Bernoulli*, 5, 3–48, 1999]

Mardia distribution (alias: **multivariate Pareto distribution**) A r.vr $\mathbf{X} = (X_1, \ldots, X_n)$ has a Mardia distribution in standard form when

$$\mathbb{P}(X > x) = \left\{1 + \sum_{r=1}^{n} \frac{x_r}{b_r}\right\}^{-a},$$

for $a > 0$, and $b_r > 0$ and $x_r > 0, 1 \leq r \leq n$. Its multivariate marginal distributions are of the same type, as are its conditional distributions. [K. Mardia, Multivariate Pareto distributions, *Ann. Math. Stat.*, 33, 1008–1015, 1962]

Mardia's multivariate kurtosis *see* kurtosis

marginal distribution Arises in its simplest form as the distribution of a discrete r.v. X that is jointly distributed with a discrete r.v. Y, so that $p(x) = \mathbb{P}(X = x) = \sum_y \mathbb{P}(X = x, Y = y) = \sum_y p(x, y)$, where $p(x, y)$ is the joint distribution of X and Y. The natural display of $p(x, y)$ as an array with $p(x)$ at one margin illustrates why $p(x)$ is then called the marginal distribution of X. More generally, for X and Y having joint distribution function $F(x, y)$, the marginal distribution function $F(x)$ of X is given by

$$F(x) = \lim_{y \to \infty} F(x, y) = F(x, \infty).$$

Yet more generally, the distribution of a r.v. (or r.vr) $\mathbf{X}$, obtained by considering it as a subset of the components of some jointly distributed larger r.vr $\mathbf{Z}$ is a (multivariate) marginal distribution. Marginal distributions are uniquely determined by the parent distribution; but the converse does not hold in general unless extra conditions are imposed, e.g. independence. [J. Kingman & S. Taylor, *Introduction to Measure and Probability*. Cambridge University Press, 1966]

Margulis (or Margulis–Russo) formula *see* Russo's formula

marked (point) process *see* point process

market model Arises in financial mathematics as any one of many random processes (in discrete or continuous time) constructed in attempting to describe real-world markets. The basic (or standard) market model comprises a random vector $\mathbf{X}(t)$ (on some stochastic basis), where all components are non-negative for all t, and at least one is strictly positive for all t (representing a risk-free bank account, or numeraire). Classic models include Black–Scholes in continuous time, and Cox–Ross–Rubinstein (or binomial) in discrete time. [M. Musiela & M. Rutkowski, *Martingale Methods in Financial Modelling*, 2nd edn, 3rd printing. Springer, 2009; M. Capiński & E. Kopp, *Discrete Models of Financial Markets*. Cambridge University Press, 2012]

marking theorem (alias: **product theorem; colouring theorem**) Arises in the context of a Poisson process carrying marks, i.e. a Poisson point process N on a measurable space S, where each point X of N carries an associated r.v. Y (called the mark of X), taking values in a measurable space M. The marks of different points of N are independent, but may depend on X. Then the marking theorem asserts that the random countable subset of $S \times M$, induced by the points of N and their corresponding marks, is also a Poisson process. [J. F. C. Kingman, *Poisson Processes*. Oxford University Press, 1993]

Markov blanket *see* Bayes network

Markov chain A random process $X(t), t \in T$ (where T is commonly either $\mathbb{Z}^+$ or $\mathbb{R}^+$), such that $X(t)$ takes values in a countable set S (conventionally $\mathbb{Z}$), and the process has a Markov property; which in its simplest basic form may be written as:

$$\mathbb{P}(X(t) = j \mid X(t_1) = i_1, \ldots, X(t_n) = i_n) = \mathbb{P}(X(t) = j \mid X(t_n) = i_n),$$

for all $j, i_1, \ldots, i_n \in S$, and any sequence $t_1 < t_2 < \cdots < t_n < t$ of times. (In discrete time, $n \leq t$; in continuous time $n \in \mathbb{N}$.)

Informally, the future is conditionally independent of the past, given the present value of the process. It is commonly assumed, without explicit mention, that the chain has stationary transition probabilities, which is to say that for all $s < t, \mathbb{P}(X(t) = j \mid X(s) = i) = p_{ij}(t - s)$, where the stochastic matrix $p_{ij}(t-s)$ is a function of $t-s$ only; such a chain may be called time-homogeneous. (When the state space is not countable, the term Markov process is commonly used to refer to processes having a more general form of the Markov property; though the term Markov chain is also used in this case when time is discrete.)

It follows that for a time-homogeneous chain in discrete time, the evolution of $X(n)$ is described by the n-step transition probabilities $p_{ij}(n) = \mathbb{P}(X(n) = j \mid X(0) = i)$, which are determined by the one-step transition probabilities denoted by the stochastic matrix $p_{ij} = p_{ij}(1)$. In particular, (p_{ij}) determines whether any given state may ever be entered by the chain from any other, and if so with what frequency. This yields various classifications of the states of S, as follows:

- A state i accesses (or leads to, or reaches) state j if $p_{ij}(n) > 0$ for some finite n, and we write $i \to j$.
- If also j accesses i, then i and j are said to communicate, and we write $i \leftrightarrow j$.
- A state that accesses no other state is called absorbing.
- A state that cannot be accessed by any other state (not even itself) is called a non-return state.
- A state i that accesses at least one other state j such that j does not access i, is called inessential (otherwise i is essential).
- A set C of states is closed if no state in C accesses a state outside C.
- The communication symbol $\leftrightarrow$, seen above, defines an equivalence relation on S, and thus partitions the state space accordingly into disjoint classes $C_k, S = C_1 \cup \cdots \cup C_r$, where each C_k is either a communicating class of states (closed or not closed), or comprises a single non-return state.
- A chain whose state space comprises exactly one (necessarily closed) equivalence class is said to be irreducible (or non-decomposable).

Then consideration of the long-run (or asymptotic) properties of the chain induces further classification, thus:

- A state i is recurrent (persistent) if the chain started in i returns to i eventually with probability 1.
- If the expected time μ_i of first return to the recurrent state i is infinite, then it is said to be null; otherwise, if $\mu_i < \infty$, it is said to be non-null (or positive).
- Any state that is not recurrent is said to be transient.

It simplifies the analysis of the chain that the communicating classes satisfy a number of solidarity theorems, thus:

- If $i \leftrightarrow j$, then:
  (a) i is transient iff j is transient;
  (b) i is null recurrent iff j is null recurrent;
  (c) the greatest common divisors d_i and d_j of $\{n : p_{ii}(n) > 0\}$ and $\{m : p_{jj}(m) > 0\}$ are equal; if $d > 1$, then i and j are said to have period d; if $d = 1$, they are aperiodic.
- An irreducible, aperiodic, non-null recurrent chain is **ergodic**.

Note that no class can be recurrent unless it is closed, and that no closed class in a finite chain can be transient.

In continuous time, $t \geq 0$, the evolution of the chain is described in principle by the family of stochastic matrices $p_{ij}(t), t \in \mathbb{R}^+$, which determine the finite dimensional distributions of $X(t)$ via the Chapman–Kolmogorov eqns, so that $p_{ij}(t)$ comprises a stochastic semi-group. However, in the most general case, the analysis becomes extremely complicated. For tractability it is convenient to make the following assumptions:

(i) $\lim_{t\to 0} p_{ij}(t) = \begin{cases} 1, & i = j, \\ 0, & i \neq j, \end{cases}$

and in this case the matrix (and semi-group) is called standard. It is then differentiable for all $t \geq 0$, and in a standard notation the right-hand derivative at $t = 0$ is

$$\left.\frac{\mathrm{d}p_{ij}}{\mathrm{d}t}\right|_{t=0} = p'_{ij}(0) = q_{ij} \geq 0, \quad i \neq j \text{ and } -p'_{ii}(0) = q_i \geq 0.$$

Some authors use the convention $q_{ii} = -q_i$. In either case the resulting array is called the q-matrix (or Q-matrix), and can be seen as playing a role in generating the process that is analogous to that played in discrete time by p_{ij}. Then it is further convenient to require that:

(ii) $q_i < \infty$ for all i, and then the states are said to be stable (when $q_i = \infty, i$ is said to be instantaneous).

(iii) $\sum_{j \in S} q_{ij} = q_i$, for all i, and then the chain is said to be conservative.

(iv) The sample paths of $X(t)$ are right-continuous; in this case the finite dimensional distributions are sufficient to give the probability of events determined by (i.e. measurable functions of) the process $X(t)$.

(v) The process is defined up to the first explosion (*see* Feller minimal process).

The simplest case arises when there is no explosion, and the process is honest, and the Q-matrix determines $p_{ij}(t)$ uniquely. There exists a discrete Markov chain Y_n, with a transition matrix R, whose entries are determined by Q, that follows the successive distinct values taken by $X(t)$; it is called the jump chain, and R may be called the jump matrix (or the routing matrix). Between jumps, $X(t)$ remains in its current state i for a holding time that is exponentially distributed with parameter q_i. Note that if $q_i = 0$, then i is said to be an absorbing state.

The class structure of $X(t)$ is then just the class structure of the jump chain, as determined by its routing matrix (as given above), with two notable exceptions:
(i) Periodicity does not arise as a useful concept, because (by the Lévy dichotomy) either $p_{ij}(t) = 0$ for all $t \geq 0$, or $p_{ij}(t) > 0$ for all $t > 0$.
(ii) The mean recurrence time of a state i (used to distinguish between positive recurrent and null recurrence states and classes) is defined thus: let J be the duration of the holding time in i of $X(t)$ started in $X(0) = i$ (i.e. J is the time of the first jump). Define the first return time $T_i = \inf\{t > J : X(t) = i\}$. Then the mean recurrence (or expected return) time is $m_i = \mathbb{E}T_i$, and a recurrent state i is positive or null recurrent according to whether $\mathbb{E}T_i < \infty$ or $\mathbb{E}T_i = \infty$.
The classification can be interpreted in terms of the Q-matrix, if desired. *See* Feller's construction and Feller–Reuter theorem.[J. Norris, *Markov Chains*. Cambridge University Press, 1997]

Markov chain ergodic theorem Arises in considering the long-run properties of a Markov chain (or, more generally, Markov process). The term most commonly refers to the following basic properties of a Markov chain $X_n, n \geq 1$, on a countable state space S, with n-step transition probabilities $p_{ij}(n)$. Then for all $i, j \in S$:
(a) The limits $\pi_{ij} = \lim_{n\to\infty} \frac{1}{n}\sum_{r=1}^{n} p_{ij}(r)$ exist, and satisfy $\pi_{ij} \geq 0$, and $\sum_{j\in S}\pi_{ij} \leq 1$, and $\pi_{ik} = \sum_{j\in S}\pi_{ij}\pi_{jk}$.
The chain is then said to be:
(i) dissipative if $\pi_{ij} = 0$ for all i, j;
(ii) non-dissipative if $\sum_{j\in S}\pi_{ij} = 1$ for all i;
(iii) semi-dissipative if $\pi_{ij} > 0$ for some i, j, but $\sum_{j\in S}\pi_{kj} < 1$ for some k.
In that (a) deals with Cesàro limits, it may be called a weak ergodic theorem. A stronger form asserts:
(b) If j is an aperiodic state of the chain then $\lim_{n\to\infty} p_{jj}(n) = \mu_j^{-1}$, where μ_j is the mean recurrence time of j. If f_{ij} denotes the probability of ever reaching j from i, $\lim_{n\to\infty} p_{ij}(n) = f_{ij}\mu_j^{-1}$. Hence, one has the strong ergodic theorem for an irreducible aperiodic chain: $\lim_{n\to\infty} p_{ij}(n) = \mu_j^{-1}$, for all $i, j, \in S$. The term also denotes the following (sometimes called the strong law of large numbers for a Markov chain): let $\boldsymbol{\pi}$ be the unique stationary distn of a non-null recurrent irreducible chain $X_n, n \geq 0$, and g a function such that $\sum_k \pi_k |g(k)| < \infty$. Then, a.s., $\lim_{n\to\infty} \frac{1}{n}\sum_{r=0}^{n} g(X_r) = \sum_k \pi_k g(k)$.

Analogous ergodic theorems exist for Markov processes in continuous time. [F. Foster, Markov chains with an enumerable number of states, *Proc. Camb. Phil. Soc.*, 48(4), 587–591, 1952]

Markov chain Monte Carlo (alias: **McMC method; random walk Monte Carlo**) An example of a Monte Carlo method in which the required sampling from a probability distribution of interest is obtained by running (i.e. simulating) a suitable Markov chain. The method arose in investigating otherwise intractable stochastic physical systems, and has since been applied in image analysis, the evaluation of integrals and solutions of partial differential equations, and optimization problems. Particular schemes include those of Metropolis and Hastings. *See also* Propp–Wilson method and Gibbs sampling. [S. Brooks et al. (eds), *Handbook of Markov Chain Monte Carlo*. Chapman & Hall/CRC Press, 2011]

Markov decision process (alias: **Markov dynamic programming (or control, or sequential optimization); MDP**) Broadly, a Markov chain $X(t)$, in discrete or continuous time $t \geq 0$, whose transition probabilities also depend on the current state of the process through actions that are chosen according to some policy of a decision maker. Typically the policy for choosing actions is intended to minimise some cost function (or maximise some set of rewards) that depends on the sequence of states of the chain. More formally, an MDP comprises:
1. A set T of times, called decision epochs; typically $T = \mathbb{N}^+$ in discrete time.
2. The state space S (assumed countable), of $X(t)$.
3. A set $A(s,t)$ of actions for all $s \in S$ and $t \in T$, together with a policy π for choosing an action, which may perhaps include randomized choices.
4. A set $C(s,a)$ of costs (or rewards, seen as negative costs), for $s \in S$ and $a \in A$, arising if action a is taken in state s.
5. A set $p_{jk}(a)$ of transition probabilities (which are stationary under reasonably mild conditions), such that transitions are conditionally independent of the past of the process given the current state and the present action arising from it.

[M. Puterman, *Markov Decision Processes*. Wiley, 1994]

Markovian fluid model *see* fluid Markov model

Markov inequality In probability theory, Markov's inequality is the assertion that for any non-negative r.v. X, and $t > 0$, we have $\mathbb{P}(X \geq t) \leq \mathbb{E}X/t$. It is often encountered in the statement that for any r.v. X and $t > 0$, $\mathbb{P}(|X| \geq t) \leq \mathbb{E}|X|/t$, of which Chebyshov's inequality is a simple corollary. More generally, for any increasing sequence $1 < t_1 < t_2 < \cdots < t_n$, and positive r.v. X,

$$\sum_{r=1}^{n} \mathbb{P}(X \geq t_r) \leq \max_{1 \leq r \leq n} \{r\mathbb{E}X/t_r\}.$$

[B. Eisenberg & B. Ghosh, A generalization of Markov's inequality, *Stat. Probab. Lett.*, 53, 59–65, 2001]

Markov matrix *see* stochastic matrix

Markov-modulated Describes a random process in which some parameter, or component of the model, is assumed to follow a Markov process rather than being some deterministic function (such as a constant). For example, the server in a queue may move between states of readiness or efficiency according to some Markov chain. Very commonly, one has a Markov-modulated Poisson process, which is a Cox process with intensity $\Lambda(t)$ that is governed by a Markov process $X(t)$. If the parameter is not observed, it may be called a **hidden Markov model.** [W. Fischer & K. Meier–Hellstern, The Markov-modulated Poisson process cook book, *Performance Evaluation*, 18(2), 149–171, 1992]

Markov moment problem *see* moment problem

Markov order A term arising in entirely different contexts. Most commonly, it denotes the order of dependence in a multiple Markov chain. Less commonly, it arises in seeking a general characterization of distributions that are maximally uncertain (in the sense of being dispersed over their support) when subject to constraints. Formally, let **a** and **b** be probability distributions on the same finite set S. Then **b** is said to be more random than **a** in Markov preorder with respect to $\boldsymbol{\pi}$ if there is an ergodic Markov chain $X(t), t \geq 0$, with stationary distribution $\boldsymbol{\pi}$, such that $X(0)$ and $X(1)$ have respective distributions **a** and **b**; the transitive closure of this preorder is the Markov order for $\boldsymbol{\pi}$. [A. Gorban et al., Entropy: the Markov ordering approach, *Entropy*, 12, 1145–1193, 2010]

Markov point process Arises in extending the concept of the Markov property to finite point processes in $\mathbb{R}^d$ (or other suitable space); it can be seen as a process analogous to the classical Markov random field. The distribution (or density) of the process is determined by the states of adjacent points, i.e. (more informally) by the interactions of neighbours. The difference from the classical format is that the concept of adjacency may be defined by a wider class of reflexive and symmetric neighbour relations, depending (in general) on the realization of the process. [B. Ripley & F. Kelly, Markov point processes, *J. Lond. Math. Soc.*, 15, 188–192, 1977]

Markov process (alias: **Markov system**) Observation of the real world suggests that many processes evolving in time have the property that their future development after any time t depends only on their current state at t, and (given that state) does not in any other way depend on the history of the process before t (i.e. the sequence of values of the process leading to its state at time t). The theory of Markov processes is a formalization of this empirical fact for processes whose evolution is random, and the lack of dependence on the past is therefore characterized by some form of conditional stochastic independence, called the Markov property. Processes having some form of Markov property (that are thus entitled to be called Markov) arise, and are defined, in several different ways. The classic elementary example is a random walk $S_n = \sum_{r=1}^{n} X_r$, $n \geq 1$, having independent steps $X_r, r \geq 1$. By the independence of the steps, any $S_r, r \leq n-1$ is conditionally independent of any $S_{n+s}, s \geq 1$, given S_n, for all n. This is a basic form of the Markov property, and the random walk is a Markov process.

Equivalently, given $S_n = x$, the subsequent trajectory of the walk is given by $S_{n+s} = x + \sum_{r=n+1}^{s} X_r, s \geq 1$, and its distribution is that of a walk started from x, at time n, using steps $\hat{X}_r, r \geq n+1$, that have the same distribution as, while being independent of, the original walk. This is another form of the Markov property. It is often assumed that a Markov process is time-homogeneous, unless explicitly stated otherwise, and we conform to that practice henceforth. In this case the steps of the walk are i.i.d., and it is important and useful that the walk after time n, conditional on $S_n = x$, is given by $S_{n+s} = x + \sum_{r=1}^{s} \hat{X}_r$, which is a very readily generalized expression of the Markov property. These ideas extend naturally to general independent increment (or Lévy) processes, which are likewise Markov. Other classic examples of Markov processes include Markov chains and diffusion processes, which may be defined by their transition probabilities or transition densities. More generally, they may take values in a measurable space (typically assumed to be a completely separable metric space), and may be defined by their transition kernels. These different contexts (and other generalizations of the Markov property) lead to various forms of the Markov property, or condition; these in turn lead to various forms of the Chapman–Kolmogorov equations. Furthermore, there are general types of Markov process according to the stringency of the Markov condition; we use a simple basic form of the condition to define these ideas: Let $X(t)$ be a random process (defined on a filtered space; the filtration $\mathcal{F}_t$ need not be that generated by $X(t)$ itself, though here, for simplicity, we assume that it is).

(A) For any fixed t, let B be any event depending only on $X(s), s \leq t$, and A any event depending only on $X(u), u \geq t$. If for all such A, B, and $t, \mathbb{P}(A \mid X(t), B) = \mathbb{P}(A \mid X(t))$, then $X(t)$ is a Markov process in the usual sense.

(Note that some few writers call this a Markov process in the strict sense, because of (C) below; but this is not to be confused with (B) below.)

(B) Let S be any stopping (or Markov) time for $X(t)$ (i.e. for the filtration $\mathcal{F}_t$ generated by $X(t)$), and let A and B be events depending respectively on $X(u), u \geq S$, and $X(s), s \leq S$. If $\mathbb{P}(A \mid X(S), B) = \mathbb{P}(A \mid X(S))$ for all such A, B and S, then $X(t)$ is called a **strong Markov process** and is said to have the strong Markov property.

(C) Suppose that $\mathbb{E}(X(t)^2) < \infty$ for all t, and for any times $0 \leq t_1 < t_2 < \cdots < t_n < t$ define $\hat{\mathbb{E}}(X(t) \mid X(t_1), \ldots, X(t_n))$ to be the linear minimum mean square estimator of $X(t)$ given $X(t_1), \ldots, X(t_n)$, i.e. the linear function of $X(t_1), \ldots, X(t_n)$, denoted by $L_n = \sum_{r=1}^{n} a_r X(t_r)$, that minimizes $\mathbb{E}[(X(t) - L_n)^2]$ over all choices of $a_1, \ldots, a_n$. If for all $t_1, \ldots, t_n$ $\hat{\mathbb{E}}(X(t) \mid X(t_1), \ldots, X(t_n)) = \hat{\mathbb{E}}(X(t) \mid X(t_n))$, then $X(t)$ is said to be a wide-sense Markov process.

Other types of Markov process may be defined by extending the idea of conditional independence to parameter sets that are not rectilinear (i.e. not time-like); *see* Markov random field. *See* Feller process; Hunt process; and Ray–Knight compactification.

[S. Karlin & H. Taylor, *A First Course in Stochastic Processes*, 2nd edn. Academic Press, 1975; *A Second Course in Stochastic Processes*. Academic Press, 1981]

Markov property (alias: **Markov condition**)

Arises as a formalization, in the theory of random processes, of the observable lack-of-memory property of many real-world phenomena evolving in time. Such a process has the property that whenever it may have started, and whatever its path to its given present state s at time t may have been, its future evolution from its given state s at time t proceeds as though it had been started afresh from s at t. Equivalently, the future values of the process after t are conditionally independent of its past values prior to t, given its value $X(t) = s$ at t. These viewpoints give rise to technically different formalizations of the Markov property in a range of different frameworks, corresponding to different types of Markov process.

The classic elementary example is $X(n)$, the Markov chain in discrete time having stationary transition probabilities $p_{ij}(n), n \geq 1; i, j \in S$; where we write $p_{ij}(1) = p_{ij}$. Then $X(n)$ is said to have the Markov property (and to be a Markov process), if for any $0 < t_1 < \cdots < t_n \leq n$, and $x_1, \ldots, x_n, x \in S$, we have

$$\mathbb{P}(X_{n+1} = x \mid X(t_n) = x_n, \ldots, X(t_1) = x_1) = \mathbb{P}(X_{n+1} = x \mid X(t_n) = x_n).$$

This is equivalent to several other similar conditions, each of which is thus also a Markov property, or condition, e.g.:

(i) $\mathbb{P}(X_{n+1} = x \mid X_0 = x_0, \ldots, X_n = x_n) = \mathbb{P}(X_{n+1} = x \mid X_n = x_n)$ for all $x_0, \ldots, x_n$, x and $n \geq 1$.

(ii) $\mathbb{P}(X_0 = x_0, \ldots, X_{r-1} = x_{r-1}, X_{r+1} = x_{r+1}, \ldots, X_n = x_n \mid X_r = x_r) = \mathbb{P}(X_0 = x_0, \ldots, X_{r-1} = x_{r-1} \mid X_r = x_r)\mathbb{P}(X_{r+1} = x_{r+1}, \ldots, X_n = x_n \mid X_r = x_r)$ for all $x_0, \ldots, x_n$, and $1 \leq r \leq n$.

(iii) $\mathbb{P}(X_0 = x_0, X_1 = x_1, \ldots, X_n = x_n) = \mathbb{P}(X_0 = x_0)\, p_{x_0 x_1} p_{x_1 x_2} \cdots p_{x_{n-1} x_n}$.

For Markov chains in continuous time there are similar, naturally analogous Markov conditions. In both discrete and continuous time this leads to suitable forms of the Chapman–Kolmogorov equations as a corollary. If the Markov condition holds at stopping times (i.e. Markov times) of the process, then one refers to the **strong Markov property**. For example, one form of the strong Markov condition asserts that for any stopping time T of a Markov chain X_n,

$$\mathbb{P}(X_{T+1} = k \mid X_0 = x_0, \ldots, X_T = j) = p_{jk}.$$

In fact any discrete time Markov process (with countable or general state space) is a strong Markov process. In continuous time additional conditions of regularity are required for this to hold.

More generally, and more formally, a commonly seen form of the Markov property is this: Let $X(t), t \in T \subseteq \mathbb{R}$, be a random process defined on a suitable probability space, where $X(t)$ is adapted to a filtration $\mathcal{F}_t$ (typically that generated by $X(t)$). Let A and B be events depending on $X(u), u \geq t$ and $X(s), s \leq t$, respectively. Then $X(t)$ is said to be a Markov process if, for all such A, B, and t, it has the Markov property

$$\mathbb{P}(A \cap B \mid X(t)) = \mathbb{P}(A \mid X(t))\mathbb{P}(B \mid X(t)).$$

This is equivalent to another popular form of the Markov property,

$$\mathbb{P}(A \mid \mathcal{F}_t) = \mathbb{P}(A \mid X(t)).$$

The Markov property may be alternatively formulated in terms of a probability distribution on the space of possible paths (or trajectories) of the process $\mathbb{P}_x(\cdot)$, as x ranges over the possible initial values x of the process. Then (omitting some technical details) the random process $X(t)$, adapted to the filtration $\mathcal{F}_t, t \geq 0$, is said to be a Markov process if it has the Markov property

$$\mathbb{P}_x(X(t+u) \in A \mid \mathcal{F}_t) = \mathbb{P}_{X(t)}(X(u) \in A),$$

with probability 1 (wrt $\mathbb{P}_x(\cdot)), t, u \geq 0$.

This can alternatively be written as a suitable expectation of the form $\mathbb{E}_x(h(X(t+u)) \mid \mathcal{F}_t) = \mathbb{E}_{X(t)}(h(X(u)), u, t \geq 0$, where $\mathbb{E}_x$ denotes the expectation wrt to the family of measures indexed by the initial values of the process, and h is any member of a sufficiently rich class of functions (e.g. all bounded, measurable, real-valued functions).

For Markov processes on a more general state space S (in discrete time, for simplicity), one defines a transition (or Markov) kernel $k(x, A)$, for $x \in S$ and A an element of the sigma-field $\mathcal{F}$ of subsets of S. It is required that:

(i) $k(x, .)$ is a probability distribution an $(S, \mathcal{F})$ for each fixed x; and

(ii) $k(., A)$ is $\mathcal{F}$-measurable for each fixed A.

In this framework, the Markov property takes the form

$$\mathbb{P}_x(X(n+1) \in A \mid \mathcal{F}_n) = k(X(n), A),$$

where $X(n)$ is adapted to the filtration $\mathcal{F}_n$; with a similar property for kernels $k_t(x, A)$ arising in continuous time. In both cases there are alternative versions, and one obtains a form of the Chapman–Kolmogorov equations as a corollary.

For processes with parameter sets that are not time-like, the Markov property is expressed differently; *see* Markov random field and Bayes network (where one speaks of the local directed Markov property on a graph). [M. Sharpe, *General Theory of Markov Processes*. Academic Press, 1988; R. Bhattacharya & E. Waymire, *Stochastic Processes with Application*. Wiley, 1992; reprinted by SIAM, 2009]

Markov random field Broadly, a collection of random variables satisfying a Markov property (i.e. a conditional independence property), where the index set is a graph; so it may be referred to as a spatial Markov process (in contrast to the usual Markov processes having a time-like index). Formally, let $G = (E, V)$ be a finite graph (i.e. a collection of vertices V, joined by the edges of E), and suppose that G contains no loops (an edge joining a vertex to itself) and no multiple edges (more than one edge joining two vertices v and w). At each vertex $v \in V$ a random variable X_v takes values in a finite set A, where for simplicity we write $A = \{0, 1\}$. Thus the collection $X_v, v \in V$ takes values in $[0, 1]^V$, and we require that the probability of every point $s \in \{0, 1\}^V$ is strictly positive. For any $A \subseteq V$, define the boundary ∂A of A to be all vertices that are outside A connected by an edge to a point inside A. Then the collection $X_v, v \in V$ (or, equivalently, their probability distribution), is said to be a Markov random field if the two collections $(X_v : v \in A)$ and $(X_w : w \in V \backslash A)$ are conditionally independent given the values of $(X_u : u \in \partial A)$; or, equivalently, given $(X_w : w \in V \backslash A)$, the distribution of $(X_v : v \in A)$ depends only on the values of $(X_u : u \in \partial A)$. This is called the **global Markov property**. The **local Markov property** is the same statement made in the case when A is a single vertex a, so that ∂A is the set N_a of neighbours of a. These are equivalent to the so-called **pair-wise Markov property**, which requires that for any non-neighbouring vertices v and w (i.e. not joined by an edge), X_v and X_w are conditionally independent given the values of $(X_u : u \in V \backslash v \backslash w)$.

A key property of such fields is given by the Hammersley–Clifford theorem: they may be represented as a Gibbs random field (which is to say that the distribution may be written as a product over the complete subgraphs- or cliques-of G). The conditional distribution of X_a, given the values of $(X_v : v \in N_a)$, may be called the local characteristic, and one may refer to a conditionally specified process (or model). *See* Bayes network. [G. R. Grimmett, *Probability on Graphs*. Cambridge University Press, 2010]

Markov random set Arises as a level set of suitable Markov processes, and takes the form of a random closed subset of $\mathbb{R}^+ = [0, \infty)$. Informally, the most natural definition is through a type of Markov property, thus: let T be a non-empty random subset of $\mathbb{R}^+$ (defined on a suitable space), and for any fixed $t \in \mathbb{R}^+$ let M_t be the largest element of T not exceeding t. If $T \cap (t, \infty)$ is conditionally independent of $T \cap [0, M_t)$, given M_t, then T may be called a Markov random set. If t can be replaced by any appropriate stopping time S, then T may be called a strong Markov random set. The basic example of such a set is given by the zero set of a Markov chain with finitely many states, where T comprises i.i.d. exponentially distributed intervals separated by i.i.d. intervals (that are not necessarily exponentially distributed). The general Markov random set arises as the zero set of a semilinear Markov process (i.e. non-negative sawtooth process); and all Markov random sets can be given as the range of a suitable subordinator X_t, i.e. $T = \{x : x = X_t \text{ for some } t > 0\}$. [J. Horowitz, Semilinear Markov processes, subordinators and renewal theory, *Probab. Theor. Rel. Fields*, 24, 167–193, 1972]

Markov recurrence *see* random mapping representation

Markov renewal process A random process that is seen both as a generalization of Markov processes with a countable state space, and as a generalization of an ordinary renewal process. It has therefore been described as a marriage of the theories of Markov chains and renewal processes; and one of its offspring is the semi-Markov process. Formally, let $X(t), t \geq 0$, taking values in a countable state space S, be a jump process, whose nth jump occurs at $S_n = \sum_{r=1}^{n} T_r$, $n \geq 1$, where $T_r > 0$, $r \geq 1$, are the times between jumps. The successive values $X(S_n) = X_n, n \geq 0$ (where the empty sum is 0), comprise a Markov chain; and, conditional on $X_{n-1} = j$ and $X_n = k$, the inter-jump time T_n is independent of all other interevent times (and values of the chain) with a distribution function F_{jk} depending only on j and k. Define $N_s(t), s \in S$, to be the number of times that X_n has taken the value s up to time t, and let $S(t)$ be the state occupied by $X(t)$ at time t. Then the term Markov renewal process denotes the vector $N(t) = (N_1(t), N_2(t), \ldots)$, $t \geq 0$, or either of the bivariate processes (X_n, S_n) and $(X_n, T_n), n \geq 0$; while $S(t)$ is called a **semi-Markov (renewal) process**. [R. Pyke, Markov renewal processes: definitions and preliminary properties, *Ann. Math. Stat.*, 32(4), 1231–1242, 1961]

Markov semi-group *see* stochastic semi-group

Markov system *see* Markov process

Markov time *see* stopping time

Markov tree process *see* tree-indexed process; Bayes network

Markov uniqueness *see* Dirichlet form

Markowitz portfolio (alias: **Markowitz mean-variance model (or principle)**) Arises in financial mathematics when an investor seeks an investment strategy (or portfolio) over a set of assets whose returns comprise a random vector $(X_1, \ldots, X_a)$ with total return $R_a = \sum_1^a w_r X_r$, where $(w_1, \ldots, w_a)$ is the portfolio weighting. The mean-variance principle asserts that the allocation should seek to maximize the expected return $\mathbb{E}R_a$, for any given fixed value of the variance var R_a where var R_a is interpreted as a measure of the level of risk the investor will accept. The set of portfolios with smaller risk than any other for which the expected return is at least as great is called the **efficient** (or **Markowitz**) **frontier**. [J. Guerard (ed.), *Handbook of Portfolio Construction*. Springer, 2010]

marriage problem *see* secretary problem; Daniel Bernoulli's marriage problem

married couples problem *see* ménage problem

Marsaglia effect Arises in the context of Lehmer's multiplicative congruential generator for a sequence of pseudo-random numbers. Such a sequence $r_n, n \geq 0$, is generated (for some multiplier μ and modulus m) by $r_{n+1} = \mu r_n$, modulo m, with suitable seed (or initial value) r_0. Set $x_n = r_n/m$. Then a sequence of N-tuples $S_a = (x_{a+1}, \ldots, x_{a+N})$, regarded as a collection of pseudo-random points in the N-cube, has the property that all of them lie in at most $(N!m)^{1/N}$ parallel hyperplanes (which G. Marsaglia called the 'crystalline' effect). [G. Marsaglia, Random numbers fall mainly in the planes, *Proc. Nat. Acad. Sci. USA*, 61(1), 25–28, 1968]

Marsaglia's method Any one of several methods for generating pseudo-random numbers from a variety of distributions. Generate an i.i.d. sequence of pairs $(U_r, V_r), r \geq 1$, of independent r.vs uniform on $(0, 1)$. Then pseudo-r.vs appear as follows:

(a) Standard normal. Inspect the sequence until $Z = (2U_r - 1)^2 + (2V_r - 1)^2 \leq 1$; then $X = (2U_r - 1)\left(\frac{-2}{Z}\log Z\right)^{1/2}$ and $Y = (2V_r - 1)\left(\frac{-2}{Z}\log Z\right)^{1/2}$ are independent $N(0, 1)$ r.vs.

(b) Uniform distribution on a sphere. For the pair (U_r, V_r) defining Z in (a), set $S = 2U_r - 1$ and $T = 2V_r - 1$, and define $P = 2S(1 - S^2 - T^2)^{½}$;

$Q = 2T(1 - S^2 - T^2)^{½}; R = 1 - 2(S^2 + T^2)$. Then (P, Q, R) is uniformly distributed on the unit sphere.

(c) Normal tail. Inspect the sequence (U_r, V_r) until $X = V_r(c^2 - 2\log U_r)^{½} \le c$, for given $c > 0$. Then X is a sample from the density $f(x) = \mathrm{e}^{-x^2/2} / \int_c^\infty \mathrm{e}^{-x^2/2}\mathrm{d}x, \quad x \ge c$.

(d) Exponential. Generate auxiliary r.vs M and N, independent of $U_r, r \ge 1$, where $M \ge 0$ is a modified geometric r.v. with $\mathbb{P}(M = m) \propto \mathrm{e}^{-cm}$, and $N \ge 1$ is a modified Poisson r.v. with $\mathbb{P}(N = n) \propto c^n/n!; c > 0$. Then $W = (M + \min\{U_1, \ldots, U_N\})$ is a sample from the density $f(x) = \mathrm{e}^{-x}, x \ge 0$.

[L. Devroye, *Non-uniform Random Variate Generation*. Springer, 1986; J. Gentle, *Random Number Generation and Monte Carlo Methods*, 2nd edn. Springer, 2003]

Marshall's inequality Arises as a maximal inequality for martingales, thus: let $X_r, r \ge 1$, satisfy $\mathbb{E}X_1 = 0$, and $\mathbb{E}(X_n \mid X_1, \ldots, X_{n-1}) = 0, n \ge 2$, and $\mathbb{E}(X_n^2) < \infty$, $n \ge 1$. Then for $t > 0$,

$$\mathbb{P}\Big(\max_{1 \le r \le n}(X_1 + X_2 + \cdots + X_r) \ge t\Big) \le s_n/(s_n + t^2),$$

where $s_n = \sum_{r=1}^n \mathbb{E}(X_r^2)$. [A. Marshall, A one-sided analog of Kolmogorov's inequality, *Ann. Math. Stat.*, 31(2), 483–487, 1960]

Marshall–Olkin distribution (alias: **multivariate exponential distribution**) A multivariate distribution having exponentially distributed marginals that is characterized by an extension of the simple lack-of-memory property to higher dimensions. It arises also as a model for lifetimes that are curtailed by competing Poisson processes. Thus, defining the joint tail (or survival function) for the r.vr $X_1, \ldots, X_n$ as

$$\overline{F}(x_1, \ldots, x_n) = \mathbb{P}(X_1 > x_1, \ldots, X_n > x_n),$$

one natural generalization of lack-of-memory is

$$\overline{F}(x_1 + t, \ldots, x_n + t) = \overline{F}(x_1, \ldots, x_n)\overline{F}(t, \ldots, t).$$

Requiring the components $X_1, \ldots, X_n$ to have lack-of-memory individually entails (in, e.g. the bivariate case) that

$$\overline{F}(x, y) = \exp\{-ax - by - c\max\{x, y\}\}$$

for $a, b, c, x, y > 0$. The distribution is a mixture of a singular component on $x = y$, and a density on $\mathbb{R}_+^2$. Furthermore, there exist independent exponential r.vs U, V, and W such that (X, Y) has the representation $X = \min(U, W); Y = \min(V, W)$.

In the multivariate case, likewise, there is a family S of independent exponential r.vs $S = [Z_i : i \in I]$ such that for each r there is a subset $S_r \subseteq S$ such that $X_r = \min[Z_i : i \in S_r]$. [A. Marshall & I. Olkin, A multivariate exponential distribution, *J. Amer. Stat. Ass.*, 62, 30–44, 1967]

Martin boundary Arose originally in potential theory; in particular in the connexion between functions harmonic in a region D and non-negative measures on the boundary ∂D of D, which may be defined for suitable D by the so-called Martin compactification. In probability theory the term Martin boundary commonly refers to the probabilistic counterpart of classical potential theory, developed by J. Doob inter alia, with a suitable metric defined on a transient Markov chain with countable state space, using the so-called Martin kernel defined in terms of the Green function of the chain. The evolution of the process yields the Martin exit boundary, and suitable time-reversal yields a Martin entrance boundary. [J. Doob, Discrete potential theory and boundaries, *J. Math. Mech.*, 8, 433–458, 1959; J. Kemeny & L. Snell, Boundary theory for recurrent Markov chains, *Trans. Amer. Math. Soc.*, 106(3), 495–520, 1963]

martingale An important type of random process that is ubiquitous in probability and its applications, and arises naturally – in its most elementary form – as a model for the fortune of a gambler making a sequence of fair bets. A bet is said to be fair if, whatever the outcomes of the previous wagers, the expected return equals the stake; yielding zero net expected return. Broadly speaking, the sequence of fortunes comprises a sequence of (finite mean) r.vs such that each is centred at its conditional mean given the previous terms in the sequence.

Formally, let $M_0, M_1, \ldots$, and $D_1, D_2, \ldots$, be sequences of finite-mean r.vs such that for $n \ge 1, M_n = D_1 + \cdots + D_n + f$, with $M_0 = f$, and $M_{n+1} - M_n = D_{n+1}$, where f is a constant that may be interpreted as the initial fortune of the gambler. Then $M_n, n \ge 0$, is called a martingale (or sometimes a true, or strict martingale) if, for all $n \ge 0, \mathbb{E}(M_{n+1} \mid M_1, \ldots, M_n) = M_n$, and $D_n, n \ge 1$, is called a **martingale difference** (or absolutely fair) sequence having the property that for $n \ge 1, \mathbb{E}(D_{n+1} \mid D_1, \ldots, D_n) = 0$, with $\mathbb{E}D_1 = 0$. Then D_n is seen as the net result of the nth bet, and M_n the resulting net fortune after the nth bet.

More generally, $X_1, X_2, \ldots$ is a martingale with respect to a sequence Y_n if $\mathbb{E}|X_n| < \infty$ and $\mathbb{E}(X_{n+1} \mid Y_1, \ldots, Y_n) = X_n$. For example, M_n above is a martingale with respect to its difference sequence $D_n, n \ge 1$.

Yet more generally, if $\mathcal{F}_n, n \geq 1$ is an increasing sequence of sigma-fields $\mathcal{F}_1 \subseteq \mathcal{F}_2 \subseteq \cdots \subseteq \mathcal{F}$, then $X_1, X_2, \ldots$ defined on some $(\Omega, \mathcal{F}, \mathbb{P})$ is called a martingale wrt $\mathcal{F}_1, \mathcal{F}_2, \ldots$, if $\mathbb{E}|X_n| < \infty$, and X_n is $\mathcal{F}_n$-measurable, and $\mathbb{E}(X_n \mid \mathcal{F}_m) = X_m$, for all $m \leq n$.

Martingales have useful properties, of which the most important are perhaps the following (subject to suitable and rather weak conditions in each case):

(a) they converge (*see* martingale convergence);

(b) the martingale property is preserved at stopping times (*see* Doob's optional stopping theorem);

(c) some natural linear transformations of the martingale sequence are also martingales (*see* martingale transform and stochastic integral);

(d) they satisfy a number of powerful inequalities (*see* Doob's maximal inequality).

Related concepts include: (i) Let $X_n, n \geq 0$ be a sequence of r.vs adapted to the filtration $\mathcal{F}_n, n \geq 0$. Then X_1, X_2, is a **submartingale** wrt $\mathcal{F}_n$ if for all n, $\mathbb{E}X_n^+ < \infty$, and $\mathbb{E}(X_{n+1} \mid \mathcal{F}_n) \geq X_n$.

Reversing the inequality above,

(ii) a sequence $Y_0, Y_1, \ldots$ likewise adapted, to $\mathcal{F}_0, \mathcal{F}_1, \ldots$, is called a **supermartingale** if for all n, $\mathbb{E}Y_n^- < \infty$ and $\mathbb{E}(Y_{n+1} \mid \mathcal{F}_n) \leq Y_n$.

(iii) If a martingale sequence $X_0, X_1, \ldots$, is uniformly integrable then it may be represented as $X_n = \mathbb{E}(Z \mid \mathcal{F}_n)$ for a suitable finite-mean r.v. Z, and an increasing sequence of sigma-fields $\mathcal{F}_n$. It may then be called a Doob (or Doob–Lévy) martingale, or a regular martingale closed by Z.

(iv) If $\mathbb{E}X_n^2 < \infty$ for all X_n of a martingale sequence, then it may be called an L_2-martingale.

For more related processes *see* backward martingale, harness, local martingale, semi-martingale, weak martingale, wide-sense martingale, and martingale generalization. All the above definitions may be extended to martingales in continuous time by analogous definitions.

Finally, a random process $X(t), t \geq 0$, such that there exists a martingale $M(t), t \geq 0$, where the distribution of $X(t)$ and $M(t)$ is the same for all t, has been called a 1-martingale. *See* Kellerer's theorem. [O. Kallenberg, *Foundations of Modern Probability*, 2nd edn. Springer, 2002]

martingale betting system (alias: **doubling, or geometric progression system**) A classic sequential betting strategy, whose basic principle is this: In any sequence of bets the next stake following a lost bet should be sufficiently large that if the bet is won the gambler will be ahead, i.e. all previous stakes in the sequence of losses will be recouped plus a surplus. More formally, consider a sequence of independent games at evens, where the probability of winning (and thus profiting by the amount of the stake) is $p \leq 1/2$. We assume, wlog, that the first stake is 1.

If this wins, the gambler is ahead by 1. If this loses, and at every consecutive following loss, the gambler doubles the stake for the next game. Then L losses followed by a win yields a net gain of $2^{L+1} - 1 - 2 - 4 - \cdots - 2^L = 1$, with probability 1. However, the expected size of this winning bet is infinite, and anyone who uses this system repeatedly is almost sure to be bankrupt in the long run. [S. Ethier, *The Doctrine of Chances*. Springer, 2010]

martingale characterization of Markov chains Let X_n be a sequence of r.vs (taking values in some measurable space S) adapted to a filtration $\mathcal{F}_n$, and let $P(x, B)$ be a Markov transition kernel on S. For any bounded measurable function f on S define $(Pf)(x) = \int_S P(x, \mathrm{d}y) f(y)$. If, for every such f, and $n \geq 0$,

$$M_n = f(X_n) = \sum_{r=1}^{n-1} (Pf - f)(X_r) + f(X_0),$$

comprises a martingale for $\mathcal{F}_n$, then X_n is the Markov chain on S with kernel $P(x, B)$ (adapted to $\mathcal{F}_n$), and conversely. *See* martingale problem. [G. Last & A. Brandt, *Marked Point Processes*, Springer, 1995]

martingale characterization of the Wiener process *see* Lévy characterization of Brownian motion

martingale convergence theorem (alias: **Doob's (sub-)martingale convergence theorem**) Any of several theorems supplying conditions for the convergence (either almost surely or in rth mean) of a martingale (or of a sub- or super-martingale), in discrete or continuous time. For example, in discrete time, the basic submartingale convergence theorem is this: let $X_n, n \geq 1$, be a submartingale (either wrt itself or wrt a filtration $\mathcal{F}_n$). If $\mathbb{E}X_n^+ \leq M < \infty$ for some M and all n, then there exists a r.v. X such that $X_n \xrightarrow{\text{a.s.}} X$ as $n \to \infty$. Furthermore, if $\mathbb{E}|X_1| < \infty$ then $\mathbb{E}|X| < \infty$, and X_n converges in mean to X if the collection $(X_n, n \geq 1)$ is uniformly integrable. In particular, if $X_n, n \geq 1$, is a martingale, then the condition, $\mathbb{E}X_n^+ \leq m < \infty$ may be replaced by $\mathbb{E}|X_n| \leq m < \infty$. Furthermore, the following are equivalent for the martingale X_n:

(a) X_n is uniformly integrable;

(b) X_n converges in mean as $n \longrightarrow \infty$;
(c) there exists a finite-mean r.v. Z such that $X_n = \mathbb{E}(Z \mid \mathcal{F}_n)$;
(d) $\mathbb{E}|X_n| \leq m < \infty$, and the limit X above is such that $X_n = \mathbb{E}(X \mid \mathcal{F}_n)$, for the filtration $\mathcal{F}_n$.

Such a martingale is said to be regular, and closed; *see* Doob–Lévy martingale. [J. Doob, *Stochastic Processes*. Wiley, 1953, chapter VII]

martingale difference sequence Let $X_n, n \geq 0$, be a martingale wrt a sequence of sigma-fields $\mathcal{F}_n$. Its difference sequence is $D_n = X_n - \mathbb{E}(X_n \mid \mathcal{F}_{n-1})$, and by the properties of conditional expectation

$$(^*)\mathbb{E}[D_{n+1}g(D_1, \ldots, D_n)] = 0,$$

for $n \geq 1$, and any real-valued continuous function $g(\cdot)$ on $\mathbb{R}^n$; in particular $\mathbb{E}(D_{n+1} \mid \mathcal{F}_n) = 0$.

Therefore any sequence of r.vs D_n, adapted to the filtration F_n, is called a martingale difference sequence, or mds, if $\mathbb{E}|D_n| < \infty$, and $\mathbb{E}(D_{n+1} \mid \mathcal{F}_n) = 0$ for all n; or equivalently, if $(^*)$ holds, which is to say that D_{n+1} and $g(D_1, \ldots, D_n)$ are orthogonal for all such $g(\cdot)$ and n. More generally, if $\mathbb{E}(D_{n+1} \mid \mathcal{F}_n) \geq 0$ (resp. ≤ 0) then D_n is called a submartingale (resp. supermartingale) difference sequence. [D. Burkholder, Boundary value problems and sharp inequalities for martingale transforms, *Ann. Probab.*, 12(3), 647–702, 1984]

martingale generalization The term generalized martingale most often denotes a sequence X_n adapted to a filtration $\mathcal{F}_n, n \geq 0$, such that the conditional expectation exists and satisfies $\mathbb{E}(X_n \mid \mathcal{F}_m) = X_m, 0 \leq m \leq n$. That is to say, the integrability condition of the martingale is weakened. The term has also been applied to generalized stochastic processes (*see* white noise) that obey an analogue of the martingale condition. Finally, another class of generalizations seek to formalize the informal idea that a sequence of r.vs that is not a martingale may nevertheless be approximately or asymptotically a martingale, in some sense. Thus, let X_n be a sequence of r.vs such that $\mathbb{E}|X_n| < \infty$, and the sequence is adapted to the filtration $\mathcal{F}_n, n \geq 0$. Then:

(a) If $\mathbb{P}(\mathbb{E}(X_n \mid \mathcal{F}_m) - X_m) > \varepsilon$ converges to 0, for every $\varepsilon > 0$, as $n, m \longrightarrow \infty$ with $n \geq m$, then the sequence X_n is called a **game fairer with time**.
(b) If the limit in (a) holds almost surely, then the sequence is called a **martingale in the limit**.
(c) If for each increasing sequence of bounded stopping times $T_r, r \geq 1$, such that $T_n \longrightarrow \infty$ a.s., the sequence $\mathbb{E}(X_{T_r})$ converges as $r \longrightarrow \infty$, then the sequence X_n is called an **amart**, or **asymptotic martingale**.
(d) If $\sum_{n=0}^{\infty}\mathbb{E} \mid \mathbb{E}(X_{n+1} - X_n \mid \mathcal{F}_n)| < \infty$, then the sequence is called a **quasimartingale**.

[G. Edgar & L. Sucheston, A class of asymptotic martingales part A, *J. Multivariate Anal.*, 6(2), 193–221, 1976; & part B, loc. cit., 6(4), 572–591, 1976; J. Tompkins, Martingale generalizations and preservation of martingale properties, *Canadian J. Stat.*, 12(2), 99–106, 1984]

martingale inequality *see* Doob's maximal inequality

martingale measure *see* risk-neutral measure

martingale problem Denotes a method of characterizing weak solutions of suitable SDEs by associating them to an appropriate martingale. Informally, if the SDE is that of a diffusion with generator A (i.e. elliptic differential operator A), then the continuous stochastic process $X(t)$ (or its distributions $\mathbb{P}_t$), is said to solve the associated martingale problem for A if (for a class of suitably differentiable functions $g(\cdot)$),

$$M(t) = g(X(t)) - g(X(0)) + \int_0^t Ag(X(s))\mathrm{d}s, \quad t > 0,$$

is a martingale. [O. Kallenberg, *Foundations of Modern Probability*, 2nd edn. Springer, 2002]

martingale representation theorem (alias: **predictable representation (or decomposition)**) Any of several theorems that supply a representation (or decomposition) of a suitable class of martingales as integrals of a predictable process against another martingale. The basic example is the **Itô representation**, thus: let $M(t), t \geq 0$, be an a.s. continuous square-integrable martingale wrt the natural filtration $\mathcal{F}_t$ of the Wiener process $W(t)$; then there is a unique square-integrable predictable process $Y(t)$ such that $M(t) - M(0) = \int_0^t Y(s)\mathrm{d}W(s)$, a.s. More generally, the **Kunita–Watanabe decomposition** asserts that if X and M are suitable continuous local martingales wrt the same filtration (satisfying the usual conditions), with $M(0) = 0$, then a.s.

$$M(t) = \int_0^t Y(s)\mathrm{d}X(s) + Z(t),$$

for some predictable Y, where $Z(t)$ is a local martingale orthogonal to X (with $Z(0) = 0$), which is to say that the cross-variation of Z and X is zero for all $t \geq 0$. [B. Øksendal, *Stochastic Differential Equations*, 6th edn. Springer, 2003; P. Protter, *Stochastic Integration and Differential Equations*, 2nd edn. Springer, 2005]

martingale system *see* martingale betting system; martingale transform; optimal stopping

martingale transform Arises naturally in financial mathematics and gambling, as the outcome of a suitable investment strategy for an asset whose price follows a martingale, or as the resulting fortune after a sequence of bets.

Let $M_n, n \geq 0$, be a martingale (for the increasing sequence of sigma-fields $\mathcal{F}_n$; or, more simply, wrt itself). Suppose that $A_n, n \geq 1$, is a sequence of r.vs such that A_n depends only on $M_1, \ldots, M_{n-1}$ (or, more generally, is $\mathcal{F}_{n-1}$-measurable, i.e. non-anticipating for $\mathcal{F}_n$). Then the transform of M_n by A_n is defined as the process $X_n, n \geq 0$, where $X_0 = 0$, and $X_n = A_1(M_1 - M_0) + A_2(M_2 - M_1) + \cdots + A_n(M_n - M_{n-1})$. If $\mathbb{E}|X_n| < \infty$ for all n, then X_n is also a martingale; this holds (for example) if the sequence A_n is bounded.

More generally, the martingale transform theorem asserts that for a sequence X_n of r.vs adapted to the filtration $\mathcal{F}_n, n \geq 0$, the following are equivalent:
(a) X_n is a martingale transform;
(b) X_n is a local martingale;
(c) X_n is a generalized martingale.
Analogously, if M_n is a submartingale and each $A_n \geq 0$, then X_n defined above is a submartingale transform; that is also a submartingale if $\mathbb{E}X_n^+ < \infty$ for all n.
[A. Shiryayev, *Probability*, 2nd edn. Springer, 1996]

Martin-Löf randomness (alias: **1-randomness**) Arises in considering whether or not a particular infinite sequence $x_1x_2x_3\ldots$ of symbols can be regarded as random (wlog we may assume the sequence to be the binary expansion of some $x = 0 : x_1x_2 \cdots \in [0, 1]$). Informally, such a number is said to be Martin-Löf random if it passes every (statistical) test in a suitably defined class of such tests (and therefore called a universal test), that we might apply to conclude that x is not random. Note that this question is not addressed by the usual idea of randomness in Kolmogorov's axiomatization, in which, e.g. for a uniform r.v. $X \in [0, 1]$ the outcome $\{X = 0\}$ (say) is treated equally with $\{X = x\}$ for any x in $[0, 1]$; but taken in isolation, the sequence 0.0000… is not seen as intrinsically random. The original definition of P. Martin-Löf showed that non-random sequences identified by the universal test comprise a null set in $[0, 1]$, the rest are Martin-Löf random. Several equivalent definitions now exist, e.g.: A sequence $x_1x_2\ldots$ is M-L random iff there is a constant $c > 0$ such that for all $n, K(x_1 \cdots x_n) \geq n - c$, where $K(x_1x_2 \cdots x_n)$ is the Kolmogorov complexity of $x_1x_2 \ldots x_n$. It is a type of algorithmic randomness. [P. Martin-Löf, The definition of random sequences, *Inf. Control*, 9, 602–619, 1966]

Marton's inequality *see* Pinsker's inequality

Maruyama's (mixing) theorem Arises in seeking necessary and sufficient conditions for a stationary infinitely divisible process $X(t), t \in \mathbb{R}$, to be mixing in the following sense: Let $X(t)$ be defined on $(\Omega, \mathcal{F}, \mathbb{P})$ with associated shift operator T; then X is called mixing if, for all A and S in $\mathcal{F}$, $\lim_{t\to\infty}\mathbb{P}(A \cap T^tB) = \mathbb{P}(A)\mathbb{P}(B)$. In its commonest form, the theorem asserts that a Gaussian stationary process $X(t)$ is mixing iff $\lim_{|t|\to\infty}\mathrm{cov}(X(t+u), X(u)) = 0$. [M. Magdziarz, A note on Maruyama's mixing theorem, *Theor. Probab. Appl. (SIAM)*, 54(2), 322–324, 2010]

mass function *see* probability mass function

master equation A term widely used in the physical sciences to denote a class of differential equations modelling processes evolving over time; very often the processes are random. Broadly speaking, a master equation takes the form $\frac{\partial f}{\partial t} = Lf$, where f is (for example) a probability or density, and the operator L may involve differences, differentials, integrals, and so on; such equations often arise from a balance or conservation principle. In particular, in the case of Markov processes, the master equation is commonly derived from a formal (Taylor) expansion of the Chapman–Kolmogorov equations, and is called the forward equation in general, or by a variety of other names in special cases. [C. Gardiner, *Stochastic Methods: A Handbook*, 4th edn. Springer, 2010]

matching problem *see* derangement problem

Matérn covariance model (alias: **Whittle–Matérn covariance (or correlation) model**) Arises in considering suitable spatial random processes as a model that includes a parameter characterizing the smoothness (i.e. differentiability in a suitable sense) of the process. For example, in the case of a second-order stationary isotropic process having spectral density

$$f(\lambda) = a(b^2 + \lambda^2)^{-c-\frac{1}{2}}; \quad a, b, c > 0,$$

the corresponding Matérn, or Whittle–Matérn, covariance function is

$$c(t) = \frac{a\sqrt{\pi}b^c|t|^c}{2^{c-1}\Gamma\left(c+\frac{1}{2}\right)b^{2c}} K_c(b|t|),$$

where K_c is a modified Bessel function. The random process is then n times differentiable in mean square for any integer $n < c$, where c may be called the smoothness parameter. [P. Guttorp & T. Gneiting, On the Whittle-Matérn correlation family, *NR. CSE Technical Report*, 80, 2005; P. Whittle, On stationary processes in the plane, *Biometrika*, 41, 434–449, 1954]

Matérn hard-core model *see* hard-sphere model

Mathias theorem Supplies a criterion for an integrable, real, continuous, even function $\phi(t), t \in \mathbb{R}$, to be a characteristic function. It asserts that such a ϕ is a chf iff $\phi(0) = 1$ and

$$(-1)^n \int_{-\infty}^{\infty} \phi(pt)\mathrm{e}^{-t^2/2} H_{2n}(t)\mathrm{d}t \geq 0,$$

for all $n = 0, 1, 2, \ldots$, and $p > 0$; where $H_{2n}(t)$ is the Hermite polynomial of degree $2n$. [E. Lukacs, *Characteristic Functions*, 2nd edn. Griffin, 1970]

matrix-exponential distribution (alias: **matrix-geometric distribution**) Comprises a class of distributions on $\mathbb{R}^+$ that generalizes both Coxian and phase-type distributions on $\mathbb{R}^+$, and coincides with the distributions on $\mathbb{R}^+$ that have an mgf (or Laplace transform) $M(\theta)$ that is a rational function of θ. Explicitly, X has a matrix exponential distn if there exists a row n-vector $\mathbf{a}$, a column n-vector $\mathbf{b}$, and an $n \times n$ matrix A (with possibly complex entries), such that X has density $f(x) = \mathbf{a}\mathrm{e}^{Ax}\mathbf{b}, x \geq 0$.

This representation is not unique: the smallest n for which there is such a representation is the minimal order or degree of f. For a non-negative integer-valued r.v. X, a distribution of the same form on $\mathbb{Z}^+$ is called matrix-geometric. [M. Bladt & B. Nielsen, Multivariate matrix-exponential distributions, *Stoch. Models*, 26(1), 1–26, 2010; M. Neuts, *Matrix-geometric Solutions in Stochastic Models*. Johns Hopkins, 1981; reprinted by Dover, 1994]

matrix process *see* pseudo-Markov model

Matsui's piling-up lemma *see* piling-up lemma

Matsumoto–Yor property Supplies a characterization of the generalized inverse Gaussian (GIG) and gamma distributions, as follows. Let X and Y be independent with the GIG and gamma densities respectively

$$f_X(x; a, b) = Ax^{-c-1}\exp(-ax - b/x), \quad x > 0;$$
$$f_Y(x, a) = Bx^{c-1}\exp(-ax), \quad x > 0;$$

where a, b, c, A, and B are non-negative constants. Define $U = (X+Y)^{-1}$ and $V = X^{-1} - (X+Y)^{-1}$. Then U and V are independent with respective densities $f_X(u; b, a)$ and $f_Y(v; b)$, and conversely. [G. Letac & J. Wesołowski, An independence property for the product of GIG and gamma laws, *Ann. Probab.*, 28(3), 1371–1383, 2000]

Matthews bound Any one of several inequalities for the expected value (or mgf) of the cover time (or partial cover times) of a Markov chain, that can be obtained by Matthews' method which uses an independent auxiliary randomization of the order in which the states are visited. One example is this: let $X(t), t \geq 0$, be an irreducible Markov chain on a finite state space $\{1, \ldots, n\}$. Let C_x be the first time at which all states have been visited starting from x, and let c be its worst-case expectation: $c = \max_x \mathbb{E}C_x$. Let T_{jk} be the first time (≥ 0) at which the chain is k starting from j, and let w be its worst case expectation $w = \max_{j,k}\mathbb{E}T_{jk}$. Then $c \leq \left\{1 + \frac{1}{2} + \cdots + \frac{1}{n}\right\}w$. [D. Levin et al., *Markov Chains and Mixing Times*. American Mathematical Society, 2009]

Matula's theorem Denotes any one of several inequalities for the distribution and moments of various functions of a sequence $X_n, n \in \mathbb{N}$, of associated random variables; and also denotes any one of several convergence theorems, including laws of large numbers and central limit theorems. [P. Matula, Limit theorems for sums of nonmonotonic functions of associated random variables, *J. Math. Sci.*, 105(6), 2590–2593, 2001]

maximal correlation (alias: **Gebelein maximal correlation coefficient**) A measure of association between jointly distributed r.vs X and Y given in terms of the ordinary (Pearson) correlation function $\rho(.,.)$ by

$$r(X, Y) = \sup \rho(f(X), g(Y)) \in [0, 1],$$

where the supremum is over all real-valued measurable f and g having non-zero finite variance. Unlike ρ, r is defined whenever neither X nor Y is degenerate, and $r = 0$ iff X and Y are independent. (Though there exist sequences $(X_n, Y_n), n \geq 1$, such that $r(X_n, Y_n) = 1$ for all n, but in the limit as $n \to \infty, X_n$ and Y_n are asymptotically independent.) If f and g are restricted to be monotone in their arguments, the supremum defined as above is the monotone correlation coefficient r_{mon}, which is also 0 iff X and Y are independent.

In particular, if (X, Y) is bivariate normal then $r = |\rho|$ (*see* Gebelein–Lancaster theorem). [N. Papadatos & T. Xifara, A simple method of obtaining the maximal correlation coefficient, and related characterizations, *J. Multivariate Anal.*, 118, 102–114, 2013]

maximal ergodic theorem (or lemma) (alias: **Wiener's (or Wiener–Yosida–Kakutani) maximal inequality**) Arises in ergodic theory, with applications to Birkhoff's ergodic theorem. Let $X_r, r \geq 0$ be a strongly stationary sequence of r.vs having finite mean; such as may be generated by a measure-preserving transformation T_r acting on a finite-mean r.v. Y, so that $X_r = Y \circ T^r, r \geq 0$. Let $M = \sup_{n\geq 1} \frac{1}{n}\sum_{r=0}^{n-1} X_r$, and for $\lambda > 0$, let $I(\lambda)$ be the indicator of the event that $M > \lambda$ a.s. Then $\lambda \mathbb{P}(M > \lambda) \leq \mathbb{E}\{X_0 I(\lambda)\}$.

More generally, this still holds when λ is an invariant function of the transformation T (such that $\mathbb{E}\lambda^+ < \infty$.), in the form $\mathbb{E}\{(Y - \lambda)I(\lambda)\} \geq 0$. [M. Keane & K. Petersen, Easy and nearly simultaneous proofs of the ergodic theorem and maximal ergodic theorem, in *Dynamics and Stochastics*, D. Denteneer et al. (eds). IMS, 2006]

maximal inequality Broadly speaking, such an inequality (sometimes called a maximum inequality) supplies a bound for the moments, or for the tail probabilities, of the supremum of a collection of random variables. In particular, the collection may be a random process, or the collection of partial sums of a random process. Examples include the inequalities of Bernstein, Burkholder–Davis–Gundy; Doob; Háyek–Rényi–Chow; Hoeffding; Kolmogorov; Lévy; Ottaviani, inter alia.

maximin *see* minimax principle

maximum entropy principle (or method) (alias: **Jayne's principle; MAXENT, or maxent**) Arises as a method for assigning a probability distribution $p(x), x \in S$, to the possible values (or states) of a random system $X \in S$ (where the true distribution is, by hypothesis, unknown; but the state space S is known). Typically, there are constraints on the expected value of various functions of X, e.g. $\mathbb{E}X > 0$; var $X = \sigma^2 < \infty$, and so on. The maximum entropy principle chooses the distribution $\hat{p}$ (or family F of distributions) that satisfies the constraints and has maximum entropy; where the entropy to be maximized is the Shannon entropy $-\sum p(x)\log_2 p(x)$ in the discrete case, or the (Jaynes) differential entropy if $p(x)$ is a density on S.

If the maxent distribution is not unique, and there exists an empirical measurement, or an a priori estimate, $q(x)$ of $p(x)$, then the so-called minimum cross-(or relative) entropy principle chooses that $\hat{p}$ in F which minimizes the relative (or cross-) entropy between $\hat{p}$ and q, i.e. it minimizes

$$d(\hat{p}, q) = \sum \hat{p}(x)\log\{\hat{p}(x)/q(x)\}.$$

[E. Jaynes, *Probability Theory: The Logic of Science*. Cambridge University Press, 2003]

maximum likelihood principle *see* likelihood

max-stable distribution *see* extreme value distributions

max-stable process Arises in extending the usual extreme value theory (*see* extreme value distributions) to sequences of i.i.d. random processes. Thus, let $Y_0, Y_1, \ldots$, be a sequence of i.i.d.r.vrs in $\mathbb{R}^d$. If there exist vectors b_n, and $a_n > 0$, such that for all $n, a_n Y_0 + b_n$ has the same distribution as $\max_{1\leq r\leq n} Y_r$ (the maximum being taken coordinatewise) then Y_0 is said to have a max-stable distribution. A random process $X(t), t \in T$, is said to be max-stable if all its finite dimensional distributions are max-stable. Now let $X_i(t)$ be i.i.d. copies of $X(t)$, and suppose that for some suitable sequences of functions $a_n(t)$ and $b_n(t)$ the process $a_n(t)^{-1}\max_{1\leq r\leq n} X_r(t) - b_n(t)$, $t \in T$, converges (as $n \to \infty$) in distribution to a non-degenerate random process $Y(t)$. Then $Y(t)$ is necessarily max-stable. [S. Resnick, *Extreme Values, Regular Variation, and Point Processes*. Springer, 1987]

Maxwell's characterization of the normal density Asserts that the distribution of the vector $(X_1, \ldots, X_d)$ is spherically symmetric in $\mathbb{R}^d, d \geq 2$, where $X_1, \ldots, X_d$ are i.i.d.r.vs, iff the Xs are zero-mean normal (i.e. Gaussian) r.vs. On J.C. Maxwell's assumption that the distribution of the velocity $V = (V_1, V_2, V_3)$ of a molecule in three dimensions has this property, it follows that the density of $|V| = (V_1^2 + V_2^2 + V_3^2)^{1/2}$ is (in standard form)

$$f(v) = \sqrt{\frac{2}{\pi}} v^2 e^{-\frac{1}{2}v^2}, \quad v > 0.$$

This is called **Maxwell's distn.** [J. Maxwell, Illustration of the dynamical theory of gases, *Phil. Mag.*, 19, 19–32, 1860]

McDiarmid's inequality (alias: **bounded differences inequality**) Let $X_1, \ldots, X_n$ be independent, but not necessarily identically distributed, and define $Z = g(X_1, \ldots, X_n)$, where g is a function having bounded differences (which can be seen as a type of Lipschitz condition). That is to say, for $1 \le r \le n$, there are constants c_r so that

$$|g(x_1, \ldots, x_n) - g(x_1, \ldots, x_{r-1}, y, x_{r+1}, \ldots, x_n)| \le c_r > 0.$$

Then for all $t > 0$,

$$\mathbb{P}(|Z - \mathbb{E}Z| > t) \le 2\exp\left\{ -2t^2 \Big/ \sum_{r=1}^{n} c_r^2 \right\}.$$

It generalizes Hoeffding's inequality. [C. McDiarmid, Concentration; in: *Probabilistic Methods for Algorithmic Discrete Mathematics*, M. Habib et al. (eds). Springer, 1998, pp. 195–248]

McKay's law (or measure) Arises as the limit of the empirical distributions of the eigenvalues of the adjacency matrices of a suitable sequence of d-regular random graphs with n vertices, as $n \to \infty$. For fixed vertex degree d, the density of the limit is

$$k(x) = \frac{d}{2\pi(d^2 - x^2)}\{4(d-1) - x^2\}^{\frac{1}{2}},\ x^2 \le 4(d-1).$$

This is also called **Kesten's** (or the **Kesten–McKay**) **measure**, having arisen earlier in the context of random walks on groups. As $d \to \infty$, a suitably scaled version of k converges to Wigner's semicircle distribution. [B. McKay, The expected eigenvalue distribution of a large regular graph, *J. Linear Alg. Appl.*, 40, 203–2161981]

McKean–Vlasov equation Arises in seeking to describe the behaviour of a plasma, in which particles have long-range interaction. Kac's model involves n interacting and diffusing particles in a line; in an appropriate limit as $n \to \infty$ (called the McKean–Vlasov limit) one obtains the so-called McKean–Vlasov eqn for the density f of the particles in the form

$$\frac{\partial f}{\partial t} = \frac{1}{2}\frac{\partial^2 f}{\partial x^2} - \frac{\partial}{\partial x}(K(x,f)f),$$

where $K(x,f) = \int a(x,y)f(t,y)\mathrm{d}y$, and $a(x,y)$ is the interaction function. [H. McKean, A class of Markov process associated with non-linear parabolic equations, *Proc. Natl. Acad. Sci. USA*, 56, 1907–1911, 1966]

McMillan's theorem *see* Shannon–McMillan theorem

McShane stochastic integral (alias: **stochastic (or Itô–) belated integral**) An integral similar to (but perhaps simpler than) the Itô integral, that agrees with the Itô integral when both exist, and with the Lebesgue integral in the non-random case, being a special case of the Kunita–Watanabe integral wrt a suitable semi-martingale. It has a useful consistency property, in that solutions of suitable differential equations driven by wide-band noise converge in their white noise limit to solutions of McShane-type SDEs. [E. McShane, *Stochastic Calculus and Stochastic Models*. Academic Press, 1974]

***m*-dependence** A sequence of r.vs $X_r, r \ge 1$, is said to be m-dependent if, for $k - j > m$, the two sets $(X_1, \ldots, X_j)$ and $(X_k, X_{k+1}, \ldots)$ are independent. Similarly it is $g(n)$-dependent if for $k - j > g(n)$, $(X_1, \ldots, X_j)$ and $(X_k, \ldots, X_n)$ are independent. [W. Hoeffding & H. Robbins, The central limit theorem for dependent random variables, *Duke Math. J.*, 15, 773–780, 1948]

mean *see* expected value; Kolmogorov–Nagumo mean

mean deviation *see* variation coefficient

mean difference (alias: **coefficient of mean difference; absolute mean difference; Gini mean difference**) A measure of the dispersion (over its support) of a distribution function $F(x)$, given by $\Delta = \int\int |x - y|\mathrm{d}F(x)\mathrm{d}F(y) = \mathbb{E}|X_1 - X_2|$, where X_1 and X_2 are i.i.d. with distn fn $F(x)$. One multiple of Δ that is scale-free (or dimensionless) is called the Gini coefficient of concentration G; there are several other representations of Δ and G; *see* Lorenz curve. [M. Kendall et al., *Advanced Theory of Statistics, Vol. 1*, 6th edn. Edward Arnold, 1999]

mean direction Arises in considering a directional distribution in $\mathbb{R}^d, d \ge 2$, typically represented by a random unit vector $\mathbf{U}$ on a circle, sphere, or hypersphere. For $d = 2$, denoting the polar angle of $\mathbf{U}$ by Θ, define μ by $\mathbb{E}\cos\Theta = r\cos\mu, \mathbb{E}\sin\Theta = r\sin\mu$, provided $r > 0$. Then the mean direction is given by the vector $(\cos\mu, \sin\mu)$ and by its polar angle μ. Likewise for $d = 3$, in spherical polars (Θ, Φ), the mean direction is (μ, υ) defined by

$$\mathbb{E}\cos\Theta\sin\Phi = r\cos\mu\sin\upsilon,$$
$$\mathbb{E}\sin\Theta\sin\Phi = r\sin\mu\sin\upsilon,$$
$$\mathbb{E}\cos\Phi = r\cos\upsilon,$$

provided that $r > 0$. For $d > 3$, the mean direction is defined similarly. [K. Mardia & P. Jupp, *Directional Statistics*. Wiley, 2000]

mean field theory A term used in many different areas of statistical physics and other sciences; it can denote various meanings depending on the context. Very roughly speaking indeed, the term is often (but not exclusively) used in dealing with non-linear interactions between many objects, when (for tractability) some random quantity is replaced by its expected value. The manoeuvre can often be interpreted in a rigorous context either as the result of taking some limit, e.g. the number of dimensions (or particles) goes to infinity, or of assuming some independence. [M. Opper & D. Saad, *Advanced Mean Field Methods*. M.I.T. Press, 2001]

mean measure *see* point process

mean-reverting square-root diffusion process *see* Cox–Ingersoll–Ross model

mean square contingency (alias: **phi (or Φ-) coefficient; Pearson's ϕ**) Arises as a measure of association (or correlation) for a bivariate (or multivariate) distribution. Thus, let r.vs X and Y have either a joint density or a joint probability mass function, denoted by $f(x,y)$, with marginals $f_X(x)$ and $f_Y(y)$. If $k(x,y) = f(x,y)/\{f_X(x)f_Y(y)\}$ is integrable, then the mean square contingency $\phi \geq 0$ is given by

$$\phi^2 = \iint (k-1)f(x,y)\mathrm{d}x\mathrm{d}y = \mathbb{E}k(X,Y) - 1$$
$$= \iint (k-1)^2 f_X(x)f_Y(y)\mathrm{d}x\mathrm{d}y,$$

(with $\sum$ replacing $\int$ in the discrete case). It is a type of f-divergence. [A. Rényi, On measures of dependence, *Acta Math. Acad. Sci. Hungary*, 10(3–4), 441–451, 1959]

mean-square convergence Let $X, X_1, X_2, \ldots$ be a sequence of finite-variance r.vs. If $\lim_{n\to\infty}\mathbb{E}(X_n - X)^2 = 0$, then X_n is said to be mean-square convergent to X, and we write $X_n \overset{\text{m.s.}}{\to} X$, or $X_n \overset{L_2}{\to} X$, or $X_n \overset{2}{\to} X$. It is equivalent to the statement $\lim_{m,n\to\infty}\mathbb{E}(X_n - X_m)^2 = 0$; which is Cauchy mean-square convergence. More generally, in $\mathbb{R}^d$, for a sequence of random vectors $X_n = (X_{n1}, \ldots, X_{nd})$, $n \geq 1$, each of whose components all have finite variance, $\mathbf{X}_n$ is said to converge in mean square to a random vector $\mathbf{X} = (X_1, \ldots, X_d)$, if $\lim_{n\to\infty}\mathbb{E}\sum_{r=1}^d (X_{nr} - X_r)^2$. This is equivalent to the ordinary mean-square convergence of each of the d sequences $X_{nr}, n \geq 1, 1 \leq r \leq d$. [S. Karlin & H. Taylor, *A First Course in Stochastic Processes*, 2nd edn. Academic Press, 1975]

mean-square error (alias: **expected quadratic loss; m-s error of prediction**) Arises in the context of jointly distributed random finite-variance variables X and Y, one of which (Y, say) is observed, so that a suitable function $h(Y)$ of Y may be used to give a prediction or estimate of X, assumed to be unobserved. The mean square error of prediction (m.s.e.p, or m-s.e.p.), is

$$\eta(X, h(Y)) = \mathbb{E}\{(X - h(Y))^2\}.$$

The minimum m-s.e. predictor is a function $h*(Y)$ that minimizes $\eta(X, h(Y))$. Here $h*(Y) = \mathbb{E}(X \mid Y)$, and the m-s.e.p. is $\mathbb{E}(X^2) - \mathbb{E}\{(\mathbb{E}(X \mid Y))^2\} = \mathbb{E}\,\mathrm{var}\,(X \mid Y)$. The best m-s.e. linear predictor is a function $\hat{h}(Y) = a + bY$ such that $\eta(X, h(Y))$ is least, for some constants a and b.

In this case $\hat{h}(Y) = \mathbb{E}X + \mathrm{cov}(X,Y)\frac{(Y - \mathbb{E}Y)}{\mathrm{var}\,Y}$, with m-s. e.p. given by $\{1 - \rho^2(X,Y)\}\mathrm{var}\,X$, where ρ is the correlation. [S. Karlin & H. Taylor, *A First Course in Stochastic Processes*, 2nd edn. Academic Press, 1975]

mean value option *see* asian option; exotic option

measure Informally, a number $m(A)$ that is assigned to any suitable set A, that indicates its size or content, broadly interpreted. The concept is essential in Kolmogorov's axiomatization of the theory of probability and random processes. Basically, a space Ω, being an arbitrary non-empty set, is equipped with a non-empty sigma algebra (or σ-field) $\mathcal{F}$ of subsets of Ω. That is to say, if A is in $\mathcal{F}$ then A^c is in $\mathcal{F}$; and, also, if $A_n \in \mathcal{F}$, $n \geq 1$, then $\cup_{n=1}^{\infty} A_n \in F$. The pair $(\Omega, \mathcal{F})$ is called a **measurable space**, with $A \in \mathcal{F}$ being a **measurable set**.

A set function $m(A) \in \mathbb{R}$, defined on all A in $\mathcal{F}$ is countably (or sigma) additive if for any family $A_n, n \geq 1$, of disjoint members of $\mathcal{F}$ having union $A = \cup_{n=1}^{\infty} A_n$, the sum $\sum_{n=1}^{\infty} m(A_n)$ exists and equals $m(A)$. A non-negative countably additive set function $m(\cdot)$ on $\mathcal{F}$, with $m(\emptyset) = 0$, is a **measure**, and $(\Omega, \mathcal{F}, m)$ is a **measure space**.

If $m(\Omega) = 1$, then it is a probability space. In this case, suppose that $X(\omega) \in \hat{\Omega}$, $\omega \in \Omega$, is a function (or mapping) from $(\Omega, \mathcal{F})$ to another measurable space $(\hat{\Omega}, \hat{\mathcal{F}})$; then it is called a measurable mapping (or said to be a measurable function) if $X^{-1}(B) \in \mathcal{F}$ for any $B \in \hat{\mathcal{F}}$ and X is then called a random element (or object) in $\hat{\Omega}$. A real-valued random element is a random variable, and an $\mathbb{R}^d$ valued random element is a random vector. If $\hat{\Omega} = \mathbb{R}^{\infty}$, then X may be called a random (or stochastic) process (or sequence); the same term is used when $\hat{\Omega}$ is a suitable space of functions. When $\hat{\Omega}$ is a class of sets or measures, then $X(\cdot)$ may be called a random set, or random measure, or a measure-valued random process, etc. [O. Kallenberg, *Foundations of Modern Probability*, 2nd edn. Springer, 2002]

measure-preserving transformation Arises in the context of strongly stationary sequences and ergodic theory. Given a probability space $(\Omega, \mathcal{F}, \mathbb{P})$, a measurable transformation (function) T, of Ω into Ω, is said to be **measure-preserving** if for every $A \in \mathcal{F}, \mathbb{P}(T^{-1}(A)) = \mathbb{P}(A)$. If $X_1(\omega)$ is a random variable on $(\Omega, \mathcal{F}, \mathbb{P})$ and $X_n(\omega) = X_1(T^{n-1}(\omega)), n \geq 2$, then the sequence $X_1, X_2, \ldots$ is stationary (strictly). Conversely, if $X_1, X_2, \ldots$ is a strongly stationary sequence, there exists a probability space $(\Omega^*, \mathcal{F}^*, \mathbb{P}^*)$, and a random variable Y_1, and a measure-preserving transformation T on it, such that $Y_1(\omega^*), Y_1(T(\omega^*)), \ldots$ has the same distribution as $X_1, X_2, \ldots$.

If T is measure-preserving, and $T^{-1}(A) = A$ for $A \in \mathcal{F}$, then A is **invariant** (or almost invariant if A and $T^{-1}(A)$ differ by a set of measure 0). If every invariant set A has either $\mathbb{P}(A) = 1$ or $\mathbb{P}(A) = 0$, then T is said to be **ergodic**, or **metrically transitive**. This occurs iff every almost invariant set A has $\mathbb{P}(A) = 0$ or $\mathbb{P}(A) = 1$.

A measure-preserving transformation T is mixing if $\lim_{n\to\infty}\mathbb{P}(A \cap T^{-n}(B)) = \mathbb{P}(A)\mathbb{P}(B)$ for $A, B \in \mathcal{F}$, and any such T is ergodic. Less strongly, if $\lim_{n\to\infty}\frac{1}{n}\sum_{r=0}^{n}\mathbb{P}(A \cap T^{-n}(B)) = \mathbb{P}(A)\mathbb{P}(B)$, then T is said to be weakly (or Cesàro) mixing. For any measure-preserving T, Poincaré's recurrence theorem asserts that for $\omega \in A \in \mathcal{F}, T^n(\omega) \in A$ for infinitely many values of n. For an ergodic T, Birkhoff's ergodic theorem asserts that when $\mathbb{E}|X| < \infty$, with probability 1

$$\lim_{n\to\infty}\frac{1}{n}\sum_{r=0}^{n-1} X(T^r(\omega)) = \mathbb{E}X.$$

[R. Ash & C. Doléans-Dade, *Probability and Measure Theory*, 2nd edn. Academic Press, 2000; S. Kalikow & R. McCutcheon, *An Outline of Ergodic Theory*. Cambridge, 2010]

measure theoretic entropy *see* Kolmogorov–Sinai entropy

Mecke's formula *see* Campbell's theorem

median Formalizes the idea of a middlemost (or central) value of a probability distribution, thus:
A median m of a distribution function $F(x)$ is any number m such that $F(m-) = \lim_{a\uparrow m}F(a) \leq 1/2 \leq F(m)$.
If X is continuous, then m is unique, and $\mathbb{P}(X \geq m) = \mathbb{P}(X \leq m) = 1/2$. In general, if $F(m) = 1/2$ for every m in an interval (a, b) then every point of $[a, b]$ is commonly termed a median (*see* quantile). A median m of X has the property of minimizing $\mathbb{E}|X - a|$ over all choices of a; furthermore $(m - \mathbb{E}X)^2 \leq \operatorname{var} X$ for any finite-mean r.v. X. More generally, the spatial median $\mathbf{m}$ of the multivariate distribution of $\mathbf{X}$ may be defined as $\mathbf{m}$ that minimizes $\mathbb{E}\|\mathbf{X} - \mathbf{a}\|$ over choices of $\mathbf{a}$, where $\|.\|$ is Euclidean distance. *See* depth function. [H. Hotelling & L. Solomons, The limits of a measure of skewness, *Ann. Math. Stat.*, 3(4), 141–142, 1932]

Mehler's expansion (or formula, or identity) (alias: **tetrachoric series; Mehler Hermite polynomial formula**) Let $\phi(x, y)$ be the standard bivariate normal density with correlation ρ, having marginal densities $\phi_X(x)$ and $\phi_Y(y)$. Then

$$\phi(x, y) = \phi_X(x)\phi_Y(y)\left\{1 + \sum_{r=1}^{\infty}\frac{\rho^n}{r!}H_r(x)H_r(y)\right\},$$

where $H_r(x)$ is the Hermite–Chebyshov polynomial of order r (and the normalized version is given by $\overline{H}_r(x) = (r!)^{-1/2}H_r(x), \quad r \geq 1$). [C. Klaassen & J. Wellner, Efficient estimation in the bivariate normal copula model, §6, *Bernoulli*, 3(1), 55–77, 1997]

Meixner distribution In its standard form (before shifting and scaling) this has density

$$f(x) = \frac{e^{bx}}{2\pi\Gamma(2a)}\left\{2\cos\left(\frac{1}{2}b\right)\right\}^{2a}|\Gamma(a + ix)|^2, \quad x \in \mathbb{R},$$

where $\Gamma(\cdot)$ is the gamma function and $i = \sqrt{-1}$. It has mean $\mathbb{E}X = a \tan\left(\frac{1}{2}b\right)$, variance $\operatorname{var} X = \frac{a}{1+\cos b}$, and characteristic function

$$\phi(t) = \left\{2\cos\left(\frac{1}{2}b\right)\right\}^{2a}\left\{\cosh\frac{1}{2}(t - ib)\right\}^{-2a}.$$

The distribution is infinitely divisible, and the associated Lévy process is called a **Meixner process**. [E. Mazzola & P. Muliere, Reviewing alternative characterizations of the Meixner process, *Probab. Surv.*, 8, 127–154, 2011]

Melamed's theorem *see* Jackson network

memory *see* long-range dependence

memoryless channel *see* channel

ménage problem (alias: **married couples problem; probléme des mènages**) A classic problem of probability, first stated as follows by E. Lucas in 1891. A circular table has $2n$ seats numbered consecutively, and n married couples are seated with men in even seats and women in odd seats, but otherwise each sex is seated at random. What is the probability p that nobody is next to their spouse? It is

$$p = \frac{1}{n!}\sum_{r=0}^{n}(-1)^r \frac{2n}{2n-r}\binom{2n-r}{r}(n-r)!, \quad n \geq 1.$$

[K. Bogart & P. Doyle, Non-sexist solution of the ménage problem, *Amer. Math. Mon.*, 93(7), 514–518, 1986]

merged processes *see* superposition

Mersenne twister A pseudo-random number generator, whose recurrence is based on a matrix multiplication, that yields outcomes distributed over the set $\{0, \ldots, 2^W - 1\}$ for some integer W, and which has period p that is a Mersenne prime (of the form $p = 2^m - 1$); hence the name. [M. Matsumoto & T. Nishimura, Mersenne twister, a 623-dimensionally equidistributed uniform pseudo-random number generator, *ACM Trans. Mod. Comp. Simul.*, 8(1), 3–30, 1998]

Merton formula *see* Black–Scholes–Merton theory

Merton model Any of several models proposed by R. C. Merton for concepts in mathematical finance. These include models for the probability of default (alias: asset value model, or credit risk model); and extensions of the Black–Scholes market model, yielding Black–Scholes–Merton theory, with its various **Black–Scholes–Merton formulas** for prices of options to be traded in that market. In particular, the **Merton price-process model** is a jump-diffusion process, in which the usual geometric Brownian motion is augmented by i.i.d. jumps at the instants of an independent Poisson process. Furthermore, the Merton short-term interest-rate model for the instantaneous short-term interest-rate $r(t)$ is the SDE $\mathrm{d}r = a\mathrm{d}t + b\mathrm{d}W(t)$, for suitable a and b and the standard Wiener process $W(t)$, under the appropriate risk-neutral measure. [R. Merton, Theory of rational option pricing, *Bell J. Econ. Manage. Sci.*, 4(1), 141–183, 1973]

Merton portfolio problem Comprises an investor whose wealth at time t is $M(t), 0 \leq t \leq L$, who must decide for all $t \in [0, L]$ how to divide $M(t)$, between either consumption, or investment in stocks and an interest yielding bond (i.e. between risky and risk-free assets). The intention is to maximize the total expected discounted utility of the investor's lifetime consumption and final fortune (or legacy) $M(L)$. Under certain conditions, the optimal trading strategy is to keep a constant fraction of wealth in the risky (and the risk-free) asset. *See* trading strategy. [R. Merton, Optimum consumption and portfolio rules in a continuous-time model, *J. Econ. Theor.*, 3, 373–413, 1971]

method of images *see* reflection principle

method of moments *see* Fréchet–Shohat theorem

method of phases (alias: **method (or device) of stages**) A device that may, in suitable cases, allow the powerful apparatus of Markov process theory to bear on a random process that, as it stands, does not have the Markov property. The basic idea was used by A. Erlang to analyze the M/G/1 queue; the non-exponential density $f_s(x)$ of any service time is replaced by an Erlang (gamma) density that is (in some sense) a good approximation to $f_s(x)$. Since this gamma density is that of a sum of n (say) i.i.d. exponential r.vs, the customer's service is envisaged as passing through a number n of phases (comprising a Markov process) up to completion.

More generally, such non-exponential distributions may be replaced by the distribution of the hitting time of some given state of a suitable finite state space Markov process. The state of this auxiliary chain is the phase (or stage) of the process. In this context of approximation of an arbitrary distribution by mixtures of sums of exponential r.vs (*see* phase-type distribution), **Schassberger's thm** is useful: let $F(x)$ be a distn on $[0, \infty)$, with $F(0) = 0$, and let X_k be an Erlang distn with parameters k and λ, so that $\mathbb{E}\mathrm{e}^{\mathrm{i}tX_k} = \left(\frac{\lambda}{\lambda - \mathrm{i}t}\right)^k$. Define

$$p_k = F\left(\frac{k}{\lambda}\right) - F\left(\frac{k-1}{\lambda}\right).$$

Then as $\lambda \to \infty$ the distribution of the r.v. $S = \sum_{k=1}^{\infty} p_k X_k$ converges to $F(x)$ (at the continuity points of $F(x)$). [M. Neuts, *Matrix-geometric Solutions in Stochastic Models: An Algorithmic Approach*. Johns Hopkins, 1981]

metrically transitive *see* measure-preserving transformation

metric entropy *see* Kolmogorov–Sinai entropy

Metropolis–Hastings method (or algorithm) The basic Markov chain Monte Carlo method, in which the required sample paths of the chain X_n, $n \geq 0$, with given stationary distribution $\pi_k, k \in S$, are generated by the transition matrix p_{ij} defined as follows: let $h_{ij}, i, j \in S$, be an arbitrary stochastic matrix (called the proposal, or candidate matrix), and define the **acceptance probabilities**

$$a_{ij} = \min\left\{1, \frac{\pi_j h_{ji}}{\pi_i h_{ij}}\right\}.$$

Supposing that $X_n = i$, generate $Y \in S$ according to the distribution $\mathbb{P}(Y = j \,|\, X_n = i) = h_{ij}$, and set

$$X_{n+1} = \begin{cases} j(= Y) & \text{with probability } a_{ij}, \\ X_n & \text{with probability } 1 - a_{ij.} \end{cases}$$

Thus $X_n, n \geq 1$, is generated recursively with transition probabilities

$$p_{ij} = \begin{cases} h_{ij} a_{ij}, & i \neq j; \\ 1 - \sum_{k:k \neq i} h_{ik} a_{ik}, & i = j. \end{cases}$$

(In particular, when h_{ij} does not depend on i, this is called the independence sampler.) The above method is a general form due to W. K. Hastings (1970) of the original Metropolis method (1953) in which $h_{ij} = h_{ji}$ for all i and j. In the special homogeneous case for which h_{ij} is a function of the distance $|i - j|$ between i and j, this may be called the random walk Metropolis method. When X_n is a random vector, and the chain updates its components one by one, using appropriate conditional distributions, it may be called the single-component M-H method; *see* Gibbs sampler. [W. Hastings, Monte Carlo sampling methods using Markov chains and their applications, *Biometrika*, 57(1), 97–109, 1970]

Metropolis shuffle *see* shuffle

Meyer–Tanaka formula *see* Tanaka's formula

migration process A type of stochastic network process (or particle system) with applications in queueing, data networks, etc. Broadly, it comprises a random vector describing a population of individuals that occupy, and move randomly between, a number of distributed sites (or colonies). In open migration processes, fresh individuals may enter the system, and those in the system may leave it forever; in closed migration systems neither of those events can occur. *See* Jackson network. [F. Kelly, *Reversibility and Stochastic Networks*. Wiley, 1979; reprinted by Cambridge University Press, 2011]

Mills's (or Mills') ratio The reciprocal of the hazard rate function of the standard normal distribution $\Phi(x)$ having density $\phi(x) = \Phi'(x)$; given by

$$R(x) = e^{x^2/2} \int_x^\infty e^{-y^2/2}\,dy = (1 - \Phi(x))/\phi(x).$$

By extension the term is occasionally used for the reciprocal hazard rate for distns other than normal. The related **Shenton's ratio** is $\overline{R}(x) = e^{x^2/2} \int_0^x e^{-y^2/2}\,dy$. There are many representations and inequalities for $R(x)$, including Birnbaum's inequality

$$\frac{1}{2}\{(x^2 + 4)^{1/2} - x\} \leq R(x) \leq 4\{3x + (x^2 + 8)^{1/2}\}^{-1},$$

and Laplace's asymptotic expansion

$$R(x) \sim \frac{1}{x} - \frac{1}{x^2} + \frac{1.3}{x^3} - \frac{1.3.5}{x^5} + \cdots,$$

where $R(x)$ lies between even and odd truncations. [L. Shenton, Inequalities for the normal integral including a new continued fraction, *Biometrika*, 41(1/2), 177–189, 1954]

Mineka coupling A coupling introduced to analyse the long-run behaviour of the partial sums $S_n = \sum_{r=1}^n X_r$ of a sequence of r.vs that are independent but not necessarily identically distributed. [J. Mineka, A criterion for tail events for sums of i.i.d. random variables, *Probab. Theor. Rel. Fields*, 25, 163–170, 1973]

minimal L_r distance *see* minimum L_p distance

minimal process *see* Feller minimal process

minimal solution *see* Feller minimal process

minimax principle Arises in making decisions about a system, or playing a game against an opponent, where it gives a general criterion (or rationale) for choosing a strategy for taking action. Let $\mathcal{A}$ be the space of actions from which one can make a (possibly random) choice a, and for any fixed a let $M(a)$ denote the maximum expected loss arising from a, where the maximum is taken over possible strategies of the opponent, or possible states of the system. The rule which calls for the action $\hat{a}$ minimizing $M(a)$ is the minimax rule. If $m(a)$ is the minimum expected loss arising from action a, then the rule calling for the action $\hat{a}$ maximising $m(a)$ is called the **maximin**, or **security level** (achieved by a **security strategy**).

Objective functions other than expected loss may be used, e.g. opportunity loss, giving **minimax regret**. [J. Von Neumann & O. Morgenstern, *Theory of Games and Economic Behavior*, 3rd edn. Princeton University Press, 1953]

minimum L_p distance A distance between probability distributions that is a a special case of a more general metric that is often termed the **Wasserstein–Kantorovich distance**, but is also known by several subsets (differing according to the value of p, generality, form, and context) of: **Dall'Aglio–earth mover's–Fortet–Fréchet–Gini–Kantorovich–Mallows–Monge-Mourier–Rubinstein–Salvemini–transportation–Wasserstein** (etc.) **distance**

In general, the Wasserstein–Kantorovich distance (or metric) supplies a metric for probability distributions on a (separable) metric space S, the distance between probability measures $\mathbb{P}$ and $\mathbb{Q}$ being denoted by $W_p(\mathbb{P}, \mathbb{Q})$, where $p > 0$ is the order (or exponent) of the metric. In the basic case, denoting the metric on S by $d(x, y)$, for r.vs X and Y having distns $\mathbb{P}$ and $\mathbb{Q}$, the first order, distance is denoted by

$$W_1(\mathbb{P}, \mathbb{Q}) = \inf\{\mathbb{E}d(X, Y)\}$$

where the infimum is taken over all joint distributions of X and Y having marginal distns $\mathbb{P}$ and $\mathbb{Q}$. By the Kantorovich–Rubinstein theorem

$$W_1(\mathbb{P}, \mathbb{Q}) = \sup\left\{\left|\int g\mathrm{d}\mathbb{P} - \int g\mathrm{d}\mathbb{Q}\right|\right\},$$

where the supremum is taken over all functions g obeying the Lipschitz condition $\sup\{|g(x) - g(y)|/d(x, y):$ $x \neq y\} \leq 1$.

More generally, with the infimum as defined above, the distance of order p is

$$W_p(\mathbb{P}, \mathbb{Q}) = \begin{cases} \{\inf \mathbb{E}(d(X, Y)^p)\}^{1/p}, & p \geq 1, \\ \inf \mathbb{E}(d(X, Y)^p), & 0 < p < 1. \end{cases}$$

In particular, when $S = \mathbb{R}$, and using the usual L_p norm, this gives the minimal L_p distance of order (or exponent) $p > 0$. (The case $p = 0$ yields a metric equivalent to total variation distance.)

In the simplest case when $p \geq 1$ and $S = \mathbb{R}$, and F and G are the distribution functions of $\mathbb{P}$ and $\mathbb{Q}$, one has the usual (so-called) minimum $\mathrm{L_p}$ distance given by:

$$\begin{aligned} W_p(\mathbb{P}, \mathbb{Q}) &= \left(\int_{-\infty}^{\infty} |F(x) - G(x)\mathrm{d}x|^p\right)^{\frac{1}{p}} \\ &= \left(\int_0^1 |F^{-1}(y) - G^{-1}(y)|^p\mathrm{d}y\right)^{\frac{1}{p}} \\ &= (\inf \mathbb{E}|X - Y|^p)^{1/p} \end{aligned}$$

where F^{-1} and G^{-1} are the usual inverse functions. The nomenclature of this family of distances is far from fixed. [E. Levina & P. Bickel, The earthmover's distance is the Mallows distance, *Proc. 8th IEEE Conf. Comp. Vision*, 251–256, 2001; S. Rachev, The Monge–Kantorovich mass transference problem and its stochastic applications, *Theor. Probab. Appl.*, 29(4), 647–676, 1984; C. Villani, *Topics in Optimal Transportation*. American Mathematical Society, 2003; D. A. Edwards, On the Kantorovich-Rubinstein theorem, *Expositiones Math.*, 29(4), 387–398, 2011]

minimum mean-square error *see* mean-square error

Minkowski's inequality In the context of random variables X and Y, this asserts that, for $r \geq 1$,

$$\{\mathbb{E}(|X + Y|^r)\}^{1/r} \leq \{\mathbb{E}(|X|^r)\}^{1/r} + \{\mathbb{E}(|Y|^r)\}^{1/r},$$

whenever both sides exist. It follows from Hölder's inequality. More generally, for $Z = g(X, Y)$, where X and Y are independent

$$(\mathbb{E}\{|\mathbb{E}(Z|X)|^r|Y\})^{1/r} \leq \mathbb{E}\{\mathbb{E}(|Z|^r|Y)^{1/r}|X)\}.$$

[S. Boucheron *et al.*, *Concentration Inequalities*, Oxford University Press, 2013]

minorization condition Arises in the context of a Markov chain on a measurable state space $(S, \mathcal{B})$ having n-step transition kernel $p^n(x, A)$. A minorization condition holds on some set $C \in \mathcal{B}$ if for some $\varepsilon > 0$, finite integer m, and probability measure $\mathbb{Q}$ on $(S, \mathcal{B})$ we have $p^m(s, A) \geq \varepsilon\mathbb{Q}(A)$ for all $s \in C$ and $A \in \mathcal{B}$; and then C may be called a small set. [J. Rosenthal, Minorization conditions and convergence rates for Markov chain Monte Carlo, *J. Amer. Stat. Ass.*, 90(430), 558–566, 1995]

mirror (or reflection) coupling *see* coupling; Kendall–Cranston coupling

Mittag–Leffler distribution A distribution defined in terms of the Mittag–Leffler fn:

$$M(z) = \sum_{r=0}^{\infty} \frac{z^r}{\Gamma(1 + c\,r)}, \quad 0 < c \leq 1.$$

The function $F(x) = 1 - M(-x^c), x \geq 0$, is a cumulative distribution function, called a Mittag–Leffler distn, and the Laplace transform of this distribution is $L(t) = (1 + t^c)^{-1}, t \geq 0$. The compound Poisson distn having parameter λX, where X has the cdf $F(x)$, and $\lambda > 0$, is called a discrete Mittag–Leffler distn. [R. Pillai, On Mittag-Leffler functions and related distributions, *Ann. Inst. Stat. Math*, 42(1), 157–161, 1990]

mixed distribution (alias: **mixture (of) distributions**) A probability density $f(x), x \in \mathbb{R}$, obtained as a suitable linear combination of a set of probability densities $f_s(x), s \in I$, or a distn $F(x)$ obtained similarly. Thus for a discrete index set I: $f(x) = \sum_{s \in I} p_s f_s(x)$, where $\sum_{s \in I} p_s = 1$; $f_s > 0$, and more generally, $f(x) = \int_{s \in I} f_s(x) \mathrm{d}F(s)$, where $F(s)$ is a cdf with support I. A mixture is a type of compound distribution; also called a composition. Yet more generally one may mix other, possibly random, objects, e.g. **a mixed strategy** in decision theory is one selected at random according to a distribution over a set I of strategies that are usually pure, i.e. not mixed. [K. Binmore, *Playing for Real*. Oxford University Press, 2007; J. Gentle, *Random Number Generation and Monte Carlo Methods*, 2nd edn. Springer, 2003]

mixed Poisson process (alias: **conditional Poisson process**) A point process $N(t)$ with stationary increments whose distribution is a mixture of Poisson distributions. Thus the number $N(s, s+t)$ of points in $(s, s+t)$ has the pgf $G_t(z) = \mathbb{E}z^{N(s,\ s+t)} = \mathbb{E}e^{\Lambda t(z-1)}$, where Λ is some positive r.v. There are several other characterizations of the process. More generally, a mixed non-homogeneous Poisson process is usually called a Cox process. Increments of the process are independent iff it is an ordinary homogeneous Poisson process. [J. Grandell, *Mixed Poisson Processes*. Chapman & Hall, 1997]

mixed process *see* autoregressive process

mixed strategy *see* mixed distribution

mixing see measure-preserving transformation

mixing The concept of mixing arises naturally in relaxing the assumption of mutual independence for a sequence of r.vs $X_n, n \in \mathbb{Z}$. Informally, the idea is that, even though dependent, r.vs that are sufficiently far apart in the sequence may be so nearly independent (as indicated by some suitable 'distance') that the sequence behaves to some extent like an independent sequence. For example, it may obey a CLT, or LLN, or have ergodic properties; in this case one may refer to asymptotic independence.

There are many possible measures of dependence, and each of them may give rise to a so-called mixing coefficient c_{mix} measuring closeness to independence. The speed with which this goes to zero as the separation increases is the mixing rate, and any such sequence is said to satisfy a mixing condition. Depending on its nature, one speaks of strong mixing, weak mixing (typically when convergence is in the Cesàro sense), absolute regularity, etc.; and various coefficients are identified by labels such as alpha-, beta-, phi-, psi-, or rho-mixing. The terminology is not entirely fixed throughout the literature. [R. Bradley, *Introduction to Strong Mixing Conditions*, vols 1, 2, & 3. Kendrick Press, 2007]

mixingale A random process that combines the concepts of martingale-type dependence and mixing. Let $X_n\ n \geq 1$ be a sequence of r.vs, and $\mathcal{F}_n, n \geq 1$ an increasing sequence of sigma-fields. It is assumed that $\|X_n\|_r < \infty$ for each n, and $1 \leq r \leq 2$, where $\|X\|_r = \{\mathbb{E}(|X|^r)\}^{1/r}$. Then the sequence is said to be an L_r-mixingale if there are constants $a_n, n \geq 0$, and $C_n, n \geq 1$, such that $\lim_{n \to \infty} a_n = 0$, and for all $m, n \geq 0$, $\|\mathbb{E}(X_n \mid \mathcal{F}_{n-m})\|_r \leq C_n\, a_m$, and $\|X_n - \mathbb{E}(X_n \mid \mathcal{F}_{n+m})\|_r \leq C_n\, a_{m+1}$. (The second condition is always satisfied if each X_n is $\mathcal{F}_n$-measurable.) When $r = 2$, it is usual simply to call X_n a mixingale. [D. Andrews, Laws of large numbers for dependent non-identically distributed random variables, *Econ. Theor.*, 4(3), 458–467, 1988]

mixing time Arises in considering the evolution of an ergodic Markov chain $X_n, n \geq 1$, towards stationarity, where the mixing time τ is a measure of how many steps the chain takes for its distribution $(\mathbb{P}(X_n = j \mid X_0 = i), j \in S) = \mathbf{p}_i(n)$ to become close (in some sense) to the stationary distn $\boldsymbol{\pi}$. The concept has natural applications in shuffling, and Monte Carlo methods, etc. Commonly the distance of $\mathbf{p}_i(n)$ from $\boldsymbol{\pi}$ is measured by the total variation distance $\|\mathbf{p}_i(n) - \pi\|$, and we set $d(n) = \max_{i \in S} \|p_i(n) - \pi\|$. Then the ε-mixing time is $\tau(\varepsilon) = \min\{n: d(n) \leq \varepsilon\}$, and the (standard) mixing time is $\tau = \tau(1/4)$. [D. Levin et al., *Markov Chains and Mixing Times*. American Mathematical Society, 2009]

mixture method *see* composition method

mode If the density $f_X(x)$ of a continuous random variable X has a single local maximum at m (which is also the global maximum), then m is said to be the mode of X, and X is said to be unimodal. If $f(x)$ has several local maxima, of which one at m is the global maximum, then m is said to be the principal (or major) mode of X, and

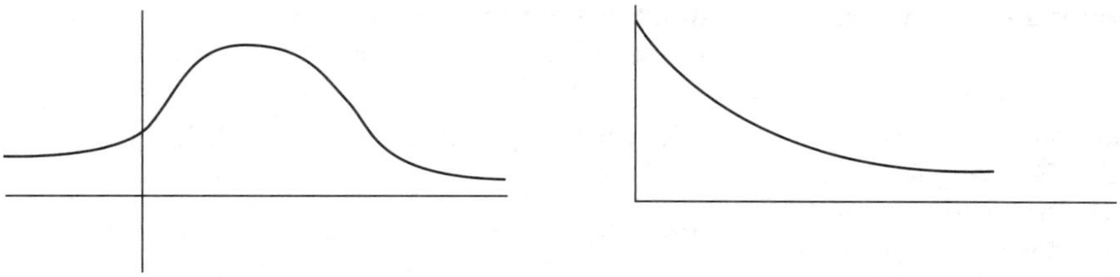

Two unimodal densities.

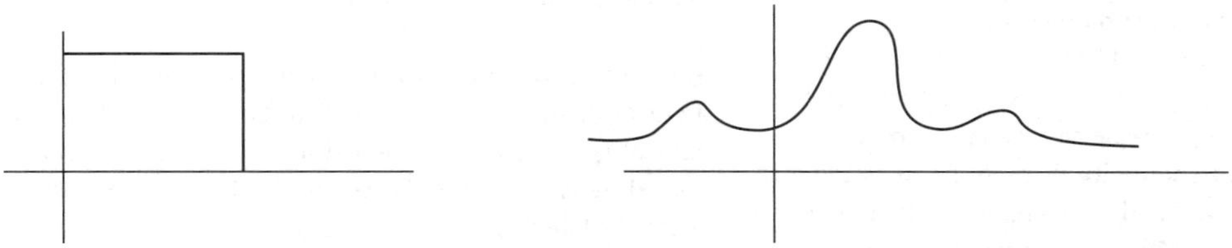

Two multimodal densities: the uniform density, and a density with a single principal mode.

X is said to be **multimodal**. If the global maximum is achieved at several points, there may be several such principal modes, and non-global maxima may also be referred to as minor modes. If $f(x)$ achieves its global maximum over an interval, then a unique mode may be assigned according to some convenient rule. If X is continuous and unimodal, with variance σ^2 and mode m, then $|m - \mathbb{E}X| < \sqrt{3}\sigma$. For discrete random variables the definition of mode is essentially similar.

modification *see* equivalent

modified distribution Broadly, a term denoting a probability distribution obtained by some amendment of a standard distribution. Note that the term is often used loosely, where the words 'truncated' or 'shifted' (for example), would be more precise. Commonly seen cases include:

1. The modified geometric distn with pgf

$$G(z) = 1 - a + a\sum_{r=1}^{\infty} pq^{r-1}z^r, \quad 0 \le a, p \le 1, p + q = 1.$$

Here $a = 1$ gives the standard geometric distn on $\{1, 2, 3, \ldots\}$, while $a = q$ gives the shifted geometric distn on $[0, 1, 2, \ldots]$.

2. The modified Poisson distn with pgf

$$G(z) = 1 - b + be^{\lambda(z-1)}, \quad \lambda > 0, 0 \le b \le \frac{e^{\lambda}}{e^{\lambda} - 1}.$$

Here $b = 1$ gives the standard Poisson distn on $[0, 1, 2, \ldots]$, while $b = \frac{e^{\lambda}}{e^{\lambda}-1}$ gives the zero-truncated Poisson distn on $[1, 2, \ldots]$, with pgf $G_0(z) = \frac{e^{\lambda z}-1}{e^{\lambda}-1}$. Intermediate forms of either distn may be called **zero-inflated** or **zero-deflated** depending on the value of the parameter. *See* zero-modified.

[N. Johnson *et al.*, *Univariate Discrete Distributions*, 3rd edn. Wiley, 2005]

Mogulskii's theorem A large deviations (or Laplace) principle for the sample paths of suitably scaled and modified random walk. [A. Dembo & O. Zeitouni, *Large Deviations: Techniques and Applications*, 2nd edn. Springer, 1998]

Molloy–Reed random graph *see* random graph

moment A quantity arising naturally in the context of random variables and their distributions. For any integer n and random variable X, the nth moment of X is the expectation $\mathbb{E}(X^n) = m_n$, when it exists. It is sometimes called the moment about the origin, to distinguish it from other moments defined below. In particular, the first moment $\mathbb{E}X$ is the mean. When n is not an integer, $\mathbb{E}X^n$ may be called a **fractional moment**.

For X with mean m_1, the nth central moment (or moment about the mean) is $\mathbb{E}(\{X - m_1\}^n) = \sigma_n$. In particular, $\sigma_1 = 0$, and $\sigma_2 = \sigma^2 = \operatorname{var} X$. Then $s_n = \mathbb{E}(\{(X - m_1)/\sigma\}^n)$ is the nth standardized moment of X. For any value of c other than $m_1, c_n = \mathbb{E}\{(X - c)^n\}$ is the nth centred moment, centred at c. When $\mathbf{X} = (X_1, \ldots, X_r)$ is a random vector, expectations of the form $\mathbb{E}(\prod_{i=1}^{r} X_i^{n_i})$ are called joint, or product moments (which may be central or centred, in the natural way). Finally, expectations of the form $\mathbb{E}(|X|^r)$, and so on, may be called absolute moments. *See* factorial moment.

moment convergence theorem *see* Fréchet–Shohat theorem

moment generating function For a random variable X the function $M(t) = \mathbb{E}e^{tX} \geq 0$, for $t \in \mathbb{R}$, is called the moment generating function (or mgf) of X. The mgf is especially useful if it is finite in an open interval including $t = 0$, for then we have $M(t) = \sum_{r=0}^{\infty} \frac{t^r}{r!}\mathbb{E}(X^r)$, and $M^{(r)}(0) = \mathbb{E}(X^r)$. Thus $M(t)$ is the exponential generating function of the moments $m_r = \mathbb{E}(X^r)$. Furthermore, in this case ($M(t) < \infty$ for $0 \leq |t| < a > 0$), the distribution $F_X(x)$ of X is uniquely determined by $M_X(t)$, using the appropriate inversion theorem; in addition, a form of continuity theorem holds (known as **Curtiss's thm**); and, finally, the characteristic function of X is $\phi_X(t) = M_X(\mathrm{i}t)$. These properties do not necessarily hold if $M_X(t)$ does not exist for $|t| < a > 0$. For a random vector $\mathbf{X}$, the joint moment generating function is $M(\mathbf{t}) = \mathbb{E}\exp(\mathbf{t}\cdot\mathbf{X})$, which has analogous properties (here $\mathbf{t}\cdot\mathbf{X} = \sum_i t_i X_i$, the scalar product). [T. Severini, *Elements of Distribution Theory*. Cambridge University Press, 2005]

moment matrix *see* covariance matrix, in covariance

moment problem (alias: **determinacy problem**) Seeks an answer to the question: 'Given a sequence of real numbers $m_0 = 1$, and $m_n, n \geq 1$, is there a distribution function $F(x), x \in I \subseteq \mathbb{R}$ such that $m_n = \int_I x^n \mathrm{d}F(x), n \geq 0$ and, if so, is $F(x)$ unique?'

If there is such a unique $F(x)$ the moment problem is said to be **determinate**; otherwise it is indeterminate. Classic special cases are:

(a) The **Hamburger moment problem**, when $I = \mathbb{R}$.

(b) The **Stieltjes moment problem**, when $I = \mathbb{R}^+$.

(c) (i) The **Hausdorff moment problem**, when $I = [0, 1]$.

(ii) The **Markov moment problem**, when $I = [0, 1]$ and $F(x)$ is required to have a bounded density.

Analogously, the **complex moment problem** seeks to characterize sequences of complex numbers $C_{nm}, n, m \geq 0$, that are the moments of a distribution $F(x, y)$, on $\mathbb{C}$, i.e. $C_{n,m} = \int z^n \bar{z}^m \mathrm{d}F$. The finite moment problem arises when the given set of moments is finite, and a generalized moment problem arises when given $s_n = \int_I x^{r_n} \mathrm{d}F(x), \quad 0 = r_0 < r_1 < r_2 < \cdots$ for an unbounded sequence $r_n, n \geq 0$. When I is the unit circle in $\mathbb{C}$, one may refer to the **trigonometric moment problem**. *See* Carleman's criteria; Heyde's example; and Krein condition. [N. Akhiezer, *The Classical Moment Problem*. Hafner, 1965]

moments lemma Asserts that for any r.v. X, any $r > 0$, and any $x > 0$, we have

$$x^r \sum_{n=1}^{\infty} \mathbb{P}(|X| \geq xn^{1/r}) \leq \mathbb{E}(|X|^r) \leq x^r \left(1 + \sum_{n=1}^{\infty} \mathbb{P}(|X| \geq xn^{1/r})\right).$$

[M. Loéve, *Probability Theory*, 4th edn. Springer, 1978]

Monge–Kantorovich distance; *see* earth mover's distance; minimum $\mathrm{L_p}$ distance

monomer–dimer model Arises in statistical physics. In its basic form a finite graph G (often a finite subset of a lattice) is randomly partitioned thus: any pair of vertices joined by an edge may be labelled a dimer, where no vertex is labelled twice; the remaining vertices are labelled monomers. The probability that there are x dimers is assumed proportional to λ^x for some $\lambda > 0$, giving the monomer–dimer distribution on G. [J. van den Berg, On the absence of phase-transition in the monomer-dimer model, in *Perplexing Problems in Probability*, M. Bramson & R. Durrett (eds). Birkhäuser, 1999]

monotone class theorem (alias: **Halmos's monotone class theorem**) A result from measure theory, related to Dynkin's π-λ theorem that finds applications in probability. It arises in the context of a collection of events E that is closed under the formation of countable monotone unions, and countable monotone intersections, which is called a monotone class. The theorem asserts that if $\mathcal{A}$ is a field of sets in E, then the smallest σ-field including $\mathcal{A}$ also lies in E. [P. Billingsley, *Probability and Measure*, 3rd edn. Wiley, 1995]

monotone convergence theorem *see* Beppo Levi theorem

monotone dependence Broadly, this denotes the propensity of two random variables X and Y to increase and decrease together, or (contrariwise) for X to decrease (resp. increase) when Y increases (resp. decreases); *see* concordance, and association. More narrowly, X and Y are monotone dependent if there exists a monotone function h such that $\mathbb{P}(X = h(Y)) = 1$.

The dependence is said to be increasing or decreasing according as h increases or decreases.

If X and Y are non-degenerate then their **monotone correlation fn** is $r(X, Y) = \sup \rho(g(X), h(Y))$, where ρ is the usual correlation function, and the supremum is over all monotone g and h such that $0 < \operatorname{var} g(X) \operatorname{var} h(Y) < \infty$.

The maximal correlation is not less than r. [G. Kimeldorf & A. Sampson, Monotone dependence, *Ann. Stat.*, 6(4), 895–903, 1978]

monotone Markov chain A Markov chain $X_n \in \mathbb{R}, n \geq 0$, is stochastically monotone if, for all fixed y, $\mathbb{P}(X_{n+1} \leq y \mid X_n = x)$ is non-increasing in x. (A chain may also be monotone in the elementary sense of being a. s. non-decreasing in n.) [D. Daley, Stochastically monotone Markov chains, *Probab. Theor. Rel. Fields*, 10, 305–317, 1968]

monotonic dependence *see* association; dependence

Monroe's theorem Asserts that a random process X is equivalent to a time change of Brownian motion (which is to say that it is embedded in a Brownian motion) iff X is a local semi-martingale. [I. Monroe, Processes that can be embedded in Brownian motion, *Ann. Probab.*, 6(1), 42–56, 1978]

Monte Carlo algorithm Broadly, a randomized algorithm (i.e. a computational procedure whose outcome is a function of (pseudo-)random numbers), whose running time is relatively short (and generally a deterministic function of the input), but whose output is incorrect with some probability that is commonly designed to be small. *See* Las Vegas algorithm. [R. Motwani & P. Raghavan, *Randomized Algorithms*. Cambridge University Press, 1995]

Monte Carlo method Broadly, a type of experimental mathematics that makes use of repeated sampling of random (or pseudo-, or quasi-random) numbers to obtain approximate solutions to a wide variety of problems that may be either deterministic or probabilistic in nature, e.g. optimization problems; random search; statistical inference; solving partial differential equations numerically; and simulating complicated systems (often with a view to computing some functional of a random process). A classic example of the method is Monte Carlo integration, thus: an integral $I = \int_0^1 f(x)\mathrm{d}x$, where $0 \leq f(x) \leq 1$, is estimated by simulating points uniformly at random in the unit square $0 \leq x, y \leq 1$. The proportion lying below $y = f(x)$ supplies the required estimate of I; this is often called **hit-or-miss Monte Carlo**.

Buffon's needle can be seen as an early example of the idea. Rayleigh noted in 1899 that a random walk would supply an approximate solution to the heat equation; both Student and E. S. Pearson used simulated data to investigate estimators; and E. Fermi (1931 onwards) used the same idea to predict experimental results. Modern Monte Carlo methods stem from ideas due to S. Ulam, developed by J. von Neumann, applied to solving a neutron diffusion problem. [R. Rubinstein & D. Kroese, *Simulation and the Monte Carlo Method*, 2nd edn. Wiley, 2008]

Montgomery–Odlyzko law (alias: **Dyson–Odlyzko–Montgomery law**) Asserts that the distribution of the gaps between non-trivial zeros of the zeta function is the same as the distribution of the gaps between eigenvalues of a Gaussian unitary ensemble (i.e. a random Hermitian matrix), when suitably normalized in an appropriate limit. [D. Rockmore, *Stalking the Riemann Hypothesis*. Random House, 2011; A. Odlyzko, On the distribution of spacings between zeros of the zeta function, *Math. Comp.*, 48(177), 273–308, 1987]

Montmort's problem *see* derangement problem

Montroll–Weiss equation *see* continuous-time random walk

Monty Hall problem (alias: **three-doors problem; car and goats problem; 'let's make a deal' problem; M. Vos Savant's problem**) A classic problem of probability, conditional probability, and psychological perception; or, more strictly, a set of such problems. It is related to the prisoner's dilemma and Bertrand's box paradox; and has its origins in a TV show, but is not the same as the game actually performed there.

It was first cast as a problem in probability by S. Selvin in 1975; and later (slightly differently, and more famously) by M. Vos Savant in 1990. Her version is this: In a game show you must choose one of three doors D1, D2, and D3, knowing that behind one door is a car, and behind the others goats. Suppose your choice is D1, and the host, who knows where the car is, opens D3 to reveal a goat. Then he asks you if you wish to switch your choice to D2. Is it to your advantage to switch?

This question has yielded a variety of answers, and a large literature, because the phraseology does not define a unique mathematical problem. In particular, it does not supply a probability distribution or a suitable probability space, and it does not say whether the host is an opponent, in which case a game theoretic formulation would be appropriate. However, both naive and more sophisticated interpretations of the problem lead to the conclusion that a policy of switching is (in general) advantageous, and that the conditional probability of winning the car is greater if you do switch than if you don't. [S. Selvin, A problem in probability, *Amer. Stat.*, 29(1), 67, 1975; S. Selvin, On the Monty Hall problem, *Amer. Stat.*, 29(3), 134, 1975: M. vos Savant, Marilyn vos Savant's reply, *Amer. Stat.*, 45(4), 347, 1991; R. Gill, The Monty Hall problem is not a probability puzzle – it is a challenge in mathematical modelling, *Statistic Neerlandica*, 65(1), 58–71, 2011; J. Rosenhouse, *The Monty Hall Problem.* Oxford University Press, 2009]

moral expectation A concept proposed by Daniel Bernoulli in 1738 to deal with problems such as the St Petersburg paradox, among others, where simple mathematical expectation is inappropriate as a measure of value, essentially because it fails to account for risk aversion; note that 'moral' meant to Bernoulli what might today be called 'subjective'. Bernoulli's expectation can be seen as a special case of the more general concept of expected utility, which supersedes it.

As an example, suppose your fortune is x, and you flip a fair coin that yields a, or loses $b(< x)$, according to the outcome, then the (Bernoulli) moral expectation of this venture is $(x+a)^{1/2}(x-b)^{1/2}/x$. [D. Bernoulli, Exposition of a new theory on the measurement of risk, Trans. from Latin to English by L. Sommer, *Econometrica*, 22(1), 23–36, 1954]

Moran dam model *see* dam process; Moran gamma process

Moran–Downton distribution (alias: **Kibble–Moran–Downton distribution; Downton's distribution**) A bivariate exponential distribution having density

$$f(x,y) = \frac{1}{1-\rho}\exp(-(x+y)/(1-\rho))I_0\left\{2(xy\rho)^{\frac{1}{2}}/(1-\rho)\right\},$$

for $x, y, \geq 0$, where I_0 is the modified Bessel function of the first kind, and ρ is the correlation. The mgf is $M(s,t) = \{(1-s)(1-t)-\rho st\}^{-1}$, which is infinitely divisible. [P. Moran, Testing for correlation between non-negative variates, *Biometrika*, 54(3&4), 385–394, 1967]

Moran gamma process (alias: **gamma process; Moran subordinator**) A random process $X(t), t \in \mathbb{R}$ (or $t \in \mathbb{R}^+$), with stationary independent increments having a gamma density; it is thus a particular Lévy process. In standard form its Lévy measure is $\lambda(\mathrm{d}x) = x^{-1}\mathrm{e}^{-x}\mathrm{d}x$, where

$$\begin{aligned}\mathbb{E}\{\exp(-\theta(X(t)-X(0)))\} &= (1+\theta)^{-t}\\ &= \exp\left\{-t\int_{(0,\infty)}(1-\mathrm{e}^{-\theta x})\lambda(\mathrm{d}x)\right\}\end{aligned}$$

is the Lévy formula. It was introduced in the Moran dam model, i.e. the stochastic recurrence $Z_{t+1} = \min\{(Z_t + X(t+1) - m)^+, c\}$, where $m > 0$ and $c > 0$ are constants. *See* dam process. [J. F. C. Kingman, *Poisson Processes.* Oxford University Press, 1993]

Moran model (or process) Any of several mathematical models in population genetics, but most commonly the following (neutral) Moran model: a population of genes of constant size M evolves as a Markov chain (in discrete or continuous time), where, at transition times, a pair of genes B and C is selected uniformly at random with replacement (so $B = C$ is possible); one splits into two copies of itself, and the other is removed. In the non-neutral generalization, genes of different types are more (or less) likely to be removed (rather than split) at each step. *See* Wright–Fisher model. [P. Moran, *Statistical Processes of Evolutionary Theory.* Oxford University Press, 1962]

Morgenstern distribution *see* Farlie–Gumbel–Morgenstern distribution

Mossin's theorem (alias: **Mossin–Smith theorem**) Arises in the theory of rational insurance. It shows that for a suitable class of insurance contracts, it is optimal for a risk-averse purchaser to be fully insured if the contract is fair (the premium equals the expected indemnity), but optimal to be under-insured if the premium is loaded in favour of the insurer. [G. Dionne (ed.), *Handbook of Insurance.* Kluwer, 2000]

Motoo's theorem Arises in large deviations theory, where it asserts that for a continuous, regular, time-homogeneous strong Markov process $X(t)$, with finite speed measure m and scale function s (i.e. for a suitable diffusion), and for any increasing positive function g, we have

$$\mathbb{P}\left(\lim_{t\to\infty}\sup X(t)/g(t) \geq 1\right) = 0 \text{ or } 1$$

according as

$$\int^{\infty}\frac{\mathrm{d}y}{s(g(y))} < \infty \text{ or is } \infty.$$

(For another theorem of Motoo, *see* combinatorial central limit theorem.) [M. Motoo, Proof of the law of the iterated logarithm through diffusion equation, *Ann. Inst Stat. Math.*, 10(1), 21–28, 1959]

move-to-front process *see* heaps process

moving average In general, a moving average (MA) forms a new random process $Y_n, n \in \mathbb{Z}$, from a given process $X_n, n \in \mathbb{Z}$, by setting $Y_n = \sum_{r=-\infty}^{\infty} a_r X_{n-r}$ for suitable real (or complex) constants $a_r, r \in \mathbb{Z}$. In continuous time the analogous formulation is $Y_t = \int_{-\infty}^{\infty} a(s)X(t-s)\mathrm{d}s$. Such moving averages (sometimes called rolling, or running averages) may be called two-sided to distinguish them from one-sided averages arising when $a_r = 0$ for either $r > 0$ or $r < 0$. In particular, if there is a finite index set I (e.g. $I = \{0, \ldots, m-1\}$) such that a_r is constant for $r \in I$, and $a_r = 0$ for $r \in I^c$, then one speaks of a simple (or uniform) moving average; if $a_r = \lambda^{|r|}$ for some $\lambda \in (-1, 1)$, then one speaks of an **exponential or geometric MA**. In the context of signal processing, estimation, prediction, etc., Y_n may be called a linear filter applied to X_n; some writers call Y_n a smoothed process, and speak of **smoothing**. In particular, the MA is most often seen when X_n is a stationary process; and even more often when X_n is a stationary sequence of uncorrelated (or independent) r.vs, in which case it is called the **innovations sequence**. Then Y_n is called a moving average process, and the sum converges (in mean) if, e.g. var $X_n = \sigma^2 < \infty$ and $\sum_{-\infty}^{\infty} |a_r|^2 < \infty$. In particular, if $Y_n = X_n + a_1 X_{n-1} + \cdots + a_n X_{n-q}$, then Y_n is called an MA process of order q, denoted by MA(q). Such processes may have a representation as an autoregressive process, and are a special case of ARMA(p, q) processes. A related process is the **integrated moving average** Z_n satisfying $(1-B)Z_n = Z_n - Z_{n-1} = \sum a_r X_{n-r}$, where B is the backward shift operator; more generally $(1-B)^i Z_n = \sum a_r X_{n-r}$ is an ith-order integrated moving average, which is a special case of an ARIMA (p, i, q) process. [P. Brockwell & R. Davis, *Time Series*, 2nd edn. Springer, 1991]

moving average representation *see* Hida–Cramér canonical representation

multiclass queue *see* queue

multidimensional *see* joint

multifractal *see* Mandelbrot multiplicative cascade; fractal

multimodal distribution *see* mode

multinomial distribution (alias: **polynomial distribution (rare usage)**) The joint distribution of the discrete random vector $(X_1, \ldots, X_k)$, where each X_r is a non-negative integer such that $X_1 + \cdots + X_k = n \geq 1$, and

$$p(x_1, \ldots, x_k) = \mathbb{P}(X_1 = x_1, \ldots, X_k = x_k)$$
$$= \frac{n!}{x_1! \cdots x_k!} p_1^{x_1} \ldots p_k^{x_k} \ldots, \qquad \text{for} \sum_{r=1}^{k} p_r = 1, \, p_r \geq 0.$$

It arises as the distribution of n independent trials, where each trial has k possible outcomes, of which the rth occurs with probability p_r. When $n = 2$ it is the binomial distribution, and for $n = 3$ the trinomial distribution. The ratio of factorials is called the **multinomial coefficient**, denoted by

$$\binom{n}{x_1, \ldots, x_k} = \frac{n!}{x_1! \cdots x_k!}.$$

The multinomial distn pgf is $\mathbb{E}(s_1^{X_1} \cdots s_k^{X_k}) = (p_1 s_1 + \cdots + p_k s_k)^n$, while $\mathbb{E}X_r = np_r$, var $X_r = np_r(1-p_r)$, and $\text{cov}(X_j, X_k) = -np_j p_k, j \neq k$.

Some writers define a generalized multinomial distn as that having joint pgf $A(s_1, \ldots, s_k)^n$, for a bilinear form (or affine polynomial) A whose coefficients are non-negative, with $A(1, \ldots, 1) = 1$.

By analogy with the negative binomial distn, a negative multinomial distn is one having a pgf of the form

$$G(s_1, \ldots, s_k) = p_0^n \left\{1 - \sum_{r=1}^{k} s_r p_r\right\}^{-n}, \sum_{r=0}^{k} p_r = 1,$$

with $p(x_1, \ldots, x_k) = \frac{\Gamma(n+x_1+\cdots+x_k)}{x_1! \ldots x_k! \Gamma(n)} p_0{}^n p_1^{x_1} \ldots p_k^{x_k}$. [W. Feller, *Introduction to Probability Theory and its Applications*, Vol. 1, 3rd edn. Wiley, 1968]

multinormal distribution (alias: **multivariate normal (or Gaussian) distribution**) The distribution of a r.vr. $\mathbf{X} = (X_1, \ldots, X_n)$ that supplies a natural generalization of the normal distribution. One straightforward definition is this: Suppose that V is a positive-definite symmetric $n \times n$ matrix, having determinant $|V|$. Then the density

$$f(\mathbf{x}) = \{(2\pi^n|V|)\}^{-\frac{1}{2}}\exp\left\{-\frac{1}{2}\mathbf{x}V^{-1}\mathbf{x}^T\right\}, \quad \mathbf{x} \in \mathbb{R}^n,$$

is the multinormal density of the r.vr. **X**, centred at the origin, with variance-covariance matrix V. The distribution may be denoted by $N(0, V)$, and then $\mathbf{Y} = \mathbf{m} + \mathbf{X}$ has distribution $N(\mathbf{m}, V)$, with mean vector **m**. Important properties include:

(a) For any $r \times n$ matrix A with rank $r \leq n, \mathbf{Z} = \mathbf{X}A$ is multinormal $N(0, A^TVA)$.

(b) **X** has joint characteristic function

$$\phi(\mathbf{t}) = \mathbb{E}e^{i\mathbf{t}\mathbf{X}^T} = \mathbb{E}e^{i(\mathbf{t}.\mathbf{X})} = \exp\left(-\frac{1}{2}\mathbf{t}V\mathbf{t}^T\right).$$

More generally, **X** is multinormal if for all $\mathbf{a} \in \mathbb{R}^n$ the r.v. $\mathbf{X}\mathbf{a}^T = a_1X_1 + \cdots + a_nX_n$ has a normal distribution. If for some such **a** there exists $b \in \mathbb{R}$ such that $\mathbb{P}(\mathbf{X}\mathbf{a}^T = b) = 1$, then **X** is said to have a **singular (or degenerate) multinormal distn**. Equivalently, the distn of **X** is multinormal if there exists an $r \times n$ matrix A, and an r-vector **Z** of independent standard normal r.vs such that $\mathbf{X} = \mathbf{m} + \mathbf{Z}A$, for some m-vector **m** of means. If $n > r$, then **X** is singular, while for $n \leq r$ it is singular iff the n equations defining **X** are linearly dependent (i.e. the rank of A is less than n). Finally, $\mathbf{Z} = \mathbf{X} + i\mathbf{Y}$ has a complex multinormal distn if $(\mathbf{X}, \mathbf{Y})$ is multinormal. [T. Severini, *Elements of Distribution Theory*. Cambridge University Press, 2005]

multiple Markov chain A sequence of r.vs $X_n, n \in \mathbb{Z}$, is said to be a multiple Markov chain of order $r \geq 1$ if the sequence $Y_n^{(k)} = (X_n, X_{n-1}, \ldots, X_{n-k+1})$ is a Markov process (in the standard sense) for $k \geq r$. (When $r = 1, X_n$ is simply a Markov process.) A similar definition holds with $n \in \mathbb{N}$, given suitable initial conditions. [J. Doob, *Stochastic Processes*. Wiley, 1953]

multiplication rule for probabilities (alias: **theorem of compound probability; chain rule for conditional probabilities**) Arises in considering so-called compound events such as $A \cap B$ and $\cap_{r=1}^n A_r$, where it uses conditional probability to write $\mathbb{P}(A \cap B) = \mathbb{P}(A \,|\, B)\mathbb{P}(B) = \mathbb{P}(B \,|\, A)\mathbb{P}(A)$; and, more generally

$$\mathbb{P}(\bigcap_{r=1}^{n} A_r) = \mathbb{P}(A_1)\mathbb{P}(A_2 \,|\, A_1)\mathbb{P}(A_3 \,|\, A_1 \cap A_2)\cdots\mathbb{P}(A_n \,|\, A_1 \cap \cdots A_{n-1}).$$

In particular, when the events are independent, $\mathbb{P}(\cap_{r=1}^n A_r) = \Pi_{r=1}^n \mathbb{P}(A_r)$, which may be called the **product rule** for independent events. The sequence of binary r. vs $I(A_r), 1 \leq r \leq n$, generates a natural binary tree of height n, whose nodes may be labeled by their corresponding probabilities; ($\mathbb{P}(A_1)$ and $\mathbb{P}(A_1^c)$ at height 1, $\mathbb{P}(A_1 \cap A_2)$ etc, at height 2, and so on). This is called a **probability tree**.

multiplicative process Most often refers to the Galton–Watson branching process; or, more generally, to any branching process; or, yet more generally, to any model for the growth of a reproducing population. *See* cascade process. Occasionally, it simply means a sequence of products of random variables; or, more generally, X_n given by the stochastic recurrence $X_{n+1} = a_nX_n + b_n$, called the **Kesten multiplicative process**.

multiplier central limit theorem Arises in the context of the empirical process of a distribution F (or the empirical measure of a probability measure $\mathbb{P}$). The empirical CLT asserts the convergence of a suitably scaled version of $\sum_{r=1}^n I(X_r \leq x)$ to a Brownian bridge, while the multiplier CLT asserts the same convergence of $\sum_{r=1}^n Y_rI(X_r \leq x)$ for a suitable sequence of i.i.d.r.vs $Y_1, \ldots, Y_n$, that are also independent of the X_r. [A. Van der Vaart & J. Wellner, *Weak Convergence and Empirical Processes*. Springer, 1996]

multivariate Chebyshov inequality *see* Olkin–Pratt inequality

multivariate Pareto distribution *see* Mardia distribution

Musiela equation Arises in financial mathematics as an SDE for forward interest rate dynamics when the rate is parameterized as a function of the time t and the time $s = T - t$ to the maturity date T. [T. Bjőrk, *Arbitrage Theory in Continuous Time*, 3rd edn. Oxford University Press, 2009]

mutual information Arises naturally in communication and information theory, where the Shannon mutual information between simple r.vs X and Y, seen as the input and output of a channel, is

$$I(X;Y) = \sum_{x,y} f(x,y)\log\frac{f(x,y)}{f_X(x)f_Y(y)} = \mathbb{E}\log\frac{f(X,Y)}{f_X(X)f_Y(Y)} = I(Y;X),$$

where f, f_X and f_Y are the joint and marginal distributions of X and Y. It is the Kullback–Leibler divergence of the distn $p = f(x,y)$ from the distn $q = f_X(x)f_Y(y)$, where I takes the value 0 iff X and Y are independent.

The conditional mutual information between X and Y given Z is defined likewise in terms of the conditional distributions $f(x, y \mid z), f_X(x \mid z)$, and $f_Y(y \mid z)$. Another type is the **Rényi mutual information**:

$$I_a(X, Y) = \frac{1}{a-1} \log \mathbb{E}\left\{ \left(\frac{f(X,Y)}{f_X(X) f_Y(Y)} \right)^{a-1} \right\}, 0 < a \neq 1.$$

[T. Cover & J. Thomas, *Elements of Information Theory*, 2nd edn. Wiley, 2006]

mutual variation *see* covariation

myopic rule *see* one-stage look-ahead policy

N

Nash's inequality *see* Poincaré inequality

Nash's theorem Arises in the theory of non-cooperative games, involving a finite number of players, each of whom may choose randomly among a finite number of pure (non-random) strategies. The theorem asserts that there is a distribution on the player's strategies such that no player's expected return from the game (that is, their pay-off, or expected utility) can be increased by a unilateral change of strategy. That is to say there is a **Nash equilibrium**. [K. Binmore, *Playing for Real*. Oxford University Press, 2007]

Nash-Williams theorem Supplies a necessary and sufficient condition for the recurrence of a reversible irreducible Markov chain that is locally finite, i.e. each row and column of the transition matrix contains only a finite number of non-zero terms. [C. Nash-Williams, Random walk and electric currents in networks, *Proc, Camb. Phil. Soc.*, 55, 181–194, 1959]

natural boundary Arises in the context of a diffusion process $X(t)$ on an interval $(a, b) \subseteq \mathbb{R}$. Informally, a natural boundary a (say) operates like the point at infinity for the standard Wiener process. Very roughly speaking, it takes the Wiener process a long time to reach anywhere near ∞, and a similarly long time to return to 0. Formally, a is a natural boundary for $X(t) \in (a, b)$ if for all $t > 0$ and $a < x < y < b$

$$\lim_{x \downarrow a} \mathbb{P}\big(T_y < t | X(0) = x\big) = 0,$$

where T_y is the hitting time (time to first-passage) of y. *See also* entrance boundary and regular boundary point. [L. Breiman, *Probability*. Addison-Wesley,1968; reprinted by SIAM, 1992]

natural increasing process *see* Doob–Meyer decomposition

natural scale *see* scale function

NBU, and NBUE *see* ageing

nearest neighbours model *see* interacting particle process

necktie problem *see* two-envelopes problem

negative association *see* association

negative binomial distribution A distribution on the non-negative integers that arises naturally in its simplest form as that of the number of tails shown before the appearance of the nth head in a sequence of independent flips of a biased coin (or, equivalently, the number of failures before the nth success in a sequence of Bernoulli trials). Thus, for $0 \le p \le 1$,

$$\mathbb{P}(X = x) = p(x) = \binom{n + x - 1}{x}(1 - p)^x p^n, x = 0, 1, 2, \ldots,$$

where X has pgf

$$G(s) = \mathbb{E}s^X = \left\{\frac{p}{1 - (1 - p)s}\right\}^n,$$

with $\mathbb{E}X = n(1 - p)/p$ and $\operatorname{var} X = n(1 - p)/p^2$. More generally, $G(s)$ is the pgf of a negative binomial r.v. for any real positive (not necessarily integral) n; this is a limiting case of the **Pólya distn**, with shape parameter n. The case when n is an integer may be called a **Pascal distn**. See multinomial distribution. [W. Feller, *Introduction to Probability Theory and its Applications*, Vol. 1, 3rd edn. Wiley, 1968]

negative multinomial distribution *see* multinomial distribution

negentropy A term used in some applications of information theory to denote the Kullback–Leibler divergence of a density f on $\mathbb{R}$ from the normal density g having the same (assumed finite) mean and variance as f, i.e.

$$\begin{aligned} d(f, g) &= \int_{\mathbb{R}} f \log \frac{f}{g} \,\mathrm{d}x = h(g) - h(f) \\ &= \frac{1}{2}\log(2\pi\sigma^2 \mathrm{e}) - h(f), \end{aligned}$$

where h denotes differential entropy. The term has also been used in a somewhat different context by physicists, notably L. Brillouin and E. Schrödinger. [E. Schrödinger, *What is Life?* Cambridge University Press, 1944]

negligible set *see* null event

Nelson's free (Markov) field *see* free Markov field

Nerman's theorem Any one of several theorems demonstrating the convergence of suitable supercritical Crump–Mode–Jagers processes. Depending on certain conditions the convergence may be in probability, or in L_1, or almost surely. [O. Nerman, On the convergence of supercritical general (CMJ) branching processes, *Probab. Theor. Rel. Fields*, 57(3), 365–395, 1981]

netput process Arises in the context of queues, dams, inventories, etc., and related fluid and diffusion approximations. It denotes an associated process arising as the difference between the input process and the output process that would arise if the server were always busy (or the buffer never empty, in other contexts). It generalizes the associated r.w. of the classic single server queue. [J. M. Harrison, *Brownian Motion and Stochastic Flow Systems*. Wiley, 1985; reprinted by Krieger, 1990]

neutral random vector This is a special form of independence. Let the random vector $\mathbf{X} = (X_1, \ldots, X_n)$ have $n \geq 2$ non-negative coordinates with $S_n = \sum_{r=1}^{n} X_r \leq 1$; and suppose that neither $1 - S_n$, nor any X_r, is degenerate at zero. Then for $r \leq n-1$, the vector $(X_1, \ldots, X_r)$ is said to be neutral in $\mathbf{X}$ if $(X_1, \ldots, X_r)$ and $(X_{r+1}/(1-S_r), X_{r+2}/(1-S_r), \ldots, X_n/(1-S_r))$ are independent. If $(X_1, \ldots, X_r)$ is neutral in $\mathbf{X}$ for all $r = 1, \ldots, n-1$, then $\mathbf{X}$ is **completely neutral**; this is equivalent to $X_1, X_2/(1-S_1), \ldots, X_n/(1-S_{n-1})$ being mutually independent. Furthermore, the following are equivalent:

(a) The distn of $\mathbf{X}$ is Dirichlet, or a limit of Dirichlet distns.
(b) X_n is neutral in $\mathbf{X}$, and also $\mathbf{X}$ is completely neutral.
(c) X_r is neutral in $\mathbf{X}$ for all $r = 1, \ldots, n$.
(d) The vector $(X_k; k \neq r)$ is neutral in $\mathbf{X}$ for all $r = 1, \ldots, n$.

[I. James & J. Mosimann, A new characterization of the Dirichlet distribution through neutrality, *Ann. Stat.*, 8(1), 183–189, 1980]

Neuts's method *see* matrix-exponential distribution

new better (or worse) than used *see* ageing

Newman–Wright theorem (alias: **Newman–Wright invariance principle, or CLT**) Arises in the context of a stationary sequence of zero-mean associated r.vs, X_r, $r \in \mathbb{N}$, where it asserts that if,

$$0 < \operatorname{var} X_1 + 2\sum_{r=2}^{\infty} \operatorname{cov}(X_1, X_r) < \infty,$$

then the sequence satisfies an invariance principle, i.e. it obeys a functional CLT. The original proof used the **Newman–Wright inequality** for associated random variables $X_1, \ldots, X_n$, i.e. $|\mathbb{E}\exp(\mathrm{i}\sum_{r=1}^{n} \lambda_r X_r) - \prod_{r=1}^{n} \mathbb{E}\exp(\mathrm{i}\lambda_r X_r)| \leq \frac{1}{2} \sum_{1 \leq r \neq} \sum_{s \leq n} |\lambda_r||\lambda_s| \operatorname{cov}(X_r X_s)$, where each $\lambda_r \in \mathbb{R}$. See Cox–Grimmett coefficient. [C. Newman & A. Wright, An invariance principle for certain dependent sequences, *Ann. Probab.*, 9(4), 671–675, 1981]

newsvendor model *see* single-period model

Newton–Pepys problem *see* Pepys' problem

Neyman–Scott process (alias: **centre-satellites process**) A type of Poisson cluster process in $\mathbb{R}^d$, in which there is an underlying Poisson process of parents; the actual process comprises the collection of random clusters of points having a parent point as origin. Conditional on the size of the cluster, the points of the cluster are i.i.d. about its origin. [J. Neyman & E. Scott, Statistical approach to problems of cosmology, *J. Roy. Stat. Soc. B*, 20(1), 1–43, 1958]

Neyman type-A process (alias: **Neyman's contagious type-A distribution**) A compound Poisson process in which the Poisson sum is of i.i.d.r.vs also having a Poisson distn. Its pgf is thus of the form

$$G(s) = \exp\{\lambda\{\exp[\mu(s-1)] - 1\}\}, \quad \lambda, \mu > 0.$$

[J. Neyman, On a new class of 'contagious' distributions, applicable in entomology and bacteriology, *Ann. Math. Stat.*, 10(1), 35–57, 1939]

no-arbitrage *see* arbitrage

no-arbitrage measure *see* risk-neutral measure

noise A concept arising in communication theory, control theory, etc.; the terms deviation, error, and residual are used to describe similar ideas in statistics and decision theory. Broadly, noise refers to any unwanted contribution to a measurement or received signal S, that corrupts some original true value (of a signal, or system) X. Commonly noise is modelled as a random variable Z (or a random process Z_n in the case of a sequence of observations S_n of a random input X_n), and S is some suitable function of X and Z. Classic examples include the noisy channel, in which noise is described by the conditional distribution of S given X; shot noise, in which quantum effects are modeled as e.g. a Poisson process; and discrete-time white noise, where (in its simplest form) the disturbance to X_n is assumed to comprise a sequence of i.i.d. (or, more weakly, orthogonal i.d.) r.vs. If each Z_n is normal, one refers to Gaussian (white) noise. When the sequence Z_n is not independent, one may speak of generalized **noise with memory**, or of **coloured noise** because the spectral density of the process is not uniform, as is the case with white noise.

In continuous time, the classic noise analogous to discrete-time white noise is formalized in terms of Hida white noise as a type of derivative of the Wiener process. More generally, given a suitable probability space (Ω, $\mathcal{F}$, $\mathbb{P}$), a noise has been defined as comprising:

1. A two-parameter family $\mathcal{F}(s,t) s \leq t$, of sub-$\sigma$-fields of $\mathcal{F}$, where s, $t \in \mathbb{R}$, such that for all $s \leq t \leq u$, $\mathcal{F}(s,t)$ and $\mathcal{F}(t,u)$ are independent and generate $\mathcal{F}(s, u)$ (this being called a continuous factorization of $\mathcal{F}$).
2. A one-parameter measurable group of automorphisms T_h, $h \in \mathbb{R}$, that is continuous at 0, and such that T_h maps $\mathcal{F}(s,t)$ to $\mathcal{F}(s+h, t+h)$.

[B. Tsirelson, Noise as a Boolean algebra of σ-fields, *Ann. Probab.*, 42(1), 311–353, 2014]

noiseless coding theorem (alias: **Shannon's source (or first) coding theorem; entropy bound theorem; data compression theorem**) Arises in the context of communication theory (and data-processing), where it defines a lower bound for the extent to which a message from a source (or other data) can be concisely and invertibly encoded (i.e. losslessly compressed). Formally, let X be a r.v. taking values in a finite alphabet A, having entropy $H(X)$, and let $c(X)$ be a lossless encoding (i.e. invertible map) taking values in the set B^* of binary strings of finite length, where $|c(x)|$ denotes the length of the string encoding x. Then the theorem asserts that $\mathbb{E}|c(X)| \geq H(X)$. More generally, for a coding alphabet of size q, $\mathbb{E}|c(X)| \geq H(X)/\log_2 q$.

Furthermore, for arbitrarily long messages from a memoryless source (and more generally) the expected codeword length per source symbol may be made arbitrarily close to the lower bound. [R. McEliece,

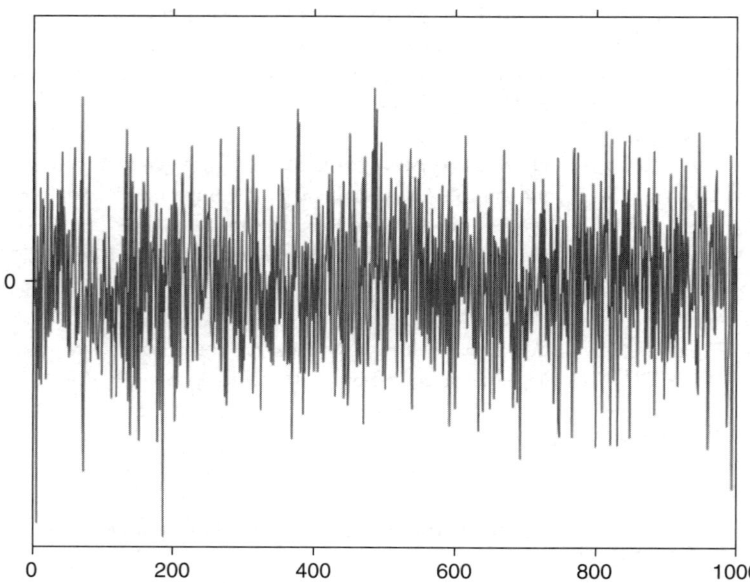

Realization of empirical noise for a communication channel in continuous time, which white noise models seek to describe.

The Theory of Information and Coding. Addison-Wesley, 1977; reprinted by Cambridge University Press, 1985]

noise sensitivity A concept arising in computer science, communication theory, and percolation, among others. One basic form is this: Let $X \in \{0,1\}^n$ be a binary string of length n, uniformly selected from all 2^n such strings, that is passed through a discrete memoryless channel with error probability p, yielding an output $Y \in \{0,1\}^n$ and let $g(\cdot)$ be a Boolean (or pseudo-Boolean) function. Then the noise-sensitivity of g at p is $S_n(p) = \mathbb{P}(g(X) \neq g(Y))$. If $\lim_{p\downarrow 0}\sup_{n\geq 0}S_n(p) = 0$, then g is (asymptotically) noise stable; if $S_n(p) \to \frac{1}{2}$ as $n \to \infty$ then it is (asymptotically) noise sensitive. [C. Garban, Oded Schramm's contributions to noise sensitivity, *Ann. Probab.*, 39(5), 1702–1767, 2011]

noisy channel *see* channel

noisy coding theorem (alias: **Shannon's channel (or 2nd) coding theorem**) Arises in the context of communication through a noisy channel, where it defines an upper bound C, called the capacity of the channel, for the rate of transmission of information (in bits per symbol) while requiring the probability of error in transmission to be arbitrarily small for sufficiently long messages (more formally, there exists a coding and decoding scheme such that the maximum probability of error in decoding the received version of any codeword sent is arbitrarily small).

Conversely, for no rate of transmission greater than C is this possible. Indeed, for any such rate the probability of error is asymptotic to 1 as the message length increases. This may be called **Wolfowitz's thm.** [J. Wolfowitz, *Coding Theorems of Information Theory*, 3rd edn. Springer, 1978]

non-anticipating process *see* adapted process

non-anticipating time *see* stopping time

non-central distribution Denotes a group of distributions that are defined in their standard form in terms of some r.v. (or r.vs) having the standard normal distribution. In their non-standard, or non-central, form, the relevant normal r.v. (or r.vs) have non-zero mean(s). Thus, e.g.:

(a) Let X and Y be independent, where X has an $N(\mu, 1)$ density, and Y a $\chi^2(n)$ density. Then $T = X/(Y/n)^{1/2}$ has a non-central t-density.

(b) Let X_r, $r \geq 1$ be independent normal $N(\mu_r, 1)$ r.vs. Then $Y_\theta = \sum_{r=1}^{n} X_r{}^2$ has a non-central chi-squared distn, $\chi^2(n, \theta)$, with non-centrality parameter $\theta = \sum_{r=1}^{n} \mu_r{}^2$.

(c) Let U and V be independent r.vs having the $\chi^2(n, \theta)$ and $\chi^2(m)$ distns respectively. Then $W = \frac{U}{U+V}$ has a non-central beta density, and $Z = (U/n)/(V/m)$ has a non-central F distribution. If V is also non-central, then W and Z may be called doubly non-central.

[H. Scheffé, *The Analysis of Variance*. Wiley, 1959]

non-central limit theorem Broadly, this denotes the convergence in distribution of normalized partial sums of a sequence of r.vs X_n, $n \geq 1$, where the limiting distributions are not Gaussian (because the sequence is not i.i.d., or suitably mixing), and the norming sequences are not of standard type. Typically, such theorems arise when the sequence has long-range memory (and dependence). [R. Dobrushin & P. Major, Non-central limit theorems for non-linear functionals of Gaussian fields, *Probab. Theor. Rel. Fields*, 50, 27–52, 1979]

non-commutative probability *see* quantum probability

non-null *see* recurrence time

non-recurrent *see* transient

non-transitivity *see* Efron's dice

norm *see* distance; Lp space; Orlicz norm

normal copula *see* copula

normal distribution (alias: **Gauss distribution, Gauss–Laplace distribution**) A most important distribution, arising throughout probability theory and its applications, that has a density $f(x), x \in \mathbb{R}$, given by,

$$f(x) = \frac{1}{\sigma\sqrt{2\pi}}\exp\left\{-\frac{1}{2}\left(\frac{x-\mu}{\sigma}\right)^2\right\},$$

for $\mu \in \mathbb{R}$ and $\sigma > 0$, and often denoted by $N(\mu, \sigma^2)$. For a random variable X having the $N(\mu, \sigma^2)$ density, $\mathbb{E}X = \mu$ and var $X = \sigma^2$. Then $Z = \frac{X-\mu}{\sigma}$ has the $N(0, 1)$ density, which is called the standard normal density, denoted by $\phi(x)$, with distn fn $\Phi(x)$; Z may be called a **standard normal** r.v., and its distribution is often called the **Z-distn**. Some authors write $Q(x) = 1 - \Phi(x)$.

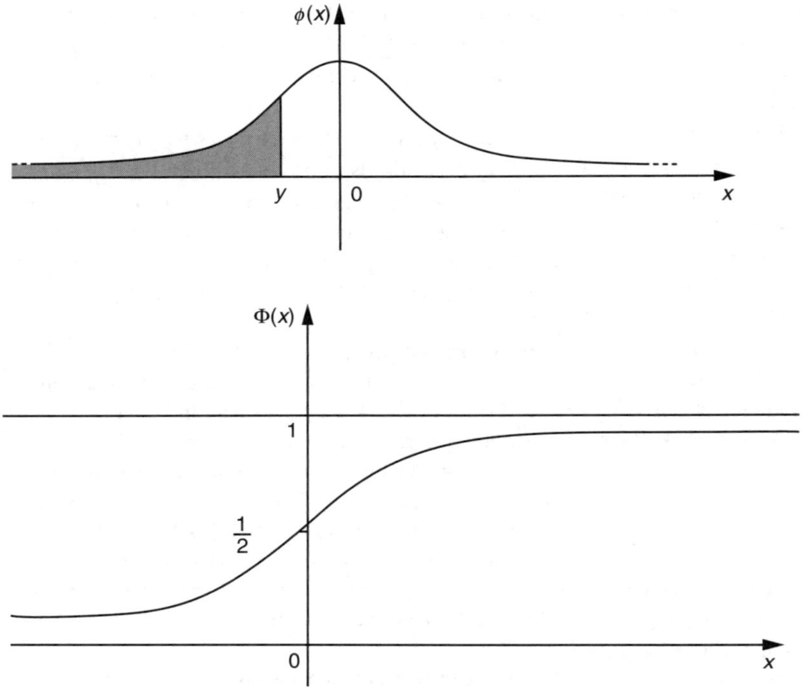

Density of an $N(0, 1)$ random variable. The shaded area is $\Phi(y) = \int_{-\infty}^{y} \phi(x)\, dx$.

The characteristic function of the general normal r.v. is

$$\varphi(t) = \mathrm{E}e^{\mathrm{i}tX} = \exp\left\{\mathrm{i}\mu t - \frac{1}{2}\sigma^2 t^2\right\}.$$

Importantly, the sum of independent normal r.vs is normal; and the normal distn is stable (with index 2), and therefore infinitely divisible.

It first appeared in the work of A. de Moivre as an approximation to the binomial distribution, and in subsequent CLTs. C. Gauss and P. Laplace used it as a model for errors. [W. Bryc, *Normal Distribution Characterizations with Applications*. Springer, 1995; posted online 2005]

normal-gamma distribution *see* variance-gamma distribution

normal-Laplace distribution Arises as the sum of a normal r.v. and an independent Laplace distribution, yielding a distribution having chf

$$\phi(t) = \frac{ab}{(a - \mathrm{i}t)(b + \mathrm{i}t)} \exp\left\{\mathrm{i}\mu t - \frac{1}{2}\sigma^2 t^2\right\}.$$

normal martingale A square integrable martingale $X(t), t \geq 0$, is called a normal martingale if $X(t)^2 - t$ is also a martingale. [N. Privault, *Stochastic Analysis in Discrete and Continuous Settings*. Springer, 2009]

normal number *see* Borel's normal number theorem

normal sample A set of i.i.d. normal r.vs, having therefore a special multinormal distribution. See Geary's theorem and Cochran's theorem.

normal scale mixture *see* Gaussian scale mixture

normal variance-mean mixture *see* variance-mean mixture

norm inequality *see* Garsia-Neveu inequality

Norton's theorem *see* flow-equivalent method

Novikov condition (or criterion) Supplies a sufficient condition for an exponential local martingale to be a martingale (i.e. a true or honest martingale). One form is this: Let $W(t), 0 \leq t \leq T$, be a Wiener process, and $X(t) \in \mathbb{R}, 0 \le t \le T$, a measurable process, both being adapted to the filtered space $(\Omega, \mathcal{F}_t, \mathbb{P})$. If $\mathbb{E}\exp\left(\frac{1}{2}\int_0^T |X(s)|^2 \mathrm{d}s\right) < \infty$, then $Z(t) = \exp\left(\int_0^t X(s) \mathrm{d}W(s) - \frac{1}{2}\int_0^t |X(s)|^2 \mathrm{d}s\right), 0 \le t \le T$, is a martingale. [I. Karatzas & S. Shreve, *Brownian Motion and Stochastic Calculus*, 2nd edn. Springer, 1998]

null array (alias: **infinitesimal array; uniformly asymptotically negligible (or u.a.n.) array; holospoudic array**) Arises in the context of the CLT and other classical limit laws, where (informally speaking) it requires that no single term of the sequence of r.vs $X(n,r), 1 \le r \le k(n)$ influences the (possibly scaled and shifted) limit of $S_n = \sum_{r=1}^{k(n)} X(n,r)$ as $n \to \infty$ and $k(n) \to \infty$. Formally, the array $X(n,r), n \ge 1, 1 \le r \le k(n)$, is said to be null, or u.a.n., if for any $\varepsilon > 0$,

$$\lim_{n\to\infty} \max_{1 \le r \le k(n)} \mathbb{P}(|X(n,r)| > \varepsilon) = 0,$$

i.e. $X(n,r)$ converges in probability to 0 as $n \to \infty$, uniformly in r. Such arrays yield infinitely divisible distributions in the limit. Often $k(n) = n$, and the array is triangular. [O. Kallenberg, *Foundations of Modern Probability*, 2nd edn. Springer, 2002]

null event An event A such that $\mathbb{P}(A) = 0$. Since A is defined as an element of $\mathcal{F}$ in a space $(\Omega, \mathcal{F}, \mathbb{P})$, it may be called $\mathbb{P}$-null. More generally, a subset B of Ω is said to be $\mathbb{P}$-null if there is a null event $A \in \mathcal{F}$ such that $B \subseteq A$. Note that a null set is not necessarily empty; but some authors call the empty set the null set, and then refer to A such that $\mathbb{P}(A) = 0$ as a negligible event.

null recurrent *see* recurrence

Nummelin splitting *see* Athreya–Ney–Nummelin splitting

O

objective method A branch of weak convergence theory that finds applications in probabilistic combinatorics. [D. Aldous & J. Steele, The objective method, in *Encyclopaedia Math. Sci.*, 110. Springer, 2004]

objective probability *see* subjective probability

observable In filtering, prediction, control, etc., some stochastic process $Z(t) = (X(t), Y(t))$ is such that only the realized values of $Y(t)$ are known to an observer (so that $Z(t)$ is partially observable) and the task is to infer (or estimate) the conditional distn (or moments) of $X(t)$ given $Y(t)$. In control theory the measure of the merits (or usefulness) of such estimates can be called **observability**.

In quantum theory an observable is a physical quantity that is represented (in the classical theory) by a self-adjoint (Hermitian) operator on a Hilbert space, where a measurement of the observable is assumed to yield an eigenvalue with a well-defined probability. [A. Bensoussan, *Stochastic Control of Partially Observable Systems*, Cambridge University Press, 1992]

occlusion model *see* dead leaves model

occupancy problem *see* urn model

occupation time (or measure) Broadly speaking, for a random process $X_n, n \in \mathbb{N}$, having state space S, the occupation time up to $T > 1$ (or the pre-T occupation time), $V_T(A)$ of a suitable subset A of S, is the number of visits of X to A before T (or the total sojourn time in A up to T). Formally, denoting by $I(X_n \in A)$ the indicator of the event that the process is in A at time n,

$$V_T(A) = \sum_{n=1}^{T-1} I(X_n \in A).$$

Then the expected value of $V_T(A)$ is the occupation measure

$$v_T(A) = \mathbb{E}V_T(A) = \sum_{n=1}^{T-1} \mathbb{P}(X_n \in A).$$

Note that some writers call $V_T(A)$ the occupation measure, and $v_T(A)$ the occupation intensity. Very often X_n is a Markov chain; thus, e.g. when X_n is a random walk, and $T = \infty$, $v_T(A)$ is a renewal measure; *see* renewal theory.

For a suitable process $X(t)$ in continuous time, similarly, $V_A(T) = \int_0^T I(X(s) \in A)\mathrm{d}s$, and $v_A(T) = \int_0^T \mathbb{P}(X(s) \in A)\mathrm{d}s$.

If, further, there exists $g(x, T)$ such that for all T and A

$$v_A(T) = \int_{x \in A} g(x, T)\mathrm{d}x$$

then g may be called an **occupation time density**. The equivalent function for $V_A(T)$ is local time. [S. Karlin & H. Taylor, *A Second Course in Stochastic Processes*. Academic Press, 1981]

Ocone martingale A class of rcll real-valued local martingales whose laws are invariant under integral transformations that preserve quadratic variation (i.e. a transform of the martingale $M(t)$ of the form $\int_0^t b(s)\mathrm{d}M(s)$, where $b(t)$ is adapted to $M(t)$ and $|b(s)| = 1$ for all s).

There are several characterizations of Ocone martingales: e.g. they have symmetric jumps and conditionally independent increments; further, when continuous, the increments are Gaussian. [D. Ocone, A symmetry characterization of conditionally independent increment martingales, *Barcelona Seminar on Stoch. Analysis, Progr, Probab.*, 32, 147–167, 1993]

odds A term arising in probability, statistics, and betting, where it has quite different meanings (though in each case it denotes a ratio). Note that in all three applications many authors use the word incorrectly. In probability, given an event A with probability $\mathbb{P}(A)$, the odds for (or on) A are $\mathbb{P}(A)/(1 - \mathbb{P}(A))$ and the odds against A (or for A^c) are $(1 - \mathbb{P}(A))/\mathbb{P}(A)$. Very commonly $\mathbb{P}(A) = \frac{a}{a+b}$ is rational (a and b being integers) and one writes a:b for the odds on A and b:a for the odds against A. For example, on the flip of a fair coin the odds for and against a head are both 1:1, and one speaks of even odds, or an evens chance.

This ratio should not be confused with the **odds ratio**, a term used in statistics to denote a relative risk of the form $\frac{\mathbb{P}(A)}{\mathbb{P}(A^c)} \Big/ \frac{\mathbb{P}(B)}{\mathbb{P}(B^c)}$ for events A and B, this being a ratio of odds (i.e. a ratio of ratios).

Most commonly the term arises in betting where it denotes the ratio of the amounts staked by two gamblers on a wager, where the winner takes both stakes. In particular, when one of the parties is a casino or bookmaker (whose stake is implicit) one speaks of pay-off odds. Such odds commonly assume a returnable unit stake from the customer, and are written as a:1 (pronounced 'a-to-one'). When the customer wins with probability p, the expected net change in fortune g is then $g = ap - (1-p) = (a+1)p - 1$.

Note that if the pay-off odds are $\frac{1}{p} - 1 : 1$, then $g = 0$, and these may be called fair pay-off odds. More generally, the fair pay-off odds for a partition A_i; with probabilities $\mathbb{P}(A_i)$, $1 \le i \le n$ where $\sum_{k=1}^n \mathbb{P}(A_i) = 1$, are a_i:1, where $a_i = \frac{1}{\mathbb{P}(A_i)} - 1$, and so $\sum_{i=1}^n \frac{1}{a_i+1} = 1$.

Finally, in lotteries there is a non-refundable (even to a winner) admission fee, so it is customary to speak of W-for-1 odds where the fee is 1 for a lottery giving one a prize size W. [T. Cover & J. Thomas, *Elements of Information Theory*, 2nd edn. Wiley, 2006]

Odlyzko–Montgomery law *see* Montgomery–Odlyzko law

Ogawa integral A type of stochastic integral that is non-adapted (or anticipative), first defined by S. Ogawa in terms of a suitable orthonormal system. [D. Nualart & M. Zakai, On the relation between the Stratonovich and Ogawa integrals, *Ann. Probab.*, 17(4), 1536–1540, 1989]

Ohlin's lemma *see* Karlin–Novikoff cut criterion

OK corral model A bivariate Markov chain (X_n, Y_n), $n \ge 1$, taking values in the non-negative quadrant of $\mathbb{Z}^2$, with transition probabilities $p(x-1, y|x, y) = \frac{y}{x+y}$ and $p(x, y-1|x, y) = \frac{x}{x+y}$.

A typical initial condition is $X_0 = Y_0 = g > 0$, and the process is ultimately absorbed, on one of the axes $x = 0$ or $y = 0$, at time T_g, when the value of the non-zero coordinate is S. It comprises a type of bivariate death process and, as $g \to \infty$, the limiting distribution of $\frac{3}{16}\frac{S^4}{g^3}$ has density $(\pi x)^{-\frac{1}{2}}\mathrm{e}^{-x}$, i.e. a gamma distribution. In particular, as $g \to \infty$,

$$\mathbb{E}S \sim \frac{2}{\pi^{1/2}} \cdot \frac{1}{3^{1/4}} \cdot \Gamma\left(\frac{3}{4}\right) g^{3/4},$$

where $\Gamma(\cdot)$ is the gamma function. The appearance of $g^{3/4}$ is a curiosity. [J. F. C. Kingman & S. Volkov, Solution to the OK corral model via decoupling of Friedman's urn, *J. Theor. Probab.*, 16(1), 267–276, 2003]

Olkin–Pratt inequality (alias: **multivariate Chebyshov inequality**) Asserts that for jointly distributed zero mean r.vs $X_1, \ldots, X_n$, having non-singular covariance matrix with var $X_r = \sigma_r{}^2$ and $\rho(X_r, X_s) = \rho_{rs}$, we have that for any positive constants $c_1, \ldots, c_n$

$$\mathbb{P}(|X_r| > c_r\sigma_r \text{ for some } r) \le \{a^{1/2} + [(nt-a)(n-1)]^{1/2}\}^2/n^2,$$

where $t = \sum_{r=1}^n c_r{}^{-2}$, and $a = \sum_r \sum_s \rho_{rs} c_r^{-1} c_s{}^{-1}$. [I. Olkin & J. Pratt, A multivariate Tchebycheff inequality, *Ann. Math. Stat.*, 29(1), 226–234, 1958]

Olkin–Shepp inequality *see* Chernoff's inequality

one-cut criterion *see* Karlin–Novikoff cut criterion

one-dependent process *see* m-dependence

one-martingale *see* martingale

one-sided Chebyshov inequality *see* Chebyshov's inequality

one-stage look-ahead policy (alias: **one-step look ahead rule; myopic rule; greedy rule**) Arises in sequential decision and/or control theory, optimal stopping, etc. Let X_n, $n \ge 1$, be some underlying process subject to actions (or decisions) of an agent, with reward function $R(X_1, X_2, \ldots)$. Then, broadly speaking, at any stage (or time) n, this rule chooses its next action $A_n = a$ (yielding X_{n+1}), to maximize $\mathbb{E}(R|X_1, \ldots, X_n, A_n = a)$ over possible values of a. For example, if $R_n(X_1, \ldots, X_n)$ is the total reward for stopping at time n, and c denotes the decision to continue (stop or continue being the only possible actions) then the one-step look-ahead rule stops at $N = \min\{n \ge 1 : R_n \ge \mathbb{E}(R_{n+1}|X_1, \ldots, X_n, A_n = c)$. [D. Bertsekas, *Dynamic Programming and Optimal Control. Athena*, Vol. I, 3rd edn, 2005; Vol. II, 4th edn. 2012; E. Porteus, *Foundations of Stochastic Inventory Theory*. Stanford University Press, 2002]

one-way function Arises in cryptography and complexity theory. Informally, such a function $y = f(x)$, where x takes values in e.g. the set of finite binary strings, is easily computable for any given x, but its inverse $f^{-1}(y)$ is (in some sense) not easily computable. More formally, it is required that $f(x)$ is computable in polynomial time for any x, while any probabilistic algorithms for inverting f has a negligible chance of finding a pre-image of $Y = f(X)$, when X is a suitable random variable taking values in the set of finite binary strings. [J. Talbot & D. Welsh, *Complexity and Cryptography*. Cambridge University Press, 2006]

operational research (alias: **operations research (in the USA); O.R.; systems analysis**) A large, varied, and imperfectly defined portfolio of methods for making optimal (or at least better) decisions about real-world problems. In general the methodology, and criteria for judging optimality, are drawn from mathematics, and related areas (rather than from legal, moral, or political arguments). Since the problems are real, probability and its applications feature prominently; as e.g. in queues and networks, stochastic programming and optimization; reliability and maintenance; storage and inventories; decision theory; game theory; simulation and Markov chain Monte Carlo; image processing; and sorting, packing, and partitioning in many contexts. [D. Heyman & M. Sobel, *Stochastic Models in Operations Research*, Vols I and II. McGraw-Hill, 1982 & 1984, reprinted by Dover, 2004]

optimal compression coefficient *see* redundancy

optimality equation (or principle) *see* Bellman equation

optimal stopping Comprises a special case of sequential decision theory in which, at any time t, there are just two possible actions: to continue or to stop (and take no more actions). Applications arise in many branches of OR, statistics, and mathematical finance; elementary examples include the parking problem and the secretary problem. Formally, in discrete time, one is given a sequence of r.vs $X_1, X_2, \ldots$, that are jointly distributed, and a sequence of functions $r_0, r_1(X_1)$, $r_2(X_1, X_2), \ldots, r_\infty(X_1, X_2, \ldots)$. For any index n you may stop and receive the reward r_n; where you get r_0 if you choose not to observe any X_s, and you get r_∞ if you never stop. The classic problem is to devise a rule (or policy) for defining a random stopping time N such that $\mathbb{E}r_N$ is maximal. [T. Ferguson, Optimal stopping and applications, available online at: www.math.ucla.edu/~tom; R. Cairoli & R. Dalang, *Sequential Stochastic Optimization*. Wiley, 1996]

option *see* contingent claim

optional process Given a complete filtered probability space $(\Omega, \mathcal{F}, \mathcal{F}_t, \mathbb{P})$, where $\mathcal{F}_t, t \geq 0$, is right-continuous, the σ-algebra on $[0, \infty) \times \Omega$ generated by all rcll $\mathcal{F}_t$-adapted processes is called the optional σ-algebra $\mathcal{G}$. Then a process that is $\mathcal{G}$-measurable is called optional. [D. Revuz, & M. Yor, *Continuous Martingales and Brownian Motion*, 3rd edn. Springer, 1999]

optional sampling *see* martingale transform; Doob's optional stopping theorem

optional skipping *see* martingale betting system; Doob's optional stopping theorem

optional stopping *see* Doob's optional stopping theorem

optional system *see* martingale transform

optional time Given a filtered measurable space $(\Omega, \mathcal{F}_t, \mathcal{F}, \mathbb{P})$ a random time T is called optional if $\{T < t\} \in \mathcal{F}_t$ for all $t \geq 0$. [I. Karatzas & S. Shreve, *Brownian Motion and Stochastic Calculus*, 2nd edn. Springer, 1998]

oracle *see* random oracle

orderliness Arises as a property of a point process $N(s,t), s < t$ on the real line $\mathbb{R}$, which is said to be orderly if, for all $t \in \mathbb{R}, \mathbb{P}(N(t, t+h) > 1) = o(h) as\, h \downarrow 0$. This implies that $N(t)$ has no multiple points (or batches) with probability 1. [D. Daley, Various concepts of orderliness for point processes, in *Stochastic Geometry*, E. Harding & D. Kendall (eds). Wiley, 1974, pp. 148–161]

order statistics If a collection $(X_1, \ldots, X_n)$ of i.i.d.r.vs are listed by increasing order of magnitude and then denoted by $X_{(1)} \leq X_{(2)} \leq \cdots \leq X_{(n)}$, then $X_{(r)}$ is called the rth order statistic of $X_1, \ldots, X_n$. Very commonly the distribution $F(x)$ of X_1 is continuous with density $f(x)$, and so with probability 1, we have: $X_{(1)} < X_{(2)} < \cdots < X_{(n)}$. Then the density of $X_{(r)}$ is given by

$$f_r(x) = nf(x)\binom{n-1}{r-1}F(x)^{r-1}\{1 - F(x)\}^{n-r}.$$

[H. David & H. Nagaraja, *Order Statistics*, 3rd edn. Wiley, 2003]

Orey's theorem Extends the classic (Kolmogorov) convergence theorem for a Markov chain $X_n, n \geq 0$, on a countable state space S, as follows: let P be the transition matrix of an irreducible, aperiodic, recurrent Markov chain on S, and let $\mathbf{p}$ and $\mathbf{q}$ be any two initial distributions of X_0. Then $\lim_{n\to\infty}\|\mathbf{p}P^n - \mathbf{q}P^n\| = 0$, where $\|\mathbf{r}\| = \sum_{i \in S}|r_i|$ for a probability distn $\mathbf{r}$ on S. [S. Orey, An ergodic theorem for Markov chains, *Probab. Theor. Rel. Fields*, 1(2), 174–176, 1962]

Orlicz norm A generalization of the L_p-norm (*see* L_p-space) that takes the following form in the probabilistic context: let $\psi(x) \geq 0$, $x \geq 0$, be convex and strictly increasing with $\psi(0) = 0$ (which may be called an Orlicz function). Then the Orlicz norm of a r.v. X is

$$\|X\| = \inf\left\{c > 0 : E\psi\left(\frac{|X|}{c}\right) \leq 1\right\}.$$

[S. van de Geer & J. Lederer, The Bernstein-Orlicz norm and deviation inequalities, *Probab. Theor. Rel. Fields*, 157(1–2), 225–250, 2013]

Ornstein coupling A coupling for aperiodic random walk $S_n = S_0 + \sum_{r=1}^{n} X_r$ on the integers that is successful. It may be sketched as follows: let $p_k = \mathbb{P}(X_r = k), k \in \mathbb{Z}$ be the distribution of the independent steps of the walk, and let W_n be an independent sequence having the same distribution as X_n. For an arbitrarily large constant c define the sequences Y_n and R_n thus:

$$Y_n = \begin{cases} X_n & \text{if } |X_n - W_n| > c \\ W_n & \text{if } |X_n - W_n| \leq c \end{cases}$$

and $R_n = R_0 + \sum_{r=1}^{n} Y_r$. It can be shown that each Y_n has distribution (p_k), and the coupling $C_n(i,j)$ for S_n defined by $C_0 = (S_0, R_0) = (i,j)$, and $C_n = (S_n, R_n)$ for $n \leq \inf \{n : S_n = R_n\} = T_D$, is successful; which is to say that $\mathbb{P}(T_D < \infty) = 1$.

This follows from the fact that each of $X_n - Y_n$ is bounded and symmetric, and non-degenerate for large enough c. [T. Lindvall, *Lectures on the Coupling Method.* Wiley, 1992; reprinted by Dover, 2002; D. Ornstein, Random walks, *Trans. Amer. Math. Soc.*, 138, 1–43 & 45–60, 1969]

Ornstein $\bar{d}$ distance (or metric) A metric arising in ergodic theory as a measure of distance between stationary processes; one basic form is this: let $\mathbf{X}_n = (X_1, \ldots, X_n)$ and $\mathbf{Y}_n = (Y_1, \ldots, Y_n)$ be random vectors (with given distributions $F_n(\mathbf{x})$ and $G_n(\mathbf{y})$), each taking values in a discrete state space A^n. Define

$$\bar{h}_n(X_1, \ldots, X_n; Y_1, \ldots, Y_n) = \inf \frac{1}{n} \sum_{r=1}^{n} \mathbb{P}(X_r \neq Y_r),$$

where the infimum is taken over all joint distns of $\mathbf{X}_n$ and $\mathbf{Y}_n$ with marginals F_n and G_n. Then the Ornstein $\bar{d}$ distance between processes X_r; $r \geq 1$, and Y_r, $r \geq 1$, is $\bar{d} = \limsup_{n \to \infty} \bar{h}_n$. [R. Gray *et al.*, A generalization of Ornstein's $\bar{d}$ distance, with applications to information theory, *Ann. Probab.*, 3(2), 315–328, 1975]

Ornstein's theorem (alias: **Kolmogorov–Ornstein theorem**) Arises in the ergodic theory of dynamical systems and the Kolmogorov–Sinai theory of entropy, where it asserts that if two such systems are isomorphic then they have the same entropy; and, conversely, that two Bernoulli systems are isomorphic if they have the same entropy. [D. Ornstein, Some new results in the Kolmogorov–Sinai theory of entropy and ergodic theory; *Bull. Amer. Math. Soc.*, 77(6), 878–890, 1971]

Ornstein–Uhlenbeck process A homogenous diffusion arising as the solution to the Langevin equation modelling the velocity $V(t)$ of a particle engaged in Brownian motion. The standard form of the equation is

$$\mathrm{d}V(t) = -aV(t) + \mathrm{d}W(t), \quad t \geq 0, a > 0,$$

where $W(t)$ is the Wiener process, having a strong solution given by

$$V(t) = V(0)\mathrm{e}^{-at} + \int_0^t \mathrm{e}^{-a(t-u)} \mathrm{d}W(u).$$

It may also be represented in terms of the Wiener process by

$$V(t) = V(0)\mathrm{e}^{-at} + (2a)^{-\frac{1}{2}}\mathrm{e}^{-at} W(\mathrm{e}^{2at} - 1),$$

and (conditional on $V(0)$) comprises a Gaussian process. In particular, $X(t) = \mathrm{e}^{-at} W(\mathrm{e}^{2at})$ is a stationary Ornstein–Uhlenbeck (Gaussian) process; with autocovariance fn $e^{-a|t|}$, and with Cauchy spectral density. [G. Uhlenbeck & L. Ornstein, On the theory of the Brownian motion, *Phys. Rev.*, 36(5), 823–841, 1930; M. Lifshits, *Lectures on Gaussian Processes.* Springer, 2012]

orthant dependence *see* association of random variables

orthant probability Broadly, the probability that a random vector $X_1, \ldots, X_n$ lies in any one of the 2^n orthants determined by an arbitrary point $(x_1, \ldots, x_n) \in \mathbb{R}^n$. In particular, $\{X_1 > x_1, \ldots, X_n > x_n\}$ is called an upper orthant and $\{X_1 > 0, \ldots, X_n > 0\}$ the positive orthant; with $\leq$ replacing $>$ one has a lower, and the negative orthant. The (unqualified) term is often used to refer to the standard multinormal orthant probabilities; *see* Sheppard's theorem.

orthogonal Two real r.vs X and Y are orthogonal if $\mathbb{E}(XY) = 0$; more generally complex-valued r.vs are orthogonal if $\mathbb{E}(X\overline{Y}) = 0$, where $\overline{Y}$ is the complex conjugate of Y. Any uncorrelated r.vs are orthogonal when centred at their means. A sequence Z_n, $n \in \mathbb{Z}$ is orthogonal if $\mathbb{E}Z_i\overline{Z_j} = 0, i \neq j$; and a random process $X(t), t \in \mathbb{R}$ has orthogonal increments if $\mathbb{E}\{(X(v) - X(u))(\overline{X}(t) - \overline{X(s)})\} = 0$ for all $u \leq v \leq s \leq t$.

Finite variance martingales $X(t)$ and $Y(t)$, $X(0) = Y(0) = 0$, are orthogonal if $X(t)Y(t)$ is a martingale, or weakly orthogonal if $\mathbb{E}(X(s)Y(t)) = 0$ for all $s, t \geq 0$. Probability measures $\mathbb{P}$ and $\mathbb{Q}$ are orthogonal, or (mutually) singular, on some Borel field $\mathcal{B}$ if there is $A \in \mathcal{B}$ such that $\mathbb{P}(A) = 0$ and $\mathbb{Q}(A) = 1$. [O. Kallenberg, *Foundations of Modern Probability*, 2nd edn. Springer, 2002]

orthogonal increments process *see* additive process

Oscar system A progressive (or sequential) betting system in which the first stake is a unit, and after any losing bet the stake remains the same. After a winning bet the subsequent stake is raised by 1 unit, unless this would lead to a total gain exceeding 1 unit, in which case the stake is left unchanged. [S. Ethier, *The Doctrine of Chances*. Springer, 2010]

oscillating random walk A time-homogeneous Markov chain with state space $\mathbb{R}^n$, that evolves thus: there is a hyperplane H dividing $\mathbb{R}^n$ into half-spaces L and R. At each step the process acts as a random walk with jump distribution (or measure) $\mathbb{P}_H, \mathbb{P}_L$, or $\mathbb{P}_R$ according as the step begins in H, L, or R, respectively. The case $\mathbb{P}_H = \mathbb{P}_L = \mathbb{P}_R$ is ordinary random walk. Oscillating Brownian motion is defined analogously. (Some authors impose the extra condition of drift away from H.) [J. Keilson & J. Wellner, Oscillating Brownian Motion, *J. Appl. Probab.*, 15(2), 300–310, 1978]

oscillatory random walk *see* Kesten's theorem

Ottaviani–Skorokhod inequality Let $S_m = \sum_{r=1}^{m} X_r$ be the partial sums of independent r.vs $X_1, \ldots, X_n$(or random vectors), $1 \leq m \leq n$. Then for $x, y > 0$, and any norm $\|.\|$

$$\mathbb{P}(\max_{m \leq n} \|S_m\| > x + y)\Big(1 - \max_{m \leq n} \mathbb{P}(\|S_n - S_m\| > x)\Big) \leq \mathbb{P}(\|S_n\| > y).$$

It is often seen with $x = y$. [A. Gut, *Probability: A Graduate Course*, 2nd edn. Springer, 2013; R. Dudley, *Real Analysis and Probability*. Cambridge University Press, 2002]

Otter–Dwass formula *see* hitting-time theorem

output process *see* departure process

P

PageRank Markov chain (alias: **random surfer model**) Arises as a method for ranking (i.e. estimating the relative significance of) a set S of pages on the Web, which comprises the state space of the chain. Its transition matrix M is defined thus: let $\mathbf{q}$ and $\mathbf{r}$ be two probability distns on S (which are often assumed to be equal), and for any $i \in S$ let L_i denote the subset (numbering $|L_i|$) of S comprising pages one step from i on outgoing links. Define the transition matrix P on S given by

$$p_{ij} = \begin{cases} |L_i|^{-1}, & j \in L_i, \quad \text{if } |L_i| \neq 0, \\ q_j, & \text{if } |L_i| = 0, \end{cases}$$

and then for some $\rho \in (0,1)$, called the reset probability, define $M = (1-\rho)P + \rho R$, where the transition matrix R has all its rows equal to $\mathbf{r}$. [S. Kamvar *et al.*, Adaptive methods for the computation of Page Rank, *Linear Algebra Appl.*, 386, 51–65, 2004]

pairwise independent *see* independent

Pakes's lemma (or condition, or theorem) Supplies a criterion for the ergodicity (also called stability in applications) of a suitable Markov chain X_n on a countable state space S. One form is this: let X_n with transition probabilities p_{ij}, be an aperiodic, irreducible, time-homogeneous Markov chain on the non-negative integers. If there exist constants $a > 0$ and $N > 0$ such that $d_i = \mathbb{E}(X_{n+1} - X_n | X_n = i) = \sum_j (j-i) p_{ij}$ satisfies $d_i < \infty$ for all $i \geq 0$, and $d_i \leq -a$ for all $i \geq N$, then the chain is ergodic.

Conversely, **Kaplan's lemma** asserts that if there are constants $a \geq 0$, and $b \in [0,1]$, and N such that for all $i \geq N, d_i > 0$, and $x^i - \sum_j p_{ij} x^j \geq -a(1-x)$ for all $x \in [b,1]$, then the chain is not ergodic. *See* Foster's condition and drift condition. [M. Kaplan, A sufficient condition for nonergodicity of a Markov chain, *IEEE Trans. Inf. Theor.*, 25(4), 470–471, 1979]

Paley–Wiener–Zygmund integral (alias: **Paley–Wiener integral; PWZ-integral**) A stochastic integral that may be denoted by $\int_0^t v(s) \mathrm{d}W(s)$, where $v(s)$ is L_2-integrable on $[0, t]$, $W(t)$ is the Wiener process (in $d \geq 1$ dimensions), and the integral is defined pathwise on the Wiener space of real-valued continuous functions on $[0,t]$, taking the value 0 at the origin. [R. Paley *et al.*, Notes on random functions, *Math. Z.*, 37(1), 647–688, 1933]

Paley–Zygmund inequality (alias: **second moment method**) In its basic form this asserts that for a finite-variance non-negative r.v. X, and $0 \leq a \leq 1, \mathbb{P}(X > a\mathbb{E}X) \geq (1-a)^2 (\mathbb{E}X)^2 / \mathbb{E}(X^2)$. Among several generalizations, one has e.g. for a non-negative r.v. X with finite rth mean, $0 \leq a \leq 1$, and $r > p > 0$,

$$\mathbb{P}\left(X \geq a(\mathbb{E}(X^p))^{1/p}\right)^{\frac{1}{p}-\frac{1}{r}} \geq (1-a^p)^{1/p} \frac{(\mathbb{E}(X^p))^{1/p}}{(\mathbb{E}(X^r))^{1/r}},$$

which has been called **Petrov's inequality**. [D. Khoshnevisan, *Probability*. American Mathematical Society, 2007; M. Weber, *Dynamical Systems and Processes*. EMS, 2009]

Paley–Zygmund theorem Arises in the context of random trigonometric (and Fourier) series. One form is this: let

$$Y(t) = \sum_{r=0}^{\infty} a_r X_r \cos(rt + b_r),$$

where X_r, $r \geq 0$ is a sequence of i.i.d.r.vs such that $\mathbb{P}(X_r = +1) = \mathbb{P}(X_r = -1) = 1/2$, and a_r and b_r, $r \geq 0$, are given real constants. If $\sum_0^\infty a_r^2 < \infty$ then the series converges a.s. for almost all t to an L_p-integrable function, $1 \leq p < \infty$; if $\sum_0^\infty a_r^2 = \infty$, then the series diverges a.s. for almost all t. [J.-P. Kahane, *Some Random Series of Functions*, 2nd edn. Cambridge University Press, 1985]

Palm's machine repair problem *see* Erlang loss system

Palm's theorem Arises in its original and simplest form in the context of the M/G/∞ queue, where it asserts that the long-run distribution of the number of customers present is Poisson. [M. Carrillo, Extensions of Palm's theorem: a review, *Manage. Sci.*, 37(6), 739–744, 1991]

Palm theory (alias: **Palm calculus**) Arises naturally in the context of stationary simple point processes, where – informally speaking – it interprets the probabilistic properties of the process from the point of view of a typical point of the process, yielding Palm distributions (and probabilities and measures). More formally, it considers suitable processes obtained as functionals of stationary processes, and the relationships between the two, which are given explicitly by so-called Palm inversion formulas. [D. Daley & D. Vere-Jones, *Introduction to the Theory of Point Processes*, 2nd edn. Springer, Vol. I, 2003; Vol. II, 2008]

Panjer recursion Supplies an iterative method for calculating the values of a family of distributions that are often used in actuarial mathematics. A non-degenerate probability distribution, p_n, $n \geq 0$, on the non-negative integers is said to be a Panjer distribution of order *r*, with parameters *a* and *b*, if for all $n \geq r$

$$p_{n+1} = \left(a + \frac{b}{n+1}\right)p_n,$$

while $p_n = 0$ for $n \leq r - 1$, and *a* and *b* are suitable real constants such that $a + b \geq 0$. The set of such distns is a **Panjer class** of order *r*. In particular, the compound Poisson case is called the Adelson recursion. [K. Hess *et al.*, An extension of Panjer's recursion, *ASTIN Bull.*, 32(2), 283–297, 2002]

Papangelou conditional intensity Arises in the theory of random point processes where, broadly speaking, it is seen as the intensity measure for the probability of finding a point of the process in an infinitesimally small neighbourhood of any point *x*, conditional on the realization of the process (i.e. the configuration of points) outside that neighbourhood. [O. Kallenberg, On conditional intensities of point processes, *Probab. Theor. Rel. Fields*, 41(3), 205–220, 1978]

paradox A term most often arising in probability to denote statements that are true but diverge from uninformed intuition; or, less often, a false but superficially plausible argument or assertion. For examples, see Berkson; Bertrand; birthdays; Borel; Carroll; Galton; inspection; Monty Hall; Parrondo; Saint Petersburg; Simpson; two aces; waiting time (or line); and two envelopes. [W. Eckhardt, *Paradoxes in Probability*, Springer, 2013]

paradox of the second ace *see* two aces problem

parallelogram rule (or law) For r.vs *X* and *Y*, the elementary fact that $\mathbb{E}\{(X-Y)^2\} + \mathbb{E}\{(X+Y)^2\} = 2\mathbb{E}(X^2) + 2\mathbb{E}(Y^2)$. More generally, the rule holds for any norm $\|.\|$ (e.g. on r.vrs) arising as an inner product, in the form $\|X-Y\|^2 + \|X+Y\|^2 = 2\|X\|^2 + 2\|Y\|^2$. [M. Capinski & E. Kopp, *Measure, Integral and Probability*, 2nd edn. Springer, 2004]

Pareto distribution (alias: **power-law distribution**) The Pareto probability density with parameter $a > 0$ is given by $f(x) = \frac{ac^a}{x^{a+1}}, x \geq c > 0$, and the corresponding distribution function is $F(x) = 1 - \left(\frac{c}{x}\right)^a, x \geq c > 0$. It arose in the context of **Pareto's law**, which asserts that incomes have approximately such a distribution. More generally, an asymptotically power-law distn has density $f(x) \propto K(x)x^{-a}$, where $K(x)$ is slowly varying. The analogously defined distribution on positive integers is called the zeta (or Zipf) distribution. [C. Kleiber & S. Kotz, *Statistical Size Distributions in Economics and Actuarial Sciences*. Wiley, 2003]

parking problem A classic problem of optimal stopping, with many variations and extensions. In its basic form you are driving along a road with randomly available parking spots, towards a destination *T*, with the intention of parking so that the expected distance to *T* is minimal. Vacancies ahead cannot be seen, and vacancies passed cannot be returned to. Variations include the possibility of circling blocks if you pass *T*. *See also* Rényi's parking constant. [J. MacQueen & R. Miller, Optimal persistence policies, *Oper. Res.*, 8(3), 362–380, 1960]

Parrondo's paradox A paradox of game theory and probability that has many formulations. An elementary version in a gambling context asserts this: it is possible to have two games that are both systematically unfavourable to the gambler when either is played repeatedly (to the exclusion of the other) but if they are played alternately then the composite game is systematically favourable to the gambler. Related examples include the Allinson mixture, and the Brownian ratchet. [G. Harmer & D. Abbott, Parrondo's paradox, *Stat. Sci.*, 14(2), 206–213, 1999]

partially observed Markov sequence *see* hidden Markov process

partition function *see* Gibbs distribution

partition structure *see* Kingman's partition structure

partition theorem (alias: **law of total probability; total probability theorem**) A term used in probability theory to denote a basic formula of conditional probability (and also of conditional expectation) thus: let the events B_n, $n \in I$, be a partition of the sample space Ω, which is to say that $\cup_{k \in I} B_k = \Omega$, and $B_i \cap B_j = \emptyset, i \neq j$. Then for any event $A \subseteq \Omega$,

$$\mathbb{P}(A) = \sum_{n \in I} \mathbb{P}(A \cap B_n) = \sum_{n \in I} \mathbb{P}(A|B_n)\mathbb{P}(B_n);$$

and for a suitable r.v. on this probability space

$$\mathbb{E}(X) = \sum_{n \in I} \mathbb{E}(XI(B_n)) = \sum_{n \in I} \mathbb{E}(X|B_n)\mathbb{P}(B_n),$$

where $I(B_n)$ is the usual indicator r.v. *See also* addition rule. [G. Grimmett & D. Welsh, *Probability: An Introduction*, 2nd edn. Oxford University Press, 2014]

Pascal distribution *see* negative binomial distribution

PASTA *see* arrival theorem

path decomposition Arises in probability theory most often in the context of Markov processes where, broadly speaking, it is a method of analysis that uses suitable conditioning of the paths of some random process so that they split into sections that are (possibly when reordered) equivalent to the paths of some other process. Such splittings commonly occur at first-passages, extreme values, and/or last exits of the path. Perhaps the best known are Williams' path decompositions. [P. Millar, A path decomposition for Markov processes, *Ann. Probab.*, 6(2), 345–348, 1978; A. Borodin & P. Salminen, *Handbook of Brownian Motion*, 2nd edn. Birkhäuser, 2002]

path-dependent process *see* reinforced processes

pathwise uniqueness Arises in considering the nature and uniqueness of solution to SDEs of the form $dX(t) = \mu(X(t))dt + \sigma(X(t))dW(t)$, where $W(t)$ is the Wiener process, $t > 0$. If $X_1(t)$ and $X_2(t)$ are two solutions on the same filtration that are indistinguishable whenever $\mathbb{P}(X_1(0) = X_2(0)) = 1$, then pathwise uniqueness holds for the SDE. More weakly, if $X_1(t)$ and $X_2(t)$ (not necessarily on the same filtration), have the same joint distributions, whenever $X_1(0)$ and $X_2(0)$ have the same distribution, then one has weak (or distributional, uniqueness, or uniqueness in law (or distribution); this is implied by pathwise uniqueness, but not conversely. [D. Revuz, & M. Yor, *Continuous Martingales and Brownian Motion*, 3rd edn. Springer, 1999]

pay-off matrix (alias: **pay-off function (or table, or array); loss (or reward) matrix, etc.**) Arises in two-person games (and in decision theory) thus: let the sets U and V comprise the pure strategies available to players A and B respectively (or, in decision theory, the non-randomized decisions – or actions – available to the decider D and the possible states of nature respectively). For any pair $(u, v) \in U \times V$ the game yields a (possibly random) reward to player A (or to the decider D) whose utility (or expected utility for a random reward) is denoted by a_{uv}; which yields the pay-off matrix. In the context of decision theory the negative of a_{uv}, is called a loss, and $-a_{uv}$ $(u, v) \in u \times v$, is the loss matrix (or function). If $-a_{uv} = b_{uv}$, where b_{uv} is B's pay-off matrix, then the game has zero sum. [K. Binmore, *Fun and Games*. D.C. Heath, 1992]

peacock *see* Kellerer's theorem

Pearson's ϕ *see* mean square contingency

Pearson's product moment coefficient *see* correlation coefficient

Pearson system (alias: **Pearson curve (or distribution)**) If a probability density function $f(x)$ satisfies the differential equation

$$\frac{\mathrm{d}f(x)}{\mathrm{d}x} = -\frac{(x+a)}{bx^2 + cx + d} f(x),$$

for some real a, b, c, and d, then it is said to be a Pearson curve (or distribution), or to be in the Pearson system. Any such density is determined by its first four moments, provided they are finite. K. Pearson identified 12 distinct types in this system, which include many important densities. In the discrete analogy, if X is non-negative, non-degenerate, and integer valued, it is said to be of **Ord–Carver type** if $p(x) = \mathbb{P}(X = x)$ satisfies

$$p(x+1) - p(x) = -\frac{x+a}{bx^2 + cx + d} p(x), \quad x \geq 0.$$

By extension, a **Pearson diffusion** (alias Kolmogorov–Pearson diffusion) is a solution X of $\mathrm{d}X = -(X + a)\mathrm{d}t + \{2(bX^2 + cX + d)\}^{1/2}\mathrm{d}W$, for suitable values of a, b, c, and d. [A. Stuart & K. Ord, *Kendall's Advanced Theory of Statistics*, Vol. I, 6th edn. Wiley, 1994]

Pearson's type III distribution *see* gamma distribution

Penney Ante (alias: **Penney's game**) A game devised by Walter Penney in 1969. A fair a-sided die bears all the letters of an alphabet Ω of size $a \geq 3$. Player A chooses a motto comprising a string of n letters from Ω, and then player B also chooses a motto of length n from Ω, different from A's motto. The die is rolled until the first appearance of either motto in n consecutive rolls, and that motto wins. It is always possible for B to choose a motto such that $\mathbb{P}(B \text{ wins the game}) > 1/2$. [D. Felix, Optimal Penney Ante strategy via correlation polynomial identities, *Elec. J. Combin.*, 13, R35, 1–15, 2006]

Penrose–Banzhaf index *see* importance

Pepys' problem (alias: **Newton–Pepys problem**) Poses the following question: if player A rolls $6n$ conventional fair dice and needs at least n sixes to win, while player B rolls $6(n+1)$ such dice and needs at least $n+1$ sixes to win, who is the more likely to win? The question was put by S. Pepys in 1693, with $n = 1$ and $n = 2$, to I. Newton, who supplied the correct answer that A is more likely to win; this answer remains true for any n (but not for unfair dice). [S. Stigler, Isaac Newton as a probabilist, *Stat. Sci.*, 21(3), 400–403, 2006]

percolation process Arises as a model for a random medium, as exemplified by this archetypal physical problem: if a porous rock is placed in a bucket of water, what is the probability that the centre of the rock becomes wet? The classic **Broadbent–Hammersley model**, or **bond percolation**, is formulated on the square lattice $\mathbb{Z}^2$, thus: Each edge is seen as a pipe along which fluid may flow if it is open, and each edge is independently open with probability p (or closed with probability $1-p$). Then for a finite connected subset of $\mathbb{Z}^2$ with a wet perimeter one asks for the probability that an interior point is wet; or on the infinite lattice $\mathbb{Z}^2$, with fluid supplied at the origin, one first asks for the probability $\theta(p)$ that an infinite number of vertices became wet, and second for the properties of the function $\theta(p), 0 < p < 1$. Alternatively, **site percolation** considers these problems when vertices are independently open or closed to the passage of fluid. If both may occur one has **mixed percolation**, and if edges permit only directed flow then one has **oriented percolation**. In **first-passage percolation** a r.v. is attached to each edge indicating the time taken for fluid to pass from one end to the other, and one studies the random process $W(t)$ of wet sites at times $t \geq 0$, arising from fluid supplied at the origin from $t = 0$. [G. Grimmett, *Percolation*, 2nd edn. Springer, 1999; D. V. Wood, Problem 5, *Amer. Math. Mon.*, 1(3),99, 1894; and solutions in subsequent issues]

perfect probability measure A probability space $(\Omega, \mathcal{F}, \mathbb{P})$ is said to be perfect, and $\mathbb{P}$ is a perfect probability measure, if for any random variable $X \in \mathbb{R}$ (defined on $(\Omega, \mathcal{F}, \mathbb{P})$), and any set $A \subseteq \mathbb{R}$ such that $X^{-1}(A) \in \mathcal{F}$, there is a subset B of A that is Borel, and $\mathbb{P}\big(X^{-1}(B)\big) = \mathbb{P}(X^{-1}(A)$. [K. Itô, *Introduction to Probability Theory*. Cambridge University Press, 1984]

perfect sampling (or simulation) *see* Propp–Wilson method

performance function (or measure) *see* reliability function

performance rate *see* capacity

periodic state (or chain) A state s of a Markov chain (which is time-homogeneous on a countable state space) is said to be periodic with period d if return to s is possible only (but not necessarily) at times $d, 2d, 3d, \ldots,$

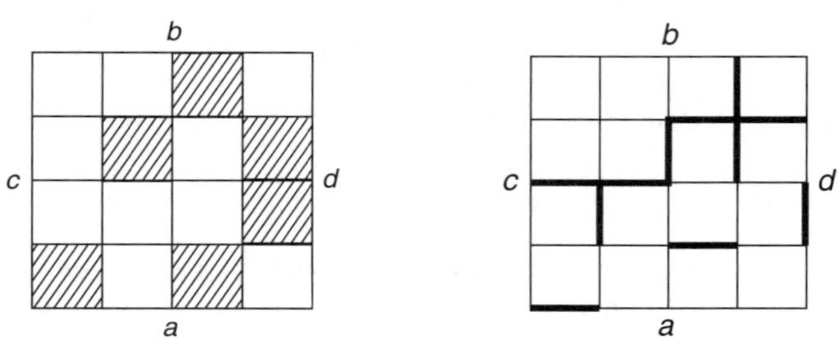

Left: **Site percolation. Fluid may pass from face a to face b, but not from face c to face d.** ***Right:*** **Bond percolation. Fluid may pass from face c to face d, but not from face a to face b.**

and $d > 1$ is the largest integer with this property. Otherwise, i.e. if $d = 1$, the state is aperiodic. This is a class property (*see* solidarity theorem) so that communicating states have the same period, as do all the states of irreducible chain, which is called periodic if $d > 1$. [W. Feller, *Introduction to Probability Theory and its Applications*, Vol. 1, 3rd edn. Wiley, 1968]

Perks's distribution Most commonly denotes a distn having a density $f(x), x \in \mathbb{R}$, proportional to $\{e^x + a + e^{-x}\}$ for some constant $a > -2$. [J. Talacko, A note about a family of Perks' distributions, *Sankhyā*, 20(3/4), 323–328, 1958]

perpetuity (alias: **perpetual annuity; stochastic perpetuity**) In financial and actuarial mathematics, a perpetuity is the present value V of a contract to pay a sum of money regularly (annually, say) forever, allowing for the force of interest. In probability theory its basic form is

$$V = X_1 + X_1X_2 + X_1X_2X_3 + \cdots,$$

where $X_r, r \geq 1$, is a sequence of i.i.d.r.vs having the distn of X. (Some authors define a perpetuity as $1 + V$, where V is as above.) This distribution of V is then the same as that of $X(1 + V)$, where X and V are independent, and V is said to be the perpetuity generated by the distn of X. In particular, if X has the distribution of $U^{1/a}$, for some $a > 0$, where U is uniformly distributed on [0, 1], one speaks of the **Vervaat family** of perpetuities.

In another formulation, a r.v. V that has the same distribution as $X\hat{V} + Y$, where the pair (X, Y) is independent of the r.v. $\hat{V}$ that has the same distn as V, one refers to the perpetuity generated by the distn of (X, Y). *See also* Dufresne's relation. [J. Fill & M. Huber, Perfect simulation of Vervaat perpetuities, *Electr. J. Probab.*, 15, 96–109, 2010]

Perrin distribution An axially symmetric distribution on the sphere which arises as the density of the position of a particle executing a spatially and temporally homogeneous diffusion process on the surface of the sphere. (Such a diffusion is essentially unique, by Yosida's theorem.) In spherical polar coordinates (θ, ϕ) when the particle starts at the pole $\theta = 0$, at $t = 0$, the density of its position at time t is

$$f(t, \theta, \phi) = \frac{1}{4\pi}\sum_{r=0}^{\infty}(2r+1)e^{\frac{-1}{2}\sigma^2 r(r+1)t}p_r(\cos\theta)\sin\theta,$$
$$0 \leq \theta \leq \pi,\ 0 \leq \phi < 2\pi,\ 0 \leq t;$$

where $p_r(x)$ is the rth Legendre polynomial, and σ^2 is the variance parameter of the diffusion. [P. Roberts & H. Ursell, Random walk on a sphere and on a Riemannian manifold, *Proc. Trans. Roy. Soc. A*, 252(1012), 317–356, 1960]

persistent *see* recurrence

personal probability *see* subjective probability

perturbed Brownian motion Arises from the Wiener process $W(t), t \geq 0$, thus: for a random process $X(t), t \geq 0$, denote $\overline{X}(t) = \sup_{0 \leq s \leq t} X(s)$, and $\underline{X}(t) = \inf_{0 \leq s \leq t} X(s)$, then for any $a < 1$ $Y(t) = W(t) + \frac{a}{1-a}\overline{W}(t)$ is called perturbed Brownian motion. It is the pathwise unique solution of $X(t) = W(t) + a\overline{X}(t)$. More generally, for $b < 1$, the solution of $X(t) = W(t) + a\overline{X}(t) + b\underline{X}(t)$is called doubly perturbed Brownian motion. [L. Chaumont & R. Doney, Some calculations for doubly perturbed Brownian motion, *Stoch. Process. Appl.*, 85(1), 61–74, 2000]

Petersburg problem (or paradox) *see* Saint Petersburg problem

Peterson bounds Arise in the context of competing risks, where they supply bounds for a joint distribution in terms of its so-called subdistribution functions, as follows: let X and Y be jointly distributed with $T = \min(X,Y)$. Then the basic forms are:

$$\mathbb{P}(X \leq x, X \leq Y) \leq \mathbb{P}(X \leq x) \leq \mathbb{P}(T \leq x), \text{ and}$$
$$\mathbb{P}(Y \leq y, Y \leq X) \leq \mathbb{P}(Y \leq y) \leq \mathbb{P}(T \leq y).$$

[T. Bedford & I. Meilijson, Characterization of marginal distributions of (possibly dependent) lifetime variables which right censor each other, *Ann. Stat.*, 25(4), 1622–1645, 1997]

petite set *see* small set

Petrov's inequality Any of several concentration inequalities, or inequalities for moderate deviations, of random walk. *See also* Paley–Zygmund inequality. [V. V. Petrov, *Limit Theorems of Probability Theory*. Oxford University Press, 1995]

***p*-function** *see* regenerative phenomenon

phase space *see* sample space

phase-type distribution (alias: **PH distribution; absorption-time distribution**) The distribution of the time of absorption of a continuous-time Markov chain on a finite state-space S, having one absorbing state and $s = |S| - 1$ transient states, which may be called the phases. Discrete phase-type distns are defined likewise in terms of discrete-time Markov chains.

An important subclass comprises the Coxian distributions, and more generally one has matrix-exponential distributions; *see also* method of phases. [M. Neuts, *Matrix-geometric Solutions in Stochastic Models*. Johns Hopkins, 1981; reprinted by Dover, 1994]

phi coefficient *see* mean square contingency

phi-irreducibility Arises in considering a Markov chain $X_n, n \geq 1$, on a general state space $(S, \mathcal{F})$, where $\mathcal{F}$ is the σ-field of subsets of S; such a concept being necessary in order to prove solidarity theorems analogous to that for chains on a countable state space. The chain X_n is said to be ϕ-irreducible if there is a non-zero sigma-finite measure $\phi(\cdot)$ on $(S, \mathcal{F})$ such that for any $B \in \mathcal{F}$ for which $\phi(B) > 0$, it is possible to reach B (in finite time) from every point of S. *See* Harris recurrence. [R. Tweedie, R-theory for Markov chains on a general state space I, *Ann. Probab.*, 2(5), 840–864, 1974]

philosophers' process A nearest-neighbour interacting-particle system (being a special type of a spin system), introduced as a stochastic (Markov process) version of a model for resource sharing. [B. Ycart, The Philosophers' process: an ergodic reversible nearest particle system, *Ann. Appl. Probab.*, 3(2), 356–363, 1993]

phi-mixing A mixing condition (i.e. a type of measure of dependence) that arises in one form thus: Let $X_r, r \geq 0$, be a sequence of r.vs, and A any event that depends only on $(X_1, \ldots, X_n)$, and B any event that depends only on $(X_{k+n}, X_{k+n+1}, \ldots)$, and set

$$\phi_n = \sup_{k\geq 1} \sup_{A,\, B: \mathbb{P}(B)>0} |\mathbb{P}(A|B) - \mathbb{P}(A)|.$$

If $\lim_{n\to\infty}\phi_n = 0$, then the sequence is said to be phi-mixing; it implies α-mixing and β-mixing. [R. Bradley, Basic properties of some strong mixing conditions, *Probab. Surv.*, 2, 107–144, 2005]

phi-recurrence *see* Harris recurrence

Pickands dependence function Arises in considering the possible limits $C(x, y)$ of bivariate copulas of componentwise maxima of i.i.d.r.vrs, called extreme-value copulas. Then $C(x, y)$ is an extreme-value copula iff there is a convex function $A(z), 0 \leq z \leq 1$, called Pickands dependence function, such that

$$C(x, y) = \exp\left\{\log(xy)A\left(\frac{\log y}{\log(xy)}\right)\right\}, \quad 0 < x, y < 1,$$

where $\max\{z, 1 - z\} \leq A(z) \leq 1$. [D. Zhang *et al.*, Nonparametric estimation of the dependence function for a multivariate extreme value distribution, *J. Multivariate Anal.*, 99(4), 577–588, 2008]

pi-lambda theorem *see* Dynkin's pi-lambda theorem

piling-up lemma (alias: **Matsui's piling-up lemma**) A term arising in cryptography, where it denotes the property of independent (not necessarily identically distributed) binary r.vs $X_r \in \{0, 1\}, r \geq 1$, that

$$\mathbb{P}\Big(\sum_{r=1}^{n} X_r(\text{modulo}\,2) = 0\Big) = \frac{1}{2} + 2^{n-1}\prod_{r=1}^{n}\Big(\mathbb{P}(X_r = 0) - \frac{1}{2}\Big).$$

[Z. Kukorelly, The piling-up lemma and dependent random variables, in *Cryptography and Coding*, M. Walker (ed). Springer, 1999, pp. 186–190]

Pinsker's inequality (alias: **Pinsker–Csiszár–Kullback inequality**) Let $\mathbb{P}$ and $\mathbb{Q}$ be probability distributions (measures) on the same (arbitrary) set Ω. Then $d_{KL}(\mathbb{P}, \mathbb{Q}) \geq 2\{d_{TV}(\mathbb{P}, \mathbb{Q})\}^2$, where d_{KL} and d_{TV} denote the Kullback–Leibler divergence and total variation distance respectively. (Note that other non-standard definitions of d_{TV} differ by a factor of 2, yielding a different apparent form of Pinsker's inequality, with ½ replacing 2.) See e.g. Vajda's inequality and Volkonskii–Rozanov inequality. [M. Reid & R. Williamson, Generalized Pinsker inequalities, *Conference on Learning Theory*, 2009, arXiv:0906.1244]

pistimetric probability *see* subjective probability

pi-system *see* Dynkin's pi-lambda theorem

Pitman's sampling formula Arises in the context of random partitions, with widespread applications. Explicitly, for parameters $\alpha \in [0, 1)$ and $\theta > -\alpha$, and non-negative integers $x_1, \ldots, x_n$ such that $\sum_1^n r x_r = n$, and $\sum_1^n x_r = k$, it is a probability distribution of the form

$$p(x_1, x_2, \ldots, x_n) = \frac{n!}{(\theta)^{\overline{n}}}\prod_{r=0}^{k-1}(\theta + r\alpha)\prod_{t=1}^{n}\left\{\frac{1}{x_t!}\left(\frac{(1-\alpha)^{\overline{t-1}}}{t!}\right)^{x_t}\right\},$$

where $(\theta)^{\overline{n}} = \theta(\theta + 1)\ldots(\theta + n - 1), n \geq 1$, is the rising factorial. The special case $\alpha = 0$ is called **Ewens'**

sampling formula. [J. Pitman, Exchangeable and partially exchangeable random partitions, *Probab. Theor. Rel. Fields*, 102 (2), 145–148, 1995]

Pitman's theorem (alias: **Pitman's 2M-X (or 2M-B) theorem; Pitman's representation theorem**) Asserts that for the standard Wiener process $W(t), t \geq 0$, with maximum function $M(t) = \sup_{0 \leq s \leq t} W(s), t \geq 0$, the process $Y(t) = 2M(t) - W(t)$ has the same fdds as the three-dimensional Bessel process. More generally, the **Pitman–Rogers thm** asserts a similar result for the drifting Wiener process $W_\mu(t)$, having drift parameter $\mu \geq 0$; if $M_\mu(t)$ is the maximum function of $W_\mu(t)$, then $Z(t) = 2M_\mu(t) - W_\mu(t)$ is a diffusion having the same fdds as W_μ suitably conditioned to stay positive. *See* Ciesielski–Taylor theorem. [B. Hambly *et al.*, Pitman's 2*M-X* theorem for skip-free random walks with Markovian increments, *Electr. Commun. Probab.*, 6, 73–77, 2001]

Pitman–Yor process *see* Dirichlet–Ferguson process

pivotal *see* importance

Plackett's distribution (alias: **contingency-type distribution**) A bivariate distribution from a one-parameter family of such distns, having prescribed marginal distn fns $F_X(x)$ and $F_Y(y)$, that is defined as the unique root P (which is a proper joint distn with those marginals), of $\left(1 - F_X(x) - F_Y(y) + P\right) P = a\left(F_X(x) - P\right)\left(F_Y(y) - P\right)$, for some fixed $a \geq 0$.

This family includes the Fréchet–Hoeffding bounds for P when the coefficient of association a takes the values 0 or ∞; and $P = F_X(x)F_Y(y)$ when $a = 1$. [K. Mardia, Some contributions to contingency-type bivariate distribution, *Biometrika*, 54(1–2), 235–249, 1967; & Correction, 55, 597]

Plackett's formula A recurrence relation supplying a reduction formula that gives a convenient method for the computation of multivariate normal probabilities. [H. Grassman, Multivariate normal probabilities: implementing an old idea of Plackett's, *J. Comp. Graph. Stat.*, 12(3), 731–752, 2003]

Plackett's problem An extremal problem that in its basic form is this: Let X and Y be i.i.d.r.vs having distribution function $F(x), x \in \mathbb{R}$; for what $F(x)$ is $\mathbb{E}(|X - Y|)$ maximal? For zero-mean unit-variance r.vs, the distn is uniform on $[-\sqrt{3}, +\sqrt{3}]$. [L. Bloomer & T. Hill, Random probability measures with given mean and variance, *J. Theor. Probab.*, 15(4), 919–937, 2002]

planar aggregation (or clustering) model *see* Hastings–Levitov model

plausibility measure Arises in the context of models for uncertainty, in a formalization similar to that of probability measure, of which it is a generalization. Thus, let $\mathcal{F}$ be a sigma-field of subsets of a universal set (or sample space) Ω; the plausibility function (or measure) $\pi(\cdot)$ is defined on the members of $\mathcal{F}$ and takes values in a partially ordered set V of plausibility values. It is further assumed that for A and B in $\mathcal{F}$ such that $A \subseteq B$, we have $\pi(A) \leq \pi(B)$; and, further, that V contains elements b and t such that for all $v \in V, \pi(\emptyset) = b \leq v \leq t = \pi(\Omega)$. [N. Freidman & J. Halpern, Plausibility measures: a user's guide, in *Proc. 11th Conference on Uncertainty in Artificial Intelligence*. Morgan Kaufmann, 1995, pp. 175–184]

$\mathbb{P}$-null process (alias: **evanescent process**) An event $A \in \mathcal{F}$ of a probability space $(\Omega, \mathcal{F}, \mathbb{P})$ may be called $\mathbb{P}$-null (or null) if $\mathbb{P}(A) = 0$. By extension, a random process $X(t, \omega)$ defined on a stochastic basis $(\Omega, \mathcal{F}, \mathcal{F}_t, \mathbb{P})$ is said to be $\mathbb{P}$-null if the paths $t \mapsto X(t, \omega)$ are identically zero a.s. wrt $\mathbb{P}(\cdot)$. Two processes X and Y on the same basis are equivalent (or indistinguishable or $\mathbb{P}$-equal) if $X(t, \omega) - Y(t, \omega)$ is $\mathbb{P}$-null. [M. Métivier, *Semimartingales*. De Gruyter, 1982]

Poincaré inequality (alias: **spectral gap inequality**) Broadly, a type of inequality that supplies a bound for the appropriate L_p-norm of a suitably defined function g, in terms of the derivative(s) of g, and a constant C that is independent of g; the optimal constant in any given case may be called the Poincaré constant. In the context of r. vs, a classic example is **Nash's Poincaré inequality**, also called the **Gaussian Poincaré inequality**, which asserts (in its basic form) that for any real-valued continuously differentiable function $g(X)$ of a standard normal r.v. X, we have var $g(X) \leq \mathbb{E}\{(g\prime(X))^2\}$, with Poincaré constant 1. [J. Nash, Continuity of solutions of parabolic and elliptic equations, *Amer. J. Math.*, 80, 931–954, 1958]

Poincaré's observation (or lemma) Asserts that for the r.vr **U** that is uniformly distributed on the surface of sphere of radius $\sqrt{n}$ in n dimensions, if n is sufficiently large then the distribution of the first r coordinates of **U** differs little in total variation distance from that of r i.i.d. standard normal r.vs. [O. Johnson, Entropy and a generalization of 'Poincaré's observation', *Math. Proc. Cambridge Phil. Soc.*, 135(2), 375–384, 2003]

Poincaré's recurrence theorem *see* measure-preserving transformation

point of increase The distribution function $F(x)$, $x \in \mathbb{R}$ has a point of increase at x if, for every $a > 0$, $F(x + a) - F(x - a) > 0$. The set of such points may be called the **spectrum** (or **support**) of F. It includes the jumps of F.

Similarly, the real-valued function $g(t), t \in \mathbb{R}$, has a point of increase at x if, for some $a > 0, g(s) \leq g(x)$, for $s \in (x - a, x)$, and $g(s) \geq g(x)$, for $s \in (x, x + a)$. It is notable that, with probability 1, the Wiener process has no points of increase. [K. Burdzy, On non-increase of Brownian motion, *Ann. Probab.*, 1990, 18(3), 978–980]

point process (alias: **random counting measure**) Broadly, a type of random process whose realizations comprise collections of points in $\mathbb{R}^n, n \geq 1$, or more generally, sets of random elements of some suitable space. Such a process may be formally characterized in several ways. Perhaps the most common is to define it as a random counting measure $N(\cdot)$ (i.e. taking values in the non-negative integers, possibly including the value ∞) defined on a suitable metric space S together with its field $\mathcal{B}$ of Borel sets.

Then for $A \in \mathcal{B}, N(A)$ is the number of points of the process in A; it is customary to consider so-called finite processes in which $N(A)$ is finite if A is (thus excluding processes with limit points, such as e.g. an explosive birth process). The function $m(A) = \mathbb{E}N(A)$ defined for each A in $\mathcal{B}$ is called the mean (or first moment) measure of N; while $\mathbb{E}\{N(A_1)N(A_2)\}, A_1, A_2 \in \mathcal{B}$, is the second moment measure, and the process is called second-order if this is finite. There are numerous types of point process, of which important examples are:

(i) If the distribution of $N(B)$ is unchanged under translations, the process is simply (or crudely) stationary.
(ii) If a.s. no two points coincide, the process is simple. (Some authors use the term 'ordinary'.)
(iii) If to each point of the process there is attached (or associated) a random variable X (or r.vr) $\mathbf{X}$, then one speaks of a **marked (point) process**, and, X (or $\mathbf{X}$) is called a mark.

[D. Daley & D. Vere-Jones, *Introduction to the Theory of Point Processes*, 2nd edn. Springer, Vol. I, 2003; Vol. II, 2008]

pointwise ergodic theorem *see* Birkhoff's ergodic theorem

Poisson approximation *see* Stein's method; law of rare events

Poisson-binomial distribution The distribution of the sum $S_n = \sum_{r=1}^{n} X_r$, where $X_1, \ldots, X_n$ are independent Bernoulli trials such that $\mathbb{E}X_r = p_r$ (which is to say that $\mathbb{P}(X_r = 1) = p_r = 1 - \mathbb{P}(X_r = 0)$). The distribution of the r.vs $(X_1, \ldots, X_n)$ conditional on the event that $S_n = k \leq n$ is called the conditional Bernoulli model. [B. Roos, On the rate of multivariate Poisson convergence, *J. Multivariate Anal.*, 69(1), 120–134, 1999]

Poisson bridge *see* pontogram

Poisson chaos *see* chaos expansion

Poisson cluster process *see* cluster process; Bartlett–Lewis model

Poisson–Dirichlet distribution A member of a family of distributions arising from random ranked (i.e. ordered by size) discrete probability distributions, having applications in genetics, combinatorics, and number theory; that is, it is the distribution of a type of random vector $(X_1, X_2, \ldots)$ such that $X_1 > X_2 > \cdots$, and $\sum_{r=1}^{\infty} X_r = 1$. The so-called one-parameter family, introduced by J. Kingman in 1975, is a subset of the general two-parameter Poisson–Dirichlet distn. *See also* Pitman's sampling formula. [S. Feng, *The Poisson–Dirichlet Distribution and Related Topics*. Springer, 2010; J. F. C. Kingman, Random discrete distributions, *J. Roy Stat. Soc. B*, 37 (1), 1–22, 1975]

Poisson distribution The distribution of a discrete r.v. X taking values in $\mathbb{Z}^+$, such that for some $\lambda > 0, p(x, \lambda) = \mathbb{P}(X = x) = \lambda^x e^{-\lambda}/x!$, where $\mathbb{E}X = \operatorname{var} X = \lambda$, and the distn is infinitely divisible with chf $\phi(t) = \exp\{\lambda(e^{it} - 1)\}$.

It is often denoted by $\operatorname{Poi}(\lambda)$ or $\operatorname{Po}(\lambda)$, or $\operatorname{Pois}(\lambda)$, etc., and the probabilities satisfy the integral relation

$$\sum_{r=0}^{n} p(r, \lambda) = 1 - \int_0^{\lambda} p(n, s)\mathrm{d}s.$$

It arises naturally in counting random occurrences in time, whose counts are independent in disjoint regions, and hence (when appropriately formulated) is an approximation to the distribution of a wider class of rare occurrences. This generalizes to the Poisson process in $\mathbb{R}$ and more general spaces. *See also* compound Poisson; Poisson's theorem; and Raikov's theorem. [W. Feller,

Introduction to Probability Theory and its Applications, Vol. 1, 3rd edn. Wiley, 1968]

Poissonization (alias: **Poisson randomization**) Broadly, the replacing of an integer-valued index n of some family of random variables by a Poisson-distributed random index N; very often N has parameter n. Further, when n is seen as time, N is often taken to have continuous parameter $t \in \mathbb{R}^+$ (so that N is the Poisson process). The concept arises in very disparate areas, with applications from statistics to the analysis of algorithms and combinatorial problems. The idea was first developed by M. Kac (1949) in replacing the sample size n of an empirical process by N. K. Athreya and S. Karlin (1968) considered the embedding of urn schemes in continuous-time processes comprising a vector of Poisson processes. The resulting applications in combinatorics make use of the Poisson transform, whose inversion is called de-Poissonization. In general, any suitable random process $X_n, n \geq 0$, may be randomized using an independent Poisson process $N(t)$ by forming $X(t) = X_{N(t)}, t \geq 0$. In this case one speaks of a subordinated, or Poisson-directed (or regulated) (or continuized), process. [A. Gnedin *et al.*, Notes on the occupancy problem with infinitely many boxes, *Probab. Surv.*, 4, 146–171, 2007]

Poisson mosaic (or tessellation) A random mosaic in (or tessellation of) the plane that comprises the so-called Poisson polygons arising from a Poisson line process, i.e. a random process of lines that is stationary and isotropic (which is to say that their distribution is invariant under translation and rotation).

In $\mathbb{R}^d$ a Poisson process of hyperplanes yields the Poisson honeycomb (or hyperplane mosaic) of random cells (or crystals). More generally, any suitable random line process yields a random division or $\mathbb{R}^2$ into random cells, often called a random tessellation (or mosaic). *See also* Poisson–Voronoi mosaic. [R. Schneider & W. Weil, *Stochastic and Integral Geometry*. Springer, 2008]

Poisson-normal distribution *see* Hermite distribution

Poisson process Arises in its basic form as a random process $N(t)$, in continuous time, for $t \geq 0$, that counts the occurrences of a sequence of random events that a.s. never coincide, and are all of the same type, where informally, the events occur independently at random while the physical circumstances causing the sequence stay constant over time. Classic examples giving rise to this model include the arrival times of telephone calls at an exchange, and the times of emission of particles from radioactive material.

There are several formalizations (i.e. different characterizations) of the process: in the time-dependent context it is natural to define it in terms of the inter-event times, which are required to be i.i.d. exponential r.vs with parameter $\lambda > 0$; or, equivalently, the numbers of events in

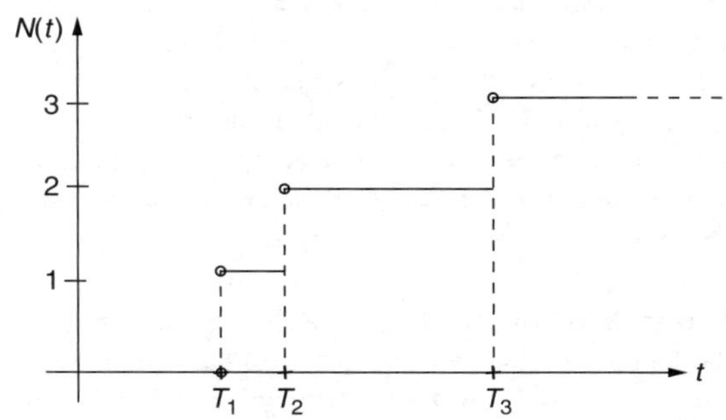

Poisson process on $\mathbb{R}$ whose first three events occur at T_1, T_2, and T_3.

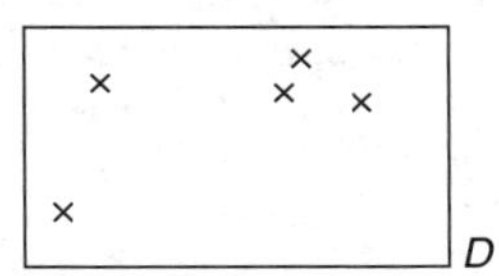

Poisson process in $\mathbb{R}^2$, showing points x that appear in region D.

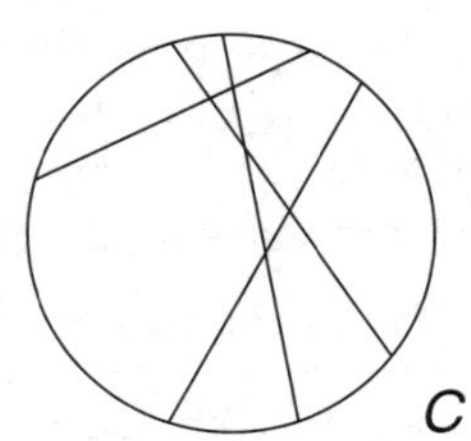

Poisson process of lines in $\mathbb{R}^2$, showing those that intersect the circle C.

disjoint intervals in $\mathbb{R}^+$, of lengths a_r, $1 \leq r \leq n$, are required to be independent Poisson r.vs with parameters λa_r respectively. For random points in space (e.g. $\mathbb{R}^d$, $d \geq 2$) it is natural to require that the counts in suitable disjoint sets

$A_1, A_2, \ldots$ are independent Poisson r.vs with parameters $\lambda|A_r|, r \geq 1$, where $|A|$ is the area (or volume) of A.

Yet more generally (in a completely separable metric space S) a Poisson process is defined as a random countable subject π of S such that for disjoint measurable subsets $A_1, \ldots A_n$ of S, the random counts $N(A_r)$ of points of π in the A_r, are independent and have a Poisson distribution with respective parameters $\lambda(A_r)$ for some suitable non-atomic measure $\lambda(\cdot)$, which may be called the mean measure. If $\lambda(\cdot)$ is constant, so that $\lambda(A_r) = \lambda|A_r|$, then one speaks of a homogeneous (or uniform) Poisson process; otherwise it is said to be non-homogeneous. *See* Rényi's theorem; superposition; Campbell's theorem; Bartlett's theorem; Cox process; and marking theorem. [J. F. C. Kingman, *Poisson Processes*. Oxford University Press, 1993]

Poisson's theorem In its simplest form this denotes the following property of the binomial distribution with parameters n and p: As $n \to \infty$ with $np = \lambda$ held fixed, the distribution converges to the Poisson distn with parameter λ. More generally, under suitable conditions, a similar result is true of the Poisson-binomial distribution. A result in sequential sampling has also been termed Poisson's thm. [S. Poisson, *Recherches sur la probabilité des jugements en matière criminelle et en matière civile*. Bachelier, 1837; K. Chung & F. AitSahlia, *Elementary Probability Theory*, 4th edn. Springer, 2003]

Poisson transform Several mathematical objects have been referred to as the Poisson transform. In a probabilistic context, it commonly denotes a Poissonized function $g(n), n \in \mathbb{Z}^+$ whose Poisson transform is given by $g^*(\lambda) = \sum_{n=0}^{\infty} g(n)\mathrm{e}^{-\lambda}\lambda^n/n!$. Much less often, the term has also been used for a Poisson compound, so that the Poisson transform of $f(x), x \in \mathbb{R}^+$, is $f^*(n) = \frac{1}{n!}\int_0^\infty f(x)x^n\mathrm{e}^{-x}\mathrm{d}x$, for suitable f. (In analysis, a Poisson transform of suitable $h(x)$, $x \in \mathbb{R}$, is

$$h^*(t) = \frac{1}{\pi}\int_{\mathbb{R}} \frac{h(x)}{1+(x-t)^2}\mathrm{d}x.\ )$$

The inversion of $g*$ is called de-Poissonization. [G. Gonnet & J. Munro, The analysis of linear probing sort by the use of a new mathematical transform, *J. Algor.*, 5(4), 451–470, 1984]

Poisson–Voronoi mosaic (or tessellation) (alias: **Poisson–Dirichlet mosaic**) A random mosaic that in its simplest form tessellates the plane as follows: given a Poisson process of points π_r, $r \in 1, 2, \ldots$ (usually homogeneous), the Poisson–Voronoi mosaic is the partition of the plane into disjoint cells C_r, where the rth cell C_r contains those points of the plane that are closer to π_r than they are to any other point of the Poisson process.

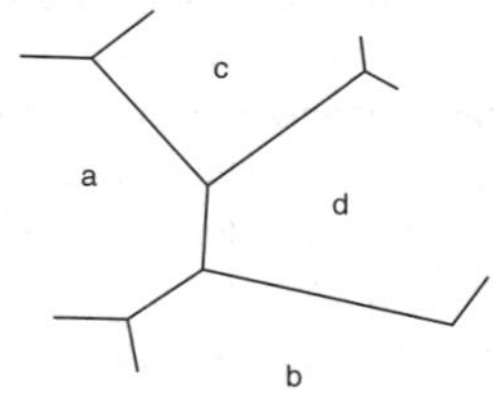

Part of a Poisson–Voronoi mosaic.

More generally, the Poisson–Voronoi mosaic in $\mathbb{R}^d$ is defined likewise; furthermore a random Voronoi mosaic is similarly defined for any (not necessarily Poisson) process of points in $\mathbb{R}^d$, whose cells are thus d-dimensional open bounded convex polytopes, called ***d*-facets**. Given a random Voronoi mosaic, its straight-line dual yields the so-called random **Delaunay mosaic** (Poisson–Delaunay in the special case). Yet more generally, mosaics may be formed by weighting the distance function used to determine the cells. Thus let a weight W_r be attached to π_r, and define the power of $y \in \mathbb{R}^d$ with respect to (π, w), in terms of the Euclidean distance $\|.\|$ by $p(\pi, w; y) = \|\pi - y\|^2 - w$, (or more generally $\|\pi - y\|^q$). Then for any π_r the set of points whose power wrt (π_r, w_r) is less than their power for any other point yields a cell in a random mosaic that is often called the random **Laguerre tessellation** (among several other names). *See also* Johnson–Mehl tessellation. [S. Chiu *et al.*, *Stochastic Geometry and its Applications*, 3rd edn. Wiley, 2013]

polar method *see* Marsaglia's method

policy improvement method *see* Howard policy improvement

Pollaczek–Khinchin formula (or equation) (**alias: Pollaczek's formula; Cramér's formula**) Any of several identities arising in the context of the single server queue of type M/G/1, which is to say that it has arrivals at the instants of a Poisson process of rate λ, and i.i.d. service times with distn fn $F(x)$ and mean μ. Similar identities arising in ruin theory of insurance mathematics are denoted likewise, or may be called **Cramér's formula**. In F. Pollaczek's original queueing result, as recast in probabilistic terms by A. Khinchin, let W be the waiting time of a customer arriving at the above M/G/1 queue in equilibrium, and let $H(x)$ be the distribution fn with density $h(x) = \mu^{-1}(1 - F(x))$, whose n-fold convolution has distribution fn $H^{*n}(x)$.

Then the Pollaczek–Khinchin formula is commonly given as

$$F_W(x) = \mathbb{P}(W \leq x) = (1 - \lambda\mu)\sum_{r=0}^{\infty}(\lambda\mu)^r H^{*r}(x).$$

If $F(x)$ and $F_W(x)$ have respective mgfs $M_s(t)$ and $M_W(t)$, then the **P-K transform formula** is

$$M_W(t) = (1 - \lambda\mu)t/\{t - \lambda(1 - M_s(t)\}.$$

If follows that $\mathbb{E}W = \frac{1}{2}\lambda\ \mathbb{E}(S^2)/(1 - \lambda\mu)$, where S is a typical service time; this is commonly called **Pollaczek's formula** (but also, by some people, the Pollaczek–Khinchin formula). [S. Asmussen, *Applied Probability and Queues*, 2nd edn. Springer, 2003]

Pollaczek–Spitzer identity (alias: **Spitzer's identity**) Arises in considering the distribution of the maxima $M_n = \max_{0\leq k\leq n} S_n$ of the partial sums $S_n = \sum_{r=1}^{n} X_r$ of i.i.d.r.vs $X_1, X_2, \ldots$ (where the empty sum is 0). F. Spitzer's classic form of the identity is

$$\sum_{n=0}^{\infty} \mathbb{E}e^{itM_n} z^n = \exp\left(\sum_{n=1}^{\infty} n^{-1}\ \mathbb{E}e^{itS_n^+} z^n\right),$$

being a key result of fluctuation theory. The equivalent result for Lévy processes may be called the **Spitzer–Rogozin–Pecherskii identity**. *See also* Sparre Andersen theorem. [P. Greenwood & J. Pitman, Fluctuation identities for Lévy processes and splitting at the maximum, *Adv. Appl. Probab.*, 12(4), 893–902, 1980; P. Greenwood & J. Pitman, Fluctuation identities for random walk by path decomposition at the maximum, *Adv. Appl. Probab.*, 12(2), 291–293, 1980; F. Spitzer, A combinatorial lemma and its application to probability theory, *Trans. Amer. Math. Soc.*, 82, 323–339, 1956; J. F. C. Kingman, Spitzer's identity and its use in probability theory, *J. Lond. Math. Soc.*, 37, 309–316, 1962]

Pólya–Aeppli distribution (alias: **Poisson-geometric distribution**) Arises as the sum $Y = \sum_{r=1}^{N} X_r$ of i.i.d.r.vs that are geometrically distributed on the non-negative integers, and independent of the Poisson r.v. N having parameter λ. Then Y has pgf $G_Y(s) = \exp\left\{\lambda\left(\frac{1-p}{1-ps} - 1\right)\right\}$, for some $0 < p < 1$. More generally, if the X_r have a negative binomial distn, the generalized Pólya–Aeppli distn (or **Poisson–Pascal distn**) has pgf $G(s) = \exp\left[\lambda\left\{\left(\frac{1-p}{1-ps}\right)^m - 1\right\}\right]$, for some $m > 0$.

This is a special case of a **generalized Poisson distn** having pgf $G(s) = \exp\{\lambda(G_X(s) - 1)\}$ for a suitable pgf $G_X(s)$. [P. Moran, *Introduction to Probability Theory*. Oxford University Press, 1968; N. Johnson *et al.*, *Univariate Discrete Distributions*, 3rd edn. Wiley, 2005]

Pólya characteristic function (alias: **Pólya's condition (or criterion, or theorem)**) Arises in seeking criteria sufficient for a given function to be the characteristic function of a probability distribution. Let $\phi(t)$ be a continuous, real-valued, even function for $t \in \mathbb{R}$. Then $\phi(t)$ is the chf of an absolutely continuous distribution $F(x)$ if it satisfies the so-called Pólya condition(s):

(a) $\phi(0) = 1$;
(b) $\lim_{t\to\infty}\phi(t) = 0$;
(c) $\phi(t)$ is convex for $t > 0$.

[E. Lukacs, *Characteristic Functions*, 2nd edn. Griffin, 1970]

Pólya's characterization theorem Arises in characterizing the normal distribution and is commonly stated thus: Let $X_1, \ldots, X_n$ be i.i.d.r.vs, and $c_1, \ldots, c_n$ nonzero real numbers such that $\sum_1^n c_i^2 = 1$. If $\sum_1^n c_r X_r$ has the same distribution as X_1, then all the r.vs are normal. *See* Skitovich–Darmois theorem. [W. Bryc, *The Normal Distribution: Characterizations with Applications*. Springer, 1995]

Pólya density (alias: **Pólya-type distribution**) A family of densities $f(x, t), x \in \mathbb{R}, t \in T \subseteq \mathbb{R}$, depending on the parameter t, is said to be of Pólya type n (or a Pólya density of order n) if for any $m \leq n$, and any $x_1 < x_2 < \cdots < x_m$, and any $t_1 < t_2 < \cdots < t_m$, we have:

$$\det\|f(x_i, t_j)\| = \begin{vmatrix} f(x_1, t_1) & \cdots & f(x_1, t_m) \\ \vdots & \ddots & \vdots \\ f(x_m, t_1) & \ldots & f(x_m, t_m) \end{vmatrix} \geq 0;$$

which is to say that f is totally positive of order m. More narrowly, the term is used for a density such that $\det\|f(x_i - y_j)\| \geq 0$, where $x_1 < x_2 < \cdots < x_m$, and $y_1 < y_2 < \cdots < y_m$. [S. Karlin & F. Proschan, Pólya type distributions of convolutions, *Ann. Math. Stat.*, 31(3), 721–736, 1960]

Pólya distribution (alias: **Pólya–Eggenberger distribution; beta-binomial distribution**) Arises as a compound distn, where the parameter p of a binomial $B(n, p)$ distribution is a r.v. having a beta distribution;

it also arises (in a special case) in the context of Pólya's urn. In one standard form, for $0 < p = 1 - q < 1$, and $a > 0$, it is written as

$$\mathbb{P}(X=r) = \binom{n}{r} \frac{p(p+a)\cdots(p+(r-1)a)q(q+a)\cdots(q+(n-r-1)a)}{(a+1)(2a+1)\cdots\{(n-1)a+1\}}, \quad 0 \le r \le n.$$

It includes many common discrete distributions, either as special cases or as limiting cases. [P. Moran, *Introduction to Probability Theory*. Oxford University Press, 1968; A. Dasgupta, *Fundamentals of Probability*. Springer, 2010]

Pólya process Arises as a continuous-time limit of Pólya's urn process, and comprises a non-stationary pure birth process where time spent in state i by $X(t), t > 0$, is exponentially distributed with parameter $\frac{a+i}{b+t}$, for suitable non-negative constants a and b. It also arises as a gamma-mixture of Poisson processes (i.e. a doubly stochastic Poisson process whose random parameter has a gamma density), in which case it is sometimes called a **Pólya–Lundberg process**. [W. Feller, *Introduction to Probability Theory and its Applications*, Vol. 1, 3rd edn. Wiley, 1968]

Pólya's theorem Any of several theorems, but in probability perhaps most often the celebrated result that a symmetric simple random walk on the integer lattice in d dimensions is recurrent (persistent) for $d \le 2$, but transient for $d \ge 3$. Less often, it may denote the early (1926) result that for a sequence of distribution functions converging to a continuous distribution function $F(x)$, $x \in \mathbb{R}$, the convergence is uniform in x. [G. Alexanderson (ed.), *The Random Walks of George Pólya*. Mathematical Association of America, 2000]

Pólya's urn process (alias: **Pólya–Eggenberger urn model**) Arises as a model for 'contagion', also called reinforcement. The basic process is classically given in terms of an urn containing red and green balls; initially there are a red and b green balls. A ball is drawn at random and returned together with s balls of the same colour as the drawn ball (where s may be negative, so that balls are subtracted). If this process is repeated n times, then the distribution of the total number of red balls drawn has a Pólya distribution. *See also* Pólya process and Friedman's urn. [G. Pólya, Sur quelques points de la théorie des probabilités, *Ann. Inst H. Poincare*, 1(2), 117–161, 1930]

Pólya–Williams approximation (alias: **Pólya normal inequalities**) For the standard normal distribution function $\Phi(x)$ this usually denotes the inequality $\Phi(x) - \Phi(-x) < (1 - \mathrm{e}^{-2x^2/\pi})^{\frac{1}{2}}$, $x > 0$, which is never in error by as much as 0.0075. This is **Williams' approximation**.

More generally, G. Pólya showed that

$$\Phi(x) - \Phi(-x) < a_n^{-1}(2\pi)^{-\frac{1}{2}}\left\{\int_0^{a_n x} nt^{n-1}\mathrm{e}^{-\frac{1}{2}t^2}\mathrm{d}t\right\}^{\frac{1}{n}}$$

for $n \ge 2$, where $a_n = 2\pi^{-\frac{1}{2}}\{\Gamma(\frac{1}{2}n+1)\}^{\frac{1}{n}}$. [G. Pólya, Remarks on computing the probability integral in one and two dimensions, *Proc. 1st Berkeley Symp. Math. Stat. Prob.*, 63–78, 1949; J. Williams, An approximation to the probability integral, *Ann. Math. Stat.*, 17(3), 363–365, 1946]

polymer model Arises in seeking to describe chemical compounds whose physical form comprises a chain (or tree) of simple monomers (of possibly different types). Its realization is necessarily self-avoiding, and classic models include the self-avoiding random walk, self-interacting random walk, and suitable random lattice polygons and trees, together with continuous versions. [G. Giacomin, *Random Polymer Models*. Imperial College Press, 2007]

polynomial chaos *see* chaos

polyspectrum Arises in the theory of stationary time series, where it extends the idea of a spectral distribution for the covariance function to the definition of a spectral distn (i.e. Fourier transform) for moment (and cumulant) functions of higher order. In particular, for the third-order moment function

$$c(u, v) = \mathbb{E}(X(t)X(t+u)X(T+v))$$

of the stationary process $X(t), t \in \mathbb{R}$, the bivariate Fourier transform may be called the **bispectrum**, or **bispectral density**, or **third-order polyspectral density**, provided that $\int\int |c(u,v)|\mathrm{d}u\mathrm{d}v < \infty$. [D. Brillinger, Introduction to polyspectra, *Ann. Math. Stat.*, 36(5), 1351–1374, 1965]

pontogram A process introduced by D. & W. Kendall in considering possible alignments of random points in the plane. Let $N(t)$ be a renewal process. Then one basic form of pontogram process is defined as

$$B_n(t) = \frac{N(nt) - tN(n)}{\sqrt{n}}, \quad n = 1, 2, 3 \ldots, 0 \le t \le 1.$$

In the special case when $N(t)$ is a Poisson process, it may be called a **Poisson bridge**. [H. Zhang, On a weighted embedding for generalized pontograms, *Stoch. Proc. Appl.*, 88(2), 213–224, 2000]

portfolio process *see* trading strategy; Merton portfolio problem

portmanteau theorem (alias: **Alexandrov's theorem**) Arises in considering the weak convergence of a sequence of probability measures, which may be defined by the assertion that $\mathbb{P}_n, n \geq 1$, converges weakly to $\mathbb{P}$ as $n \to \infty$ if $\lim_{n\to\infty} \int g\mathrm{d}\mathbb{P}_n = \int g\mathrm{d}\mathbb{P}$ for all continuous bounded real functions $g(\cdot)$ defined on the same metric space as $\mathbb{P}_n$ and $\mathbb{P}$. The portmanteau theorem gives several equivalent characterizations of such weak convergence (hence the use of this name for the result). *See also* Mann–Wald theorem. [P. Billingsley, *Convergence of Probability Measures*, 2nd edn. Wiley, 1999]

positive dependence *see* attractive event

posterior distribution *see* inverse probability; prior probability distribution

posterior risk *see* expected posterior loss

potential theory In describing phenomena such as gravitation and heat flow, the laws of physics give rise to a function $\phi(\mathbf{x}), \mathbf{x} \in \mathbb{R}^d$, called a potential function, that satisfies Laplace's eqn $\nabla^2\phi = 0$ (or, more generally, Poisson's equation $\nabla^2\phi = g(\mathbf{x})$, for a suitable function g).

By Green's theorems, solutions to Laplace's eqn also satisfy a type of integral equation that expresses a local averaging property. It is the case that various functions (and functionals) arising from the Wiener process (such as hitting probabilities and expected hitting times) either have this averaging property and satisfy Laplace's equation, or satisfy some form of Poisson's equation. Furthermore, functionals of Markov chains (seen as random walks on their associated graph) have similar averaging properties in discrete form, that give rise to interpretations in terms of current flow and potentials in electrical networks.

These observations lead to **probabilistic potential theory**, which explores and exploits this duality between the properties of Markov processes (with others having suitable averaging properties, such as martingales) and numerous classical results of harmonic function theory. Some vocabulary is transferred in this process; for example, the **Green fn** (which serves as the kernel of the so-called potential operator) for the Wiener process in three dimensions with transition probability density $f(t, r \mid x_0)$ is $\int_0^\infty f(u, r \mid x_0)\mathrm{d}u$, and for a homogeneous Markov chain with transition matrix P, the **Greenian** (or Green function) is $G = \sum_{r=0}^{\infty} P^r$. [J. Doob, *Classical Potential Theory and its Probabilistic Counterpart*. Springer, 1984; S. Port & C. Stone, *Brownian Motion and Classical Potential Theory*. Academic Press, 1978]

potlatch process *see* smoothing process

power constraint *see* Shannon–Hartley formula

power divergence (alias: **density power divergence**) A divergence between densities f and g defined for $0 < a$ by $d_a(f, g) = \int \{f^{a+1}(x) - (1 + \frac{1}{a})f^a(x)g(x) + \frac{1}{a}g^{a+1}(x)\}\mathrm{d}x$, and where for $a = 0$ the divergence is defined to be the Kullback–Leibler divergence. [A. Basu *et al.*, Robust and efficient estimation by minimising a density power divergence, *Biometrika*, 85(3), 549–559, 1998]

power entropy *see* entropy

power-law probability distribution A term used somewhat loosely to denote a probability distribution whose density is either exactly, or asymptotically as $x \to \infty$, of the form $f(x) \propto x^{-a}, x \geq x_0 > 0$, for some a, called the **scaling parameter** (or **exponent**). Some authors use the term to denote a density of the form $f(x) = ax^c$, for x in a finite interval, and suitable constants a and c. *See* Pareto distribution; zeta distribution; and regular variation. [A. Clauset *et al.*, Power-Law distributions, *SIAM Review*, 51(4), 661–703, 2009]

power mixture A special case of compounding, thus: let $\phi(t)$ be the infinitely divisible characteristic fn of a r.v. X, and $G(x)$ a distn on $[0, \infty)$. Then $(\phi(t))^x$ is a chf for all $x > 0$, and the chf $\psi(t) = \int_0^\infty (\phi(t))^x \mathrm{d}G(x)$ is called the power mixture of X (or, equivalently, of its distn $F_X(x)$). [D. Pommeret, Polynomial expansions of density of power mixtures, *ESAIM Probab. Stat.*, 11, 248–263, 2007]

power series distribution Let $g(z) = \sum_{x=0}^{\infty} c_x z^x$ be a power series, with non-negative coefficients c_x, that converges for $|z| < r > 0$. Then the probability distribution defined by $\mathbb{P}(X = x) = c_x z^x / g(z), 0 < z < r$, is said to be a power series distribution. [A. Noack, A class of random variables with discrete distributions, *Ann. Math. Stat.*, 21(1), 127–132, 1950]

predictability (alias: **previsibility**) A concept arising naturally in the context of functions of a random process that is evolving in time; a classic example is supplied by a martingale as a model of a fair game, where any betting system (i.e. martingale transform) cannot depend on future values, and thus is certainly known (i.e. is predictable) at any stage, given the values of the process to date. More formally, the sequence X_n, $n \geq 0$ is predictable with respect to the filtration $\mathcal{F}_n$, $n \geq 0$, if X_n is $\mathcal{F}_{n-1}$-measurable for all $n \geq 1$ (or one says that the pair $(X_n, F_n), n \geq 0$, is predictable).

In continuous time, given a filtration $\mathcal{F}_t$, $t \geq 0$, on a measurable space $(\Omega, \mathcal{F})$, the predictable σ-field $\mathcal{F}_p$ is defined on $\Omega \times [0, \infty)$ as the smallest σ-field such that any process $X(t)$ that is bounded, left-continuous, and adapted to $\mathcal{F}_t$ is also $\mathcal{F}_p$-measurable. Then a process that is $\mathcal{F}_p$-measurable is said to be predictable. (Furthermore, note that the σ-field on $\Omega \times [0, \infty)$ generated by all bounded right-continuous process adapted to $\mathcal{F}_t$ is called the **optional σ-field**.)

A positive stopping time T is predictable if it is the limit of a non-decreasing sequence of stopping times $T_1 \leq T_2 \leq \cdots$. This sequence may be said to announce T, which may also be called an **announceable time**. [R. Bass, *Stochastic Processes*. Cambridge University Press, 2011]

predictable representation *see* martingale representation theorem

prediction problem (or theory) *see* filtering

preemptive service *see* queueing theory

preferential attachment model *see* Barabási–Albert model

Prékopa–Leindler inequality Arises in proving that certain probability measures are log-concave, and that convolutions of such log-concave distributions are also log-concave (with natural applications in optimization); it is related to the isoperimetric and Brunn–Minkowski inequalities of geometry. It is commonly stated thus: let f, g, and h be integrable non-negative functions such that for some $a \in (0, 1)$, and all $s, t \in \mathbb{R}^n, h\big((1-a)s + at\big) \geq \big(f\ (s)\big)^{1-\,a}\big(g(t)\big)^a$. Then

$$\int_{\mathbb{R}^n} h(x)\mathrm{d}x \geq \left(\int_{\mathbb{R}^n} f(x)\mathrm{d}x\right)^{1-a} \left(\int_{\mathbb{R}^n} g(x)\mathrm{d}x\right)^a.$$

Any of the several corollaries demonstrating the log-concavity of some distribution may be called Prékopa's thm.

Note that the term Prékopa's thm is occasionally used to denote a result that supplies a characterization of the Poisson process. [P. Billingsley, *Probability and Measure*, 3rd edn. Wiley, 1995; A. Prékopa, on logarithmic concave measures and functions, *Acta Sci. Math. Szeged*, 34, 335–343, 1973]

Prékopa's theorem *see* log-concave distribution

premium principle Arises in insurance mathematics, in which a risk to be insured is denoted by a r.v. X with distribution $F(x)$; the insurer seeks to determine a premium (i.e. fee to cover the risk), which is a real-valued (including $+\infty$) function of X denoted by $\pi(X)$ (or by $\pi(F)$ in the simplest case when the premium depends only on the distn of X).

Any rule that determines $\pi(X)$ is called a premium principle. Such principles may be required to satisfy a number of natural constraints, yielding various classes of premium principle. Typically, $\pi(X)$ will depend on a subset of the following: $\mathbb{E}X$; var X; the mgf of X; the median of X; a suitable utility function; or other risks insured by the premium taker. In particular, X is said to be **insurable** if $\mathbb{E}X < \infty$, and the so-called **safety loading** of $\pi(X)$ is $\pi(X) - \mathbb{E}X$. [T. Rolski *et al.*, *Stochastic Processes for Insurance and Finance*. Wiley, 1999]

pre-stopping time *see* stopping time

previsible *see* predictability

principal component analysis *see* Karhunen–Loève expansion

principle of indifference *see* indifference principle

principle of insufficient reason *see* indifference principle

priority queue *see* queueing theory

prior probability distribution Arises naturally in the application of Bayes theorem to decision theory and inference, where some parameter θ (or other undetermined quantity) is interpreted as a r.v. denoted by Θ. The distribution $\pi(\theta)$ of Θ, postulated in advance of the result X of some experiment, is called the prior distribution. In cases of interest the distribution $p(x|\theta)$ of X depends on θ, and then the conditional distribution $\pi(\theta|x)$ of Θ given $X = x$ is called the posterior distn. By Bayes theorem (e.g. in the discrete case),

$$\pi(\theta|x) = p(x\,|\,\theta)\pi(\theta)\Big/\Big\{\sum_{\theta} p(x\,|\,\theta)\pi(\theta)\Big\},$$

where $p(x\,|\,\theta)$ is called the likelihood function in this context. In an extension, $\pi(\theta)$ may be improper (so that Θ is not a r.v.) which is to say that $\sum_{\theta}\pi(\theta)$ diverges. [E. Jaynes, *Probability Theory: The Logic of Science*. Cambridge University Press, 2003]

prisoners' dilemma (or paradox, or problem) May denote either a classic problem of game theory, or a classic variant of the Monty Hall problem. In the former, partners in crime are interrogated separately: if both are silent, penalties are light; if both confess, penalties are heavy. If only one confesses, the confessor is penalized very lightly while the non-confessor is penalized severely. The dominating strategy (confession) is optimal for neither. [K. Binmore, *Playing for Real*. Oxford University Press, 2007; F. Mosteller, *50 Challenging Problems in Probability*. Addison-Wesley, 1965; reprinted by Dover, 1987]

probabilistic *see* randomization

probabilistic method A technique, popularized by P. Erdös, for demonstrating the existence of some suitable types of mathematical entity; its most frequent application is to combinatorial problems. The essential idea is to assign a probability measure to some wider class of objects, and prove that the probability of the desired type of entity is greater than zero. [N. Alon & J. Spencer, *The Probabilistic Method*, 3rd edn. Wiley, 2008]

probabilistic programming *see* stochastic programming

probability A term arising in the presence of uncertainty about some event's occurrence, or about the truth of some statement; in either case it may be used in a broad everyday sense, or used more narrowly to denote a formalization of the concept. Broadly speaking, 'probability' denotes an assessment of the likelihood that the event will occur, or that the statement is true, where the method of assessment will depend on any of several concepts (or interpretations) of probability, of which there is a debatable number of types. The simplest taxonomy of types is this:

1. Probability is an objective (or physical) property of real world events, that is measurable by suitable experiments, and most notably observable in the stability of statistical ratios. One speaks of aleatory, or frequentist, or propensity, etc., probability.
2. Probability is a subjective (or epistemic) property of some agent, that is arrived at by consideration of the available information and evidence; such consideration may (at one extreme) employ the methods of formal logic, or (at another extreme) arise simply from the prejudices and ideas of the agent, or it may comprise some intermediate combination of reason and belief. One speaks of personal, or logical, or classical, etc., probability; or, in the words of A. de Morgan: 'Probability is a feeling of the mind.'

More narrowly, in the usual mathematical axiomatization, probability is a number $p \in [0, 1]$, whose value is given by a suitable function $\mathbb{P}(\cdot)$. In this case, formally, $\mathbb{P}(\cdot)$ is defined on an algebra (or sigma-field), $\mathcal{F}$ of subsets (called events) of a universal set Ω; thus $p = \mathbb{P}(A)$ for some event $A \in \mathcal{F}$.

In the standard (Kolmogorov) version of this definition, $\mathbb{P}(\cdot)$ is countably additive, and is often called a probability measure. In other (less often seen) axiomatizations, probability may be e.g. finitely additive, or non-additive, or an interval-valued function, or a binary relation of partial ordering; such alternatives are often (but not always) careful to use different nomenclature, e.g. belief function, plausibility, possibility, upper and lower probability, and so on. [J. van Plato, *Creating Modern Probability*. Cambridge University Press, 1994; A. N. Kolmogorov, *Foundations of the Theory of Probability*, 2nd English edn. Chelsea, 1956; T. Fine, *Theories of Probability*. Academic Press, 1973; D. Mellor, *Probability: A Philosophical Introduction*. Routledge, 2003]

probability density *see* density

probability distribution *see* distribution

probability generating function (alias: **pgf; Z-transform**) Arises naturally as the ordinary generating function of the probability distribution $p_r = \mathbb{P}(X = r), r \geq 0$ of any non-negative integer-valued r.v. X:

$$G_X(s) = \sum_{r=0}^{\infty} s^r \mathbb{P}(X = r) = \sum_{r=o}^{\infty} s^r p_r = \mathbb{E}s^X.$$

This power series has radius of convergence no less than 1, and converges absolutely at least for $|s| \leq 1$. Furthermore, if $\mathbb{E}|X| < \infty$ then $\mathbb{E}X = \lim_{s\uparrow 1} \frac{\mathrm{d}}{\mathrm{d}s} G_X(s)$, with corresponding expressions for higher moments that exist. More generally, the joint pgf of the vector $\mathbf{X}$ of integer-valued jointly distributed r.vs $(X_1, \ldots, X_n) = X$ is $G_X(s) = \mathbb{E}s_1^{X_1} \cdots s_n^{X_n}$. For non-integer-valued r.vs X, or for r.vs taking negative values, the pgf is often called the Z-transform: $G(z) = \mathbb{E}z^X$. [W. Feller, *Introduction to Probability Theory and its Applications*, Vol. 1, 3rd edn. Wiley, 1968]

probability generating functional *see* generating functional

probability integral transform (alias: **quantile function; (cumulative) distribution function transform**) Arises naturally in the inverse transform method, and has several applications in statistics. Its simplest form is this: let $F(x)$ be the strictly increasing and continuous (as a function of x) distribution function of a r.v. X. Then the r.v. $Y = F(X)$ is uniformly distributed on (0,1), and called the probability integral transform; furthermore, given a r.v. U that is uniform on (0,1), and denoting the well-defined inverse of F by $F^{-1}(\cdot)$, the r.v. $Z = F^{-1}(U)$ has the distribution of X.

More generally, allowing F to be discontinuous and not necessarily strictly increasing, the function defined as

$$F^{-1}(y) = \inf\{x : F(x) \geq y, \quad 0 < y < 1\}$$

still has the property that $F^{-1}(U)$ has distribution function $F(x)$; this has been called the **quantile fn thm**, and $F^{-1}(y)$ is the quantile function. In such cases, the generalized distribution transform of X is defined in terms of the function

$$T(x, u) = \mathbb{P}(X < x) + u\mathbb{P}(X = x).$$

For a r.v. U that is independent of X, and uniformly distributed on (0,1), this yields the transform $Y = T(X, U)$, which is uniformly distributed on (0,1).

Furthermore, with $F_X^{-1}(.)$ as defined above, $F_X^{-1}(Y)$ has distribution function F_X. *See also* Rosenblatt's transform. [T. Ferguson, *Mathematical Statistics*. Academic Press, 1967, p. 216; J. Angus, The probability integral transform and related results; *SIAM Rev.*, 36(4), 652–654, 1994; L. Rüschendorf & V. de Valk, On regression representations of stochastic processes, *Stoch. Proc. Appl.*, 46, 183–198, 1993]

probability mass function A term used to denote the probability distribution $\mathbb{P}(X = r) = p_r, r \in \mathbb{Z}$, of a discrete r.v. X. The term density is likewise used for the distribution of an absolutely continuous r.v. [H. Cramér, *Mathematical Methods of Statistics*. Princeton, 1946, p. 57ff]

probability measure *see* measure; probability

probability metric *see* distance

probability space Arises as a natural setting for the usual mathematical theory of probability, where it comprises the triple $(\Omega, \mathcal{F}, \mathbb{P})$ consisting of a universal set Ω (called the sample space), with a sigma-field (or algebra) $\mathcal{F}$ of subsets of Ω, and a probability function (or probability measure) $\mathbb{P}(\cdot)$ on $\mathcal{F}$, with $\mathbb{P}(\Omega) = 1$. Note that such a measure is countably (or completely) additive, and non-negative.

This space supports the subsequent definitions of independence, random variables, expectation, conditional probability, expectation, etc. [A. N. Kolmogorov, *Foundations of the Theory of Probability*, 2nd English edn. Chelsea, 1956; A. A. Borovkov, *Probability Theory*. Springer, 2013]

probability tree *see* multiplication rule

problem des rencontres *see* derangement problem

problem of the points (alias: **problem of the unfinished game; division problem**) A classic problem of elementary probability (arising in Italian manuscripts of the 14th century) whose first correct solution was given by B. Pascal in 1654 for the case of a fair game (the problem having been posed to him by de Méré; *see* de Méré's problem). The usual version is this: two players play a series of independent games, where each is won by A with probability p, or by B with probability $q = 1 - p$ (where $p = q = ½$ in the case of a fair game), until one has won t games and is then awarded the prize (or stake) of size S. The game is stopped when A has won $a < t$ games, and B has won $b < t$ games. How should S be fairly divided between A and B?

Following his solution, B. Pascal initiated a celebrated correspondence with P. Fermat, that included the extension of the fair division problem to three or more players. The extension to the case when $p \neq q$ was made independently by A. de Moivre and J. Bernoulli. [A. Hald, *History of Probability and Statistics and their Applications Before 1750*. Wiley, 1990]

product probability measure (or distribution, or space) Arises naturally when seeking a probability space appropriate as a setting for the joint description of independent random variables. Let $(\Omega_1, \mathcal{F}_1, \mathbb{P}_1)$ and $(\Omega_2, \mathcal{F}_2, \mathbb{P}_2)$ be two probability spaces, and define the joint (product) sample space $\Omega = \Omega_1 \times \Omega_2 = \{(\omega_1, \omega_2) : \omega_1 \in \Omega, \omega_2 \in \Omega_2\}$. Then there exists a unique smallest σ-field $\mathcal{F}$ generated by the so-called rectangles

$\mathcal{F}_1 \times \mathcal{F}_2 = \{A \times B; A \in \mathcal{F}_1, B \in \mathcal{F}_2\}$; and the product probability defined on the rectangles by $\mathbb{P}(A \times B) = \mathbb{P}_1(A)\mathbb{P}_2(B)$ can be extended to $\mathcal{F}$, and called the product measure on the product space (Ω, $\mathcal{F}$, $\mathbb{P}$). Note that dependent experiments yield other measures; the converse operation of writing a suitable joint measure as a product is called **disintegration**. [K. Itô, *Introduction to Probability Theory*. Cambridge University Press, 1984]

product theorem *see* marking theorem

program size complexity *see* Kolmogorov complexity

progressively measurable (alias: **progressive**) Arises in the context of a random process $X(t), t \geq 0$, as a natural condition ensuring that the value $X(T)$ of process at a stopping time T is a suitably measurable r.v. Formally, $X(t, \omega), t \geq 0, \omega \in \Omega$, is progressive (or progressively measurable) for a filtration $\mathcal{F}_t, t \geq 0$, if $X(t, \omega)$ is measurable on the product space $\mathcal{B}_t \times \mathcal{F}_t$, where $\mathcal{B}_t$ is the sigma-field of Borel sets on [0, t]. Such a process is adapted; conversely, a right- or left-continuous adapted process is progressive. [D. Revuz & M. Yor, *Continuous Martingales and Brownian Motion*, 3rd edn. Springer, 1999]

projected normal distribution (alias: **angular normal, or offset normal, or spherically projected normal distribution**) Arises in considering directional distributions, thus: Let $\mathbf{X}$ have the multivariate normal distn in d dimensions with mean vector $\mathbf{m}$, and covariance matrix V. Then the random unit vector $\mathbf{u} = \mathbf{X}/|\mathbf{X}|$, where $|\mathbf{X}|$ is the length of $\mathbf{X}$, has the projected normal density on the unit sphere. In the special case $V = I$ (the identity matrix), it is the displaced normal. [F. Wang & A. Gelfand, Directional data analysis under the general projected normal distribution, *Stat. Meth.*, 10(1), 113–127, 2013]

projection *see* lumpability

Prokhorov's inequality An inequality for the tail of the distribution of the sum $S_n = \sum_{r=1}^{n} X_r$ of independent r.vs; it is related to Bernstein's and Hoeffding's inequalities. Explicitly, suppose that $\mathbb{E}X_r = 0, r \geq 1$, and that $0 < s^2 = \operatorname{var} S_n < \infty$, and that $\mathbb{P}(|X_r| \leq b) = 1$ for some $b < \infty$. Then for $x > 0$, $\mathbb{P}(S_n > x) \leq \exp\left(-\frac{x}{2b}\sinh^{-1}\frac{bx}{2s^2}\right)$ where $\sinh^{-1}y = \operatorname{arc\,sinh} y$ is the inverse hyperbolic sine. [Y. V. Prokhorov, An extremal problem in probability theory, *Theor. Probab. Appl.*, 4(2), 201–203, 1959]

Prokhorov metric *see* Lévy metric

Prokhorov's theorem (or criterion) Arises in considering the convergence of a sequence of probability measures, where it supplies a condition for such a sequence to have a weakly convergent subsequence. Formally, let S be a metric space (equipped with its Borel field of open sets), and $\mathbb{P}_n, n \geq 1$, a collection of probability measures on S. Then $\mathbb{P}_n$ is said to be tight, or to satisfy Prokhorov's condition, if for any $t > 0$ there is a compact set $C(t)$ such that

$$\sup_{n \geq 1} \mathbb{P}_n\left(C(t)^{\mathrm{c}}\right) \leq t.$$

Then Prokhorov's theorem asserts that whenever a family $\mathbb{P}_n$ is tight, it has a weakly convergent subsequence. If S is separable and complete, then a converse theorem holds.

Prokhorov's criterion, together with an extension due to L. Le Cam, comprise the **Prokhorov–Le Cam compactness criteria**. [L. Le Cam, Convergence in distribution of stochastic processes, *Univ. California Publ. Stat.*, 2, 207–236, 1957; R. Bass, *Stochastic Processes*. Cambridge University Press, 2011]

propensity probability An interpretation of probability as an expression of a physical property of chance experiments. It is a type of objective probability, and is rarely invoked by probabilists. [K. Popper, The propensity interpretation of probability, *Brit. J. Phil. Sci.*, 10(37), 25–42, 1959]

proper A random variable X is said to be proper if $\mathbb{P}(-\infty < X < \infty) = 1$. A distribution function $F(x)$, $-\infty < x < \infty$ is said to be proper if $\lim_{x \to -\infty} F(x) = 0$, and $\lim_{x \to +\infty} F(x) = 1$. Likewise, a probability measure $\mathbb{P}$ on $\{\Omega, \mathcal{F}\}$ is proper if $\mathbb{P}(\Omega) = 1$.

A distribution function that is not proper may be called **defective**, or a **sub-probability distn**, or an **improper distn**, or (rarely) a **pseudo-distn**. [W. Feller, *Introduction to Probability Theory and its Applications*, Vol. 2, 2nd edn. Wiley, 1971]

proper convergence Arises in considering weak convergence of probability measures, where it is used when a sequence of distributions converges (at continuity points of $F(x)$) to a distribution $F(x)$ that is proper. [S. Resnick, *A Probability Path*. Birkhäuser, 1999]

proper orthogonal decomposition *see* Karhunen–Loève expansion

prophet inequality Arises in optimal stopping problems of sequential decision theory, with applications to the theory of algorithms. Let $X_1, X_2, \ldots$ be a sequence of r.vs with finite expectations, and define $M_n = \mathbb{E}(\max_{1 \le r \le n} X_r)$. Let T_n be the set of stopping times (for the sequence $X_1, X_2, \ldots$) that must stop at or before time n (the stopping rule being based only on past and present values of the sequence), and define $V_n = \max_{T \in T_n} \mathbb{E}X_T$. Then an inequality of the form $M_n \le g(V_n)$, for a suitable function $g(\cdot)$ is called a prophet inequality; the name arises because M_n is the reward available to a 'prophet' who can see all values of the sequence, in contrast to V_n which is the reward available to a real-time decision maker who only sees the past and present. The classic prophet inequality of U. Krengel, L. Sucheston, & D. Garling asserts that for i.i.d. non-negative r.vs $X_1, X_2, \ldots$, we have $M_n \le 2V_n$. [T. Hill & R. Kertz, A survey of prophet inequalities in optimal stopping theory, *Contemp. Math.*, 125, 191–207, 1992]

Propp–Wilson method (or algorithm) (alias: **coupling from the past; backward coupling; perfect (or exact) simulation**) A type of Markov chain Monte Carlo method for sampling from a probability distribution $\boldsymbol{\pi}$ that is the stationary distribution of a suitably constructed Markov chain on a state space S. The key attribute of the method is that it runs successive sets of $|S|$ copies of the chain (started from the points of S), each set starting at a receding sequence of times $0 > -n_1 > -n_2 > -n_3 > \cdots$, up to time 0. If a set is coupled at time 0, with value X_0, then X_0 is an exact sample from $\boldsymbol{\pi}$; such coupling eventually occurs a.s. under weak conditions on the chain.

This differs from the usual McMC method in stopping automatically (at 0) with an exact (or perfect) simulation of a r.v. with distribution $\boldsymbol{\pi}$. [O. Häggström, *Finite Markov Chains and Algorithmic Applications*. Cambridge University Press, 2002]

proportional betting *see* Kelly betting

prosecutor's fallacy (alias: **fallacy of the transposed conditional**) Arises in seeking to use evidence E of a probabilistic nature to support the hypothesis G that some accused is guilty. The error amounts to the assertion that $\mathbb{P}(G|E) = \mathbb{P}(E|G)$; by Bayes' rule this is equivalent to the claim that $\mathbb{P}(E) = \mathbb{P}(G)$, which is (in general) an entirely implausible numerical coincidence.

Another error arising in similar contexts is the so-called **defender's fallacy,** which amounts to the assertion for a defendant of some identifiable type T, that $\mathbb{P}(G \cap T|E) = \mathbb{P}(G \cap T)$. *See also* base-rate fallacy. [D. Balding, *Weight of Evidence for Forensic DNA Profiles*. Wiley, 2005]

pseudo-Boolean function *see* Boolean function

pseudo-differential operator An operator D on suitable functions $g(x), x \in \mathbb{R}^d$ whose effect is defined by an integral relation of the form $f(x, D)g(x) = (2\pi)^{-d} \int\int e^{i(x-y)t} f(x,t)g(y)dydt$.

Such operators have applications in Feller's theory of Markov processes. [D. Applebaum, Pseudo-differential operators and Markov semigroups on compact Lie groups, *J. Math. Anal. Appl.*, 384(2)), 331–348, 2011]

pseudo-distribution *see* defective; proper

pseudo-Markov model (alias: **matrix process**) A generalization of the concept of a Markov process that finds applications in communication and coding theory, and also in hidden Markov models. A random process $X_n, n \ge 0$, on a finite state space S, with $|S| = s$ is said to be pseudo-Markov (or a matrix process) if there exist s-vectors $\boldsymbol{\alpha}$ and $\boldsymbol{\beta}$ and an $s \times s$ matrix $M = (m_{ij}), i, j \in S$, having possibly complex entries (and where M is in general stochastic) such that for all n, and $x_0, \ldots, x_n \in S$, $\mathbb{P}(X_0 = x_0, X_1 = x_1, \ldots, X_n = x_n) = \alpha_{x_0} m_{x_0 x_1} \cdots m_{x_{n-1} x_n} \beta_{x_n}$.

The term 'pseudo-Markov' has also been used to denote other processes that share properties with, or are derived from, some Markov process. [W. Turin, *Performance Analysis and Modelling of Digital Transmission Systems*. Kluwer, 2004]

pseudo-moment *see* Zolotarev pseudomoment

pseudo-Poisson process Let $Z_n, n \ge 0$, be a discrete time Markov process taking values in some space S, and let $N(t), t \ge 0$, be a Poisson process of rate λ that is independent of Z_n. Then the continuous-time process $X(t) = Z_{N(t)}, t \ge 0$, is a pseudo-Poisson process; it is a Markov process. In the special case when Z_n is a suitable random walk, $X(t)$ is a compound Poisson process. [W. Feller, *Introduction to Probability Theory and its Applications*, Vol. 2, 2nd edn. Wiley, 1971]

pseudo-process *see* pseudo-Markov model; pseudo-random

pseudo-random (alias: **simulated; computer (or algorithmically) generated; resampled; pseudo-stochastic**) A term arising when the random variables (or other random objects) in the theoretical version of some procedure (or theorem) are replaced in the implementation of the procedure by numbers obtained by means of a computation initiated with an essentially non-random input (that is often called a **seed**). The program defining the computation is a random number generator, and the results are called pseudo-random numbers (or pseudo-random fields, graphs, keys processes, samples, variable, etc., depending on the context) because they cannot be truly random in any generally accepted sense of the concept; the output from any fixed seed being the same for each of a sequence of such computations. However, a sequence of pseudo-random numbers may have the property that it cannot be substantially (or efficiently, in some suitable sense) distinguished from a realization of some truly random sequence, i.e. the sequence will pass any reasonable statistical test for randomness. The archetype of this is a sequence of the symbols H and T that cannot be effectively distinguished from what could be obtained by flipping a real coin. Such sequences are widely used in Monte Carlo methods, cryptography, simulation, and statistics; and producing such a number is called sampling from the appropriate distribution $F(x)$, or simulating a value with distribution $F(x)$. *See also* quasi-random and algorithmic randomness. [O. Goldreich, *A Primer on Pseudorandom Generators*. American Mathematical Society, 2010; RAND Corporation, *A Million Random Digits with 100,000 Normal Deviates*. Free Press, 1955; reprinted by RAND, 2001]

pseudo-stochastic *see* pseudo-Markov model; pseudo-random

pseudo-stopping time Let T be a random time defined on $(\Omega, \mathcal{F})$, where $(\Omega, \mathcal{F}, \mathcal{F}_t, \mathbb{P})$, $t > 0$, is a stochastic basis. Then T is an $(\mathcal{F}_t-)$ pseudo-stopping time if $\mathbb{E}X(T) = \mathbb{E}X(0)$ for every bounded rcll martingale (on the above basis). [A. Nikeghbali & M. Yor, A definition and some characteristic properties of pseudo-stopping times, *Ann. Probab.*, 33(5), 1804–1824, 2005]

pull-through property (alias: **pull-out property; product (or factor) rule; conditional determinism**) The basic property of conditional expectation that for r.vs X and Y satisfying $\mathbb{E}|X| < \infty$ and $\mathbb{E}|XY| < \infty$, we have $\mathbb{E}(XY \mid X) = X\mathbb{E}(Y \mid X)$, a.s. [O. Kallenberg, *Foundations of Modern Probability*, 2nd edn. Springer, 2002]

pure jump process *see* jump process

push-forward measure *see* induced probability

push forward probability measure *see* induced probability

put option *see* option

Q

q-distribution Broadly, a q-deformation of a probability distribution $\mathbb{P}$ is a family $\mathbb{P}_q$ depending on a parameter q such that (conventionally) $\mathbb{P}_1 = \mathbb{P}$, or $\lim_{q\to 1}\mathbb{P}_q = \mathbb{P}$, and then $\mathbb{P}_q$ is called a q-distribution. Of several such families, the so-called **Tsallis *q*-distns** are perhaps most often seen, being defined in terms of the ***q*-deformed exponential fn**,

$$d_q(x) = \{1 + (1-q)x\}_+^{\frac{1}{1-q}}, \quad x \in \mathbb{R},\, q \neq 1;$$

with inverse q-logarithm,

$$\log_q x = (x^{1-q} - 1)/(1-q), \quad x > 0,\, q \neq 1.$$

(The shorter term q-exponential commonly denotes another mathematical object in F. Jackson's q-calculus.) Then for $\lambda > 0$, the q-exponential density may be defined as

$$f(x) = \lambda(2-q)d_q(-\lambda x), \begin{cases} x > 0, & 1 \leq q < 2, \\ 0 \leq x < \dfrac{1}{\lambda(1-q)}, & q < 1. \end{cases}$$

Its survival function is proportional to another deformed exponential function, and this leads some authors to define the $\hat{q}$-exponential distribution function as $F(x) = 1 - d_{\hat{q}}(-\lambda x)$. In general, a Tsallis q-deformed distribution is one that maximizes the Havrda–Charvát–Tsallis q-entropy; examples include the q-deformed gamma distribution, with density proportional to $x^r d_q(-\lambda x)$, and the q-deformed Gaussian (or normal) distribution, which has density proportional to $d_q(-ax^2)$, for some $a > 0$, and $-\infty < q < 3$; the support of the distribution is $\mathbb{R}$ for $1 \leq q < 3$, but only a finite interval for $q < 1$. As noted above, such families arise naturally as maximizing the q-entropy that appears in non-extensive statistical mechanics. Other physical applications give rise to the Božejko–Speicher and Diaz–Ternel families of q-distributions. In each case, the limit as $q \to 1$ yields a corresponding standard probability density. [R. Diaz & E. Parignan, On the Gaussian q-distribution, *J. Math. Anal. Appl.*, 358(1), 1–9, 2009; M. Gell-Mann & C. Tsallis (eds), *Nonextensive Entropy: Interdisciplinary Applications*. Oxford University Press, 2004]

q-entropy *see* entropy

Q-function *see* normal distribution

q-Gaussian (or q-normal) *see* q-distribution

Q-matrix (alias: ***q*-matrix; transition rate matrix, or intensity matrix, or (infinitesimal) generator matrix**) Arises as the derivative at 0, $Q = P'(0)$, of the matrix $P(t)$ of transition probabilities of a continuous-time Markov chain $X(t), t \geq 0$, on a countable state space S; and appears naturally in the construction of a version of $X(t)$ where the chain evolves by successively occupying states for an exponentially distributed holding time and moves between successive states according to the stochastic routing (or jump) matrix determining the jump chain. Formally, in the simplest case, $Q = (q_{jk});\, j, k \in S$, is a Q-matrix if:

(a) $0 \leq q_{jk} < \infty$ for all $j \neq k$;
(b) $\sum_{k \neq j} q_{jk} = 0$.

Informally, q_{jk} is the rate (or intensity) of transitions from j to k, so that $-q_{jj}$ (where we often write $-q_{jj} = q_j$), is the rate of departure from j (being the parameter of the exponentially distributed holding time in j) (*see* Feller's construction). More generally, a Q-matrix satisfies:

(i) $0 \leq q_{jk} < \infty, \quad j \neq k$;
(ii) $\sum_{k \neq j} q_{jk} \leq q_j \leq \infty, \quad$ for all $j \in S$.

If $q_j < \infty$ for all $j \in S$, then Q is said to be **stable**, otherwise it is **unstable**; if $\sum_{k \neq j} q_{jk} = q_j$, then Q is said to be **conservative**, otherwise it is **non-conservative**. Those Q-matrices defined by (a) and (b) are thus stable and conservative, and are what is usually understood by the unadorned term Q-matrix.

As noted above, the matrix Q also arises as the right-hand derivative at zero of the transition matrix $P(t) = (p_{jk}(t))$ *of* $X(t)$, on the usual condition that $P(t)$ is standard. Conversely, for any Q-matrix (given by (a) and (b)) there is at least one $P(t)$ with $P'(0) = Q$; if $P(t)$ is unique it is said to be regular. When not unique, any of the possible processes corresponding to Q may be called a Q-process, and its $P(t)$ may be called a Q-function. (*See* Feller minimal process.)

The jump (or routing) matrix $R = (r_{jk})$ corresponding to a given Q-matrix (q_{jk}) is given by

$$r_{jk} = \begin{cases} q_{jk}/q_j, & \text{if } k \neq j \text{ and } q_j \neq 0 \\ 0, & \text{if } k \neq j \text{ and } q_j = 0 \end{cases}$$

$$r_{jj} = \begin{cases} 1 & \text{if } q_j = 0 \\ 0 & \text{if } q_j \neq 0 \end{cases}.$$

Note that a state j is absorbing if $q_j = 0$; and it may be called instantaneous if $q_j = \infty$. [W. Anderson, *Continuous-time Markov Chains*. Springer, 1991]

quadrant dependence *see* association of random variables; dependence

quadratic entropy (alias: **quadratic diversity index (or index of diversity); quadratic similarity index**) Any one of several measures (or indices), to be denoted by Q, of the spread of a probability distribution $(p_1, \ldots, p_n)$, where Q is expressed as a function of $S = \sum_{r=1}^{n} p_r^2$, whose extreme values are $S = 1$ and $S = n^{-1}$, corresponding to minimal and maximal spread respectively. More generally, Q is a function of $R = \sum_{r,s} d_{rs}\, p_r p_s$, where d_{rs} is a weighting function that is interpreted as 'distance' or 'difference', and R may be called **Rao's quadratic entropy**.

In disparate applications, 'spread' may be interpreted as diversity, or dispersion, or dissimilarity, or heterogeneity, etc; and, conversely, S and Q yield measures of concentration, dominance, similarity, homogeneity, etc. The wide field of applicability has given rise to several indices of this type, almost all of which are known by a variety of different names in their separate fields, such as:

(i) S is **Simpson's** (concentration, or dominance) **index**, or the Herfindahl index.
(ii) $2(1 - S)$ is the Havrda–Charvát quadratic entropy, or identification entropy.
(iii) $1 - S$ is Tsallis quadratic entropy, also called the **diversity index**, or **Gini–Simpson index**.
(iv) $\sqrt{S}$ is the Hirschman index.
(v) $-\log S$ is the Rényi quadratic entropy, or **collision entropy**.
(vi) S/p_r, $1 \le r \le n$, is Weaver's surprise index.

Where S has also been called the repeat rate, and the index of coincidence, and L_2 is the categorical diversity; $1 - S$ has also been called the logical entropy, and the concentration measure, and the diversity index.

Usage not being definitive, one arbitrary convention is this: $Q_E = 1 - S$ denotes the quadratic entropy; while $Q_c = S$ denotes the concentration or coincidence index. [D. Aldous, In search of a missing word: entropy and ???, *Bernoulli News*, 18, 1, 2011; I. J. Good, Diversity as a concept and its measurement: Comment, *J. Amer. Stat. Ass.*, 77(379), 561–563, 1982; C. R. Rao, Quadratic entropy and analysis of diversity, *Sankhyā*, 72-A(1), 70–80, 2010; R. Ahlswede & N. Cai, An interpretation of identification entropy, *IEEE Trans. Inf. Theor.*, 52(9), 4198–4207, 2006]

quadratic index *see* quadratic entropy

quadratic variation Broadly, in the context of random processes (especially martingales) a variation function (or process) measures the oscillatory properties of the process, i.e. the irregularity – or spikiness – of its paths. Thus, for the Wiener process $W(t)$, whose ordinary (total) variation is infinite (by its continuity and the independence of its increments), nevertheless the quadratic variation $Q(a, b)$, defined as follows, is finite.

For a partition $a = t_0(n) < t_1(n) < \cdots < t_m(n) = b$, define

$$Q_n(a, b) = \sum\nolimits_r \{W(t_{r+1}(n)) - W(t_r(n))\}^2.$$

Then $Q_n(a, b)$ converges to $Q(a, b) = b - a$ in mean square, as $n \to \infty$ with $\max_r |t_{r+1}(n) - t_r(n)| \to 0$. Alternatively, the predictable quadratic variation of $W(t)$ is the unique continuous increasing adapted process $<W>_t$ such that $W(t)^2 - <W>_t$ is a martingale; and $<W>_t = t$. More generally, for arbitrary L_2 local martingales (not necessarily continuous) the basic and predictable quadratic variations need not coincide. *See* brackets process. [P. Protter, *Stochastic Integration and Differential Equations*, 2nd edn. Springer, 2005]

quadratic variation process *see* brackets process

qualitative probability An axiomatization of the concept of comparative probabilities, in which the σ-field $\mathcal{F}$ of subsets of some finite set Ω supports a binary relation $\le$, such that for $A, B \in \mathcal{F}$, one of these holds:

$A \le B \equiv A$ is no more probable than B;
$A \ge B \equiv B$ is no more probable than A.

Furthermore, in one formulation, for all $A, B, C \in \mathcal{F}$:
(a) $A \le B$ and $B \le C \Rightarrow A \le C$;
(b) $A \ge \emptyset$ and $\Omega \le {\sim}\emptyset$;
(c) If $A \cap C = \emptyset = B \cap C$, then $A \le B$ iff $A \cup C \le B \cup C$.

A central problem is to determine conditions for the existence of a probability function $\mathbb{P}(\cdot) \in [0, 1]$ on the members of A, such that $A \le B$ iff $\mathbb{P}(A) \le \mathbb{P}(B)$, and further conditions for $\mathbb{P}(\cdot)$ to be additive. One version of such conditions is supplied by **Fine's thm**. [T. Fine, *Theories of Probability*. Academic Press, 1973]

quantile Arises in describing the location of a probability distribution; in statistics the same term is used in denoting a function that describes the location of a random sample. Let $F(x)$ be the distn function of a r.v. X; then a pth quantile of X (or of $F(x)$), is any $x \in \mathbb{R}$ such that $F(x) \geq p$, and $\lim_{y\uparrow x} F(y) \leq p$. In particular:

(a) if $p = \frac{1}{2}$, then one speaks of a median;

(b) if $p = \frac{r}{4}$, $1 \leq r \leq 3$, then one speaks of the rth quartile;

(c) if $p = \frac{r}{100}$, $1 \leq r \leq 99$, then one speaks of the rth centile, and so on (with quintiles, octiles, deciles, etc.).

For suitable F, the pth quantile may be unique; more generally, for any p, the set of pth quantiles is a closed interval of $\mathbb{R}$, one of which may be deemed to be the pth quantile. In particular, for any $F(x)$, the so-called quantile function is given by

$$F^{-1}(y) = \inf\{x;\ F(x) \geq y, \quad 0 < y < 1\},$$

which is left-continuous, as F is right-continuous. *See* probability integral transform, and inverse transform method.

quantile function *see* probability integral transform; quantile

quantile of Brownian motion Is defined in terms of Brownian motion $W(t), t \geq 0$, for $0 \leq a < 1$, as $Q(a) = \inf\{x : \int_0^1 I(W(y) \leq x)dy > a\}$, where $I(\cdot)$ is the usual indicator function. [M. Yor, The distribution of Brownian quantiles, *J. Appl. Probab.*, 32(2), 405–416, 1995]

quantile process Arises in relation to the empirical distribution function as follows: Let the i.i.d.r.vs $X_1, \ldots, X_n$, with order statistics $X_{(r)}, 1 \leq r \leq n$, have distribution function $F(x)$, whose inverse (or quantile) function is $Q(y) = F^{-1}(y) = \inf\{x : F(x) \geq y\}$. The empirical inverse (or quantile) function is

$$Q_n(y) = F_n{}^{-1}(y) = X_{(r)} \text{ if } \frac{r-1}{n} \leq y < \frac{r}{n},$$

for some $1 \leq r \leq n$, and then the quantile process is defined to be $\sqrt{n}(F_n^{-1}(y) - F^{-1}(y)), 0 \leq y \leq 1$. [M. Csörgö, *Quantile Processes with Statistical Applications*. SIAM, 1983]

quantize Arises in communication theory, and other applications of probability, where it commonly describes the action of a function $f(\cdot)$ taking discrete values, whose argument is a random variable X taking values in $\mathbb{R}$; very often X is continuous and $f(X)$ is simple. The function $f(\cdot)$ may be called a quantizer. [S. Graf & H. Luschgy, *Foundations of Quantization for Probability Distributions*. Springer, 2000]

quantum probability (alias: **non-commutative probability; algebraic probability**) Arises naturally in the context of quantum mechanics, and can be seen as a non-commutative generalization of classical probability. It was first given an axiomatic development by J. von Neumann, which can be seen as analogous to A. Kolmogorov's axiomatization of classical probability. It postulates a family of non-commutative linear operators on a suitable Hilbert space; these correspond to observables, and may be e.g. square matrices. Analogues of various concepts of classical probability (such as expectation) can be defined, including several different types of independence (*see* free probability). Formally, a quantum probability space (also called an algebraic probability space), denoted by $(\mathcal{A}, \phi)$, comprises a von Neumann algebra $\mathcal{A}$ of operators on a suitable Hilbert space, together with a non-negative linear function ϕ, often called a state, on $\mathcal{A}$. The orthogonal projections π in $\mathcal{A}$ may be called events, and $\phi(\pi)$ may be called the probability of π. In this framework, a suitable mapping of $\mathcal{A}$ can be seen as a quantum communication channel. [J. von Neumann, *Mathematical Foundations of Quantum Mechanics*. Princeton, 1955; P. A. Meyer, *Quantum Probability for Probabilists*, 2nd edn. Springer, 1995]

quartile *see* quantile

quasi-arithmetic mean *see* Kolmogorov–Nagumo mean

quasi-birth–death process A Markov chain in continuous time whose Q-matrix is block tri-diagonal, which is to say that it is of the form

$$Q = \begin{pmatrix} A & U & & 0 \\ L & D & U & \\ & L & D & U \\ 0 & \ddots & \ddots & \ddots \end{pmatrix}$$

for suitable matrices A, D, L, and U. [G. Latouche & V. Ramaswami, *Introduction to Matrix Analytic Methods in Stochastic Modelling*. SIAM, 1999]

quasi-diffusion process (alias: **gap diffusion; generalized diffusion**) An extension of the standard theory of diffusions that can be accomplished in several equivalent ways, e.g. via generalized differential operators, or by relaxing the requirement that speed measure is strictly positive, or by a suitable time change of Brownian motion, or via Krein's spectral theory of strings. [A. M. G. Cox *et al.*, Time-homogeneous diffusions with a given marginal at a random time, *ESAIM Probab. Stat.*, 15, Suppl., 11–24, 2011]

quasi-left-continuous Arises in considering stopping times in a general theory of stochastic processes, with applications in the context of Markov processes. Given a stochastic basis (or filtered probability space) $(\Omega, \mathcal{F}, \mathcal{F}_t, \mathbb{P})t \geq 0$, the filtration $\mathcal{F}_t$, or an rcll process $X(t)$ on the basis, is said to be quasi-left-continuous if for every predictable stopping time T, we have (respectively) on the event $\{T < \infty\}$, $\mathcal{F}_{T-} = \mathcal{F}_T$ or $X(T-) = X(T)$ a.s.

When a strong Markov process is quasi-left-continuous it is called a Hunt process. [O. Kallenberg, *Foundations of Modern Probability*, 2nd edn. Springer, 2002]

quasi-limiting distribution *see* Yaglom's theorem

quasilinear mean *see* Kolmogorov–Nagumo mean

quasi-Markov process Arises in the context of a regenerative phenomenon. Formally, a quasi-Markov process of order m, denoted by $X(t), t \geq 0$, is a random process taking values in a set $\{0, 1, \ldots, m\} = S$, such that for all n and all $0 = t_0 < t_1 < \cdots < t_n$, and all $s_0, \ldots, s_n \in S \backslash \{0\} = \{1, 2, \ldots, m\}$, we have

$$\mathbb{P}(X(t_r) = s_r, 1 \leq r \leq n \,|\, X(t_0) = s_0) = p_{s_0 s_1}(t_1 - t_0) \ldots p_{s_{n-1} s_n}(t_n - t_{n-1}),$$

for some so-called p-matrix, $\mathbf{p}(t) = (p_{rs}(t))$ of functions, called the p-functions. It should be noted that this does not necessarily hold if $s_r = 0$ for some r. The chain (and $\mathbf{p}$) is said to be standard if $\lim_{t \to 0} \mathbf{p}(t)$ is the $m \times m$ identity matrix. [J. F. C. Kingman, *Regenerative Phenomena*. Wiley, 1972]

quasi-randomness Arises in Monte Carlo methods for integration, optimization, and simulation, etc., where a quasi-random set of points in the unit hypercube is designed to be much more uniformly distributed than would be a collection of independent uniformly distributed points. The intention is to avoid clustering, i.e. to achieve (as far as possible) similar proportions of points in uniform dissections of the cube. They may be called low-discrepancy (pseudo-random) numbers. [J. Dick & F. Pillichshammer, *Digital Nets and Sequences*. Cambridge University Press, 2010]

quasi-reversibility A concept arising in queueing theory, with applications to networks of queues in equilibrium. Formally, the stationary Markov process $X(t), t \geq 0$, representing the state of a queue at time t, is said to be quasi-reversible if, for all t, $X(t)$ is independent of both the arrival times of customers after time t and the departure times of customers before time t (where customers may be of different types or classes).

This property is not implied by, nor does it imply, reversibility in the ordinary sense. [S. Asmussen, *Applied Probability and Queues*, 2nd edn. Springer, 2003]

quasi-stationarity Arises in the context of a Markov process $X(t), t \geq 0$ (in discrete or continuous time), whose state space S includes a set of absorbing (or forbidden, or taboo, or trap) states Δ; the process is stopped (or absorbed, or killed, or trapped) at its entry time T to Δ, where $\mathbb{P}(T < \infty) = 1$. Then a distribution $\mathbb{Q}(\cdot)$ on $S \backslash \Delta$ such that $\mathbb{P}(X(t) \in A \,|\, T > t, X(0) \notin \Delta$ has distribution $\mathbb{Q}) = \mathbb{Q}(A)$ is called quasi-stationary.

The **Yaglom limit** (*see* Yaglom's theorem), or quasi-limiting distribution, $\lim_{t \to \infty} \mathbb{P}(X(t) \in A \,|\, T > t,$ $X(0) = x \in S \backslash \Delta)$, is quasi-stationary when it exists and is proper. In the non-Markov case, one speaks of **taboo-stationarity**. [P. Collet *et al.*, *Quasi-stationary Distributions*. Springer, 2013]

quenched regime A term arising naturally in the context of a random process $X(t)$ evolving in (i.e. being a function of) a random environment or medium, denoted by $M(\omega)$, where ω takes values in some sample space Ω. If $X(t)$ is analysed for some fixed ω, then the process is said to be in the quenched (or frozen) regime (or setting).

The term was coined, and is popular, in physics literature, and there, if M is replaced by $\mathbb{E}M$, they refer to this as the annealed regime, or annealing. Confusingly, in the mathematical literature the annealed regime refers to the expectation of X (and functions of X) taken wrt the distribution of M. [D. A. Dawson *et al.* (eds), *Probability and Mathematical Physics*. American Mathematical Society, 2007]

queueing theory Broadly, the analysis of systems that exist to deal with (i.e. provide a service to) a sequence of entities (which may be objects, or signals etc., but are usually called **customers**) that arrive at random and make random demands of some resource, e. g. time, bandwidth, space, processing capacity, etc., from one or more so-called **servers**. There is a huge variety of queueing models, but basic concepts are most readily defined in terms of the fundamental single server queue, whose structure may be described in terms of the arrivals process, the customers' demands, and the queue discipline, thus:

(A) In most cases, the **arrivals process** constitutes a counting process (or point process in time); commonly customers arrive one at a time, otherwise one speaks of **batch arrivals**. The process may be seen as a marked point process, in that with each customer is associated some potential demand on the system (*see* (D), following); furthermore, customers may be of different types, entailing different treatment (*see* (C), following). The arrivals process is generally assumed to be independent of the state of the queue and of their demands and types.

(B) In the simplest single server queueing model each arriving customer requires a service time from the server; these are assumed to be i.i.d.r.vs that are also independent of the arrivals process. (*See* Kendall's notation.)

(C) The **queue discipline** is the procedure that determines how the server deals with the sequence of arriving customers (and, in more complicated models, how customers react to their experience in the system). Some of the more commonly seen rules, given a single server, are the following:
   (i) **First in, first out** (or FIFO).
   (ii) **First come, first served** (or FCFS).
   (iii) **Last in, first out** (or LIFO)
   (iv) **Last come, first served** (or LCFS).
   (v) **Priority service**: the server deals first with those highest in order of precedence.
   (vi) **Preemptive service**: some customers can (on arrival) terminate or interrupt the server in dealing with a customer being served, and be served first.
   (vii) **Batch service**: the server attends to more than one customer simultaneously. One form of this is gated service, when a random batch of customers accumulates, e.g. while the server is absent for some reason (often referred to as being on vacation).
   (viii) Among many other possibilities, one has:
   - **balking**, where customers may leave, or refuse to join a queue before being served;
   - **reneging**, where the server may eject the customer before completing service;
   - queues with finite **waiting rooms** (or **buffers**), or no buffer at all, so that customers are lost;
   - **retrial queues**, where customers may rejoin the same queue after service;
   - **quorum queue**, where the server will not work until a given number of customers are present;
   - **hysteretic queue**, being a quorum queue where the server stops work when the number of customers waiting drops below a given threshold – these thresholds may also be based on the total service needed;
   - **random service,** where the server picks the next to be served randomly from the queue.

(D) Finally, the interaction between arriving customers and the server yields a great many functions and processes describing the behaviour of the system, including the following:
   (i) The number of customers present and waiting to begin service.
   (ii) The **number in system**, being customers waiting for service plus those being served. Also called **queue size**, **congestion process**, or **buffer content**.
   (iii) The **idle period** of the server, being the interval from an instant when a departing customer leaves no waiting customer, until the server begins service again. If idleness is transient, one speaks of **congestion**, and of heavy traffic, as traffic intensity approaches congestion.
   (iv) A **busy period** is an interval between idle periods.
   (v) The **waiting time before service** of an arriving customer, i.e. the interval between arrival and beginning to be served; also called the delay.
   (vi) The **waiting time in the system**, i.e. the interval between arrival and completing service (not including any retrial); also called the sojourn time in the system.
   (vii) **Virtual waiting time**; which is the wait for service that would be experienced by a customer if they were to join the queue. This being the time for the server to clear all those in front in a FIFO queue, it is also called the workload.

Various other more complicated disciplines and functionals arise when one has a network of queues with multiple servers.

[S. Asmussen, *Applied Probability and Queues*, 2nd edn. Springer, 2003]

quickest detection problem *see* change-point problem

quizmaster problem *see* Monty Hall problem

R

Rabin–Miller test *see* Solovay–Strassen test

Rabin primality test *see* Solovay–Strassen test

Rademacher chaos Arises from a sequence of i.i.d.r.vs $X_1, X_2, \ldots$, each having the symmetric distribution $\mathbb{P}(X_r = 1) = \mathbb{P}(X_r = -1) = \frac{1}{2}$; *see* Rademacher random variable. For $n \geq 1$, let $c(r_1, \ldots, r_n) \in \mathbb{C}$, and then the (homogeneous) Rademacher chaos of order n is

$$R_n = \sum_{r_1 < r_2 < \cdots < r_n} c(r_1, \ldots, r_n) X_{r_1} \cdots X_{r_n}.$$

In particular, the chaos of order 1 is also called a **Rademacher process** (or series): $R_1 = \sum_r c(r) X_r$. [R. Blei & S. Janson, Rademacher chaos: tail estimates vs limit theorems, *Ark. Mat.*, 42, 13–29, 2004]

Rademacher random variable Has the so-called Rademacher distribution $\mathbb{P}(X = 1) = \mathbb{P}(X = -1) = \frac{1}{2}$, and may be called a **random sign**. The simple Rademacher series $R(n) = \sum_{r=1}^{n} X_r$, where the X_r are i.i.d.r.vs with the Rademacher distribution, obeys $\mathbb{P}(|R(n)| \leq \sqrt{n}) \geq \frac{1}{2}$. [I. Pinelis, An asymptotically Gaussian bound on the Rademacher tails, *Electron. J. Probab.*, 2012, 17(35), 1–22]

radial Brownian motion *see* Bessel process

radiation boundary condition *see* elastic boundary

Radon–Nikodym theorem A result in the general theory of measure and integration that has important applications in probability, e.g. in the foundations of conditional expectation, and in results such as the Cameron–Martin–Girsanov theorem. One form is this: let $\mathbb{P}$ be a probability measure on $(\Omega, \mathcal{F})$, and let $\mathbb{Q}$ be a finite measure on $(\Omega, \mathcal{F})$ that is absolutely continuous wrt $\mathbb{P}$. Then there is a non-negative, integrable, $\mathcal{F}$-measurable function g (which is a.s. unique wrt $\mathbb{P}$), such that for all $A \in \mathcal{F}$

$$\mathbb{Q}(A) = \int_A g \, d\mathbb{P}.$$

The notation $g = \frac{d\mathbb{Q}}{d\mathbb{P}}$ is often used, and g is referred to as the Radon–Nikodym derivative. The conflation of this and the Lebesgue decomposition may be called the **Lebesgue–Radon–Nikodym thm**. [E. Çinlar, *Probability and Stochastics*. Springer, 2011]

Raikov's theorem Asserts that for independent non-degenerate non-negative r.vs X and Y such that $X + Y$ has a Poisson distribution, both X and Y must have Poisson distributions. *See* Cramér–Lévy theorem.
A generalization of these results to sums of independent Poisson and normal r.vs is called Linnik's theorem. [P. Moran, *Introduction to Probability Theory*. Oxford University Press, 1968]

Rajski's distance Supplies a measure of the difference (or association, from another point of view) between discrete probability distributions $\mathbb{P}$ and $\mathbb{Q}$, having joint Shannon entropy $H(\mathbb{P}, \mathbb{Q})$ and mutual Shannon information $I(\mathbb{P}; \mathbb{Q})$, as

$$d(\mathbb{P}, \mathbb{Q}) = 1 - \frac{I(\mathbb{P}; \mathbb{Q})}{H(\mathbb{P}, \mathbb{Q})}.$$

It is a normalized form of the simple Shannon (or information, or entropy) metric

$$d_s(\mathbb{P}, \mathbb{Q}) = H(\mathbb{P}, \mathbb{Q}) - I(\mathbb{P}; \mathbb{Q}).$$

[Y. Horibe, Entropy and correlation, *IEEE Trans. Syst. Man Cyber.*, SMC-15(5), 641–642, 1985]

random allocation *see* urn

random-cluster model (alias: **Fortuin–Kasteleyn random-cluster model**) Arises in statistical mechanics as a model for phenomena such as phase transitions that are seen in suitable physical systems on lattices or graphs. Thus, a random-cluster model is one of a family, indexed by a parameter $q \in (0, \infty)$, of probability distributions on the subsets of the edges of the lattice or graph in question. Particular values of q yield classic special cases, such as percolation and the Ising model. [P. Kastelyn & C. Fortuin,

Phase transitions in lattice systems with random local properties, *J. Phys. Soc. Jap.*, 26, Suppl., 11–14, 1969; G. Grimmett, *The Random-cluster Model*. Springer, 2006]

random counting measure *see* point process

random directions algorithm *see* hit-and-run algorithm

random element Arises when considering functions $g(\omega)$ (or maps), defined on the sample space Ω of a probability space $(\Omega, \mathcal{F}, \mathbb{P})$. When $g(\omega)$, $\omega \in \Omega$, is an $\mathcal{F}$-measurable function that takes values in a measurable space $(S, \mathcal{G})$, it is called a random element. In particular, e.g. a random element is called a:

- random variable when $S \subseteq \mathbb{R}$;
- random vector when $S \subseteq \mathbb{R}^n$, $n > 1$;
- random set when S is a suitable class of sets;
- random process when S is a suitable class of functions, and so on.

Some writers use the term random object for random element, but this term is used otherwise in computer science where the word 'object' has a particular meaning. [O. Kallenberg, *Foundations of Modern Probability*, 2nd edn. Springer, 2002]

random environment (alias: **random medium**) Arises in considering non-homogeneous random processes in which the non-homogeneity is given by a random process. The classic model is the random walk in a random environment (or medium, or RWRE, or RWRM) thus: Let $H_n, n \in \mathbb{Z}$, be a sequence of i.i.d.r.vs taking values in [0, 1], and let $X_t, t \geq 0$, be the position of a particle which moves on the integers from $X_0 = 0$; at each integer time $t > 0$ the particle moves to $X_t + 1$ (respectively $X_t - 1$), with probability H_{X_t}, (respectively $1 - H_{X_t}$), given the value of X_t. This may be called a site-random RWRE to distinguish it from the so-called bond-random RWRE, in which $H(n, n+1)$ and $H(n, n-1)$, $n \in \mathbb{Z}$, comprise two families of non-negative r.vs, and the walk steps to $X_t + 1$ (resp. $X_t - 1$) with probability $p = H(X(t), X(t)+1)/\{H(X(t), X(t)+1) + H(X(t), X(t)-1)\}$ (resp. $1 - p$). For the site-random RWRE, **Solomon's thm** assets that the origin is a.e. recurrent if $\mathbb{E}\log\left(\frac{H_0}{1-H_0}\right) = 0$; and **Sinai's thm** asserts for such a recurrent walk, if also $0 < \mathbb{E}\left[\left(\log \frac{H_0}{1-H_0}\right)^2\right] < \infty$, then $X_n/(\log n)^2$ converges in distribution, with a non-degenerate limiting probability measure $K(x)$, that may be called **Kesten's distn**.

Similarly, a branching process in a random environment entails the selection of random family size pgfs from a suitable class of such pgfs; in the i.i.d. case, this is the **Smith–Wilkinson model**.

The RWRE is not be confused with a random walk in a random scenery, or RWRS, *see* Kesten–Spitzer random walk in random scenery. [H. Kesten, The limit distribution of Sinai's random walk in random environment, *Physica A*, 138A, 299–309, 1986; B. Hughes, *Random Walks and Random Environments*, Vols I & II. Oxford University Press, 1995]

random evolution *see* random velocity model

random field *see* stochastic process

random flight (alias: **random directions process**) Arises naturally in modelling diffusion, and also as an idealized model of a polymer (seen as a random chain). In its simplest form it comprises the sequence of random vectors $S_n = \sum_{r=1}^{n} U_r$, where $U_r, r \geq 1$, are i.i.d.r.vrs of unit length in $\mathbb{R}^d, d \geq 1$, that are often assumed to be uniformly distributed on the unit sphere. In one dimension this is simple random walk; in three dimensions it is known as **Rayleigh's random flight**.

More generally, the U_r may have i.i.d. random lengths, as well as random directions; if each is exponentially distributed one has a Poissonized (or randomized) flight, if each is normally distributed it is called a Gaussian flight; if each has a stable distribution, then it may be referred to as a **Lévy flight**. [W. Feller, *Introduction to Probability Theory and its Applications*, Vol. 2, 2nd edn. Wiley, 1971]

random graph A concept arising naturally in considering the reliability of a system, and in several combinatorial problems on networks. Very broadly, given a sample space Ω of graphs, and a probability space $(\Omega, \mathcal{F}, \mathbb{P})$, any realization ω, with probability $\mathbb{P}(\omega)$ is a random graph. In practice, two types are of the greatest importance:

1. The **binomial random graph**. Given n vertices, each of the $\binom{n}{2}$ pairs of vertices is connected by an edge independently with probability p. The resulting graph is denoted as $G_{n,p}$. It was introduced for $p = \frac{1}{2}$ by P. Erdös, and for general p by E. Gilbert, and might be called the Erdös–Gilbert graph; many writers refer to it as an Erdös–Rényi graph.

2. The **uniform random graph**. Given n vertices, and a non-negative integer $m \leq \binom{n}{2}$, one of the $\binom{\binom{n}{2}}{m}$ graphs having exactly m edges joining pairs of these n vertices is selected with equal probability $\binom{\binom{n}{2}}{m}^{-1}$ of choosing any. The resulting graph is denoted by $G_{n,m}$. It was introduced by P. Erdös, and extensively studied by P. Erdös and A. Rényi, and might be called the Erdös–Rényi graph (many authors do so).

A large variety of other random graphs may be defined e.g. by deleting vertices rather than edges; and by requiring the graph to be planar, or bipartite, or regular, etc. In particular, the **Molloy–Reed random graph** is constructed to have a given distribution of vertex degree. [S. Janson *et al.*, *Random Graphs*. Wiley, 2000; B. Bollobás, *Random Graphs*, 2nd edn. Cambridge University Press, 2001]

randomization A term arising when a set S of objects is selected (or ordered) randomly, i.e. more formally, when a probability distribution is assigned to the set S, which is typically, not itself intrinsically random. Such randomization is usually designed to achieve fairness (or lack of bias), or efficiency, in some task, with applications in statistics, cryptography, optimization, game theory, etc. Examples include:

- Lotteries – a lottery for an indivisible prize (or penalty) induces fairness, in the sense of equal chances.
- Random allocation – in statistics this reduces various types of bias
- Randomized algorithms – supply efficient procedures in many contexts
- Randomized decision rules, and randomized strategies – supply optimal solutions in game and decision theory.
- Randomized (or uniformized, or Poissonized) process – the process obtained from a discrete parameter process $X_n, n \geq 0$ (commonly a Markov chain) by imbedding in the continuous-time process $Y(t) = X_{N(t)}, t \geq 0$, where $N(t)$ is a Poisson process.
- The probabilistic method.

In this context, any procedure or program that yields a suitable random (or pseudo-random) outcome from a probability distribution of interest is called a **randomizer**. Early randomizers included coins, dice, cards, and tickets drawn from urns; more recently, computers are predominant in this.

[P. Diaconis *et al.*, Dynamical bias in the coin toss, *SIAM Review*, 49(2), 211–235, 2007; M. Mitzenmacher & E. Upfal, *Probability and Computing: Randomized Algorithms and Probabilistic Analysis*. Cambridge University Press, 2005]

random mapping representation (alias: **canonical (or white noise) construction (or representation); Markov recurrence**) Arises in simulating a Markov chain on a finite state space S. Let $Z_n, n \geq 0$, be a sequence of i.i.d.r.vs, and let $g(.,.)$ be a function taking values in S such that $\mathbb{P}(g(j, Z_0) = k) = p_{jk}, j, k \in S$. Then the sequence of r.vs $X_n, n \geq 0$, taking values in S defined by $X_n = g(X_{n-1}, Z_n), n \geq 1$, with $\mathbb{P}(X_0 = s) = p_0(s), s \in S$, comprises a Markov chain with initial distribution $p_0(\cdot)$, and transition matrix $P = (p_{jk})$. The pair (g, Z) is said to give a random mapping representation of P, and of the chain; and every P (and chain), on a finite state space has such a representation. *See* Rosenblatt's transformation. [D. Levin *et al.*, *Markov Chains and Mixing Times*. American Mathematical Society, 2009]

random matrix A matrix M whose entries are random variables that are real-valued, or complex-valued, or matrix-valued, e.g. Pauli matrices (or quaternions). In the basic case when M is square with i.i.d. zero-mean complex-valued entries X_{ij}, $1 \leq i, j \leq n$, one has the circular law conjecture, thus: If $0 < \text{var}\, X_{ij} < \infty$ then the suitably normalized eigenvalue density of M is uniform on the disk $|z| < 1$ of the complex plane in the limit as $n \to \infty$. It has been proved on the additional condition that $\mathbb{E}X_{ij}^{2+\delta}$ exists for some $\delta > 0$.

In many applications M is symmetric (or, more generally, Hermitian), and the term **Wigner matrix** is often used when X_{ij} are i.i.d. for $i < j$, and $X_{ij} = \overline{X}_{ji}$ for all i, j, and $X_{jj}, 1 \leq j \leq n$, are also i.i.d. real-valued r.vs (that may have a distribution different from off-diagonal elements). It is usual to assume that $\mathbb{E}X_{ij} = 0$, and that the entries are non-degenerate with finite variance. Then one has Wigner's semicircle law, thus: for a Wigner matrix the suitably normalized density of the eigenvalues of M converges (weakly, in probability), as $n \to \infty$, to $S(x) = (2\pi)^{-1}(4 - x^2)^{1/2}, |x| \leq 2$. In particular, when $X_{ij}, 1 \leq i, j \leq n$ are i.i.d. $N(0, 1)$ r.vs, the matrix G, given by $G_1 = \frac{1}{2}(M + M^{\mathrm{T}})$, where T denotes transpose, is called the **Gaussian orthogonal ensemble**, or GOE, with Dyson parameter (or index) $b = 1$. When $X_{ij}, 1 \leq i, j \leq n$ are i.i.d. standard complex valued normal r.vs, then $G_2 = \frac{1}{2}(M + M^{\mathrm{T}})$, where T denotes Hermitian transpose, is called the **Gaussian unitary ensemble**, or GUE, with Dyson parameter $b = 2$. When X_{ij} are suitably defined Pauli (quaternion) matrices, then $G_4 = \frac{1}{2}(M + M^{\mathrm{T}})$, where T denotes the dual transpose, is the **Gaussian symplectic ensemble**, or GSE, with Dyson parameter $b = 4$. Several characteristics of these Gaussian random matrices are expressed using the Dyson parameter b, most notably the joint density $f(\lambda)$ of their eigenvalues, which is such that (suitably scaled)

$$f(\lambda) \propto \prod_{1\le j<k\le n} |\lambda_j - \lambda_k|^b \exp\left\{-\sum_{r=1}^{n} \lambda_r^2/2\right\};$$

in this context one often speaks of b-ensembles, or **beta-ensembles**. *See also* Wishart distribution; Tracy–Widom distribution; Marčenko–Pastur law; and Wigner's semicircle law. [G. Anderson *et al.*, *Introduction to Random Matrices*. Cambridge University Press, 2010]

random measure A random object (or element) $M(\cdot)$ taking values in a suitable space of measures, such that $M(B)$ is a non-negative r.v. for each measurable set B, and $M(\cdot)$ is countably additive, with $M(\emptyset) = 0$. When $M(B)$ is integer-valued for each B, the random measure is called a point process; and when $M(B_1), \ldots, M(B_n)$ are independent for all disjoint $B_1, \ldots, B_n$, the measure is said to be **independently scattered**, or to have independent increments, or to be **completely random**. An independently scattered point process such that $M(B)$ has a finite-mean Poisson distribution, for all B, is called a **Poisson random measure** (or process). [O. Kallenberg, *Random Measures*, 3rd edn. Academic Press, 1983; G. Last, Modern random measures: Palm theory and related models, in: *New Perspectives in Stochastic Geometry*, W. Kendall & I. Molchanov (eds). Oxford University Press, 2009]

random mosaic (or tessellation) (alias: **random cell model; random tiling**) Broadly, any locally finite random division of space, usually $\mathbb{R}^d$, into closed d-dimensional sets that have pairwise disjoint interiors, and cover $\mathbb{R}^d$ or some given subset of it. The subsets are commonly called cells (or crystals), and in many cases are required to be convex bounded polytopes. Such a random division may be constructed directly from a suitable random process of geometric objects, such as hyperplanes or hyperspheres. *See* Johnson–Mehl tessellation; Poisson–Voronoi mosaic. [J. Møller, *Lectures on Random Voronoi Tessellations*, Springer, 1994]

randomness (alias: **chance**) The property of unpredictability possessed by many realizable events, for which probability theory supplies a model (or description) in terms of random variables and stochastic processes. Randomness commonly arises from a lack of knowledge of some system, or the universe (in which case it is called an epistemological property); it may conceivably be intrinsic to the universe (in which case it is said to be an ontological property), but this possibility is contested by several authors. Related concepts include:

- pseudo-randomness, where a string of symbols (e.g. numbers) that has been deterministically generated will nevertheless pass any reasonable (statistical) test for randomness;
- algorithmic randomness, where a string of symbols is shorter than any set of instructions for reproducing it; *see* Kolmogorov complexity and Martin-Löf randomness.

[G. Chaitin, *Exploring Randomness*. Springer, 2001]

randomness deficiency Arises, and is formally defined, in the context of Kolmogorov complexity. Informally, the randomness deficiency $r(x)$ of $x \in S$, for some set S, is a measure of how far x is from acceptability as a random sample from S; where the selection is to be either uniform or, more generally, according to some distribution $\mathbb{P}$ on S. [M. Li & P. Vitányi, *An Introduction to Kolmogorov Complexity and its Applications*, 3rd edn. Springer, 2008]

random number *see* pseudo-random

random oracle Arises in cryptography, where cryptographic systems may be proved secure on the assumption that a random oracle exists; this is called the random oracle model. The oracle itself is a random function $X(q)$ from the set of binary strings $Q = \{0,1\}^m$ to the set $A = \{0,1\}^L$, where members of Q are interpreted as questions put to the oracle, and A comprises its answers. [J. Katz & Y. Lindell, *Introduction to Modern Cryptography*. Chapman & Hall/CRC Press, 2008]

random power series *see* random series

random process *see* stochastic process

random sample from a distribution Denotes a collection of i.i.d.r.vs $X_1, \ldots, X_n$; whose order statistics $X_{(1)} \le X_{(2)} \le \ldots \le X_{(n)}$ may be called an ordered sample. When each X_r is normally distributed, one speaks of a **normal sample**. [S. Silvey, *Statistical Inference*. Penguin, 1970; reprinted by Chapman & Hall, 1975]

random sample from a population Arises when a subset $s_{a_1}, \ldots, s_{a_r}$ of r symbols is selected (or drawn) randomly from a population $S_1, \ldots, S_n$ of n symbols. If each of the n^r possible (including repetitions) sequences of length r is equally likely, one speaks of a random sample with replacement. If each of the $\binom{n}{r}$ possible choices of distinct subsets of size $r \leq n$ is equally likely (or $n(n-1), \ldots, (n-r+1)$ sequences, when ordered), one speaks of a random sample without replacement. [W. Feller, *Introduction to Probability Theory and its Applications*, Vol. 1, 3rd edn. Wiley, 1968]

random scenery *see* random environment, and Kesten–Spitzer random walk in random scenery

random series In its basic form, the sum $S_n = \sum_{r=1}^{n} X_r$ of a sequence $X_1, X_2, \ldots,$ of r.vs; considerable interest arises in the question of convergence of S_n, as $n \to \infty$, for suitable random sequences $X_r, r \geq 1$. More generally, given a sequence of random processes (i.e. random functions) $X_1(t), X_2(t), \ldots, t \in \mathbb{R}$ (say), a random series of functions denotes $S_n(t) = \sum_{r=1}^{n} X_r(t)$. More narrowly, for a series of the form $\sum_n = \sum_{r=1}^{n} X_r c_r(t)$, where the $c_r(t)$ are commonly not random, one speaks of a **Rademacher series** when the X_r are i.i.d. Rademacher r.vs, or a normal series when the X_r are i.i.d. normally distributed, or a **Steinhaus series** when the X_r are i.i.d. uniform on [0, 1}, etc. When $c_r = t^r$, one has a random power series. *See* Billard's theorem and Paley–Zygmund theorem. [J.-P. Kahane, *Some Random Series of Functions*, 2nd edn. Cambridge University Press, 1985]

random speed process *see* random velocity model

random string A term with two principal meanings:
1. A sequence of symbols (i.e. elements of some set A) drawn according to some probability distribution. The set A is often called the alphabet, and its elements called letters; a string may be called a **word** in computational applications, or a **message** in cryptography and communication theory.
2. A sequence of adjacent points chosen randomly in some suitable space.

In both cases, strings may evolve randomly in time. [R. Durrett, On the shape of a random string, *Ann. Probab.*, 7(6), 1014–1027, 1979; T. Funaki, Random motion of strings, and related stochastic evolution equations, *Nagoya Math. J.*, 89, 129–193, 1983; A. Gairat *et al.*, Classification of Markov chains describing the evolution of random strings, *Russian Math. Surveys*, 50(2), 237–255, 1995]

random surfer model *see* PageRank Markov chain

random target lemma *see* Kemeny's constant

random tessellation *see* random mosaic

random time change theorem A theorem of P. Meyer, which asserts that for any member of a suitable class of multivariate point processes there exists a random time change transforming it to a collection of independent Poisson processes of rate 1. [T. Brown & M. Nair, A simple proof of the multivariate random time change theorem for point processes, *J. Appl. Probab.*, 25(1), 210–214, 1988]

random turn game Any game in which at a given stage the player entitled to make the next move is chosen at random. [Y. Peres *et al.*, Random-turn Hex and other selection games, *Amer. Math. Mon.*, 114(5), 373–387, 2007]

random variable (alias: **chance (or stochastic) variable; variate**) Arises in mathematical probability as a description of unpredictable measurements (or other numerical results) produced by experiments, together with a measure of their relative likelihoods.

A random variable is thus a real-valued function defined on the sample space Ω of outcomes, and is denoted by $X(\omega), \omega \in \Omega$. Furthermore, to fulfil its stated purpose, a random variable must be a function such that a probability distribution (on the values of $X(\cdot)$ in $\mathbb{R}$) can be defined; it is therefore also required of $X(\omega)$ that the set $A(x) = \{\omega : X(\omega) \leq x\}$ is an event as a subset of Ω. Hence, for a probability distribution $\mathbb{P}(\cdot)$ defined on the events in Ω, X has a distribution function $F(x) = \mathbb{P}(A(x)) = \mathbb{P}_X(X \leq x)$, where $\mathbb{P}_X(\cdot)$ is the probability distribution on $\mathbb{R}$ induced by X.

Note that a **complex-valued r.v.** Z is of the form $Z = X + iY$, where X and Y are real r.vs, and a (jointly distributed) set $V = (X_1, \ldots, X_n)$ of r.vs may be called a **random vector**. More formally, given a probability space $\{\Omega, \mathcal{F}, \mathbb{P}\}$, a random variable $X(\cdot)$ is a real-valued function on Ω that is measurable with respect to $\mathcal{F}$ (which is to say that $A(x) \in \mathcal{F}$ for all x).

More generally, if $X(\cdot)$ is $\mathcal{F}$-measurable, and takes values in a measurable space other than $\mathbb{R}$, one speaks of a random element. *See* continuous random variable; discrete distribution; and simple random variable. [W. Feller, *Introduction to Probability Theory and its Applications*, Vol. 2, 2nd edn. Wiley, 1971]

random vector *see* random variable

random velocity model (or process) Any one of several random processes in which the basic model envisages a particle in random motion that has a random velocity on each of a sequence of linear segments comprising its path (which is usually, but not in every model, continuous). One basic model is the telegraph process of M. Kac, in which a particle moves on a line at speed $+v$ or at speed $-v$, changing from one direction to the other at the instants of a Poisson process. More generally, the particle may change among a set S of speeds according to a Markov chain.

Yet more generally, let G be the generator of a process taking values in a suitable linear space S; the process obtained by switching from G to $-G$ at the instants of a Poisson process (or among a family of generators), is called a **random evolution**.

Another type of random velocity process arises when the members of a family of lines in the plane (say) are labelled with random velocities, and a particle moves along lines at the corresponding velocity; it may move from one line to another (at random) at intersections, or it may jump randomly from line to line. [R. Hersh, The birth of random evolutions, *Math. Intell.*, 25(1), 53–60, 2003]

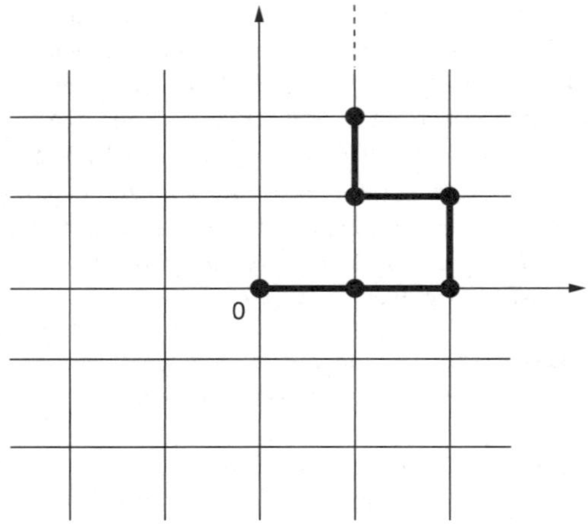

First five steps of a random walk on $\mathbb{Z}^2$.

random walk (alias: **r.w.**) Arises in almost every branch of applied probability as a model for e.g. the unpredictable motion of particles (or indices, or animals), or the fortune of a gambler making a sequence of i.i.d. wagers; this last example being the classical first appearance of random walk in probability theory.

The basic random walk or r.w., started from s_0, comprises the partial sums $S_n = s_0 + \sum_{r=1}^{n} X_r$ of a sequence $X_1, X_2, \ldots$ of i.i.d.r.vs that are often called the steps of the walk. Very often $s_0 = 0$, and $X_r \in \mathbb{Z}$, giving a r.w. on the integers. In particular, when $X_r \in \{-1, 1\}$ (or more generally $X_n \in \{-1, 0, 1\}$), one has **simple random walk** (also called **binomial random walk**). When $X_r \in \mathbb{Z}^d$ one has a r.w. on the integer lattice which may be called simple r.w. if $|S_{n+1} - S_n| = 1$ for all n. More generally, a r.w. (in discrete or continuous time) comprises a sequence of steps, in some suitable state space W, having the property that the step X_{n+1} from S_n to S_{n+1} is conditionally independent of its previous steps given that $S_n = s \in W$. This is a form of the Markov property, and a Markov chain can be seen as executing a r.w. on its state space. Examples include:

- $W = \mathbb{R}^d$ and $X_r \in \mathbb{R}^d$, and S_n is a r.w. in $\mathbb{R}^d$ that may be called a random flight.
- $W = G$, where G is a graph, and the walk steps from vertex to vertex according to a probability distribution on edges from its current vertex.
- $W = G$, where G is a group with an associative operation, and identity, and an inverse denoted by x^{-1}. For a given distribution $p(\cdot)$ on the elements of G a random walk is developed by left-multiplying the current state by an element randomly selected according to the distribution $p(\cdot)$; this is a (right-invariant) **random walk on a group**, with transition probabilities $p(j,k) = p(kj^{-1}), j, k \in G$
- The (left-invariant) random walk on a group, with transition probabilities $p(j,k) = p(j^{-1}k)$ is developed by right-multiplying the current state by an element of G independently selected according to the distn $p(\cdot)$.

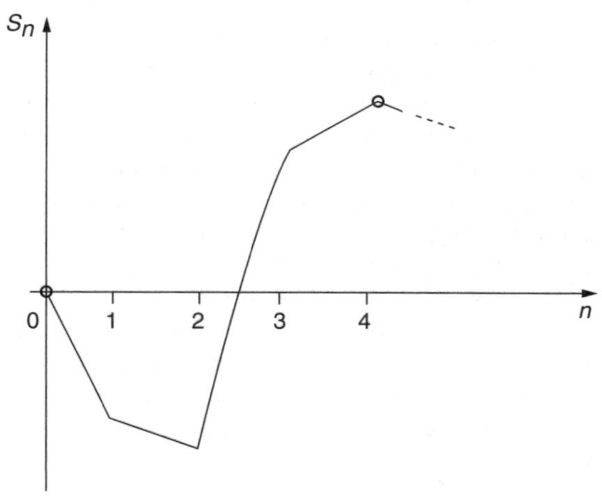

First four steps of a random walk on $\mathbb{R}$.

The term 'random walk' was introduced by K. Pearson in a letter to *Nature* in 1905 posing a r.w. problem; following Rayleigh's solution Pearson remarked that: '... in open country the most likely place to find a drunken man is somewhere near his starting point'.

Many extensions exist, *see* loop-erased random walk; gambler's ruin; random environment; self-interacting random walk; and Kesten–Spitzer random walk in random scenery. [G. Lawler & V. Limic, *Random Walk: A Modern Introduction*. Cambridge University Press, 2010; P. Diaconis, Random walk on groups, in *Groups St Andrews 2001 in Oxford*, C. Campbell *et al.* (eds). Cambridge University Press, 2003]

Rao's quadratic entropy *see* quadratic entropy

rapid variation A probability distribution function $F(x), x > 0$, is said to have a rapidly varying tail if

$$\lim_{\theta\to\infty} \frac{1-F(\theta x)}{1-F(\theta)} = \begin{cases} 0 & \text{for } x > 1 \\ \infty & \text{for } 0 < x < 1 \end{cases}.$$

[S. Resnick, *Extreme Values, Regular Variation, and Point Processes*. Springer, 1987]

rare event *see* law of rare events

rate distortion theory Arises in communication theory when a message is to be encoded for later decoding; in practice there are two competing costs in such a system, viz:

- the cost of encoding at a higher rate, i.e. committing more bits per source symbol to ensure error-free decoding;
- the cost of differences between the decoded (received) signal and the original message, i.e. distortion arising from an insufficiency of bits per symbol.

In such systems, each cost can be reduced at the expense of increasing the other, and rate distortion theory quantifies this trade-off. That is to say it shows the minimum distortion possible at a given rate; or, conversely, it seeks the minimum rate required for a given permissible distortion. In the latter case one speaks of lossy coding for a given fidelity criterion. The distortion may be either unavoidable, or accepted on grounds of economy, according to the nature of the system.
[R. McEleice, *Information and Coding*, 2nd edn. Cambridge University Press, 2002]

rate function *see* large deviations; Cramér's transform

ratio of uniforms method A technique for sampling from a distribution having a suitable density $f(x)$; it is implemented as a type of rejection method. In its simplest form, assume that $f(x) \le \min\{1, x^{-2}\}$ for $x \in \mathbb{R}$, and let U and V be independent and uniform on (0, 1). Then the r.v. $R = U/V$, conditional on the event $A = \left\{V \le \left(f(U/V)\right)^{\frac{1}{2}}\right\}$, has density $f(x)$. For example, conditional on $U \ge 0$ and $U^2 + V^2 \le 1$, the r.v. U/V has a Cauchy density. [A. Kinderman & J. Monahan, Computer generation of random variables using the ratio of uniform deviates, *ACM Trans. Math. Software*, 3(3), 257–260, 1977]

Ray–Knight compactification Arises in considering whether there is a modification of a Markov process on a suitable state space S that is regular, i.e. is right-continuous, and has the strong Markov property; under appropriate conditions this may be done by embedding S in some compact metrizable space, and this is the Ray–Knight compactification. [H. Kunita & T. Watanabe, Some theorems concerning resolvents over locally compact spaces, *Proc. 5th Berkeley Symp. Math. Stat. Probab.*, 2(2), 131–164, 1967]

Ray–Knight theorem Any of several theorems arising in the context of local time of Brownian motion $W(t)$ denoted by $L(x,t)$; and more generally similar theorems for a wider class of local times. They assert that, at suitable stopping times T, the process $L(x,T)$ is also a Markov process; such stopping times include: (a) an exponentially distributed T that is independent of $W(t)$; (b) the first-passage time of $W(t)$ to 1 (say); and (c) the first time that L exceeds 1. Further, the latter processes are squared Bessel processes. [D. Revuz, & M. Yor, *Continuous Martingales and Brownian Motion*, 3rd edn. Springer, 1999; R. Mansuy & M. Yor, *Aspects of Brownian Motion*. Springer, 2008]

Rayleigh distribution Arises as the distribution of the length $R = (X^2 + Y^2)^{1/2}$ of the random vector (X, Y) when X and Y are i.i.d. zero-mean normal r.vs, $N(0, \sigma^2)$. The density of R is

$$f(r) = \frac{r}{\sigma^2}\exp\left(-r^2/(2\sigma^2)\right), \quad r \ge 0,$$

with mean $\sigma\,(\pi/2)^{1/2}$ and variance $(4-\pi)\sigma^2/2$.

More generally $R = (X_1^2 + \cdots + X_n^2)^{1/2}$, for i.i.d. $N(0,\sigma^2)$r.vs $X_1, \ldots, X_n$, may be called a Rayleigh distn of order n. *See* Rician distribution. [M. Simon, *Probability Distributions Involving Gaussian Random Variables*. Springer, 2006].

Rayleigh flight *see* random flight

Rayleigh process Refers either to a certain Markov process on the space of real compact trees; or to a diffusion process with SDE: $\mathrm{d}X = (\alpha/x + \mu x)\mathrm{d}t + \sigma^2 > \mathrm{d}W(t)$, where $W(t)$ is the Wiener process, $\alpha \in \mathbb{R}$, $\mu \neq 0$, $\sigma^2 > 0$. [S. Evans *et al.*, Rayleigh processes, real trees, and root growth with re-grafting, *Probab. Theor. Rel. Fields*, 134, 81–126, 2006; V. Giorno *et al.*, Some remarks on the Rayleigh process, *J. Appl. Probab.*, 23(2), 398–408, 1986]

rcll process (alias: **RCLL, or Skorokhod, or regular, or càdlàg, or corlol process**) A process $X(t), t \in \mathbb{R}$ (or $t \in \mathbb{R}^+$, etc.), that for all t is continuous on the right at t, and whose left limit exists at t, i.e.:
(1) for any sequence $t_1, t_2, \ldots$ decreasing to t, $\lim_{n\to\infty} X(t_n) = X(t)$;
(2) for any sequence $s_1, s_2, \ldots$increasing to t, $\lim_{n\to\infty} X(s_n) = X(t-)$ exists.
The space of all such functions may be called **Skorokhod space**. This property arises both in seeking usefully regular versions of random processes in continuous time and also in the context of probability distribution functions. The terms ***R*-process** and **right process** are also used, but '*r*-process' has another meaning in physics, and D. Stroock has noted that 'right process' is ambiguous: 'the term "right" having been applied to various processes in much the same spirit as it used to be applied to certain whales'.

Processes that are left continuous with right limits may be called lcrl, or càglàd, or collor, or left processes. [O. Kallenberg, *Foundations of Modern Probability*, 2nd edn. Springer, 2002]

realization (alias: **sample (or observed) value**) Informally, a realization of a random variable X is that particular value (from its set of possible values in $\mathbb{R}$) that it is observed to take after the experiment (from which it arises) takes place. Formally, $X(\cdot)$ is a (measurable) function on some space Ω, and a realization of X is $X(\omega) = x$ for any fixed $\omega \in \Omega$.

Similarly, for a random process $X(t, .), t \in I$, defined on a space Ω, the collection of random variables $X(t, \omega)$, for any fixed $\omega \in \Omega$, is called a realization of the process, or a **trajectory**, or a **sample path**, or a path (or sample) function. [G. R. Grimmett & D. R. Stirzaker, *Probability and Random Processes*, 3rd edn. Oxford University Press, 2001]

reciprocal process (alias: **Bernstein (reciprocal) process**) One that has a conditional independence property that generalizes the Markov property. Formally, the random process $X(t), 0 \leq t \leq 1$, is a Bernstein or reciprocal process if for all $u \in (s, t)$ and all $v \in [0, s) \cup (t, 1], X(u)$ and $X(v)$ are conditionally independent given $X(s)$ and $X(t)$, for any $0 \leq s \leq t \leq 1$. It is the special case of a (Lévy) Markov random field in one dimension, and any Markov process $X(t)$ is reciprocal. [S. Bernstein, Sur les liaisons entre les grandeurs aléatoires, *Proc. Int. Cong. Math. Zürich*, 1, 288–309, 1932]

record value Arises in considering a wide variety of random sequences, e.g. financial and economic indexes, weather data, reliability testing, etc. For the usual basic form, let $X_1, X_2, X_3, \ldots$ be a sequence of i.i.d.r.vs; then X_n is said to be an upper (resp. lower) record value (at the record time n) if $X_j < X_n$ (resp. $X_j > X_n$) for all $j = 1, 2, \ldots, n-1$. (Here X_1 may be defined by convention either as not a record value or as both an upper and a lower record value.) More generally, X_n is an upper (resp. lower) k-record value if $X_r \geq X_n$ (resp. $X_r \leq X_n$) for exactly k of $X_1, \ldots, X_n$; the case $k = 1$ gives the simple record values above.

When X_1, has a density, records are a.s. not repeated; when X_1 is e.g. integer valued they may be repeated with positive probability; in this case a weak (upper) record value X_n is such that $X_n \geq X_j$ for $j < n$. *See* Ignatov's theorem and ladder process. [V. Nevzorov, *Records: Mathematical Theory*. American Mathematical Society, 2001]

rectangular distribution *see* uniform distribution

recurrence (alias: **persistence**) A concept arising in physical systems that are evolving randomly, where it is often important to know whether the initial state of the process ever recurs; and, if so, how often, and with what distribution of the time to do so (which is often called the **recurrence time** or interval). In the non-random case the first question is answered by the Poincaré recurrence theorem; for the stochastic case in its simplest formulation, the random process $X_n, n \in \mathbb{Z}$, is assumed to have a discrete state space S, and there are various analogues of Poincaré's theorem. For example, when X_n is stationary one has Kac's lemma (or Kac's ergodic theorem), asserting that the expected return time of a state s (i.e. mean recurrence time), or of a finite sequence of states **s**, is the reciprocal of the probability of s, or of **s**. Very often X_n is a Markov chain and then a state is said to be recurrent (or persistent) if the chain started in s returns to s in a finite number of steps with probability 1, and thus returns to s infinitely often; otherwise the state is called transient. If the mean recurrence time of s is finite, then s is called **positive** (or **non-null**); if the mean recurrence time is infinite (but s is not transient) then s may be called either **recurrent null** or **persistent null**. When the state space is uncountable the same underlying ideas may be formalized in several different ways, e.g. for a random walk $S_n = \sum_{r=1}^{n} X_r$ having i.i.d. steps X_r with density $f(x)$, the state $s \in \mathbb{R}$ is called a **possible value** if for any $h > 0$, there is some $n < \infty$ such that $\mathbb{P}(|S_n - s| < h) > 0$; then s is called recurrent if for any $h > 0, \mathbb{P}(|S_n - s| < h$ infinitely often$) = 1$. (*See* Chung–Fuchs theorem.) More generally, for a Markov chain on a general state space, *see* Harris recurrence. Likewise, for a Markov chain $X(t)$ in continuous time with a countable state space, the recurrence (or first return) time of s is $T_s = \inf(t > J_s : X(t) = s \,|\, X(0) = s)$, where $J_s = \inf(t > 0 : X(t) \neq s \,|\, X(0) = s)$, and the above definition for recurrence on an infinite state space likewise extends to diffusions.

In the related area of renewal theory, the intervals between renewals may be called **recurrence times**, while the time until the renewal next after t and that the most recent renewal, may be called the **forward and backward recurrence times** respectively. [W. Anderson, *Continuous-time Markov Chains*. Springer, 1991; P. Baxendale, T.E. Harris's contributions to recurrent Markov process and stochastic flows, *Ann. Probab.*, 39(2), 417–428, 2011; R. Bhattacharya, Criteria for recurrence and existence of invariant measures for multidimensional diffusions, *Ann. Probab.*, 6(4), 541–553, 1978]

recurrent event process A term used in statistics, in its everyday sense, to distinguish e.g. reliability models in which a component has a random lifetime and then fails just once, from e.g. accident or medical models in which entities such as cars and people may experience the same type of event repeatedly at random intervals.

The term has also been used to denote an attribute or property A that may arise at random times in a given sequence $A_r, r \geq 1$, of events, and such that the occurrence times of this attribute form a discrete renewal process. *See also* regenerative process and renewal theory. [W. Feller, *Introduction to Probability Theory and its Applications*, Vol. 1, 3rd edn. Wiley, 1968]

red and black *see* bold play

redundancy In the sense of unnecessary surplus, this term has two meanings in communication theory:

1. If a source S (or more extensively, a language) has alphabet A, and (Shannon) entropy H, then the redundancy of S (or the language), using A is $R = 1 - H/\log_2|A|$.
2. If a suitable memoryless source S, having entropy H and distribution $\mathbf{p}$ on its alphabet A, is encoded using a uniquely decipherable symbol code $c(x), x \in A$, then the redundancy of the code is $r = \sum_{x \in A} p_x c(x) - H/\log_2|A|$.

The ratio $H/\log_2|A|$ is sometimes called the **optimal compression coefficient** (it was called 'relative entropy' by Shannon, but this term usually now denotes Kullback–Leibler divergence).

This redundancy (as a suitable minimax redundancy over a family of distributions) arises in the redundancy-capacity theorem, connecting universal coding and channel capacity. More generally, one speaks of **redundancy in a channel** whenever the input distribution fails to achieve the channel capacity. [T. Cover & J. Thomas, *Elements of Information Theory*, 2nd edn. Wiley, 2006]

redundancy-capacity theorem *see* redundancy

Reed–Frost epidemic model A classic discrete-time model for a simple epidemic in a population of X_n susceptible, Y_n infectious, and $a - X_n - Y_n$ immune individuals at time n. For some $p \in (0, 1)$ given X_n and Y_n, X_{n+1} is a binomial r.v. with parameters X_n and p^{Y_n}, and $Y_{n+1} = X_n - X_{n+1}$; the epidemic stops when there are no infectives. Because of its method of construction; this is called a **chain-binomial model**. [D. Daley & J. Gani, *Epidemic Modelling*. Cambridge University Press, 1999]

reference space *see* sample space

reflected process *see* reflecting barrier

reflected Wiener process *see* Brownian motion

reflecting barrier (or boundary) Arises naturally in describing real-world processes whose paths, evolving in some region D (and seen or interpreted as the motion of a particle), may arrive at a point a in the boundary ∂D of D, which boundary they are not permitted to cross. For example, the contents of a dam cannot exceed its capacity or fall below 0. If the path does not stop at a (see absorbing), but is returned to D according to some rule, then the barrier is said to be retaining, or of reflecting type and the process a **reflected process**. The return rule may involve jumps into D, and it may include time spent at a. For example, a simple random walk on the non-negative

integers may have a reflecting barrier at 0 with transitions from 0 according to one of these probability distributions:
1. $p_{01} = 1$;
2. $p_{00} = q, p_{01} = 1 - q$;
3. $p_{0r} = q_r, r = 0, 1, 2, \ldots$, where $\sum_r q_r = 1$.
If a particle spends no time at a, one speaks of fast reflection; and, roughly speaking, if the incident and subsequent directions of the particle follow the rules of reflected light one has the usual simple reflecting boundary, otherwise one has oblique reflection. If a particle spends time at a one speaks of slow reflection, and there are several types, e.g. an elastic boundary, which may support a number of reflections but then traps the particle; or a **sticky boundary,** which holds the particle according to some random process between returns to the interior of D. Formally, these and other types are distinguished by the nature of the local time process at a; and they may be expressed in terms of the density function of the process and its derivatives at a, or by other functionals of the process. [S. Karlin & H. Taylor, *A Second Course in Stochastic Processes*. Academic Press, 1981]

reflection principle Arises naturally in considering simple symmetric random walk, and more generally, any symmetric r.w., and also for the Wiener process. Its basic form is this: let $X(t)$ denote either a symmetric r.w. for $t = 0, 1, 2, \ldots$, or the Wiener process (Brownian motion) for $t \in [0, \infty)$, and let T be a stopping time for $X(t)$. (Typically, T is the first-passage time of X to $a > 0$.) Then $X(t), t \geq 0$, has the same joint distributions as

$$Y(t) = X(t \wedge T) - (X(t) - X(t \wedge T)) = \begin{cases} X(t), & t \leq T \\ 2X(T) - X(t), & t \geq T \end{cases}, \quad t \geq 0.$$

This follows from the observation that for $t > T$, by symmetry, $X(t) - X(T)$ has the same distributions as $X(T) - X(t)$, together with the strong Markov property of these process. The name 'reflection principle' arises from the more informal observation that each path of $X(t) - X(T), t \geq T$ reflected in the level $y = X(T)$ gives a path of $-\,(X(t) - X(T))$ having the same probability, and vice versa.

For the Wiener process $X(t)$, defining $M(t) = \sup_{s \leq t} X(s)$, it follows that for $0 < a > b, \mathbb{P}(M(t) > a) = 2\mathbb{P}(X(t) > a)$, and $\mathbb{P}(M(t) > a, X(t) < b) = \mathbb{P}(X(t) > 2b - a)$, both of which are sometimes referred to as the reflection principle. Similar but more complicated expressions hold for symmetric r.w.

This reflection principle is effectively a special case of the **method of images** employed in solving related

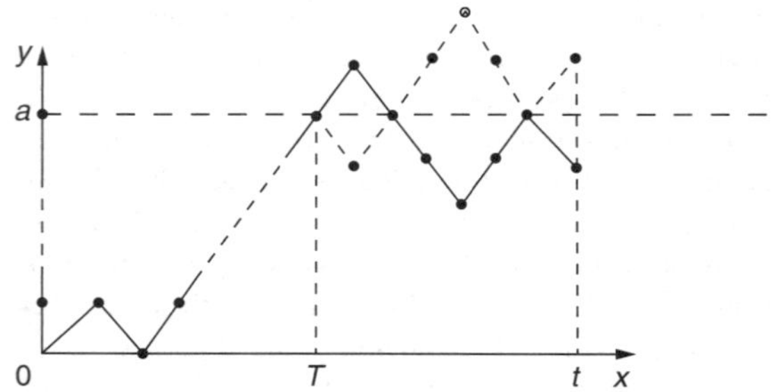

Reflection for a random walk. The solid line is the path of the walk; the dashed line is the reflection in *y* = *a* of the part of the walk after its first visit to *a*.

differential equations and difference equations, which amounts to writing the solution as a suitable linear superposition of solutions that can be interpreted as an image of some fundamental solution reflected or inverted in a boundary; sometimes called **Kelvin's method of images**. [W. Feller, *Introduction to Probability Theory and its Applications*, Vol. 1, 3rd edn. Wiley, 1968; Vol. 2, 2nd edn. Wiley, 1971]

regeneration epoch (or point) *see* renewal

regenerative phenomenon (alias: **p-function theory; regenerative event**) Arises in considering a continuous-time Markov process $X(t), t \geq 0$, on a state space S, where $X(0) = s$, and $\mathbb{P}(X(t) = s \mid X(0) = s) = p(s)$. Define $R(t)$ to be the indicator of the event that $X(t) = s$, taking the value $R(t) = 1$ if $X(t) = s$, and $R(t) = 0$ otherwise. Then for any $0 < t_1 < t_2 < \cdots < t_n$,

$$\mathbb{P}(R(t_r) = 1, 1 \leq r \leq n) = p(t_1)p(t_2 - t_1)\ldots p(t_n - t_{n-1}).$$

Any random process $R(t)$ satisfying this equation is said to be a regenerative phenomenon with $p(t)$ as its p function, which is called standard if $\lim_{t \downarrow 0} p(t) = 1$. [J. F. C. Kingman, Progress and problems in the theory of regenerative phenomena, *Bull Lond. Math. Soc.*, 38(6), 881–896, 2006; J. F. C. Kingman, *Regenerative Phenomena*. Wiley, 1972]

regenerative process Informally, a random process $X(t), t \geq 0$, that evolves through a sequence of independent cycles (or tours), that are also identically distributed except (perhaps) for the first. Thus the beginning of each cycle can be seen as a point at which the process regenerates itself independently of its past. A classic example is supplied by the basic time-homogeneous queue (i.e. $GI/G/s$ in Kendall's notation) in which a natural cycle begins when an arriving customer finds all s servers free, and the length of a cycle is the resulting busy period of the queue added to the subsequent idle period.

Formally, let $T_n, n \geq 0$ be a (possibly general, or delayed) renewal process, on the same probability space as $X(t)$. Then $X(t)$ is said to be regenerative on the regeneration points T_n if the segments $S_n = (T_{n+1} - T_n, \{X(T_n + s), 0 \leq s < T_{n+1} - T_n\}), n \geq 0$, are i.i.d. Very commonly the renewal times T_n are determined by $X(t)$ itself, being e.g. visits to some special set of states; the possible values of $X(t)$ at regeneration points may be called the **homing set**. It is usual to consider the case when $\mathbb{E}(T_{n+1} - T_n) < \infty$.

Such a process can be seen as a generalization of the recurrent event processes associated with discrete renewal theory, and further generalizations arise with the names loosely regenerative, and wide-sense regenerative. More weakly, a process $X(t)$ is said to be **crudely regenerative** at $Z > 0$ if, for all $t \geq 0, \mathbb{E}(X(Z+t)\,|\,Z) = \mathbb{E}X(t) < \infty$ (where Z and $X(t)$ are both adapted to the same filtration $\mathcal{F}_t$).

So-called regeneration theorems characterize the properties of $W(t) = \int_0^t f(X(s))\mathrm{d}s$ for suitable functions $f(\cdot)$ of a regenerative process $X(t)$ as $t \to \infty$, e.g. on weak conditions it is shown by W. Smith that $\mathbb{E}W(t) = At + \mathrm{o}(t)$ and the explicit representation of A is called **Smith's formula**. [W.L. Smith, Regenerative stochastic processes, *Proc. Roy. Soc. Lond. A*, 232(1188), 6–31, 1955; R. Serfozo, Applications of the key renewal theorem: crudely regenerative processes, *J. Appl. Probab.*, 29(2), 384–395, 1992]

regression dependence A measure of dependence between jointly distributed random variables X and Y, thus: Y is said to be positively regression dependent on X if $\mathbb{P}(Y \leq y\,|\,X = x)$ is non-increasing in x. (The expression Y is stochastically increasing on X denotes the same property.) Likewise, if $\mathbb{P}(Y \leq y\,|\,X = x)$ is non-decreasing in x then Y is negatively regression dependent (or stochastically decreasing on X). *See* dependence and likelihood ratio dependence. [D. Mari & S. Kotz, *Correlation and Dependence*. Imperial College Press, 2001]

regular *see* harmonic

regular boundary point Arises in considering a diffusion process on an interval I of $\mathbb{R}$; broadly speaking, an end point x of I is called regular if the process can both leave I at x, and enter I from x, in the sense that for any point of I (arbitrarily close to x) the process can reach any other point of I, in finite time. The idea may be more formally defined using the speed function. In higher dimensions, for the Wiener process in a region D of $\mathbb{R}^d$, with boundary ∂D, a point $x \in \partial D$ is said to be regular if the process started at x hits D^{c} immediately. If all points of ∂D are regular then D is called a regular domain. A classic example of a non-regular domain is one with an inward-pointing Lebesgue thorn (or spine). [J. Doob, *Classical Potential Theory and its Probabilistic Counterpart.* Springer, 1984; L. Breiman, *Probability.* Addison-Wesley, 1968; reprinted by SIAM, 1992; A. Knopp, Regular boundary points in Markov chains, *Proc. Amer. Math. Soc.*, 17(2), 435–4401966]

regular conditional probability Arises when seeking to extend the elementary definition of conditional probability so that one may consider e.g. the conditional distribution and expectation of one random element given another. Given a probability space $(\Omega, \mathcal{F}, \mathbb{P})$ and a sub-σ-field $\mathcal{G} \subseteq \mathcal{F}$, a conditional probability function $\mathbb{P}_{\mathcal{G}}(A, \omega), A \in \mathcal{F}, \omega \in \Omega$, is defined by the conditional expectation of the indicator I_A of A given $\mathcal{G}$. But this is not necessarily a probability measure on $\mathcal{F}$ for all ω (or even for almost all ω), in which case it cannot be used in the usual integral calculations of conditional expectation. Therefore, one defines a regular conditional probability which is such a measure, and can be so used, as follows: $\mathbb{P}_{\mathcal{G}}(A, \omega) \in [0, 1]$ is a regular conditional probability if:

(i) for almost every $\omega \in \Omega, \mathbb{P}_{\mathcal{G}}(A, \omega)$ is a probability measure on $\mathcal{F}$;
(ii) for every $A \in \mathcal{F}, \mathbb{P}_{\mathcal{G}}(A, .)$ is $\mathcal{G}$-measurable;
(iii) for every $A \in \mathcal{F}$ and $B \in \mathcal{G}$, $\int_{\mathrm{B}} \mathbb{P}_{\mathcal{G}}(A, \omega)\mathbb{P}(d\omega) = \mathbb{P}(A \cap B)$.

For example, if Ω is a complete separable metric space, then such a $\mathbb{P}_{\mathcal{G}}(A, \omega)$ exists.
[L. Breiman, *Probability.* Addison-Wesley, 1968; reprinted by SIAM, 1992]

regular domain *see* regular boundary point

regular process Loosely and broadly, any process satisfying a regularity (or 'good behaviour') condition may be called regular, and several such conditions have given rise to the term. Examples include:

1. An rcll process may be called regular.
2. A Markov chain with transition matrix P is called regular if for some $n < \infty$ all the entries in P^n are positive.
3. A diffusion on an interval I is called regular if, for any points x and y in the interior of I, the first-passage time from x to y is finite with positive probability.

4. A Markov process may be called regular if the conditional probabilities in its definition have regular versions.
5. An honest process may be called regular.
6. An orderly point process may be called regular.

regular transition matrix *see* regular process

regular variation (alias: **Karamata theory**) A property of functions arising in the theory of extreme values of suitable sequences of random variables, and with other applications in probability. Informally, a regularly varying function $u(x), x > 0$, is asymptotic to a power-law function x^r, as $x \to \infty$, for some r. Formally, let $u(x) > 0, x > 0$, satisfy

$$\lim_{t\to\infty} \frac{u(tx)}{u(t)} = v(x) \le \infty, \quad \text{for all } x > 0,$$

for some function $v(x)$. Under weak conditions on $u(x)$ (such as monotonicity, or measurability), the function $v(x) = x^\rho$, for $-\infty \le \rho \le \infty$; then $u(x)$ is said to vary regularly at infinity (or to be regularly varying at infinity), and ρ is called the **exponent (or index) of variation**, or **tail index**. If $\rho = 0$, u is said to be **slowly varying**, and it has the so-called **Karamata integral representation**:

$$u(x) = c(x)\exp\{\int_1^x t^{-1}b(t)\mathrm{d}t\}, \quad x \ge 1,$$

for functions $b(x)$ and $c(x)$ such that as $x \to \infty, b(x) \to 0$ and $c(x) \to c > 0$. In the regularly varying case, $b(x) \to b \neq 0$. [S. Resnick, *Extreme Values, Regular Variation, and Point Processes*. Springer, 2008; N. Bingham *et al.*, *Regular Variation*. Cambridge University Press, 1989]

Reimer's theorem *see* van den Berg–Kesten inequality

reinforced processes (alias: **history-dependent process; self-interacting process; path-dependent process**) A class of random processes whose evolution at any time depends on its history up to that time, i.e. its current sample path. Classic examples include self-avoiding and self-interacting random walks. [R. Pemantle, A survey of random processes with reinforcement, *Probab. Surv.*, 4, 1–79, 2007]

rejection method *see* acceptance-rejection

relative entropy A term now used to denote the Kullback–Leibler divergence, though it was first introduced by Shannon to denote the quantity $H(X)/M(X)$, where $H(X)$ is the entropy of the r.v. X, and $M(X)$ is the maximum value of the entropies of the collection of r.vs having the same alphabet as X (or, more generally, the same support as X). Then $1 - H(X)/M(X)$ is the redundancy of X. [C. Shannon, A mathematical theory of communication, *Bell Syst. Tech. J.*, 27(3), 379–423, 1948]

relative frequency *see* frequency probability

relative pointwise distance (alias: L_∞ **distance**) A measure of the discrepancy (or divergence) between two probability distributions $\pi_r, q_r, r \ge 1$, given by

$$d(\pi, q) = \max_r \left|1 - \frac{\pi_r}{q_r}\right|,$$

a two-sided form of the so-called **separation distance**

$$d_s(\pi, q) = \max_r \left(1 - \frac{\pi_r}{q_r}\right),$$

and related to a form of the L_p distance, given by.

$$d_p(\pi, q) = \left\{\sum_r q_r \left|1 - \frac{\pi_r}{q_r}\right|^p\right\}^{1/p}.$$

It finds applications in considering the mixing rate of a Markov chain. [J. Fill & J. Kahn, Comparison inequalities and fastest-mixing Markov chains, *Ann. Appl. Probab.*, 23(5), 1778–1816, 2013; D. Aldous & P. Diaconis, Strong uniform times and finite random walks, *Adv. Appl. Math.*, 8, 69–97, 1987]

relative probability A term used in several different ways:

- In logical probability it denotes the confirmation function, i.e. the probability (or degree of belief) that some hypothesis is true given the evidence (both being statements from some propositional language).
- It sometimes denotes A. Rényi's axiomatization of probability theory in which conditional probability is the fundamental concept.
- It may refer to odds.
- It may be used in its everyday non-technical sense of a ratio of probabilities.

[A. Rényi, On a new axiomatic theory of probability, *Acta Math. Acad. Sci. Hungarica*, 6(3–4), 285–335, 1955]

relative randomness Arises in algorithmic (i.e. descriptive, or computational) randomness, when seeking to quantify the degree of randomness processed by a sequence x. For example, a sequence x that passes stronger (or more) tests for randomness than a sequence y, can be seen as having a higher degree of randomness, and is said to be relatively more random; or a sequence x that has a shortest description (in terms of a prefix-free computer program, *see* Kolmogorov complexity) longer than that of y, can be seen as relatively more random than y. [R. Downey & D. Hirschfeldt, *Algorithmic Randomness and Complexity*. Springer, 2010]

relative stability A sequence of r.vs $X_n, n \geq 1$, is called relatively stable if there exists a sequence of numbers $c_n, n \geq 1$, such that X_n/c_n converges to 1 in probability; then c_n may be called a norming sequence. More strongly, if the convergence is almost surely (or in rth mean) one speaks of a.s. (or rth mean) relative stability.

The sequence X_n often comprises the running maximum of a sequence $Z_n, n \geq 1$, or the partial sums $X_n = \sum_{r=1}^{n} Y_r$ of a sequence of i.i.d.r.vs $Y_n, n \geq 1$; in the latter case the distribution of Y_1, may be called relatively stable.

relaxation time Arises in the context of finite state-space ergodic Markov chains, where it supplies a bound on the rate of convergence of the chain to equilibrium (i.e. the rate of mixing). Formally, let P be a reversible transition matrix with eigenvalues λ_r such that $-1 \leq \lambda_{|s|} \leq \cdots \leq \lambda_2 < \lambda_1 = 1$, where $|s|$ is the size of the state space. Then the relaxation time is $(1 - \lambda_2)^{-1}$; where $1 - \lambda_2$ is called the **absolute spectral gap**. [D. Levin *et al.*, *Markov Chains and Mixing Times*. American Mathematical Society, 2009]

reliability chain *see* success run chain; two-state process

reliability function *see* structure function

reliability theory Arises in considering the performance (i.e. working, impaired, or failed) of a system S comprising a number n of components that may fail (or be impaired) after some random lifetime of operation. The analysis of such systems is strongly linked with the concept of ageing. The basic model consists of a collection of indicator random processes $X_r(t), 1 \leq r \leq n, t \geq 0$, where $X_r(t) = 1$ if the rth component C_r is working, and $X_r = 0$ otherwise; initially $X_r(0) = 1$, and then $X_r(t) = 0$ for $t \geq T_r$, where T_r is the random failure time of C_r. The vector $(X_1, \ldots, X_n)$ is called the **state process**, and the state of the system is some Boolean function (indicator random process) of the state vector $A(t) = \phi(X_1(t), \ldots, X_n(t)) \in \{0, 1\}$, where ϕ is called the structure function, and $A(t)$ is called the **availability fn**. The structure function is usually required to satisfy natural properties of monotonicity and coherency (e.g. better performance by components entails better performance by the system) and then the reliability (or system survivor function) is $R(t) = \mathbb{E}A(t) = \mathbb{P}(\inf\{s > 0 : A(s) = 0\} > t)$.

More generally, in non-binary (or non-Boolean) systems each component may degrade gradually with degradation level at time t denoted by $D_r(t)$, yielding a performance (or capacity) function (or process), $C(t) = \psi(D_1(t), \ldots, D_n(t))$, for a multilevel structure function ψ. Yet more generally, systems may be repaired, so that $D_r(t)$ is not necessarily monotone decreasing. [A. Høyland & M. Rausand, *System Reliability Theory*, 2nd edn. Wiley, 2004]

renewal equation *see* renewal theory

renewal function *see* renewal theory

renewal process *see* renewal theory

renewal reward process *see* renewal theory

renewal theory Arises naturally in (and is named after) a classic model for the replacement (i.e. renewal), immediately upon their failure, of a sequence of components having i.i.d. random lifetimes $X_r, r \geq 1$. Then the nth renewal takes place at time $S_n = \sum_{r=1}^{n} X_r$, and at time $t > 0$ the number of renewals that have occurred up to (and including) t is denoted by $N(t)$. Thus $N(t) = \max\{n : S_n \leq t\}$, and $N(t)$ is called a **renewal process**; many authors also call the sequence $X_1, X_2, \ldots$, a renewal process. Note that to avoid tiresome trivialities it is usual to require that $\mathbb{P}(X_r > 0) = 1$; that is, no new component is broken on installation. Also, it is often the case that $\mathbb{P}(X_1 < \infty) = 1$; otherwise (when the distribution of X_i is defective) we have a **terminating renewal process**. Sometimes the time of first renewal is not required to be 0, but occurs at a time X_0 that is different from X_1; this is called a **general** (or a **delayed** or **modified**) **renewal process**. In discrete time, a renewal process is also called a **recurrent event process.** For suitable X_0, it may be that $\mathbb{E}N(t) = ct$ for some constant c, and one speaks of a **stationary renewal process**. In all cases interest centres on the so-called **renewal fn** $m(t) = \mathbb{E}N(t)$; in particular on its asymptotic behaviour as $t \to \infty$. For example the **elementary renewal thm** asserts that $\lim_{t\to\infty} m(t)/t = (\mathbb{E}X_1)^{-1}$, where the RHS is taken as 0 when $\mathbb{E}X_1 = \infty$.

More generally, the **key renewal thm** (also called **Smith's renewal thm**) asserts that for a non-increasing, non-negative function $g(x)$ such that $\int_0^\infty g(t)\mathrm{d}t < \infty$, and non-arithmetic X_1, we have

$$\lim_{t\to\infty}\int_0^t g(t-u)\mathrm{d}m(u) = \frac{1}{\mathbb{E}X_1}\int_0^\infty g(t)\mathrm{d}t.$$

(*See also* Blackwell's theorem, to which it is equivalent.) A similar result holds in the arithmetic case.

The renewal function satisfies the **renewal eqn**,

$$m(x) = F(x) + \int_0^x m(x-t)\mathrm{d}F(t),$$

and more generally renewal theory comprises the study of renewal-type equations of the form

$$m(x) = G(x) + \int_0^x m(x-t)\mathrm{d}F(t), \quad x \geq 0,$$

where $G(x)$ is a uniformly bounded function. Such equations arise in many contexts other than simple renewal as described above. Renewal theory has been generalized in many directions, e.g.:

- when $X_{2n}, n \geq 1$ are i.i.d. with distn F, and $X_{2n-1}, n \geq 1$, are i.i.d. with distn G, one has an **alternating renewal process**;
- when $(X_n, R_n), n \geq 1$, is a sequence of jointly distributed i.i.d.r.vrs, then $C(t) = \sum_{r=1}^{N(t)} R_r$ is a **renewal reward** (or cumulative) **process**;
- when $X_n, n \geq 1$, are conditionally independent given a Markov chain that determines their successive distributions one has a Markov renewal process;
- for any interval $I \in \mathbb{R}$, the renewal function satisfies $m(I) = \sum_{r=0}^{\infty}\mathbb{P}(S_r \in I)$, and one speaks of m as a **renewal measure** on $\mathbb{R}^+$.

More generally, for X_n not necessarily positive, and any sequence $c_n, n \geq 1$, of non-negative numbers, the generalized renewal measure is

$$\mu(A) = \sum_{r=0}^{\infty} c_r\mathbb{P}(S_n \in A).$$

In particular, for $c_r = \frac{1}{r}$, one has **harmonic renewal measure.** [D. R. Cox, *Renewal Theory*. Methuen, 1962; W. Smith, Renewal theory and its ramifications, *J. Roy. Stat. Soc. B*, 20(2), 243–302, 1958]

Rényi axioms A term used to denote either Rényi's axioms for entropy, or Rényi's axioms for conditional probability (*see* relative probability), or Rényi's postulates for measures of dependence. [A. Rényi, On measures of entropy and information, *Proc. 4th Berkeley Symp. Math. Stat. Probab.*, 1, 547–561, 1961; A. Rényi, On a new axiomatic theory of probability, *Acta Math. Acad. Sci. Hungarica*, 6(3–4), 285–335, 1955; A. Rényi, On measures of dependence, *Acta Math. Acad. Sci. Hungarica*, 10(3–4), 441–451, 1959]

Rényi constant *see* Rényi parking constant

Rényi correlation *see* maximal correlation

Rényi distance (or divergence) (alias: ***a*-divergence; Rényi's information gain of order *a***) A measure of the distance (or divergence, or difference) between two probability distributions, arising from the Rényi entropy, and generalizing the Kullback–Leibler divergence. For $0 < a \neq 1$, and probability distributions $p_1, \ldots, p_n$ and $q_1, \ldots, q_n$, it is given by $R_a(p,q) = \frac{1}{a-1}\log_2\sum_{r=1}^{n} p_r^a q_r^{1-a}$.

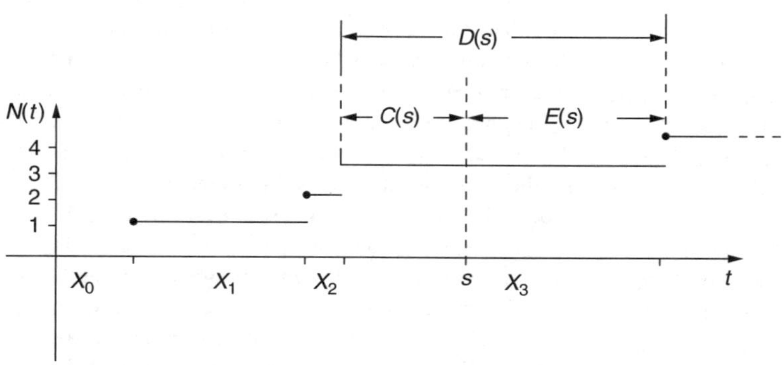

General renewal process. At time s, the current life is $C(s)$, the residual (or excess) life is $E(s)$, and the total life is $D(s) = C(s) + E(s)$. Here $N(s) = 3$.

As $a \to 1$, this yields the Kullback–Leibler divergence. [A. Rényi, On measures of entropy and information, *Proc. 4th Berkeley Symp. Math. Stat. Probab.*, 1, 547–561, 1961]

Rényi entropy *see* entropy; Rényi distance

Rényi–Epstein–Sobel theorem *see* Sukhatme's theorem

Rényi information *see* Rényi distance; entropy

Rényi parking (or packing) constant Arises in Rényi's one-dimensional packing (or parking or space-filling) problem, in which a sequence of cars of unit length attempt to park at random (uniformly) in the remaining space available in an interval of length $l > 1$. The expected total number of cars that succeed in finding space is denoted by $r(l)$, which satisfies $r(l) = \frac{2}{l-1}\int_0^{l-1} r(u)\mathrm{d}u$, and $\lim_{l\to\infty}\frac{r(l)}{l} \doteqdot 0.7475979\ldots$ is **Rényi's (parking) constant.** [H. Solomon & H. Weiner, A review of the packing problem, *Commun. Stat. Theor. Method.*, 15, 2571–260, 1986]

Rényi rate The entropy rate of a random process where the Rényi entropy is used rather than the Shannon entropy.

Rényi's theorem Any of several results. In probability the term commonly arises in connexion with characterizations of the Poisson process and associated corollaries. For the first, consider a renewal process whose points are independently deleted with probability $q \in (0, 1)$, and scaled in position by $1 - q$. The resulting process is a copy (or version) of the original renewal process iff it is a Poisson process. If this thinning and scaling is applied to an arbitrary renewal process, then as $q \to 1$ the resulting point process converges to a Poisson process. This is equivalent to the following, also called Rényi's theorem: let $X_r, r \geq 1$ be non-negative i.i.d.r.vs with unit mean, and N an independent geometric r.v. with parameter $p = 1 - q \in (0, 1)$. Then as $q \to 1$, the distribution of $S(p) = p\sum_{r=1}^{N} X_r$ converges to the exponential distn with parameter 1.

For the second, let π be a random countable subset of $\mathbb{R}^d$ such that for any finite union of boxes $A \subseteq \mathbb{R}^d, \mathbb{P}(\pi \cap A = \emptyset) = \exp\{-m(A)\}$, where for some $f(\mathbf{r}) \geq 0$, $m(A) = \int_A f(\mathbf{r})\mathrm{d}\mathbf{r} < \infty$. Then π is a Poisson process with intensity $f(\mathbf{r})$. [A. Rényi, A characterization of Poisson processes, *M.T.A. Mat. Kutato. Int. Közl.*, 1, 519–527, 1956; A. Rényi, Remarks on the Poisson process, *Studia Sci. Math. Hungarica*, 2, 119–123, 1967; J. F. C. Kingman, *Poisson Processes.* Oxford University Press, 1993]

repairman problem *see* machine interference problem

replicating portfolio (or strategy) *see* contingent claim

reproductive property A family of distributions $f(x, c)$ depending on a (possibly vector) parameter $c \in C$ is said to have the reproductive property if the sum of independent r.vs having distributions $f(x, c_1)$ and $f(x, c_2)$ has distribution $f(x, c_3)$ for some $c_3 \in C$.

reset time *see* access time

residence time *see* holding time

residual allocation model (alias: **random alms model; stick-breaking model**) A random discrete probability distribution $P_1, P_2, \ldots$, is called a residual allocation model if it can be written as $P_1 = X_1$, and $P_r = X_r(1 - X_{r-1}) \cdots (1 - X_1), r \geq 2$, where the X_r are i.i.d.r.vs, each distributed on (0,1). [P. Halmos, Random alms, *Ann. Math. Stat.*, 15(2), 182–189, 1944; J. Bertoin, *Random Fragmentation and Coagulation Processes.* Cambridge University Press, 2006]

residual waiting time *see* excess life

resolvent Arises in the context of continuous-time Markov processes that are Feller processes, with transition kernels (or Feller operators) denoted by $P_t, t \geq 0$. Then the resolvent is the family of operators $R(\lambda)$ such that

$$R(\lambda)f = \int_0^\infty \mathrm{e}^{-\lambda t}(P_t f)\mathrm{d}t, \quad \lambda > 0,$$

for all suitable functions f (e.g. continuous and vanishing at infinity). It is often convenient to determine the generator of the Feller semi-group through its resolvent. In particular, the so-called **resolvent eqn**,

$$R(\lambda) - R(\mu) = (\mu - \lambda)R(\lambda)R(\mu), \quad \lambda, \mu > 0,$$

corresponds to the semi-group property of P_t. When P_t is standard, $\lim_{\lambda\to\infty}\lambda R(\lambda) = I$, the identity. [O. Kallenberg, *Foundations of Modern Probability*, 2nd edn. Springer, 2002]

resurrected process A Markov process that is subject to killing (either by entering a cemetery set of absorbing states, or by an independent killing process) but having the property that it spends an exponentially distributed interval in the cemetery before being returned to its state space according to some so-called ascent distribution. [A. Pakes, Killing and resurrection of Markov processes, *Commun. Stat. Stoch. Models*, 13(2), 255–269, 1997]

retaining *see* reflecting

retrial *see* queueing theory

Reuter's criterion (or condition) *see* Feller–Reuter theorem

reverse martingale *see* backward martingale

reversibility (alias: **time-reversibility**) A random process $X(t)$ is said to be reversible (in equilibrium) if, for all $t_1, t_2, \ldots, t_n$, and T, the joint distributions of $X(t_1), \ldots, X(t_n)$ and $X(T - t_1), \ldots, X(T - t_n)$ are the same; $X(T - t)$ is called the reversed process. Such a process must be stationary, and the concept is often applied in the context of Markov chains, where it is characterized thus: an irreducible time-homogeneous Markov chain $X(n), n \in \mathbb{Z}^+$, with transition probabilities p_{jk} is reversible in equilibrium iff there exists $\boldsymbol{\pi} = (\pi_1, \pi_2, \ldots) > 0$ such that $\sum \pi_j = 1$, and $\pi_j p_{jk} = \pi_k p_{kj}$, for all j and k; and then $\boldsymbol{\pi}$ is the stationary distribution of X. These are the **detailed balance eqns**. In continuous time for an irreducible time-homogeneous chain with Q-matrix (q_{jk}) the condition is $\pi_j q_{jk} = \pi_k q_{kj}$, for all j, k.

Alternatively, reversibility is characterized by **Kolmogorov's criteria**, also called **Kolmogorov's cycle conditions** A stationary irreducible Markov chain is reversible if for any finite n, and sequence $i_1, \ldots, i_n$, its transition probabilities obey $p_{i_1 i_2} p_{i_2 i_3} \cdots p_{i_n i_1} = p_{i_1 i_n} \cdots p_{i_2 i_1}$; in continuous time the same result holds with the transition rates (Q-matrix entries) q_{ij} replacing p_{ij}.

Some authors call this property strong reversibility, allowing the term reversible to be used if there is a measure $\mathbf{m} = m_1, m_2, \ldots$ on the state space of X such that $m_i p_{ij} = m_j p_{ji}$, where $\mathbf{m}$ is not necessarily a probability distribution (i.e.; $\sum_i m_i = \infty$).

For a time-homogeneous diffusion $X \in \mathbb{R}^d$ with SDE $\mathrm{d}X(t) = \mu(X)\mathrm{d}t + \mathrm{d}W(t)$, where $W(t)$ is the Wiener process in d dimensions, **Kolmogorov's thm** asserts that X is reversible in equilibrium iff the vector μ is the gradient of some scalar field (which is to say that μ is a conservative vector field). In the context of stationary time-series, any stationary Gaussian process is reversible in equilibrium; furthermore if a causal autoregressive moving-average process with i.i.d. innovations (or noise) is reversible then it is Gaussian. This may be called **Weiss's thm.** [F, Kelly, *Reversibility and Stochastic Networks*, Wiley, 1979; reprinted by Cambridge University Press, 2011; A. N. Kolmogorov, Zur Umkehrbarkeit der statistischen Naturgesetze, *Math. Annalen*, 113, 766–772, 1937]

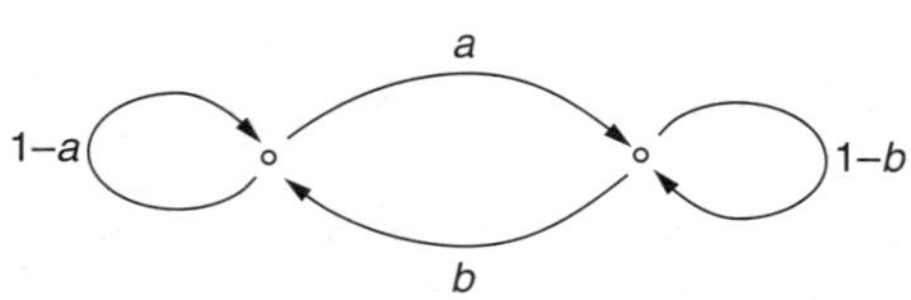

Transition diagram for 2-state Markov chain. When $a + b > 0$ the chain is irreducible and reversible in equilibrium.

Rice's distribution *see* Rician distribution

Rice's formula Arises in considering the rate (or intensity) $r(l)$ of crossings of a fixed level l by a suitable continuous stationary random process $X(t) \in \mathbb{R}, t \in \mathbb{R}$, that is mean-square differentiable in t. A standard form is this:

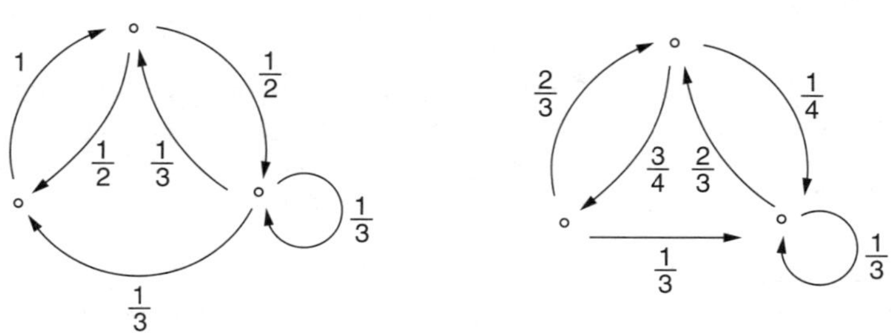

Left: **Diagram for 3-state Markov chain.** ***Right:*** **Diagram for this chain reversed. It is not reversible in equilibrium.**

$$r(l) = \int |y| f(l, y) \mathrm{d}y,$$

where $f(x, \dot{x})$ is the joint density of $X(t)$ and its derivative $\dot{X}(t)$. The term also denotes various generalizations and special cases of this result; perhaps most often when $X(t)$ is Gaussian with autocorrelation function $c(t), t \in \mathbb{R}$, and then Rice's formula is

$$r(0) = \frac{1}{\pi}\{-c''(0)\}^{1/2}.$$

[S. Rice, Mathematical analysis of random noise, in *Selected Papers on Noise and Stochastic Processes*, N. Wax (ed.). Dover, 1954; reprinted 2003]

Richardson model *see* Eden–Richardson model

Rician distribution Arises in signalling theory as the distribution of $X = (Y^2 + Z^2)^{\frac{1}{2}}$, where Y and Z are independent normal r.vs having densities $N(a, \sigma^2)$ and $N(b, \sigma^2)$ respectively. Then Rice's distribution has density

$$f(x) = \frac{x}{\sigma^2} I_0\left(\frac{xc}{\sigma^2}\right) \exp\left\{-\frac{(x^2 + c^2)}{2\sigma^2}\right\},$$

where $c = \sqrt{a^2 + b^2}$, and I_0 is a Bessel function (modified, of the first kind, and of order zero).

More generally, the square root of the sum of n squared independent normal r.vs having the densities $N(a_r, \sigma^2), 1 \le r \le n$, has the generalized Rayleigh distribution

$$f(x) = \frac{c}{\sigma^2}\left(\frac{x}{c}\right)^{n/2} I_{\frac{1}{2}(n-2)}\left(\frac{xc}{\sigma^2}\right) \exp\left\{-\frac{(x^2 + c^2)}{2\sigma^2}\right\},$$

where $c = +(a_1^2 + \cdots + a_n^2)^{1/2}$. [M. Simon, *Probability Distributions Involving Gaussian Random Variables*. Springer, 2006]

Riemann–Liouville process For $a > 0$, this may be defined as

$$X(t) = \int_0^t (t - s)^{a - \frac{1}{2}} \mathrm{d}W(s), \quad t \ge 0,$$

where $W(s), s \ge 0$, is the Wiener process. It is a self-similar zero-mean a.s. continuous Gaussian process. The name arises from an alternative view of the definition as a Riemann–Liouville fractional integral of a white noise process. [X. Chen *et al.*, Large deviations for local times of fractional Brownian motion and Riemann-Liouville processes, *Ann. Probab.*, 39(2), 729–778, 2011]

Riemann zeta distribution *see* zeta distribution

Riesz–Bessel distribution A probability distribution on $\mathbb{R}^d$ having a characteristic function of the form $\phi(t) = \exp\{-c|t|^{2a}(1 + |t|^2)^b\}, t \in \mathbb{R}^d$, where $c \ge 0, 0 < a \le 1$, and $0 \le a + b \le 1$. [V. Anh *et al.*, Estimation and simulation of the Riesz-Bessel distribution, *Commun. Stat. Theor. Method.*, 34, 1881–1897, 2005]

right-averaging principle *see* Kemeney's constant

right process *see* rcll process

right spread order (alias: **excess wealth order**) An ordering of probability distributions that arises in seeking to compare their variability (or spread). Formally, for r.vs X and Y with distn fns F_X and F_Y, one says that Y is greater than X in the right spread ordering if, for all $0 < t < 1$,

$$\mathbb{E}\left[\left(Y - F_Y{}^{-1}(t)\right)^+\right] \ge \mathbb{E}\left[\left(X - F_X^{-1}(t)\right)^+\right];$$

or equivalently

$$\int_{F_Y^{-1}(t)}^{\infty} (1 - F_Y(x))dx \ge \int_{F_X^{-1}(t)}^{\infty} (1 - F_X(x))dx,$$

where $F_X^{-1}(t) = \inf\{x : F_X(x) \ge t\}$ is the quantile function. This property implies that var $Y \ge$ var X, and is implied when Y is greater than X in dispersive order. [M. Shaked & J. G. Shanthikumar, *Stochastic Orders*. Springer, 2007]

Rio's inequality Arises in the context of strongly mixing processes; the basic form is this: let X and Y be jointly distributed real random variables and define the quantile function $Q_X(t) = \inf\{x : \mathbb{P}(|X| > x) \le t\}$, with $Q_Y(t)$ defined similarly. Then the covariance $\mathrm{cov}(X, Y)$ satisfies

$$|\mathrm{cov}(X, Y)| \le 2\int_0^{2\alpha} Q_X(t) Q_Y(t) \mathrm{d}t,$$

where α is the strong mixing coefficient:

$$\alpha = \sup_{A, B} |\mathbb{P}(A \cap B) - \mathbb{P}(A)\mathbb{P}(B)| \le \frac{1}{4},$$

and the supremum is taken over all events A and B such that A and B are in the sigma-fields generated by X and Y, respectively. [E. Rio, Covariance inequalities for strongly mixing processes, *Ann. de L'I.H.P.*, Section B, 29(4), 587–597, 1993]

risk Broadly, a term arising when there is a possibility (i.e. non-zero probability) of the occurrence of an event that is in some way undesirable. It appears in the technical jargon of various disciplines with different (incompatible) meanings:

- In medicine a risk is commonly the probability of an event, e.g. infection.
- In actuarial and insurance mathematics a risk is a random variable; *see* risk process.
- In statistical decision theory risk is the expected value of the loss function; *see* also utility function.
- In financial mathematics risk is the variability (or spread) of some random variable (such as a stock price), which may be measured by any one of a number of risk measures of which the variance is the simplest. Associated usages include risk-free for a replicating portfolio; and risk-neutral to denote the no-arbitrage value of a contingent claim.
- In utility theory, **Bell's risk measure** of a financial gamble X is $R(X) = \mathbb{E}\exp\{-a(X - \mathbb{E}X)\}$, where the parameter a depends on the agent.

[D. E. Bell, One-switch utility functions and a measure of risk, *Manage. Sci.*, 34(12), 1416–1424, 1988]

risk measure Arises in financial mathematics, where the prices of assets may fall as well as rise; it is natural to seek a measure of the risk entailed in owning such an asset. Typically, assets are assumed to be random variables, and a risk measure $r(x)$ is defined on the space of such r.vs, and required to satisfy appropriate constraints. For example, $r(\cdot)$ is said to be a **coherent risk measure** if:

(a) $r(X) \geq r(Y)$ whenever $X \leq Y$;
(b) $r(X + c) = r(X) - c$ for constant c;
(c) $r(0) = 0$, and $r(cX) = cr(X)$ for $c > 0$;
(d) $r(X + Y) \leq r(X) + r(Y)$.

Under suitable conditions, such risk measures may correspond to deviation measures. [S. Rachev *et al.*, *A Probability Metrics Approach to Financial Risk Measures*. Wiley, 2011]

risk-neutral measure (alias: **(equivalent) martingale measure; no-arbitrage measure**) A concept of considerable importance in financial mathematics, defined thus: let $X(t)$ be a suitable random process (on a stochastic basis $(\Omega, \mathcal{F}, \mathcal{F}_t, \mathbb{P})$), then a risk-neutral (or martingale) measure $\mathbb{Q}$ is a probability measure on Ω such that $X(t)$ is a martingale on $(\Omega, \mathcal{F}, \mathcal{F}_t, \mathbb{Q})$, and the Radon–Nikodym derivative $\frac{d\mathbb{Q}}{d\mathbb{P}}$ exists. In general $\mathbb{Q}$ is not unique, yielding a class of so-called equivalent martingale measures. [N. Bingham & R. Kiesel, *Risk-neutral Valuation*, 2nd edn. Springer, 2004]

risk process (alias **Cramér–Lundberg (or v.v.) model; compound Poisson model**) Arises in insurance mathematics in considering the so-called collective risk problem (or model). Broadly, this seeks to describe the surplus $R(t)$ of money held at time t by an insurer issuing many policies over time; in particular one is interested in the probability of ruin (where ruin is defined as $R(t) < 0$).

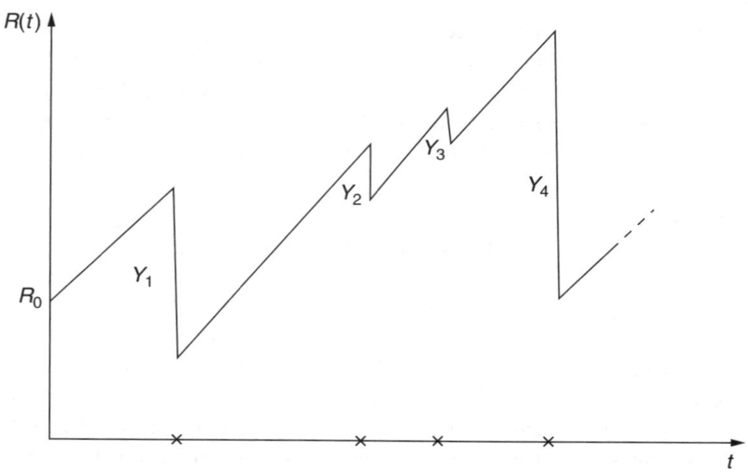

Risk process. The slope of the line segments is the rate of premium income, and the points x denote the Poisson process of successive claims, having sizes $Y_1, Y_2, \ldots$.

The surplus $R(t)$ is also called the **risk-reserve**, or **current capital**. The classic model for $R(t)$, also called the **Lundberg model**, or **Lundberg–Cramér–Andersen model**, assumes that

$$R(t) = R_0 - \sum_{r=1}^{N(t)} Y_r + ct,$$

where $\sum_1^{N(t)} Y_r$ is called the **claims process**, and $N(t)$ is the Poisson (or renewal) process counting the number of claims by time *t*, the i.i.d.r.vs Y_r denote the costs of the claims, R_0 is the initial reserve, and $c > 0$ is the (risk) premium rate. Let T_r be the interval between the *r*th claim and its predecessor. If there is a number *L*, called the **Lundberg exponent**, satisfying the **Lundberg eqn**, $\mathbb{E}\{\exp(L(Y_1 - cT_1))\} = 1$, then we have the **Lundberg inequality (or Lundberg bound):**

$$\mathbb{P}(\text{ultimate ruin}) = \mathbb{P}(R(t) < 0, \text{some } t < \infty) \leq \exp(-LR_0),$$

and the **Cramér–Lundberg approximation,**

$$\lim_{R_0 \to \infty} \{e^{LR_0} \mathbb{P}\,(\text{ultimate ruin})\} = C,$$

where *C* is the **Cramér–Lundberg constant (or exponent)**, or **adjustment coefficient**.
[D. Dickson, *Insurance Risk and Ruin*. Cambridge University Press, 2005]

Robbins improvement *see* Stirling's formula

Robbins–Munro method A classic example of an iterative stochastic approximation procedure (converging in probability) that seeks the root *r* of a suitable equation $m(x) = a$, where $m(x)$ is the unknown mean function of a family of uniformly bounded r.vs $Y(x)$, whose distributions are also unknown. [H. Robbins & S. Munro, A stochastic approximation method, *Ann. Math. Stat.*, 22(3), 400–407, 1951]

Robbins' problem A classic problem of optimal stopping, thus: Let $X_1, X_2, \ldots, X_n$ be i.i.d.r.vs, uniformly distributed on (0,1), that are inspected in order without recall. The problem is to determine the stopping rule that minimizes the expected rank of the r.v. selected at the stopping time. [F. Bruss, What is known about Robbins' problem?, *J. Appl. Probab.*, 42(1), 108–120, 2005]

Robbins–Siegmund theorem Arises in considering the almost sure convergence of random processes that are almost supermartingales, in a technical sense that they make precise. Formally, let $\mathcal{F}_n, n \geq 1$, be an increasing sequence of σ fields (i.e. a filtration) and let W_n, X_n, Y_n, Z_n, be $\mathcal{F}_n$-measurable non-negative r.vs, such that $\mathbb{E}(Z_{n+1}|\mathcal{F}_n| \leq (1 + W_n)Z_n + X_n - Y_n$. Then with probability 1, as $n \to \infty, Z_n$ and $\sum_{r=1}^{n} Y_r$ converge to finite limits on the event $\{\sum_{r=1}^{\infty} W_r < \infty$, $\sum_{r=1}^{\infty} X_r < \infty\}$. [H. Robbins & D. Siegmund, A convergence theorem for non-negative almost supermartingales, and some applications, in *Herbert Robbins Selected Papers*. Springer, 1985]

Rogozin's factorization *see* Pollaczek–Spitzer identity

Rosenblatt's theorem Any one of several results, but most often this CLT: Let $X_n, n \in \mathbb{Z}$, be a zero-mean strictly stationary process that is strongly mixing (i.e. α-mixing) then under suitable moment conditions it obeys a central limit theorem. [M. Rosenblatt, *Selected Works of Murray Rosenblatt*, R. Davis *et al.* (eds). Springer, 2011]

Rosenblatt's transformation In its original form, a transformation of an absolutely continuous *n*-dimensional distribution into the uniform distribution on the *n*-dimensional unit cube $(0, 1)^n$. More generally, there is a similar transformation for an arbitrary *n*-dimensional distribution (not necessarily having a joint density). One important corollary asserts that for a stationary irreducible aperiodic Markov chain $X_n, n \geq 0$, with countable state space *S*, there is a function $g(\cdot)$: $(0, 1)^{\mathbb{Z}} \to S$ such that for a sequence of i.i.d.r.vs $U_n, n \geq 0$, each uniform on (0,1), the sequence $Y_0 = X_0$, and $Y_n = g(U_n, U_{n-1}, \ldots), n \geq 1$, has the same joint distns as X_n. *See* probability integral transform and random mapping representation.
[M. Rosenblatt, Stationary Markov chains and independent random variables, *J. Math. Mech.*, 9, 945–950, 1960;
M. Rosenblatt, Remarks on a multivariate transformation, *Ann. Math. Stat.*, 23, 470–472, 1952]

Rosenthal inequality Any one of several closely related inequalities, of which the most often seen is this: Let $X_1, \ldots, X_n$ be independent zero-mean r.vs with $\mathbb{E}|X_k|^r < \infty$ for some $r > 2$, and set

$$R = \max\left\{\sum_{k=1}^{n} \mathbb{E}|X_k|^r, \left(\sum_{r=1}^{n} \mathbb{E}X_k^2\right)^{r/2}\right\}.$$

Then $\mathbb{E}|\sum_{r=1}^{n} X_k|^r \leq c(r)R$, where $c(r)$ is a constant depending only on *r*. [R. Ibragimov & Sh. Sharakhmetov, The exact constant in the Rosenthal inequality for random variables with mean zero, *Theor. Probab. Appl.*, 46(1), 127–132, 2002]

rough path theory A theory of integration enabling the integration of suitable differential forms against paths (i.e. functions) that are of finite p-variation for some $p \geq 1$. It has applications in stochastic integration where e.g. the Wiener process is of finite p-variation for all $p > 2$. [A. Lejay, An introduction to rough paths, in *Séminaire de Probabilités XXXVII, Lecture Notes Math.*, 1832, 1–59, 2003; A. Lejang & N. Victoir, On (p, q)-rough paths, *J. Diff. Eq.*, 225 (1), 103–133, 2006]

routing matrix *see* Q-matrix

R-transform Arises in free probability, where it is analogous to the ordinary cumulant generating function (see cumulants). Formally denote the Stieltjes (alias Cauchy) transform of a r.v. X on a suitable non-commutative (free) probability space, with moments $m_n, n \geq 1$, by

$$C(Z) = \frac{1}{Z} + \sum_{n=1}^{\infty} \frac{m_n}{Z^{n+1}}.$$

Then the R-transform $R_X(\cdot)$ is defined by $C\left(\frac{1}{Z} + R_X(Z)\right) = Z$; or equivalently, $\frac{1}{C(Z)} + R_X(C(Z)) = Z$.

Voilescu's thm asserts that r.vs X and Y are free iff $R_{X+Y}(Z) = R_X(Z) + R_Y(Z)$. Note that in the expansion $R_X(Z) = \sum_{r=1}^{\infty} k_r Z^{r-1}$, the coefficients k_r may be called free cumulants. [R. Speicher, Free probability theory, Chapter 22 in *The Oxford Handbook of Random Matrix Theory*, G. Akemann *et al.* (eds). Oxford University Press, 2011]

Rubin's construction *see* Athreya–Ney embedding

Ruelle probability cascade A model arising in statistical mechanics that comprises a cascade (i.e. iteratively generated sequence) of Poisson point processes and functionals of these. [E. Bolthausen & A.-S. Sznitman, On Ruelle's probability cascades, *Commun. Math. Phys.*, 197, 247–276, 1998]

ruin problem Arises in insurance mathematics when considering the risks run by an insurer. In one classic model the risk reserve process is

$$R(t) = u + ct - \sum_{i=1}^{N(t)} X_i, \quad t \geq 0,$$

where c is the rate at which premiums are received, and $N(t)$ is the point process of claims whose respective sizes are $X_i, i \geq 1$. Then the ruin time T is the instant at which $R(t)$ first becomes negative, whose distribution in general is given (in terms of a convolution integral) by what is called **Seal's formula**; and $\psi(u) = \mathbb{P}(T < \infty)$ is the probability of ruin, also called the **ruin fn**, whose evaluation is the basic ruin problem. In simple cases $\psi(u)$ has an explicit representation; more generally one seeks ruin probability approximations, e.g. by using the method of phases (or stages). (Among several others one has the **Beekman–Bowers**, the **Cramér–Lundberg**, and the **de Vylder approximations**.) Given a discount factor $r > 0$ (for the value of money) the expected discounted cost of ruin (insolvency) is

$$c(u) = \mathbb{E}\{e^{-rT} I\{T < \infty\} R(T)\};$$

or more generally one has e.g. the (**Powers–**)**Gerber–Shiu ruin penalty fn,** $k(u) = \mathbb{E}\{e^{-rT} I(T < \infty) g(R(T), R(T-))\}$, for a suitable function g. *See* risk process. [S. Asmussen & H. Albrecher, *Ruin Probabilities*, 2nd edn. World Scientific, 2010]

rumour model Any of several random processes describing the spread of a rumour in a population; such processes are related to epidemic models. Early versions include the **Daley–Kendall model**, where the population is perfectly mixing and comprises ignorants, spreaders and stiflers; a variant of this is the **Maki–Thompson model**. Later versions incorporate the social topology of the population, leading to rumour models on graphs (i.e. social networks); they may also allow scepticism and denial in the population. [V. Junior *et al.*, Rumour processes on $\mathbb{N}$, *J. Appl. Probab.*, 48(3), 624–636, 2011; F. Chierichetti *et al.*, Rumour spreading and graph conductance, in *Proc. 21st Annual ACM-SIAM Symposium on Discrete Algorithms*. SIAM, 2010, pp. 1657–1663]

Russian option *see* exotic option; Shepp–Shiryaev problem

Russo's formula (or lemma) (alias: **Margulis–Russo formula (or identity)**) Arises in the context of Boolean functions, with application in percolation and allied fields. Formally, suppose that f is a function from $\{0,1\}^n$ to $\{0,1\}$ that is monotone, i.e. it cannot decrease if any 0 in its set of arguments is replaced by 1 and let $\mathbf{X} = (X_1, \ldots, X_n)$ be i.i.d.r.vs with $\mathbb{P}(X_r = 1) = p = 1 - \mathbb{P}(X_r = 0)$. Define $m(p) = \mathbb{P}(f(\mathbf{X}) = 1)$, and for $\mathbf{x} \in \{0,1\}^n$ let $n(\mathbf{x})$ be the number of vectors $\mathbf{y} \in \{0,1\}^n$ that differ from $\mathbf{x}$ in exactly one argument and such that $f(\mathbf{y}) \neq f(\mathbf{x})$. Then Russo's formula asserts that

$$\frac{\mathrm{d}m}{\mathrm{d}p} = \mathbb{E}n(\mathbf{X}),$$

where $\mathbb{E}n(\mathbf{X})$ may be called the (total) influence of f, and denoted by $I(p,f)$. [G. R. Grimmett, *Probability on Graphs*. Cambridge University Press, 2010]

Russo–Seymour–Welsh theorem (or inequality) Arises in percolation theory, where it relates the probability of a path crossing a box to the probability that there exists a circuit in an associated annulus. [G. Grimmett, *Percolation*, 2nd edn. Springer, 1999]

Ryll–Nardzewski theorem *see* exchangeable

S

safety loading *see* premium principle

Saint Petersburg paradox (or problem) (alias: **Petersburg game**) Arises in considering a gamble (or lottery) such that the possession of a ticket yields a random prize X ducats whose distribution is

$$\mathbb{P}(X = 2^r) = 2^{-r}, \quad r = 1, 2, 3, \ldots.$$

The question, first posed in a letter from N. Bernoulli to P. de Montmort in 1713, asks what is a price for this ticket that is in some sense fair?

For a finite lottery the fair price of a ticket is often taken to be the expected value of the resulting prize, but here $\mathbb{E}X = \infty$, which is clearly absurd as a ticket price. Daniel Bernoulli discussed the problem in a paper published by the Imperial Academy of Petersburg in 1738, where he introduced moral expectation, essentially equivalent to utility, and observed that if your utility of money is of logarithmic type then your fair price for the ticket could be $\exp(\mathbb{E}\log X) < \infty$. However, this example does not resolve the paradox for general infinite lotteries such as the **super St. Petersburg game** Y, where $\mathbb{P}(Y = 2^{2^r}) = 2^{-r}$, $r \geq 1$, for which $\mathbb{E}\log Y = \infty$, and which yields a fair price if your utility is e.g. iterated-logarithmic.

Several further resolutions of the paradox have been proposed, most assuming that one can participate in an i.i.d. sequence of such lotteries, so that one can use appropriate limit theorems to fix a fair price (in some suitable sense). [D. Bernoulli, *Econometrica*, 22(1), 23–36, 1954, translated from the Latin by L. Sommer; T. Cover, On the St. Petersburg paradox, *Proc. IEEE International Symposium on Information Theory*, 2011, pp. 1758–1761; A. Martin-Löf, A limit theorem which clarifies the St. Petersburg problem, *J. Appl. Probab.*, 22(3), 634–643, 1985; O. Peters, The time resolution of the St Petersburg paradox, *Phil. Trans. Roy. Soc. A*, 369, 4913–4931, 2011]

saltus *see* jump

sample *see* random sample

sample space (alias: **universal set; reference space; set of outcomes**) Probability arises from ignorance of the actual state (or outcome) of a system S that may be in one of several states (or outcomes), e.g. S may be the weather tomorrow, or someone's DNA, or the face shown by a flipped coin, or the path of a molecule in a fluid, etc. The generic term for such a system is a (random) experiment, and the set of all possible (and indecomposable) outcomes of S is called the sample space and denoted by Ω. It is essential in defining a probability space.

The elements ω of Ω are called sample points, and are the domain values of a random variable defined on any associated probability space. Any ω that is a member of the σ-field of events (subsets of Ω) may be called an **elementary event**. [G. R. Grimmett & D. R. Stirzaker, *Probability and Random Processes*, 3rd edn. Oxford University Press, 2001]

sample (or observed) value *see* realization

sampling *see* pseudo-random

sampling formula A term used in two distinct contexts, thus:

1. Broadly, a multivariate distribution arising when a random sample is taken from a population of individuals of various types; in particular, Ewens sampling formula (*see* Pitman's sampling formula) giving the probability that, for a random sample of genes at a given locus, exactly a_r alleles appear exactly r times, $r \geq 1$.
2. Arises in the sampling theorem, denoting an inversion formula that reconstructs a band limited signal from suitable sample values.

sampling theorem (alias: **Shannon–Nyquist (or vice versa) theorem; Whittaker–Shannon (or v.v.) theorem**) Arises in communication theory, where it asserts that a signal $X(t), t \in \mathbb{R}$, whose Fourier transform contains no frequencies higher than B, can be reconstructed exactly by sampling its values at a sequence of suitable discrete times $t_n, n \in \mathbb{Z}$, where the minimum rate of sampling is two for each period of highest frequency. Furthermore, $X(t)$ is given by the **sampling formula**. C. Shannon's form is this:

$$X(t) = \sum_{n=-\infty}^{\infty} X\left(\frac{n}{2B}\right) \frac{\sin\{\pi(2Bt - n)\}}{\pi(2Bt - n)}.$$

Sampling theory originated with E. T. Whittaker, and there are many other sampling theorems, in many areas of mathematics. [A. Zayed, *Advances in Shannon's Sampling Theorem*. CRC Press, 1993]

Samuels' theorem Arises in the classical gambler's ruin problem when the two players have equal initial fortunes, where it asserts that the duration of the game is independent of who wins. The theorem was extended by J. Wendel, who also further showed that, for the n-dimensional Wiener process with constant drift, the interarrival times between successive concentric spheres are independent of the point at which the process hits the outermost sphere. [J. Wendel, An independence property of Brownian motion with drift, *Ann. Probab.*, 8(3), 600–601, 1980]

sandpile automaton *see* Abelian sandpile

sandpile model *see* Abelian sandpile

Sanov's theorem Arises in considering the empirical distributions

$$\hat{p}(A) = \frac{1}{n}\sum_{r=1}^{n} I(X_r \in A), \quad n \geq 1,$$

of a sequence $X_1, X_2, \ldots$ of i.i.d.r.vs, each having probability distribution $p(\cdot)$; where the theorem addresses the behaviour of the limiting probability (as $n \to \infty$) that $\hat{p}(\cdot)$ lies in a suitable set S of distributions, and also supplies an expression for the rate of convergence. For an appropriate choice of S it is an example of a large deviations principle, and it has applications in information theory. Formally, S is said to have the Sanov property if,

$$\lim_{n\to\infty} \frac{1}{n} \mathbb{P}(\hat{p} \in S) = -\inf_{\mathbb{Q}\in S} d(\mathbb{Q}; p),$$

where $d(\mathbb{Q}, p)$ is the Kullback–Leibler divergence. Any theorem giving conditions for this to hold is commonly called Sanov's theorem. [A. Dembo & O. Zeitouni, *Large Deviations*, 2nd edn. Springer, 1998]

Sargan density A probability density used in econometrics, and having the form: $f(x) \propto \exp(-a|x|)\sum_{r=0}^{s} c_r a^r |x|^r, x \in \mathbb{R}$, where $a > 0$, $c_r \geq 0$, and $c_0 = 1$. [Y. Tse, A note on Sargan densities, *J. Econ.*, 34, 349–354, 1987]

Sarmanov's distribution A bivariate density having the form: $f(x, y) = g(x)\, h(y)\{1 + au(x)\, v(y)\}$, where g and h are probability densities, and u and v are bounded functions such that $\int u(x)\, g(x)\, dx = \int v(x)\, h(x)\, \mathrm{d}x = 0$, and a is such that f is non-negative. [M. Shubina & M. Lee, On maximum attainable correlation and other measures of dependence for the Sarmonov family of bivariate distns, *Commun. Stat. Theor. Method.*, 35(5), 1031–1052, 2004]

Sato's theorem *see* self-decomposable

scale function (alias: **canonical scale**) Arises in considering a regular diffusion process $X(t) \in \mathbb{R}, t \geq 0$, on an interval I; there exists a strictly increasing continuous function $s(x), x \in [a, b] \subseteq I$, such that the probability that $X(t)$ started at x hits a before b is $\frac{s(b)-s(x)}{s(b)-s(a)}$. Then $s(x)$ is called the scale function, and in particular when $s(x) = \frac{b-x}{b-a}$ the diffusion X is said to be on its **natural scale**; it follows that $s(X(t))$ is a diffusion on its natural scale (and is a local martingale). The existence of such an $s(\cdot)$ may be called the **Feller–Dynkin thm**. For an Itô diffusion given by the SDE $\mathrm{d}X(t) = \mu(X)\mathrm{d}t + \sigma(X)\mathrm{d}W(t)$, the scale function is the solution $s(x)$ of

$$\frac{1}{2}\sigma^2(x)s'' + \mu(x)s' = 0.$$

For example, the drifting standard Wiener process $D(t)$ with drift m has $s(x) = \mathrm{e}^{-2mx}$, and $\exp(-2mD(t))$ is a martingale. *See* speed function. [R. Bass, *Stochastic Processes*. Cambridge University Press, 2011]

scale mixture Denotes a transformation of the r.v. X, having the form $Y = X/V$, where V is a positive r.v. that is independent of X. (*See* normal variance-mean mixture.) The concept extends to scale mixtures of a random vector in the natural way. [T. Eltoft *et al.*, Multivariate scale mixture of Gaussians modeling, in: Independent component analysis and blind signal separation, *Lecture Notes Comp. Sci.*, 3889, 799–806, 2006]

scale parameter *see* location parameter

scaling property *see* self-similar process

scenery reconstruction problem Arises from a random walk in a random scenery, and asks if one can determine (reconstruct) the full scenery, given only the scenery observed by the walker, but not the path taken. [F. den Hollander & J. Steif, Random walk in random scenery: a survey of recent results, *IMS Lecture Notes Dynam. Stoch.*, 48, 53–65, 2006]

Schassberger's theorem *see* method of phases

scheduling problem *see* allocation problem

Scheffé's theorem Arises in considering a sequence $f_1(x), f_2(x), \ldots$ of proper densities (i.e. $\int f_r(x)\mathrm{d}x = 1$) on $\mathbb{R}^n$, and its basic form is this: let $f(x)$ be a proper density; then as $n \to \infty$, $f_n(x) \to f(x)$ almost everywhere in $\mathbb{R}^n$ iff $\int |f_n(x) - f(x)|\mathrm{d}x \to 0$. [H. Scheffé, A useful convergence theorem for probability distributions, *Ann. Math. Stat.*, 13(3), 434–438, 1947]

Scher–Lax equation *see* continuous-time random walk

Schilder's theorem Arises in considering the large deviations (from the mean) of a suitably rescaled Wiener process $W(t)$ in n-dimensions. More formally, it supplies a large deviations principle in the form of bounds for the lim inf and lim sup of $\varepsilon^2 \log \mathbb{P}(\varepsilon W(t) \in A), 0 \le t \le 1$, as $\varepsilon \to 0$, for any Borel set A in the space of continuous functions in $\mathbb{R}^n$, over $[0, 1]$, started at the origin. [O. Kallenberg, *Foundations of Modern Probability*, 2nd edn. Springer, 2002]

Schnorr's theorem (alias: **Levin–Schnorr theorem**) Arises in descriptive (i.e. algorithmic) randomness where it establishes the equivalence of Martin-Löf randomness and Kolmogorov–Levin–Chaitin randomness (*see* Kolmogorov complexity) for infinite strings. [C. P. Schnorr, Process complexity and effective random tests, *J. Comp. Syst. Sci.*, 7(4), 376–388, 1973]

Schoenberg's theorem A term used in probability theory to denote either of the following assertions:

1. A function $\phi(\mathbf{t}), \mathbf{t} \in \mathbb{R}^n$ is a radial characteristic function (which is to say that it depends only on $\|\mathbf{t}\|$ iff it has the form $\phi(\mathbf{t}) = \int_0^\infty \Omega_n(\|\mathbf{t}\|y)\mathrm{d}F(y)$, for some distribution function F on $[0, \infty)$, where

$$\Omega_n(x) = \left(\frac{2}{x}\right)^{\frac{1}{2}n-1} \Gamma\left(\frac{n}{2}\right) J_{\frac{1}{2}n-1}(x), \quad x \ge 0,$$

and where J is a Bessel function. Note that $\Omega_n(\|\mathbf{t}\|)$ is the chf of a random unit vector uniformly distributed in $n-1$ dimensions

2. A characteristic function $\phi(t), t \in \mathbb{R}$ is such that $\phi(\|\mathbf{t}\|)$, $\mathbf{t} \in \mathbb{R}^n$, is a chf for all n iff it has the form $\phi(t) = \int_0^\infty e^{-t^2y^2/2}\mathrm{d}F(y)$, for some distribution function F.

[I. Schoenberg, Metric spaces and completely monotone functions, *Ann. Math.*, 39(4), 811–841, 1938]

Schramm–Loewner evolution *see* harmonic explorer; *see also* Schramm–Löwner evolution.

Schramm–Löwner evolution (alias: **SLE; stochastic Löwner evolution**)

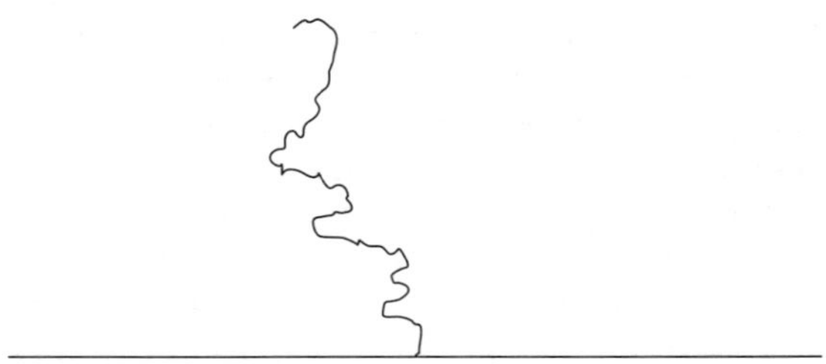

Chordal SLE.

Arises in considering suitably scaled limits of non-crossing random paths on planar lattices that appear in classic problems of statistical mechanics, e.g. percolation, the Ising model, diffusion limited aggregation, and self-avoiding random walks. An SLE is a random curve in either the upper half-plane $\mathbb{H}$ giving **chordal SLE**, or in the unit disc $\mathbb{U}$ giving **radial SLE**. The process is constructed through the stochastic (or Schramm) Loewner equation in a region D of the complex plane:

$$\frac{\partial g}{\partial t}(t, z) = \frac{2}{g(t, z) - W(Kt)},$$

which supplies an association between the driving Wiener process $W(Kt)$ on the boundary of D (being $\mathbb{H}$ or $\mathbb{U}$), and a simple random curve $g(t)$ that evolves from a point on the boundary of D, depending on the value of the parameter K. [I. Benjamin & O. Häggström (eds), *Selected works of Oded Schramm*, Vol. I. Springer, 2011]

Schuette–Nesbitt formula A generalization of **Jordan's identity** (or formula), thus: let $A_1, \ldots, A_n$ be n events, and let $S_r = \sum_i \mathbb{P}(A_{i_1} \cap A_{i_2} \cap \ldots \cap A_{i_r})$ be the sum of the probabilities that r given events $A_{i_1}, \ldots, A_{i_r}$ occur for all choices of $\mathbf{i} = (i_1, \ldots, i_r)$. The probability that exactly r of the n events occur is $p_{[r]}$. Then the Schuette–Nesbitt formula asserts that for arbitrary numbers $a_0, a_1, \ldots, a_n$

$$\sum_{r=0}^{n} a_r p_{[r]} = \sum_{r=0}^{n} S_r \sum_{j=0}^{r} \binom{r}{j} (-1)^{r-j} a_j = \sum_{r=0}^{n} S_r \Delta^r a_0,$$

where $\Delta a_k = a_{k+1} - a_k$ is the difference operator. *See also* inclusion-exclusion bounds. [H. Gerber, A proof of the

Schuette-Nesbitt formula of dependent events, *Actuarial Res. Clearing House*, 1, 9–10, 1979]

Schwarz inequality *see* Cauchy–Schwarz inequality

score function (alias: **informant; efficient score**) Arises in defining the Fisher information of a suitably differentiable density $f(x,\theta), x \in \mathbb{R}$, of a r.v. X, where $\theta \in \Theta$ is a parameter. Then the score (function) is $u(\theta, x) = \frac{\partial}{\partial\theta}\log f(\theta,x)$; in the case when θ is a vector of parameters the score is the gradient (wrt θ) of $\log f(\theta,x)$. In regular cases (when f is a nice function), $\mathbb{E}u(\theta,X) = 0$ and var $u(\theta,X) = I(\theta)$, where I is Fisher's information. When θ is a location parameter, $f(\theta,x) = f(x+\theta)$, and $u(\theta,x) = \frac{\partial}{\partial x}\log f(\theta,x)$, and so the function $f'(x)/f(x)$ is commonly called the score function of X; in the multivariate case the score of $f(\mathbf{x})$ is $\nabla \log f \equiv \text{grad} \log f$. [O. Johnson, *Information Theory and the Central Limit Theorem*. Imperial College Press, 2004]

scoring rule Arises in prediction and forecasting as a measure of the quality (or uncertainty) in a forecast. Formally, a scoring rule $S(x,\mathbb{Q})$ is the penalty (or loss) incurred by having chosen $\mathbb{Q}$ to be the distribution of some r.v. X having an unknown distribution, when x is the observed (realized) value of X. In applications a natural constraint is that when X has distribution $\mathbb{P}$, the expected score $s(\mathbb{P},\mathbb{Q}) = \mathbb{E}S(X,\mathbb{Q})$ is minimized over $\mathbb{Q}$ by $\mathbb{Q} = \mathbb{P}$, and then the rule $S(x,\mathbb{Q})$ is called proper. A basic example is the log score for a discrete distribution $\mathbb{Q} = (q_x), x \in \mathbb{Z}, S(x,\mathbb{Q}) = -\log q_x, x \in \mathbb{Z}$, which is proper by Gibbs inequality. [M. Parry *et al.*, Proper local scoring rules, *Ann. Stat.*, 40(1), 561–592, 2012]

Seal's formula *see* ruin problem

search problem Any of a class of problems in which an object T is located at random in some region R, and it is required to minimize the expected time (or the length of path) needed to find it. A classic example is **Bellman's linear search problem**, in which T is a random point of $\mathbb{R}$. More generally, two agents S and T, initially placed at random in R, seek to meet in the shortest expected time. Conversely, T may seek to evade S. [S. Alpern & S. Gal, *Theory of Search Games and Rendezvous*. Kluwer, 2003]

sech-squared distribution *see* hyperbolic secant distribution

second moment method *see* Paley–Zygmund inequality

second-order process *see* stationarity

second-tail distribution *see* integrated tail distribution

second theorem of extreme value theory *see* Balkema–De Haan–Pickands theorem

secrecy system Broadly, a mathematical model of some cryptographic (coding–decoding) process for preserving the secrecy of a message from an observer (or enemy) by means of encryption. More narrowly, the term commonly refers specifically to **Shannon's secrecy model** (or **cryptosystem**) and extensions of it; this is also called classical cryptography, or private key cryptography. The essential components comprise a message M (or plain text, i.e. a finite string of letters), encrypted by means of a key K (also a finite string of symbols) to yield a cryptogram $C = e(M,K)$; these, together with a decryption function $d(C,K) = M$, form the system. Note that Shannon's scheme assumes that the enemy can read C without effort, and that M and K are seen as essentially random. The conditional entropies $H(K \mid C)$ and $H(M \mid C)$ are the key equivocation and message equivocation respectively. A system has perfect secrecy if $H(M) = H(M \mid C)$, and for this it is necessary that $H(K) \geq H(M)$, called **Shannon's secrecy bound** (or **thm**), where H denotes entropy; in the uniform case this requires that there be at least as many keys as there are messages.

Finally, Shannon notes that an effective secrecy system must have a high work characteristic, which is to say that any decryption scheme must have high computational complexity. [C. Shannon, Communication theory of secrecy systems, *Bell Syst. Tech. J.*, 28(4), 656–715, 1949]

secretary problem (alias: **interview problem; marriage problem; fiancé(e) problem; googol game; best-choice problem; dowry problem; etc.**) A classic problem of optimal stopping that has given rise to many variants and generalizations, and which has many names. The basic version is this:

(i) There is one available job, with a known fixed number n of candidates.
(ii) The candidates are strictly ranked (no ties) according to some suitable criteria.
(iii) They are interviewed (or assessed) in random order, all orders being equally probable, and the accept/reject decision is made for each at once knowing only the relative ranks of previously seen candidates.

(iv) A rejected candidate cannot be accepted later on.
(v) You seek the best (i.e. top-ranked) candidate B, and require a stopping rule that maximizes your probability of accepting B.

The solution asserts that the optimal rule rejects the first r candidates, and then accepts the first subsequent candidate who is better than all predecessors, where

$$r = \left[\left(n - \frac{1}{2}\right)e^{-1} + \frac{1}{2}\right],$$
$$\simeq n/e \text{ for large } n.$$

The probability that this rule succeeds in picking the best candidate is approximately e^{-1}, for large enough n. *See also* Bruss odds rule.
[T. Ferguson, Who solved the secretary problem?, *Stat. Sci.*, 4(3), 282–296, 1989]

security strategy *see* minimax principle

seek time *see* Kemeney's constant

Selberg's inequality A Chebyshov-type inequality for a r.v. X, that supplies a lower bound for $\mathbb{P}(a < X < b)$ in terms of the variance of X, whenever $a < \mathbb{E}X < b$. [H. Godwin, On generalizations of Tchebychef's inequality, *J. Amer. Stat. Ass.*, 50(271), 923–945, 1955]

selection expectation *see* Aumann expectation

selection property A term arising in two distinct contexts; first in considering a sequence of distribution functions (*see* Helly's theorem); and second in considering a sequence of r.vs $X_1, X_2, \ldots$, which is said to have the selection property if for every $1 \leq j_1 < j_2 < \ldots < j_k$, the joint distribution of $X_{j_1}, \ldots, X_{j_k}$ is that of $X_1, \ldots, X_k$. The **Dacunha–Castelle thm** asserts that such an infinite sequence is exchangeable iff it has the selection property. [J. F. C. Kingman, Uses of exchangeability, *Ann. Probab.*, 6(2), 183–197, 1978]

self-affine process *see* self-similar process

self-averaging Arises in statistical mechanics and interacting particle systems when considering some random function (or functional) $X_n, n \geq 1$, of a random graph of size n, or a lattice of n interacting particles, or a random field on n points, etc. If a suitably scaled version of X_n converges to a constant in some sense (e.g. in mean square) then the sequence is said to be self-averaging; otherwise it is not self-averaging, which typically arises in the presence of long-range correlations. For a toy example, for a strictly stationary sequence $Y_n, n \in \mathbb{Z}$ with finite mean, $X_n = \frac{1}{n}\sum_{r=1}^{n} Y_r$ converges (by the ergodic theorem) to a r.v. X; if X is a.s. constant, then the sequence is self-averaging. [O. Riordan & L. Warnke, Achlioptas processes are not always self-averaging, *Phys. Rev. E*, 86(1), 01129, 2013]

self-avoiding random walk *see* self-interacting random walk

self-correcting point process Loosely speaking, a random point process that attempts to generate realizations that are nearly deterministic, in the sense of nearly having λt points in $[0, t]$ for all t and some λ. More formally, it has a counting process $N(t), t \geq 0$, with a conditional intensity function given the process up to time t that may be expressed as

$$\lambda(t \mid \mathcal{F}_t) = \lambda(t \mid N(t)) = cu(ct - N(t))$$

for some $c > 0$, and a suitable function $u(\cdot)$, where $\mathcal{F}_t$ is the σ-field generated by $N(s), 0 \leq s \leq t$. [T. Hayashi, Laws of large numbers in self-correcting point processes, *Stoch. Proc. Appl.*, 23(2), 319–326, 1986; V. Isham & M. Westcott, A self-correcting point process, *Stoch. Proc. Appl.*, 8, 335–347, 1979]

self-decomposable A r.vr $\mathbf{X} \in \mathbb{R}^d$ (or its distn), is said to be self-decomposable if for all $a \in (0, 1)$ there is a r.vr $\mathbf{Y}_a$, independent of $\mathbf{X}$, such that $a\mathbf{X} + \mathbf{Y}_a$ has the same distn as $\mathbf{X}$. All stable distns are self-decomposable, and all self-decomposable distns are infinitely divisible. All non-degenerate self-decomposable distns have a density (i.e. are absolutely continuous); this, along with some other related results, may be called **Sato's thm.**

A non-negative, integer-valued r.v. X is called discrete self-decomposable if for all $a \in (0, 1)$ there is a r.v. Y_a with generating function $G_Y(z)$ such that Y_a is independent of X and

$$G_X(z) = \mathbb{E}z^X = G_X(1 - a + az)G_Y(z).$$

Then for some $b > 0$ and pgf $H(z)$, with $H(0) = 0$,

$$G_X(z) = \exp\left\{-b\int_z^1 \frac{1 - H(y)}{1 - y} dy\right\}.$$

[K. Sato, *Lévy Processes and Infinitely Divisible Distributions*. Cambridge University Press, 1999]

self-exciting process *see* Hawkes process

self-financing portfolio Arises in financial mathematics, where it describes a holding of stocks and bonds over time that is subject neither to withdrawals nor injections of cash (or numeraire). That is to say, its value $V(t)$ at time t is the initial holding plus the accrued changes arising from market movements and reallocation of resources among the portfolio's assets. More formally, if,

$$V(t) = a(t)S(t) + b(t)M(t),$$

where $S(t)$ denotes the stock, and $M(t)$ denotes cash (or riskless bond, etc.), the portfolio (or the strategy determining the holdings) is self-financing if

$$dV(t) = a(t)dS(t) + b(t)dM(t).$$

[T. Björk, *Arbitrage Theory in Continuous Time*, 3rd edn. Oxford University Press, 2009]

self-information *see* entropy

self-interacting diffusion *see* self-interacting random walk

self-interacting process *see* reinforced process; self-interacting random walk

self-interacting random walk A classic example (and archetype) of a reinforced random process (also called a path-(or history-)dependent process), in which the distribution of any step of the path depends on the past path up to that step. Of the many varieties, important examples are these:

1. **Excited random walk**, in which the distribution of a step from a vertex v depends on the number of previous visits to v. This is also called called a **cookie random walk**.
2. **Edge-reinforced random walk** (or ERRW), in which the probability of stepping along any edge E is a function (weight function) of the number of times $N(E)$ that the path has previously traversed E; in the basic case of a linear weight function (proportional to $N(E) + 1$), such a walk on a finite graph converges to a random stationary distribution.
3. **Vertex-reinforced r.w.** (or VRRW), in which the probability of next stepping to any vertex V is given by a function $S(V)$ of the number of previous visits to V (the occupation or sojourn time). When this is linear and increasing in $S(V)$ such walks tend to get trapped in a subset of vertices, e.g. for simple VRRW on $\mathbb{Z}$, with linear weight function, the walk is ultimately trapped on five consecutive points.
4. When the weight function in VRRW is decreasing in $S(V)$, one speaks of self-repelling walk, and the extreme case of this is **self-avoiding r.w.** (or SARW), in which at no step can the walk revisit any vertex that it has previously occupied. A classic example of this is simple SARW on the integer lattice $\mathbb{Z}^d$ in d dimensions; denoting the number of possible paths of length n (i.e. having n steps) by $p_d(n)$, one has **Hammersley's thm** asserting that $\lim_{n\to\infty}(p_d(n))^{\frac{1}{n}} = b_d > 0$ for some b_d called the **connective constant**; and **Kesten's thm** asserts that $\lim_{n\to\infty}p_d(n+2)/p_d(n) = b_d^2$.

The concept extends naturally to a diffusion process $X(t)$ whose instantaneous evolution is a function of the local time $L(t,x)$ of the process. Such a process may be specified as the solution to an SDE of the form $dX(t) = g(X(t), L(t, X(t))dt + dW(t)$, where one speaks of self-attracting or self-repelling diffusion, respectively, according as the function $g(.,.)$ pushes the process towards or away from points with the largest local time. [J. Norris *et al.*, Self-avoiding random walk: a Brownian motion model with local time drift, *Probab. Theor. Rel. Fields*, 74(2), 271–287, 1987; R. Pemantle, A survey of random processes with reinforcement, *Probab. Surv.*, 4, 1–79, 2007; B. Tóth, Self-interacting random motions: a survey, in *Random Walks: A Collection of Surveys, Bolyai Soc. Math. Studies*, 9, 349–384, 1999; M. Benaïm & O. Raimond, Self-interacting diffusions, lll: symmetric interactions, *Ann. Probab.*, 33(5), 1716–1759, 2005; O. Raimond & B. Schapira, Excited Bramian motions, *ALEA Latin Amer. J. Probab. Math. Stat.*, 8, 19–41, 2011]

self-normalized process (alias: **Studentized process**) Arises in generalizing classical limit theorems, and also in statistical distribution theory. Broadly, given a sequence $X_1, X_2, \ldots$ of r.vs, an important class of limit theorems postulates sequences of constants a_n and $b_n, n \geq 1$, such that $\lim_{n\to\infty}(X_n - a_n)/b_n$ exists appropriately; a corresponding self-normalized (or Studentized) process (writing $\mathbf{X}_n = (X_1, \ldots, X_n)$) is of the form $S_n = (X_n - A_n(\mathbf{X}_n)/(B_n(\mathbf{X}_n)$, for suitable functions $A_n(\cdot)$ and $B_n(\cdot)$, where S_n has the same (or similar) limiting properties as $n \to \infty$. For example, the classic CLT asserts that for i.i.d. zero-mean r.vs $Z_r, r \geq 1$, with unit variance $\sum_{r=1}^{n} Z_r/\sqrt{n}$ converges in distribution to the standard normal; generalizing this, the **Giné–Götze–Mason self-normalized CLT** asserts that for i.i.d.r.vs $Y_1, Y_2, \ldots, \sum_{r=1}^{n} Y_r/(\sum_{r=1}^{n} Y_r^2)^{\frac{1}{2}}$, as $n \to \infty$, has asymptotically a standard normal distribution iff $\mathbb{E}Y_r = 0$ and Y_r is in the domain of attraction of the normal law. [V. de la Peña *et al.*, *Self-normalized Processes*. Springer, 2009]

self-similarity exponent *see* Hurst exponent

self-similar process Arises in seeking to characterize those random processes that can occur as limits when a random process is suitably scaled, repeatedly, in time and space. The basic definition is this: $X(t), t \geq 0$, is called self-similar if for any $a \geq 0$ there exists $b(a)$ such that $X(at)$ has the same joint distributions (fdds) as $bX(t),\ t \geq 0$.

J. Lamperti introduced such processes under the name semi-stable (now used in a related but different context), and proved an equivalent definition: $X(t), t \geq 0$, is self-similar if for any $a \geq 0$ there exists $H > 0$ such that $X(at)$ has the same fdds as $a^H X(t)$. Then H is called the **self-similarity** (or **Hurst**) **parameter** (or index, or exponent). There are other characterizations of self-similarity.

More weakly, $X(t), t \geq 0$, is semi-self-similar if for some $0 < a \neq 1$ there exists $b(a)$ such that $X(at)$ has the same fdds as $bX(t), t \geq 0$. A Lévy process $X(t), t \geq 0$, is self-similar iff $X(1)$ is strictly stable, and semi-self-similar iff $X(1)$ is semi-stable.

More generally, given affine transformations $u(\cdot)$ from $\mathbb{R}^d$ to $\mathbb{R}^n$, and $v(\cdot)$ from $\mathbb{R}^n$ to $\mathbb{R}^d$, $X(t)$ is called (u, v)-self-affine if $X(u(t))$ has the same fdds as $v(X(t))$. Such processes arise typically as a vector of self-similar process with different Hurst parameters. [K. Sato, *Lévy Processes and Infinitely Divisible Distributions*. Cambridge University Press, 1999]

semi-group *see* stochastic semi-group; Feller property

semi-invariants *see* cumulants

semi-Markov process *see* Markov renewal process

semi-martingale Arises as the most natural class of integrators $X(t)$ in stochastic integrals of the usual form $\int Y(t)\mathrm{d}X(t)$, under reasonable tractability requirements. Formally, a process $S(t), t \geq 0$, adapted to a filtration $\mathcal{F}_t, t \geq 0$, is called a semi-martingale if it is rcll, and can be written (or decomposed) as $S(t) = S(0) + M(t) + A(t)$, where $S(0)$ is a.s. finite, $A(t)$ is a process of bounded variation on $[0, t]$, and $M(t)$ is a local martingale. If $A(t)$ is $\mathcal{F}_t$-predictable, then the decomposition is unique (up to stochastic equivalence) and $S(t)$ may be called a special semi-martingale.

If $S(t) = \int_0^t B(t)\mathrm{d}N(t)$, for predictable $B(t)$ and some local martingale $N(t)$ then $S(t)$ may be called a **sigma-martingale**. Note that some early authors call a submartingale a semi-martingale. [P. Protter, *Stochastic Integration and Differential Equations*, 2nd edn. Springer, 2005]

semi-reversible *see* reversibility

semi-stable *see* stable

semi-stationary *see* stationarity

semivariance *see* variance

Seneta–Heyde theorem (alias: **Seneta's theorem; Seneta–Heyde norming**) Arises in considering the limiting behaviour of a super-critical ordinary Galton–Watson branching process $Z_0 = 1, Z_1, Z_2, \ldots$, with probability of extinction $\eta \in [0, 1)$, and such that Z_1 is non-degenerate, while $1 < \mathbb{E}Z_1 < \infty$. The theorem asserts that there exists a sequence of positive numbers $S_n, n \geq 1$, called the **Seneta constants**, such that $\lim_{n\to\infty} S_n = \infty$, and $\lim_{n\to\infty} S_n{}^{-1} S_{n+1} = \mathbb{E}Z_1$ and $\lim_{n\to\infty} S_n{}^{-1} Z_n = W$ a.s., where $\mathbb{P}(W = 0) = \eta$ and W has a density on $(0, \infty)$. *See* Kesten–Stigum theorem. [C. C. Heyde, Extension of a result of Seneta for the super-critical Galton–Watson process, *Ann. Math. Stat.*, 41(2), 739–742, 1970]

separability Arises in considering a random process $X(t, \omega), \omega \in \Omega, t \in T \subseteq \mathbb{R}$, as a regularity condition ensuring that suitable sets of sample paths are measurable. Broadly speaking, a separable process $X(t)$ is such that its paths are effectively determined (i.e. up to null sets) by a knowledge of the values $X(t_i), t_i \in S$, where S is a countable subset of T. For every system of joint distributions (fdds) that are consistent (*see* Kolmogorov consistency) there exists a separable version of the process having those fdds. [P. Billingsley, *Probability and Measure*, 3rd edn. Wiley, 1995]

separation distance *see* relative pointwise distance

sequencing problem *see* allocation problem

sequential decision problem *see* decision theory

Serfling's inequality Any of several concentration inequalities for the sample mean in sampling without replacement from a set of numbers. [R. Serfling, Probability inequalities for the sum in sampling without replacement, *Ann. Statist.*, 2(1), 39–48, 1974]

Sevastyanov branching process A special case of a Crump–Mode–Jagers process, in which all offspring replace their parent instantly at the same time, and where the parent's life-span and family-size may be dependent.

[P. Jagers & S. Sagitov, General branching processes in discrete time as random trees, *Bernoulli*, 14(4), 949–962, 2008]

Shannon (channel) coding theorem *see* Shannon's theorem; noiseless coding theorem; noisy coding theorem

Shannon divergence (or distance) *see* Jensen–Shannon divergence

Shannon entropy Introduced by C. Shannon in the context of communication theory as a measure of the information produced (or uncertainty resolved) when a message (or point) is given (at random) from the set of all possible messages (or sample space). Formally, let there be n possible messages, whose respective probabilities are $p_1, \ldots, p_n$; then the Shannon entropy is.

$$H(p_1, \ldots, p_n) = -\sum_{r=1}^{n} p_r \log p_r.$$

More generally, for a r.v. (or r.vs) having density f with respect to a measure $M(x)$, the (generalized, or differential) Shannon entropy is $H = -\int f \log f M(\mathrm{d}x)$. When g is also a density wrt $M(x)$, the relative Shannon entropy between f and g (or Kullback–Leibler divergence, especially when M is Lebesgue measure), is

$$d = \int f \log \frac{f}{g} M(\mathrm{d}x).$$

This has also been called the cross-entropy between f and g, but note that the function $\int f \log g\, M(\mathrm{d}x)$ is also called the cross-entropy by a number of authors, *see* Kerridge's inaccuracy. *See also* entropy and differential entropy. [R. Rubinstein & D. Kroese, *The Cross-Entropy Method*. Springer, 2004; S. Verdú, Fifty years of Shannon entropy, *IEEE Trans. Inf. Theor.*, 44(6), 2057–2078, 1998; C. Shannon, A mathematical theory of communication, *Bell Syst. Tech. J.*, 27(3), 379–423, 1948]

Shannon's entropy power inequality *see* Shannon–Stam inequality

Shannon–Hartley formula (or theorem) (alias: **Shannon's formula**) Arises in communication theory when the input X to a channel is subject to additive Gaussian noise, i.e. the output is given by $Y = X + Z$, where Z is normally distributed as $N(0, \sigma^2)$ and independent of X. It is assumed that the channel is power-limited, so that $\mathbb{E}X^2 \leq p < \infty$, and then Shannon's capacity theorem asserts that this **Gaussian channel** has capacity at most $C = \frac{1}{2}\log_2(1 + p/\sigma^2)$, which is the **Shannon–Hartley formula**. In continuous time the output signal is given by $Y(t) = X(t) + Z(t)$, where $Z(t)$ is Gaussian white noise independent of $X(t)$, and the input is assumed to be band-limited with bandwidth W, as well as power-limited. Then the capacity is $C = W\log_2(1 + p/\sigma^2)$, in bits per unit time, where p/σ^2 is the **signal to noise ratio**. In each case, capacity is achieved by a Gaussian input. [C. Shannon, A mathematical theory of communication, *Bell Syst. Tech. J.*, 21(3), 623–656, 1948; R. Ash, *Information Theory*. Wiley, 1965; reprinted by Dover, 1990]

Shannon index *see* Shannon entropy

Shannon's inequality *see* Shannon–Stam inequality; Gibbs inequality

Shannon–McMillan (–Breiman–Chung) theorem Generalizes Shannon's asymptotic equipartition property for i.i.d.r.vs to suitable stationary processes X_n, $n \in \mathbb{Z}$, having joint distributions denoted by $f(x_1, \ldots, x_n)$. One form is this: Suppose that H is the entropy of the sequence X_n, where $H(X_0) < \infty$, and define $H_n = -\frac{1}{n}\log f(X_1, \ldots, X_n), n > 1$. Then there is some r.v. Z, with $\mathbb{E}Z = H$, such that as $n \to \infty$, H_n converges to Z a.s. and in mean; and if the sequence is ergodic then $\mathbb{P}(Z = H) = 1$. [R. Ash & C. Doléans-Dade, *Probability and Measure Theory*, 2nd edn. Academic Press, 2000]

Shannon–Nyquist theorem *see* sampling theorem

Shannon power inequality *see* entropy power inequality

Shannon's problem Arises in the context of the differential entropy $h(\cdot)$ of a suitably normalized sum of finite variance i.i.d.r.vs $X_1, \ldots, X_n$, denoted by $h_n = h\left(\frac{1}{\sqrt{n}}(X_1 + \cdots + X_n)\right)$. C. Shannon and A. Stam showed that $h_2 \geq h_1$, and the conjecture that h_n is a non-decreasing function of n is Shannon's problem; the question was resolved affirmatively by a theorem of S. Artstein, K. Ball, F. Barthe, & A. Naor. *See also* entropy power inequality and Shannon–Stam inequality. [S. Artstein *et al.*, Solution of Shannon's problem on the monotonicity of entropy, *J. Amer. Math. Soc.*, 17(4), 925–982, 2004]

Shannon secrecy problem *see* secrecy system

Shannon's source coding theorem *see* noiseless coding theorem

Shannon–Stam inequality *see* entropy power inequality; Shannon's problem; Stam's inequality

Shannon's theorem Any of a large number of results in the theory of communication and information and cryptography, of which these are perhaps most often seen:

1. **Shannon's first thm**, or **noiseless coding thm**, or **source coding thm**, or **data-compression thm**, or **entropy bound thm**. This asserts that in encoding the output X from a source, having entropy $H(X)$, the expected codeword length per source symbol, if the code is to be uniquely decodable, is bounded below by $H(X)$. Conversely, **Shannon's asymptotic equipartition thm** (and its generalization, the **Shannon–McMillan–Breiman–Chung thm** for general sources) show that for arbitrarily long messages the bound can be arbitrarily closely approached.
2. **Shannon's second thm**, or **noisy coding thm**, or **channel coding thm.** This asserts that for a noisy channel there is a number C called the capacity such that the rate R of information arriving at the channel does not exceed C. Messages may be passed through the channel with arbitrarily small probability of error as their length increases, but if $R > C$ the probability of error is close to 1 in this case.
3. Shannon–Hartley formula, which gives the capacity of a continuous alphabet power-limited Gaussian channel. (Sometimes called Shannon's third thm, but see below.)
4. **Shannon's rate distortion thm**, which gives the smallest achievable distortion for a given rate of transmission; or, equivalently, the largest rate that does not exceed a given distortion. (This is also sometimes called Shannon's third thm, see above.)
5. Shannon–Nyquist thm; *see* sampling theorem.
6. Shannon's secrecy thm; *see* secrecy system.

[C. Shannon, Collected Papers, 1993, IEEE]

Shannon transform The Shannon transform of a non-negative r.v. X is defined for $z \geq 0$ by $S(z) = \mathbb{E}\log(1 + zX)$. It is related to the Stieltjes transform $T(z) = \mathbb{E}\left(\frac{1}{X-z}\right)$ by the relationship

$$\frac{z}{\log \mathrm{e}} \frac{\mathrm{d}S}{\mathrm{d}z} = 1 - \frac{1}{z} T\left(-\frac{1}{z}\right),$$

where $\frac{1}{z}T\left(-\frac{1}{z}\right)$ is sometimes called the $\boldsymbol{\eta}$**-transform**.
[R. Couillet & M. Debbah, *Random Matrix Methods*. Cambridge University Press, 2011]

shape theorem *see* Eden–Richardson model

Shapley–Shubik index *see* importance measure

Shearer's lemma (or inequality) Establishes an entropy inequality with applications in combinatorics, thus: Let $\mathbf{X}_1 = (X_1, \ldots, X_n)$ be a r.vr, and let $S = (S_k)_{k \in I}$ be a family of subsets of $\{1, \ldots, n\}$ such that each integer $r \in \{1, \ldots, n\}$ is a member of at least m of the subsets in S. For each $S_k = (i_1, \ldots, i_{n_k})$ denote the r.vr $X_{i_1}, \ldots, X_{i_{n_k}}$ by X_{S_k}. Then the lemma asserts that

$$mH(X_1, \ldots, X_n) \leq \sum_{k \in I} H(X_{S_k}),$$

where $H(\cdot)$ denotes the Shannon entropy. *See* Han's inequalities. [P. Balister & B. Bollobás, Projections, entropy and sumsets, *Combinatorica*, 32(2), 125–141, 2012]

Sheppard's theorem (or formula) Gives an expression for the probability that a standard bivariate normal r.vr (X, Y) has both variables positive (i.e. lies in the positive quadrant) thus:

$$\mathbb{P}(X > 0,\ Y > 0) = \frac{1}{4} + \frac{1}{2\pi} \sin^{-1}\rho,$$

where ρ is the correlation. For a standard trivariate normal density with correlation coefficients $\rho_{12}, \rho_{23}, \rho_{31}$,

$$\mathbb{P}(X > 0, Y > 0, Z > 0) = \frac{1}{8} + \frac{1}{4\pi}(\sin^{-1}\rho_{12} + \sin^{-1}\rho_{23} + \sin^{-1}\rho_{31}).$$

[A. Stuart & J. Ord, *Kendall's Advanced Theory of Statistics*, 6th edn. Wiley, 2009]

Shepp's covering theorem Supplies a solution to the so-called **Dvoretzky problem**: let $1 > a_1 \geq a_2 \geq \cdots \geq 0$ be the lengths of a sequence of arcs placed independently and uniformly on a circle with unit circumference; when is the circle covered by arcs with probability 1? Shepp's theorem asserts that this holds iff $\sum_{n=1}^{\infty} \frac{1}{n^2} \exp(a_1 + \cdots + a_n) = \infty$.

The Fitzsimmons–Fristedt–Shepp thm further addresses the nature of the uncovered set when this fails. [P. Fitzsimmons *et al.*, The set of real numbers left uncovered by random covering intervals, *Probab. Theor. Rel. Fields*, 70, 175–189, 1985]

Shepp's law *see* Erdös–Rényi law of large numbers

Shepp's local limit theorem Supplies asymptotic estimates for

$$\mathbb{P}\left(a \leq \sum_{r=1}^{n} X_r \leq b\right), \quad a < b,$$

when $X_r, r \geq 1$, is a sequence of finite-variance, zero-mean i.i.d.r.vs that may be discrete or continuous. [L. Shepp, A local limit theorem, *Ann. Math. Stat.*, 35(1), 419–423, 1964]

Shepp–Shiryaev problem (or game) An optimal stopping problem for a novel (and theoretical, i.e. untraded) type of financial option (that the original proposers called a **Russian option**) in which the owner receives the (discounted) supremum of prices that the option has ever achieved between the owner buying it and exercising it. [L. Shepp & A. Shiryaev, The Russian option: reduced regret, *Ann. Appl. Probab.*, 3(3), 631–640, 1993]

Shepp's unimodality criterion *see* Khinchin's criterion

Sherrington–Kirkpatrick model Arises in statistical mechanics as a disordered spin system, i.e. a random field whose state space is $\Omega = \{-1, 1\}^n$, whose elements $\boldsymbol{\omega} = (\omega_1, \ldots, \omega_n) \in \Omega$ are called configurations. The distribution on Ω is taken to be a Gibbs distribution of the form

$$\mathbb{P}(\boldsymbol{\omega}) = \exp\left\{-\frac{\beta}{\sqrt{n}} \sum_{1\leq i<j\leq n} b_{ij}\omega_i\omega_j - h\sum_i a_i\omega_i\right\}.$$

[M. Talagrand, *Spin Glasses: A Challenge for Mathematicians*. Springer, 2003]

shift coupling Broadly, this refers to the construction of two processes $\hat{X}_n$ and $\hat{Y}_n, n \in I$, on a single probability space so that they have the same distributions as a given pair of processes (not necessarily on the same space), but modulo a random shift in time index. [D. Aldous & H. Thorisson, Shift coupling, *Stoch. Proc. Appl.*, 44(1), 1–14, 1993]

shock model (alias: **cumulative damage model**) Arises in reliability theory to describe the survival of components subject to failure by degradation; that is, they fail when a cumulative process (of 'shocks') exceeds some threshold (that may be random). [T. Nakagawa, *Shock and Damage Models in Reliability Theory*. Springer, 2007]

Shor code An analogue of the error-correcting codes arising in classical communication theory that plays a similar role in quantum computation by reducing the decoherence in some superposition of states. [P. Shor, Scheme for reducing decoherence in quantum computer memory, *Phys. Rev. A*, 52(4), R2493–R2496, 1995]

shot noise process (alias: **functional of point process**) Arises naturally in modelling the behaviour and effects of electrons moving in a charged field (which may be distinguished in the physics literature from Johnson, or thermal, noise, arising from e.g. Brownian motion of particles in a resistor), and in several other contexts. The basic model assumes that particles arrive at the instants $S_r, r \in \mathbb{Z}(\text{ or } \mathbb{N})$, of a Poisson process, and that the residual effect of a particle is given by a response function $r(\cdot)$, and that effects are additive. Then the shot noise process is

$$S(t) = \sum_{S_r \leq t} r(t - S_r), \quad t \in \mathbb{R}, \ (\text{or } t \geq 0).$$

Typically (on physical grounds), $r(t) = 0$ for $t < 0$, and it is assumed to be decreasing, bounded, and suitably integrable. (Otherwise one may speak of explosive shot noise.) In particular, when $r(t) = \delta(t)$, the Dirac delta, one has white noise. More generally, for a point process Π in $\mathbb{R}^n$, a general shot noise process is

$$S(x) = \sum_{\pi} R(\pi, x),$$

where the random processes $R(\pi, x), \mathbb{R}^n \times \mathbb{R}^n \to [0, \infty)$ comprise a collection of i.i.d. response functions (over $\pi \in \Pi$). [D. Daley & D. Vere-Jones, *Introduction to the Theory of Point Processes*, 2nd edn. Springer, Vol. I, 2003; Vol. II, 2008]

shuffle A mixing or rearrangement of a pack of cards (and, by extension, other objects), designed to provide some randomization of the order. A key property of an effective shuffle is that repetition should provide reasonably speedy convergence to the uniform distribution over all permutations of the pack. A natural question is how many shuffles are necessary to achieve this?

For the riffle shuffle, D. Bayer & P. Diaconis show that 7 is necessary and sufficient for a pack of 52 cards.

Examples of shuffling rules include:

1. **Riffle shuffle**. There are many types of riffle shuffle (also called the dovetail); the basic procedure randomly divides the pack into two parts, and then drops cards (singly or in groups) alternately from each subpack to create a newly ordered pack. That most often seen is the **Gilbert–Shannon–Reeds shuffle**, in which the division has a binomial distribution (with $p = ½$) and the cards are dropped from each subpile with probability equal to the proportion in each.
2. **Fisher–Yates–(Durstenfeld–Knuth) shuffle**. For a pack of n cards, for $r = 0, 1, \ldots, n-1$ successively, select a card at random from the first $n - r$ positions, and swap it with the card in the $(n - r)$th place. The final order is uniform on the $n!$ permutations of the pack.
3. **Top to random shuffle**. The top m cards are inserted one at a time uniformly at random in the residual pile (including the successive newcomers). It takes approximately $\frac{n}{m}(\log n + c)$ shuffles for the distribution to approach the uniform.
4. **Transposition shuffle**. Take U and V independently uniform on $[1, \ldots, n]$, and swap the cards in those positions. Approximate uniformity needs around $\frac{1}{2} n \log n$ repetitions. A variant of this is **Metropolis (or adjacent) transposition shuffling**, in which a pair of adjacent cards is picked uniformly at random. With probability p they are placed in increasing order, or in decreasing order with probability $1 - p$. If $p = ½$, repetition yields an approximately uniform distribution, otherwise $(p \neq ½)$ the limit is not uniform and may be called **Metropolis biased**.

[P. Diaconis, Mathematical developments from the analysis of riffle shuffling, in: *Groups, Combinatorics & Geometry*, A. Ivanov *et al.* (eds). World Scientific, 2003, pp. 73–99; L. & L. Trefethen, How many shuffles to randomize a deck of cards? *Proc. R. Soc. Lond. A*, 456, 2561–2568, 2000]

Sibuya distribution A probability distribution on the positive integers having pgf $G(z) = 1 - (1 - z)^a$, $0 < a \leq 1$. More generally, a r.v. on the non-negative integers with pgf $G(z) = 1 - b(1 - z)^a, 0 < a, b \leq 1$, is said to have a scaled Sibuya distribution. [G. Christoph & K. Schreiber, Scaled Sibuya distribution and discrete self-decomposability, *Stat. Probab. Lett.*, 48, 181–187, 2000]

Sichel distribution *see* inverse Gaussian distribution

Šidak's inequality Arises in considering a zero-mean multivariate normal random vector $\mathbf{X} = (X_1, \ldots, X_n)$, where it asserts that for any positive numbers $x_1, \ldots, x_n$,

$$\mathbb{P}(|X_1| \leq x_1, \ldots, |X_n| \leq x_n) \geq \prod_{r=1}^{n} \mathbb{P}(|X_r| \leq x_r).$$

More generally, the term also encompasses this assertion: if the correlation matrix ρ_{jk} of $\mathbf{X}$ takes the form $\rho_{jk} = \lambda_j \lambda_k r_{jk}$, where r_{jk} is any correlation matrix, and $0 \leq \lambda_r \leq 1$, $1 \leq r \leq n$, then $\mathbb{P}(|X_1| \leq x_1, \ldots, |X_n| \leq x_n)$ is a non-decreasing function of every λ_r. Similar results hold for multivariate Student and F distributions. *See* Khatri–Šidak inequality. [Z. Šidak, On probabilities of rectangles in multivariate student distributions: their dependence on correlations, *Ann. Math. Stat.*, 42(1), 169–175, 1971]

sigma-algebra *see* sigma-field

sigma-field (alias: **σ-field; sigma-(or σ-) algebra; Borel field**) A class (or collection) $\mathcal{F}$ of subsets of a space (or universal set) Ω, such that:

1. $\mathcal{F}$ is closed under complements, which is to say that if $A \in \mathcal{F}$, then $A^c \in \mathcal{F}$;
2. $\Omega \in \mathcal{F}$ (or equivalently, $\emptyset \in \mathcal{F}$; or $\mathcal{F}$ is non-empty);
3. $\mathcal{F}$ is closed under countable unions, which is to say that if $A_1, A_2, \ldots \in \mathcal{F}$, then $\cup_{r=1}^{\infty} A_r \in \mathcal{F}$.

The standard definition of mathematical probability is a measure $\mathbb{P}(\cdot)$ on a sigma-field $\mathcal{F}$ of subsets of the certain event Ω (*see* probability space). If $\mathcal{F} = \{\emptyset, \Omega\}$, it is called trivial. When Ω is a topological space, the σ-field of open sets may be called the **Borel field** of **Borel sets**. [D. Pollard, *A User's Guide to Measure Theoretic Probability*. Cambridge University Press, 2002]

sigma-finite Arises in measure theory as a weaker property than finiteness. A measure is sigma-finite if every measurable set is a countable union of sets, each of which has finite measure. Its importance in probability theory is largely through the Radon–Nikodym theorem, and extension theorems. [D. Pollard, *A User's Guide to Measure Theoretic Probability*. Cambridge University Press, 2002]

sigma-martingale *see* semi-martingale

signal-to-noise ratio *see* Shannon–Hartley formula

significant-digit law *see* Benford's distribution

Simon Newcomb's problem A classic problem of combinatorial probability. In its basic form a pack of n cards bearing n distinct numbers is inspected in random order, and one asks for the distribution of the number N of occasions when any number is less than its immediate predecessor (i.e. an increasing run stops). Then P. MacMahon showed that

$$\mathbb{P}(N \le x) = \frac{1}{n!}\sum_{j=0}^{x}\binom{n}{j}(-1)^j(x+1-j)^n.$$

This problem is so named to distinguish it from **Newcomb's problem**, which is a paradox of the philosophy of decision theory. [D. Barton & C. Mallows, Some aspects of the random sequence, *Ann. Math. Stat.*, 36(1), 236–260, 1965; F. Drane & D. Roselle, The extended Simon Newcomb problem, *J. Comb. Theor. A*, 24, 348–356, 1978]

simple random variable *see* discrete distribution

simple random walk *see* random walk

Simpson's paradox *see* Yule–Simpson paradox

simulated annealing *see* annealing method

simulation *see* pseudo-random; Monte Carlo method

Sinai's problem (or theorem) *see* random environment

Singh–Maddala distribution *see* Burr distribution

single-period model (alias: **S-s inventory policy; newsvendor model**) The key model of stochastic inventory theory, which supplies the basis for analysing many multiperiod models. There is a random demand $D \ge 0$, with continuous distn $F(x)$, for units (or items) to be sold by the vendor in some single period. The vendor has an initial stock i, and must decide how many further units to order to minimize the expected total costs. Costs accumulate thus (where profits are negative costs):

A set up cost K if any extra units are ordered.

A cost C per ordered unit.

A cost C_d for each unit of unmet demand, where $C_d > C$.

A cost C_n for each unit of unsold stock, where $C + C_n > 0$.

The total expected cost of a day's trading, when $i = 0$ and x units are ordered is thus $K + k(x)$, for some function $k(\cdot)$.

The optimal policy is shown to be: make no order if $i \ge s$, but bring the stock up to S if $i < s$, where $S = F^{-1}\left(\frac{C_d - C}{C_d + C_n}\right)$, and s is the smallest value of x such that $k(x) = K + k(S)$. If there is no set-up cost for an order, i.e. $K = 0$, then $s = S$ and it is optimal to bring the stock up to S if $i < S$, but otherwise do not order. The ratio $(C_d - C)/(C_d + C_n)$ may be called the **critical fractile**. [K. Arrow *et al.*, Optimal inventory policy, *Econometrica*, 19(3), 250–272, 1951; T.-M. Choi (ed.), *Handbook of Newsvendor Problems*. Springer, 2012; E. Porteus, *Foundations of Stochastic Inventory Theory*. Stanford University Press, 2002]

singular *see* degenerate distribution

singular distribution (or measure) Broadly, singularity is a property of probability distributions that do not have a density, i.e. are not absolutely continuous. Formally, a probability measure $\mathbb{P}(\cdot)$ on $\mathbb{R}^n$ is singular with respect to Lebesgue measure on $\mathbb{R}^n$ if there is a set A of zero Lebesgue measure (informally: volume) such that $\mathbb{P}(A^c) = 0$. Important examples include:

1. The uniform (or other) distribution on a sphere of radius $a > 0$.
2. The singular multivariate normal distribution, which is the distribution of $\mathbf{Y} = A\mathbf{X} \in \mathbb{R}^m$, where $\mathbf{X}$ is random n-vector having a multinomial distribution, $N(0, V)$, and A is an $m \times n$ matrix with rank $r < m$. Then $\mathbf{Y}$ lies in a linear subspace of $\mathbb{R}^m$, having dimension r, and the distribution of $\mathbf{Y}$ is denoted by $N(0, AVA^T)$.

On this definition, any atomic (i.e. discrete) probability distribution is singular wrt Lebesgue measure; and distributions that are singular but attach zero probability to any point set are called singular continuous. Lebesgue's decomposition thm asserts that for a probability measure $\mathbb{P} = \mathbb{P}_a + \mathbb{P}_d + \mathbb{P}_s$, where $\mathbb{P}_a$ is absolutely continuous, $\mathbb{P}_d$ is discrete, and $\mathbb{P}_s$ is singular continuous. Very often, especially in $\mathbb{R}$, singularity is defined more narrowly to exclude discrete distributions. In this case, the Cantor distribution is an example of a singular distn.

More generally, given measures $m(\cdot)$ and $v(\cdot)$ on a measurable space $(\Omega, \mathcal{F})$, m is said to be singular wrt v (or, equivalently, m and v are said to be mutually singular, or orthogonal), if there is a set A in $\mathcal{F}$ such that $m(A) = 0$ and $v(A^c) = v(\Omega \backslash A) = 0$. One writes: $m \perp v$. [R. Ash & C. Doléans-Dade, *Probability & Measure Theory*, 2nd edn. Academic Press, 2000]

singular multinormal distribution *see* degenerate normal distribution

sinusoidal limit theorem (alias: **Slutsky's (sinusoidal limit) theorem; Slutsky's effect**) Arises in considering the repeated smoothing (or filtering, i.e. forming moving averages and differencing) of stationary time series, a process which leads to series that are arbitrarily close to sine waves over finite segments. Formally, let $Z_r, r \in \mathbb{Z}$ be a sequence of zero-mean i.i.d.r.vs with unit variance (i.e. white noise); then take the moving average $Z_r + Z_{r+1}, r \in \mathbb{Z}, n$ times, and then take the mth difference of the result. Any finite segment of this smoothed sequence converges in probability to a sine wave as $n \to \infty$ and $m \to \infty$ with $\lambda = m/n \in (0,1)$ held fixed; where the period s of the sine curve is given by $\cos(2\pi s^{-1}) = (1-\lambda)/(1+\lambda)$. *See* Slutsky–Yule effect. [P. Moran, The spectral theory of discrete stochastic processes, *Biometrika*, 36(1–2), 63–70, 1949]

size-biased distribution (alias: **length-biased distribution**) Arises naturally in random sampling of individuals that comprise n groups of different sizes $x_1, \ldots, x_n$, so $x_i \neq x_j$ when $i \neq j$. Let Y be the size of the group when a group is picked uniformly at random from the n, and Z the size of the group containing an individual picked uniformly at random from the $\sum_1^n x_r$ individuals. Then by definition $\mathbb{P}(Y = x_r) = \frac{1}{n}$, and $\mathbb{P}(Z = x_r) = \frac{x_r}{\mathbb{E}Y}\,\mathbb{P}(Y = x_r)$. The distribution of Z is that of Y biased by the values of Y, and is called the size biased distribution for Y. The fact that $\mathbb{E}Z = \mathbb{E}Y^2/\mathbb{E}Y \geq \mathbb{E}Y$ is called the **class-size paradox**, among other similar names. In general for a non-negative finite mean r.v. X, having distribution function $F(x)$, the size-biased distribution function is that of a r.v. X_b, and is given explicitly by

$$F_b(x) = \frac{1}{\mathbb{E}X}\int_0^x y\,\mathrm{d}F(y),$$

or implicitly (for arbitrary g) by $\mathbb{E}[Xg(X)] = \mathbb{E}X\mathbb{E}(g(X_b))$, so that the size biased density is $f_b(x) = \frac{x}{\mathbb{E}X} f_X(x)$, when X has density $f_X(x)$. In the context of renewal theory, the above paradox appears (suitably formulated) as the **waiting time paradox** (with length-biasing), and $F_b(x)$ also arises there as the equilibrium distn yielding stationary renewal. *See* inspection paradox. [L. Chen *et al.*, *Normal Approximation*. Springer, 2011]

Skellam's distribution (alias: **Irwin–Skellam distribution**) The distribution of the difference between independent Poisson random variables having different parameters λ and μ respectively; it has the form,

$$p(x) = \mathrm{e}^{-(\lambda+\mu)}\left(\frac{\lambda}{\mu}\right)^{\frac{1}{2}x} I_{|x|}(2\sqrt{\lambda\mu}), \quad x \in \mathbb{Z}$$

where $I_x(\cdot)$ is a modified Bessel fn, of the 1st kind. The case $\lambda = \mu$ may be called **Irwin's distn**. *See* Kendall's taxicab problem. [J. Skellam, The frequency distribution of the difference between two Poisson variates belonging to different populations, *J. Roy. Stat. Soc. A*, 109(3), 296, 1946]

skew Brownian motion *see* Brownian motion

skew distribution The distribution of a random variable that is not symmetric; the term skewness may be used to denote any measure of departure from symmetry of a r.v. X, most commonly its third central moment scaled by the cube of the non-zero standard deviation σ, i.e.

$$\text{skw}\,X = \mathbb{E}(X - \text{E}X)^3/\sigma^3.$$

Other quantities proposed as measures of the skewness of X include, using the *mode* of X: sko $X = \{\mathbb{E}X - \text{mode}(X)\}/\sigma$; or using the *median* of X: ske $X = \{\mathbb{E}X - \text{median}(X)\}/\sigma$; or when $\mathbb{E}(X - \mathbb{E}X)^3 = 0$, the first non-zero odd moment of X. But note that the asymmetric distribution with density

$$f = \frac{1}{48}\left\{1 - \text{sgn}\,x\,\sin|x|^{\frac{1}{4}}\right\}\exp\left(-|x|^{\frac{1}{4}}\right)$$

has zero odd moments of all orders. [E. Churchill, Information given by odd moments, *Ann. Math. Stat.*, 7(2), 244–246, 1946; K. Mardia, Measures of multivariate skewness and kurtosis with applications, *Biometrika*, 57(3), 519–530, 1970]

skew-normal distribution Has density $f(x), x \in \mathbb{R}$, given in terms of the standard normal density ϕ and cdf Φ by $f(x) = 2\phi(x)\,\Phi(\lambda x)$, where $\lambda \in \mathbb{R}$ is arbitrary. Given independent standard normal r.vs U and V, a skew normal r.v. X can be written as $X = (1+\lambda^2)^{-\frac{1}{2}}U + \lambda(1+\lambda^2)^{-\frac{1}{2}}|V|$.

More generally, the extended skew normal has density

$$f(x) = \phi(x)\Phi(\lambda x + a)/\Phi(a(1+\lambda^2)^{\frac{1}{2}}),$$

where $a \in \mathbb{R}$ is also arbitrary. The multivariate skew normal has density $f(\mathbf{x}) = 2g(\mathbf{x})\,\Phi(\mathbf{a}^T\mathbf{x})$, where g is the density of a standard non degenerate multinormal r.v., and $\mathbf{a}$ is a suitable vector. [A. Azzalini, *The Skew-normal and Related Families*. Cambridge University Press, 2013]

skew-symmetric distribution Arises in its simplest form thus: let $f(x)$ be a probability density on $\mathbb{R}$ that is symmetric about 0, and $G(x)$ a distribution function on $\mathbb{R}$ having a density $g(x)$ that is symmetric about 0. Then $h(x) = 2f(x)\,G(\lambda x)$ is a so-called skew-symmetric density, where $\lambda \in \mathbb{R}$ is arbitrary and called the **skewing**

parameter. When $f = g = \phi$ one has the skew-normal distribution.

More generally, let $f(\mathbf{x})$ be a probability density on $\mathbb{R}^n$ that is symmetric about 0 (which is to say that $f(\mathbf{x}) = f(-\mathbf{x})$), and $S(\mathbf{x}) : \mathbb{R}^n \to [0, 1]$ is such that $S(\mathbf{x}) + S(-\mathbf{x}) = 1$ with $0 \leq S(\mathbf{x}) \leq 1$; then $S(\mathbf{x})$ is called a **skewing fn**, and $h(\mathbf{x}) = 2f(\mathbf{x})S(\mathbf{x})$ is a skew-symmetric density function (having skew component S and symmetric component f). If f is e.g. a Cauchy density, then one may speak of h as a **skew-Cauchy** density. [J. Wang *et al.*, A skew-symmetric representation of multivariate distributions, *Stat. Sinica*, 14(4), 1259–1270, 2004]

skip-free A concept arising naturally in the context of a random walk (or Markov chain)$X(t)$ on the integers $\mathbb{Z}$. If the transition probabilities p_{ij} (or generator, for $t \in \mathbb{R}$), of X have the property that $p_{ij} = 0$ for $j \leq i - 2$, then X is said to be skip-free to the left (or downwards), or left-continuous. Likewise, if $p_{ij} = 0$ for $j \geq i - 2$, X is skip-free to the right (or upwards), or right-continuous. If p_{ij} has both these properties, then X is simply skip-free, and may be called a simple random walk; or a birth–death process if the state space is $\mathbb{Z}^+$.

More generally, a Markov chain in discrete-time on a more general state-space may be called skip-free if the steps $X(n + 1) - X(n)$ are uniformly bounded (in an appropriate norm) by a constant, a.s. for all n, and all initial conditions. [M. Brown *et al.*, Some results for skip-free random walk, *Probab. Eng. Inf. Sci.*, 24(4), 491–507, 2010]

Skitovich–Darmois theorem (alias: **Skitovič–Darmois theorem**) A characterization of the normal distribution, thus: Let $X_1, \ldots, X_n, n \geq 2$, be independent; if, for some non-zero constants $a_1, \ldots, a_n, b_1, \ldots, b_n$, $U = \sum_{r=1}^n a_r X_r$ is independent of $V = \sum_{r=1}^n b_r X_r$, then each X_r is normally distributed. The case $n = 2$ is Bernstein's (or the Bernstein–Kac) thm; the case $n = 4$ implies a related thm of H. Cramér. More generally, for random vectors $\mathbf{X}_1, \ldots, \mathbf{X}_n$ of dimension d, the independence of $\sum_r A_r \mathbf{X}_r^{\mathrm{T}}$ and $\sum_r B_r \mathbf{X}_r^{\mathrm{T}}$ for non-singular square matrices A_r and B_r $1 \leq r \leq n$, entails the multivariate normality of each $\mathbf{X}_r$. A related result is **Heyde's thm**, which asserts that if $a_r/b_r + a_s/b_s \neq 0$ for any r and s, and the conditional distribution of U given V is symmetric, then each X_r is normal. An extension to certain algebraic structures is more often called the **Darmois–Skitovich thm.** [M. Quine & E. Seneta, The generalization of the Kac–Bernstein theorem, *Probab. Math. Stat.*, 19(2), 441–452, 1999]

Sklar's theorem *see* copula

Skorokhod embedding *see* Skorokhod imbedding

Skorokhod's equation (alias: **Skorokhod's lemma (or problem)**) Arises in the context of reflecting Brownian motion $X(t) = |W(t) + z|, z \geq 0$, where $W(t)$, $t \geq 0$, is the Wiener process. Then $X(t)$ has the fdds of the solution of $X(t) = X(0) + Z(t) + L(t)$, where $Z(t)$ is a Wiener process independent of $X(0)$, and $L(t)$ is a non-decreasing process that is flat unless $X(t) = 0$, i.e., $\int_0^t I(X(s) \neq 0) \mathrm{d}L(s) = 0$, and $L(0) = 0$. Furthermore, $L(t) = \max[0, \max_{0 \leq s \leq t}\{-Z(s) + X(0)\}]$ is often called Skorokhod's lemma.

More generally, Skorokhod's problem and equation denote the solution of similar problems of reflected BM in $\mathbb{R}^d, d \geq 2$. [I. Karatzas & S. Shreve, *Brownian Motion and Stochastic Calculus*, 2nd edn. Springer, 1998]

Skorokhod imbedding (or embedding) (alias: **Skorokhod representation**) In its basic form this comprises a Wiener process $W(t)$, with filtration $\mathcal{F}_t$, and a stopping time T such that $W(T)$ has the distribution of an arbitrary zero-mean probability distribution on $\mathbb{R}$, where $W(T \wedge t)$ is a uniformly integrable martingale. Skorokhod's extended imbedding gives a sequence of stopping times $T_1, T_2, \ldots$ such that $W(T_r)$, $r \geq 1$ have the joint distributions of a random walk $S_n = \sum_{r=1}^n X_r$. Even more generally, one imbeds a martingale $(M_n), M_0 = 0$, in $W(\cdot)$ with stopping times $T_1, T_2, \ldots$ such that $W(T_n)$, $n \geq 1$ has the same distribution as M_n, $n \geq 1$, and so on. [J. Obłój, The Skorokhod embedding problem and its offspring, *Probab. Surv.*, I, 321–329, 2004]

Skorokhod's inequality Let $S_n = \sum_{r=1}^n X_r$ be the partial sums of a sequence $X_1, X_2, \ldots$ of independent r.vs having the property that for some $c \in (0, 1)$, $\sup_{r \leq n} \mathbb{P}(|S_n - S_r| > x) = c$, for some $x \in (0, \infty)$.

Then $\mathbb{P}(\sup_{r \leq n} |S_r| > 2x) \leq \frac{1}{1-c} \mathbb{P}(|S_n| > x)$. [L. Breiman, *Probability*. Addison-Wesley, 1968; reprinted by SIAM, 1992]

Skorokhod integral (alias: **Hitsuda integral**) A stochastic integral that is essentially the Itô stochastic integral in the case when the integrand is not adapted to the Wiener space (i.e. not adapted to the filtration generated by the Wiener process $W(t), t \geq 0$, against which it is integrated). [C. Tudor, Martingale-type stochastic calculus for anticipating integral processes, *Bernoulli*, 10(2), 313–325, 2004]

Skorokhod metric Arises in metrizing the Skorokhod space, i.e. the set of all rcll real-valued functions defined on $[0, \infty)$ (or some subset such as $[0, 1]$, which defines the Skorokhod topology. **Billingsley's metric** is equivalent, and makes the Skorokhod metric space complete as well as separable. [P. Billingsley, *Convergence of Probability Measures*, 2nd edn. Wiley, 1999]

Skorokhod's theorem (alias: **Skorokhod coupling (or representation)**) In its basic form this represents the convergence of a sequence of probability distributions $F_n(x)$, $n \geq 1$, on $\mathbb{R}$ in terms of an a.s. convergent sequence of r.vs X_n, $n \geq 1$. Suppose that $\lim_{n\to\infty} F_n(x) = F(x)$ at the continuity points of F; then there exist X_n, $n \geq 1$, and X defined on the same space $(\Omega, \mathcal{F}, \mathbb{P})$, such that $\lim_{n\to\infty} X_n = X$ with probability one, and X_n and X have distribution functions F_n and F respectively. More generally, a similar result holds for a sequence of probability measures converging weakly on a separable metric space. The result supplies attractive proofs of results such as the Mann–Wald, or continuous mapping, theorem. Note that the representation says nothing about any joint distributions of either sequence. [R. Bass, *Stochastic Processes*. Cambridge University Press, 2011]

Slepian's inequality (or lemma) Any one of several inequalities (called **Gaussian comparison inqualities**) comparing the distributions of maxima, and intervals of non-negativity, for Gaussian processes having different covariance functions; as well as related inequalities for multinormal random vectors. One form is this: let $X_1, \ldots, X_n$ be a zero-mean multivariate normal r.vr, with covariance matrix $V = (v_{ij})$. Then both $\mathbb{P}(\min_r X_r \geq x)$ and $\mathbb{P}(\max_r X_r \leq x)$ are increasing in v_{ij} for all $i \neq j$. A related result is **Pitt's thm**, which asserts that $X_1, \ldots, X_n$ are associated if $v_{ij} \geq 0$ for all i and j. More strongly, let $h(x_1, \ldots, x_n)$ be a sufficiently smooth and well-behaved function such that $\frac{\partial^2 h}{\partial x_i \, \partial x_j} \geq 0$ everywhere for some $i \leq j$. Then $\mathbb{E}h(X_1, \ldots, X_n)$ is increasing in v_{ij}. An extension to Gaussian random fields, under weaker conditions, is called the **Sudakov–Fernique inequality**. [K. Joag-Dev *et al.*, Association of normal random variables and Slepian's inequality, *Ann. Probab.*, 11(2), 451–455, 1983; D. Slepian, The one-sided barrier problem for Gaussian noise, *Bell Syst. Tech. J.*, 41, 463–501, 1962]

Slepian–Wolf theorem (or coding) Arises in communication theory in seeking the most efficient encoding of the output (X, Y) of two sources, where (X, Y) are jointly distributed, i.e. the data-compression problem for correlated sources. The theorem asserts that separate (i.e. distributed) coding is no less efficient than the optimal joint encoding, i.e. both achieve the entropy lower bound $H(X, Y)$. [T. Cover & J. Thomas, *Elements of Information Theory*, 2nd edn. Wiley, 2006]

Slivnyak's theorem This is a characterization of the Poisson process, in $\mathbb{R}^d$ or in some more general Borel space. Informally, a simple point process $N(\mathbf{x})$ is a Poisson point process if and only if its properties, as seen from any point $\mathbf{x} \in \mathbb{R}^d$, are the same whether or not we consider them conditional on the event that $\mathbf{x}$ is a point of the process; provided only that the point $\mathbf{x}$ is not itself considered in the property in question. More formally, a simple point process with finite positive intensity is a Poisson process if and only if its reduced Palm measure at $\mathbf{x}$ (i.e. the measure conditional on $\mathbf{x}$ being a point of the process, but discounting $\mathbf{x}$ itself) is the same as the original distribution. A related integral form is called the Slivnyak–Mecke formula (or eqn, or thm). [O. Kallenberg, Iterated Palm conditioning and some Slivnyak-type theorems for Cox and cluster processes, *J. Theor. Probab.*, 24, 875–893, 2011]

SLLN *see* strong law of large numbers

slow variation *see* regular variation

Slutsky–Fréchet theorem (alias: **Slutsky's convergence theorem**) Addresses the question of convergence in distribution for a suitably continuous function f of a sequence of random vectors $(X_{n1}, \ldots, X_{nd}), = \mathbf{X}_n; n \geq 1$, where each component converges in distribution as $n \to \infty$. Formally, if $\mathbf{X}_n \to^{d} \mathbf{X}$ as $n \to \infty$, and, with probability 1, $\mathbf{X}$ is in the set of points at which f is continuous, then $f(\mathbf{X}_n) \to^{d} f(\mathbf{X})$ as $n \to \infty$. The theorem is most often seen for rational functions f, especially, e.g. $f(x, y) = xy$, $f(x, y) = x/y$, and $f(x, y) = x + y$. [J. Hammersley, An extension of the Slutsky–Fréchet theorem, *Acta Mathematica*, 87(1), 243–257, 1952]

Slutsky's theorem *see* sinusoidal limit theorem; Slutsky–Fréchet theorem

Slutsky–Yule effect (alias: **Slutsky effect**) Arises in the analysis of stationary time series where it refers to the observable (empirical) fact that repeatedly filtering or smoothing such a series by a moving average may yield a sequence exhibiting oscillatory (cyclical) behaviour, even when the original sequence is white noise. *See* sinusoidal limit theorem. [G. Yule, Why do we sometimes get nonsense-correlations between time series?, *J. Roy. Stat. Soc.*, 89(1), 1–63, 1926]

small ball probability (alias: **small value (or deviation) probability; small value (or ball) problem**) Broadly, given a sequence of non-negative r.vs Y_n, $n \geq 1$, and a sequence of non-negative constants c_n, such that $\lim_{n\to\infty} c_n = 0 = \lim_{n\to\infty}\mathbb{P}(Y_n \leq c_n)$, the small ball problem seeks to describe the asymptotic rate at which the small ball probability $\mathbb{P}(Y_n \leq c_n)$ tends to zero as $n \longrightarrow \infty$. [W. Li, & Q.-M. Shao, Gaussian processes: inequalities, small ball probabilities and applications, in *Stochastic Processes: Theory and Methods, Handbook of Statistics*, Vol. 19, C. Rao & D. Shanbhag (eds). Elsevier, 2001]

small set Arises in considering a Markov chain $X_n, n \geq 1$, on a general state space, where (roughly speaking) small sets fulfil the same function that point sets (singletons) perform in analysing recurrence and transience for a chain on a discrete (countable) state space. Formally, denoting the n-step transition kernel of the chain by $p^n(x, B)$, a subset S of the state space is small if for all $x \in S$ and all Borel sets B, $p^n(x, B) \geq m(B)$ (called a **minorization condition**), for some positive measure $m(\cdot)$, and some $n < \infty$. More generally, for a transition kernel $q(x, B) = p^T(x, B)$ obtained by a suitable random sampling at time T of the transition kernels of X_n, a set S satisfying the corresponding minorization condition is called a **petite set**. Every small set is petite, and petite sets are small for aperiodic irreducible chains. [S. Meyn & R. Tweedie, *Markov Chains and Stochastic Stability*, 2nd edn. Cambridge University Press, 2009]

small-world network *see* Watts–Strogatz small-world model

Smirnov's theorem *see* Cardy's formula

Smith–Wilkinson model *see* random environment

Smoluchowski equation Arises in Smoluchowski's general model for coagulation (*see* coalescence) where it comprises an integrodifferential equation for the evolution of the expected number $m(s, t)$ of clumps of size s at time t. *See also* Chapman–Kolmogorov equation. [D. Aldous, Deterministic and stochastic models for coalescence (aggregation and coagulation), *Bernoulli*, 5(1), 3–48, 1999]

Smoluchowski process A strictly stationary vector random process describing the occupation numbers of disjoint regions of some space by particles executing i.i.d. random motions from an initial Poisson point process. [P. McDunnough, Some aspects of the Smoluchowski process, *J. Appl. Probab.*, 15(4), 663–674, 1978]

smoothing inequality (alias: **Berry, or Esseen, or Berry–Esseen inequality**) A type of inequality that supplies a useful bound for the uniform (or sup norm, or Kolmogorov, or Chebyshov) distance $d(F, G) = \sup_{x\in\mathbb{R}}|F(x) - G(x)|$ between a distribution function F and a function $G(x)$ of bounded variation with $G(-\infty) = 0$ and $G(\infty) = 1$. One form is this: Let $\sup_x|G'(x)| \leq c < \infty$, then for $0 < T < \infty$

$$d(F, G) \leq \frac{1}{\pi}\int_{-T}^{T} \left|\frac{\phi(t) - \psi(t)}{t}\right| \mathrm{d}t + \frac{24c}{\pi T},$$

where ϕ and ψ are the Fourier–Stieltjes transforms of F and G respectively (i.e. the chf in the case of F).

More generally, a bound involving the so-called smoothed difference

$$\sup_x \left| \int_{\mathbb{R}} (F(t) - G(t))v(x - t)\mathrm{d}t \right|$$

for a suitable function v may be called a smoothing inequality. For sufficiently nice $v(\cdot)$, this may in turn be bounded by an integral of $|\phi - \psi|$, as above. [W. Feller, *Introduction to Probability Theory and its Applications*, Vol. 2, 2nd edn. Wiley, 1971; C. Esseen, Fourier analysis of distribution functions, *Acta Math.*, 77, 1–125, 1945; C. Esseen, A moment inequality, *Skand. Actuar.*, 3–4, 160–170, 1956]

smoothing process An interacting particle system introduced by F. Spitzer, where transitions are defined in terms of suitable weighted averages of the current configuration. A related process, which can be seen as a dual of the smoothing process, is the **potlatch process**. [F. Spitzer, Infinite systems with locally interacting components, *Ann. Probab.*, 9, 349–364, 1981; R. Holley & T. Liggett, Generalized Potlatch and smoothing processes, *Probab. Theor. Rel. Fields*, 55, 165–195, 1981]

smoothing property *see* tower property

smoothing transform A concept arising in branching processes and related fields, that is defined as follows: let $(A, B_1, B_2, \ldots)$ be a sequence of r.vs such that each B_r is non-negative, and $\mathbb{P}(\sup\{r \geq 1 : B_r > 0\} < \infty) = 1$; for any sequence $X_r, r \geq 1$ of i.i.d.r.vs with distribution $\mathbb{P}_X(\cdot)$, and independent of $(A, B_1, B_2, \ldots)$, the smoothing transform is the r.v. Y, with distribution $\mathbb{Q}(\cdot)$, given by

$$Y = A + \sum_{r\geq 1} B_r X_r.$$

Interest centres on fixed points of this when $\mathbb{P}_X = \mathbb{Q}$. [G. Alsmeyer & M. Meiners, Fixed points of the smoothing transform; two-sided solutions, *Probab. Theor. Rel. Fields*, 155, 165–199, 2013]

Snedecor's F distribution *see* F distribution

Snell envelope Arises in optimal stopping problems, in particular in Snell's problem, which seeks the best time for a gambler to quit; and also in pricing American options. Given a random process X_n, adapted to a filtration $\mathcal{F}_n$, $n \geq 0$, the Snell envelope, S_n of X_n, is the smallest supermartingale, adapted to the filtration $\mathcal{F}_n$, that dominates (or majorizes) X_n, i.e. $S_n \geq X_n$ for all n with probability 1. The envelope may be defined by a suitable backward induction on any finite interval. [J. Neveu, *Discrete-parameter Martingales*. North-Holland, 1975]

soft-sphere model *see* hard-sphere model

sojourn time A term used to denote several quantities:

1. In queueing theory the sojourn time of a customer is the time elapsing between first arrival for service and final departure from the system.
2. The holding time of a Markov chain between entering a state and first departure.
3. The occupation time of a class of random processes, i.e. the total time spent in a state (or subset of states) over some interval. *see* local time.
4. The time spent by a random process in a subset B of its state space during a single visit, i.e. the length of the interval (in continuous time) or the number of steps (in discrete time) between entering B and its first subsequent exit from B.

Ambiguity is possible in some contexts.

solidarity theorem Arises in considering e.g. Markov chains and Markov renewal processes that are irreducible. A theorem that asserts, of a given property that any state of the state space may have, that either all states have that property or none do is called a solidarity theorem. Thus, for a discrete-time countable state space irreducible chain, recurrence, transience, and aperiodicity are solidarity properties. Such theorems also hold for geometric (or exponential) ergodicity, e.g. those of D. Kendall, D. Vere-Jones, J. F. C. Kingman *et al.* [J. Teugels, Exponential ergodicity in Markov renewal processes, *J. Appl. Probab.*, 5(2), 387–400, 1968]

Solomonoff probability *see* algorithmic probability; Kolmogorov complexity

Solovay–Strassen test A randomized (i.e. probabilistic, or Monte Carlo) procedure for testing whether a given (arbitrary) number n is prime, if n is prime the test gives a correct decision, if n is composite the probability of an incorrect decision does not exceed ½. A related (more efficient) result is the so-called **Rabin–Miller test**, having similar properties. [M. Rabin, Probabilistic algorithm for testing primality, *J. Num. Theor.*, 12(1), 128–138, 1980]

Song's measure A measure of the shape of the distribution of a r.v X (or r.vr $\mathbf{X}$), having density $f(x)$ (or joint density $f(\mathbf{x})$, given by $S(f) = \text{var} \log f(X)$. It can be seen as considerably extending the idea of kurtosis to heavy tailed distributions with few (or no) moments. [K. Song, Rényi information, log likelihood and an intrinsic distribution measure, *J. Stat. Plan. Inf.*, 93(1–2), 51–69, 2001]

sortilege *see* lottery

source In a variety of network and queueing models, a vertex from which customers (or other objects) may enter the system is a source; while a vertex which is accessible, but has no links to any other vertex, is a **sink**. In communication theory, a source is a random process; in particular, in discrete time, a sequence of i.i.d.rvs taking values in an alphabet A is called a **discrete memoryless source**, while a sequence of i.i.d.r.vs having a normal density is a memoryless Gaussian source.

source–channel separation theorem (alias: **information transmission theorem; Shannon's separation theorem**) Arises in communication theory, where its basic form asserts that (asymptotically, as the length of the message increases to any arbitrarily long length), optimality in transmitting information from a source through a noisy channel is still achievable even if the operations of compressing the data from the source (compact coding) and coding for error corrections (channel coding) are performed separately in sequence. [C. Shannon, Coding theorems for a discrete source with a fidelity criterion, *IRE Nat. Conv. Rec.*, 7(4), 142–163, 1959]

space-fractional diffusion *see* Lévy–Feller diffusion

Sparre Andersen's model (or process) (alias: **Andersen's model**) Generalizes the classical risk-reserve process seen in the ruin problem in permitting the point process of claim times to be a renewal process, rather than a Poisson process. [T. Rolski *et al.*, *Stochastic Processes for Insurance and Finance*. Wiley, 1999]

Sparre Andersen's theorem (or identity) (alias: **Andersen's theorem (or identity)**) Arises in considering the partial sums $S_k = \sum_{r=1}^{k} X_r, 1 \leq k \leq n$, of a collection of exchangeable random variables $X_1, \ldots, X_n$, and the term denotes any of several related results in the fluctuation theory of random walk. Thus, let N_n be the number of strictly positive elements in $S_1, \ldots, S_n$, then N_n has the same distribution as the time of the first maximum in $0, S_1, \ldots, S_n$. It follows that for $0 \leq r \leq n, \mathbb{P}(N_n = r) = \mathbb{P}(N_r = r)\mathbb{P}(N_{n-r} = 0)$, which is also called Sparre Andersen's identity. [W. Feller, *Introduction to Probability Theory and its Applications*, Vol. 2, 2nd edn. Wiley, 1971]

Spearman's rho *see* grade correlation

spectral distribution Arises in the spectral analysis (or decomposition, or representation) of stationary time series (essentially, Fourier analysis). Thus, if $X_n,\ n \in \mathbb{Z}$, is a (possibly complex-valued) second-order stationary process with autocorrelation function $r(n)$, then Herglotz' theorem asserts that $r(n) = \int_{-\pi}^{\pi} \mathrm{e}^{\mathrm{i}n\lambda} \mathrm{d}F(\lambda)$, for some distribution function $F(\lambda),\ -\pi \leq \lambda \leq \pi$, called the spectral distribution function. When $f(\lambda) = \frac{\mathrm{d}F}{\mathrm{d}\lambda}$ exists, it is called the **spectral density**.

If X_n is real, then $r(n) = 2\int_0^{\pi} \cos n\lambda \mathrm{d}F(\lambda)$. Similarly, for a continuous-time stationary process $X(t)$, with autocorrelation $r(t)$, $r(t) = \int_{-\infty}^{\infty} \mathrm{e}^{\mathrm{i}t\lambda} \mathrm{d}F(\lambda)$. The set of pure jumps of $F(\lambda)$ is called the **spectrum**; and, conversely, the spectral distribution $F(\lambda)$ may be given in terms of the autocorrelation function. Furthermore, there is a direct **spectral representation** of X_n as $X_n = \int_{-\pi}^{\pi} \mathrm{e}^{\mathrm{i}t\lambda} \mathrm{d}Z(\lambda)$, (with a similar expression for $X(t)$ in continuous time) for some orthogonal-increments random process $Z(\lambda)$ defined on $[-\pi, \pi]$, called the **spectral process**. This may be called the **Cramér**, or **Cramér–Loève, representation**. [P. Brockwell & R. Davis, *Time Series*, 2nd edn. Springer, 1991]

spectral gap inequality *see* Poincaré inequality

speed function (or density) (alias: **canonical measure**) Arises in considering linear diffusion processes, where it can be seen as governing the rate at which the diffusion traverses its path. Thus, if $X(t) \in [a, b], t \geq 0$, is a regular diffusion on its natural scale, and $I = (x - \varepsilon,\ x + \varepsilon) \leq [a, b]$, then the expected time that $X(t)$, started at x, spends in I before exiting the interval is $m(x)\varepsilon^2 + \mathrm{o}(\varepsilon^2)$ as $\varepsilon \downarrow 0$, where the function $m(x)$ is called the speed density, and $M(A) = \int_A m(x)\mathrm{d}x$ determines the canonical (or speed) measure $M(\cdot)$. Informally, $m(x)$ shows how fast the 'clock' of the process runs at x. Formally, for a regular diffusion with instantaneous variance $\sigma^2(x)$, and scale function $s(x)$, the speed density is

$$m(x) = \frac{1}{s'(x)\sigma^2(x)};$$

for the Wiener process this is Lebesgue measure. Then, in general, the expected exit time $e(a, b)$ of X, started at x in I, from I, is given by $e(a, b) = \int_a^b G_{ab}(x, y)M(\mathrm{d}y)$, where G_{ab} is called the Green function of the diffusion. For a diffusion on its natural scale, a suitable time-change by means of the speed function yields the Wiener process. [R. Bass, *Stochastic Processes*. Cambridge University Press, 2011]

spent waiting time *see* age

sphere walk *see* ball walk

spherical Brownian motion *see* Brownian motion

spherical distribution (alias: **hyperspherical distribution; directional distribution**) Arises naturally in random experiments whose outcomes are unit vectors (signed or unsigned) $\mathbf{V}$ representing a direction in d dimensions, $d \geq 2$, so that the sample space (and support of the distribution) is a $(d-1)$-dimensional sphere, being a circle when $d = 2$. When the vector is unsigned (so that $-\mathbf{V}$ is the same sample point as $\mathbf{V}$) one speaks of an axial spherical distribution. In three dimensions $(d = 3)$ the analogue of the bivariate normal distribution on the plane is called Fisher's distribution. *See* Bingham distribution. [J. Kent, The Fisher–Bingham distribution on the sphere, *J. Roy. Stat. Soc. B*, 44(1), 71–80, 1982; K. Mardia & P. Jupp, *Directional Statistics*. Wiley, 2000]

spherical matrix distribution An $r \times n$ random matrix X is called spherical if its distribution is invariant both under pre- and post-multiplication by a suitable orthogonal matrix. If only the former (or the latter) holds, it is said to be left-(or right-) spherical, respectively. [A. P. Dawid, Spherical matrix distributions and a multivariate model, *J. Roy. Stat. Soc. B*, 39(2), 254–261, 1977]

spine decomposition (or technique) A method for analysing the Galton–Watson branching process (and, more generally, branching diffusion or branching r.w., etc.), which identifies a path of descent in successive generations that is called the spine. It is used to construct a different measure on the natural space of trees. [R. Lyons *et al.*, Conceptual proofs of L log L criteria for mean behavior of branching processes, *Ann. Probab.*, 23(3), 1125–1138, 1995]

spin-glass model (or system) Arises in statistical mechanics when seeking to extend the basic theory of ferromagnetism (*see* Ising model) to dilute magnetic alloys (called spin glasses). The classic formulation is the **Edwards–Anderson model,** in which the Gibbs distribution of configurations **S** on a suitable lattice or graph is proportional to exp $\left(-b\sum_{i<j} J_{ij}s_is_j\right)$, where the interaction function J_{ij} comprises i.i.d. symmetric r.vs in $\mathbb{R}$. *See also* Sherrington–Kirkpatrick model. [M. Talagrand, *Spin Glasses: A Challenge for Mathematicians.* Springer, 2003]

Spitzer's condition *see* arcsine law

Spitzer's identity (or formula) *see* Pollaczek–Spitzer identity

Spitzer's law Arises in planar Brownian motion (i.e. a two-dimensional Wiener process), where it considers the winding number $N(t)$ of the process about the origin, and asserts that as $t \to \infty$, for $x \in \mathbb{R}$,

$$\mathbb{P}\Big(2N(t)/\log t \le x\Big) \to \int_{-\infty}^{x} \frac{\mathrm{d}u}{\pi(1+u^2)},$$

which is the Cauchy distribution. [P. Mörters & Y. Peres, *Brownian Motion.* Cambridge University Press, 2010]

Spitzer–Rogozin criterion (or theorem) *see* Pollaczek–Spitzer identity

Spitzer's theorem Any of several results in the theories of random walk, Brownian motion, or interacting particle systems, inter alia; but perhaps most often (other than the above) this law of large numbers: let X_r, $r \ge 1$, be i.i.d.r.vs, and set $S_n = \sum_{r=1}^{n} X_r$. Then $\sum_{n\ge1} \frac{1}{n}\mathbb{P}(|S_n| > an) < \infty$ for all $a > 0$ iff $\mathbb{E}X_1 = 0$ and $\mathbb{E}(X_1) < \infty$. [A. Spătaru, Precise asymptotics in Spitzer's law of large numbers, *J. Theor. Probab.*, 12(3), 811–819, 1999; C. Tudor, Hsu-Robbins' and Spitzer's theorems for the variations of fractional Brownian motion, *Electr. Commun. Probab.*, 14, 278–289, 2009]

splitting *see* Athreya–Ney–Nummelin pseudo-atom construction

spreadability A type of distributional symmetry that is weaker than exchangeability. A sequence of r.vs (or random elements) $X_1, X_2, \ldots$ is said to be spreadable (or contractable, or spreading invariant) if for any increasing subsequence $i_1, \ldots, i_n$, $n < \infty$, $X_{i_1}, \ldots, X_{i_n}$ has the same joint distributions as $X_1, \ldots, X_n$. [O. Kallenberg, *Probabilistic Symmetries and Invariance Principles.* Springer, 2005]

spread option *see* exotic option

spread-out A distribution function $F(x)$ is said to be spread out if for some finite n the n-fold convolution F^{*n} of F with itself is not singular wrt Lebesgue measure (i.e. it has an absolutely continuous component). A spread-out measure on some space is defined similarly. If F has no discrete component it is called **diffuse**; *see* atomic. [S. Asmussen, *Applied Probability and Queues*, 2nd edn. Springer, 2003]

squared Bessel process *see* Bessel process

square-root diffusion *see* Cox–Ingersoll–Ross model

square-root trick Arises in the context of percolation and related problems, where an event A in a suitable probability space $(\Omega, \mathcal{F}, \mathbb{P})$ is called increasing if $I_A(\omega) \le I_A(\omega')$ whenever $\omega \le \omega'$; (I_A denotes the indicator of A). If A and B are increasing, and $C = A \cup B$, and $\mathbb{P}(A) = \mathbb{P}(B)$, then it follows (using the Harris–FKG inequality) that

$$\mathbb{P}(A) \ge 1 - (1 - \mathbb{P}(C))^{1/2},$$

and this result is called the square root trick. More generally, if $A_1, \ldots, A_n$ are all increasing, and $C = \cup_1^n A_r$, and $\mathbb{P}(A_r) = \mathbb{P}(A_s)$, all r, s, then $\mathbb{P}(A) \ge 1 - \{1 - \mathbb{P}(C)\}^{1/n}$, which may be called the ***n*th root trick**. [J. Cox & R. Durrett, Limit theorems for the spread of epidemics and forest fires, *Stoch. Proc. Appl.*, 30, 171–191, 1988]

squeezing method *see* envelope method

S-shaped theorem Arises in reliability theory where it asserts that for a wide class of networks having coherent structure functions, the reliability function (graphed as a function of component reliability p) crosses the diagonal at most once and always from below, i.e. it is S-shaped. [J. Esary & F. Proschan, Coherent structures of nonidentical components, *Technometrics*, 5(2), 191–209, 1963]

S-s policy *see* single-period model

stability of statistical ratios Refers to the empirical observation that in a long sequence of N similar trials (i.e. experiments conducted under essentially the same conditions) of which $N(A)$ yield some fixed event A, the ratio $N(A)/N$ almost always seems to settle around a number p in $[0,1]$ as N increases. A classic example, first noted by J. Graunt in 1662, is the sex ratio in infants. This natural law supports the frequency interpretation of probability, and K. Pearson has noted that its importance parallels the fundamental principle of the exact sciences that the same causes produce the same effects. [K. Pearson, *The History of Statistics in the 17th and 18th Centuries*, E. Pearson (ed.). Griffin, 1978]

stable age distribution Arises in population models when the proportion of the population in any fixed set (or interval) of ages converges to (or is) a constant. In particular, for the classic age-dependent (or Bellman–Harris) branching process with lifetime distribution function $G(x) = 1 - \overline{G}(x)$, the expected proportion of the population of age less than a converges to the stable age distribution

$$\int_0^a \mathrm{e}^{-mx}\overline{G}(x)\mathrm{d}x \Big/ \int_0^\infty \mathrm{e}^{-mx}\overline{G}(x)\mathrm{d}x,$$

where m is the Malthusian parameter. [P. Haccou *et al.*, *Branching Processes*. Cambridge University Press, 2005; F. Hoppensteadt, *Mathematical Theory of Populations*. SIAM, 1975]

stable distribution (alias: **Lévy-(or alpha-) stable distribution**) A class of distributions that arises when seeking random variables X whose distributions are stable in the sense that the sum of independent copies of X has the same distribution as X after linear rescaling; this class coincides with those distributions that can appear as the limiting distribution of a sequence $a_n^{-1}\sum_{r=1}^{n} X_r - b_n$, for i.i.d.r.vs $X_1, X_2, \ldots$, and suitable constants $a_1, a_2, \ldots, b_1, b_2, \ldots$; *see* domain of attraction. (Note that such distns may be called **sum-stable**, to distinguish them from those that are max-stable.) Equivalent definitions of stable include:

1. A non-degenerate random variable X (or its distribution F) is said to be stable (or broad-sense, or weakly, or quasi-stable) if for any integer n there exist $c_n > 0$ and d_n such that the sum $S_n = X_1 + \cdots + X_n$ of i.i.d. copies of X has the same distribution as $c_nX + d_n$. If $d_n = 0$ for all n, then X (or F) is stable in the strict (or narrow) sense.
2. The r.v. X is stable if for independent copies X_1 and X_2 of X, and any positive constants a and b, there exist $c > 0$ and d such that $aX_1 + bX_2$ has the same distribution as $cX + d$. If $d = 0$ for all a and b, then X is strictly (or narrowly) stable.
4. A r.v. X is stable if its chf $\phi(t)$ is given by log $\phi(t) = imt - c|t|^\alpha\{1 - ib\,\mathrm{sign}(t)\,h(t,\alpha)\}$ where $m \in \mathbb{R}, 0 < \alpha \le 2, c > 0, |b| \le 1$, and

$$h(t,\alpha) = \begin{cases} \tan(\alpha\pi/2), & \alpha \ne 1 \\ -(2/\pi)\log|t|, & \alpha = 1; \end{cases}$$

and sign (t) is the usual sign function. It is then said to be α-stable, or stable of order α, or stable with (characteristic) exponent α; or index α. Stable distributions have a unimodal and infinitely divisible density whose support is either $\mathbb{R}$ or a half-line. Pleasant closed-form representations for the density exist in three cases, with chfs:

(i) $\phi = \exp\{imt - ct^2\}$ – normal density;
(ii) $\phi = \exp\{imt - c|t|\}$ – a Cauchy density;
(iii) $\phi = \exp[imt - c|t|^{\frac{1}{2}}\{1 - \mathrm{i}\,\mathrm{sign}(t)\}]$; – a Lévy (or first-passage) density;

these being α-stable with $\alpha = 2, \alpha = 1$, and $\alpha = \frac{1}{2}$ respectively. Stable distributions with $\alpha = \frac{3}{2}$ include the Airy and Holtsmark distributions.

When $\phi(t) = \exp\{-c|t|^\alpha\}$, X is said to have a symmetric α-stable distn. P. Lévy also defined a class of **semi-stable distns**: a r.v. or its distn is (weakly) semi-stable if for some $0 < a \ne 1$ there exists $b > 0$ and c such that its chf $\phi(t)$ satisfies $\phi(t)^a = \phi(bt)\mathrm{e}^{ict}$. If $c = 0$, it is strictly (or narrowly) semi-stable. [P. Hall, A comedy of errors: the canonical form for a stable characteristic function, *Bull. Lond. Math. Soc.*, 13, 23–27, 1981; T. Huillet et al., On Lévy stable and semistable distributions, *Fractals*, 9(3), 347–364, 2001]

stable process Arises in the context of a Lévy process (i.e. a stochastically continuous process $X(t)$, $t \geq 0$, starting at 0, with independent stationary increments). Such a process is called a stable (resp. strongly stable) process if $X(1)$ is a stable (resp. strongly stable) r.v. Likewise, $X(t)$ is semi-stable if it is a Lévy process such that $X(1)$ has a semi-stable distribution. (Note that such processes may be called sum-stable to distinguish them from those that are max-stable.)

More generally, a **stable random field** has a multidimensional time parameter $\mathbf{t} \in \mathbb{R}^n$ (or $\mathbf{t} \in \mathbb{Z}^n$), and has the property that for all $\mathbf{t}_1, \ldots, \mathbf{t}_m$, and $m \geq 1 (X(\mathbf{t}_1), \ldots, X(\mathbf{t}_m))$ has a stable distribution. In the case of strictly stable, or symmetrically stable fields, an equivalent definition is that all linear combination $\sum_{j=1}^{m} a_j X(\mathbf{t}_j)$ are strictly stable, or symmetrically stable, where $a_1, \ldots, a_m$, and $m \geq 1$, are arbitrary. This may be called a sum-stable random field to distinguish it from a max-stable random field in which the maximum of arbitrary linear combinations has a scaled max-stable distribution. [K. Sato, *Lévy Processes and Infinitely Divisible Distributions*. Cambridge University Press, 1999; J. Rosiński, Decomposition of stationary α-stable random fields; *Ann. Probab.*, 28(4), 1797–1813, 2000; Y. Wang et al., Ergodic properties of sum- and max-stable random fields via null and positive group actions, *Ann. Probab*, 41(1), 206–228, 2013; M. Schlather, Models for stationary max-stable random fields, *Extremes*, 5(1), 33–44, 2002]

stable queue A queue, or more generally a queueing network, is said to be stable if the Markov process of queue lengths (at the nodes of the network) is non-null recurrent, or (more generally) positive Harris recurrent. For the simplest queues with a single server the queue is stable if the traffic intensity is less than unity; in queueing networks some counter-intuitive results are possible. [M. Bramson, Stability of queueing networks, *Probab. Surv.*, 5, 169–345, 2008]

stable random field *see* stable process

stable state A state s of a Markov chain $X(t)$, $t \geq 0$, on a countable state space is said to be stable if the holding time of X in s is a.s. strictly positive. In terms of the Q-matrix this is equivalent to $q_s < \infty$; when $q_s = \infty$ the state is called instantaneous (and when $q_s = 0$ the state is absorbing). [W. Anderson, *Continuous-time Markov Chains*. Springer, 1991]

stable tail dependence *see* tail dependence measure

Stam's inequality Any one of several inequalities satisfied by Fisher's information and the (Shannon) entropy function, perhaps most often the term denotes these: Let $Z = X + Y$, where X, Y, and Z are r.vs each having a density (i.e. are absolutely continuous), and write the Fisher information of X in terms of its density f_X as $I_X = \int_{\mathbb{R}} f\prime_X^2 / f_X \mathrm{d}x$, etc. Then for independent X and Y, $1/I_X + 1/I_Y \leq 1/I_{X+Y}$; with equality iff X and Y are normally distributed.

The equivalent assertion that for any constants a and b $(a+b)^2 I_Z \leq a^2 I_X + b^2 I_Y$, is called **Blackman's** or the **Blackman–Stam inequality**; the latter term also being used for the multivariate form of the **entropy power (or Shannon–Stam) inequality**, $\mathrm{e}^{2H(\mathbf{X}+\mathbf{Y})/n} \geq \mathrm{e}^{2H(\mathbf{X})/n} + \mathrm{e}^{2H(\mathbf{Y})/n}$, where $H(\cdot)$ is the usual Shannon entropy, with equality iff the independent r.vrs $\mathbf{X}$ and $\mathbf{Y}$ (of length n), are normally distributed, having proportional covariance matrices. [N. Blackman, The convolution inequality for entropy powers, *IEEE Trans. Inf. Theor.*, 2, 267–271, 1965]

standard Broadly this term often serves to identify one among a family of objects as typical or representative, or convenient for purposes of reference, or from which the entire family can be generated. Thus the standard normal density has mean 0 and variance 1; and for any $N(\mu, \sigma^2)$ normal r.v. $X, Y = (X - \mu)/\sigma$ is the standard normal $N(0, 1)$. A standard normal r.v. is often denoted by Z, its distribution fn by Φ, and its density by ϕ; some writers use the term Z distribution. By extension, for any r.v. X with finite variance σ^2 and mean μ, the r.v. $Y = (X - \mu)/|\sigma|$ is the standardized form of X, and $|\sigma|$ is the **standard deviation**. More generally, the distribution of a r.v. may be called standard if it is centred at its mean, or median, etc., and its scale parameter is set to a standard value, such as 1.

Similarly, the standard Wiener process $W(t)$ is a version of the Wiener process that is a.s. continuous, with $W(0) = 0 = \mathbb{E}W(t)$, and var $W(t) = t$. The concept extends to random vectors in a natural way, so that e.g. the standard multinormal distribution has a mean vector of zeros, and unit entries on the diagonal of its covariance matrix. In another context, a standard spherical distribution has its axis at $\theta = 0$, in spherical polar coordinates. Furthermore, the transition probabilities $p_{ij}(t)$ of a homogeneous Markov chain on a countable state space are said to be standard (or comprise a standard semi group) if for all i: $\lim_{t \downarrow 0} p_{ii}(t) = 1$, and $\lim_{t \downarrow 0} p_{ij}(t) = 0$, $i \neq j$.

standard probability space A class of probability spaces that includes almost all such spaces that appear in practice, and whose members have attractive analytical properties. Thus, if Ω is a complete separable metric space, any probability measure $\mathbb{P}$ on Ω is standard if $\mathbb{P}$ is regular (which is to say that it is the completion of the measure on the Borel field, i.e. the Lebesgue extension). Equivalently, a probability space is standard if it is suitably isomorphic to a regular probability distribution on $\mathbb{R}$, or a subset such as $[0, 1]$ of $\mathbb{R}$. (Note that some writers exclude atoms, while some permit them.) [K. Itô, *Introduction to Probability Theory*. Cambridge University Press, 1984]

state space The set S of possible values or states of a random process $X(t), t \in T$; or, more precisely, the measurable space $(S, \mathcal{F})$, where $\mathcal{F}$ is some σ-field of subsets of S. In statistical mechanics S is often called the **phase space**. And in control theory the state space is the set of variables sufficient to determine the system at any time. [P. Billingsley, *Probability and Measure*, 3rd edn. Wiley, 1995]

stationarity Broadly, a term applied to some characteristic of a family of r.vs, or of a random process, that denotes the property of being invariant under suitable shifts in the parameter (or index), particularly when the index is seen as time; ergodic theorems are among the most important outcomes of the assumption of stationarity. More narrowly there are various specific definitions depending on the type of process, the particular characteristic considered, and the nature of the shifts. In the basic one-parameter case, $X(t), t \in T$ is said to be:

(a) **strongly (or strictly) stationary** if its finite-dimensional distributions are invariant under any time-shift, which is to say that for any $s, x_1, \ldots, x_n, t_1, \ldots, t_n$, and $n < \infty, \mathbb{P}(X(t_1 + s) \leq x_1, \ldots, X(t_n + s) \leq x_n)$ does not depend on s. The distribution of $X(t)$, being the same for all t, may be called the stationary distribution; in some contexts it is called the equilibrium distribution, and $X(t)$ may be said to be in statistical equilibrium (or simply in equilibrium).

(b) **weakly (or wide-sense, or second-order) stationary** if $\mathbb{E}X(t) = m < \infty$, for some constant m, and the covariance $c(s) = \text{cov}(X(t), X(t + s))$ depends only on s.

If $\int_0^\infty c(s)\text{d}s = \infty$, the process is said to have a **long memory**.

(c) **conditionally stationary** if its conditional fdds are invariant under time-shifts, so (a.s.) for any $s, x, t_1, \ldots, t_n$, and $n < \infty$,

$$\begin{aligned}&\mathbb{P}(X(t_n + s) \leq x \,|\, X(t_{n-1} + s), \ldots, X(t_1 + s))\\ &\quad = \mathbb{P}(X(t_n) \leq x \,|\, X(t_{n-1}), \ldots, X(t_1)).\end{aligned}$$

In the special case when $X(t)$ is a Markov process it is said to have stationary transition probabilities.

(d) **cycle-stationary** if there is a sequence of cycle points (i.e. random times) $S_n, n \in \mathbb{Z}$ that are strictly increasing in n, with $S_{-1} < 0 \leq S_0$, and such that the cycles (i.e. segments of $X(t)$ between consecutive cycle points), comprise a stationary sequence of random elements. This property is known under several other names, including event, occurrence-, and point-stationarity. (And is not to be confused with **cyclostationary** processes having periodic covariance.)

Other uses of the term include **interval-stationary**, which applies to a simple point process on $\mathbb{R}$ such that the intervals between successive points comprise a strongly stationary sequence; *see also* crudely stationary. And a random process $X(t)$ such that $X(t + s) - X(s)$ has the same distribution as $X(t) - X(0)$ for all s and t is said to have **stationary increments**. [H. Thorisson, On time- and cycle-stationarity, *Stoch. Proc. Appl.*, 1995, 55, 183–209; S. Karlin & H. Taylor, *A First Course in Stochastic Processes*, 2nd edn. Academic Press, 1975; O. Kallenberg, *Foundations of Modern Probability*, 2nd edn. Springer, 2002]

stationary distribution *see* invariant distribution

stationary increments *see* stationarity

stationary measure *see* invariant distribution

stationary policy *see* Markov decision process

stationary process *see* stationarity

stationary time Arises in considering mixing of a Markov chain $X_n, n \geq 0$, with invariant distribution π. A stationary time T for X is a stopping time (possibly randomized) such that $\mathbb{P}(X_T = j \,|\, X_0 = i) = \pi_j, j \in S$.

Thus the distribution of X_T, for the chain started at i, is the stationary distribution. If X_T has distribution π and is independent of T, then T is called a strong stationary time. [D. Levin *et al.*, *Markov Chains and Mixing Times*. American Mathematical Society, 2009]

stationary transition probabilities *see* stationarity

statistical decision theory Broadly, that branch of decision theory (including game theory) in which decisions are made (or strategies formulated) conditional on the observation of a random variable (or vector) X, that is seen (or is interpreted) as representing the state of nature. In particular, the term often refers to that branch of statistical inference in which the decision amounts to drawing conclusions about the unknown distribution of X, e.g. its type; or more precisely, its parameters. [J. Berger, *Statistical Decision Theory and Bayesian Analysis*, 2nd edn. Springer, 1985]

statistical independence *see* independence

statistical mechanics (or physics) Broadly, the branch of physics that uses the techniques of probability to describe the large scale properties of systems comprising a large number of small entities (such as molecules) whose individual properties must be treated as random. From its origins in thermodynamics, the subject has extended to include a number of problems addressed in the probabilistic literature, see Abelian sandpile, Ising model, Gibbs distribution, percolation, random cluster model, and spin-glass model. [A. Bovier, *Statistical Mechanics of Disordered Systems: A Mathematical Perspective*. Cambridge University Press, 2006; S. Sheffield & T. Spencer (eds), *Statistical Mechanics*. American Mathematical Society, 2009]

statistical probability *see* frequency probability

Steele's inequality Asserts that for an arbitrary real-valued function g of n variables, and i.i.d.r.vs $X_1, \ldots, X_n, Y_1, \ldots, Y_n$,

$$\operatorname{var} g(X_1, \ldots, X_n) \leq \frac{1}{2}\sum_{r=1}^{n} \mathbb{E}[(g - g_r)^2],$$

where $g_r = g(X_1, \ldots, X_{r-1}, Y_r, X_{r+1}, \ldots, X_n)$, $1 \leq r \leq n$. [J. M. Steele, An Efron–Stein inequality for nonsymmetric statistics, *Ann. Stat.*, 14(12), 753–758, 1986]

Stein–Chen method *see* Stein's method

Stein's identity *see* Stein's lemma

Stein's lemma Arises as a property of the normal distribution, and more generally of the exponential family of distns. In its basic form it asserts that a r.v. X is normal (with unit variance and mean m) iff for all differentiable real-valued functions $g(x)$, $x \in \mathbb{R}$, such that $\mathbb{E}|g'(X)| < \infty$, it is the case that

$$\mathbb{E}\{(X - m)g(X)\} = \mathbb{E}\{g'(X)\}.$$

This equality is often called **Stein's identity**, as is also the elementary identity of integration by parts that the score function $s(x) = f'(x)/f(x)$ for a rv. X with density $f(x)$ satisfies

$$\mathbb{E}\{f(X)s(X)\} = -\mathbb{E}f'(X),$$

provided only that f is suitably differentiable, well-behaved at $\pm\infty$, and the expectations exist.

More generally, H. Hudson showed that for a continuous random variable with density f in the exponential family $f(x) = b(x)\exp\{\theta x - a(\theta)\}$, it is the case that for $g(x)$ as above, and $t(x) = -b'(x)/b(x)$,

$$\mathbb{E}[(t(X) - \theta)g(X)] = \mathbb{E}g'(X).$$

B. Prakasa Rao established a converse, showing that this **generalized Stein identity** characterized the exponential family. [B. L. S. Prakasa Rao, Characterization of distributions through some identities, *J. Appl. Probab.*, 16(4), 903–909, 1979]

Stein's method A general procedure for approximating the distribution of one random variable (or random vector) by another, first introduced by C. Stein as a technique for normal approximation in the context of estimating the rate of convergence in CLTs. The special case of Poisson approximation is called the **Stein–Chen method**, in which the following lemma arises: the discrete r.v. X has a Poisson (λ) distribution iff for all real-valued functions $g(\cdot)$ such that both sides exist:

$$\lambda\mathbb{E}g(X + 1) = \mathbb{E}(Xg(X)).$$

This equality is called the **Stein–Chen identity**. A classic example of the method arises when Y is the sum of n independent Bernoulli (p_i) r.vs such that $\sum_i p_i = \lambda$, and X is Poisson (λ). Then the total variation distance between X and Y is bounded above by $\lambda^{-1}(1 - e^{-\lambda})\sum_{i=1}^{n} p_i^2$. When a bound in this method includes a parameter of X as here, it may be called a **magic** (or **Stein**) **factor**. A related technique using characteristic functions is the **Stein–Tikhomirov method**. [A. Barbour & L. Chen (eds), *Introduction to Stein's Method*. Singapore University Press & W.S.P., 2005; A. Barbour et al., *Poisson Approximation*. Oxford University Press, 1992]

stepping stone model Arises in stochastic population genetics when considering dispersed (or divided) populations. In **Wright's island model**, individuals migrate (i.e. alleles are exchanged) symmetrically and randomly between a number of colonies of equal size; in the stepping stone model the rates of exchange(or migration, or mutation) depend on the colonies according to a function representing a suitable distance between them. The special case when populations are 0 or 1 is a type of voter model. [J. T. Cox & R. Durrett, The stepping stone model: new formulas expose old myths, *Ann. Appl. Probab.*, 12(4), 1348–1377, 2002]

STER distribution *see* Bissinger's distribution

stick breaking model (or process) *see* residual allocation model

sticky boundary *see* reflecting barrier

sticky Brownian motion Informally, a reflecting Brownian motion that pauses at the origin in such a way that it does not remain there over any time-interval, but the time spent there has positive Lebesgue measure, proportional to the local time at zero of reflecting Brownian motion. More formally, it is a continuous non-negative solution $X(t), t \geq 0$, of the SDE

$$X(t) = x_0 + a\int_0^t I\{X(s) = 0\}\,\mathrm{d}s + \int_0^t I(X(s) > 0)\,\mathrm{d}W(s),$$

where $a > 0$ is the parameter of the process, and $W(t)$ is the Weiner process, and $x_0 \geq 0$. [J. Warren, The noise made by a Poisson snake, *Electr. J. Probab.*, 7(21), 1–21, 2002]

Stieltjes class A family of probability distributions having the same moments, given in terms of some distribution function $F(x)$ (where $F(x) = 0, x < 0$), that is not determined by its moments. *See* moment problem. [A. Pakes, Structure of Stieltjes classes of moment equivalent probability laws, *J. Math. Anal Appl.*, 326, 1268–1290, 2007]

Stieltjes transform (alias: **Cauchy, or Cauchy–Stieltjes transform**) Defined for a probability distribution function $F(x)$, $x \in \mathbb{R}$, by

$$S_X(z) = \int_{\mathbb{R}} \frac{1}{z-x}\mathrm{d}F(x), \ \mathrm{Im}\,(z) \neq 0.$$

It finds application in random matrices and free probability. More generally, for $a > 0$, the generalized (Cauchy–)Stieltjes transform is

$$C_X(z,a) = \int_{\mathbb{R}} \frac{1}{(z-x)^a}\mathrm{d}F(x).$$

See also Shannon transform. [G. Anderson *et al.*, *Introduction to Random Matrices*. Cambridge University Press, 2010]

Stienen model A random set process that falls within the general class of hard-sphere models. In its basic form, each point Q of a d-dimensional Poisson process Π ($d \geq 2$), is the centre of a sphere whose diameter is the distance from Q to the nearest point of Π. Any two such spheres touch iff they are nearest neighbours of each other. The fractional volume (i.e. proportion) of space occupied by spheres is 2^{-d}. *See also* dead leaves model and lily-pond model. [S. Chiu *et al.*, *Stochastic Geometry and its Applications*, 3rd edn. Wiley, 2013]

stiff Markov chain A term arising in the numerical analysis of the transient properties of a Markov chain (e.g. in the context of reliability theory), where a chain whose finite Q-matrix has eigenvalues $\lambda_1, \ldots, \lambda_n$, is said to be stiff when $\max_r|\mathrm{Re}(\lambda_r)|$ is much greater than $\min_r|\mathrm{Re}(\lambda_r)|$. [M. Malhotra, Stiffness tolerant methods for transient analysis of stiff Markov chains, *Microelectron. Reliab.*, 34(11), 1825–1841, 1994]

Stirling's formula (or approximation) An asymptotic approximation to the factorial function $n!$, as $n \to \infty$. In its basic form

$$\lim_{n\to\infty} \frac{n!}{\sqrt{2\pi n}}\left(\frac{\mathrm{e}}{n}\right)^n = 1$$

it was first used by A. de Moivre in approximating terms of the binomial distribution for a CLT. **Robbins improvement** asserts that, for $n \geq 1$,

$$\frac{1}{12n+1} < \log\frac{n!}{\sqrt{2\pi n}}\left(\frac{\mathrm{e}}{n}\right)^n < \frac{1}{12n}.$$

Numerous other approximations have been derived, with a great variety of proofs. [H. Lou, A short proof of Stirling's formula, *Amer. Math. Mon.*, 121(2), 154–157, 2014]

stirring process (alias: **random stirring process**) Arising in statistical mechanics, this comprises a random process on the group of permutations of the vertices of a graph. [O. Angel, Random infinite permutations and the cyclic time random walk, in *Proc. Discrete Random Walks, Paris*, 2003, pp. 9–16].

stochastically larger (or smaller) *see* stochastic order

stochastically monotone *see* stochastic monotonicity

stochastic approximation Broadly, a collection of methods that use stochastic difference equations (or recurrence relations) to establish some property of a function $g(x)$, such as the position of a zero, or the position and value of a maximum (or minimum); where the function g is unknown but the values of a sequence of r.vs that depend on $g(x)$ will become known. Archetypes of such techniques are the Robbins–Munro method and the Kiefer–Wolfowitz algorithm. [H. Kushner & G. Yin, *Stochastic Approximation and Recursive Algorithms and Applications*, 2nd edn. Springer, 2003]

stochastic basis (or base) *see* filtration

stochastic billiards *see* Knudsen process

stochastic calculus (or analysis) Broadly, the extension of several concepts of the usual calculus of real functions (i.e. the study of various limiting procedures) to functions $X(t),\ t \in S \subseteq \mathbb{R}$ that are random, and are called stochastic processes. More generally, stochastic analysis deals with a much wider range of functions and spaces that lie in the conjunction of analysis and probability theory.

At the elementary level, a function $X(t)$ is stochastically bounded if for any $\varepsilon > 0$ there exists $a \in (0, \infty)$ such that $\mathbb{P}(|X(t)| > a) < \varepsilon$ for all t; more strongly $X(t)$ is a.s. bounded if its sample paths are bounded with probability 1, i.e. $\mathbb{P}(|X(t)| < c$ for all $t) = 1$, for some $c \in (0, \infty)$. A function $X(t) \in \mathbb{R}$ is stochastically continuous (or continuous in probability) at t if $\lim_{h\to 0}\mathbb{P}(|X(t+h) - X(t)|>\varepsilon) = 0$, for all $\varepsilon > 0$, or continuous a.s. if $\mathbb{P}(\lim_{h\to 0}|X(t+h) - X(t)| = 0) = 1$. *See* continuous random process. (Some authors use the term stochastically continuous to denote mean square continuity, i.e. $\lim_{h\to 0}\mathbb{E}(X(t+h) - X(t))^2 = 0$.)

Likewise, $X(t)$ is a.s., or stochastically, or mean square differentiable if $\lim_{h\to 0}h^{-1}\{X(t+h) - X(t)\}$ exists appropriately.

More narrowly (and more usually), the term stochastic calculus is used in the context of stochastic integrals, and stochastic differential equations. *See* Itô's integral and Itô's formula. For random processes having finite variance, the term **second order calculus** has been used for calculus in mean square. [M. Loève, *Probability Theory*, 4th edn. Springer, 1978]

stochastic calculus of variations *see* Malliavin calculus

stochastic differential equation (or SDE) *see* stochastic integral

stochastic dominance (or domination) *see* stochastic order

stochastic dynamic programming *see* Bellman's equation

stochastic equivalence *see* equivalence

stochastic exponential *see* stochastic logarithm

stochastic flow Arises in its simplest form as the strong solution $X(t) \in \mathbb{R}^d,\ t \geq 0$ (that is jointly continuous on $\mathbb{R}^+ \times \mathbb{R}^d$) of an Itô stochastic differential equation. More generally, a random field $\mathbf{R}(s, t; \mathbf{x}) \in \mathbb{R}^d,\ \mathbf{x} \in \mathbb{R}^d,\ 0 \leq s \leq t \leq T$, is called a stochastic flow if it is jointly continuous in s, t, and $\mathbf{x}$, and:

(1) $\mathbf{R}(t, t; \mathbf{x}) = \mathbf{x}$, for all t;
(2) $\mathbf{R}(t, u; \mathbf{R}(s, t; \mathbf{x})) = \mathbf{R}(s, u; \mathbf{x})$, for all $\mathbf{x}$, and $s \leq t \leq u$;
(3) the function $\mathbf{R}(s, t; \mathbf{x})$ of $\mathbf{x}$ is a homeomorphism for all s and t.

If it has independent increments on the intervals defined by $0 \leq t_1 < t_2 < \cdots < t_n < T$, it is called a **Brownian flow**.[H. Kunita, *Stochastic Flows and Stochastic Differential Equations*. Cambridge University Press, 1990]

stochastic geometry *see* geometrical probability

stochastic independence *see* independence

stochastic integral Arises in stochastic calculus when seeking to extend the elementary integral of classic calculus, written as $h(x) = \int_a^x f(t)\mathrm{d}g(t)$ in standard notation, to the case when g and f are suitable stochastic processes on $\mathbb{R}$. As in the non-random case there are various stochastic integrals whose definition and nature depends on the field in which they arise, and the theorems of interest in that area which might be proved using a well-defined integral. In an important case, $g(t) = W(t)$, where $W(t)$ is the Wiener process; because W is nowhere differentiable the integral cannot be defined in the usual way. When $f(t)$ is not random, the Paley–Wiener integral can be defined as a limit, using an application of integration by parts and a suitable isometry. The Itô integral extends this similarly to functions $f(t)$ that are non-anticipating functionals of the paths of the Wiener process (i.e. measurable wrt the natural filtration); in particular, functions of the form $f(t, W(t))$, for suitable $f(\cdot,\cdot)$.

In this framework, the integral equation

$$X(t) = X(0) + \int_0^t a(s, X(s))\mathrm{d}s + \int_0^t b(s, X(s))\mathrm{d}W(s)$$

may be written as a **stochastic differential eqn** or **SDE**

$$\mathrm{d}X = a(t, X)\mathrm{d}t + b(t, X)\mathrm{d}W, \quad t \geq 0.$$

For suitable a and b this has a solution $X(t)$, with continuous sample paths, called an **Itô diffusion (or process)**. Similar procedures, using different limiting procedures yield other stochastic integrals, e.g. McShane's integral and the Fisk–Stratonovich integral.

More generally, the integrator $g(t)$ can be a martingale, in which case the integral may be seen as the equivalent in continuous time of the martingale transform. Most generally, for an integral required to have certain desirable properties, the integrator may be a semi-martingale; *see* Bichteler–Dellacherie theorem.

If these requirements are relaxed, then one may integrate against, e.g. fractional Brownian motion, which is not in general a semi-martingale. Also, among many other variants, one may allow $f(t)$ to be an anticipating functional, and stochastic integrals may be formulated in the framework of white noise theory. There is an analogous theory for **backward stochastic differential eqns**, or BSDEs, of the form

$$X(t) = x + \int_t^T a(S, X(S))\mathrm{d}S + \int_t^T b(S, X(S))\mathrm{d}W(S), \quad 0 \leq t \leq T.$$

Stochastic partial differential equations, as well as stochastic multiple integrals, may also be defined likewise in various frameworks, at the expense of some technical complications. [K. L. Chung & R. Williams, *Introduction to Stochastic Integration*, 2nd edn. Birkhäuser, 1990; P. Protter, *Stochastic Integration and Differential Equations*, 2nd edn. Springer, 2005; H. Holden et al., *Stochastic Partial Differential Equations: A Modeling, White Noise Functional Approach*, 2nd edn. Springer, 2010; B. Øksendal, *Stochastic Differential Equations*, 6th edn. Springer, 2003]

stochastic interval *see* stopping time

stochastic logarithm Arises naturally as the inverse function of the stochastic exponential; *see* Doléans-Dade exponential. Thus for a semi-martingale $Z(t)$ such that for all $t, Z(t) \neq 0 \neq Z(t-)$, the stochastic logarithm is

$$Y(t) = \int_0^t \frac{1}{Z(s-)}\mathrm{d}Z(s),$$

while the stochastic (Doléans-Dade) exponential is the solution of the SDE,

$$Z(t) = 1 + \int_0^t Z(s-)\mathrm{d}Y(s),$$

for a given semi-martingale Y. [J. Kallsen & A. Shiryaev, The cumulant process and Esscher's change of measure, *Finan. Stoch.*, 6(4), 397–428, 2002]

stochastic Löwner evolution *see* Schramm–Löwner evolution

stochastic majorization *see* stochastic order; majorization

stochastic matrix (alias: **transition probability matrix; Markov matrix**) Arises naturally as the array of one-step transition probabilities of a Markov chain on a countable state space, i.e. a square (possibly infinite) matrix of non-negative terms each of whose rows sums to unity. If each column also has unit sum, then the matrix is said to be **doubly stochastic**; *see* Birkhoff's theorem. If all row sums are ≤ 1, it is called **substochastic**, or **sub-Markov**.

More narrowly, some authors reserve the term **Markov matrix** for a stochastic matrix that has at least one strictly positive column. And a stochastic matrix (p_{rs}) with the weaker property that for any i and j there exists a k such that p_{ik} and p_{jk} are both strictly positive is called **scrambling**. (Note that a matrix of random variables is called a random matrix.) [E. Seneta, *Non-negative Matrices and Markov Chains*; revised printing. Springer, 2006]

stochastic monotonicity Arises naturally in considering a family of random variables having an index set I that is ordered (or, at least, partially ordered) by a relation $\prec$. If $\prec_{\mathrm{st}}$ denotes some stochastic order (maybe partial) on random variables and their distributions, then the set $\{X_i; i \in I\}$ is called stochastically monotone if for any i and j in I such that $i \prec j$ it follows that $X_i \prec_{\mathrm{st}} X_j$.

The concept is perhaps most often met in the context of Markov processes, thus: let $X(t) \in \mathbb{R}, t \geq 0$, be a Markov chain, then it is stochastically monotone if $\mathbb{P}(X(t) \geq y \,|\, X(0) = x)$ is non-decreasing in x for any fixed t and y. More generally, a transition kernel $k(x, A)$ on a state space S with order relation $\prec$ is stochastically monotone if $k(x, .) \prec_{\text{st}} k(y, .)$ whenever $x \prec y$, where $\prec_{\text{st}}$ is the usual stochastic order, and $\prec$ is an order on S. Several alternate formulations exist; for example, a Markov process on $\mathbb{R}$ is stochastically monotone if its Markov semi-group preserves the set of non-decreasing functions. [W. Anderson, *Continuous-time Markov Chains*. Springer, 1991; T. Lindvall, *Lectures on the Coupling Method*. Wiley, 1992; reprinted by Dover, 2002; I. Kadi et al., Different monotonicity definitions in stochastic modelling, in *Analytical and Stochastic Modelling Techniques and Applications, Lecture Notes Comp. Sci.*, 5513, 144–158, 2009]

stochastic order (or ordering) Broadly, any of several relationships between a given class of probability distributions that allows some (possibly all) of them to be placed either in linear order or in partial order; and hence yield an associated (partial) ordering of r.vs, r.vrs, or other random objects. Such orders have widespread applications in many areas, and it follows that there are many stochastic orders depending on which property of distributions is chosen for comparison; most often one sees ordering either by location, or by variability (i.e. dispersion). In the former case some authors speak of **stochastic dominance**, and in the latter case some authors speak of **stochastic majorization**, but nomenclature is not fixed.

What is called 'the usual stochastic order' is this: let $F(x)$ and $G(x), x \in \mathbb{R}$, be the distribution functions of X and Y respectively, If, for all x,

$$1 - F(x) = \mathbb{P}(X > x) \leq \mathbb{P}(Y > x) = 1 - G(x),$$

then one says any of:

- Y (or G) is stochastically larger than X (or F);
- Y stochastically dominates X;
- Y is larger than X in the usual stochastic order;
- $Y \geq_{\text{st}} X$.

Note that the condition is equivalent to:

- $\mathbb{P}(X \geq x) \leq \mathbb{P}(Y \geq x)$ for all x;
- $\mathbb{E}g(X) \leq \mathbb{E}g(Y)$ for all non-decreasing functions $g \in \mathbb{R}$ such that both sides exist;
- there is a probability space and two r.vs $\tilde{X}$ and $\tilde{Y}$ defined on it, having respective distns F and G, such that $\mathbb{P}(\tilde{X} \leq \tilde{Y}) = 1$; this last is often called Strassen's theorem.

In the same vein, Y is said to dominate X in second order (or to have **second order dominance** over X, etc.) if $\int_x^\infty F(y)\mathrm{d}y \leq \int_x^\infty G(y)\mathrm{d}y$, and further integrations yield third-order, etc., dominance.

In the multivariate case $\mathbf{X}$ is said to be smaller than $\mathbf{Y}$ in the usual stochastic order if $g(\mathbf{X}) \leq_{\text{st}} g(\mathbf{Y})$, for all functions $g(x_1, \ldots, x_n) \in \mathbb{R}$ that are increasing in each x_r while the other arguments are held fixed. Other orders may be defined using the likelihood ratio, or hazard rates.

The principal order based on dispersion or variability is this: If, for r.vs X and $Y, \mathbb{E}g(X) \leq \mathbb{E}g(Y)$, for all convex functions $g \in \mathbb{R}$ such that both sides exist, then Y is said to be greater than X in convex order (or, to majorize X); one writes $X \leq_{cx} Y$.

Note (by choosing $g(x) = x$ and then $g(x) = -x$) that this implies $\mathbb{E}X = \mathbb{E}Y$. The convex order is equivalent to the stop-loss order with the extra assumption that $\mathbb{E}X = \mathbb{E}Y$. And convex ordering can be extended to the multivariate case in a natural way.

More generally, if $\mathbb{E}g(X) \leq \mathbb{E}g(Y)$ for all g in some class C of functions, then one speaks of the C-order, e.g. increasing concave order, or supermodular order, and so on. *See* right spread order and majorization. [M. Shaked & J. Shanthikumar, *Stochastic Orders*. Springer, 2007; A. Marshall et al., *Inequalities: Theory of Majorization and its Applications*, 2nd edn. Springer, 2011]

stochastic partial differential equation *see* stochastic integral

stochastic pin-ball *see* Lorentz gas model

stochastic process (alias: **random process**) Arises naturally in generalizing the concept of a random sequence $X_n, n \in \mathbb{Z}^+$ (or $n \in \mathbb{Z}$), to processes with continuous parameter that are often seen as describing (i.e. modelling) the random evolution of some system in time. Broadly, a stochastic process is a collection of random variables on some probability space $(\Omega, \mathcal{F}, \mathbb{P})$, denoted by $X(t, \omega), \omega \in \Omega, t \in I$, where the index set I is commonly an interval subset of $\mathbb{R}$, but may be any linearly ordered set. When the index set is not so ordered, e.g. $\mathbb{R} \times \mathbb{R}$, one has a **random field**, though the term stochastic process is still sometimes used in such a case.

More formally, a stochastic process is a map from $\Omega \times I$ to $\mathbb{R}$, where each $X(t, .)$ is $\mathcal{F}$-measurable. (More generally, a complex-valued stochastic process takes values in $\mathbb{C}$.) For many practical purposes it is often necessary to require further measurability properties; *see* adapted process and filtration.

Alternatively, given a probability space $(\Omega, \mathcal{F}, \mathbb{P})$, it is often convenient (though perhaps less intuitively natural) to define a stochastic process as a measurable mapping to (or a measurable function taking values in) a measurable space of functions on I, that then comprises the set of possible paths (as realizations, or trajectories, or sample functions) of the process as t runs over I. In this framework, some writers see $X(t, \omega)$ as an infinite-dimensional random vector, and call the process a **random fn**. [R. Bass, *Stochastic Processes*. Cambridge University Press, 2011; S. Karlin & H. Taylor, *A First Course & A Second Course in Stochastic Processes*. Academic Press, 1975 & 1981]

stochastic programming (or optimization) (alias: **probabilistic programming**) Broadly, a class of optimization and decision problems (including e.g. resource allocation and portfolio management) where actions are constrained, and the parameters of the problem may be random variables, e.g. the constraints, costs, pay-offs, and objective functions. Many of the problems can be seen as generalizing non-random linear, quadratic, or dynamic programming problems; hence the name. *See* Bellman's equation, and single-period model. [J. Birge & F. Louveaux, *Introduction to Stochastic Programming*, 2nd edn. Springer, 2011]

stochastic semi-group (alias: **Markov semi-group**) Arises in the context of Markov processes, where the Markov property entails the Chapman–Kolmogorov equations, which exhibit a semi-group property. (A semi-group is a set that is closed under some associative binary operation; the set includes a unit element.)

Thus, for example, let $P(t) = (p_{ij}(t))$, $i, j \in S$, $t \geq 0$, be the transition probabilities of a Markov chain, so that the Chapman–Kolmogorov equations take the form $p_{ik}(s+t) = \sum_j p_{ij}(s)p_{jk}(t)$, $s, t \geq 0$, or in matrix notation $P(s+t) = P(s)P(t)$, exhibiting the semi-group property, with identity (or unit) $P(0)$. Furthermore, both $P(t)$, and – more generally – a Markov transition kernel $k_t(x, A)$ on S, are associated with a semi-group of linear transition operators $T_t, t \geq 0$, on S, thus: for a suitable space G of functions on S, define for $g \in G$ either $T_t\, g(x) = \sum_y g(y)p_{xy}(t)$, or $T_t\, g(x) = \int g(y)k_t(x, \mathrm{d}y), x \in S$.

Then in each case $T_{s+t}g(x) = T_s T_t g(x)$, so that $T_t, t \geq 0$, comprise a stochastic (or Markov) semi-group with unit (or identity) T_0. If P or k is **substochastic**, one has a substochastic (or sub Markov) semi-group. *See* Feller property. [A. Bobrowski, *Functional Analysis for Probability and Stochastic Processes: An Introduction*. Cambridge University Press, 2005]

stochastic string model Arises in financial mathematics in seeking to describe the term structure of forward interest rates, where it comprises a type of random field. [P. Santa-Clara & D. Sornette, Dynamics of forward interest rate curve with stochastic string shocks, *Rev. Financ. Stud.*, 14(1), 149–185, 2001]

stochastic transition kernel *see* transition function

stochastic volatility model Arises in financial mathematics in seeking to explain the empirical observation that different options have different implied volatility in the standard models (*see* Black–Scholes–Merton theory). Thus, e.g. if the stock price S is assumed to obey the SDE, $\mathrm{d}S = u(t)\,S\mathrm{d}t + v^p S\mathrm{d}W_1$, where v is the volatility, a stochastic volatility model further assumes that $v(t)$ obeys an SDE of the form $\mathrm{d}v = a(t, v, S)\mathrm{d}t + b(t, v, S)\mathrm{d}W_2$, for suitable a and b, where $W_1(t)$ and $W_2(t)$ are jointly distributed Wiener processes. *See* Heston's model. [J.-P. Fouque *et al.*, *Derivatives in Markets with Stochastic Volatility*. Cambridge University Press, 2000]

Stone's decomposition Arises in renewal theory when the interarrival distribution F is such that some n-fold convolution $F^{*(n)}$ of F has a non-trivial absolutely continuous component (i.e. is spread out). In this case the renewal measure $U = \sum_{n=0}^{\infty} F^{*(n)}$ is decomposed into two parts, one of which is finite and has the same tail behaviour as F, while the other part is absolutely continuous and has the tail behaviour of a renewal density (i.e. it is nearly stationary at ∞). [C. Stone, On absolutely continuous components and renewal theory, *Ann. Math. Stat.*, 37(1), 271–275, 1966]

Stone's local limit theorem Any of several local limit theorems, but most often that arising in several dimensions for non-lattice random variables. [C. Stone, On local and ratio limit theorems, *Proc. 5th Berkeley Symp. Math. Stat. Probab.*, 2(2), 217–224, 1967]

stop-loss ordering (alias: **increasing convex ordering**) A type of stochastic order that arises naturally in actuarial and insurance mathematics, thus: if random variables X and Y satisfy

$$S_X(a) = \mathbb{E}\{(X-a)^+\} \le \mathbb{E}\{(Y-a)^+\} = S_Y(a),\ \ a \in \mathbb{R},$$

then X is said to be less than Y in stop-loss order.

This holds iff $\mathbb{E}g(X) \le \mathbb{E}g(Y)$ for all convex non-decreasing functions $g(\cdot)$, so that X is less than Y in increasing convex order. One writes $X \le_{\text{sl}} Y$, or $X \le_{\text{icx}} Y$. (*See* Karlin–Novikoff cut criterion.)

The function $S_X(a)$, $a \in \mathbb{R}$, is called the **stop-loss transformation**, or **stop-loss premium**, for X. [T. Rolski *et al.*, *Stochastic Processes for Insurance and Finance*. Wiley, 1999]

stopped process *see* stopping time

stopping a success run *see* burglar problem

stopping rule *see* stopping time

stopping set *see* stopping time

stopping time (alias: **Markov time; non-anticipating time; random variable independent of the future**) Informally, in its simplest form, a stopping time arises in considering the sequence X_n, $n \ge 0$, of r.vs representing the fortune of a gambler A playing a sequence of games (or trials) against B, being another gambler, or a casino, etc. At some time $T \in \{0, 1, 2, \ldots\}$ either A or B runs out of cash and the game stops (if this never happens, $T = \infty$). Clearly the occurrence of the event $\{T \le n\}$ is determined by inspection of X_0, $X_1, \ldots, X_n$.

In practice, A may also decide to stop playing (or, alternatively e.g. double the stake) after any trial, at time T, according to some rule (or betting, or gambling system); but any such system or rule for deciding must have the property that the event $\{T \le n]$ depends only on $X_0, X_1, \ldots, X_n$.

Formally, therefore, a non-negative integer-valued r.v. T is a stopping time for $X_n, n \ge 0$, if the indicator $I(T \le n)$ is a function only of $X_0, X_1, \ldots, X_n$ (and thus does not depend on any of $X_{n+r}, r \ge 1$).

Alternatively, given an increasing sequence of σ-fields $\mathcal{F}_n, n \ge 0$, a r.v. $T \in \{0, 1, 2, \ldots\} \cup \{\infty\}$ is a stopping time for $\mathcal{F}_n$ if $\{T = n\} \in \mathcal{F}_n$ for $n \ge 0$, which is to say that also $\{T \le n\} \in \mathcal{F}_n$ for all n. Some authors require $\mathbb{P}(T < \infty) = 1$ as a property of a stopping time, others call such times proper, and allow improper times such that $\mathbb{P}(T = \infty) > 0$, yet others call a.s. finite stopping times **stopping rules**.

In continuous time likewise, let $\mathcal{F}_t, t \ge 0$, be a filtration on a suitable space $(\Omega, \mathcal{F}, \mathbb{P})$, then a non-negative r.v. on the same space is a stopping time for $\mathcal{F}_t$ if $\{T \le t\} \in \mathcal{F}_t$, (i.e. is $\mathcal{F}_t$-measurable), for all $t \ge 0$.

In the case when $\{T < t\} \in \mathcal{F}_t$ for all t, T is called an **optional time** for $\mathcal{F}_t$; stopping times are always optional and an optional time is a stopping time if $\mathcal{F}_t$ is right-continuous.

Reasons for the importance of stopping times include the preservation of the Markov property, and also the martingale property, at such times; in these contexts one speaks of the **stopped process** defined as

$$X(t \wedge T) = \begin{cases} X(t), & t < T \\ X(T), & t \ge T. \end{cases}$$

Stopping times are classified as follows: if T is the limit of an increasing sequence of stopping times $T_n, n \ge 1$, such that $T_n < T$ whenever $T > 0$, then T is called a **predictable stopping time** (or a **previsible stopping time**), and the sequence T_n is said to **announce** T. A stopping time T is called **accessible** if there is a sequence of predictable stopping times $T_n, n \ge 1$, such that for all $\omega \in \Omega$ there is an n such that $T(\omega) = T_n(\omega)$. Otherwise T is called **totally inaccessible**, and in this case $\mathbb{P}(T = \tau < \infty) = 0$ for every predictable stopping time τ.

Furthermore, $T \ge 0$ may be called a **randomized stopping time** for $X(t)$ if $\{T \le t\}$, depends not only on $X(s), s \le t$, but also on an auxiliary r.v. Y (or random sequence Y_t). *See also* stationary time, Wald's identity, and strong Markov property.

Finally, given stopping times S and T for $\mathcal{F}_t$, the subset of $\mathbb{R}^+ \times \Omega$ given by $I(S, T) = \{(t, \omega) \colon S(\omega) \le t \le T(\omega)\}$ is called a **stochastic interval**, and then $\sum(\omega) = \{t : (\omega, t) \in I(S, T)\}$ may be called a **stopping set**, particularly when $S = 0$. Note that in the case of a set-induced filtration, the term stopping-set will mean not this, but the analogue of a stopping time; neither should be confused with the stopping sets of coding theory. [R. Bass, *Stochastic Processes*. Cambridge University Press, 2011]

storage process *see* dam process

S-transform A term used to denote a number of different integral transforms. In random matrix theory (and free probability generally) it denotes an operator analagous to the Mellin–Fourier transform, thus: let $m_n, n \ge 1$, be the moments of some r.v. (or, equivalently, of its distribution) where it is assumed that $m_1 \ne 0$. Define $m(z) = \sum_{r=1}^{\infty} m_r z^r$, with inverse function $m^{-1}(\cdot)$; then the associated S-transform is defined by $S(z) = m^{-1}(z)(1+z)/z$.

In signalling theory, the S-transform of a suitably integrable function $g(t)$ is defined as

$$S(a,b) = \int_{\mathbb{R}} g(x)\frac{|a|}{\sqrt{2\pi}}\exp\left\{-2\pi i a x - \frac{1}{2}a^2(x-b)^2\right\}\mathrm{d}x.$$

In white noise theory, and much more widely, it denotes an appropriate version of the **Segal–Bargmann transform**, which for finite dimensional linear spaces takes the form

$$S(g) = \int_{\mathbb{R}^n} g(x+z)d\Phi(x), \quad x \in \mathbb{R}^n,\ z \in \mathbb{C}^n,$$

where Φ is the standard multinormal distribution. [D. Voiculescu, Multiplication of certain non-commuting random variables, *J. Op. Theor.*, 18, 223–235, 1987; R. Stockwell et al., Localization of the complex spectrum: the S transform, *IEEE Trans. Signal Proc.*, 44(4), 998–1001, 1996; B. Hall, The Segal–Bargmann transform and the Gross ergodicity theorem, in *Contemporary Mathematics*, Vol. 317, H.-H. Kuo & A. Sengupta (eds). American Mathematical Society, 2001]

Strassen–Dudley theorem *see* Strassen's theorem

Strassen's invariance principle (alias: **Strassen's law of the iterated logarithm; Strassen's theorem; Strassen's functional LIL**) Denotes a collection of results yielding a functional law of the iterated logarithm, with its corollaries and extensions. In its original form the principal theorem considers the continuous paths obtained by linear interpolation of the sequence $(2n\log\log n)^{-1/2}S_r, 0 \le r \le n$, at r/n, $n \ge 3$, where $S_r = \sum_{k=1}^{r} X_k$ are the partial sums of a sequence $X_k, k \ge 1$, of i.i.d. zero-mean r.vs with unit variance. Then as $n \to \infty$, the set of limiting paths of these piecewise linear graphs on [0, 1] comprises the absolutely continuous functions $c(\cdot)$ on [0, 1], such that $c(0) = 0$ and $\int_0^1 \dot{c}(t)^2 \mathrm{d}t \le 1$.

This follows from preliminary results, any of which may be referred to as Strassen's theorem, thus: Let $W(t)$, $t \ge 0$, be the Wiener process and set $X(n,t) = W(nt)/(2n \log\log n)^{1/2}, n \ge 3, 0 \le t \le 1$; then as $n \to \infty$ the limiting functions of $X(n,t)$ comprise the set defined above a.s.

Furthermore, with probability 1, there exists a Wiener process $W(t)$ defined on the same probability space as $S_r, r \ge 1$, such that with probability 1:

$$\lim_{t\to\infty} \frac{W(t) - S_{[t]}}{(2t\log\log t)^{\frac{1}{2}}} = 0.$$

[V. Strassen, An invariance principle for the law of the iterated logarithm, *Probab. Theor. Rel. Fields*, 3(3), 211–226, 1964]

Strassen's theorem Any of several results that arise in considering the existence of joint probability distributions with given marginals; in particular the term is perhaps most often applied to denote special cases of the general theorems. For example, in applications to stochastic dominance, Strassen's theorem may refer to these:

1. Let $F(x)$ and $G(x)$ be distribution functions such that $\int g(x)\mathrm{d}F(x) \le \int g(x)\mathrm{d}G(x)$ for all increasing bounded functions $g(\cdot)$. Then there is a joint distn $H(x,y)$, with marginals F and G, such that
$$\iint_{x\le y} \mathrm{d}H(x,y) = 1.$$
2. Let $\mathbb{E}u(X) \le \mathbb{E}u(Y)$ for all convex functions $u(\cdot)$ such that both sides exist. Then there is a probability space on which $\tilde{X}$ and $\tilde{Y}$ are defined, and having the distributions of X and Y respectively, such that $\mathbb{E}(\tilde{Y}\,|\,\tilde{X}) = \tilde{X}$, (a.s.). (If $\mathbb{E}u(X) \le \mathbb{E}u(Y)$ for all increasing convex functions $u(\cdot)$, then the corresponding result is $\mathbb{E}(\tilde{Y}\,|\,\tilde{X}) \ge \tilde{X}$ (a.s.).)

In another manifestation, Strassen's theorem shows that for random variables X and Y (not necessarily on the same space) that are near to each other in the Lévy (i.e. Lévy–Prokhorov) distance, then there is a probability space and r.vs $\tilde{X}$ and $\tilde{Y}$ having the distns of X and Y, such that $\tilde{X}$ and $\tilde{Y}$ are equally near to each other in the Ky Fan distance. An extension that supplies a constructive method for finding r.vs that are nearest in this sense was given by R. Dudley, so results in this group are often termed as a **Strassen–Dudley thm.** [R. Dudley, *Real Analysis and Probability*. Cambridge University Press, 2002; V. Strassen, The existence of probability measures with given marginals, *Ann. Math. Stat.*, 36(2), 423–439, 1965; T. Lindvall, On Strassen's theorem on stochastic domination, *Electr. Commun. Probab.*, 4, 51–59, 1999]

Stratonovich integral *see* Fisk–Stratonovich integral

Strauss process A point process with pairwise interaction depending on a parameter $r > 0$ called the interaction radius (or range). For any configuration (i.e. set) Z of points, of size $|Z|$, let

$$n(Z) = \sum_{x,\,y\,\in\,Z} I(\|x-y\| \le r)$$

be the number of points of Z that are no more than a distance r apart. Then the density f of the process is proportional to $a^{|Z|}b^{n(Z)}$, for some $0 \le b \le 1$, where $0^0 = 1$. When $b = 1$, this is a Poisson process, when $b = 0$ it is a **hard-sphere process**. The density being a

Gibbs distribution, this is an example of a **Gibbs point process**.

In a **generalized Strauss process** $f \propto \prod_{x,y} g(\|x-y\|)$, for some suitable function $g(\cdot)$. [A. Baddeley & G. Nair, Fast approximation of the intensity of Gibbs point processes, *Electr. J. Stat.*, 6, 1155–1169, 2012]

stretched exponential *see* exponential distribution

Stricker's theorem Let M_t be a semi-martingale for a filtration $\mathcal{G}_t$, that is also adapted to a filtration $\mathcal{F}_t$, where $\mathcal{F}_t \subseteq \mathcal{G}_t$. Then Stricker's theorem asserts that M_t is also a semi-martingale for F_t. [H. Föllmer & P. Protter, Local martingales and filtration shrinkage, *ESAIM Probab. Stat.*, 15, 25–38, 2010]

strike price *see* contingent claim

strong independence axiom *see* sure thing principle

strong law of large numbers (alias: **SLLN**) Arises naturally in considering the long-run properties of the empirical average $\frac{1}{n}S_n = \frac{1}{n}\sum_{r=1}^{n} X_r$, $n \geq 1$, of a sequence $X_r, r \geq 1$, of r.vs. Such averages are of importance in many applications, and are also crucial to intuitive notions of probability and expectation. In its basic form, the SLLN (also called Kolmogorov's SLLN), assumes the X_r to be i.i.d.r.vs, and asserts this:

(i) If $\mathbb{E}|X_1| < \infty$, then with probability 1, $\lim_{n\to\infty} \frac{1}{n}S_n = \mathbb{E}X_1$.

(ii) If $\mathbb{E}|X_1| = \infty$, then $\lim\sup_{n\to\infty} |\frac{1}{n}S_n| = \infty$, a.s.

Earlier strong laws established (i) under more restrictive conditions, e.g. Cantelli's finite fourth-moment SLLN. More generally, another theorem of A. Kolmogorov for independent but not necessarily identically distributed r.vs, shows that (i) holds if $\sum_{r=1}^{\infty} \frac{1}{r^2} \operatorname{var} X_r < \infty$,. Yet more generally, the assumption of independence may be relaxed to e.g. pairwise independent, or exchangeable r.vs, or martingale differences, etc.

Finally, one may speak of a (generalized) strong law of large numbers if there exist sequences a_n, and $b_n, n \geq 1$ (where $0 < b_n \uparrow \infty$), such that $\frac{1}{b_n}\sum_{r=1}^{n}(X_r - a_r)$ converges to 0 wp. 1, as $n \to \infty$. *See* Markov chain ergodic theorem. [Y. Chow & H Teicher, *Probability Theory*, 3rd edn. Springer, 1997]

strongly non-lattice condition *see* Cramér's condition

strong Markov property *see* Markov property

strong mixing *see* alpha-mixing

strong solution Arises in considering the nature of solutions to an SDE of the form $\mathrm{d}X = a(t,X)\mathrm{d}t + b(t,X)\mathrm{d}W(t)$, with initial condition $\mathbb{P}(X(0) = Z) = 1$, and where a and b satisfy mild regularity conditions. A strong solution $X(t)$ is a continuous process that a.s. satisfies the SDE and the initial condition, and is adapted to the natural complete filtration generated by Z and the Wiener process $W(t), t \geq 0$; thus $X(t)$ is given a.s. for all t by a measurable functional of Z and $W(s), 0 \leq s \leq t$.

By contrast, a **weak solution** comprises a pair of processes X and W, each adapted to a stochastic basis, that satisfy the SDE, and such that $X(0)$ has the distribution of Z; but the filtration of the basis is not necessarily that generated by Z and the Wiener process, so $X(t)$ is not necessarily a functional of these. *See* uniqueness in law. [F. Klebaner, *Introduction to Stochastic Calculus with Applications*, 3rd edn. Imperial College Press, 2012]

Stroock's formula Supplies a representation for the kernels k_n, in the Wiener–Itô (or Wiener) chaos decomposition of a square integrable functional in terms of nth-order differential operators (also called Malliavin derivatives) on Wiener space. [D. Stroock, Homogeneous chaos revisited, *Séminaire de Probabilités XXI*, 1247, 1–7, 1987]

Stroock–Varadhan theorem Any of several theorems arising in the theory of diffusions and SDEs, but perhaps most often their solution of the martingale problem, i.e. given a suitable linear operator L on a space of appropriate functions, is there a probability measure on the space of continuous functions X such that $g(X(t)) - g(X(0)) - \int_0^t Lg(X)\mathrm{d}s$ is a local martingale for all test functions g? The theorem supplies conditions for the existence and uniqueness of such a measure, in which case the problem is said to be well-posed, and further proves that L characterizes a strongly Markov Feller process. [D. Stroock & S. Varadhan, *Multidimensional Diffusion Processes*. Springer, 1979]

structure equation *see* Emery's structure equation

structure function An indicator function $\phi(\mathbf{X}) \in \{0,1\}$ of a vector of indicator r.vs $\mathbf{X} = (X_1, \ldots, X_n)$ (called the **status vector**), where $X_r \in \{0,1\}$ indicates whether the rth component of some system is working, and $\phi(\mathbf{X})$ indicates whether the system is working. It arises in reliability theory, where important types are:

1. Parallel structure: $\phi(\mathbf{x}) = \max\{x_1, \ldots, x_n\} = 1 - \prod_{r=1}^{n}(1 - x_r)$;
2. Series structure: $\phi(x) = \min\{x_1, \ldots, x_n\} = \prod_{r=1}^{n} x_r$;
3. k-out-of-n structure: $\phi(x) = \begin{cases} 1 & \text{if } \sum_1^n x_r \geq k, \\ 0 & \text{otherwise;} \end{cases}$
4. Monotone structure: $\phi(\mathbf{x}) \geq \phi(\mathbf{y})$ for any $\mathbf{x}$ and $\mathbf{y}$ such that $x_r \geq y_r, 1 \leq r \leq n$;
5. Coherent structure: if $\phi(\mathbf{x}) = \phi(\mathbf{y})$ for any $\mathbf{x}$ and $\mathbf{y}$ such that $x_r = y_r$ for all $r \neq i$ (i fixed), then the ith component x_i is said to be **irrelevant** to the system. If the system is monotone and has no irrelevant components it is said to be **coherent**.

Denoting $\mathbb{E}X_r = p_r$, the function $r(p_1, \ldots, p_n) = r(\mathbf{p}) = \mathbb{E}\phi(\mathbf{X}) = \mathbb{P}(\phi(\mathbf{X}) = 1)$ is called the **reliability** (of the system).

(Note that this is not to be confused with A. Kolmogorov's structure function of his complexity theory.) [A. Høyland & M. Rausand, *System Reliability Theory*, 2nd edn. Wiley, 2004]

structure function of a random process A characteristic of a random process $X(t)$ first introduced by A. Kolmogorov in turbulence. For $X(t) \in \mathbb{R}, t \in \mathbb{R}$, the structure function is $S(u, v) = \mathbb{E}|X(u) - X(v)|^2$; more generally, the nth-order structure function is $S_n(u, v) = \mathbb{E}|X(u) - X(v)|^n$. For complex-valued processes the (second-order) structure function is.

$$k(t; u, v) = \mathbb{E}\Big\{(X(u) - X(t))\Big(\overline{X(v) - X(t)}\Big)\Big\}.$$

[A. Yaglom, *An Introduction to the Theory of Stationary Random Functions*. Prentice-Hall, 1962; reprinted by Dover, 2004]

Studentized process *see* self-normalized process

Student's *t*-distribution The probability distribution on $\mathbb{R}$ with density

$$f_n(x) = \frac{\Gamma\left(\frac{1}{2}(n+1)\right)}{\sqrt{n\pi}\,\Gamma\left(\frac{1}{2}n\right)}\left(1 + \frac{x^2}{n}\right)^{-\frac{1}{2}(n+1)},$$

where the parameter $n > 0$ is called the **number of degrees of freedom**, and is commonly an integer.

The density arises naturally in the context of a normal, $N(\mu, \sigma^2)$, random sample $(X_1, \ldots, X_n)$, with sample mean $\overline{X} = \frac{1}{n}\sum_{r=1}^{n} X_r$, for which

$$T = \sqrt{n}(\overline{X} - \mu)\Big/\left\{\sum_{r=1}^{n}(X_r - \overline{X})^2\right\}^{1/2}$$

has density function $f_{n-1}(x), n \geq 2$. This does not depend on σ, and the sample is said to have been Studentized; the term is sometimes used of any function of a sample (statistic) whose distribution does not depend on one or more parameters of the parent distn. *See* non-central distribution. [B. Grigelionis, *Student's t-distribution and Related Stochastic Processes*. Springer, 2013]

subadditive ergodic theorem *see* Kingman's subadditive ergodic theorem

subcritical branching process For the basic case of an ordinary Galton–Watson process, $Z_n, n \geq 0$, with $Z_0 = 1$, the process is said to be subcritical if $\mathbb{E}Z_1 < 1$. In this case the **Heathcote–Seneta–Vere–Jones thm** asserts that $\mathbb{E}(Z_n \mid Z_n > 0)$ converges to a finite limit C, or diverges, according as $\mathbb{E}Z_1 \log^+ Z_1$ is finite or infinite; and, further, $G_n(s) = \mathbb{E}(s^{Z_n} \mid Z_n > 0)$ converges to $G(s)$, the pgf of a proper distribution, such that $1 - G(\mathbb{E}s^{Z_1}) = \{1 - G(s)\}\mathbb{E}Z_1$ (and G is the unique solution of this functional equation that is a pgf with $G(0) = 0$), while $G'(1) = C$ if C is finite, otherwise $G'(1)$ is infinite. [C. Heathcote *et al.*, A refinement of two theorems in the theory of branching processes, *Theor. Probab. Appl.*, 12(2), 297–301, 1967]

subdiffusion *see* anomalous diffusion

subexponential distribution Broadly, a distribution whose tail decays more slowly than any exponential (i.e. the tail is heavy), and has a number of regularity properties, thus: let $F(x) < 1, 0 \leq x < \infty$ be the distribution function of independent positive r.vs $X_1, X_2, \ldots$. Then F is called subexponential if, for some $n \geq 2$,

$$\lim_{x\to\infty} \frac{\mathbb{P}(X_1 + \cdots + X_n > x)}{\mathbb{P}(\max\{X_1, \ldots, X_n\} > x)} = 1.$$

If this holds for any such n, then it holds for all such n, in particular for $n = 2$; thus, equivalently, F is subexponential if

$$\lim_{x\to\infty} \frac{\mathbb{P}(X_1 + X_2 > x)}{\mathbb{P}(X_1 > x)} = 2$$

or

$$\lim_{x\to\infty} \mathbb{P}(X_1 > x \mid X_1 + X_2 > x) = \frac{1}{2}.$$

More generally, a distribution $F(x) < 1$ on $\mathbb{R}$ is subexponential if it is long tailed and the above conditions hold. [S. Foss *et al.*, *Introduction to Heavy-tailed and Subexponential Distributions*, 2nd edn. Springer, 2013]

subfair *see* bold play

subfair primitive casino Dubins–Savage casino

sub-Gaussian distribution A r.v. X is said to have a sub-Gaussian distn (or to be sub-Gaussian) if there exists some $a > 0$ such that $\mathbb{E}e^{tX} \le e^{a^2t^2/2}$, for all $t \in \mathbb{R}$. It follows that $\mathbb{E}X = 0$, and var $X \le a^2$. For any zero mean r.v. X, the number $S = \inf\{a \ge 0\colon \mathbb{E}e^{tX} \le e^{a^2t^2/2}$, for all $t \in \mathbb{R}\}$ is called the **subgaussian shift**, or standard, or moment, of X.

Equivalently, X is sub-Gaussian if for some $b > 0$ and all $x > 0$, $\mathbb{P}(|X| \ge x) \le 2e^{-bx^2}$. Analogously, a random process $X(t), t \in T$, is said to be sub-Gaussian if, for all $x > 0$, and $s, t \in T$,

$$\mathbb{P}(|X(s) - X(t)| > x) \le 2\exp\left\{-\frac{1}{2}x^2/m^2(s,t)\right\},$$

where $m(s,t)$ is a semimetric. [V. Buldygin & Yu. Kozachenko, Sub-Gaussian random variables, *Ukrainian Math. J.*, 32(6), 483–489, 1980]

subharmonic *see* harmonic

subjective probability (alias: **personal probability; Bayesian probability; pistimetric probability**) Broadly an interpretation of probability that is often encapsulated in the maxim 'Probability is a degree of belief', or in A. de Morgan's words, 'Probability is a feeling of the mind'. In its widest sense, it defines the probability $\mathbb{P}(A)$ of an event A occurring (or some statement A being true), as a measure (or degree) of one's belief that A will occur (or turn out to be true). So $\mathbb{P}(A)$ arises from an individual's experience, and the evidence available, and the form of reasoning applied to these factors. This contrasts with so-called **objective probability**, as exemplified by the frequentist interpretation (which arises from long-run relative frequencies in repeated trials). This general definition has been refined into various separate, narrower forms; one result of that has been the proliferation of an inconsistent and confusing terminology.

Subjective probability is especially convenient in Bayesian inference and methods, in which unknown quantities may be treated in some respects as though they were random variables, in particular to the extent of being assigned (possibly improper) probability distributions. At one extreme, orthodox (or unconstrained) Bayesians allow any degrees of belief (probabilities) about events; at another extreme the so-called rational (or logical, or objective) Bayesians require beliefs (probabilities) that are consistent in some sense, e.g. obeying the rules of some formal logic, or being coherent in choosing from a set of offered bets, or satisfying some justifiable principle. [R. Jeffrey, *Subjective Probability*. Cambridge University Press, 2004]

submartingale *see* martingale; martingale convergence theorem

subordinated Markov process *see* derived Markov process

subordination *see* change of time

subordinator (alias: **randomized operational time; directing process; time-change process; stochastic clock (or chronometer); Lévy time**) Arises as a change of time for a random process $X(t)$, the time-changed process being $X(T(t)), t \ge 0$, where $T(t)$ is a suitable random process, not necessarily independent of X. Most often in applications $T(t)$ is a subordinator, which is to say that it is non-decreasing from $T(0) = 0$, with independent stationary increments, i.e. it is a Lévy process. Importantly, a Lévy process time-changed by an independent subordinator yields a Lévy process.

More precisely, a subordinator that is not independent of X is designated a wide-sense subordinator, and more generally a subordinator subjected to independent exponential killing is called a generalized subordinator. [D. Applebaum, *Lévy Processes and Stochastic Calculus*, 2nd edn. Cambridge University Press, 2009]

sub-probability *see* defective distribution

sub-stochastic matrix *see* stochastic matrix; transition function

success-run chain A Markov chain on the non-negative integers that, at each step, increases its value by unity or drops to zero. [S. Resnick, *Adventures in Stochastic Processes*, Birkhäuser, 1992]

Sudakov–Fernique inequality *see* Slepian's inequality

Sudakov's minoration (or inequality) In the context of a Gaussian process $X(t), t \in T$, one form is this: let distance be: $d(u, v) = \{\mathbb{E}(X(v) - X(u))^2\}^{\frac{1}{2}}$, $u, v \in T$, and let $C(\delta)$ be the smallest number of open balls having radius $\delta > 0$ required to cover the metric (or pseudo-metric) space T. Then there exists $a > 0$ such that

$$\mathbb{E} \sup_{t \in T} X(t) \geq a\delta\{\log C(\delta)\}^{\frac{1}{2}},$$

where $\log C(\delta)$ may be called the metric (or $\delta-$) entropy of T, with entropy numbers $C(\delta)$. [M. Ledoux & M. Talagrand, *Probability in Banach Spaces*. Springer, 1991]

Suen's (correlation) inequality *see* Janson's inequality

Suen–Janson inequality *see* Janson's inequality

Sukhatme's theorem (alias: **Rényi–Epstein–Sobel theorem (and subsets)**) Denotes the property of a random sample $X_1, \ldots, X_n$ from the exponential $\text{Exp}(\gamma)$ distribution, with order statistics $X_{(1)} < X_{(2)} < \cdots < X_{(n)}$, that the following are also i.i.d. $\text{Exp}(\gamma)$ r.vs:

$$nX_{(1)}, (n-1)\big(X_{(2)} - X_{(1)}\big), \ldots, X_{(n)} - X_{(n-1)}.$$

[P. Sukhatme, Tests of significance for samples of the χ^2-population with two degrees of freedom, *Ann. Human Genetics (or Eugenics)*, 8(1), 52–56, 1937]

sum–product method *see* belief propagation

sum-the-odds theorem *see* Bruss's odds rule

superBrownian motion *see* superprocess

superconvolutive sequence A sequence of (possibly improper) distribution functions $F_n(x), n \geq 1, x \in \mathbb{R}$, is superconvolutive if for all r and s

$$F_{r+s}(x) \geq \int_{\mathbb{R}} F_r(x - y)\mathrm{d}F_s(y),$$

for all x; and then their corresponding cumulant generating functions $K_n(t)$ are superadditive, i.e.

$$K_{m+n}(t) \geq K_m(t) + K_n(t).$$

A theorem of J. Hammersley establishes a limit theorem for such sequences which generalizes the WLLN, among other results. [J. Hammersley, Postulates for subadditive processes, *Ann. Probab.*, 2(4), 652–680, 1974]

superharmonic *see* harmonic

supermarket model A classic many-server queueing system in which arriving customers join the shortest line (or the queue with the smallest workload). [M. Luczak & C. McDiarmid, On the maximum queue length in the supermarket model, *Ann. Probab.*, 34(2), 493–527, 2006]

supermartingale *see* martingale

superposition Given two point processes N and M in $\mathbb{R}$, their superposition is the union of the points in N and M. The concept arises naturally in applications, where N and M may be (say) the arrival times of different types of particle or customer. Important theorems about such processes include:

1. If the superposition of two independent non-degenerate stationary renewal processes is a stationary renewal process, then all three processes must be Poisson processes.
2. A result sometimes called **Grigelionis's thm** asserts that if $N(t)$ is a finite-intensity (being a constant c) simple point process on $\mathbb{R}$, and $X(t)$ is the point process obtained by superposing m independent copies of $N(t)$ (and changing the scale by a factor m), then as $m \to \infty, X(t)$ converges weakly to a Poisson process of rate c.

[E. Çinlar, Superposition of point processes, in: *Stochastic Point Processes*. Wiley, 1972, pp. 549–606]

superprocess (alias: **measure-valued (or continuous state) branching process; branching measure-valued process (BMVP)**) Broadly, a random process taking values in a space of finite measures over a suitable measurable space (e.g. $\mathbb{R}^n$, with the Borel σ-field); such processes arise in connexion with non-linear pdes, and as solutions of stochastic pdes, and as scaling limits of interacting particle systems, etc., but perhaps most naturally as limits of spatial branching processes. They thus have the branching property, i.e. for a measure $m = m_X + m_Y$, the process started with m has the same distribution at time t as the sum of independent processes started with m_X and m_Y respectively.

The classic example is **superBrownian motion**, also called the **Dawson–Watanabe process**, arising as a limit of the process in which particles follow independent Brownian motions until branching according to the pgf, or branching mechanism, $G(s) = \frac{1}{2}(1 + s^2)$. More generally, the particles paths may be diffusions, or Lévy processes, and the branching mechanism may take the form $a^{-1}(1-s)^a + s, \quad 1 < a \le 2$. Other famous examples include the **Fleming–Viot model** for mutation, describing the frequencies of alleles in some population, and the **Dawson–Perkins process** (or historical superprocess) in which the space includes the paths of particles, not just their locations. [A. Etheridge, *An Introduction to Superprocesses*. American Mathematical Society, 2000]

superregular *see* harmonic

supersmooth distribution A concept arising in density estimation, also called deconvolution, where a distribution having chf $\phi(t)$ is called supersmooth of order c if, as $t \to \infty$,

$$a_1|t|^{b_1}\exp(-|t|^c/d) \le |\phi(t)| \le a_2|t|^{b_2}\exp(-|t|^c/d),$$

where a_1, a_2, c, and d are positive constants, with b_1 and b_2 constant. If $\alpha|t|^{-\beta} \le |\phi(t)| \le \alpha|t|^{-\beta}$ as $t \to \infty$, for positive constants α, β, γ, then the distribution is called **ordinary smooth** of order β. [J. Fan, Deconvolution with supersmooth distributions, *Canadian J. Stat.*, 20(2), 155–169, 1992]

supplementary variable *see* augmented state space

sure thing principle (alias: **strong independence axiom**) Arises in the context of decision theory where it seeks to formalize the intuitively natural idea that for some given decision there are events such that your knowledge of their occurrence (or non-occurrence) is irrelevant to your preference. More formally, if you prefer a to b, either knowing c happened or knowing that c didn't happen, and also the distribution of c is independent of your preference (i.e. what you decide), then you surely prefer a to b.

Note that if the distribution of the event c depends on your preference, then the Yule–Simpson paradox yields situations where the principle doesn't hold. [J. Pearl, *Causality*. Cambridge University Press, 2000; P. Fishburn & P. Wakker, The invention of the independence condition for preferences, *Manage. Sci.*, 41(7), 1130–1144, 1995]

surfer process *see* PageRank Markov chain

surprise A term sometimes used in the context of Bayesian inference and methods to denote either of the Kullback–Leibler divergences between a posterior distribution $p(y \mid x)$ given an observation x, and the corresponding prior distribution $p(y)$, i.e., either $d_{KL}(p(y), p(y \mid x))$ or $d_{KL}(p(y \mid x), p(y))$, which is interpreted as the information supplied by the observation x. *See* Lindley's information measure. [P. Baldi & L. Itti, Of bits and wows, *Neural Networks*, 23, 649–666, 2010]

surprise index Any one of a number of functions suggested as a measure of relative unexpectedness of an event with probability p_i, $1 \le i \le n$, occurring, when the possible mutually exclusive outcomes have probabilities $p_1, \ldots, p_n$. Examples include **Weaver's surprise index**, $W_i = p_i^{-1}\sum_{r=1}^n p_r^2$, which is a special case of **Good's surprise index**

$$g_i = p_i^{-1}\left(\sum_1^n p_r^{a+1}\right)^{1/a}, \quad a \in \mathbb{R}.$$

Good's logarithmic surprise index is $\gamma_i = -\log p_i + \sum_r p_r \log p_r$.

See surprisingness. [I.J. Good, Antisurprise, *J. Stat. Comp. Simul.*, 17(1), 69–71, 1983]

surprisingness (alias: **surprise index**) Let $X_1, \ldots, X_r$ be non-negative integer-valued r.vs, with $\sum_{j=1}^r X_j = n$, having joint pmf $f(x_1, \ldots, x_r)$, that has a maximum at $(m_1, \ldots, m_r)$. Then K. Jordan defined the surprise index (or surprisingness) of the observation $(x_1, \ldots, x_r) = \mathbf{x}$, as

$$S(\mathbf{x}) = 1 - \frac{f(x_1, \ldots, x_r)}{f(m_1, \ldots, m_r)}.$$

[E. Csáki, Mathematical statistics, in: *A Panorama of Hungarian Mathematicians in the 20th Century*, J. Horváth (ed.). Springer, 2006]

survival function *see* distribution function

sweeping measure *see* balayage

Swendsen–Wang method *see* Gibbs sampling

Swiss army formula Arises in the context of stationary point processes, in particular in considering the ergodic properties of suitable functionals of the process; the Swiss army formula is a general identity which yields a number of well-known identities appearing in the theory. [P. Brémaud, A Swiss army formula of Palm calculus, *J. Appl. Probab.*, 30(1), 40–51, 1993]

Sylvester–Kac matrix *see* Kac matrix

Sylvester's problem (alias: **Sylvester's four point problem**) This asks for the probability $\mathbb{P}(\Delta)$ that four points picked independently at random in a finite convex region C of the plane have triangular convex hull, i.e. the probability that one of them lies inside the triangle formed by the other three. **Blaschke's inequality** asserts that

$$\frac{35}{12\pi^2} \leq \mathbb{P}(\Delta) \leq \frac{1}{3},$$

where the left-hand side is attained for C any ellipse, and the right-hand side is attained for C any triangle. [R. Pfiefer, The historical development of J.J. Sylvester's four point problem, *Math. Mag.*, 62(5), 309–317, 1989]

symbiotic branching process A quite general continuous-time two-type spatial branching diffusion process, where the branching rates are functions of both types in the population and are driven by correlated Wiener processes. Formally, the defining stochastic partial differential equation may be written, for $i \in \{1, 2\}$, as $\mathrm{d}X_i(t,x) = a\frac{\partial^2}{\partial x^2}X_i(t,x)\mathrm{d}t + b(X_1X_2)^{\frac{1}{2}}\mathrm{d}W_i(t,x)$, where a and b are positive, $x \in \mathbb{R}, t \geq 0$, and for $|\rho| \leq 1$, $\mathbb{E}\{W_1(s,\mathrm{d}x)W_2(t,\mathrm{d}y)\} = \rho s \wedge t\delta(x-y)\mathrm{d}x\mathrm{d}y$. [J. Blath *et al.*, On the moments and the interface of the symbiotic branching model, *Ann. Probab.*, 39(1), 252–290, 2011]

symmetrically dependent *see* exchangeable

symmetric alpha-stable distribution A subset of the class of stable distributions in which the characteristic function $\phi(t)$, in standard form, is given by $\phi(t) = \exp\{-c|t|^\alpha\}$, for $c > 0$ and $0 < \alpha \leq 2$. [J. Lamperti, *Probability*, 2nd edn. Wiley, 1996]

symmetric channel In the context of communication theory, a channel matrix $M = (m_{ij})$ is said to be symmetric (or strongly symmetric) if each row is a permutation of every other row, and each column is a permutation of every other column. The special case when M is a square $n \times n$ matrix with $m_{ii} = p, m_{ij} = q, i \neq j$, and $p + (n-1)q = 1$, may be called a **very strongly** (or strongly) **symmetric channel**. More generally, if the rows of M are permutations of each other, and the column sums are all equal then M is said to be **weakly symmetric** (or symmetric). Yet more generally, if the columns of M can be reordered in such a way that they can then be partitioned into submatrices that are symmetric in any of the above senses, then M is weakly (or very weakly) symmetric. Thus the binary symmetric channel is (strongly) symmetric, and the binary symmetric erasure channel,

$$M = \begin{pmatrix} 1-q-c & q & c \\ q & 1-q-c & c \end{pmatrix},$$

is weakly symmetric. The capacity of any such channel is achieved by a uniform distribution on the input alphabet. [R. McEliece, *The Theory of Information and Coding*. Addison-Wesley, 1977; reprinted by Cambridge University Press, 1985]

symmetric distribution A random variable X is said to have a symmetric distribution if $-X$ has the same distribution as X. More generally, a random vector $\mathbf{X}$ is symmetric if $-\mathbf{X}$ has the same distribution as $\mathbf{X}$. If $\mathbf{X} = (X_1, \ldots, X_n)$ has the same distribution as $(u_1X_1, \ldots, u_nX_n)$ for all $(u_1, \ldots, u_n) \in \{-1, 1\}^n$, then $\mathbf{X}$ is called **coordinate symmetric** (or **supersymmetric**). If the distribution of $\mathbf{X}$ is invariant under all rotations, then it is called **spherically** (or **radially**) **symmetric**. *See* elliptical distribution. [W. Schindler, *Measures with Symmetry*. Springer, 2003]

symmetric geometric-stable distribution *see* Linnik distribution

symmetric queue Commonly denotes a particular type of ordered queue in which the fraction of the server's effort that is expended on the customer in the rth place equals the probability that an arriving customer is assigned the rth place in the queue. [F. Kelly, *Reversibility and Stochastic Networks*, Wiley, 1979; reprinted by Cambridge University Press, 2011]

symmetrization of random variables Given a r.v. X with distn fn $F(x)$, it is often useful or necessary to find a closely related (in some suitable sense) r.v. Y, whose distribution function $G(y)$ is symmetric, which is to say that $G(y) = 1 - G(-y)$ at all continuity points of $G(y), y \in \mathbb{R}$. This process is called symmetrization, and there are several methods, including:

1. Let $\hat{X}$ be independent of X with the same distn; then $Y = X - \hat{X}$ is symmetric, with chf $\phi_Y(t) = |\phi_X(t)|^2$.
2. Let $Y = +X$ or $Y = -X$ with probability ½ each. Then Y is symmetric with chf $\phi_Y(t) = \frac{1}{2}\{\phi_X(t) + \overline{\phi_X(t)}\}$.

3. If X has density $f(x)$, let Y have density proportional to $\{f(x)f(-x)\}^{\frac{1}{2}}$; then Y is symmetric.
4. If Z is independent of X, and $Y = X + Z$ is symmetric, then Z is said to be a **symmetrizer** for X.

[C. Withers & S. Nadarajah, Methods for symmetrizing random variables, *PEIS*, 24(4), 549–559, 2010]

symmetrization inequality (alias: **Lévy's symmetrization inequality**) Any of several inequalities linking the properties of some r.v. X with a symmetrization Z of X. Perhaps most often this relates to the symmetrized r.v. $Z = X - \hat{X}$, where $\hat{X}$ is independent of X and has the same distn as X. In this case, for $t > 0$, and $a \in \mathbb{R}$, e.g. $\mathbb{P}(X - \text{med}(X) \geq t) \leq 2\mathbb{P}(Z \geq t)$, where $med(X)$ is a median of X; also

$$\frac{1}{2}\mathbb{P}\Big(|X-\text{med}(X)|\geq t\Big) \leq \mathbb{P}(|Z|\geq t) \leq 2\mathbb{P}\left(|X-a|\geq \frac{1}{2}t\right).$$

[M. Loève, *Probability Theory*, 4th edn. Springer, 1977]

symmetrized stochastic integral *see* Fisk–Stratonovich integral

synchronous process A suitable random process $X(t), t \geq 0$, is said to be synchronous for a sequence of random times $0 \leq T_0 < T_1 < T_2 < \cdots$, where $\lim_{n\to\infty} T_n = \infty$, a.s., if $X_n(u) = X(T_{n-1} + u)$, $0 \leq u < T_n - T_{n-1}$, is a stationary sequence. [P. Glynn & K. Sigman, Uniform Cesàro limit theorems for synchronous processes with applications to queues, *Stoch. Proc. Appl.*, 40(1), 29–43, 1992]

synonymous process Arises as a weak property of similarity or equivalence of random processes $X(t)$ and $Y(t)$, that is stronger than having the same finite-dimensional distributions. Let $X(t)$ adapted to the filtration $\mathcal{F}_t$, and $Y_{(t)}$ adapted to the filtration $\mathcal{G}_t, t \geq 0$, be defined on possibly different stochastic bases. They are said to be synonymous if for any $n \geq 1$, and non-negative $s_1, s_2, \ldots, s_n, t_1, \ldots, t_n$, and bounded measurable functions $g, g_1, g_2, \ldots, g_n$, we have

$$\begin{aligned}
&\mathbb{E}\Big[g\Big(\mathbb{E}\Big(g_1\Big(X(s_1),..,X(s_n)\Big)\Big|\mathcal{F}_{t_1}\Big)\Big), \ldots,\\
&\quad \mathbb{E}\Big(g_n\Big(X(s_1), \ldots, X(s_n)\Big)\Big|\mathcal{F}_{t_n}\Big)\Big]\\
&\quad = \mathbb{E}\Big[g\Big(\mathbb{E}\Big(g_1\Big(Y(s_1), \ldots, Y(s_n)\Big)\Big|\mathcal{G}_{t_1}\Big)\Big), \ldots,\\
&\quad \mathbb{E}\Big(g_n\Big(Y(s_1), \ldots, Y(s_n)\Big)\Big|\mathcal{G}_{t_n}\Big)\Big].
\end{aligned}$$

A yet stronger equivalence arises if $X(t)$ and $Y(t)$ have the same adapted distributions. [D. Hower & H. Keisler, Adapted probability distributions, *Trans. Amer. Math. Soc.*, 286(1), 159–201, 1984]

system *see* betting system; martingale transform

systems analysis *see* operational research

Szegö–Kolmogorov formula *see* Kolmogorov's formula

T

taboo probability (alias: **tabu probability; transitions with prohibition**) Arises in the theory of Markov processes, especially in the long-run behaviour of transient and null-recurrent chains, as exemplified by ratio limit theorems and conditioned limit laws. A taboo (or tabu) probability gives the n-step (in discrete time) transition probability from one state to another on the event that some state (or set of states) is not visited at any intermediate step. Formally, if T_{iB} is the first-passage time from the state i to the set B (for some homogeneous Markov chain), then ${}_Bp_{ij}(n) = \mathbb{P}(X_n = j; T_{iB} > n \,|\, X_0 = i)$ is a taboo probability for the **taboo set** B. If $B = j$, then ${}_jp_{ij}(n)$ is the familiar first-passage probability from i to j. If $B = i$, then ${}_ip_{ij}(n)$ is the last exit probability from i to j. If B is empty, then ${}_\emptyset p_{ij}(n) = p_{ij}(n)$, the n-step probabilities. A similar definition holds in continuous time, yielding taboo Q and resolvent matrices, and taboo stochastic semi groups. [R. Syski, *Passage Times for Markov Chains*. IOS Press, 1992; K. L. Chung, *Markov Chains*, 2nd edn. Springer, 2012]

tail The distribution function $F(x)$ of a r.v. X may have a right (or upper) tail, given by $1 - F(x) = \mathbb{P}(X > x)$, if $F(x) < 1$ for all $x \in \mathbb{R}$, and a left (or lower) tail, $F(x) = \mathbb{P}(X \le x)$, if $F(x) > 0$ for all x. It is double tailed if $0 < F < 1$, or single tailed if, for some x, $F(x)$ takes one of the values 0 or 1. Strictly, X has no tails if X is a.s. finite, but it is natural to speak loosely of the tails of the binomial distn, when considering $\mathbb{P}\left(|\frac{X}{n} - p| > x\right)$ for a binomial $\mathrm{B}(n,p)$ r.v. X, as $n \to \infty$. For the multivariate distn of a r.vr $\mathbf{X}$, the tail is understood as $\mathbb{P}(\|\mathbf{X}\| > x)$.

Tails are classified according to the amount of probability in them as $x \to \pm\infty$. Thus $F(x)$, or X, is said to be:

1. **long-tailed** (to the right) if, for $0 < t < \infty$,

   $$\lim_{x\to\infty} \mathbb{P}(X > x + t)/\mathbb{P}(X > x) = 1;$$

2. **heavy-tailed** (to the right) if, for all $t > 0$, $\mathbb{E}e^{tX} = \infty$;
3. **light-tailed** (to the right) if $\mathbb{P}(X > x) \le ae^{-bx}$ for some positive a and b, and all $x > 0$;

which is to say that $1 - F(x)$ is exponentially bounded and X^+ has finite moments of all orders. More precisely, for $a > 0$, $\mathbb{E}|X|^a < \infty$ iff $\int_0^\infty x^{a-1}\mathbb{P}(|X| > x)\mathrm{d}x < \infty$; thus when only a finite number of moments of X are finite, some writers say that X has very heavy tails. *See also* regular variation; subexponential distribution; and tail integral formula.

The term **fat-tailed** is used to denote several properties; some authors call X fat-tailed if X is leptokurtic, i.e.

$$\frac{\mathbb{E}(X - \mathbb{E}X)^4}{(\mathrm{var}X)^2} > 3,$$

and call platykurtic distributions, with the inequality reversed, thin-tailed. Other authors use fat-tailed as a synonym for heavy-tailed; yet others use it to denote power-law distributions. *See* regular variation. [S. Foss *et al.*, *Introduction to Heavy-tailed and Subexponential Distributions*, 2nd edn. Springer, 2013]

tail conditional expectation (alias: **conditional tail expectation**) For a r.v. X, the tail conditional expectation at level c is $\mathbb{E}(X \,|\, X > c)$; this function finds applications in actuarial and financial risk theory. [P. Artzner *et al.*, Coherent measures of risk, *Math. Finan.*, 9(3), 203–228, 1999]

tail dependence measure Broadly, any of a number of functions (or coefficients, or parameters) that arise in describing how far r.vs X and Y, with joint distn fn $F(x, y)$, are from being independent for large values of x and y, according to some suitable criterion that may be called a **tail dependence function**. The general task is usually formulated in terms of a joint distn whose marginals are standardized in some way, e.g. by using a copula with uniform marginals, or – in the context of extreme value theory – marginals having an extreme value distn such as Fréchet's with $F(x) = \exp\left(-\frac{1}{x}\right)$, $x > 0$. Thus in the copula framework the so-called dependence function of X and Y is the copula $c(x, y) = F\left(F_X^{-1}(x), F_Y^{-1}(y)\right)$, where F_X^{-1} and F_Y^{-1} are the generalized inverses of the marginal distn fns. Then the upper tail dependence parameter (or coefficient) is

$$\tau_u = \lim_{t \downarrow 0} \mathbb{P}\big(F_X(X) > 1 - t \,|\, F_Y(Y) > 1 - t\big)$$
$$= \lim_{t \downarrow 0} \frac{1}{t} \bar{c}(1 - t, 1 - t),$$

where $\bar{c}(x, y) = 1 - x - y + c(x, y)$ is the copula survivor function; the lower tail dependence parameter is $\tau_l = \lim_{t\downarrow 0}\mathbb{P}(F_X(X) \le t|F_Y(Y) \le t) = \lim_{t\downarrow 0}\frac{1}{t}c(t, t)$.

When these are zero, one speaks of **asymptotic independence**. By convention (and in applications), interest is usually confined to the upper tail dependence where the **stable tail dependence fn** is

$$\tau(x, y) = \lim_{t \downarrow 0} \frac{1}{t} \mathbb{P}\big(F_X(X) \ge 1 - tx, F_Y(Y) \ge 1 - ty\big),$$

when this exists for all $x, y \ge 0$, and then $\tau(1, 1)$ is the upper tail dependence parameter. The alternative definition

$$\sigma(x,y) = \lim_{t \downarrow 0} \frac{1}{t} \mathbb{P}\big\{\{F_X(X) \ge 1 - tx\} \cup \{F_Y(Y) \ge 1 - tY\}\big\}$$

is also seen, where $\sigma(x, y) + \tau(x, y) = x + y$. More complicated tail dependence functions seek to capture more delicate dependence between X and Y as $x, y \to \infty$. [J. Einmahl *et al.*, A method of moments estimator of tail dependence, *Bernoulli*, 14(4), 1003–1026, 2008; J. Heffernan, A directory of coefficients of tail dependence, *Extremes*, 3(3), 279–290, 2000]

tail equivalence Used in two senses:

1. Two distribution functions are tail equivalent if $s = \inf\{x\colon F(x) = 1\} = \inf\{x\colon G(x) = 1\}$ (where $s = \infty$ if $FG < 1$ for all x), and

   $$\lim_{x \uparrow s} \frac{1 - F(x)}{1 - G(x)} = c > 0,$$

   for some constant c, which is often required to be 1.
2. Two sequences of random variables X_n and Y_n, $n \ge 1$, are tail (or Khinchin) equivalent if they differ for only a finite number of terms, a.s.; more formally, if $\sum_n \mathbb{P}(X_n \ne Y_n) < \infty$. In general, the probability of any tail event is the same for both, i.e. they converge a.s. (or diverge a.s.) together, and they both satisfy (or both fail to satisfy) the same laws of large numbers.

[P. Embrechts *et al.*, *Modelling Extremal Events*. Springer, 1997; S. Resnick, *A Probability Path*. Birkhäuser, 1999]

tail event Arises in the context of a random process (or sequence) X_n, $n \ge 1$, and any event A defined in terms of the sequence $X_n(\omega)$, $\omega \in \Omega$. Such an event A whose occurrence is not determined by any finite number of the $X_n(\omega)$ (or equivalently, changing a finite number of values of the sequence does not affect whether or not $\omega \in A$), is called a tail event. More formally, let $\sigma(X_n, X_{n+1}, \ldots)$ be the smallest sigma-field for which X_m, $m \ge n$ are all measurable; which can be thought of as the future from n. Then $\mathcal{F}_\infty = \cap_n \sigma(X_n, X_{n+1}, \ldots)$ is called the **tail sigma-field**, and contains all tail events; it can be thought of as the remote future. Examples of tail events are:

$$A = \{\omega : \textstyle\sum_n X_n(\omega) \text{ converges}\}$$
$$B = \{\omega : n^{-1}\textstyle\sum_1^n X_r(\omega) \text{ converges as } n \to \infty\}$$
$$C = \{\omega : X_n(\omega) \in B_n \text{ infinitely often}\},$$

where B_n is a sequence of Borel sets in $\mathbb{R}$. A random variable measurable on $\mathcal{F}_\infty$ is a **tail random variable**. If the X_n are independent, then all tail events have probability 0 or 1, by Kolmogorov's zero-one law. *See also* Hewitt–Savage zero-one law. [R. Ash & C. Doléans-Dade, *Probability and Measure Theory*, 2nd edn. Academic Press, 2000]

tail index *see* regular variation

tail inequality *see* concentration inequality

tail integral formula Denotes this expression for the rth moment of a non-negative r.v. X with distn fn $F(x)$:

$$\mathbb{E}X^r = \mathbb{E}\int_0^\infty I(X^r > x)\mathrm{d}x = \int_0^\infty \mathbb{P}(X^r > x)\mathrm{d}x$$
$$= \int_o^\infty rx^{r-1}\big(1 - F(x)\big)\mathrm{d}x,$$

where I is the usual indicator. In particular, for $r = 1$, and integer-valued X

$$\mathbb{E}X = \sum_{r=0}^{\infty} \mathbb{P}(X > r) = \sum_{r=1}^{\infty} \mathbb{P}(X \ge r).$$

Consequently, when $\mathbb{E}X$ is finite, the distn fn

$$G(x) = (\mathbb{E}X)^{-1}\int_0^x \{1 - F(y)\}\mathrm{d}y$$

is called the equilibrium or stationary excess distn of $F(x)$. *See* renewal theory; equilibrium; and integrated tail distribution. [S. Foss *et al.*, *Introduction to Heavy-tailed and Subexponential Distributions*, 2nd edn. Springer, 2013]

tail Markov chain The chain with state space $\mathbb{R}^+$ defined in terms of a positive r.v. with distn fn $F(x)$ and continuous positive density $f(x)$, $x > 0$, by the transition kernel

$$k(x, A) = \int_A \frac{f(x+y)}{1 - F(x)} \mathrm{d}y.$$

[J. Hobert *et al.*, Stability of the tail Markov chain, *Bernoulli*, 10(3), 549–564, 2004]

tail sigma-field *see* tail event

Takács equation Arises in the context of the classic single server queue with Poisson arrivals (of rate λ, possibly time-dependent) and general independent service times (i.e. M/G/1 in Kendall's system). If the workload (or virtual waiting time, or Takács) process at time t is $V(t)$, with distn fn $F(x, t)$, then F obeys Takács equation

$$\frac{\partial F}{\partial t} = \frac{\partial F}{\partial x} - \lambda \frac{\partial F}{\partial t} + \lambda \int_0^x F_S(x-y) \mathrm{d}_y F(y, t),$$

where F_S is the service-time distribution. [D. Cox & H. Miller, *Theory of Stochastic Processes*. Methuen, 1965; reprinted by Chapman & Hall, 2001]

Takács formula Any of several identities arising in various areas of mathematics from combinatorics to queueing theory; perhaps most often it denotes expressions for waiting times and other properties of the single server queue, but it also denotes an expression for occupation times of the drifting Wiener process. [R. Doney & M. Yor, On a formula of Takács for Brownian motion with drift, *J. Appl. Probab.*, 35, 272–280, 1998]

Takács theorem A generalization of Whitworth's classical (1878) ballot theorem, thus: let $X_1, X_2, \ldots, X_{n+1}$ be non-negative, integer-valued, exchangeable r.vs, and set $S_r = X_1 + \cdots + X_r$. Let $N_{n+1}(c)$ be the number of indices $r, 1 \leq r \leq n+1$, such that $S_r - r < c$, where c is a non-negative integer. Then for $1 \leq r \leq n+1$, $\mathbb{P}(N_{n+1}(0) = r \mid S_{n+1} = n) = (n+1)^{-1}$, and

$$\mathbb{P}(N_{n+1}(c) = n+1 \mid S_{n+1} = n) = 1 - \sum_{j=1}^{n-c} \frac{c+1}{n+1-j} \mathbb{P}(S_j - j = c \mid S_{n+1} = n),$$

whenever these conditional probabilities exist. [L. Takács, *Combinatorial Methods in the Theory of Stochastic Processes*. Wiley, 1967; M. Dwass, Review of 'Combinatorial Methods in the Theory of Stochastic Processes', *SIAM Rev.*, 12(1), 158–159, 1970]

Talagrand inequality (alias: **T. isoperimetric inequality; T. theorem; T. deviation inequality; convex distance inequality**) Arises in the context of measure concentration, and is seen in several forms, depending on the application (of which there are many). Classically, it may be stated in terms of Talagrand's convex distance, defined thus: let $\Omega = \prod_1^n \Omega_i$ be a product space (each Ω_i being suitably measurable), and let $X_i \in \Omega_i$ be the ith coordinate of a r.v. $X \in \Omega$, having the product probability measure $\prod_1^n \mathbb{P}_i$. For X and $A \in \Omega$, the Talagrand convex distance of X from A is (for real numbers $a_1, \ldots, a_n$), given by $d_T(X, A) = \sup_{a \in B} \inf_{y \in A} \sum_i a_i I(X_i \neq y_i)$, where the supremum is over all $a_1, \ldots, a_n$ such that $\sum_i a_i^2 \leq 1$ (the unit ball B_1), and the indicator $I(X_i \neq Y_i) = 1$ if $X_i \neq Y_i$, and is zero otherwise. Then Talagrand's inequality is

$$\mathbb{P}(X \in A)\mathbb{E}\left(\exp\left\{\frac{1}{4} d_T(X, A)^2\right\}\right) \leq 1.$$

In particular, for $B \subseteq \Omega$, define $d_T(A, B) = \min_{x \in B} d_T(x, A)$. Then

$$\mathbb{P}(A)\mathbb{P}(B) \leq \exp\left\{-\frac{1}{4} d_T(A, B)^2\right\}.$$

Likewise, M. Talagrand defines the enlargement (or t-fattening, or convex extension) of A to be A_t, where $\mathbf{x} \in A_t$ iff for any vector $(a_1, \ldots, a_n)$ of reals there is $\mathbf{y} \in A$ such that

$$\sum_i^n a_i I(x_i \neq y_i) \leq t\left(\sum_1^n a_i^2\right)^{\frac{1}{2}}.$$

Then $\mathbb{P}(A)(1 - \mathbb{P}(A_t)) \leq \exp(-t^2/4)$. [M. Talagrand, *Concentration of Measure and Isoperimetric Inequalities in Product Spaces*. Publications Mathématiques de L'HES/ Springer, 1995]

Tanaka's construction (alias: **Tanaka's decomposition**) Arises in the context of a random walk $S_n = \sum_1^n X_r$, starting at $S_0 = 0$, that does not drift to $-\infty$. Let C_n be S_n conditioned to stay positive; that is, the distribution of the nth step Z_n of C_n is the distribution of X_n conditional on S_n never becoming negative. Then Tanaka's construction (or decomposition) takes any path of $S_n, n \geq 0$, and reassembles it into a path of $C_n, n \geq 0$. That is to say the reassembled sequence $R_n, n \geq 1$, has the same joint distributions as $C_n, n \geq 1$. Informally, the steps of segments between successive ladder points of S_n are taken in reverse order. More formally, let (L_k, H_k), $k \geq 0$ be the (strict) increasing ladder process of the walk S_n, with $L_0 = H_0 = 0$. Then the reassembled sequence is given by $R_0 = 0$ and for $k \geq 0$,

$$R_{T_k+j} = H_k + X_{T_{k+1}} + X_{T_{k+1}-1} + \cdots + X_{T_{k+1}-j+1},$$
$$1 \le j \le T_{k+1} - T_k.$$

[R. Doney, Tanaka's construction for random walks and Lévy processes, in *Séminaire de Probabilités XXXVIII, Lecture Notes in Mathematics*, Vol. 1857. Springer, 2005, pp. 1–4]

Tanaka's formula (alias: **Tanaka's equation**) This relates the Wiener process to its local time, in the context of stochastic calculus. Formally, there is a continuous increasing process $L(t)$ such that for the Wiener process $W(t)$

$$|W(t)| = \int_0^t \operatorname{sign} W(s) \mathrm{d}W(s) + L(t), \quad t \ge 0,$$

where sign $x = 1$ for $x > 0$, and sign $x = -1$ for $x \le 0$. The process $L(t)$ is called the local time of $W(t)$ at (or in) 0, and is effectively an occupation density. The equation gives the Doob–Meyer decomposition of $|W(t)|$, and there is a similar formula for $W(t)^+ = \max\{0, W(t)\}$, namely

$$W(t)^+ = \int_0^t I\Big(W(s) > 0\Big) \mathrm{d}W(s) + \frac{1}{2}L(t), \quad t \ge 0,$$

where $I(W(s) > 0)$ is 1 when $W(s)$ is positive, or 0 otherwise. [P. Mörters & Y. Peres, *Brownian Motion*. Cambridge University Press, 2010]

Tanaka–Meyer formula (alias: **Itô–Tanaka–Meyer formula; Itô–Tanaka formula**) Arises in considering the local times of a continuous semi-martingale $X(t)$, where it supplies an integral representation (or decomposition) of any convex function $u(X(t))$ of $X(t)$ in terms of the local times of $X(t)$ and the first and second derivatives of u. [D. Revuz, & M. Yor, *Continuous Martingales and Brownian Motion*, 3rd edn. Springer, 1999]

tangent process *see* first variation process

tangent sequence Two sequences X_n and Y_n, $n \ge 1$, of r.vs adapted to an increasing sequence of σ-fields $\mathcal{F}_n$, $n \ge 1$, on a complete probability space, are said to be tangent sequences if for all $n \ge 1$, and all Borel sets B, $\mathbb{P}(X_n \in B \mid \mathcal{F}_{n-1}) = \mathbb{P}(Y_n \in B \mid \mathcal{F}_{n-1})$. [S. Cox & M. Veraar, Some remarks on tangent martingale difference sequences in L^1 spaces, *Electr. Commun. Probab.*, 12, 421–433, 2007]

Tao's inequality Arises in the context of Szemerédi's lemma, and asserts this: for jointly distributed discrete r.vs X, Y, and Z, where $X \in [-1, 1]$ and Z is a (non-random) function of Y, we have

$$\mathbb{E}\,|\,\mathbb{E}(X \mid Y) - \mathbb{E}(X \mid Z)| \le \{2\log_e 2I(X; Y \mid Z)\}^{\frac{1}{2}},$$

where $I(X; Y \mid Z)$ is the mutual information between X and Y, conditional on Z. [R. Ahlswede, The final form of Tao's inequality, *Adv. Math. Commun.*, 1(2), 239–242, 2007]

target time *see* random target lemma

Tausworthe generator A popular multiplicative recursive generator of pseudo-random numbers (it is of linear congruential type). [J. Gentle, *Random Number Generation and Monte Carlo Methods*, 2nd edn. Springer, 2003]

Taylor's formula Let $D(t)$ be drifting Brownian motion, i.e. $D(t) = \mu t + W(t)$ where $W(t)$, $t \ge 0$, is the Wiener process, and let $M(t) = \max_{0 \le s \le t} D(s)$. Then for $T = \inf\{t: M(t) - D(t) = a\}$, we have

$$\mathbb{E}\exp\Big(\theta D(T) - \psi T\Big) = \frac{c\exp\{-(\mu + \theta)a\}}{c\cosh(ac) - (\mu + \theta)\sinh ac},$$

for $\theta > 0$ and $\psi < c\coth(ac) - \mu \ge 0$, where $c = (2\psi + \mu^2)^{\frac{1}{2}}$. [D. Williams, On a stopped Brownian motion formula of H. M. Taylor, in *Séminaire de Probabilités X, Université de Strasbourg*. Springer, 1976]

Taylor's theorem Asserts that for Brownian motion in n dimensions the Hausdorff (and box) dimension of the process is a.s. 2 for $n \ge 2$, and 3/2 for $n = 1$. [S. J. Taylor, The Hausdorff α-measure of Brownian paths in n-space, *Math. Proc. Camb. Phil. Soc.*, 49(1), 31–39, 1953; P. Mörters & Y. Peres, *Brownian Motion*. Cambridge University Press, 2010]

tax problem Arises in the context of optimal scheduling of jobs, and denotes the case when unprocessed jobs give rise to a cost (tax) until they are out of the system. [P. Whittle, Tax problems in the undiscounted case, *J. Appl. Probab.*, 42(3), 754–765, 2005]

***t*-distribution** *see* Student's t-distribution

Teicher's theorem Any of a number of unique factorization (or decomposition) theorems for mixtures of distributions; perhaps most often the assertion that the class of mixtures of a one-parameter additively closed family of distributions $F(x, a)$, $a \in \mathcal{A}$, is identifiable, which is to say that the equality of the mixtures formed using mixing distribution functions $M_1(x)$ and $M_2(x)$ entails the equality of $M_1(x)$ and $M_2(x)$.

Formally:

$$\int_{a\in A} F(x,a)\mathrm{d}M_1(a) = \int_{a\in A} F(x,a)\mathrm{d}M_2(a) \Rightarrow M_1 \equiv M_2.$$

[H. Teicher, Identifiability of finite mixtures, *Ann. Math. Stat.*, 34(4), 1265–1269, 1963]

Telecom process Arises as a model for traffic intensity in networks, and is obtained explicitly as a limit of suitably scaled renewal-reward processes. It is a self-similar symmetric α-stable process; where the parameter $\alpha \in [1,2]$, and for $\alpha = 2$ the process is fractional Brownian motion. [I. Kaj & M. Taqqu, Convergence to fractional Brownian motion and to the Telecom process: the integral representation approach, in *In and Out of Equilibrium 2*, V. Sidoravicius & M. Vares (eds). Birkhäuser, 2008]

telegraph process In its basic form, a telegraph process $X(t) \in \{0,1\}$, $t \geq 0$, is the symmetrical form of the two-state Markov process, and can be written in terms of a Poisson process $N(t)$ of intensity λ as follows $X(t) = X(0)(-1)^{N(t)}$.

The term is also used to denote the integrated form $K(t) = \int_u^t X(0)(-1)^{N(s)}\mathrm{d}s$, where $X(t)$ is seen as the velocity, and $K(t)$ as the position, of a randomly moving particle; *see* Goldstein–Kac process. More generally, if the particle is allowed to rest, so that $X(t) \in \{-1,0,1\}$, one has a three-state (or valued) telegraph process.

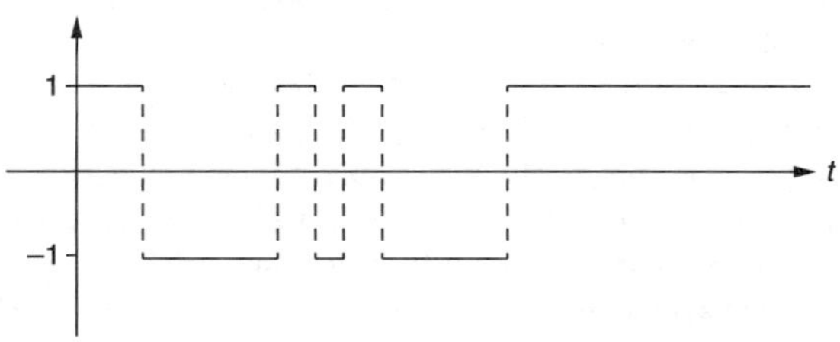

Basic telegraph process.

[A. Di Crescenzo *et al.*, Generalized telegraph process with random jumps, *J. Appl. Probab.*, 50(2), 450–463, 2013]

terminal time Given a suitably regular Markov process $X(t)$, $t \geq 0$, a stopping time T for X is called a terminal time if for any stopping time S, with probability 1 on the event $\{S < T\}, T = S + T \circ \theta_s$ where θ_s is the usual shift operator. Some authors speak of a weak terminal time in the special case when S is degenerate at t.

[M. Sharpe, *General Theory of Markov Processes*. Academic Press, 1988]

term structure model *see* bond pricing; interest rate model

test function methods *see* Foster's criteria

test paradox *see* base-rate fallacy test

tetrachoric series *see* Mehler's expansion

Teugels' martingale A family of martingales first defined in the context of a Lévy process $X(t)$, $t \geq 0$, thus: let $\Delta X(t) = X(t) - X(t-)$ be the jump at t, and define the so-called power jump process for $n \geq 2$

$$Z^{(n)}(t) = \sum_{0<s\leq t} (\Delta X(s))^n.$$

Then the Teugels martingale of order n is

$$M(t) = Z^{(n)}(t) - \mathbb{E}Z^{(n)}(t).$$

[D. Nualart & W. Schoutens, Chaotic and predictable representations for Lévy processes, *Stoch. Proc. Appl.*, 90(1), 109–122, 2000]

theorem of compound probability *see* multiplication rule for probabilities

theory of belief functions *see* Dempster–Shafer theory

Thiele's equation A classic result of insurance mathematics, which gives the reserve required to insure a single random (excess) life T. During $[0,T]$ the insured pays premiums at a constant rate r, and at time T the insurer pays out S. At any time t after time 0 (when this contract is agreed), let $v(t)$ be the expected value of the future payments discounted at rate δ, less the expected future premiums to be received. If T has density $f(t)$ and distribution fn $F(t)$, then the backward equation for $v(t)$ is Thiele's equation

$$\frac{\mathrm{d}v}{\mathrm{d}t} = r + \delta v - (S - v)f(t)/(1 - F(t)).$$

[T. Rolski *et al.*, *Stochastic Processes for Insurance and Finance*. Wiley, 1998]

thinning (alias: **random erasure or deletion; Rényi thinning**) A formalization of the idea that points of a counting process (or point process) may be omitted, or struck out, or fail to register, at random. Thus if X is a point process on $\mathbb{R}^m$ then the process obtained by deleting each point X independently with probability p is called the p-thinning of X; it was introduced by A. Rényi in a characterization of the Poisson process by invariance under thinning and scaling. More generally, the erasure probability may be $p(x)$, a function of the position of the point; and $p(x)$ may be random. The thinning procedure can be seen as a special case of colouring, with only two colours.

Analogously, if X is a non-negative integer-valued r.v., and Z_r, $r \geq 1$, comprise i.i.d. Bernoulli (p) r.vs, then $T_n = \sum_{r=1}^{X} Z_r$ is called the p-thinning of X (or of the distn of X). [P. Harremoës *et al.*, Thinning, entropy, and the law of thin numbers, *IEEE Trans. Inf. Theor.*, 56(9), 4228–4244, 2010]

Thomasian's theorem Supplies conditions for the equivalence of various modes of convergence on a probability space $\{\Omega, \mathcal{F}, \mathbb{P}\}$. Thus the following are equivalent:

(a) Ω is a countable union of disjoint atoms;
(b) there is a metric for r.vs on $\{\Omega, \mathcal{F}, \mathbb{P}\}$ such that convergence a.s. is equivalent to convergence in the metric;
(c) For any sequence X_n, $n \geq 1$ of r.vs, $X_n \xrightarrow{\mathrm{P}} X \Rightarrow X_n \xrightarrow{\text{a.s.}} X$.

These are also equivalent:

(i) Ω is a finite union of disjoint atoms;
(ii) there is a norm for r.vs on $\{\Omega, \mathcal{F}, \mathbb{P}\}$ such that convergence in the norm is equivalent to convergence in probability (and also to a.s., rth mean, and complete convergence).

[R. Tomkins, On the equivalence of modes of convergence, *Canadian Math. Bull.*, 16(4), 571–575, 1973]

Thorin class A family of generalized gamma distributions comprising distributions on $[0, \infty)$ that arise as weak limits of finite convolutions of gamma densities; they may be characterized in several other ways. In this context **Thorin's thm**(s) assert the infinite divisibility of the lognormal distn and of the Pareto distn. [L. Bondesson, *Generalized Gamma Convolutions and Related Classes of Distributions and Densities*. Springer, 1992]

three-series theorem (alias: **Kolmogorov's three-series theorem**) Asserts that for a sequence $X_n, n \geq 1$, of independent r.vs (or r.vrs), the series $S_n = \sum_{r=1}^{n} X_r$ converges a.s. iff these three series converge, where

$$Y_r = \begin{cases} X_r & \text{if } |X_r| \leq 1 \quad (\text{or } \|X_r\| \leq 1 \text{ for r.vrs}) \\ 0 & \text{if } |X_r| > 1 \quad (\text{or } \|X_r\| > 1 \text{ for r.vrs}) \end{cases},$$

(i) $\sum_{r=1}^{\infty} \mathbb{P}(X_r \neq Y_r) = \sum_{r=1}^{\infty} \mathbb{P}(|X_r| > 1)$
(ii) $\sum_{r=1}^{\infty} \mathbb{E} Y_r$
(iii) $\sum_{r=1}^{\infty} \mathrm{var} Y_r$.

More generally, the assumption of independence may be dropped, in which case (ii) and (iii) are replaced by a sum of conditional expectations and a sum of conditional variances respectively. [B. Brown, A general three-series theorem, *Proc. Amer. Math. Soc.*, 28(2), 573–577, 1971]

threshold phenomenon *see* cut-off phenomenon

tied-down Brownian motion (or Wiener process) *see* Brownian bridge

tightness A concept that is used in the theory and applications of weak convergence of probability measures. Let Ω be a metric (or, more generally, a topological) space, and $\mathcal{B}$ the sigma-field generated by the open sets of Ω. A probability measure $\mathbb{P}(\cdot)$ on $\mathcal{B}$ is tight if, for any $\delta > 0$, there exists a compact set C such that $\mathbb{P}(C) > 1 - \delta$.

If Ω is Euclidean space, then any (proper) probability measure is tight. More generally, if Ω is complete and separable, then any probability measure on $(\Omega, \mathcal{B})$ is tight. Further, a collection of probability measures $\mathbb{P}_n$, $n \in I$, is (uniformly) tight if for any $\delta > 0$, there is a compact set C such that $\mathbb{P}_n(\mathrm{C}) > 1 - \delta$, $n \in I$.

Note that the word 'tight' has other meanings; for example in asymptotic analysis the sequence $g(n)$ is said to be an asymptotically tight bound for the sequence $f(n)$, as $n \to \infty$, if $f(n) = \Theta(g(n))$, which is to say that $a|g(n)| \leq |f(n)| \leq b|g(n)|$ for some finite positive constants a and b, and all n. [P. Billingsley, *Convergence of Probability Measures*, 2nd edn. Wiley, 1999]

tilted distribution *see* Esscher transform

time change *see* change of time

time on test transform (alias: **total time on test transform**) Let $F(x)$ be the distn fn, with inverse $F^{-1}(x)$, of a non-negative r.v. X. Then the (total) time on test transform of X is

$$T_F(t) = \int_0^{F^{-1}(t)} (1 - F(x))\mathrm{d}x, \quad 0 < t < 1.$$

More generally, for a distn fn $G(x)$ having density $g(x)$, with $g(x) = 0$ for $x < 0$, the generalized time on test transform is

$$T_F(t, G) = \int_0^{F^{-1}(t)} g\big(G^{-1}F(x)\big)\mathrm{d}x,$$

which reduces to $T_F(t)$ when $g(x) = \mathrm{e}^{-x}$, $x \geq 0$. [X. Li & M. Shaked, A general family of univariate stochastic orders, *J. Stat. Plan. Inf.*, 137(11), 3601–3610, 2007]

time series Originally a sequence X_n, $n \geq 0$, of measurements or observations over time, often called an **index sequence** in economic contexts. By analogy, the term denotes a sequence $X(t)$ of r.vs, where the index t is seen as time and may take values in $\mathbb{Z}$, or $\mathbb{Z}^+$, or $\mathbb{R}$, or $\mathbb{R}^+$, etc. [P. Brockwell & R. Davis, *Time Series*, 2nd edn. Springer, 1991]

time-space harmonic polynomial Given a suitable random process $X(t)$, $t \geq 0$, with its natural filtration $\mathcal{F}_t$, a family of polynomials $P(t,x)$, $x \in \mathbb{R}$, $t \geq 0$, is said to be space-time harmonic for X if $P(t, X(t))$ is a martingale for $\mathcal{F}_t$, i.e. for all $s \leq t$, $\mathbb{E}\{P(t,X(t)) \mid \mathcal{F}_s\} = P(s,X(s))$. Classic examples include: Hermite polynomials for the Wiener process; Charlier polynomials for the Poisson process; and Laguerre polynomials for the gamma process. [E. Di Nardo & I. Oliva, A new family of time-space harmonic polynomials with respect to Lévy processes, *Ann. di Matematica*, 192(5), 917–929, 2013]

time substitution *see* change of time

T-martingale (alias: **Kahane's T-martingale**) Arises in the context of multiplicative cascades, and comprises a sequence of non-negative random functions $Q_n(\cdot)$, $n \geq 1$, on a suitable metric space, adapted to the filtration $\mathcal{F}_n$, and such that $Q_n(t)$ is a martingale for $\mathcal{F}_n$, for each fixed t. [E. Waymire & S. Williams, Cascade decomposition theory with applications to Markov and exchangeable cascades, *Trans. Amer. Math. Soc.*, 348(2), 585–632, 1996]

Tonelli's theorem *see* Fubini's theorem

tontine Arises in actuarial mathematics as a type of annuity that can be seen as the reverse of the usual life insurance context. Subscribers to a tontine pay a lump sum to the company, and receive an annuity. On the death of any subscriber their annuity continues to be distributed among all survivors (in proportion to their share). When the set of survivors is empty, the company takes the fund. [A. Hald, *History of Probability and Statistics and their Applications Before 1750*. Wiley, 1990]

Toom's theorem (alias: **Toom's stability theorem**) Arises in the context of a class of probabilistic cellular automata that may be called Toom processes; the theorem supplies a necessary and sufficient condition (called Toom's eroder condition) for the nonergodicity of such processes. [J. Gravner & D. Griffeath, Random growth models with polygonal shapes, *Ann. Probab.*, 34(1), 181–218, 2006]

Topchii–Vatutin inequality *see* von Bahr–Esseen inequality

total expectation rule (or law, or theorem, etc.) *see* partition theorem

total probability rule (or law, or theorem, etc.) *see* partition theorem; addition rule

total variance law *see* conditional variance theorem

total variation distance (alias: **the statistical distance; trace distance; $\mathbf{L_1}$ distance**) For any two probability distributions (i.e. measures) $\mathbb{P}(\cdot)$ and $\mathbb{Q}(\cdot)$ on a sample space Ω the total variation distance $d(\mathbb{P}, \mathbb{Q})$ is the supremum of the absolute difference between $\mathbb{P}(A)$ and $\mathbb{Q}(A)$ over all possible events A in Ω. That is:

$$\begin{aligned} d(\mathbb{P},\mathbb{Q}) &= \sup_{A\in\mathcal{F}}|\mathbb{P}(A)-\mathbb{Q}(A)|, \text{ where } \mathcal{F} \text{ is the } \sigma\text{-field of events,} \\ &= \frac{1}{2}\max_{|u|\leq 1}\left|\int u\mathrm{d}\mathbb{P}-\int u\mathrm{d}\mathbb{Q}\right| \in [0,1], \end{aligned}$$

where the maximum is over all suitable functions u: $\Omega \to \mathbb{R}$, such that $|u| \leq 1$.

In particular, when Ω is countable

$$\begin{aligned} d(\mathbb{P},\mathbb{Q}) &= \frac{1}{2}\sum_{x\in\Omega}|\mathbb{P}(x)-\mathbb{Q}(x)| = \sum_{x:\mathbb{P}(x)>\mathbb{Q}(x)}\{\mathbb{P}(x)-\mathbb{Q}(x)\} \\ &= \inf \mathbb{P}(X \neq Y), \end{aligned}$$

where the infimum is over all possible couplings of r.vs X and Y whose marginal distributions are $\mathbb{P}$ and $\mathbb{Q}$.

Similarly, if $\mathbb{P}$ and $\mathbb{Q}$ are absolutely continuous wrt some positive measure m, having corresponding densities (Radon–Nikodym derivatives wrt m) f and g, then $d(\mathbb{P}, \mathbb{Q}) = \frac{1}{2}\int |f - g|\mathrm{d}m$. Some authors take this as the primary definition with the omission of the factor ½, leading to the **simple variation distance**

$\mathcal{D}(\mathbb{P}, \mathbb{Q}) = 2d(\mathbb{P}, \mathbb{Q}) \in [0, 2]$. The **affinity** between $\mathbb{P}$ and $\mathbb{Q}$ is $a(\mathbb{P}, \mathbb{Q}) = 1 - d = 1 - 1/2\mathcal{D}$. [D. Levin *et al.*, *Markov Chains and Mixing Times*. American Mathematical Society, 2009; A. Gibbs & F. Su, Choosing and bounding probability metrics, *Int. Stat. Rev.*, 70(3), 419–435, 2002]

tower (or smoothing) property (alias: **chain rule for conditional expectation**) A fundamental property of conditional expectation which asserts that for r.vs X and Y defined on $(\Omega, \mathcal{F}, \mathbb{P},)$ we have $\mathbb{E}(\mathbb{E}(X \mid Y)) = \mathbb{E}X$; and, similarly for $\mathcal{G}$ any sub-sigma-field of $\mathcal{F}, \mathbb{E}(\mathbb{E}(X \mid \mathcal{G}) = \mathbb{E}X$. [D. Williams, *Probability with Martingales*. Cambridge University Press, 1991]

Tracy–Widom distribution A probability distribution that arose in the context of a class of random $n \times n$ matrices, having suitable normally distributed entries, as the limiting distribution of the largest eigenvalue (appropriately centred and scaled) as $n \to \infty$. [J. Baik *et al.*, On the distribution of the length of the largest increasing subsequence of random permutations, *J. Amer. Math. Soc.*, 12(4), 1119–1178, 1999; C. Tracy & H. Widom, The distributions of random matrix theory and their applications, in *New Trends in Mathematical Physics*, V. Sidoravičius (ed.). Springer, 2009]

trading strategy (alias: **portfolio process**) Arises in financial mathematics where there is a market comprising a vector $S_1(t), \ldots, S_d(t), t \geq 0$, of asset prices (comprising stocks, bonds, etc.); typically this is assumed to be driven by a suitable random process and adapted to the resulting natural filtration $\mathcal{F}_t, t \geq 0$. Then a trading strategy, or portfolio process, is an $\mathcal{F}_t$-measurable vector $\tau_1(t), \ldots, \tau_d(t)$, taking values in $\mathbb{R}^d$, with a resulting wealth process

$$M(t) = \sum_r \tau_r(t) S_r(t).$$

If wealth is consumed according to an $\mathcal{F}_t$-measurable consumption process $c(t) \geq 0, t \geq 0$, then the portfolio (or strategy) with consumption $c(t)$ is called self-financing if

$$M(t) = M(0) + \sum_r \int_0^t \tau_r(s) \mathrm{d}S_r(s) - \int_0^t c(u) \mathrm{d}u,$$

which may be called the **budget constraint**. The trading problem, or portfolio selection problem, is that of choosing the processes $\tau_r(t)$ and $c(t)$ to maximize some suitable reward function. For example, the optimal consumption problem seeks to maximize the agents expected utility over an interval $[0, T]$, where T is a stopping time.

In the absence of consumption, the hedging problem seeks a portfolio to replicate the value of an option; while the trading schedule problem seeks to maximize some functional of the wealth process, e.g. utility. *See* Merton portfolio problem. [M. Musiela & M. Rutkowski, *Martingale Methods in Financial Modelling*, 2nd edn. Springer, 2005]

traffic intensity (alias: **offered** (or **traffic**) **load; utilization**) Arises in queuing theory where, broadly, the traffic intensity ρ for a system is the ratio of two rates; namely, the rate at which customers arrive, and the rate at which the system is able to deal with them. (Here, rates are time averages if the system is not in equilibrium.) In general, $\rho > 1$ corresponds to congestion because the system cannot meet demand, and $\rho < 1$ offers the possibility of an equilibrium in which queues are finite. For the single server queue G/G/1, the more formal definition is

$$\rho = \frac{\text{mean service time}}{\text{mean inter}-\text{arrival time}}.$$

It is traditionally measured in **Erlangs**. *See* Heavy traffic. [S. Asmussen, *Applied Probability and Queues*, 2nd edn. Springer, 2003]

trajectory *see* realization

transformation of time *see* change of time

transient (alias: **non-recurrent**) Broadly, a transient state (or set of states) of a random process evolving in time is one that occurs (i.e. is visited by the process) for only a finite amount of time. Otherwise the state (or set of states) is occupied for an infinite amount of time and is called recurrent (or persistent). The concept is most often encountered in the context of the Markov process $X(t)$, $t \geq 0$, on a countable state space S, with transition probabilities $p_{ij}(t)$. Then a state j, whose recurrence time is $T_j > 0$, is said to be transient if $\mathbb{P}(T_j < \infty \mid X(0) = j) < 1$. Alternatively, j is transient if $\int_0^\infty p_{jj}(t)\mathrm{d}t$ (or $\sum_{n=1}^\infty p_{jj}(n) < \infty$, for a chain in discrete time), so that j is recurrent when $\int_0^\infty p_{jj}(t)\mathrm{d}t = \infty$, (or $\sum_n p_{jj}(n) = \infty$). A chain (or process) is transient if all its states (or sets of states) are transient.

On general state spaces, the same basic concepts may be formalized in several different ways. For example, a suitably well-behaved diffusion $X(t)$ on $\mathbb{R}^n$ is transient if it goes to infinity as $t \to \infty$, or is recurrent if it hits every

ε-ball infinitely often; more generally a diffusion on a domain D in $\mathbb{R}^n$ is transient if it revisits any proper subdomain of D only finitely often.

More generally, a (Borel) set A of the state space of a Markov process $X(t)$ is transient if the probability of A being hit infinitely often by X is 0; formally (where I is an indicator),

$$\mathbb{P}\Big(\limsup_{t\to\infty} I(X(t) \in A) = 1\Big) = 0.$$

If this probability is 1, then A is recurrent, but some technical conditions are required to ensure that any A is either transient or recurrent. For example, any regular Feller process is either Harris recurrent or every compact subset C of the state space has an expected visiting time that is uniformly bounded, i.e. bounded by $a < \infty$ independently of $X(0)$, i.e.

$$\mathbb{E}\int_0^\infty I(X \in C)\mathrm{d}t \leq a < \infty.$$

[S. Meyn & R. Tweedie, *Markov Chains and Stochastic Stability*, 2nd edn. Cambridge University Press, 2009; R. Pinsky, *Positive Harmonic Functions and Diffusion.* Cambridge University Press, 1995; E. Nummelin, *General Irreducible Markov Chains and Non-negative Operators.* Cambridge University Press, 1984]

transient time *see* burn-in time

transition function (alias: **transition kernel; Markov transition distribution; stochastic transition function (or kernel, or distribution)**) A natural generalization of the stochastic transition matrix arising for Markov chains in discrete time (*see* Markov property). Let $(\Omega, \mathcal{F})$ be a space Ω equipped with sigma-field $\mathcal{F}$ of subsets. A (stochastic) transition function $p(x, A)$, $x \in \Omega$, $A \in \mathcal{F}$, is such that:

(a) for fixed x, $p(x, A)$ is a probability measure on Ω, as A varies;

(b) for fixed A, $p(x, A)$ is $\mathcal{F}$-measurable as a function of x.

When $p(x, \Omega) = 1$, the kernel is called stochastic, or Markov; if $p(x, \Omega) < 1$ for any x, the kernel is called **sub-Markov**, or **substochastic**.

In continuous time, it is further assumed that the analogous kernel $p(s, x; t, A)$, $0 \leq s \leq t < \infty$, satisfies the Chapman–Kolmogorov equation

$$p(s, x; t, A) = \int_\Omega p(s, x; v, \mathrm{d}y)p(v, y; t, A), \quad s < v < t.$$

It is usual to assume also that as $t \downarrow s$, $p(s, x; t, \{x\}) \to 1$ so the kernel is standard. [O. Kallenberg, *Foundations of Modern Probability*, 2nd edn. Springer, 2002]

transition matrix *see* Markov property; stochastic matrix

transition probability *see* Markov property

transition semi-group *see* stochastic semi-group

transitive Markov chain A Markov chain on a countable state space S, with transition matrix $p(i, j)$, is transitive if for any two states x and y there is a permutation $\pi(\cdot)$ of the state space that takes x to y while the transition rates are left invariant, i.e. $\pi(x) = y$, and $p(i, j) = p(\pi(i), \pi(j))$ for all i, j. Simple random walk on a vertex-transitive graph has this property. [D. Levin *et al.*, *Markov Chains and Mixing Times.* American Mathematical Society, 2009]

translated Poisson distribution Arises in Poisson approximation, where $X \in \mathbb{Z}$ has the translated Poisson distribution, with parameters μ and σ^2, and denoted by $\mathrm{Poiss}(\mu, \sigma^2)$, if the distribution of $X - \mu + \sigma^2 + f$ is that of a standard Poisson distribution with parameter $\sigma^2 + f$, where $f = \mu - \sigma^2 - \lfloor \mu - \sigma^2 \rfloor$ is the fractional part of $\mu - \sigma^2$. Then $\mathbb{E}X = \mu$, and $\sigma^2 \leq \operatorname{var} X \leq \sigma^2 + f$. [A. Röllin, Approximation of sums of conditionally independent variables by the translated Poisson distribution, *Bernoulli*, 11(6), 1115–1128, 2005]

transmission rate *see* capacity

trap *see* absorbing

tree-indexed process Broadly, a random process whose index set comprises the vertices of a possibly infinite (but locally finite) possibly random, rooted (directed) tree T. For example, tree-indexed r.w. attaches a collection of i.i.d.r.vs to the edges of T, one to each edge, and each vertex v of T is labeled with the sum of the r.vs on the path from r to the root. A tree-indexed Markov chain runs a Markov process from the root along branches of the tree, where the state of any vertex depends only on its parent vertex. [R. Pemantle, Tree-indexed processes, *Stat. Sci.*, 10(2), 200–231, 1995]

triangular distribution

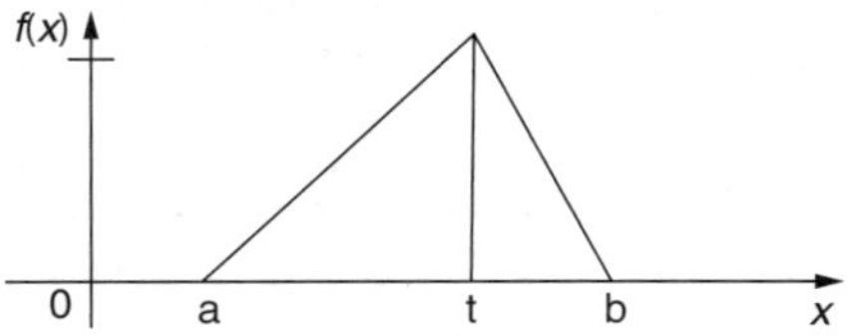

The density of the general (asymmetric) triangular distribution over [a, b]. Special cases include:

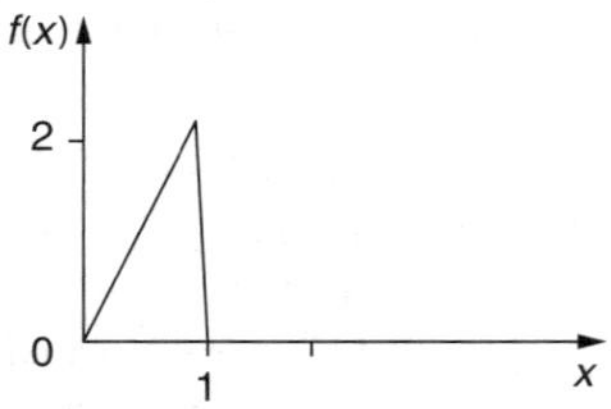

The density of $U^{1/2}$, where U is uniform on [0, 1], and:

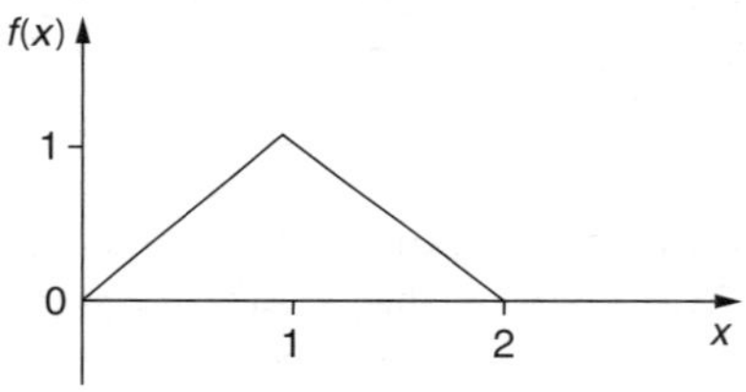

The density of $U + V$, where U and V are independently uniform on [0, 1].

trichotomy theorem *see* Kesten's theorem

trinomial distribution Arises in considering a sequence of n i.i.d. trials (i.e. experiments) each of which yields one of three outcomes with respective probabilities p, q, and $1-p-q$, where the totals in each category are X, Y, and $n-X-Y$. Then (X, Y) has joint distribution

$$f(x,y) = \frac{n!}{x!y!(n-x-y)!}p^x q^y (1-p-q)^{n-x-y},$$

for $x, y \geq 0$ and $x+y \leq n$. The distribution has binomial marginals, and $\text{cov}(X, Y) = -npq$. *See* multinomial distribution. The distribution with pgf $G(s,t) = \left(\frac{1-q}{1-q(ps+(1-p)t)}\right)^n, 0 < p, q < 1$, has been called a **negative trinomial distn**. [P. Consul & F. Famoye, *Lagrangian Probability Distributions*, Birkhäuser, 2006]

Trotter's theorem Arises in considering the local time of the Wiener process $W(t), t \geq 0$, where it asserts that there is a process $L(t, x)$ which is jointly continuous (a.s.) in $t \geq 0$ and $x \in \mathbb{R}$, such that for any measurable bounded function $u(\cdot)$

$$\int_{-\infty}^{\infty} u(x)L(t,x)\mathrm{d}x = \int_0^t u(W(s))\mathrm{d}s.$$

[H. Trotter, A property of Brownian motion paths, *Illinois J. Math.*, 2(3), 425–433, 1958]

truncation A random variable X is said to be truncated to $[b, a]$, with corresponding truncated form $\hat{X}$, when one sets

$$\begin{aligned} \hat{X} &= \begin{cases} X & \text{if } b \leq X \leq a \\ c & \text{otherwise, for some constant } c \end{cases} \\ &= XI(b \leq X \leq a) + cI^c. \end{aligned}$$

Very often $b = -a$, and $c = 0$. In a statistical context, one often defines

$$X' = \begin{cases} X & \text{if } b \leq X \leq a \\ a & \text{if } X \geq a \\ b & \text{if } X \leq b \end{cases}$$

and then X' is said to be X **censored** by $[b, a]$, although some writers also refer to this as truncation. Furthermore, the conditional distribution of X, conditional on the event that $a \leq X \leq b$, is sometimes called the truncated distribution of X truncated to $[a, b]$. [A. Gut, *Probability: A Graduate Course*, 2nd edn. Springer, 2013]

truncation inequality Broadly, any inequality arising in the context of truncated random variables; but most commonly the term refers to these: let X have distribution function $F(x)$ and chf $\phi(t)$. Then for $a > 0$,

$$\int_{|x| \geq 1/a} \mathrm{d}F(x) \leq \frac{7}{a}\int_0^a \{\phi(0) - \mathrm{Re}\phi(t)\}\mathrm{d}t$$

and

$$\int_{|x| < 1/a} x^2 \mathrm{d}F(x) \leq \frac{3}{a^2}\{\phi(0) - \mathrm{Re}\phi(a)\}.$$

[R. Ash & C. Doléans-Dade, *Probability and Measure Theory*, 2nd edn. Academic Press, 2000]

truncation of a Markov chain A Markov chain on a state space S with transition matrix p_{ij} (or Q matrix q_{ij}), $i, j \in S$, is said to be truncated to $A \in S$ if p_{ij} (resp q_{ij}) is replaced by 0 for $i \in A, j \in S \backslash A$, and the new process is irreducible on A (with the appropriate modification of p_{ii}

and q_{ii}). [F, Kelly, *Reversibility and Stochastic Networks*, Wiley, 1979; reprinted by Cambridge University Press, 2011]

Tsallis q-distribution *see* q-distribution

Tsetlin library *see* heaps process

Tukey's distribution (alias: **Tukey's lambda distribution**) A family of distributions $F_\lambda(x)$, $\lambda \in \mathbb{R}$, $x \in \mathbb{R}$, each symmetric about zero. If U is a uniform r.v. on $[0, 1]$, then for $\lambda \neq 0$, the distribution $F_\lambda(x)$ is that of the r.v. $X_\lambda = \lambda^{-1}\{U^\lambda - (1-U)^\lambda\}$; and for $\lambda = 0$, $F_0(x)$ is the distribution of $X_0 = \log U - \log(1-U)$.

Alternatively, in terms of the quantile functions,

$$F_0^{-1}(x) = \log x - \log(1-x), \quad 0 < x < 1,$$
$$F_\lambda^{-1}(x) = \frac{1}{\lambda}\{x^\lambda - (1-x)^\lambda\}, \quad \lambda \neq 0,\ 0 < x < 1.$$

There are a number of generalizations, e.g. the **Wakeby distns**, where

$$X = k - a(1-U)^b + c(1-U)^{-d},$$

for suitable values of a, b, c, d, and k. [N. Johnson, Extended and multivariate Tukey lambda distributions, *Biometrika*, 60(3), 655–661, 1973]

Tukey lambda distribution *see* Tukey's distribution

Tusnády's inequality (or lemma) *see* Komlós–Major–Tusnády approximation

Tweedie distribution *see* exponential family

Tweedie variates *see* inverse variates

two aces problem (or paradox) (alias: **paradox of the second ace**) A classic problem in probability that appears in several forms; one standard version is this: a pack of four cards contains the two red aces, one club, and one spade. A hand of two cards is dealt at random to a truthful agent A. In the following two cases, what is the probability $\mathbb{P}(S)$ that the hand contains both aces?

(i) You ask: 'Do you hold the ace of hearts?', and A replies: 'Yes'.

(ii) You say: 'Name an ace in your hand', and A replies 'Ace of hearts'.

The fact that the respective values of $\mathbb{P}(S)$ are 1/3 and 1/5 in (i) and (ii) is seen by some as paradoxical. The apparent paradox is resolved by noting the different ways in which the knowledge of the ace is conveyed from A to you; or, more concisely, the different protocols that determine the probability distribution. It is straightforward to construct other protocols for a conversation such that $\mathbb{P}(S)$ may take any desired value in $[0, 1]$. [N. Gridgeman, Letter to the editor, *Amer. Stat.*, 21(3), 38–39, 1967]

two-envelopes problem (or paradox) (alias: **exchange paradox; two-boxes problem; wallet game; necktie paradox**) A classic paradox that has generated a large class of related problems known under the same name, which have been analysed in the contexts of probability theory, decision theory, game theory, Bayesian inference, and logic. Martin Gardner's version of the basic problem is the wallet game: two players (of similar means) put their wallets on the nail and agree that the owner of the wallet with the smaller amount of money takes all the money in the other. Each player argues that 'I may lose what I have but equally I may win more than I have, so the game is advantageous for me.' But a zero-sum game to the advantage of both players is paradoxical. A popular variant postulates two envelopes, one of which contains twice as much money as the other. You are given one at random and offered the chance – before opening it – to switch to the other. An argument similar to the above suggests that it is advantageous to switch; but then the same argument suggests that you should switch back, and so on. In a further variant, you are invited to switch after opening the envelope and counting the contents.

The early elementary forms of the paradox result from misleading labelling and/or descriptions of the quantities involved and are readily resolved by being properly posed. More advanced versions require more subtle arguments to develop a resolution, or to show in what way the problem was improperly posed. This family of paradoxes continues to grow. [M. McDonnell & D. Abbott, Randomized switching in the two-envelopes problem, *Proc. Roy. Soc. A*, 465, 3309–3322, 2009]

two-parameter Markov process A random process $X(s,t)$, where $(s,t) \in \mathbb{R}_+^2$, (or $(s,t) \in \mathbb{Z}_+^2$), where the index set is equipped with the partial order $(s_1, t_1) \ll (s_2, t_2)$ iff $s_1 < s_2$ and $t_1 < t_2$; the Markov property is then assigned to X in the usual way in terms of this ordering. [G. Mazziotto, Two-parameter Hunt processes and a potential theory, *Ann. Probab.*, 16(2), 600–619, 1988]

two-parameter martingale An extension of the concept of a martingale to a process $M(s,t), (s,t) \in \mathbb{R}_+^2$, which may be accomplished in several ways. The basic form requires the usual partial order: $(s_1,t_1) \ll (s_2,t_2)$ iff $s_1 < s_2$ and $t_1 < t_2$ and then M is a martingale wrt a filtration $\mathcal{F}(s,t)$ if $\mathbb{E}|M| < \infty$ for all (s,t), and for $(s_1,t_1) \ll (s_2,t_2)$, $\mathbb{E}(M(s_2,t_2) \mid \mathcal{F}(s_1,t_1)) = M(s_1,t_1)$. [R. Cairoli & J. Walsh, Stochastic integrals in the plane, *Acta Math.*, 134, 111–183, 1975; E. Wong & Zakai, Weak martingales and stochastic integrals in the plane, *Ann. Probab.*, 4, 570–586, 1976]

two-parameter Wiener process *see* Brownian sheet

two-sex branching process (alias: **Daley's two-sex (or bisexual) branching process**) A population comprises two disjoint classes of M_n males and F_n females at time $n \geq 0$ (i.e. at the nth stage) and families arise only from a pair consisting of one from each class, the ith such pair yielding $Y_{n,i}$ males and $X_{n,i}$ females, independently of other pairs. Then Daley's branching process sets $(M_{n+1}, F_{n+1}) = \sum_{r=1}^{C(M_n,F_n)} (Y_{n,i}, X_{n,i})$, where $C(M_n, F_n)$ is the number of pairs formed from (M_n, F_n), and $(Y_{n,i}, X_{n,i}), n \geq 0, 1 \leq i \leq C(M_n, F_n)$, are i.i.d.r.vrs. [M. Molina, Two-sex branching process literature, in *Workshop on Branching Processes and their Applications*, M. Velasco *et al.* (eds). Springer, 2010, pp. 279–293]

two-state process (alias: **flip-flop process; dichotomous process; switch process; telegraph (or telegrapher's) process; reliability process**) A random process having just two possible states, which typically may be denoted by $\{0,1\}$ or $\{-1,1\}$, that arises naturally in models for systems that may be on or off, or molecules with a spin that may be up or down, etc. Commonly the process is assumed to change state at the instants of an alternating renewal process; when the interchange times are independent exponential (or geometric) r.vs the process is Markov. In particular, if $N(t)$ is a Poisson process $X(t) = X(0)(-1)^{N(t)}$ is called the Poisson telegraph (or switch, or flip flop, etc.) process. This process is often compounded in some way; *see* e.g. Goldstein–Kac process and telegraph process.

type G process *see* variance-mean mixture

typicality property *see* asymptotic equipartition property

U

u.a.n. *see* null array

Ulam–Harris framework (alias: **Ulam–Harris convention; construction; device; family history; formulation; labels; notation; sample space, etc.**) Arises in the classical theory of branching processes, where the Ulam–Harris system identifies each individual with a string of positive integers, so that $\mathbf{c} = (c_1, c_2, \ldots, c_n)$ is the c_nth child of the c_{n-1}th child ... of the c_1th child of the initiator of that line of descent; the initiator is denoted, if necessary, by the empty set symbol $\emptyset$. [P. Jagers & A. Lagerås, General branching processes conditioned on extinction are still branching processes, *Electr. Commun. Probab.*, 13, 540–547, 2008]

Ulam's problem Any of a large number of problems either posed or highlighted by S. Ulam; but perhaps most often this one: you think of a number between 1 and $10^6 (< 2^{20})$, and another person is allowed to ask you 20 questions to which you must answer 'yes' or 'no'. If you are allowed to lie once or twice, how many questions would one need to determine the selected number? [J. Spencer, Ulam's searching game with a fixed number of lies, *Theor. Comp. Sci.*, 95(2), 307–321, 1992; S. Ulam, *Adventures of a Mathematician*. Scribner, 1976; S. Ulam, *A Collection of Mathematical Problems*. Wiley, 1960; reprinted as *Problems in Modern Mathematics*. Dover, 2004]

Ulam's random adding sequence A random integer sequence $X_n, n \geq 1$, generated by a history-dependent stochastic recurrence as follows: $X_{n+1} = X_{U(n)} + X_{V(n)}, n \geq r$, where $X_1 = x_1, \ldots, X_r = x_r$ are given, and $(U(n), V(n))$ comprise i.i.d.r.vrs such that $U(n)$ and $V(n)$ are each uniformly distributed on $\{1, \ldots, n\}$.

S. Ulam noted only that $\mathbb{E}X_n = \frac{1}{3}(x_1 + x_2)n, n \geq 3$, in the case $r = 2$. But, setting $S_n = \sum_{r=1}^{n} X_r$, it is seen that $S_n/(n^2 + n), n \geq r$, is a martingale wrt the natural filtration, so that $n^{-2}S_n$ converges both a.s. and in mean-square to a non-degenerate r.v. as $n \to \infty$. And, in particular, for $r = 1$ and $x_1 = 1$,

$$\lim_{n\to\infty} n^{-2}\mathbb{E}X_n{}^2 = \frac{1}{2\pi}\sinh(\pi).$$

[W. Beyer *et al.*, *Computer Studies of Some History-dependent Random Processes, LA4246*. Los Alamos, 1969; reprinted in A. Bednarek & F. Ulam, *Analogies Between Analogies*. University of California Press, 1990; P. Clifford & D. Stirzaker, *Ulam's Random Adding Sequence*, unpublished, 2008]

Ulam's random permutation problem Arises in considering the length L_n of the longest increasing subsequence of a random permutation of $\{1, \ldots, n\}$. The conjecture that $\frac{1}{\sqrt{n}}L_n$ converges in probability to a constant c is Ulam's problem. [J. Baik *et al.*, On the distribution of the length of the longest increasing subsequence of random permutations, *J. Amer. Math. Soc.*, 12(4), 1119–1178, 1999]

Ulam's redistribution of energy problem Arises in considering repeated random collisions between pairs of energetic particles in which the energy is redistributed between the pair according to some suitable distribution F. S. Ulam conjectured that under iterated collisions among a large number of particles the energy distribution of each particle converges to a limiting **'collision transform'** of F. [D. Blackwell & R. Mauldin, Ulam's redistribution of energy problem: collision transformations, *Lett. Math. Phys.*, 10(2–3), 149–153, 1985]

unbiased In some contexts this term denotes that outcomes are equally likely; thus an unbiased coin (resp. die, or lottery) is equally likely to yield either face (resp. any side, or any ticket). It is used differently in statistics, where e.g. a sample mean $\overline{X} = \frac{1}{n}\sum_{r=1}^{n} X_r$ is said to be an unbiased estimator of $\mu = \mathbb{E}X_1 < \infty$, because $\mathbb{E}\overline{X} = \mu$. *See* fairness.

uncertainty A quality attributed to a quantity that is undetermined by reason of chance, doubt, or ignorance. The term is used in different ways (according to some mixture of the three factors) in probability and its applications. For example:

(a) The uncertainty in a random variable X is its Shannon entropy $H(X)$.

(b) In decision theory, analysis under uncertainty entails making decisions in the absence of any (given or assumed) probability distribution on the decision space.

(c) In quantum mechanics the uncertainty principle is informally rendered as the assertion that for a number of pairs of physical quantities (such as position and momentum) you cannot know their exact values simultaneously (nor can you obtain arbitrarily good approximations). Uncertainty is unavoidable.

In particular, there are a number of inequalities having applications in physics and information theory, inter alia, that are termed **uncertainty principles**. Simple examples include these: suppose that for $x \in \mathbb{R}, u(x) \in \mathbb{C}$ has Fourier transform $v(x) \in \mathbb{C}$, while $\int_{\mathbb{R}} |u(x)|^2 \mathrm{d}x = 1 = \int_{\mathbb{R}} |v(x)|^2 \mathrm{d}x$. Then the density functions $f = |u|^2$ and $g = |v|^2$ on $\mathbb{R}$ are said to be conjugate, and we denote their respective variances by V_f and V_g, and their respective Shannon differential entropies by H_f and H_g. Then the **Heisenberg–Pauli–Weyl uncertainty principle** asserts that $V_f V_g \geq (16\pi^2)^{-1}$, while **Hirschman's (or Hirschman–Beckner) uncertainty principle** asserts that $H_f + H_g \geq \log\left(\frac{e}{2}\right)$, with equality in each case when $|f|^2$ and $|g|^2$ are normal distributions. [A. Dembo *et al.*, Information theoretic inequalities, *IEEE Trans. Inf. Theor.*, 37(6), 1501–1518, 1991; I. Hirschman, A note on entropy, *Amer. J. Math.*, 79(1), 152–156, 1957]

uncorrelated *see* correlation

unessential state *see* essential state

uniform distribution A discrete r.v. X is said to have the (discrete) uniform distn on $\{1, \ldots, n\}$ when $\mathbb{P}(X = k) = n^{-1}, 1 \leq k \leq n$. A continuous r.v. Y is said to have the uniform (or **rectangular**) distn on (a, b) (or on $[a, b]$), when its density is $f_X(x) = (b - a)^{-1}, a < x < b$, (or $a \leq x \leq b$), and $f_X(x)$ vanishes outside the interval. It may be denoted by $\mathrm{U}(a, b)$, or by $\mathrm{Unif}(a, b)$ or Uniform (a, b). The uniform distn $\mathrm{U}(0, 1)$ arises naturally, e.g. as the limiting distn of $n^{-1}X$ as $n \to \infty$ (for X above); and also as the distn of $\sum_{r=1}^{\infty} Z_r/2^r$, where the Z_r are i.i.d.r.vs with $\mathbb{P}(Z_r = 0) = \mathbb{P}(Z_r = 1) = \frac{1}{2}$.

More generally, a r.v. X taking values in a finite subset S of $\mathbb{R}^n$ is said to be uniformly distributed on S (or uniform on S), if its density is constant on S; informally one speaks of point chosen uniformly at random in S. For example $f(\mathbf{x}) = 1$ for $\mathbf{x} \in (0, 1)^n$, when $\mathbf{X}$ is uniform in the unit cube; and $f(\mathbf{x}) = \frac{1}{2}\Gamma\left(\frac{1}{2}n\right)\pi^{\frac{-1}{2}n}$ for $|\mathbf{x}| = 1$, when $\mathbf{X}$ is uniform on the surface of the unit sphere in n dimensions. Similarly $f(\mathbf{x}) = 2^{-n}$ for $\mathbf{x} \in \{0, 1\}^n$, when $\mathbf{X}$ is uniformly distributed on the vertices of the unit cube (i.e. the set of all strings of 0s and 1s of length n).

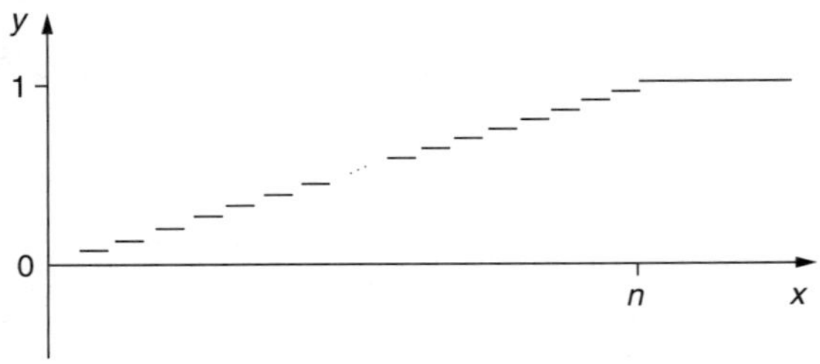

Distribution of a discrete uniform random variable.

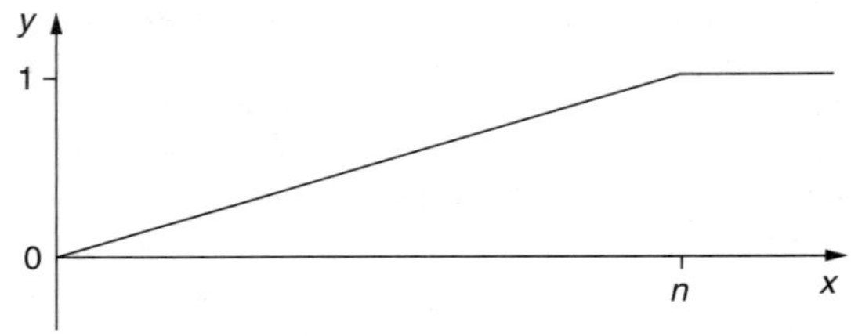

Distribution of a U(0, *n*) continuous random variable.

[Y. Suhov & M. Kelbert, *Probability and Statistics by Example*. Cambridge University Press, 2005]

uniform distribution of sequences A concept arising in analysis with a theorem of H. Weyl in 1916, having applications in many other branches of mathematics including ergodic theory, and pseudo-randomness, inter alia. A sequence $s = (x_r), r = 1, 2, \ldots$ of real numbers is said to be uniformly distributed modulo 1 if for any $0 \leq a < b \leq 1$,

$$\lim_{n \to \infty} \frac{1}{n} N_n([a, b)) = b - a,$$

where $N_n([a, b))$ is the number of terms of $x_1, \ldots, x_n$ whose fractional part $x_n - [x_n]$ lies in $[a, b)$. **Weyl's thm** asserts that this occurs iff, for all integer $k \neq 0$, $\lim_{n \to \infty} \frac{1}{n} \sum_{r=1}^{n} \exp\{2\pi i k x_r\} = 0$. *See* discrepancy and van der Corput sequence. [L. Kuipers & H. Niederreiter, *Uniform Distribution of Sequences*. Wiley, 1974; reprinted by Dover, 2006; P. Diaconis, The distribution of leading digits, and uniform distribution mod 1, *Ann. Probab.*, 5(1), 72–81, 1977]

uniformization (alias: **Jensen's method; randomization of time**) For a suitable continuous-time Markov chain $X(t), t \geq 0$, on a countable state space S, with Q-matrix $q_{ij}, i, j \in S$, the uniformization method defines a Markov chain $\hat{X}(t)$ having the same fdds as $X(t)$ (i.e. $\hat{X}$ is a version of X), that facilitates the computation of the transition probabilities $p_{ij}(t), i, j \in S$, of X, where $P(t) = P(0)\exp(Qt)$. Thus, let $\max_i |q_{ii}| \le \lambda < \infty$ and let the stochastic matrix $R = Q/\lambda + I$ be the transition matrix of a Markov chain $Y_n, n \ge 0$, on S. Then define $\hat{X}(t) = Y_{N(t)}$, where $N(t)$ is a Poisson process of rate λ, which is to say that $\hat{X}$ is the process obtained when Y_n makes its jumps at the instants of $N(t)$. [H. Tijms, *A First Course in Stochastic Models*. Wiley, 2003]

uniformly asymptotically negligible *see* null array

uniformly integrable (alias: **equi-integrable**) A sequence $X_n, n \ge 1$, of r.vs is uniformly integrable if

$$\sup_n \mathbb{E}(|X_n| I\{|X_n| \ge a\}) \to 0, \quad \text{as } a \to \infty;$$

where $I = 1$ if $|X_n| \ge a$, or $I = 0$ otherwise. Equivalently $X_n, n \ge 1$, is a uniformly integrable sequence iff $\sup_n \mathbb{E}|X_n| < \infty$ and for all $\varepsilon > 0$ there exists $\delta > 0$ such that for any event A with $\mathbb{P}(A) < \delta$, and all $n, \mathbb{E}\{|X_n|\, I(A)\} < \varepsilon$, where $I(A) = 1$ if A occurs, or $I = 0$ otherwise. [O. Kallenberg, *Foundations of Modern Probability*, 2nd edn. Springer, 2002]

uniform metric *see* Kolmogorov's distance

uniform stochastic order Let $F(x)$ and $G(x)$ be the distribution functions of r.vs X and Y respectively, with survival fns $\overline{F} = 1 - F$, and $\overline{G} = 1 - G$ then X is said to be uniformly smaller than Y in the negative direction

if $\begin{vmatrix} F(x) & F(y) \\ G(x) & G(y) \end{vmatrix} \ge 0$ for all $x \le y$;

and X is uniformly smaller than Y in the positive direction

if $\begin{vmatrix} \overline{F}(x) & \overline{F}(y) \\ \overline{G}(x) & \overline{G}(y) \end{vmatrix} \ge 0$ for all $x \le y$.

[J. Keilson & U. Sumita, Uniform stochastic ordering and related inequalities, *Canadian J. Stat.*, 10(3), 181–198, 1982]

unimodal Let $F(x)$ be the distribution function of a random variable $X \in \mathbb{R}$. Then $F(x)$ (or X), is said to be unimodal if, for some m, $F(x)$ is convex for $x < m$ and concave for $x > m$. The number m is said to be the *mode* of $F(x)$.

A distribution function is **strongly unimodal** if its convolution with any other unimodal distribution is unimodal. In this context Ibragimov's theorem asserts that the unimodal distribution F of X is strongly unimodal iff either F is degenerate (which is to say $\mathbb{P}(X = a) = 1$ for some constant a) or F is absolutely continuous (which is to say that X has a density f) and its density f is log-concave. And **Yamazato's thm** asserts that all self-decomposable distns on $\mathbb{R}$ are unimodal. *See* Khinchin's criterion; Gauss–Winckler inequality; and mode. [E. Lukacs, *Characteristic Functions*, 2nd edn. Griffin, 1970; K-I. Sato, *Lévy Processes and Infinitely Divisible Distns*. Cambridge University Press, 1999]

uniqueness in law (alias: **weak uniqueness**) Arises in considering weak solutions of stochastic differential equations, where uniqueness in law holds if, for any two weak solutions of such an SDE with the same initial distribution, the solutions have the same fdds. If, more strongly, the solutions are indistinguishable, then one speaks of pathwise uniqueness. *See* strong solution. [D. Revuz & M. Yor, *Continuous Martingales and Brownian Motion*, 3rd edn, 1999]

uniqueness theorem for chfs *see* characteristic function

universal generating function (alias: ***u*-function**) A methodology that is used in the computational analysis of multi-state reliability systems, where the so-called *u*-function is defined in terms of an appropriate composition operator acting on the usual probability (or moment) generating functions. [G. Levitin, *The Universal Generating Function in Reliability Analysis and Optimization*. Springer, 2005]

universal law *see* Doeblin's universal laws

universal portfolio (alias: **Cover's universal portfolio**) A trading strategy (i.e. a portfolio selection process) that is predictable (i.e. non-anticipating, so that each rebalancing of holdings in the market is a function only of past values), which performs at least as well as the optimal strategy when the future evolution of the market is known (i.e. the strategies have the same exponential rate of growth). [T. Cover, Universal portfolios, *Math. Finan.*, 1(1), 1–29, 1991]

universal probability *see* algorithmic probability

universal probability space *see* universal sigma-field

universal set *see* sample space

universal sigma-field (alias **universal σ field**) Given a measurable space $(\Omega, \mathcal{F})$, we denote the completion of $\mathcal{F}$ with respect to a probability measure $\mathbb{P}$ by $\mathcal{F}_{\mathbb{P}}$. Then the universal sigma-field is the intersection over all probability measures $\mathbb{P}$, denoted by $\cap_{\mathbb{P}} \mathcal{F}_{\mathbb{P}} = \overline{\mathcal{F}}$. A r.v. measurable wrt $\overline{\mathcal{F}}$ is called universally measurable. $\mathcal{F}$ may also be called universally complete, and yields a universal probability space. Likewise if $\mathcal{F}(t)$ is a filtration, and $\mathcal{F}_{\mathbb{P}}(t)$ the completion for $\mathbb{P}$, (or augmented filtration), then $\cap_{\mathbb{P}} \mathcal{F}_{\mathbb{P}}(t) = \overline{\mathcal{F}}(t)$ may be called a **universal filtration**, and a process adapted to $\overline{\mathcal{F}}(t)$ is said to be **universally adapted**. [O. Kallenberg, *Foundations of Modern Probability*, 2nd edn. Springer, 2002]

unstable state *see* instantaneous state

upcrossing inequality (or lemma) Arises in a proof of the (sub) martingale convergence theorem, and asserts this: if $M_n, n \geq 1$, is a submartingale, and $U_n(a, b)$ is the number of times that a sample path $M_1, M_2, \ldots, M_n$ traverses the interval (a, b) from below a to above b, then

$$\mathbb{E}U_n(a,b) \leq \frac{\mathbb{E}(M_n - a)^+ - \mathbb{E}(M_1 - a)^+}{b - a} \leq \frac{\mathbb{E}|M_n| + |a|}{b - a}.$$

When applied to martingales, it may be termed **Doob's upcrossing lemma**; when applied to submartingales it may be termed **Snell's upcrossing lemma**. [J. Doob, *Stochastic Processes*. Wiley, 1953]

upper class function (alias: **upper function**) Given a random sequence $X_n, n \geq 1$, a numerical sequence a_n is said to be upper (or an upper function, or of the upper class) for X_n if $\mathbb{P}(X_n > a_n$ infinitely often$) = 0$, and lower (or a lower function, or of the lower class) if this probability is 1. Similarly for a random process $X(t)$, a function $a(t)$ is upper (or lower) for $X(t)$ according as the events

$$U_n = \left\{ \sup_{s \geq t_n} \Big(X(s) - a(s) \Big) > 0 \right\}$$

occur finitely (or infinitely) often with probability 1, for some sequence $t_n \uparrow \infty$. In the context of the law of the iterated logarithm for partial sums $S_n, n \geq 1$, one often sets $a_n = \sqrt{n} f_n$, for some increasing sequence f_n, which is said to be upper or lower according as $\mathbb{P}(S_n / \sqrt{n} > f_n \text{ i.o.}) = 0$ or 1, with an analogous definition for $\sqrt{t} f(t)$ in continuous time. *See* Kolmogorov–Petrov–Erdös–Feller test. [A. A. Borovkov, *Probability Theory*, 5th edn. Springer, 2013]

urn model (alias: **urn-sampling problem; urn process; balls in boxes (or bins)**) A class of random processes (or experiments) in which an integer-valued r.v. (or vector, or process) is visualized in terms of discrete objects (e.g. balls, items, tickets, etc.) being placed in, or removed from, or moved among, a set of containers (e.g. urns, bins, boxes, etc.); such models are found in almost every branch of science. Urn models in which balls are assigned randomly, but not removed, are commonly called **occupancy problems**. When balls are removed as well as replaced, one often speaks of a **sampling problem** (or scheme); without or with replacement depending on whether balls may not or may be returned to the urns. In physical applications there are further characterizations depending on whether balls are distinguishable or not, or are of various types (often denoted by colour), or whether urns have a fixed capacity or not, etc. *See* Ehrenfest; Friedman's; and Pólya's urns. [H. Mahmoud, *Pólya Urn Models*. Chapman & Hall, 2009; N. Balakrishnan (ed), *Advances in Combinatorial Models and Applications to Probability and Statistics*, 203–286, 1997, Birkhäuser]

user impatience bias Arises naturally in simulation schemes and Monte Carlo methods when the outcome is not independent of the run-time of the procedure; as in, e.g. the Propp–Wilson method. If the time to coupling (or coalescence) is so long that the run is abandoned for external reasons, the distribution of the result is no longer the perfect sample intended, but favours those values that are produced most swiftly. [E. Thönnes, Perfect simulation of some point processes for the impatient user, *Adv. Appl. Probab.*, 31(1), 69–87, 1999]

U-shaped distribution A term applied (somewhat informally) to a distribution that is convex over an interval with (at least approximately) equal maxima at the end points of the interval, and thus resembles the letter U, e.g. the arcsine density

$$f(x) = \frac{1}{\pi\{x(1-x)\}^{\frac{1}{2}}}, \quad 0 \leq x \leq 1.$$

If, more weakly, the distribution $f(x)$ simply decreases in x to a unique minimum, and increases in x thereafter, it may be called **bath tub-shaped**. *See* J-shaped distribution.

U-statistic (alias: **Hoeffding U-statistic**) A generalization of the concept of sample mean to higher-order symmetric functions, thus: Let $X_1, \ldots, X_n$

be i.i.d.r.vs (or, more generally, exchangeable random vectors), and let $s(x_1, \ldots, x_m)$ be a real-valued function (often called the kernel) that is symmetric in its arguments. Then the sample function

$$U = \sum s(X_{k(1)}, \ldots, X_{k(m)}) / \{n(n-1) \cdots (n-m+1)\}$$

where the sum is over all permutations of m different selections $k(1), \ldots, k(m)$, is called Hoeffding's U-statistic. [W. Hoeffding, A class of statistics with asymptotically normal distribution, *Ann. Math. Stat.*, 19(3), 293–325, 1948]

usual conditions *see* filtration

utility function Arises in utility theory as a function $u(x) \in \mathbb{R}$ agreeing with an agent's preferences among financial gambles. Plausible assumptions about agents (i.e. people) entail that u is increasing and continuous in x, and also concave for risk-averse agents. Popular utilities include, for $a > 0, u = -\exp(-ax); u = (1 - x^{-a})/a$; $u = \log(a + x)$; and D. E. Bell's **one-switch utilities**, such as $u = ax + be^{cx}$, for $b, c < 0$, which reflects an agent's change to riskier preferences as wealth increases, with ultimate risk neutrality. [D. E. Bell, One-switch utility functions and a measure of risk, *Manage. Stud.*, 34(12), 1416–1424, 1988; P. Gärdenfors & N.-E. Sahlin (eds), *Decision, Probability and Utility*. Cambridge University Press, 1988]

utility theory In economics, game theory, decision theory, and related fields, the concept of utility seeks to represent the collection of an agent's preferences over a set R of rewards (or decisions, or risks, etc.) by a real-valued function $u(\cdot)$ defined on R. In the basic case where rewards are certain, $u(\cdot)$ is a utility function when for any x and y in $R, u(x) \geq u(y)$, iff x is preferred to y; one seeks conditions (on R and the preference relation) for the existence and possible uniqueness of $u(\cdot)$. When rewards are random (as in choosing between lotteries, etc.), $u(\cdot)$ is a utility function when, for all X and Y in $R, \mathbb{E}u(X) \geq \mathbb{E}u(Y)$ iff X is preferred to Y. The von Neumann–Morgenstern theorem supplies conditions for such a utility function in their framework; one speaks of the **expected utility hypothesis**. When rewards are monetary, an agent is said to be **risk averse** or **risk seeking** according as $u(\cdot)$ is concave or convex.

Actual agents often fail to satisfy the conditions of this and other similar theorems, giving rise to a range of alternatives that are sometimes called non-expected utility. *See also* behavioural decision theory. [P. Gärdenfors & N.-E. Sahlin (eds), *Decision, Probability and Utility*. Cambridge University Press, 1988]

V

vacation In queueing theory, a server that is temporarily unavailable (or inactive) is said to be on vacation; this is to be distinguished from an idle server that is available for work but has no customers. Vacations arise from breakdowns, repairs, priority commitment to other tasks, etc., or by mistake. There are many vacation disciplines, each characterized by their quit-work and back-to-work rules. [N. Tian & Z. Zhang, *Vacation Queueing Models*. Springer, 2006]

vague convergence Arises in considering sequences of distribution functions (or probability measures, or distns) that are convergent, but whose limit is not necessarily proper, e.g. the distn fn is that of a defective r.v., or the limiting measure is such that $\mathbb{P}(\Omega) < 1$. Informally, one speaks of probability mass escaping to infinity. When the limit is proper, vague convergence is equivalent to weak convergence (or convergence in distn).

Formally, in the basic case, a sequence $F_n(x), n \geq 1$, $x \in \mathbb{R}$, of distn fns converges vaguely to a possibly improper $(G(\infty) < 1)$ limit $G(x)$ if, for all $a < b$ such that $G(\cdot)$ is continuous at a and b, $\lim_{n\to\infty}\{F_n(b) - F_n(a)\} = G(b) - G(a)$.

More generally, a sequence of probability measures $\mathbb{P}_n$ on $\mathbb{R}^d$ (or other suitable topological space) is vaguely convergent to $\mathbb{P}$ if, for all continuous functions $u(\cdot)$ having compact support (or for all continuous functions $u(\cdot)$ vanishing at infinity) $\lim_{n\to\infty}\int u\,d\mathbb{P}_n \to \int u\,d\mathbb{P}$, where $\mathbb{P}$ may be improper, i.e. a subprobability measure, or pseudo-distribution. *See* convergence; tightness; and weak convergence of probability measures. [K. L. Chung, *Course in Probability Theory*, 2nd edn revised. Academic Press, 2001]

Vajda's inequality Let **p** and **q** be probability distns on the same space, with Kullback–Leibler (or information) divergence $D = D(\mathbf{p}, \mathbf{q})(= \sum_i p_i \log(p_i/q_i)$ in the discrete case), and simple total variation distance $V = V(\mathbf{p}, \mathbf{q})(= \sum_i |p_i - q_i|$; in the discrete case). Then Vajda's inequality asserts that

$$D \geq \log\left(\frac{2+V}{2-V}\right) - \frac{2V}{2+V} = i(V).$$

The function $L(v) = \inf\{D: V = v\}, v \in (-2, 2)$, may be called Vajda's (tight) lower bound, and another Vajda inequality asserts that $L(v) \leq i(v) + \frac{2v^3}{(2+v)^2}$.

Related inequalities include Pinsker's; and that of Volkonskii–Rozanov, which asserts that $D \geq V - \log(1 + V)$. [G. Gilardoni, On Pinsker's and Vajda's type inequalities for Csiszár's *f*-divergences, *IEEE Trans. Inf. Theor.*, 56(10), 5377–5386, 2010]

valuation (alias: **additive set functional**) Arises in geometrical probability where e.g. a real-valued function $v(\cdot)$ on the space C of convex compact subsets of $\mathbb{R}^n$ is called a valuation if $v(\emptyset) = 0$ and for $A, B, A \cup B$ and $A \cap B$ in C, $v(A \cap B) + v(A \cup B) = v(A) + v(B)$. More generally, $v(\cdot)$ is defined likewise on a distributive lattice. [D. Klein & G.-C. Rota, *Introduction to Geometric Probability*. Cambridge University Press, 1997]

value at risk (alias: **VaR**) A term arising in financial mathematics when an agent seeks to assess the monetary risk entailed by a market portfolio (comprising stocks, bonds, assets, etc., and contingent claims on these).

There is a large literature, but in the simplest basic form of VaR it is assumed that some r.v. L represents a potential loss on the portfolio, which may be determined at some fixed future time T, or over the interval up to T, where L has distribution function $F(x)$. Then for $\lambda \in (0, 1)$, the VaR at level λ is the quantile fn (or generalized inverse distn fn)

$$R(\lambda) = \inf\{x : F(x) \geq \lambda\}.$$

A related function is the so-called **conditional value at risk**, being the conditional expectation of L, given that $L \geq R(\lambda)$. [R. Rockafellar & S. Uryasev, Conditional value-at-risk for general loss distributions, *J. Banking Finan.*, 26(7), 1443–1471, 2002]

value iteration *see* Bellman equation

van Dantzig's collective marks *see* collective marks

van den Berg–Kesten inequality (alias: **BK (or B–K) inequality**) Arises in the context of percolation and random graphs, thus: Let $H^n = \{0,1\}^n$ (which can be seen as the discrete cube in n dimensions), and assign the uniform (or independent product uniform) distribution to each point (i.e. $\mathbb{P}(\omega) = 2^{-n}, \omega \in H^n$). The events A and B are called disjoint, and one writes $A \circ B$, if they occur on disjoint subsets of the coordinates of ω. Then the B–K inequality asserts that for events A and B that are each closed under the action of increasing any 0 to 1 (called up-events, or up-sets),

$$\mathbb{P}(A \circ B) \le \mathbb{P}(A)\mathbb{P}(B).$$

The van den Berg–Kesten conjecture is that this holds without the restriction to up-sets; **Reimer's thm** asserts that this B–K conjecture is true; it is also called the **B–K–Reimer inequality**. [D. Reimer, Proof of the van den Berg-Kesten conjecture, *Combin. Probab. Comp.*, 9(1), 22–32, 2000]

van der Corput sequence A quasi-random sequence of numbers that is equidistributed over the unit interval. The nth term c_n of the binary version is constructed by taking its binary expansion after the binary radix point to be the binary expansion of n, reversed. Thus, the sequence begins: $0.1, 0.01, 0.11, 0.001, \ldots$ [A. del Junco & J. M. Steele, Hammersley's law for the van der Corput sequence: an instance of probability theory for pseudorandom numbers, *Ann. Probab.*, 7(2), 267–275, 1979]

Vapnik–Chervonenkis theory Arises in the branch of statistical inference known as statistical learning (or probabilistic computational learning) theory, where – broadly speaking – it extends various laws of large numbers to spaces of functions. In particular, the Vapnik–Chervonenkis theorem establishes a Glivenko–Cantelli type result for the convergence of a sequence of empirical probability measures $m_n(\cdot)$, arising from a probability measure $m(\cdot)$, thus: for a suitable class of sets V, such that $\sup_{A\in V}|m_n(A) - m(A)|$ is measurable, and any $\varepsilon > 0$ and $n < \infty$, the original form of the V–C theorem (or inequality) is,

$$\mathbb{P}\Big(\sup_{A\in V}|m_n(A) - m(A)| > \varepsilon\Big) \le 4c(V, 2n)\mathrm{e}^{-n\varepsilon^2/8},$$

where $c(V, n)$ is the maximum over choices of $\mathbf{x} = (x_1, \ldots, x_n)$ of $|\{\mathbf{x} \cap A; A \in V\}|$, and $c(V, n)$ is called the **shatter coefficient** (or growth function) of the class V. The largest finite integer n such that $c(V, n) = 2^n$ is called the Vapnik–Chervonenkis dimension (or index) of V; and, when that V–C index is finite, V is called a Vapnik–Chervonenkis class.

In this context, note that for a sequence of i.i.d.r.vrs $\{X_1, X_2, X_3, \ldots, X_n\} = \mathbf{X}_n$, the function

$$H(V) = \lim_{n\to\infty}\frac{1}{n}\log\mathbb{E}|\{\mathbf{X}_n \cap A; A \in V\}|$$

is called the **Vapnik–Chervonenkis entropy** of V for $X_r, r \ge 1$. [G. Lugosi, Improved upper bounds for probability of uniform deviations, *Stat. Probab. Lett.*, 25(1), 71–77, 1995; J. M. Steele, Empirical discrepancies and subadditive processes, *Ann. Probab.*, 6(1), 118–127, 1978]

Varadhan's lemma (or theorem) Arises in large deviations theory, where it can be seen as generalizing Laplace's method for the asymptotic analysis of suitable integrals, e.g. the basic result of Laplace's method shows that for continuous bounded $g(x), 0 \le x \le 1$,

$$\lim_{n\to\infty}\frac{1}{n}\log\int_0^1 \exp\{ng(x)\}\mathrm{d}x = \max_{0\le x\le 1} g(x).$$

Varadhan's lemma asserts that for a sequence of r.vs X_n (or, more generally, random elements in a Polish space), satisfying a large deviations principle with rate function I, we have for bounded continuous $g(x)$,

$$\lim_{n\to\infty}\frac{1}{n}\log\mathbb{E}\Big(\exp\{-ng(X_n)\}\Big) = -\inf_x\{I(x) + g(x)\};$$

a limiting property that may be called a **Laplace principle**. Conversely, **Bryc's thm** asserts that if

$$\lim_{n\to\infty}\frac{1}{n}\log\mathbb{E}\exp\{-ng(X_n)\} = b(g)$$

exists for all continuous bounded g, and if the sequence X_n is exponentially tight (which is to say that for any $K < \infty$ there is a compact set C such that $\lim_{n\to\infty}\sup\frac{1}{n}\log\mathbb{P}(X_n \notin C) \le -K$), then $X_n, n \ge 1$, obeys a large deviations principle. [P. Dupuis & R. Ellis, *A Weak Convergence Approach to the Theory of Large Deviations*. Wiley, 1997; F. den Hollander, *Large Deviations*. American Mathematical Society, 2000; A. Dembo & O. Zeitouni, *Large Deviations Techniques and Applications*, 2nd edn. Springer, 1998]

variance (alias: **dispersion**) A random variable X having finite mean $\mathbb{E}X$ has variance var X (possibly infinite), given by

$$\operatorname{var} X = \mathbb{E}\{(X - \mathbb{E}X^2)\} = \mathbb{E}X^2 - (\mathbb{E}X)^2 = \sigma^2,$$

where $|\sigma|$ is called the **standard deviation** of X. Denoting the distribution function of X by $F(x)$, $\operatorname{var}X = \int_{-\infty}^{\infty}(x - \mathbb{E}X)^2\mathrm{d}F(x)$, and var X may also be referred to as the variance of F, or of its density (when it exists); it is the second central moment of X centred at $\mathbb{E}X$. The corresponding **sample variance** is $S^2 = c\sum_{r=1}^{n}\left(X_r - \frac{1}{n}\sum_{i=1}^{n}X_i\right)^2$, $n \geq 2$ where $X_1, \ldots, X_n$ is a random sample with the distribution of X, and c is either n^{-1} or $(n-1)^{-1}$ in two different commonly seen conventions.

The generalized variance of a random vector with variance-covariance matrix V is det $V = |V|$, the determinant of V; the generalized sample variance is defined analogously with the single variable case.

The ratio var $X/\mathbb{E}X$ may be called the coefficient (or index) of dispersion, or variance-mean ratio (VMR); *see* variation coefficient. The **semivariance** (or downside, or left-semivariance) of X with distn fn F is

$$sv^- = \int_{-\infty}^{\mathbb{E}X} (x - \mathbb{E}X)^2 \mathrm{d}F(x),$$

with upside, or right-semivariance

$$sv^+ = \int_{\mathbb{E}X}^{\infty} (x - \mathbb{E}X)^2 \mathrm{d}F(x),$$

(with a suitable convention about intervals of integration); more generally for $t \in \mathbb{R}$, the target semivariance (or downside risk), is

$$sv^t = \int_{-\infty}^{t} (x - t)^2 \mathrm{d}F(x),$$

where $\sqrt{sv^t}$ is also called the downside deviation (or downside risk). Yet more generally,

$$R_a^t = \int_{-\infty}^{t} (x - t)^a \mathrm{d}F(x)$$

is the a-t downside risk, or lower partial moment. *See* covariance. [P. Fishburn, Mean-risk analysis associated with below-target returns, *Amer. Econ. Rev.*, 67(2), 116–126, 1977]

variance-covariance matrix *see* covariance

variance-gamma distribution (alias: **normal-gamma distribution**) Arises naturally as a normal variance-mean mixture in which the mixing parameter M has a gamma density. That is to say, given independent r.vs M and X where M has a gamma distribution and X has a normal distn (or, more generally, a multinormal distn), the variance-gamma r.v. Y is given by

$$Y = a + bM + c\sqrt{M}X,$$

for real a, b, and positive c (or, more generally for vector X and Y; real vectors $\mathbf{a}$ and $\mathbf{b}$; and where C is the square root of a positive-definite symmetric matrix).

The variance-gamma process $X(t), t \geq 0$, is a Lévy process whose value at $t = 1$ has a variance-gamma distn; it is obtained as a drifting Wiener process time-changed by a gamma process, or as the difference between two suitable gamma processes. It is sometimes called **Laplace** (or **Brownian-Laplace**) **motion**. [H. Geman *et al.*, Time changes for Lévy processes, *Math. Finan.*, 11(1), 77–96, 2001]

variance-mean mixture (or distribution)
A d-dimensional random vector $\mathbf{X}$ is said to be a variance-mean mixture with mixing distribution $F(\cdot)$ if it can be represented as

$$\mathbf{X} = \mathbf{a} + \mathbf{b}M + \sqrt{M}C\mathbf{Y},$$

where M is a non-negative r.v., with distn fn $F(\cdot)$, and M is independent of the d-dimensional r.vr $\mathbf{Y}$, with $\mathbf{a}$ and $\mathbf{b}$ being real d-vectors, and C the square root of a $d \times d$ positive-definite symmetric matrix. When $\mathbf{b} = 0$, one speaks of a variance mixture.

Most commonly $\mathbf{Y}$ is multinormal, yielding a normal variance-mean distn. In particular, if $\mathbf{b} = 0$ and M is infinitely divisible, and Y is univariate normal, then X is said to be a **type G distn**, or Gaussian transform. [O. Barndorff-Nielsen & V. Pérez-Abreu, Extensions of type G and marginal infinite divisibility, *Theor. Probab Appl*, 2002, 47(2), 202–218; O. Barndorff-Nielsen *et al.*, Normal variance-mean mixtures and Z distributions, *Int. Stat. Rev.* 1982, 50(2), 145–159]

variance ratio distribution *see* F distribution

variance reduction Arises in Monte Carlo methods of simulation and integration that use pseudo-random sampling to give an estimate X of some quantity m, where X is often unbiased, so that $\mathbb{E}X = m$. Any modification of the procedure that increases its efficiency by decreasing the variance of X may be called a variance reduction technique. The subclass of so-called correlation methods is perhaps the most useful and most often seen; the basic idea is to introduce an auxiliary pseudo-random number Y that is suitably jointly distributed with X. Then there are functions $g(X, Y)$ such that $\mathbb{E}g = m$, and var $g <$ var X. *See* antithetic variable; control variate; importance sampling. [R. Rubinstein & D. Kroese, *Simulation and the Monte Carlo Method*, 2nd edn. Wiley, 2008]

variate The term random variate is now most commonly used to denote a simulated (or pseudo-random) number from a probability distribution. In some older texts the word variate is synonymous with random variable. Some writers have used variate to denote a realization (or outcome or sample value), of a random variable; this is a rare usage. [W. Hörmann *et al.*, *Automatic Nonuniform Random Variate Generation*. Springer, 2004]

variation coefficient (alias: **coefficient of variation**) Arises as a dimensionless measure of the variability in the distribution of a positive r.v., and finds applications in e.g. queueing theory. The variation coefficient of X is given by $(\text{var}X)^{\frac{1}{2}}/\mathbb{E}X$, i.e. the standard deviation divided by the expected value.

variation distance *see* total variation distance

variation threshold time (alias: **total variation threshold time**) An example of a mixing time for random processes that is commonly seen in the context of a suitable Markov chain $X(t), t > 0$, on a discrete state space S, having transition probabilities $p_{ij}(t), i, j \in S$. Then the variation threshold time $T(\varepsilon)$, for given $\varepsilon \in (0, 1)$, is

$$T(\varepsilon) = \inf\left\{ t > 0;\ \max_{i,j\in S}\left[\frac{1}{2}\sum_k |p_{ik}(t) - p_{jk}(t)|\right] < \varepsilon \right\}.$$

A common convention is to set $\varepsilon = \mathrm{e}^{-1}$; and the total variation distance in the above definition is sometimes replaced by distance from equilibrium:

$$\hat{T}(\varepsilon) = \inf\left\{ t > 0 : \max_{i\in S}\left[\frac{1}{2}\sum_k |p_{ik}(t) - \pi_k|\right] < \varepsilon \right\}$$

where π is the equilibrium distribution of X. [D. Aldous & J. Fill, *Reversible Markov Chains and Random Walks on Graphs*. Cambridge University Press, to appear; see www.stat.berkeley.edu/~aldous/RWG/book.html]

Varn code Arises in the efficient source coding (or data compression) problem of communication theory when symbols of a discrete memoryless source have different costs. For a source uniformly distributed on its alphabet, Varn coding supplies a prefix code that has minimum expected cost. [S. Savari, Some notes on Varn coding, *IEEE Trans. Inf. Theor.*, 40(1), 181–186, 1994]

Vasershtein coupling *see* basic coupling

Vasicek model A term arising in financial mathematics, where it commonly denotes either of these concepts:

1. A model for the term structure of interest rates, in which it is assumed that the spot (instantaneous) rate follows an Ornstein–Uhlenbeck process with SDE of the form $\mathrm{d}R = a(b - R)\mathrm{d}t + c\mathrm{d}W$, where $a > 0$, and $W(t)$ is the Wiener process. (*See also* Hull–White model.)
2. A model that describes the distribution of losses (and hence potential defaults) in a portfolio of mutually dependent assets (or firms), where the dependence is described in terms of a Gaussian copula, i.e. to the rth asset (or firm) there is associated a r.v. $X_r, 1 \le r \le m$, and it is assumed that for some $\rho \in (0, 1)$, $X_r = \rho^{\frac{1}{2}}Y + \sqrt{1-\rho}Z_r$ where $Y, Z_r, 1 \le r \le m$, are i.i.d. standard normal r.vs. The r.v. Y is called the systematic (or market) risk.

[O. Vasicek, An equilibrium characterization of the term structure, *J. Finan. Econ.*, 5(2), 177–188, 1977]

Vedder's approximation A closed form, and invertible, approximation to the normal distribution function $\Phi(x)$, given by:

$$V(x) = \{1 + \exp(-ax - bx^3)\}^{-1}, \quad x \in \mathbb{R},$$

where $a = \left(\frac{8}{\pi}\right)^{\frac{1}{2}}$, and $b = \frac{4-\pi}{3\pi}\left(\frac{2}{\pi}\right)^{\frac{1}{2}}$; the maximum error being $\pm 0.000\,313$ at $x = \pm 2.29$. [J. Vedder, An invertible approximation to the normal distribution function, *Comp. Stat. Data Anal.*, 16(1), 119–123, 1993]

Venn diagram A graphical device for representing classes (or sets, or events) and their relationships (via logical propositions, or Boolean algebra) by means of regions of the plane. It finds applications in showing relationships between events in a probability space; and in depicting error-correcting codes (e.g. Hamming and Gourlay codes); and in illustrating identities satisfied by Shannon entropy and mutual information; *see* Hu correspondence. [A.W.F. Edwards, *Cogwheels of the Mind*. Johns Hopkins, 2004; R. McEliece, The reliability of computer memories, *Sci. Amer.*, 252(1), 88–95, 1985]

Ventcel (or Ven'cel) *see* Wentzel

Veraverbeke's theorem Supplies the limiting (asymptotic) distribution of the maximum S of a random walk $S_n = \sum_{r=1}^{n} X_r$, $n \ge 0$, where $X_1, X_2, \ldots$ are i.i.d.r.vs with negative mean $\mathbb{E}X_1 < 0$, having common heavy-tailed distribution $F(x)$. Let $S = \sup_{n\ge 0}\max_{0\le r\le n} S_r$, and define the integrated (or second-) tail distribution $G(x)$ by

$$1 - G(x) = \min\left\{1, \int_x^\infty \left(1 - F(y)\right)\mathrm{d}y\right\}, \quad x \in \mathbb{R}.$$

Then the theorem asserts that if $G(x)$ is a subexponential distribution then

$$\lim_{x\to\infty} \frac{\mathbb{P}(S > x)}{1 - G(x)} = \frac{1}{\mathbb{E}X_1};$$

more weakly, if $G(x)$ is long-tailed, then

$$\liminf_{x\to\infty} \frac{\mathbb{P}(S > x)}{1 - G(x)} \geq \frac{1}{\mathbb{E}X_1}.$$

[S. Zachary, A note on Veraverbeke's theorem, *Queue. Sys.*, 46(1–2), 9–14, 2004]

Vernam cypher (alias: **XOR cypher**) Arises in cryptography, where it supplies an example of one-time-pad encryption of a binary message $X_1, X_2, \ldots$. It requires a sequence of (independent) pseudo-random (or random) binary digits $K_1, K_2, \ldots$, called the key, and then the rth symbol of the encrypted message is $X_r + K_r$ modulo 2. [K. Martin, *Everyday Cryptography*. Oxford University Press, 2012]

version *see* indistinguishable; equivalent; conditional probability

Vervaat's transformation (or identity) Asserts that the path transformation of the standard Brownian bridge $A(t), 0 \leq t \leq 1$, that is obtained by placing the origin at the minimum of $A(t), 0 \leq t \leq 1$, has the same distribution as the positive Brownian excursion, $E(t), 0 \leq t \leq 1$. That is to say, if $T = \inf\{t \geq 0 : A(t) = \inf_{0\leq s\leq 1} A(s)\}$, then $E(t)$ has the same joint distns as

$$A(T + t(\text{modulo } 1)) - A(T), \quad 0 \leq t \leq 1.$$

[W. Vervaat, A relation between Brownian bridge and Brownian excursion, *Ann. Probab.*, 7(1), 143–149, 1979]

vicinity measure *see* local time

vicious random walk (alias: **non-intersecting random walk**) Denotes the random process generated by p random walks on the integers $\mathbb{Z}$ (or the line $\mathbb{R}$), whose positions at time t are $X_1(t), \ldots, X_p(t)$, and whose paths evolve on the condition that for all t $X_1(t) < X_2(t) < \cdots X_p(t)$, i.e. they never meet. *See* Dyson's theorem. [M. Fisher, Walks, walls, wetting, and melting, *J. Stat. Phys.*, 34(5–6), 667–729, 1984]

Ville's theorem Arises in the context of descriptive (or algorithmic) randomness, i.e. the task of deciding whether some given infinite binary sequence $b = b_0 b_1 b_2 \cdots$ can be called random (it is assumed that the ratio of 1s to 0s in $b_0 b_1 \cdots b_n$ tends to 1 as $n \to \infty$). R. von Mises (followed by A. Wald and A. Church) suggested that the decision might be made on the basis of a suitable countable family of functions selecting infinite subsequences of b, such that b could be called random if every such subsequence had the property that the ratio of 1s to 0s converged to 1. Ville's theorem undermines this proposal by showing that for any countable family of such selection functions there is a sequence that would be called random but is such that the proportion of 1s to 0s in $b_0 b_1 \cdots b_n$ is never less than 1. [R. Downey & D. Hirschfeldt, *Algorithmic Randomness and Complexity*. Springer, 2010]

vine copula (or structure, or model) Arises in modeling and simulating multivariate distributions in several dimensions, where a vine structure is a graphical description of the decomposition of a multivariate density as a product of its marginal densities and a suitable set of pair-copula conditional densities. The vine is a nested set of trees, in which the edges of one tree comprise the nodes of the next tree in the sequence, each tree having the maximum possible number of edges. [T. Bedford & R. Cooke, Probability density decomposition for conditionally dependent random variables modelled by vines, *Ann. Math. Artif. Intell.*, 32(1–4), 245–268, 2001]

virtual waiting time *see* queueing theory

visiting time *see* occupation time

Viswanath's constant Let $X_n, n \geq 1$, be the random Fibonacci sequence defined by the random recurrence

$$X_{n+2} = R_n X_{n+1} + X_n, \quad n \geq 1,$$

where $X_1 = X_2 = 1$, and $R_n, n \geq 1$, comprises a sequence of i.i.d.r.vs such that $\mathbb{P}(R_n = +1) = \mathbb{P}(R_n = -1) = 1/2$. Then a.s. as $n \to \infty$, $|X_n|^{1/n} \to V \simeq 1.13198824\ldots$, where V is called Viswanath's constant. [E. Makover & J. McGowan, An elementary proof that random Fibonacci sequences grow exponentially, *J. Num. Theor.*, 121(1), 40–44, 2006]

Vitale's random Brunn–Minkowski inequality A generalization of the Brunn–Minkowski inequality for compact convex sets, extending it to random sets thus:

Let $X \subseteq \mathbb{R}^d$ be a random compact set with Lebesgue volume $V(X)$, then

$$\{V(\mathbb{E}X)\}^{1/d} \geq \mathbb{E}(\{V(X)\}^{1/d}),$$

provided that $\mathbb{E}\max\{\|x\| : x \in X\} < \infty$, and where $\mathbb{E}X$ denotes the set-valued expectation of X using the Aumann integral (i.e. Aumann expectation). [R. Vitale, The Brunn–Minkowski inequality for random sets, *J. Multivariate Anal.*, 33(2), 286–293, 1990]

Vitali–Hahn–Saks theorem An important theorem from measure theory, of which one form is this: let $\mathbb{P}_n, n \geq 1$, be a sequence of probability measures on a measurable space $\{\Omega, \mathcal{F}\}$ such that $\lim_{n\to\infty}\mathbb{P}_n(B) = \mathbb{P}(B)$ for all $B \in \mathcal{F}$, and some function $\mathbb{P}$. Then $\mathbb{P}$ is a probability measure on the σ-field $\mathcal{F}$; and if $B_m, m \geq 1$ is a sequence of events in $\mathcal{F}$ that decreases to $\emptyset$, then $\lim_{m\to\infty}\sup_n\mathbb{P}_n(B_m) = 0$. [R. Ash & C. Doléans-Dade, *Probability & Measure Theory*, 2nd edn, 2000, Academic Press]

Vitali's theorem A theorem from measure and integration that has this form in the context of probability: Let $X_n, n \geq 1$, be a sequence of r.vs (with finite rth moment $1 \leq r < \infty$) converging in probability to X. Then the following are equivalent:
(a) X_n converges in rth mean to X.
(b) $\lim_{n\to\infty} \mathbb{E}|X_n|^r = \mathbb{E}|X|^r$.
(c) The sequence $|X_n|^r, n \geq 1$, is uniformly integrable.
(d) $\lim\sup \mathbb{E}|X_n|^r \leq \mathbb{E}|X|^r < \infty$.
(Note that other forms exist, and that other theorems named for G. Vitali include one exhibiting a non-Lebesgue-measurable set of $[0, 1]$; and a proof that there is no non-trivial non-negative set function $V(\cdot)$ on the power set of $\mathbb{R}$ such that $V(\{x\}) = 0$ for points x, $V([0, 1]) = 1$, $V(\cdot)$ is translation invariant, and V is σ-additive.) [G. Shorack, *Probability for Statisticians*. Springer, 2000]

Viterbi decoding algorithm (alias: **min-sum method**) Arose in communication theory as a maximum likelihood decoder of the output of a noisy channel whose input, having been encoded using a convolutional error-correcting code, comprises a Markov chain. The underlying graph of transitions in the state space yields a corresponding graph for multistage transitions that is called a **trellis diagram**. The basic principle of the method can be seen as determining the most likely path through the diagram by a form of dynamic programming. It finds applications in several other areas of inference under uncertainty, e.g. hidden Markov models. [W. Huffman & V. Pless, *Fundamentals of Error Correcting Codes*. Cambridge University Press, 2003]

volatility Arises in financial mathematics where the term may be used broadly (and loosely) to denote some measure of the random variation about its mean (over time) of the price of an asset (e.g. a stock, or contingent claim) in a market of such objects. More narrowly in mathematical models, volatility commonly denotes the standard deviation of some return on (or change of value in), some financial instrument. More precisely, in the basic model for the evolution of the price $S(t) > 0, t \geq 0$, of such an asset, it is assumed that

$$\frac{\mathrm{d}S(t)}{S(t)} = \mu\mathrm{d}t + \sigma\mathrm{d}N(t),$$

where $N(t)$ represents random effects on the process (and can be interpreted as noise). Typically $N(t)$ is the Wiener process, or a suitable Lévy process, or a semi-martingale. Then the parameter $\sigma > 0$ measuring the level of noise is called the volatility; it can be seen as the standard deviation of the instantaneous rate of return $\mathrm{d}S/S$, or equivalently as the square root of the instantaneous variance of the process $R(t) = \log S(t)$. When σ itself obeys a suitable SDE, one speaks of stochastic volatility.

When σ is estimated by statistical inference from the past, one speaks of historic volatility; when σ is estimated from current contracts for contingent claims, one speaks of implied volatility (or benchmarking).

The coefficient μ measures the non-random influences on the evolution of $S(t)$; it may be called the drift, or stock drift, or local mean rate of return, or appreciation/ depreciation rate, or rate of growth. [N. Shephard, *Stochastic Volatility: Selected Readings*. Oxford University Press, 2005]

Volkonskii–Rozanov inequality *see* Vajda's inequality

Volterra Gaussian process A Gaussian process $X(t), t \geq 0$, is said to be Volterra if there is a Volterra integral kernel $k(s,t)$ such that for a Wiener process $W(t), t \geq 0$, $X(t) = \int_0^t k(s,t)\mathrm{d}W(s)$, a.s. for $t \geq 0$. [F. Baudoin & D. Nualart, Equivalence of Volterra processes, *Stoch. Proc. Appl.*, 107(2), 327–350, 2003 & corrigendum, 115(4), 701–703, 2005]

von Bahr–Esseen inequality Any one of a number of inequalities for the sum $S_n = \sum_{r=1}^n X_r$ of a sequence $X_1, X_2, \ldots$ of r.vs, of the form

$$\mathbb{E}|S_n|^p \leq c(p,n)\sum_{r=1}^{n}\mathbb{E}|X_r|^p, \quad 1 \leq p \leq 2,$$

where $c(p,n)$ is a bounded function of p and n; and in particular one can write $c(p,n) = 2$ if $\mathbb{E}(X_{n+1} \mid S_n) = 0$; and further one can write $c(p,n) = 2$ if the distribution of X_{n+1} is symmetric given S_n.

An extension to even convex functions $f(x)$, $f(0) = 0$, with concave derivative $f'(x)$, $x > 0$, of the form

$$\mathbb{E}f(S_n) \leq 2\sum_{r=1}^{n}\mathbb{E}f(X_r),$$

is called the **Topchii–Vatutin inequality**. [B. von Bahr & C.-G. Esseen, Inequalities for the *r*th absolute moment of a sum of random variables, $1 \leq r \leq 2$, *Ann. Math. Stat.*, 36(1), 299–303, 1965]

von Kries paradox Arises as a criticism of naive application of the principle of insufficient reason (*see* indifference principle), or PIR, in assigning a uniform probability measure to an empirical random quantity X; there are many formulations of the paradox, of which Bertrand's paradox is the best known. For example, suppose it is known only that $X \in [1,2]$, then the PIR suggests that $f_X(x) = 1, 1 \leq x \leq 2$; and then $Y = X^{-1}$ has density $f_Y(y) = y^{-2}, \frac{1}{2} \leq y \leq 1$. But in the given circumstances one also initially knows only that $Y \in \left[\frac{1}{2}, 1\right]$, so that the PIR suggests $f_Y(y) = 2$, $\frac{1}{2} \leq y \leq 1$; a contradiction. [M. Heidelberger, Origins of the logical theory of probability: von Kries, Wittgenstein, Waisman, *Int. Stud. Phil. Sci.*, 15(2), 177–188, 2001]

von Mises condition In extreme value theory this is a sufficient condition for a distribution function $F(x)$ to be in the (maximum) domain of attraction of an extreme value distribution $G_b(x)$ (for necessary and sufficient conditions *see* Gnedenko–de Haan theorem). In a unified form where $G_b(x) = \exp\{-(1+bx)^{-1/b}\}, 1 + bx > 0$ (and $G_0(x) = \exp(-e^{-x}), x \in \mathbb{R}$), it is this: let $s(F) = \sup\{x : F(x) < 1\}$, and suppose that for some $a > 0$ we have $f(x) = \frac{dF}{dx} > 0$ for $a \leq x \leq s(F)$.

If for some $c > 0$ and real b

$$\lim_{x \uparrow s(F)} \frac{(1+bx)f(x)}{1-F(x)} = c,$$

then F is in the maximum domain of attraction of $G_{b/c}(x)$. Furthermore, if f is differentiable and $\lim_{x\uparrow s} \frac{d}{dx}\frac{1-F(x)}{f(x)} = 0$, then F is in the domain of attraction of $G_0(x)$. [M. Falk et al., *Laws of Small Numbers: Extremes and Rare Events*, 3rd edn. Birkhäuser, 2011]

von Mises distribution (alias: **circular normal distribution**) A distribution on the circle (or directional distn in the plane) having density

$$f(\theta) = \frac{1}{2\pi I_0(k)}\exp\{k\cos(\theta - a)\}, \quad 0 < \theta \leq 2\pi,$$

for constants a and k, where $I_0(k) = \sum_{r=0}^{\infty}\frac{k^{2r}}{2^{2r}(r!)^2}$ is the modified Bessel function of order zero. This density arises naturally in considering $X = R\cos\Theta$ and $Y = R\sin\Theta$, where X and Y are independent $N(\mu,\sigma^2)$ and $N(\nu,\sigma^2)$ r.vs respectively; the conditional density of the direction Θ of (X,Y) given $R = r$ obeys $f(\theta \mid r) \propto \exp(\{\mu\cos\theta + \nu\sin\theta\}r/\sigma^2)$, and thus has a von Mises distn. *See* Fisher's distribution. [K. Mardia & P. Jupp, *Directional Statistics*. Wiley, 2000]

von Mises inequality Supplies upper bounds on $\mathbb{P}(X \geq x)$ for a r.v. X whose distribution function $F(x)$ satisfies $F(x) \leq F(t) + m(x-t)$ for some $m > 0$, some $0 \leq s < t$, and $s \leq x < \infty$, which is to say that $F(x)$ lies below the given line through $(t, F(t))$ (i.e. F is quasi-convex). For $r > 0$, let $z > t$ be the root of $s^{r+1}z^{-r} + rz - (r+1)t = 0$, and $y > s$ the root of $y^{r+1} - s^{r+1} - (r+1)(y-s)\mathbb{E}|X|^r = 0$. When $y > z$, $\mathbb{P}(X \geq x) \leq \frac{y-x}{y-s} < z^{-r}\mathbb{E}|X|^r$. [T. Sellke, Generalized Gauss–Chebyshev inequalities for unimodal distributions, *Metrika*, 43(1), 107–121, 1996]

von Mises lattice distribution inequality Asserts that for a lattice distn with span h, and $m \geq 1, h\mathbb{E}|X - \mathbb{E}X|^m \leq 2\mathbb{E}|X - \mathbb{E}X|^{m+1}$. This has been improved. [R. von Mises, An inequality for the moments of a discontinuous distribution, *Scand. Actuarial J.*, 1, 32–36, 1939; I. Shevtsova, Moment-type estimates with asymptotically optimal structure for the accuracy of the normal approximation, *Annal. Math. Inform.*, 39, 241–307, 2012]

von Mises theorem Asserts that for the partial sums $S_n = \sum_{r=1}^{k(n)}X_{n,r}$ of an array of independent Bernoulli r.vs $X_{n,r}, 1 \leq r \leq k(n), n \geq 1$, with $\mathbb{P}(X_{n,r} = 1) = p_{n,r} = 1 - \mathbb{P}(X_{n,r} = 0)$, the distribution of S_n converges to that of a Poisson(λ) r.v. iff , as $n \to \infty$, $\sum_{r=1}^{k(n)}p_{n,r} \to \lambda$, and $\max_{1 \leq r \leq k(n)}p_{n,r} \to 0$. [J. Mačys, Bernoulli trials with variable probabilities, *Lithuanian Math. J.*, 19(4), 533–537, 1979]

von Mises–Wald–Church randomness Arises in deciding whether a given infinite binary sequence is effectively (or descriptively or recursively, or algorithmically) random. It is said to be von Mises–Wald–Church random (or stochastic) if every subsequence selected by a suitable (admissible) fixed countable set of recursive (or partially computable) rules obeys the law of large numbers, i.e. the proportion of 1s in the selected sequence converges to 1/2. *See* Ville's theorem. [M. Li & P. Vitányi, *An Introduction to Kolmogorov Complexity and its Applications*, 3rd edn. Springer, 2008]

von Neumann's comparison method (alias: **Forsythe's method; Forsythe–von Neumann method**) An acceptance-rejection method for sampling from a density of the form $f(x) \propto \exp\{-B(x)\}, x \geq 0$, where $B(x)$ is increasing in x. It can be seen as a generalization of von Neumann's exponential. [G. Forsythe, von Neumann's comparison method for random sampling from the normal and other distributions, *Math. Comp.*, 26(120), 817–826, 1972]

von Neumann corrector (or correction) (alias: **von Neumann extractor; von Neumann unbiasing (or deskewing)**) Originally suggested as a method for reduction of the effects of rounding errors (or bias) in extensive computations, the idea now finds application when seeking (e.g. in cryptography) (pseudo-) random binary sequences that are unbiased, i.e. the ratio of 1s to 0s converges to 1 in long strings. In particular, let $X_n, n \geq 1$, be an i.i.d. sequence of Bernoulli r.vs with parameter p; the von Neumann corrector takes successive pairs and deletes all those that comprise $(0, 0)$ or $(1, 1)$, while passing the first digit in unlike pairs $(0, 1)$ or $(1, 0)$. The output is an unbiased sequence. This method may also be applied to suitable stationary Bernoulli sequences. *See* XOR. [Q. Stout & B. Warren, Tree algorithms for unbiased coin tossing with a biased coin, *Ann. Probab.*, 12(1), 212–222, 1984]

von Neumann entropy Arises in von Neumann's theory of quantum mechanics, where the entropy of a quantum state (or operator) ρ, having eigenvalues $\lambda_i, i \geq 1$, is

$$S_{(\rho)} = -\text{Trace}(\rho \log \rho) = -\sum_{i=1}^{\infty} \lambda_i \log \lambda_i.$$

It may be interpreted as the information in the state or system. [G. Jaeger, *Quantum Information*. Springer, 2007]

von Neumann ergodic theorem (alias: **mean (or statistical) ergodic theorem**) In its general form this asserts that for an isometric linear operator T on a Hilbert space H, for $f \in H$, the sequence $\frac{1}{n}\sum_{k=0}^{n-1} T^k f$ converges (in the norm of H) to some $\bar{f} \in H$.

A commonly seen special case arises for the space of square-integrable random variables, and the operator takes the form $Tf(x) = f(Sx)$, where S is a measure-preserving homomorphism of the space into itself. In the simplest (classic) case, S shifts the time along a covariance-stationary sequence of finite-variance random variables $X_n, n \in \mathbb{Z}$ and one obtains the weak ergodic theorem that $\frac{1}{n}\sum_{k=1}^{n} X_k$ converges in mean square to some r.v. X (where $\mathbb{E}X = \mathbb{E}X_1$). *See* Birkhoff's theorem. [K. Petersen, *Ergodic Theory*. Cambridge University Press, 1983]

von Neumann's exponential The original special case of the Forsythe–von Neumann comparison method. It generates a sample from an exponential density (a pseudo-random exponential r.v. Y) by means of a sequence $X_n, n \geq 0$, of i.i.d.r.vs each uniform on $[0, 1]$, thus: Set $R = 0$, then:

(i) Generate $X_0 > X_1 > \cdots$ until the first n such that $X_n \geq X_{n-1}$.

(ii) If n is even increase the current R by 1 and go to (i); otherwise if n is odd set Y as the sum of X_0 in the final run and the current value of R.

Then Y has the exponential density $f(y) = \mathrm{e}^{-y}$, $y \geq 0$.

A more efficient implementation of the basic idea is known as the **von Neumann–Knuth–Yao method.** [P. Flajolet & N. Saheb, The complexity of generating an exponentially distributed variate, *J. Algor.*, 7(4), 463–488, 1986]

von Neumann's method *see* acceptance-rejection

von Neumann's minimax theorem A key result in game theory, where it asserts that for any zero-sum two-person game each player has a mixed (i.e. possibly randomized) strategy yielding an expected pay-off v, which is optimal for both in the sense that the minimax and maximin values of the pay-off function are equal to v, which is called the **value** of the game. [K. Binmore, *Playing for Real*. Oxford University Press, 2007]

von Neumann–Morgenstern utility theorem A central task of utility theory is the numerical representation of an agent's preferences among decisions leading to (possibly incommensurate) random rewards. That is to say, one seeks a real-valued function $U(r)$ of decisions with random rewards denoted by r, such that r is preferred to s iff $\mathbb{E}U(r) \geq \mathbb{E}U(s)$. This is called the arithmetization of preferences, and $u(\cdot)$ is said to be a utility function if it exists. The von Neumann–Morgenstern utility representation theorem supplies a set of four postulates characterizing so-called rational choices among risky rewards, and proves that a utility function exists for an agent whose preferences are described by those postulates. [J. von Neumann & O. Morgenstern, *Theory of Games and Economic Behaviour*, 3rd edn. Princeton, 1953 (reprinted 2004)]

Vorob'ev expectation A set-valued mean of a random compact set X, defined thus in $\mathbb{R}^n$: the probability $p_X(y) = \mathbb{E}I(y \in X)$ is called the coverage function (where $I(\cdot)$ is the usual indicator function). Then the Vorob'ev expectation of X is the set $V = \{y \in \mathbb{R}^n : p_X(y) \geq t\}$, where t is the smallest t such that $m(y \in \mathbb{R}^n : p_X(y) \geq t) = \mathbb{E}m(X)$, and $m(\cdot)$ denotes Lebesgue measure. [I. Molchanov, *Theory of Random Sets*. Springer, 2005]

Voronoi–Poisson tiling *see* random mosaic

voter model An interacting particle system comprising a Markov process, whose basic form is this: each vertex of the integer lattice $\mathbb{Z}^d$ is labeled either 0 or 1 (so the process takes values in $\{0, 1\}^{\mathbb{Z}^d}$). From any time t, the process evolves according to this rule, each vertex waits (independently of the others) for random interval that is exponentially distributed with parameter 1, and then gives its own value (0 or 1) to a neighbour vertex picked uniformly at random. The fundamental theorem asserts that such a process clusters when $d = 1$ or 2 (i.e. the probability that any finite set contains entirely 1s or entirely 0s tends to 1), but arrives at non-trivial stable equilibrium for $d \geq 3$. [P. Clifford & A. Sudbury, A model for spatial conflict, *Biometrika*, 60(1), 581–588, 1973; T. Liggett, *Stochastic Interacting Systems*. Springer, 1999]

Vysochanskij–Petunin inequality Supplies a bound for the probability in the tails of a unimodal probability density $f(x)$ on $\mathbb{R}$, thus: let $a \in \mathbb{R}$ be arbitrary, and let the r.v. X have density $f(x)$ with a single mode m; then

$$\mathbb{P}(|X - a| \geq x) \leq \frac{4}{9x^2}\mathbb{E}\{(X - a)^2\}, \quad x \geq \sqrt{\frac{8}{3}\mathbb{E}[(X - a)^2]}.$$

[F. Pukelsheim, The three sigma rule, *Amer. Stat.*, 48(2), 88–91, 1994]

W

waiting time in queue *see* queueing theory

waiting time in state *see* holding time

waiting–time paradox *see* inspection paradox

Wald distribution *see* inverse Gaussian distribution

Waldegrave's problem A group of n players play each other in circular order, with equal probability of each winning, and the winner continues to play the next. The game ends when one player beats all the rest in succession. Several early probabilists considered the distribution of the duration of the game, and the chance of winning for each player. [A. Hald, *History of Probability and Statistics and their Applications Before 1750*. Wiley, 1990]

Wald's identity (or equation) (alias: **Wald's fundamental identity, or theorem, or lemma**) Any of several equations that arise in considering the moments (and the mgf) of the random partial sums $S_T = \sum_{r=1}^{T} X_r$ of a sequence of independent r.vs $X_1, X_2, \ldots$, having common finite mean (that we here take to be 0, wlog), and T is a r.v. such that the value of the indicator $I(\{T = k\})$ is determined by $X_1, \ldots, X_k$ (i.e. a stopping time). Under suitable conditions on the distributions of T and the Xs it is the case that $\mathbb{E}S_T = 0$, and this is called **Wald's eqn** (or identity, etc.). A classic example is the **Blackwell–Wald lemma**, which asserts that Wald's equation holds for zero-mean i.i.d.r.vs $X_1, X_2, \ldots$, whenever $\mathbb{E}T < \infty$.

More generally, for a sequence $X_1, X_2, \ldots$, of zero-mean r.vs that are not necessarily identically distributed, and a (possibly randomized) stopping time T for that sequence, and writing $k_n = \mathbb{E}|\sum_{r=1}^{n} X_r|$, **Klass's thm** asserts that $\mathbb{E}k_T < \infty$ is sufficient for Wald's equation to hold.

If also the zero-mean Xs have common finite variance σ^2, then under suitable conditions one has $\mathbb{E}(S_T^2) = \sigma^2 \mathbb{E}T$, which is called **Wald's second eqn**; in the i.i.d. case, for this to hold, it is sufficient that $\mathbb{E}T < \infty$.

For a sequence of i.i.d.r.vs having an mgf such that $1 \leq M(t) = \mathbb{E}\exp(tX_1) < \infty$ for some t, **Wald's identity** (or third equation) asserts that under suitable conditions, $\mathbb{E}\{M(t)^{-T}\exp(tS_T)\} = 1$. It holds e.g. if $\mathbb{E}T < \infty$, and $|S_n|$ is uniformly bounded for $n < T$; while differentiating wrt t, and setting $t = 0$, yields the sequence of **Wald eqns**. A generalization to a sum $S_n = \sum_{r=1}^{n} \mathbf{X}_r$ of random vectors $\mathbf{X}_1, \mathbf{X}_2, \ldots$, may be called the **Blackwell–Girshick thm**; an extension to non-i.i.d.r.vs may be called **Blom's thm**.

The sequence $W_n = M(t)^{-n}\exp(tS_n)$ is sometimes called **Wald's martingale**; and, by extension, the random process

$$X(t) = \exp\left\{\theta W(t) - \frac{1}{2}t\theta^2\right\}, \quad t \geq 0, \theta \in \mathbb{R},$$

where $W(t)$ is the Wiener process, has also been referred to as Wald's martingale. [M. Klass, A best possible improvement of Wald's equation, *Ann. Probab.*, 16(2), 840–853, 1988; D. Blackwell & M. Girshick, On functions of sequences of independent chance vectors, *Ann. Math. Stat.*, 17(3), 310–317, 1946]

Walker's method *see* aliasing

Wallace's method A procedure for generating pseudo-random normally distributed r.vs (and also exponentially distributed r.vs) that is atypical in not relying essentially on a generator of uniform pseudo-random numbers. The method exploits the fact that orthogonal transformations of a vector of independent standard normal r.vs remain independent normal r.vs. [R. Brent, Some comments on C. S. Wallace's random number generators, *Comp. J.*, 51(5), 579–584, 2008.]

Wallenius distribution *see* hypergeometric distribution

wallet problem *see* two-envelopes problem

Walsh's Brownian motion *see* Brownian spider

Waring's distribution The r.v. X has simple Waring distribution with parameter $a > 0$ and $b > 0$ if X takes values in the non-negative integers with

$$\mathbb{P}(X = x) = \frac{a}{a+b}\prod_{r=1}^{x}\frac{b+r-1}{b+a+r}, \quad x \geq 0.$$

More generally, the standard Waring distn has pgf

$$G_X(z) = \frac{a^{[c]}}{(a+b)^{[c]}}F(b, c, a+b+c, z),$$

where $a^{[c]}$ is the rising factorial for $c \geq 0$, F is the hypergeometric series, and $c = 1$ yields the simple Waring distn. Several generalized Waring distns exist. *See* factorial distribution and hypergeometric distribution. [J. Irwin, The generalized Waring distribution: parts I, II, & III, *J. Roy. Stat. Soc. A*, 138(1), 18–31; (2), 204–227; (3), 374–304, 1975]

Wasserstein coupling *see* basic coupling

Wasserstein distance (or metric) *see* minimum L_p distance

Watanabe's theorem Supplies a characterization of Poisson processes in the context of martingale theory, thus: let $N(t), t \geq 0$, be a simple point process (i.e. a counting process); then $N(t)$ is a Poisson process of rate (or parameter) $\lambda > 0$ iff $N(t) - \lambda t$ is a martingale wrt $N(t)$ (or, equivalently, wrt the natural filtration $\mathcal{F}_t$ generated by $N(t)$. A similar result holds for a non-homogeneous Poisson process with associated martingale $N(t) - \Lambda(t)$; and more generally if the compensator $C(t)$ of $N(t)$ is continuous and suitably measurable, and $N(t) - C(t)$ is a martingale for $\mathcal{F}_t$, then $N(t)$ is a Cox process directed by $C(t)$. *See also* Lévy's characterization of Brownian motion. [D. Daley & D. Vere-Jones, *Introduction to the Theory of Point Processes*, 2nd edn. Springer, Vol. I, 2003; Vol. II, 2008]

watched process *see* censored process

Watson–Williams distribution *see* Fisher's distribution

Watts–Strogatz small-world model A type of random graph that exhibits small expected shortest paths together with strong local clustering. It is generated from a regular ring graph by random replacement of initial edges by randomly chosen shortcuts across the ring (provided that this does not duplicate an existing shortcut). [M. Newman *et al.*, *Structure and Dynamics of Networks*. Princeton, 2006]

weak convergence of probability measures Denotes the extension of the concept of weak convergence of distribution functions to the weak convergence of probability measures on a metric space $(\Omega, \mathcal{B})$, where $\mathcal{B}$ is the class of Borel sets in Ω. If a sequence $\mathbb{P}, \mathbb{P}_1, \mathbb{P}_2, \ldots$ of probability measures on $\mathcal{B}$ satisfies $\lim_{n\to\infty}\int g\mathrm{d}\mathbb{P}_n = \int g\mathrm{d}\mathbb{P}$ for all real-valued bounded continuous function g on Ω, then $\mathbb{P}_n$ is said to converge weakly to $\mathbb{P}$ as $n \to \infty$. If $\mathbb{P}, \mathbb{P}_1, \mathbb{P}_2$, are the distributions of some sequence $X, X_1, X_2, \ldots$ of random objects taking values in Ω, then the sequence of objects is said to converge in distribution.

The portmanteau theorem asserts the equivalence of several other characterizations of weak convergence, e.g. $\mathbb{P}_n$ converges weakly to $\mathbb{P}$ if $\lim_{n\to\infty}\mathbb{P}_n(B) = \mathbb{P}(B)$ for all $B \in \mathcal{B}$ such that the boundary ∂B of B has $\mathbb{P}(\partial B) = 0$. [P. Billingsley, *Convergence of Probability Measures*, 2nd edn. Wiley, 1999]

weak ergodicity Consider a non-homogeneous Markov chain on a finite state space whose transition matrix at time t is $P(t)$ with elements $p_{jk}(t)$, and write $P(s,t) = \prod_{r=s+1}^{t} P(r), \quad s \leq t$, with elements $p_{ik}(s,t)$. The chain is weakly ergodic if, as $t \to \infty$, for any $s \geq 0$, $|p_{ik}(s,t) - p_{jk}(s,t)| \to 0$, for all i, j, k.

If also $p_{ik}(s,t)$ converges for all i and k as $t \to \infty$ then the chain is (strongly) ergodic. Several theorems supply conditions for weak ergodicity, e.g. **Markov's sufficient condition** that $\sum_t \max_k \min_j p_{jk}(t) = \infty$; **Hajnal's thm** gives conditions when the transition matrices are scrambling (i.e. for any i, j there is k such that $p_{ik}(t) > 0$ and $p_{jk}(t) > 0$); and the **Wolfowitz condition** is that all finite products of possible transition matrices P are aperiodic and indecomposable (i.e. the limit $L = \lim_{n\to\infty} P^n$ exists and has all its rows the same). [E. Seneta, On the historical development of the theory of finite inhomogeneous Markov chains, *Math. Proc. Camb. Philos. Soc.*, 74(3), 507–513, 1973]

weak law of large numbers (alias: **WLLN**) Arises naturally in considering the sequence of running averages $M_n = \frac{1}{n}\sum_{r=1}^{n} X_r$ of the i.i.d.rvs $X_1, X_2, X_3, \ldots$, having common finite mean $\mu = \mathbb{E}X_1$. Empirically it is found that M_n is likely to be near μ in the long run; this is the so-called law of averages, of which the phenomenon of the stability of statistical ratios is a special case.

Formally, the WLLN asserts that for such a sequence $M_n, n \geq 1$, and any $t > 0$,

$$\lim_{n\to\infty} \mathbb{P}(|M_n - \mu| > t) = 0.$$

This form may be called **Khinchin's WLLN**; if $\mathbb{E}X_1^2 < \infty$, one may speak of **Chebyshov's WLLN**. More generally, the generalized WLLN asserts that there is a sequence of constants $c_n, n > 1$, such that $\lim_{n\to\infty}\mathbb{P}(|M_n - c_n| > t) = 0$, iff $\lim_{n\to\infty}\{\mathbb{P}(X_1 > x) + \mathbb{P}(X_1 \leq -x)\}/x = 0$; in this case one may choose $c_n = \int_{-n}^{n} x\mathrm{d}F(x)$, where $F(x)$ is the distribution function of X_1.

A further extension to the convergence in probability of the normalized sums of the rows of a triangular array (where the elements of a row are not necessarily identically distributed) may be called **Kolmogorov's WLLN**. [W. Feller, *Introduction to Probability Theory and its Applications*, Vol. 1, 3rd edn. Wiley, 1968; Vol. 2, 2nd edn. Wiley, 1971; R. Durrett, *Probability: Theory and Examples*, 4th edn. Cambridge University Press, 2010]

weak Markov property *see* Markov property

weak martingale A sequence $X_n, n \geq 1$, of r.vs having finite means is said to be a weak martingale (resp. submartingale) if for all $1 \leq m \leq n, \mathbb{E}(X_n \,|\, X_m) = X_m$, a.s. (resp. $\mathbb{E}(X_n \,|\, X_m) \geq X_m$, a.s.).

More weakly, if for all $n \geq 1$ one has $\mathbb{E}(X_{n+1} \,|\, X_n) = X_n$(resp. $\geq X_n$), then the sequence is called a very weak martingale (resp. very weak submartingale).

The term weak martingale has also been used to denote a random process $X(t), t \geq 0$, adapted to the filtration $\mathcal{F}_t$, such that for all m there is a uniformly integrable, right-continuous martingale $Y_m(t), m \geq 1$, such that $X(t) = Y_m(t)$ for $0 \leq t \leq T_m$, where $T_m \stackrel{\text{a.s.}}{\to} \infty$ is an increasing sequence of stopping times for $\mathcal{F}_t$. [L. Chen, A martingale inequality for the square and maximal functions, *Ann. Probab.*, 7(6), 1051–1055, 1979]

weak stationarity *see* stationarity

weak symmetry *see* balance equations

Weibull distribution (alias: **stretched exponential distribution**) A non-negative r.v. X is said to have a Weibull distribution (in standard form) when its distribution function is $F(x) = 1 - \exp(-x^a)$, $x > 0, a > 0$. Therefore, the distribution function

$$F(x) = \begin{cases} \exp(-(-x)^a), & x \leq 0, \quad a > 0 \\ 1, & x > 0, \end{cases}$$

is called the **reversed Weibull distn**. Because of its significance in extreme value theory, it is also called the extremal-Weibull distn, or extreme value distn of type III. For negative values of the parameter a, the corresponding distribution $F(x) = \exp(-x^a)$, $x > 0$, is called the Fréchet distn; the remaining type being the Gumbel distn. *See* extreme value distributions. [L. de Haan & A. Ferreira, *Extreme Value Theory*. Springer, 2006]

Weierstrass–Mandelbrot process A fractal random process defined by the series

$$W(t) = \sum_{n=-\infty}^{\infty} (1 - \exp(\mathrm{i}tm^n))\mathrm{e}^{\mathrm{i}U_n} m^{-nH}, \quad t \geq 0,$$

for $m \geq 1$ and $0 \leq H \leq 1$, where U_n is (in the simplest case) a sequence of i.i.d.r.vs each uniformly distributed on $(0, 2\pi)$. The limit $m \to 1$, for suitably scaled $W(t)$, yields complex fractional Brownian motion. [J. Szulga, Hausdorff dimension of Weierstrass-Mandelbrot process, *Stat. Probab. Lett.*, 56(3), 301–307, 2002]

Weierstrass random walk A random walk in one dimension whose independent steps have the common probability distribution given by $\mathbb{P}(X = 0) = (a-1)/a$, and $\mathbb{P}(X = b^r) = \frac{1}{2}\frac{a-1}{a}\left(\frac{1}{a}\right)^r = \mathbb{P}(X = -b^r)$, $r = 1, 2, 3, \ldots$, for $1 < a < b$. The characteristic function of steps

$$\phi(t) = \frac{a-1}{a}\sum_{r=0}^{\infty}\frac{1}{a^r}\cos(tb^r),$$

is a Weierstrass function (continuous and non-differentiable) which gives the name of the walk. It is a Lévy flight that can exhibit superdiffusive and fractal behaviour. [W. Paul & J. Baschnagel, *Stochastic Processes*, 2nd edn. Springer, 2013]

weight of evidence Appears in a formalization of the process of assessing how far evidence E confirms the choice of one hypothesis H_1, in preference to another hypothesis H_2, in some given set of circumstances. In the simplest statistical context of Bayesian hypothesis testing, the weight of the evidence is the logarithm of the likelihood ratio: $\log\{\mathbb{P}(E \,|\, H_1)/\mathbb{P}(E \,|\, H_2)\}$. [I. J. Good, Weight of evidence, corroboration, explanatory powers, information, and the utility of experiments, *J. Roy. Stat. Soc. B*, 22(2), 319–331, 1960]

Wendel's theorem Asserts that for n points that are independently and uniformly placed on the surface of the unit sphere in d dimensions, the probability that they all lie in the same hemisphere is

$$h = 2^{-n+1}\sum_{r=0}^{d-1}\binom{n-1}{r}.$$

[J. Wendel, A problem in geometric probability, *Math. Scand.*, 11, 109–111, 1962]

Wentzell boundary condition (alias: **Venttsel' (or Ventcel') conditions**) A class of linear second-order boundary conditions that arise as the most general boundary conditions appropriate for the (second-order) elliptic operator corresponding to a class of diffusion processes (i.e. homogeneous Feller processes) in suitable compact regions. [A. Venttsel'*, On boundary conditions for multidimensional diffusion processes, *Theor. Probab. Appl.*, 4(2), 164–177, 1959. *Now more commonly transliterated as Wentzell]

Weyl's uncertainty principle (or inequality) *see* uncertainty

white noise A model for the essentially unpredictable, natural (i.e. empirical), noise that appears as an (unwanted) environmental contribution $Z(t)$ to the measurement of some quantity $D(t)$, yielding some corrupted observation $Y(t)$, e.g. one often assumes $Y(t) = D(t) + Z(t)$.

The simplest form of noise is in discrete time, so that a stationary sequence $Z_n, n \in \mathbb{Z}$, of i.d. orthogonal or uncorrelated r.vs is often called discrete white noise. More strongly, if the Z_n are i.i.d., one may speak of **strictly white noise**. In both cases the autocorrelation function $r(n)$ is zero for $n \neq 0$, with a spectral density that is uniform on $[-\pi, \pi]$. Given a spectral density that is only approximately uniform, some writers refer to **coloured noise**.

For random signals in continuous time, a stationary random process $Z(t)$ whose spectral density is uniform over an interval $[-b, b]$, and zero elsewhere, is often called (especially in applications) white noise. However, its autocorrelation function obeys,

$$r(\theta) \propto \frac{\sin(2\pi b\theta)}{2\pi\theta},$$

so that $Z(s)$ and $Z(t)$ are not necessarily uncorrelated (still less independent) for $s \neq t$. When Z_n(or $Z(t)$), are normally distributed one has Gaussian white noise.

Continuous-time white noise such that $Z(s)$ and $Z(t)$ are independent for all $s \neq t$ is not defined within the usual framework of random processes as a family of random variables indexed by a parameter $t \in \mathbb{R}$. Instead, it exists as a generalized stochastic process, defined in terms of its action on a suitable space of test functions which, for the satisfactory development and application of the theory, is required to be infinite dimensional. In particular, **Hida's white noise theory** defines a probability space $(\mathcal{S}', \mathcal{B}, m)$ comprising the space $\mathcal{S}'$ of tempered distributions on $\mathbb{R}^d$, with its Borel field $\mathcal{B}$ of subsets, and a Gaussian measure m on $\mathcal{B}$ called white noise measure (whose existence is guaranteed by the Bochner–Minlos theorem). In this framework it is possible to justify the naive interpretation of a white noise process $Z(t), t \in \mathbb{R}$, as a type of generalized derivative of the Wiener process. [H. Holden *et al.*, *Stochastic Partial Differential Equations: A Modeling, White Noise Functional Approach*, 2nd edn. Springer, 2010]

white screen problem A covering time problem for a simple random walk on the square lattice on a torus, i.e. for a simple random walk stepping to neighbours on the graph of points $\{(i,j) : 0 \leq i \leq r-1, 0 \leq j \leq s-1\}$, the position of the particle taken modulo r and s respectively for the two coordinates; one asks for properties of the (cover) time taken to visit all rs points, $v(r,s)$, e.g. for $r = s$, $\frac{v(r,r)}{r^2\log^2 r}$ converges in probability to $\frac{4}{\pi}$ as $r \to \infty$. [F. Comets *et al.*, Large deviations for the cover time of two-dimensional torus, *Electr. J. Probab.*, 96, 1–18, 2013]

Whittaker–Shannon theorem *see* sampling theorem

Whittle–Matérn covariance *see* Matérn covariance model

Whitworth's formula Let $0 \leq X_{(1)} < X_{(2)} < \cdots < X_{(n-1)} \leq 1$ be the order statistics of $n-1$ i.i.d.r.vs, each uniform on $[0, 1]$, and let M_n be the largest interval: $M_n = \max\{X_{(1)}, X_{(2)} - X_{(1)}, \ldots, 1 - X_{(n-1)}\}$. Then Whitworth's formula is for the distribution:

$$\mathbb{P}(M_n \leq x) = \sum_{r=0}^{n}\binom{n}{r}(-1)^r(1-rx)_+^{n-1}, \quad \text{where } a_+ = \max\{a, 0\}.$$

[L. Holst, On the lengths of the pieces of a stick broken at random, *J. Appl. Probab.*, 12(3), 623–634, 1980]

Wick calculus (alias: **Wick integral; Wick product**) Stems from a renormalization (or orthogonalization) method in quantum physics due to G. C. Wick, that has subsequently been applied in stochastic analysis, especially to SDEs and white noise theory. The basic concept is the Wick product of random elements, whose definition is variously formalized according to the nature of the elements involved, e.g. operators in quantum theory, random functionals in white noise theory, or fractional Brownian motion in stochastic integration; leading to the Wick (or Wick–Itô) integral whose integrator need not be a semi-martingale (unlike the classic stochastic integral). There is a corresponding variety of notation.

In its simplest form, for zero mean Gaussian r.vs X, Y, Z, the first three Wick products (denoted by $:\cdot:$) are $:X: = X$, $:XY: = XY - \mathbb{E}(XY)$, and $:XYZ: = XYZ - X\mathbb{E}YZ - Y\mathbb{E}ZX - Z\mathbb{E}XY$.

The general formula may be given recursively, or in terms of Feynman diagrams. [S. Janson, *Gaussian Hilbert Spaces*. Cambridge University Press, 1991; Y. Hu & J. Yan, Wick calculus for nonlinear Gaussian functionals, *Acta Math. Appl. Sinica, English Ser.*, 25(3), 399–414, 2009; T. Duncan *et al.*, Stochastic calculus for fractional Brownian motion, *SIAM J. Control Optim.*, 38(2), 582–612, 2006; H. Holden *et al.*, *Stochastic Partial Differential Equations: A Modeling, White Noise Functional Approach*, 2nd edn. Springer, 2010]

Wick's formula *see* Isserlis's theorem

Wick's theorem *see* Isserlis's theorem

wide-sense Markov process *see* Markov process

wide-sense martingale A random process $X(t), t \in T$, such that $\mathbb{E}\,|X(t)|^2 < \infty$ and for $n \geq 1$, $t_1 < t_2 < \cdots < t_{n+1}$

$$\hat{\mathbb{E}}(X(t_{n+1})\,|\,X(t_n), \ldots, X(t_1)) = X(t_n),$$

where $\hat{\mathbb{E}}\ (u\,|\,\mathbf{v})$ denotes the minimum variance estimator (or predictor) of u that is linear in the components of $\mathbf{v}$. [J. Doob, *Stochastic Processes*. Wiley, 1953]

wide-sense stationary *see* stationarity

wide-sense subordinator *see* subordinator

Widom–Rowlinson model An interacting particle system arising as a model for the liquid–vapour phase transition. In one version it is defined in terms of particles of two types, and the appropriate probability distribution on countable subsets of some bounded set $S \subseteq \mathbb{R}^d$ is that of two independent Poisson processes conditioned to have no points of one process within unit distance of any particle of the other process; there is an analogous formulation on a lattice. [J. Lebowitz & G. Gallavotti, Phase transitions in binary lattice gases, *J. Math. Phys.*, 12, 1129–1133, 1971; D. Ruelle, Existence of phase transition in a continuous classical system, *Phys. Rev. Lett.*, 26, 303–304, 1971]

Wiener–Bachelier process *see* Brownian motion

Wiener causality Arises in considering stochastic systems that yield two (or more) stationary time series, where the Wiener causality is a measure quantifying the effect (or influence) that one system has on another, i.e. the extent to which it may be said to cause the other. Thus, let X_n and $Y_n, n \in \mathbb{Z}$, be two stationary time series and let Z_n be the (suitably normalized) prediction error (also called the innovation) of the best linear unbiased estimator (or BLUE, also called the autoregressive predictor) of X_n given $X_r, r \leq n-1$ (i.e. the past of X); likewise let W_n be the prediction error of the BLUE of X_n given both $Y_r, r \leq n-1$ and $X_r, r \leq n-1$. The reduction in the variance of the prediction error arising from knowledge of the past of Y, $\operatorname{var} Z_n - \operatorname{var} W_n$, is seen as a measure of the causality effect of Y on X, and called Wiener causality in this basic case. *See also* Granger causality. [N. Wiener, The theory of prediction, in *Modern Mathematics for the Engineer: First Series*, F. Beckenbach (ed.). McGraw–Hill, 1956, pp. 165–190; reprinted by Dover, 2013]

Wiener chaos expansion (alias: **Wiener–Itô chaos expansion**) Broadly, a type of orthogonal expansion for functionals of suitable random processes. In its basic form the process is Gaussian (e.g. the Wiener process), and the expansion (or representation) is commonly seen in two forms:

1. In terms of Hermite polynomials $H_n, n \geq 0$, when one often refers to the **Wiener chaos expansion**.
2. In terms of a series of multiple Itô integrals against Wiener processes

$$I_n f = \int_0^\infty f_1(t_1)\mathrm{d}W(t_1)\int_0^{t_1} f_2(t_2)\mathrm{d}W(t_2)\cdots\int_0^{t_{n-1}} f_n(t_n)\mathrm{d}W(t_n),$$

where the set of all such n-iterated Itô integrals for suitable f is called the ***n*th homogeneous chaos**. In this case one often refers to the **Itô (or Wiener–Itô) chaos expansion**.

[O. Kallenberg, *Foundations of Modern Probability*, 2nd edn. Springer, 2002]

Wiener–Chentsov random field *see* Brownian sheet

Wiener covariance *see* Brownian distance covariance

Wiener filter *see* Wiener–Kolmogorov filter

Wiener football (or soccer) A model for the game of football in terms of some Brownian motion (Wiener process) on a suitable surface, e.g. the complex plane, or a cylinder; one team scores against another when the path of the process crosses either of two line segments representing goals, and the state of play at time t is given by an appropriate winding number $N(t), t \geq 0$. Typically, $N(t)/\sqrt{\log t}$ converges in probability to some symmetric distribution as $t \to \infty$, e.g. symmetric exponential, or normal. [Y. Baryshnikov, Wiener soccer and its generalization, *Electr. Commun. Probab.*, 3, 1–11, 1997]

Wiener's homogeneous chaos Arises as a model for a variety of systems in statistical physics, including homogeneous turbulence, and homogeneous liquids or gases. It comprises a translation invariant random measure on $(\mathbb{R}^n, \mathcal{B})$ (where $\mathcal{B}$ is the Borel field of subsets of $\mathbb{R}^n$); a classic example constructed by Wiener is a form of white noise. [L. Gross, The mathematical ramifications of Wiener's program in statistical physics, in *Proc. Norbert Wiener Centenary Congress, 1994*. American Mathematical Society, 1997]

Wiener–Hopf factorization Broadly, this is a type of decomposition of a suitable function $k(\cdot)$, or an equivalent representation of an integral transform of $k(\cdot)$, that was first developed for the analysis of integral equations of convolution type with kernel k. In probability theory the term generally arises in the context of a random walk with steps having distribution function $F(x)$ (taking the role of k), whose Wiener–Hopf factorization is given in terms of the (possibly defective) distribution functions of the strict ascending (and weak descending) ladder height distributions H_+ (and H_-). That is to say, H_+ is the distn fn of the point of first entry of the walk into $(0, \infty)$, and H_- is the distn fn of the point of first entry into $(-\infty, 0]$. Then the fact that $F = H_+ + H_- - H_-{}^*H_+$, where $*$ denotes convolution, is the Wiener–Hopf factorization for random walk; the name is also given to the equivalent form using integral transforms (e.g. mgf, or chf, or Fourier-Stieltjes transform). The term is occasionally further applied to related identities in the fluctuation theory of random walks. [J. Kennedy, Understanding the Wiener-Hopf factorization for the simple random walk, *J. Appl. Probab.*, 31(2), 561–563, 1994; A. Kuznetsov, Analytic proof of Pecherskii-Rogozin identity and Wiener-Hopf factorization, *Theor. Probab. Appl.*, 55(3), 432–443, 2011]

Wiener integral *see* Paley–Wiener–Zygmund integral

Wiener–Itô expansion *see* chaos expansion

Wiener–Itô integral *see* Itô integral

Wiener–Itô theorem (or expansion) *see* Wiener chaos expansion (or representation)

Wiener–Khinchin theorem (or formula) (alias: **subsets of Einstein–Khinchin–Kolmogorov–Wiener–Wold (often Khinchin–Kolmogorov)**) Arises in considering a zero-mean weakly (i.e. second-order) stationary process $X(t), t \in \mathbb{R}$ (or $t \in \mathbb{Z}$), where it asserts that (under suitable conditions) the autocorrelation function $r(s) = \rho(X(t), X((t+s))$ and the (so-called) **power spectral density,**

$$s(\lambda) = \lim_{T \to \infty} \left(\mathbb{E} \frac{1}{2T} \left| \int_{-T}^{T} X(t) \mathrm{e}^{-2\pi\lambda i t} \, \mathrm{d}t \right|^2 \right),$$

essentially comprise a Fourier transform pair. However, the literature has two conventions. In physical applications $s(\lambda)$ is an object of empirical interest, and the Wiener–Khinchin theorem is the assertion that $s(\lambda)$ has a representation in terms of the autocorrelation

$$s(\lambda) = \int_{-\infty}^{\infty} \mathrm{e}^{-2\pi\lambda i s} r(s) \mathrm{d}s.$$

Conversely, in the mathematical literature, the Wiener–Khinchin theorem is the assertion that the autocorrelation function enjoys a representation as the chf of a spectral distribution function $F(\lambda)$,

$$r(s) = \int_{-\infty}^{\infty} \mathrm{e}^{\mathrm{i}s\lambda} \mathrm{d}F(\lambda).$$

When $\frac{\mathrm{d}F}{\mathrm{d}\lambda} = f(\lambda), \lambda \in \mathbb{R}$, the spectral density $f(\lambda)$ is proportional to the power spectral density $s(\lambda)$. (When F has jumps, they are called the spectrum.) Analogous Wiener–Khinchin theorems hold in discrete time; *see* spectral analysis. [L. Koopmans, *Spectral Analysis of Time Series*, 2nd edn. Academic Press, 1995]

Wiener–Kolmogorov filter (alias: **Kolmogorov–Wiener filter**) Arises in the problem of linear prediction (or extrapolation, or estimation, or interpolation) of a zero-mean, weakly (i.e. second-order) stationary sequence of r.vs $X(t), t \in \mathbb{R}$(or $t \in \mathbb{Z}$). A linear combination $\hat{X}(t)$ of the values of the sequence prior to t, that minimizes the mean squared error

$s^2 = \mathbb{E}\{(X(t+m) - \hat{X}(t))^2\}, m \geq 0$, is generally called a Wiener–Kolmogorov (or v.v.) filter; more narrowly one speaks of a Kolmogorov filter in discrete time ($t \in \mathbb{Z}$), and of a Wiener filter in continuous time ($t \in \mathbb{R}$). *See also* Kalman filter. [A. N. Kolmogorov, Interpolation and extrapolation of stationary random sequences, in *Selected Works of A.N. Kolmogorov*, Vol. II, A. N. Shiryayev (ed.). Kluwer, pp. 272–280; P. Whittle, *Prediction and Regulation by Linear Least-square Methods*, 2nd edn. Blackwell, 1984]

Wiener–Lévy process *see* Brownian motion

Wiener's maximal inequality *see* maximal ergodic theorem

Wiener measure A probability measure on the space C_0 of all $\mathbb{R}^n$–valued continuous functions $w(t) \in C_0$ on the interval $0 \leq t \leq T$, such that $w(0) = 0$; commonly $T = 1$ or $T = \infty$, and this is often called classical Wiener space. The Wiener measure is natural and appropriate for describing Brownian motion, in that the random process X given by $X(t, \omega) = W(t)$ is (under Wiener measure) the Brownian motion (or Wiener process); that is, for each fixed t, X has a normal $N(0, t)$ distn, and X has independent increments on $[0, T\}$. *See* Wiener's theorem.

Wiener process *see* Brownian motion

Wiener representation (or expansion, or formula) (alias: **Paley–Wiener representation**) A Fourier series expansion of the Wiener process $W(t), 0 \leq t \leq 1$, that has a number of forms depending on the particular orthonormal basis; these are often seen, in which $Z_r, r \geq 0$, is a sequence of i.i.d. normal $N(0, 1)$ r.vs:

$$W(t) = Z_0 t + \sqrt{2}\sum_{r=1}^{\infty} Z_r \frac{\sin(r\pi t)}{r\pi};$$

$$W(t) = Z_0 t + \sum_{r=1}^{\infty} Z_{2r} \frac{\sin(2r\pi t)}{\pi r\sqrt{2}} + \sum_{r=1}^{\infty} Z_{2r-1} \frac{(1 - \cos(2r\pi t))}{\pi r\sqrt{2}}.$$

[M. Lifshits, *Lectures on Gaussian Processes*. Springer, 2012]

Wiener sausage *see* Brownian sausage

Wiener's test A test for the recurrence or transience of a random walk (or suitable Markov process), which is so named because it arises as the probabilistic counterpart of Wiener's test (in classical potential theory) for the regularity of a boundary point in the Dirichlet problem for Laplace's equation in a domain. [K. Uchiyama, Wiener's test for random walks with mean zero and finite variance, *Ann. Probab.*, 26(1), 368–376, 1998]

Wiener's theorem Asserts that Wiener measure exists and is essentially unique. It may be constructed in a variety of ways. Classically, one considers the limit of a sequence of continuous linear interpolations of a suitable Gaussian process evaluated at the dyadic rationals, which establishes uniform convergence a.s., and then makes a continuous extension to obtain the required continuous Wiener process. [P. Billingsley, *Convergence of Probability Measures*, 2nd edn. Wiley, 1999; I. Karatzas & S. Shreve, *Brownian Motion and Stochastic Calculus*, 2nd edn. Springer, 1998]

Wigner's distribution *see* Wigner's semicircle law

Wigner's inequality (alias: **Bell–Wigner (or v.v) inequality**) The assertion that for r.vs X, Y, and Z, each taking values in $\{-1, 1\}$, and defined on the same space, we have

$$\mathbb{P}(X=1, Y=1) + \mathbb{P}(Y=-1, Z=1) \geq \mathbb{P}(X=1, Z=1).$$

It is a version of the Bell inequality, of interest in quantum mechanics. [E. Wigner, On hidden variables and quantum mechanical probabilities, *Amer. J. Phys.*, 38(8), 1005–1009, 1970]

Wigner matrix *see* Wigner's semicircle law

Wigner's semicircle law In its basic form, this arises in considering a sequence $W_n, n \geq 1$, of $n \times n$ Wigner matrices, which is to say that W_n is the nth principal submatrix $W_{ij}, 1 \leq i, j \leq n$, of some infinite symmetric matrix whose entries are i.d. r.vs having common variance σ^2, that are also independent apart from the constraint of symmetry (more generally, the matrix may be Hermitian or unitary). The theorem asserts that (under some moment conditions) the suitably scaled empirical distribution of the (necessarily real) eigenvalues of W_n converges, as $n \to \infty$, to a distribution having density

$$f(x) = \frac{1}{2\pi\sigma^2}(4\sigma^2 - x^2)^{1/2}, \quad x^2 \leq 4\sigma^2,$$

called Wigner's (semicircle) distribution. [G. Anderson *et al.*, *An Introduction to Random Matrices*. Cambridge University Press, 2010]

Wilks's lambda distribution A probability distribution on $\mathbb{R}^+$ that is seen as the generalization of the F distribution to a similar multivariate statistical analysis, arising thus: let X and Y be independent Wishart matrices of order p, whose Wishart distributions have covariance matrix $\sum$ and respective degrees of freedom $r \geq p$ and $s \geq p$. Then $\Lambda = \frac{|X|}{|X|+|Y|}$, where $|X|$ is the determinant of X, is a r.v. having Wilks's distribution that does not depend on the $p \times p$ covariance matrix $\sum$. As $r \to \infty$, $-(r - \frac{1}{2}(p+s+1))\log\Lambda$ has asymptotically a distn that is $\chi^2(ps)$. [N. Timm, *Applied Multivariate Analysis*. Springer, 2002]

Williams–Bjerknes model (alias: **biased voter model**) An interacting particle system in two dimensions that arises as a classic stochastic model for the spread of abnormal cells among normal cells, when abnormal cells have the advantage of a faster rate of division. [M. Bramson & D. Griffeath, On the Williams-Bjerknes tumour growth model: I, *Ann. Probab.*, 9(2), 173–185, 1981; & II, *Math. Proc. Camb. Phil. Soc.*, 88(2), 339–357, 1980]

Williams inequality *see* Polya–Williams approximation

Williams path decomposition Any of several relationships between diffusions and associated conditioned processes, but perhaps most often between the Wiener process (i.e. Brownian motion on $\mathbb{R}$) and the three-dimensional Bessel process (i.e. the radial part of Brownian motion in $\mathbb{R}^3$), whose underlying common idea is that some diffusion $X(t)$ of interest may be constructed by joining together appropriate segments of suitably conditioned auxiliary diffusions. The method has a variety of generalizations and applications. One classic example is this: let the Wiener process starting at $b > 0$, denoted by $W(t), t \geq 0$, and the Bessel process $R(t), t \geq 0$, and the r.v. U uniformly distributed on $[0, b]$ be independent. Write $H(t) = \inf\{t : W(t) = U\}$, and set

$$X(t) = \begin{cases} W(t), & t < H \\ R(t-H) + U, & t \geq H; \end{cases}$$

then X is a Bessel process started at b. [L.C.G. Rogers & D. Williams, *Diffusions, Markov Processes and Martingales*, Vols. 1 and 2, 2nd edn. Wiley, 1994; reprinted by Cambridge University Press, 2000]

Wilson's method A procedure for generating a (pseudo-) random spanning tree of a (possibly directed) weighted graph G on n vertices (with non-negative edge weights), where the probability of any given tree is proportional to its weight. [D. Wilson, Generating random spanning trees more quickly than the cover time, *STOC 96, Proc. 28th ACM Symposium on the Theory of Computing*. ACM, 1996, pp. 296–303]

Wishart distribution A multivariate distribution that can be seen as a generalization of the chi-squared distribution. In standard form it is defined in terms of i.i.d.r.vs $X_1, \ldots, X_n$, each having a multivariate normal $N(0, \Sigma)$ distribution with $p \times p$ covariance matrix Σ. Then the symmetric $p \times p$ matrix $S = \sum_{r=1}^{n} \mathbf{X}_r \mathbf{X}_r^T$ is called a Wishart matrix, and its distribution is the Wishart distribution, which has density f such that

$$f \propto |S|^{\frac{1}{2}(n-p-1)} \exp\left\{-\frac{1}{2}\mathrm{trace}{\sum}^{-1} S\right\}.$$

[M. Eaton, Multivariate statistics: a vector space approach, *IMS Lecture Notes*, 302–333, 2007]

Wishart process A random process $S(t), t \geq 0$ in the space of real symmetric non-negative $p \times p$ matrices, which can be seen as a multivariate generalization of the squared Bessel process. It is defined as a (strong) solution of the matrix SDE (for suitable c),

$$\mathrm{d}S(t) = \sqrt{S}\mathrm{d}WA + A^{\mathrm{T}}\mathrm{d}W^{\mathrm{T}}\sqrt{S} + \{SB + B^{\mathrm{T}}S + cA^{\mathrm{T}}A\}\mathrm{d}t,$$

where $c \geq 0$, W is a $p \times p$ matrix Wiener process, while A and B are arbitrary $p \times p$ real matrices. [M.-F. Bru, Wishart processes, *J. Theor. Probab.*, 4(4), 725–751, 1991]

with probability 1 *see* almost surely

Witten–Sander model *see* diffusion limited aggregation

Wold's process A point process on $\mathbb{R}^+$, starting at 0, such that intervals between successive points comprise a time-homogeneous Markov chain. [K. Athreya *et al.*, Asymptotic behavior of point processes with Markov-dependent intervals, *Math. Nachr.*, 99(1), 301–313, 1980]

Wold's theorem Arises in considering a weakly (i.e. covariance, or second-order) stationary process $X(t), t \in \mathbb{Z}$, where it comprises **Wold's decomposition** and **Wold's representation**, thus:

1. There is a unique decomposition $X(t) = Y(t) + D(t)$, into uncorrelated stationary processes $Y(t)$ and $D(t)$, where $D(t)$ is deterministic, which is to say that it is

linearly predictable given $D(t-j)$, $j \geq 1$, with zero error.
2. There is a representation $Y(t) = \sum_{j=0}^{\infty} a_j Z(t-j)$, where $a_0 = 1, \sum_j |a_j|^2 < \infty$, and $Z(t), t \in \mathbb{Z}$ is a sequence of uncorrelated r.vs.

(Note that the term Wold's theorem is sometimes used to denote a form of the Wiener–Khinchin theorem in discrete time.)

[P. Brockwell & R. Davis, *Time Series*, 2nd edn. Springer, 1991]

Wolfowitz theorem *see* noisy coding theorem; weak ergodicity

Wong–Zakai theorem Supplies a connection between SDEs and ODEs by comparing the solution of the SDE $dX(t) = f(X(t), t)dW(t)$ (for suitable functions f, where $W(t)$ denotes the Wiener process), with the sequence of solutions of the ODEs $dX_n(t) = f(X_n(t), t)\, dW_n(t)$, where $W_n(t)$ converges appropriately to $W(t)$ as $n \to \infty$. Under certain conditions the limit of solutions is the Stratonovich integral of the SDE, not the Itô integral; the difference is commonly called the **Wong–Zakai correction**. [E. Wong & M. Zakai, On the relation between ordinary and stochastic differential equations, *Int. J. Eng. Sci.*, 3 (2), 213–229, 1965]

worst-case *see* adversarial input

wrapped distribution Obtained from the density $f(x)$ of a r.v. $X \in \mathbb{R}$ by defining the r.v $Y = X$ modulo 2π, which has a density $g(y), 0 \leq y \leq 2\pi$, and is thus a circular r.v. (seen as $f(x)$ wrapped around a circle of unit radius). Those most often seen are:

(i) The wrapped normal density in which the density can be shown to have the representation

$$g(y) = \frac{1}{2\pi}\left(1 + 2\sum_{n=1}^{\infty} \rho^{n^2} \cos\{n(y-\mu)\}\right),$$

where $\rho = \exp\left(-\frac{1}{2}\sigma^2\right)$ for X an $N(\mu, \sigma^2)$ r.v.

(ii) The wrapped Cauchy density, which in standard form has $g(y) = \frac{1}{2\pi}\frac{\sinh\lambda}{\cosh\lambda - \cos y}$, where λ is a scale parameter.

[N. Fisher, *Statistical Analysis of Circular Data.* Cambridge University Press, 1993]

Wright–Fisher model (alias: **Fisher's model; Fisher–Wright model**) A stochastic model, proposed by R. Fisher and subsequently analyzed by S. Wright, for gene frequencies in successive generations of monoecious (hermaphrodite) individuals. The size of the population is fixed (at M if haploid, or $2N$ if diploid) and the simplest model considers a single locus with two alleles. Generations are assumed not to overlap, and the gene frequency $X(t), t \geq 1$, at the locus in the tth generation is obtained by binomial sampling from the previous generation. Thus $X(t)$ comprises a Markov chain on $\{0, \ldots, M\}$ with transition probabilities

$$p_{jk} = \binom{M}{k}\left(\frac{j}{M}\right)^k\left(1 - \frac{j}{M}\right)^{M-k}.$$

More generally, given mutation rates a and b between the two alleles

$$p_{jk} = \binom{M}{k} q_j^k (1-q_j)^{M-k}$$

where $q_j = (1-a)\frac{j}{M} + b\left(1 - \frac{j}{M}\right)$. Diffusion approximations to these discrete chains may be called **Fisher–Wright diffusions**, sometimes called **Jacobi diffusions**, with corresponding SDE of the form

$$dX(t) = -a(X-m)dt + b\{(X-c)(d-X)\}^{\frac{1}{2}}dW(t),$$

for suitable values of a, b, c, d, and m, with $W(t)$ the Wiener process. The eigenfunctions may be expressed in terms of Jacobi polynomials; in the classic (no mutation) simplest Fisher–Wright diffusion one has $a = c = 0$, $b = \frac{1}{2}$, and $d = 1$. These models naturally extend to many alleles by multinomial sampling, and there are many further generalizations. *See also* Moran model. [W. Ewens, *Mathematical Population Genetics*, 2nd edn. Springer, 2004; R. Griffiths & D. Spanò, Diffusion processes and coalescent trees, in *Probability and Mathematical Genetics*, N. Bingham & C. Goldie (eds). Cambridge University Press, 2010, pp. 358–379]

Wyner's wire-tap channel A classic secrecy system addressing the problem of confidential transmission of information when the (noisy) channel broadcasts two outputs, one to the intended (legitimate) receiver, and another to an unintended (enemy) receiver. Wyner's theorem established the existence of a secrecy capacity for the channel. Transmission below this rate ensures reliable transmission to the intended receiver while keeping the uncertainly (equivocation) of the unintended receiver arbitrarily close to its maximum. This cannot be achieved at rates above Wyner's secrecy capacity.

[A. Wyner, The wire-tap channel, *Bell Syst. Tech. J.*, 54(8), 1355–1387, 1975]

Wyner–Ziv problem *see* Slepian–Wolf theorem

X

XOR cipher *see* Vernam cypher

XOR corrector (alias: **exclusive-or corrector**) Used in the so-called post-processing (or debiasing) of a sequence of physically (i.e. hardware-based) random binary digits $X_n \in \{0, 1\}, n \geq 0$. The XOR-corrector applied to this sequence produces $Y_n = X_{2n} + X_{2n+1}$, modulo 2, $n \geq 0$.

On the assumption that X_n comprises a sequence of i.i.d.r.vs with $\mathbb{E}X_n = \frac{1}{2} + \varepsilon$, the output is a sequence of i.i.d.r.vs with $\mathbb{E}Y_n = \frac{1}{2} - 2\varepsilon^2$. See von Neumann corrector. [M. Dichtl, Bad and good ways of post-processing biased physical random numbers, in *Proc. FSE'70*. Springer, 2007, pp. 137–152]

XYZ inequality (or theorem) (alias: **Fishburn–Shepp inequality**) A correlation inequality for finite partially ordered sets (i.e. posets). One version asserts that if X, Y, and Z are distinct points in a poset, then

$$\mathbb{P}(X > Y)\mathbb{P}(X > Z) \leq \mathbb{P}(X > Y; X > Z);$$

this may be called **Shepp's thm.** [G. Brightwell & W. Trotter, A combinatorial approach to correlation inequalities, *Discrete Math.*, 257(2–3), 311–327, 2002]

Y

Yaglom's theorem Denotes either of two results in the theory of Galton–Watson branching processes. Let Z_n, $n \geq 1$ with $Z_0 = 1$, be such an ordinary branching process with non-degenerate family distribution $\mathbb{P}(Z_1 = r) = p_r$, $r \geq 0$, having pgf $G(s) = \mathbb{E}s^{Z_1}$, and mean $m = G'(1)$. When $m < 1$, Yaglom's theorem asserts that $\lim_{n\to\infty}\mathbb{P}(Z_n = r \mid Z_n > 0) = a_r$, where the pgf $y(s) = \sum_{r=1}^{\infty} a_r s^r$ of the Yaglom distribution $a_r, r \geq 1$, satisfies $y(G(s)) = my(s) + 1 - m$. When $m = 1$, then Yaglom's exponential limit law asserts (when $G'''(1) < \infty$) that

$$\lim_{n\to\infty} \mathbb{P}\Big(2Z_n/\{nG''(1)\} > x \mid Z_n > 0\Big) = \mathrm{e}^{-x}, \quad x \geq 0.$$

[K. Athreya & P. Ney, *Branching Processes*. Springer, 1972; reprinted by Dover, 2004]

Yamada–Watanabe theorem Supplies conditions on the coefficients a and b of the SDE $\mathrm{d}X = a(t, X)\mathrm{d}t + b(t, X)\mathrm{d}W(t)$, $t \geq 0$, $X(0) = c \in \mathbb{R}^n$, that are sufficient for the SDE to have a strong solution that is pathwise unique. [T. Yamada & S. Watanabe, On the uniqueness of solutions of stochastic differential equations I & II, *J. Math. Kyoto Univ.*, 11(1), 155–167, & 11(3), 533–563, 1971]

Yao's minimax principle (or theorem) Arises in computational complexity, as an application of von Neumann's minimax theorem, where it supplies a lower bound for the expected cost of the randomized algorithm in the worst case of the problem. [A. Yao, Probabilistic computations: toward a unified measure of complexity, in *Proc. SFCS '77. IEEE*, 1977, pp. 222–227]

Yates's correction *see* continuity correction

Yor's theorem *see* Kunita–Watanabe theorem

Yule distribution (alias: **Yule–Simon distribution (or process)**) A r.v. X on the positive integers has the Yule distribution with parameter $a > 0$ when

$$\mathbb{P}(X = r) = \frac{a\Gamma(1 + a)\Gamma(r)}{\Gamma(n + 1 + a)}, \quad r = 1, 2, \ldots,$$

where $\Gamma(\cdot)$ is the gamma function. [D. Aldous, Stochastic models and descriptive statistics for phylogenetic trees, from Yule to today, *Stat. Sci.*, 16(1), 23–34, 2001]

Yule's equation The second-order autoregressive equation $X(t) = aX(t - 1) + bX(t - 2) + Z(t), t \geq 2$, where $Z(t)$ is the discrete white noise (i.e. an uncorrelated sequence of zero-mean unit variance r.vs), introduced by G. Yule in seeking to explain sunspot data. [G. Yule, On a method of investigating periodicities in disturbed series with special reference to Wolfer's sunspot numbers, *Phil. Trans. Roy. Soc., A*, 226, 267–298, 1927]

Yule–Furry process *see* birth process

Yule's paradox *see* Yule–Simpson paradox

Yule–Simon distribution *see* Yule distribution

Yule–Simpson paradox (alias: **Simpson's paradox; amalgamation paradox**) A property of probability distributions which commonly arises in practice in statistical contexts, where two (or more) data-sets that each (separately) support some hypothesis but, when combined, support an entirely contrary hypothesis. (This seems to be counterintuitive, and actual examples have caused controversy and confusion.) In probabilistic terms the phenomenon is displayed by the observation that for any three events A, B, and C, it is possible to allocate probabilities $\mathbb{P}(\cdot)$ such that

$$\mathbb{P}(A \mid B \cap C^c) > \mathbb{P}(A \mid B^c \cap C^c), \text{ and}$$
$$\mathbb{P}(A \mid B \cap C) > \mathbb{P}(A \mid B^c \cap C),$$

but nevertheless we have $\mathbb{P}(A \mid B) < \mathbb{P}(A \mid B^c)$.

The paradox was first noted by G.U. Yule in 1903 under the heading: 'On the fallacies that may be caused by the mixing of distinct records.' [G. U. Yule, Notes on the theory of association of attributes in statistics,

Biometrika, 2, 121–134, 1903; I. J. Good & Y. Mittal, The amalgamation and geometry of two-by-two contingency tables, *Ann. Stat.*, 15(2), 694–711, 1987; J. Pearl, *Causality*, 2nd edn. Cambridge University Press, 2009]

Yule–Walker equations A set of linear equations relating the coefficients defining an autoregressive sequence, and the autocorrelation function of the process. [P. Brockwell & R. Davis, *Time Series*, 2nd edn. Springer, 1991]

Z

Zakai equation *see* filtering

Z distribution *see* F distribution; normal distribution

Zermelo's method *see* backward induction

zero-biased distribution For a zero-mean finite-variance r.v. X, the r.v. Z has the X-zero-biased distn if, for all differentiable functions g such that both sides exist, $\mathbb{E}(Xg(X)) = \text{var}(X)\mathbb{E}g'(Z)$. [L. Goldstein & G. Reinert, Stein's method and the zero bias transformation, *Ann. Appl. Probab.*, 7(4), 935–952, 1997]

zero-modified distribution The distribution $\mathbb{P}(X = x), x \in \mathbb{Z}$, of an integer-valued r.v. X (usually non-negative) is zero modified by forming the new (mixed) distribution $p_M(x), x \in \mathbb{Z}$, for some parameter q, thus:

$$p_M(0) = q + (1 - q)\mathbb{P}(X = 0),$$
$$p_M(x) = (1 - q)\mathbb{P}(X = x),\ x \neq 0,$$

subject to the constraint that $0 \leq q + (1 - q)\mathbb{P}(X = 0) \leq 1$. Then:

when $q > 0$, the distribution is **zero-inflated**;
when $q < 0$, the distribution **zero-deflated**;
when $p_M(0) = 0$, the distribution is **zero-truncated**.

See modified distribution. [N. Johnson *et al.*, *Univariate Discrete Distributions*, 3rd edn. Wiley, 2005]

zero–one law (alias: **zero or one law; 0–1 law**) Broadly, the assertion that for some event A, or for any event A in some set of events, it is the case that either $\mathbb{P}(A) = 0$ or $\mathbb{P}(A) = 1$. A basic example is this: for a sequence $A_r, r \geq 1$, of events let $\mathcal{F}$ be the smallest sigma-field containing all of them; if $A \in \mathcal{F}$ is independent of $[A_1, \ldots, A_n]$ for all n, then either $\mathbb{P}(A) = 0$ or $\mathbb{P}(A) = 1$. The Borel–Cantelli lemmas supply another classic $0 - 1$ law. See also the Blumenthal, Hewitt-Savage, Lévy, and Kolmogorov zero–one laws. [E. Çinlar, *Probability and Stochastics*. Springer, 2011]

zero-range process An interacting particle process on $\mathbb{Z}^d$ in which particles interact only with those particles at the same site (i.e. points of $\mathbb{Z}^d$), thus: If there are r particles at the point x then at a rate $j(r)\,p(x, y)$ one of them jumps to the point y, independently of all the particles at all the other points, where $p(x, y)$ is a homogeneous (i.e. translation invariant) probability distribution, and $j(r) \geq 0$ (with $j(0) = 0$), is the so-called jump rate. [T. Liggett, *Interacting Particle Systems*. Springer, 2005]

zero-sum game A game such that the sum of the pay-offs of all the players is zero, most often in the case of two players, in which the pay-offs amount to the loser paying the winner. Since the game is then defined by the pay-offs (a_{ij}) of player A, it is commonly called a matrix game. *See* von Neumann's minimax theorem and game theory. [A. Washburn, *Two-Person Zero-Sum games*, 4th edn. Springer, 2014]

zeta distribution (alias: **discrete power-law (or Pareto) distribution; Zipf distribution; Riemann zeta distribution**) The r.v. $X \in \{1, 2, 3, \ldots\}$ has the zeta distribution with parameter $a > 1$ when

$$\mathbb{P}(X = x) = \frac{1}{\zeta(a)x^a}; \quad x = 1, 2, \ldots,$$

where $\zeta(a)$ is the Riemann zeta function. The moments of X are given for $r < a - 1$ by $\mathbb{E}X^r = \zeta(a - r)/\zeta(a)$, and are infinite for $r \geq a - 1$.

The distribution of the r.v. $Y = -\log X$, with support $-\log n, n \geq 1$, has also been called the Riemann zeta distn; it has infinitely divisible chf

$$\phi_Y(t) = \frac{\zeta(a + it)}{\zeta(a)}, \quad t \in \mathbb{R}.$$

See Zipf's law. [G. Lin & C.-Y. Hu, The Riemann zeta distribution, *Bernoulli*, 7(5), 817–828, 2001]

Zhang's sandpile *see* Abelian sandpile

Zipf's law Is said to hold when the empirical frequency of outcomes (or observations) of type k (or rank, or class, or size k, etc.) follows approximately a power-law distribution, $f \propto k^{-a}$, for some $a > 0$, where k is some (usually finite) subset of the positive integers.

Originally, and classically, Zipf proposed the law $f \propto \frac{1}{k}$ for the frequency of words of rank k in some body of English text. [M. E. J. Newman, Power laws, Pareto distributions, and Zipf's law, *Contemp. Phys.*, 46(5), 323–351, 2005]

Zolotarev ideal metric In its basic form, for two r.vs X and Y, this metric is given in terms of the distn fns F_X and F_Y by

$$m(X,Y) = \int_{-\infty}^{\infty} |Z_X(t) - Z_Y(t)|\mathrm{d}t,$$

where $Z_X(t) = \int_{-\infty}^{t} \frac{1}{r!}(t-s)^r \mathrm{d}F_X(s)$, $r \geq 0$, and $Z_Y(t)$ is defined likewise. [J. Stoyanov *et al.*, Probability metrics with applications in finance, *J. Stat. Theor. Prac.*, 2(2), 253–277, 2008]

Zolotarev pseudomoment A numerical property of a pair of r.vs X and Y that combines some properties of moments and metrics. The simplest examples, in terms of the distn fns F_X and F_Y, are $m_n(X,Y) = \int_{-\infty}^{\infty} t^n \mathrm{d}(f_X(t) - F_Y(t))$ and $z_n(X,Y) = \int_{-\infty}^{\infty} |t|^n |\mathrm{d}F_X(t) - \mathrm{d}F_Y(t)|$. [V. M. Zolotarev, *Modern Theory of Summation of Random Variables*. De Gruyter, 1997]

Z transform *see* probability generating function

Table of distributions

Part A – discrete

Name of distribution	pmf $p(x)$	Parameter(s)	Mean $\mathbb{E}X$	Variance var X
Bernoulli	$p^x(1-p)^{1-x}$, $x \in \{0,1\}$	$0 \le p \le 1$	p	$p(1-p)$
beta binomial	$\binom{n}{x}\frac{\beta(a+x,n+b-x)}{\beta(a,b)}$ $x \in \{0,1,2,\ldots,n\}$	$a>0, b>0$ $n \in \mathbb{N}$	$\frac{na}{a+b}$	$\frac{n(a+b+n)ab}{(a+b+1)(a+b)^2}$
binomial	$\binom{n}{x}p^x(1-p)^{n-x}$ $x \in \{0,1,\ldots,n\}$	$0 \le p \le 1$ $n \in \mathbb{N}$	np	$np(1-p)$
discrete uniform (or rectangular)	$\frac{1}{b-a+1}$ $x \in \{a,a+1,\ldots,b\}$	$a \le b$ both integers	$\frac{1}{2}(a+b)$	$\frac{(b-a+1)^2-1}{12}$
geometric	$(1-p)p^x$ $x \in \{0,1,2,\ldots\}$	$0<p<1$	$\frac{p}{1-p}$	$\frac{p}{(1-p)^2}$
shifted geometric	$(1-p)p^{x-1}$ $x \in \{1,2,\ldots\}$	$0<p<1$	$\frac{1}{1-p}$	$\frac{p}{(1-p)^2}$
hypergeometric	$\frac{\binom{b}{x}\binom{N-b}{n-x}}{\binom{N}{n}}$ $x \in \{\max\{0,b+n-N\},$ $\ldots,\min\{b,n\}\}$	$1 \le b \le N$ $1 \le n \le N$	$\frac{nb}{N}$	$\frac{nb(N-b)(N-n)}{(N-1)N^2}$
logarithmic (or log series)	$\frac{-p^x}{x\log(1-p)}$ $x \in \{1,2,3,\ldots\}$	$0<p<1$	$\frac{-p}{(1-p)\log(1-p)}$	$\frac{-p}{(1-p)^2\log(1-p)}$ $\times\left\{1+\frac{p}{\log(1-p)}\right\}$
negative binomial	$\binom{x+m-1}{m-1}p^x(1-p)^m$ $=\frac{\Gamma(x+m)}{\Gamma(m)x!}p^x(1-p)^m$ $x \in \{0,1,2,\ldots\}$	$0<p<1$ $m>0$	$\frac{pm}{1-p}$	$\frac{pm}{(1-p)^2}$
shifted negative binomial	$\binom{x-1}{m-1}p^{x-m}(1-p)^m$ $x \in \{m,m+1,\ldots\}$	$0<p<1$ $m>0$	$\frac{m}{1-p}$	$\frac{pm}{(1-p)^2}$
Poisson	$\frac{\lambda^x e^{-\lambda}}{x!}$ $x \in \{0,1,2,\ldots\}$	$\lambda>0$	λ	λ

pgf $G(z) = \mathbb{E}z^X$	**Remarks**
$1 - p + pz$	Often denoted by Ber(p), or by Bernoulli(p), and called a Bernoulli trial. The case $p = 1/2$ is often referred to as a fair (or true, or symmetric) coin.
${}_2F_1(-n, a; a+b, 1-z)$	Here $\beta(a,b)$ is the beta function, and ${}_2F_1$ is Gauss' hypergeometric function (or series). The beta binomial arises when the parameter p of a binomial $\mathrm{B}(n,p)$ distn has a beta distribution Beta (a,b).
$(1 - p + pz)^n$	Commonly denoted by $\mathrm{B}(n,p)$. It is the distribution of the number of successes in n independent Bernoulli trials, which is the same as saying the distn of the sum of n i.i.d. Bernoulli (p) r.vs. In particular $\mathrm{B}(1,p)$ is Bernoulli (p).
$\dfrac{z^a - z^{1+b}}{(b-a+1)(1-z)}$	Often denoted by $\mathrm{U}(a,b)$ or Unif (a,b) or Uniform (a,b). In particular, for $a = 0,\ b = 1$ it is $\mathrm{Ber}(\frac{1}{2})$, (or a fair coin); for $a = 1, b = 6$, it is a fair die. More generally, a uniform distribution exists on any finite set of points $x_1, \ldots, x_n$, with $p(x_r) = n^{-1}, 1 \leq r \leq n$; this uniform distribution is *not* called rectangular.
$\dfrac{1-p}{1-pz}$	Often denoted by Geo(p), or Geometric(p), it arises as the distribution of the number of failures (with $\mathbb{P}(F) = p$) before the first success (with $\mathbb{P}(S) = 1 - p$) in a sequence of i.i.d. Bernoulli trials. In applications it is typically more natural to count the number of trials up to and including the first success, yielding the shifted geometric.
$\dfrac{(1-p)z}{1-pz}$	
$\dfrac{\binom{N-b}{n}}{\binom{N}{n}} {}_2F_1(-n, -b, N-b-n+1; z)$	Here ${}_2F_1$ is Gauss' hypergeometric function. The distn arises as the probability of obtaining exactly x marked items in n draws without replacement from a population of N items of which b are marked. It is sometimes denoted by Hyp(N, n, b), or similar variants.
$\dfrac{\log(1-pz)}{\log(1-p)}$	Arises as the limiting distribution obtained from the negative binomial $\mathrm{NB}(m,p)$ distn conditioned to be positive (i.e. zero-truncated) as $m \downarrow 0$. It may also be obtained as a compound geometric distribution.
$\left(\dfrac{1-p}{1-pz}\right)^m$	This may be denoted by $\mathrm{NB}(m,p)$; it arises as a compound Poisson distribution whose parameter is a r.v. Λ having a gamma distn $\Gamma(m, \frac{1}{p} - 1)$ (where m is the shape parameter); in this context it is also called a gamma Poisson distn (and is not to be confused with the Poisson gamma distn of the sum of a Poisson distributed number of i.i.d. gamma r. vs). In the special case when m is an integer, the name Pascal distribution is used; it arises as the distn of the sum of m i.i.d. geometric Geo(p) r.vs, and the shifted form appears when the geometric r.vs take values in $\{1, 2, 3, \ldots\}$.
$\left\{\dfrac{(1-p)z}{1-pz}\right\}^m$	
$e^{\lambda(z-1)}$	Commonly denoted by Poi(λ), or by Poisson (λ). It arises as the limiting distribution of the number of points (or events) in an interval of time (or subset of some space) of a suitably regular random process of points (or events) having independent increments (or independent counts for disjoint subsets). It also arises as the limiting distribution of a binomial r.v., with parameters n and p, as $n \to \infty$ with $np = \lambda$ held fixed.

Part B – continuous

Name of distribution	pdf $f(x)$	Parameter(s)	Mean	Variance
arcsine	$\frac{1}{\pi\sqrt{x(1-x)}}$, $0 < x < 1$	–	$\frac{1}{2}$	$\frac{1}{8}$
Bessel	$mx^{-1}e^{-x}I_m(x)$ $x > 0$	$m > 0$	∞	–
beta	$\frac{\Gamma(a+b)}{\Gamma(a)\Gamma(b)}x^{a-1}(1-x)^{b-1}$ $0 < x < 1$	$a > 0$ $b > 0$	$\frac{a}{a+b}$	$\frac{ab(a+b)^2}{a+b+1}$
Cauchy	$\frac{1}{\pi(1+x^2)}$ $x \in \mathbb{R}$	–	–	–
exponential	$\lambda e^{-\lambda x}$, $x > 0$	$\lambda > 0$	λ^{-1}	λ^{-2}
extreme value, or doubly exponential, or Gumbel (or Fisher–Tippett–Gumbel)	$\exp(-x - e^{-x})$, $x \in \mathbb{R}$	–	γ	$\frac{1}{6}\pi^2$
F (or Fisher's)	$\frac{m^{\frac{1}{2}m}n^{\frac{1}{2}n}}{\beta(\frac{1}{2}m, \frac{1}{2}n)} \frac{x^{\frac{1}{2}m-1}}{(n+mx)^{\frac{1}{2}(m+n)}}$ $x > 0$	$m > 0$ $n > 0$	$\frac{n}{n-2}, n > 2$	$\frac{2n^2(m+n-2)}{m(n-4)(n-2)^2}$ $n > 4$
gamma	$\frac{\lambda^m}{\Gamma(m)}x^{m-1}e^{-\lambda x}$ $x > 0$	$\lambda > 0$ $m > 0$	$\frac{m}{\lambda}$	$\frac{m}{\lambda^2}$

chf $\phi(t) = \mathbb{E}e^{itX}$	**Remarks**
$e^{it/2}\sum_{r=0}^{\infty}\frac{(-1)^r}{(r!)^2}\left(\frac{1}{2}t\right)^{2r} = e^{it/2}J_0\left(\frac{1}{2}t\right)$	Arises in the arcsine law; here J_0 is the ordinary Bessel function.
$\left\{1 - it + [(1-it)^2 - 1]^{\frac{1}{2}}\right\}^m$	Here $I_m(x) = \sum_{r=0}^{\infty}\frac{1}{r!\Gamma(r+m+1)}\left(\frac{1}{2}x\right)^{2r+m}$ is a modified Bessel function. For integer m, the density is that of the first-passage time to m of simple, symmetric random walk in continuous time (sometimes called randomized SSRW).
$M(a, a+b, it)$	Here $M(a, a+b, it)$, is a confluent hypergeometric function. The density is denoted by Beta (a, b) and named after the beta function $\beta(a,b) = \Gamma(a)\Gamma(b)/\Gamma(a+b)$. It arises as the distribution of the ratio $X/(X+Y)$ where X and Y are independent Gamma (a, λ) and Gamma (b, λ) random variables respectively. In particular, $a = b = ½$ is the arcsine distribution, and $a = b = 1$ is the uniform $U(0, 1)$ distn.
$e^{-\lvert t\rvert}$	Arises as the distribution of the ratio of two independent standard normal r.vs; it has no mean, and often appears in counterexamples.
$\frac{\lambda}{\lambda - it}$	This is often denoted by $\mathrm{Exp}(\lambda)$; it may also be parameterized as $\mu^{-1}\exp(-x/\mu)$, and denoted in this case by $\mathrm{Exp}(\mu)$. The exponential distribution is noted for, and characterized by, its lack-of-memory (or memoryless) property. It arises as a natural continuous limit of the geometric distn. More generally, the density with distn fn $F(x) = 1 - \exp\left(-(\lambda x)^b\right), b > 0, \lambda > 0, x > 0,$ is called the Weibull, or stretched exponential distn.
$\Gamma(1 - it)$	Here γ is Euler's constant. The generalized extreme value distribution has distn fn $F_b(x) = \exp\left\{-[1+bx]^{-1/b}\right\}$, for $0 < 1 + bx, b \in \mathbb{R}$, where $b = 0$ is the doubly exponential (or Gumbel) distn as given. For $b > 0, x > -1/b, F_b(x)$ is called the Fréchet distn; for $b < 0, x < -1/b, F_b(x)$ is called the reversed Weibull distn (to distinguish it from the right-tailed ordinary Weibull distn, defined in the preceding entry as a stretched exponential). The chf of $F_b(x)$ for $b \neq 0$ may be written in terms of special functions, e.g. Wright's generalized confluent hypergeometric fn and Fox's H function. For $b \neq 0$, the mean of $F_b(x)$ exists only for $b < 1$, and then it is $\{\Gamma(1-b) - 1\}/b$.
$\frac{\Gamma(\frac{1}{2}(m+n))}{\Gamma(\frac{1}{2}n)}\Psi\left(\frac{1}{2}m, 1 - \frac{1}{2}n, \frac{-n}{m}it\right)$	Here Ψ is a confluent hypergeometric fn of the second kind; β is the beta fn. It is often denoted by $F(m, n)$, where m and n are called degrees of freedom; it may be termed the central F distn, to distinguish it from the non-central F. It arises as the distn of the scaled ratio $\frac{X/m}{Y/n}$ of independent chi-squared r.vs X and Y with m and n degrees of freedom.
$\left(\frac{\lambda}{\lambda - it}\right)^m$	Very often denoted by $\Gamma(m, \lambda)$ or Gamma (m, λ), but note that many authors reverse the order of the parameters in this: m is called the shape parameter, λ the scale parameter. For integer m this is the distribution of the sum of m i.i.d. exponential $\mathrm{Exp}(\lambda)$ r.vs, and is often called the Erlang distn in this case. When $m = \frac{1}{2}n$ (for integer n), and $\lambda = \frac{1}{2}$, the distn is that of the sum of the squares of n i. i.d. standard normal r.vs, and is called chi-squared with n degrees of freedom; when $n = 3$ this may be called the Maxwell distn.

Part B (*cont.*)

Name of distribution	pdf $f(x)$	Parameter(s)	Mean	Variance
inverse Gaussian	$\left(\frac{\lambda}{2\pi x^3}\right)^{\frac{1}{2}}\exp\left\{-\frac{\lambda(x-\mu)^2}{2x\mu^2}\right\}$ $x > 0$	$\lambda > 0$ $\mu > 0$	μ	μ^3/λ
first passage density	$\frac{a}{c}\frac{1}{(2\pi x^3)^{\frac{1}{2}}}\exp\left\{\frac{-(a-bx)^2}{2xc^2}\right\}$ $x > 0$	$a > 0$ $b > 0$ $c > 0$	$\frac{a}{b}$	$\frac{ac^2}{b^3}$
Lévy first-passage density	$\frac{a}{c}\frac{1}{(2\pi x^3)^{\frac{1}{3}}}\exp\left(\frac{-a^2}{2xc^2}\right)$ $x > 0$	$a > 0$ $c > 0$	-	-
Laplace	$\frac{1}{2}\lambda e^{-\lambda\|x\|}$ $x \in \mathbb{R}$	$\lambda > 0$	0	$2\lambda^{-2}$
generalized Laplace	$\frac{\lambda\mu}{\lambda+\mu}e^{-\lambda x},\ x \geq 0$ $\frac{\lambda\mu}{\lambda+\mu}e^{\mu x},\ x < 0$	$\lambda > 0$ $\mu > 0$	$\frac{1}{\lambda} - \frac{1}{\mu}$	$\frac{1}{\lambda^2} + \frac{1}{\mu^2}$
logistic	$\frac{e^x}{(1+e^x)^2}$ $x \in \mathbb{R}$	–	0	$\frac{1}{3}\pi^2$
lognormal	$\frac{1}{x\sigma\sqrt{2\pi}}\exp\left\{-\frac{(\mu-\log x)^2}{2\sigma^2}\right\}$ $x > 0$	$\sigma > 0$ $\mu \in \mathbb{R}$	$\exp\{\mu + \frac{1}{2}\sigma^2\}$	$(e^{\sigma^2} - 1)e^{2\mu+\sigma^2}$
normal (or Gaussian)	$\frac{1}{\sigma\sqrt{2\pi}}\exp\left\{\frac{-(x-\mu)^2}{2\sigma^2}\right\}$ $x \in \mathbb{R}$	$\mu \in \mathbb{R}$ $\sigma > 0$	μ	σ^2
Pareto	$\left(\frac{a}{c}\right)\left(\frac{c}{x+c}\right)^{a+1}$ $x > 0$	$a > 0$ $c > 0$	$\frac{c}{a-1}, a > 1$	$\frac{ac^2}{(a-2)(a-1)^2}, a > 2$
shifted Pareto	$\left(\frac{a}{c}\right)\left(\frac{c}{x}\right)^{a+1}$ $x > c$	$a > 0$ $c > 0$	$\frac{ac}{a-1}, a > 1$	$\frac{ac^2}{(a-2)(a-1)^2}, a > 2$
Rayleigh	$xe^{-\frac{1}{2}x^2}$ $x > 0$	–	$\sqrt{\frac{\pi}{2}}$	$2 - \frac{\pi}{2}$
sech or inverse cosh or hyperbolic cosine	$\text{sech}(\pi x)$ $x \in \mathbb{R}$	–	0	$\frac{1}{4}$

chf $\phi(t) = \mathbb{E}e^{itX}$	**Remarks**
$\exp\left\{\frac{\lambda}{\mu}\left(1 - [1 - 2\mu^2 \mathrm{i}t/\lambda]^{\frac{1}{2}}\right)\right\}$	Often denoted by $IG(\mu, \lambda)$ in the first (standard) parameterization; there are several other parameterizations listed by M. Tweedie. In particular, the second parameterization arises as the density of the first-passage time to $a > 0$ of a drifting Brownian motion $X = bt + cW(t)$, where $W(t)$ is the Wiener process. When the drift is zero, $b = 0$, the distn has no moments, and may be called the Lévy first-passage density.
$\exp\left\{\frac{ab}{c^2}\left(1 - \left[1 - \frac{2c^2}{b^2}\mathrm{i}t\right]^{\frac{1}{2}}\right)\right\}$	
$\exp\left(\frac{-a}{c}\sqrt{-2\mathrm{i}t}\right)$	
$\frac{\lambda^2}{\lambda^2 + t^2}$	Arises as the distribution of the difference $U - V$ between independent exponentially distributed r.vs U and V, where U and V are i.i.d. $\mathrm{Exp}(\lambda)$ r.vs in the symmetric case, or independent $\mathrm{Exp}(\lambda)$ and $\mathrm{Exp}(\mu)$ in the generalized (asymmetric) case.
$\frac{\lambda\mu}{(\lambda - \mathrm{i}t)(\mu + \mathrm{i}t)}$	
$\frac{\pi t}{\sinh(\pi t)}$ $= \Gamma(1 + \mathrm{i}t)\Gamma(1 - \mathrm{i}t)$	Arises as the distribution of the difference between two i.i.d.r.vs having the doubly exponential (or Fisher–Tippett–Gumbel) extreme value density $e^{-x-e^{-x}}$; and also as the distn of $\log\{U/(1-U)\}$, where U is uniform on $(0, 1)$.
ϕ does not exist in the lower half-plane	Arises as the distribution of e^Z where Z is a normal $N(\mu, \sigma^2)$ r.v. This distn is not determined by its moments, and no closed-form expression has been given for its chf ϕ; there are approximate representations: e.g. in terms of Lambert's W function.
$\exp\{\mathrm{i}\mu t - \frac{1}{2}\sigma^2 t^2\}$	Arises in the central limit thm; and as the density on $\mathbb{R}$ having maximum differential entropy for a given variance $\sigma^2 < \infty$.
$aU(1, 1 - a, -\mathrm{i}tc)$	Here U is the second type of confluent hypergeometric function. The distn arises as a continuous analogue of the discrete Zeta distn; and is the density of $c(e^Y - 1)$, where Y is exponential $\mathrm{Exp}(a)$. The Pareto distn may be called a power law distn; it can be seen as a special case of the beta-prime distn, having density $\frac{x^{b-1}(1+x)^{-a-b}}{\beta(a, b)}$.
$a e^{\mathrm{i}tc} U(1, 1 - a, -\mathrm{i}tc)$	
$1 + \sqrt{2\pi}\,\mathrm{i}t\left(1 - \Phi(-\mathrm{i}t)\right)e^{-\frac{1}{2}t^2}$	Arises as the density of $(U^2 + V^2)^{\frac{1}{2}}$ when U and V are i.i.d. standard normal r.vs; it is also the density of $(2\lambda W)^{\frac{1}{2}}$ when W has an exponential $\mathrm{Exp}(\lambda)$ density. As usual, Φ denotes the standard normal distn function
$\mathrm{sech}\left(\frac{1}{2}t\right)$	The pdf and chf are a self-reciprocal pair as are the pdf and chf of the normal density.

Part B (*cont.*)

Name of distribution	pdf $f(x)$	Parameter(s)	Mean	Variance
t or Student's t-distribution	$\frac{\Gamma\left(\frac{1}{2}(n+1)\right)}{\sqrt{n\pi}\Gamma\left(\frac{1}{2}n\right)}\left(1+{x^2}/{n}\right)^{-\frac{1}{2}(n+1)}$ $x \in \mathbb{R}$	$n > 0$	0, for $n > 1$	$\frac{n}{n-2}$, for $n > 2$
triangular	$\frac{1}{a}\left(1-\frac{1}{a}\lvert x\rvert\right)$ $\lvert x\rvert < a$	$a > 0$	0	$\frac{1}{6}a^2$
generalized triangular	$\frac{2(x-a)}{(b-a)(\tau-a)}, a \le x \le \tau$ $\frac{2(b-x)}{(b-a)(b-\tau)}, \tau \le x \le b$	$a \in \mathbb{R}$ $b \in \mathbb{R}$ $\tau \in \mathbb{R}$	$\frac{1}{3}(a+b+\tau)$	$\frac{1}{18}(a^2+b^2+\tau^2-$ $ab-b\tau-\tau a)$
uniform	$(b-a)^{-1}$ $a \le x \le b$	$a, b \in \mathbb{R}$	$\frac{1}{2}(a+b)$	$\frac{1}{12}(a-b)^2$

Part C – multivariate

Name of distn	pdf $f(x)$	Parameters	mean vector	covariance matrix
non-singular multivariate normal, or multinormal	$\frac{1}{\sqrt{(2\pi)^n\lvert\mathbf{V}\rvert}}\exp\{-\frac{1}{2}(\mathbf{x}-\boldsymbol{\mu})\mathbf{V}^{-1}(\mathbf{x}-\boldsymbol{\mu})^T\}$ $\mathbf{x} \in \mathbb{R}^n$	$\boldsymbol{\mu} \in \mathbb{R}^n$ $\mathbf{V}$ a positive definite symmetric $n \times n$ matrix	$\boldsymbol{\mu}$	$\mathbf{V}$

chf $\phi(t) = \mathbb{E}e^{itX}$	**Remarks**
$\dfrac{K_{\frac{1}{2}n}(\sqrt{n}\lvert t\rvert)(\sqrt{n}\lvert t\rvert)^{\frac{1}{2}n}}{\Gamma(\frac{1}{2}n)2^{\frac{1}{2}n-1}}$	Here K is the second kind of modified Bessel function; and n is called the number of degrees of freedom. For $n > 1$, and i.i.d. normal $N(\mu, \sigma^2)$ r.vs $X_1, \ldots, X_n$, the r.v. $T = \dfrac{\left\{\frac{1}{n}\sum_{r=1}^{n} X_r - \mu\right\}\left(\frac{n}{n-1}\right)^{\frac{1}{2}}}{\left(\sum_{r=1}^{n}\left(X_r - \frac{1}{n}\sum_{k=1}^{n} X_k\right)^2\right)^{\frac{1}{2}}}$ is called Student's T, and has a t-distribution with $n-1$ degrees of freedom. As $n \to \infty$ the distribution converges to the standard normal.
$\dfrac{2(1-\cos(at))}{a^2t^2} = \left(\dfrac{2}{at}\sin\dfrac{at}{2}\right)^2$	When U and V are i.i.d. Uniform $\mathrm{U}(-\frac{1}{2}a, +\frac{1}{2}a)$ r.vs, $U + V$ has the symmetric triangular density.
$2\dfrac{(b-a)e^{i\tau t}+(a-\tau)e^{ibt}+(\tau-b)e^{iat}}{(b-a)(b-\tau)(\tau-a)t^2}$	
$\dfrac{e^{ibt} - e^{iat}}{it(b-a)}$	Commonly denoted by $\mathrm{U}(a, b)$, or Uniform (a, b), and sometimes called a rectangular density. When X is $\mathrm{U}(a, b)$, $U = \frac{X-a}{b-a}$ is $\mathrm{U}(0, 1)$, called the standard uniform distn. When Y is $\mathrm{Exp}(\lambda)$, $\exp(-\lambda Y)$ is $\mathrm{U}(0, 1)$, as is also $1 - e^{-\lambda Y}$.

chf $\phi(\mathbf{t}) = \mathbb{E}e^{i\mathbf{t}\mathbf{X}^T}$	**Remarks**
$\exp\{i\mathbf{t}\boldsymbol{\mu}^T - \frac{1}{2}\mathbf{t}\mathbf{V}\mathbf{t}^T\}$	Commonly denoted by $N(\boldsymbol{\mu}, \mathbf{V})$. More generally, $X_1 \ldots, X_n$ has a multinormal distn if for all $\mathbf{a} \in \mathbb{R}^n$, $\mathbf{X}\mathbf{a}^T = \sum_{r=1}^{n} a_r X_r$, has a normal distn. In the singular case this normal distn is permitted to be degenerate for some $\mathbf{a}$, and the covariance matrix $\mathbf{V}$ is singular and has no inverse.